W9-ABB-712

Essentials of Physical Geography

Sixth Edition

Robert E. Gabler, *Emeritus*
Western Illinois University

Robert J. Sager
Pierce College, Washington

Daniel L. Wise
Western Illinois University

James F. Petersen
Southwest Texas State University

Contributing Authors

David Butler
Southwest Texas State University

Norman Meek
California State University—San Bernardino

Saunders Golden Sunburst Series

SAUNDERS COLLEGE PUBLISHING
Harcourt Brace College Publishers

Fort Worth • Philadelphia • San Diego • New York • Orlando • Austin
San Antonio • Toronto • Montreal • London • Sydney • Tokyo

Publisher: Emily Barrosse
Executive Editor: John Vondeling
Acquisitions Editor: Jennifer Bortel
Product Manager: Erik Fahlgren
Developmental Editors: Sarah Fitz-Hugh, David Robson
Project Editor: Anne Gibby
Production Manager: Alicia Jackson
Art Directors: Cara Castiglio, Carol Bleistine
Text Designer: Kathleen Flanagan
Cover Designer: Cara Castiglio

Cover Credit: Alpine Light, Wind River Range, WY (Doug Remien Photography)

ESSENTIALS OF PHYSICAL GEOGRAPHY, sixth edition
ISBN: 0-03-022909-X

Library of Congress Catalog Card Number: 98-87953

Address for domestic orders: Saunders College Publishing, 6277 Sea Harbor Drive, Orlando, FL 32887-6777, 1-800-782-4479

Address for international orders: International Customer Service, Harcourt Brace & Company, 6277 Sea Harbor Drive, Orlando, FL 32887-6777, (407) 345-3800, Fax (407) 345-4060, email hbintl@harcourtbrace.com

Address for editorial correspondence: Saunders College Publishing, 150 S. Independence Mall West, Philadelphia, PA 19106-341

Web site address: http://www.hbcollege.com

Printed in the United States of America

9012345678 041 10 987654321

Preface

Earth, the life-support system for humankind, is a wondrous and complex place, yet it is also enigmatic. Our planet is in many ways robust and adaptable to environmental change, but in other ways it is fragile and threatened. In our modern day-to-day life, our use of technology has a tendency to insulate us from fully experiencing our environment, and we have become lulled into forgetting about our direct reliance on Earth's natural resources. Sometimes it is hard to imagine that as you read this you are also moving through space on a life-giving planetary oasis surrounded by the vastness of space—empty and, as far as we know, devoid of life. Yet a knowledge of our planet, its environments, and how they operate is as critical as it has ever been for human society.

For as long as humans have existed on Earth, the resources provided by their physical environment have been the key to their survival. Primitive, preindustrial societies, such as those dependent on hunting and gathering or small-scale agriculture, tended to have small populations and exerted relatively little impact on their natural surroundings. In contrast, today's industrialized societies have large populations, demand huge quantities of natural resources, and can influence or cause environmental change, not only locally, but also globally. As the human population has increased, so have the scale, degree, and cumulative effects of its impacts on the environment. Unfortunately, many of these impacts are detrimental. We have polluted the air and water. We have used up tremendous amounts of nonrenewable resources and have altered the natural landscape without fully assessing the consequences. Too often, we have failed to respect the power of Earth's natural forces when constructing our homes and cities or while pursuing our economic activities. As we enter the twenty-first century, it is evident that if we fail to comprehend Earth's potential and respect its limitations as a human habitat, we may be putting ourselves and future generations at risk. Despite the many differences between our current lifestyles and those of early humans, one aspect has remained the same: Our use of our physical environment in all its diversity is still the key to our survival.

Today, this important message is getting through. We understand that Earth has limits and does not offer boundless natural resources. Broadcast and print media have expanded their coverage of the environment and human/environment relationships. Politicians are drafting legislation designed to address environmental problems. Scientists and statesmen from around the world meet to discuss environmental issues that increasingly cross international boundaries. Humanitarian organizations, funded by governments as well as private citizens, struggle to alleviate the suffering that results when human beings cannot cope with the limitations of their physical surroundings or when they wittingly or unwittingly degrade the environments in which they live.

The study of geography is essential to developing a thorough knowledge, understanding, and appreciation of Earth as well as the people who occupy, alter, and rely upon our planet for their continued existence. Long a highly regarded subject in most nations of the world, geography in the last two decades of the twentieth century has undergone a renaissance in the United States. More and more people are realizing the importance of knowing and respecting the people, cultural contributions, and resources of other nations as well as those of the United States. National standards have been written to ensure that a high quality geographic curriculum is offered to students in American elementary, secondary, and postsecondary schools. Employers across the nation are recognizing the value and importance of geographic knowledge, skills, and techniques in the workplace. Geography as an applied field makes use of computer-assisted and space-age technologies, such as interpretation of satellite images, geographic information systems (GIS), computer cartography, and the global positioning system. At the collegiate level, physical geography offers an introduction to the concerns, ideas, knowledge, and tools that are necessary for further study of our planet. More than ever before, physical geography is being recognized as an ideal science course for general-education students—students who will make decisions that weigh human needs and desires against environmental limits and possibilities. It is for these students that *Essentials of Physical Geography* has been written.

Features

Comprehensive View of the Earth System.
Essentials of Physical Geography introduces all major aspects of the Earth system, identifying physical phenomena and stressing their distribution and relationships. It covers a wide range of topics, including the atmosphere, oceans and other water bodies, the solid Earth, and the patterns of life on our planet.

Clear Explanation.
The text uses an understandable and often narrative style to explain the origins, development, significance, and distribution of processes, physical features, and events that occur within, on, or above the surface of the Earth. The authors' writing style is targeted toward rapid reader comprehension and toward making the study of physical geography an enjoyable undertaking.

Introduction to the Geographer's Tools.
The space age and the computer have revolutionized the ways that we can study the environments of our planet. The text provides a full chapter devoted to the tools used by geographers, and illustrations throughout the book include many examples of images gathered from space, accompanied by interpretations of those aspects of the environment that the scenes illustrate. There are also introductory discussions of techniques used by geographers in analyzing or displaying locational and environmental aspects of Earth, including remote sensing, geographic information systems, computer-assisted cartography, and the global positioning system.

Focus on Student Interaction.
The text provides numerous ways to promote continuous interaction between students and the subject matter of physical geography. The Consider and Respond activities at the end of each chapter are designed to encourage students to apply their newly acquired knowledge in different and, when possible, problem-solving situations. Answering the queries at the end of the environmental essays requires critical thinking skills, and the legend questions challenge students to react to the illustrations they are examining.

"The Environment" Essays.
A series of essays on the environment is included in the text as supplementary reading. The essays deal with current issues such as alternative energy sources, desertification, global warming, rainforest destruction, protection of wetlands, and environmental hazards, including earthquakes, mudflows, and tsunamis.

Map Interpretation Series.
The authors consider the learning of map-interpretation skills a major priority in a physical geography course. To meet the needs of students who do not have the opportunity to interact with physical geography in a laboratory setting, the text includes 11 map exercises with accompanying explanations, illustrations, and interpretation questions. This dynamic feature gives all students the opportunity to learn and practice valuable map skills.

Objectives

Since the first edition, the authors of this book have sought to accomplish several major objectives:

To Address the Academic Needs of the Student.
Those instructors who are familiar with the style and content of *Essentials of Physical Geography* know that this is one textbook written specifically for the student. It has been designed to help meet the major purposes of a liberal education by providing students with sufficient knowledge and understanding to make informed decisions involving the physical environments with which they will interact throughout their lives. The text assumes little or no background in

physical geography or other Earth sciences on the part of introductory-level students. The authors include numerous examples from throughout the world to illustrate difficult concepts and help students bridge the gap between scientific theory and its practical application.

To Integrate the Illustration Program with the Text. Numerous photographs, block diagrams, and other line drawings have been carefully chosen to illustrate each of the complex concepts in physical geography the text addresses. Text discussions of these concepts often contain repeated references to the illustrations, and the student is able to examine in model form, as well as mentally reconstruct, the physical processes and phenomena involved. Particularly good examples of the integration of text and illustration include the sections on seasons (Chapter 3), heat energy budget (Chapter 4), surface wind systems (Chapter 6), atmospheric disturbances (Chapter 8), plate tectonics (Chapter 14), and ice sheets (Chapter 19).

To Communicate the Nature of Geography. The nature of geography as a physical science is discussed at length in Chapter 1, and several underlying themes are identified to illustrate the spatial focus of the discipline. In subsequent chapters, these themes are used as the organizational bases for the presentation of chapter content. For example, location is the major theme of Chapter 2 and remains an important ingredient throughout the text; spatial distributions are emphasized as each of the climatic elements are discussed in Chapters 4 to 7; a changing Earth system is the central consideration in Chapter 9; characteristics of places constitute the content of Chapters 10 and 11; spatial interactions are frequently demonstrated in discussions of weather systems (Chapter 8) and tectonic activity (Chapters 14 and 15); and human–environment interaction is given serious consideration whenever justified. As a part of the course summary, all of the themes are revisited in Chapter 23.

To Fulfill the Major Requirements of Introductory Physical Science College Courses. *Essentials of Physical Geography* offers a full chapter on the important tools and methodologies of physical geography. Throughout the book, the physical processes that are responsible for the location, distribution, and spatial relationships of physical phenomena beneath, at, and above Earth's surface are examined in detail. Scientific method, hypothesis, theory, and explanation are continually stressed. Models and systems are frequently cited in the discussion of important concepts, and scientific classification is presented in several chapters—specifically, air masses (Chapter 8), climates (Chapter 9 and the Appendix), biomes (Chapter 12), soils (Chapter 13), and coasts (Chapter 22).

Sixth Edition Revision

The challenge of revising *Essentials of Physical Geography* for a sixth edition turned out to be the most difficult the authors have faced in over 20 years. Several reviewers recommended increased numbers of chapters of varying lengths to be focused more directly on specific topics. Others urged expanded coverage in critical areas of physical geography that have recently sparked increased interest on the part of students. The authors had already agreed that the time had come to clearly demonstrate that physical geography could open the door to an interesting, exciting, and highly rewarding future for those individuals who decide to take advantage of its career opportunities. The challenge was to accomplish all of this in addition to thoroughly revising the text; preparing new graphs, maps, and diagrams; selecting dozens of new photographs; and updating the numerous examples of worldwide environmental events that keep physical geography current and topical. What follows is a brief review of major changes incorporated in the sixth edition in an attempt to meet the challenge.

New Coauthor The sixth edition of *Essentials of Physical Geography* is privileged to have James F. Petersen of Southwest Texas State University as a new coauthor. Jim's expertise in geographic education, geomorphology, and applied physical geography has been an invaluable asset to this project. His work throughout the book has allowed us to expand in a number of areas, which is sure to benefit students and teachers alike.

Chapter Reorganization. The number of chapters has been increased from 17 to 23. The fundamentals of Earth–sun relationships are covered in a separate new chapter preceding the discussion of solar energy and temperature. Wind systems and ocean currents have been combined in a chapter on global circulation. Climate classification and climate change are treated together in a new chapter. Groundwater is now covered in a separate chapter, and the erosional processes associated with streams in both humid and arid regions are now discussed together in a chapter entirely devoted to fluvial geomorphology. Wind as an erosional agent receives individual treatment, and content dealing with the world ocean has been subdivided into separate chapters—one dealing with general characteristics and another with processes that produce coastal landforms. In addition, a summary chapter has been added to the textbook to help students reflect on their experiences with a course in physical geography. The authors believe that these changes will provide increased course flexibility and that they have been accomplished without altering significantly the original sequence of topics or forcing instructors to make major changes in syllabi.

New and Revised Text. In addition to the text revision that was essential to chapter reorganization, major new textual material has been added on a variety of topics. These include models and systems (Chapter 1), modern technology and remote sensing (Chapter 2), the solar system (Chapter 3), El Niño (Chapter 6), climate classification and climate change (Chapter 9), paleogeography (Chapter 14), fluvial geomorphology (Chapter 18), and physical geography in the future (Chapter 23). Space was provided for these additions through the condensation of previous text, the removal of redundancy, and the elimination of some supplementary reading.

Enhanced Illustrations. The illustration program has undergone thorough revision for the sixth edition. The new text required new figures, which resulted in the drafting of many graphs and line drawings, the addition of numerous impressive photographs and satellite images, and the development of new maps treating subjects such as climate, soils, and landforms. In addition, over fifty previous figures have been replaced by new photographs and new or revised line drawings. Perhaps most challenging of all, the world distribution maps of population, precipitation, climate, and natural vegetation were redrafted to show new boundaries.

Increased Focus on the Discipline. Several changes in the text have been introduced in the sixth edition to provide students with a better understanding and appreciation of geography as a discipline worthy of continued study and meriting serious consideration as a career choice when choosing a career. The definition of geography, the discipline's tools and methodologies, selected themes in physical geography, and the practical applications of the discipline are all introduced in Chapter 1. In addition, the authors believe there are few better ways to understand an academic discipline than to learn firsthand from its practitioners. To that end, the Career Vision series has been included in the sixth edition to provide a close look at geographers in the workplace, how they earn their living, and what educational programs prepared them for their occupations. These brief interviews serve to inform students of the essential knowledge and skills associated with the exciting new career opportunities offered those who choose physical geography as their preferred field of study.

A Look at the Future. The addition of Chapter 23 to the sixth edition has provided authors with more than just a chance to summarize the important knowledge and valuable perspectives regarding our planet gained from a course in physical geography. It also provides students with an opportunity to imagine the future role of the subject in their lives as they make their way through the twenty-first century. No student should put *Essentials of Physical Geography* aside without reading the last chapter.

Ancillaries

Instructors and students alike will greatly benefit from the invaluable ancillary package that accompanies this text.

Study Mate. This student study guide includes specific recommendations focusing its contents as aids to learning. The body of Study Mate is organized to complement the textbook, and for each chapter it gives a chapter outline, a chapter summary, the Define and Recall terminology, Discuss and Review study questions keyed to textbook pages, supplementary reading lists, and a comprehensive pretest that students can use for self-evaluation.

Earth Systems CD-ROM. This interactive CD-ROM's video clips, photos, and animations introduce the study of plate tectonics and surface processes in a way the reader will never forget. By presenting the basics of Earth science in the context of today's technology, it serves as an invaluable aid to instructors and students alike.

Rand-McNally Atlas. This excellent addition to the textbook is available at minimal cost to student purchasers of the sixth edition of *Essentials of Physical Geography.* Its clear, comprehensive maps make it an invaluable tool for students and a practical complement to the text.

Instructor's Resource Manual and Test Bank. The manual is divided into three sections: The first section provides suggestions for using the many other ancillaries that accompany the text; the second section contains sources for additional instructional aids; and the third section includes recommendations for using the text as a major tool in reaching an instructor's individual course goals and objectives. This final section focuses on course flexibility, textbook features, and chapter-by-chapter responses to the critical-thought questions that appear in several places throughout the textbook. A test bank with a number of questions for each chapter is also included in this volume.

Computerized Test Bank. Test banks contain hundreds of multiple-choice, short-answer, and true-and-false test items that may be used for chapter or unit evaluation. The computerized test bank is available in IBM or Macintosh formats.

Lab-Pack. Lab-Pack offers a complete and fully integrated laboratory experience. It includes one or more exercises related to each textbook chapter and over 30 additional learning activities that can help students better understand the important concepts in physical geography. Some of the exercises and learning activities have been modified and included in the text to increase student involvement in the content of physical geography.

Overhead Transparencies and Slides. Over 100 carefully chosen color illustrations and photographs are available for use with overhead projectors or slide projectors for classroom presentation.

Saunders Geography Web Site. This exciting, up-to-date resource provides students and instructors with a number of intriguing opportunities. Web surfers can get the latest news in the field, go on virtual field trips, work on end-of-chapter quizzes, or find out about other geography textbooks published by Saunders College Publishing.

Instructor's Resource CD-ROM. This is a presentation CD-ROM that contains numerous images from this and other texts, as well as animations of earth processes.

Mapmaking Video. A unique 30-minute video, "Mapmaking Today and Tomorrow," focuses on mapmaking history, the work of today's cartographers, and the role maps will play in the future.

Videodisks. Also included in the ancillary package are two Saunders videodisks on Earth sciences and geography. These disks contain hundreds of still images as well as live-action and animated footage on such topics as storms, the origins of oceans, and the Big Bang Theory.

Saunders College Publishing may provide complimentary instructional aids and supplements or supplement packages to those adopters qualified under our adoption policy. Please contact your sales representative for more information. If, as an adopter or potential user, you receive supplements you do not need, please return them to your sales representative or send them to

Attn: Returns Department
Troy Warehouse
465 South Lincoln Drive
Troy, MO 63379

Acknowledgments

This edition of *Essentials of Physical Geography* would not have been possible without the encouragement and assistance of editors, friends, and colleagues from throughout the country.

We are grateful for the excellent work of our two contributing authors, David Butler of Southwest Texas State University and Norman Meek of California State University—San Bernardino, and for the skilled cartography of Scott Miner, Western Illinois University, who revised the world distribution maps developed on the Western Paragraphic Projection.

Special thanks must go to the splendid staff members of Saunders College Publishing. These include John Vondeling, Publisher; Jennifer Bortel, Acquisitions Editor; Erik Fahlgren, Marketing Strategist; Sarah Fitz-Hugh and David Robson, Developmental Editors; Alicia Jackson, Production Manager; Carol Bleistine, Manager of Art and Design; Cara Castiglio, Art Director; Anne Gibby, Senior Project Editor; Ellen Sklar, Associate Project Editor; Kim Menning, Illustration Supervisor; Kathleen Flanagan, Designer; Jane Sanders, Photo Researcher; and Mary Beth Smith, Editorial Assistant.

Listed below are the names of those colleagues who reviewed the plans and manuscript for the sixth edition of this text.

Roberto Garza, San Antonio College

Jeffrey Lee, Texas Tech University

John Lyman, Bakersfield College

Joyce Quinn, California State University, Fresno

Colin Thorn, University of Illinois at Urbana-Champaign

Thomas Wikle, Oklahoma State University, Stillwater

Craig ZumBrunnen, University of Washington, Redmond

Fifth Edition reviewers were:

Brock Brown, Southwest Texas State University

Perry Hardin, Brigham Young University

David Helgren, San Jose State University

Charles Martin, Kansas State University

James R. Powers, Pasadena City College

George A. Schnell, State University of New York, New Paltz

Ann Wyman, University of Nevada, Las Vegas

Our sincere thanks also go to each and every individual who continues to provide comments, assistance, and objective classroom feedback year after year.

D. Robert Altschul, University of Arizona

Sheryl Luzzadder Beach, University of Minnesota

Randall S. Cerverny, Arizona State University

Richard A. Crooker, Kutztown University

Mark Francek, Central Michigan University

Ned Greenwood, San Diego State University

M. Richard Hackett, Oklahoma State University

Kenneth M. Hinkel, University of Cincinnati

Robert B. Howard

The detailed comments and suggestions of all of the above individuals have been instrumental in bringing about the many changes and improvements incorporated in this latest revision of the text. Countless others, both known and unknown, deserve heartfelt thanks for their interest and support over the years.

Despite the painstaking efforts of all reviewers, there will always be questions of content, approach, and opinion associated with the text. The authors wish to make it clear that they accept full responsibility for all that is included in the sixth edition of *Essentials*.

Robert E. Gabler
Robert J. Sager
Daniel L. Wise
James F. Petersen
October 1998

Foreword to the Student

Why Study Geography?

In this global age, the study of geography is absolutely essential to an educated citizenry of a nation whose influence extends throughout the world. Geography deals with location, and a good sense of where things are, especially in relation to other things in the world, is an invaluable asset whether you are traveling, conducting international business, or sitting at home reading the newspaper.

Geography examines the characteristics of all the various places on Earth and their relationships. Most important in this regard, geography provides special insights into the relationships between humans and their environments. If all the world's people had one goal in common, it should be to better understand the physical environment and protect it for the generations to come.

Geography provides essential information about the distribution of things and the interconnections of places. The distribution pattern of Earth's volcanoes, for example, provides an excellent indication of where Earth's great crustal plates come in contact with one another; and the violent thunderstorms that plague Illinois on a given day may be directly associated with the low-pressure system spawned in Texas two days before. Geography, through a study of regions, provides a focus and a level of generalization that allows people to examine and understand the immensely varied characteristics of Earth.

As you will note when reading Chapter 1, there are many approaches to the study of geography. Some courses are regional in nature; they may include an examination of one or all of the world's political, cultural, economic, or physical regions. Some courses are topical or systematic in nature, dealing with human geography, physical geography, or one of the major subfields of the two.

The great advantage to the study of a general course in physical geography is the permanence of the knowledge learned. Although change is constant and is often sudden and dramatic in the human aspects of geography, alterations of the physical environment on a global scale are exceedingly slow when not influenced by human intervention. Theories and explanations may differ, but the broad patterns of atmospheric and oceanic circulation and of world climates, landforms, soils, natural vegetation, and physical landscapes will be the same tomorrow as they are today.

Keys to Successful Study

Good study habits are essential if you are to master science courses such as physical geography, where the topics, explanations, and terminology are often complex and unfamiliar. To help you succeed in the course in which you are currently enrolled, we offer the following suggestions.

Reading Assignments

- Read the assignments before the material contained therein is covered in class by the instructor.

- Compare what you have read with the instructor's presentation in class. Pay particular attention if the instructor introduces new examples or course content not included in the reading assignments.

- Do not be afraid to ask questions in class and seek a full understanding of material that may have been a problem during your first reading of the assignment.

- Reread the assignment as soon after class as possible, concentrating on those areas that were emphasized in class. Highlight only those items or phrases that you *now* consider to be important, and pass over lightly those sections that are already mastered.

- Add to your class notes important terms, your own comments, and summarized information from each reading assignment.

Understanding Vocabulary

Mastery of the basic vocabulary often becomes the critical issue in the success or failure of the student in the beginning science course.

- Focus on the terms that appear in boldface type in your reading assignments. Do not overlook any additional terms that the instructor may introduce in class.

- Develop your own definition of each term or phrase and associate it with other terms in physical geography.

- Identify any physical processes associated with the term. Knowing the process helps to define the term.

- Whenever possible, associate terms with location.

- Consider the significance to humans of terms you are defining. Recognizing the significance of terms and phrases can make them relevant and easier to recall.

Learning Earth Locations

A good knowledge of place names and of the relative locations of physical and cultural phenomena on Earth is fundamental to the study of geography.

- Take personal responsibility for learning locations on Earth. Your instructor may identify important physical features and place names, but you must learn their locations for yourself.

- Thoroughly understand latitude, longitude, and the Earth grid. They are fundamental to location on maps as well as on a globe. Practice locating features by their latitude and longitude until you are entirely comfortable using the system.

- Develop a general knowledge of the world political map. The most common way of expressing the location of physical features is by identifying the political unit (state, country, or region) in which it can be found.

- Make liberal use of outline maps. They are the key to learning the names of states and countries and they can be used to learn the locations of specific physical features. Personally placing features correctly on an outline map is still the best way to learn location.

- Cultivate the atlas habit. The atlas does for the individual who encounters place names or the features they represent what the dictionary does for the individual who encounters a new vocabulary word.

Utilizing Textbook Illustrations

The secret to making good use of maps, diagrams, and photographs lies in understanding why the illustration has been included in the text or incorporated as part of your instructor's presentation.

- Concentrate on the instructor's discussion of slides, overhead transparencies, and illustrations and take notes that will allow you to follow the same line of thought at a later date.

- Study all textbook illustrations on your own and be sure to note which were the focus of considerable classroom attention. Do not quit your examination of an illustration until it makes sense to you, until you can read the map or graph, or until you can recognize what a diagram or photograph has been selected to explain.

- Hand-copy important diagrams and graphs. Few of us are graphic artists, but you might be surprised at how much better you understand a graph or line drawing after you reproduce it yourself.

- Read the captions of photos and illustrations thoroughly and thoughtfully. If the information is included, be certain to note where a photograph was taken and in what way it is representative. What does it tell you about the region or site being illustrated?

- Attempt to place the principle being illustrated in new situations. Seek other opportunities to test your skills at interpreting similar maps, graphs, and photographs and think of other examples that support the text being illustrated.

- Remember that all illustrations are reference tools, particularly tables, graphs, and diagrams. Refer to them as often as you need to.

Taking Class Notes

The password to a good set of class notes is selectivity. You simply cannot and, indeed, you should not try to write down every word uttered by your classroom instructor.

- Learn to paraphrase. With the exception of specific quotations or definitions, put the instructor's ideas, explanations, and comments into your own words. You will understand them better when you read them over at a later time.

- Be succinct. Never use a sentence when a phrase will do, and never use a phrase when a word will do. Start your recall process with your note-taking by forcing yourself to rebuild an image, an explanation, or a concept from a few words.

- Outline where possible to discern the logical organization of information. As you take notes, organize them under main headings and subheadings.

- Take the instructor at his or her word. If the instructor takes the time to make a list, then you should do so too. If he or she writes something on the board, it should be in your notes. If the instructor's voice indicates special concern, take special notes.

- Come to class and take your own notes. Notes trigger the memory, but only if they are *your* notes.

Doing Well on Tests

Follow these important study techniques to make the most of your time and effort preparing for tests.

- Practice distillation. Do not try to reread but skim the assignments carefully, taking notes in your own words that record as economically as possible the important definitions, descriptions, and explanations. Do the same with any supplementary readings, handouts, and laboratory exercises. It takes practice to use this technique, but it is a lot easier to remember a few key phrases that lead to ever increasing amounts of organized information than it is to memorize all of your notes. And the act of distillation in itself is a splendid memory device.

- Combine and reorganize. Merge all your notes into a coherent study outline.

- Become familiar with the type of questions that will be asked. Knowing whether the questions will be objective, short-answer, essay, or related to diagrams and other illustrations can help in your preparation. Some instructors place old tests on file where you can examine them or will forewarn you of their evaluation styles if you inquire. If not, then turn to former students; there are usually some around the department or residence halls who have already experienced the instructor's tests.

- Anticipate the actual questions that will likely be on the test. The really successful students almost seem to be able to predict the test items before they appear. Take your educated guesses and turn them into real questions.

- Try cooperative study. This can best be described as role playing and consists very simply of serving temporarily as the instructor. So go ahead and teach. If you can demonstrate a technique, illustrate an idea, or explain a process or theory to another student so that he or she can understand it, there is little doubt that you can answer test questions over the same material.

- Avoid the "all-nighter." Use the early evening hours the night before the test for a final unhurried review of your study outline. Then get a good night's sleep.

The Importance of Maps

Like graphs, tables, and diagrams, maps are an excellent reference tool. Familiarize yourself with the maps in your textbook in order to better judge when it is appropriate to seek information from these important sources.

Maps are especially useful for comparison purposes and to illustrate relationships or possible associations of things. But the map reader must beware. Only a small portion of the apparent associations of phenomena in space (areal associations) are actually cause-and-effect relationships. In some instances the similarities in distribution are a result of a third factor that has not been mapped. For instance, a map of worldwide volcano distribution is almost exactly congruent with one of incidence of earthquakes, yet volcanoes are not the cause of earthquakes, nor is the obverse true. A third factor, the location of tectonic plate boundaries, explains the first two phenomena.

Finally, remember that the map is the most important statement of the professional geographer. It is useful to all natural and social scientists, engineers, politicians, military planners, road builders, farmers, and countless others, but it is the essential expression of the geographer's primary concern with location, distribution, and spatial interaction.

About Your Textbook

This textbook has been written for you, the student. It has been written so that the text can be read and understood easily. Explanations are as clear, concise, and uncomplicated as possible. Illustrations have been designed to complement the text and to help you visualize the processes, places, and phenomena being discussed. In addition, the authors do not believe it is sufficient to offer you a textbook that simply provides information to pass a course. We urge you to think critically about what you read in the textbook and hear in class.

As you learn about the physical aspects of Earth environments, ask yourself what they mean to you and to your fellow human beings throughout the world. Make an honest attempt to consider how what you are learning in your course relates to the problems and issues of today and tomorrow. Practice using your geographic skills and knowledge in new situations so that you will continue to use them in the years ahead. Your textbook includes several special features that will encourage you to go beyond memorization and reason geographically.

- *Consider and Respond.* At the end of each chapter, Consider and Respond questions require you to go well beyond routine chapter review. The questions are designed specifically so that you may apply your knowledge of physical geography and on occasion personally respond to critical issues in society today. Check with your instructor for answers to the problems.

- *Legend Questions.* With almost every illustration and photo in your textbook a legend (caption) links the image with the chapter text it supports. Read each caption carefully because it explains the illustration and may also contain new information. Wherever appropriate, questions at the ends of captions have been designed to help you seize the opportunity to consider your own personal reaction to the subject under consideration.

- *Map Interpretation Series.* It is a major goal of your textbook to help you become an adept map reader, and the Map Interpretation Series in your text has been designed to help you reach that goal.

- *Environmental Systems Diagrams.* Viewing Earth as a system comprising many subsystems is a fundamental concept in physical geography for researchers and instructors alike. The concept is introduced in Chapter 1 and reappears frequently throughout your textbook. The interrelationships and dependencies among the variables or components of Earth systems are so important that a series of special diagrams (see, for example, Figure 1.4) have been included with the text to help you visualize how the systems work. Each diagram depicts the system and its variables and also demonstrates their interdependence and the movement or exchanges that occur within each system. The diagrams are designed to help you understand how human activity can affect the delicate balance that exists within many Earth systems.

As authors of your textbook, we wish you well in your studies. It is our fond hope that you will become better informed about Earth and its varied environments and that you will enjoy the study of physical geography.

Contents
Overview

Contents

5 Atmospheric Pressure and Wind 117

6 Circulation Patterns 133

7 Moisture, Condensation, and Precipitation 153

8 Air Masses and Atmospheric Disturbances 183

14 Earth's Interior, Earth's Crust, and Plate Tectonics   375

15 Landforms and Tectonic Processes   399

16 Gradation, Weathering, and Mass Movement   423

17 Underground Water and Karst Landforms   445

23 Dynamic Physical Geography 593

List of Major Maps

List of Environmental Systems Series

CHAPTER

1

THE EARTH SYSTEM AND PHYSICAL GEOGRAPHY

CHAPTER PREVIEW

▶ Earth is a dynamic system that provides the exact combination of interrelated components to support life as we know it at this point in time.

In what ways is the Earth system dynamic?
How can changes within the system affect life-forms?
Can humans be responsible for significant changes in the Earth system or Earth subsystems?

▶ The use of models and the analysis of various Earth systems are important research and educational techniques for geographers and other scientists.

How can models help scientists better understand the real world?
In what ways can systems analysis lead to a more quantitative approach to research?
What precautions should be exercised when applying models and systems in the examination of a problem?

▶ The study of ecosystems provides scientists with excellent opportunities to observe how changes in either an organic or an inorganic component of a system can affect change in other components of the system.

Why are physical geographers especially interested in examining physical environments as parts of ecosystems?
In what way is the role of humans in an ecosystem so different from the roles of all other components of the system?

▶ Unlike some other physical sciences, physical geography places a special emphasis on human–environment relationships.

How does the nature of geography explain this emphasis?
Why is such an emphasis so important in the study of the physical sciences today?

▶ Every physical environment offers an array of advantages as well as challenges and potential problems or hazards to the human residents of that location.

How does the physical environment affect the way you live in your region? What adaptations are necessary for humans to live in your area?
What environmental advantages do you enjoy where you live?

▶ As a comprehensive and integrating science, geography is closely related to all of the systematic sciences, both physical and social.

How does geography help to integrate the systematic sciences?
Although geographers study phenomena also studied by scientists from other disciplines. How is the geographic approach different from others?
In what ways does the nature of geography affect opportunities for employment among those who major in the discipline?

▶ A collection of major themes outlines the discipline of physical geography, supported by tools, methods, and techniques.

How are each of these themes related to the concept of geography as a "spatial science"?
Which themes interest you the most?

▲ A space walk outside the orbiting Space Shuttle presents an unrestricted view of Earth. Astronaut Kathryn D. Sullivan has called the experience of orbiting Earth "the ultimate field trip."

Imagine that you are an astronaut like the one pictured above— on a space "walk" from your orbiting space shuttle. From this vantage point you have an unparalleled view of our planet, and you are intrigued by the striking environmental variation on Earth's surface. A part of your training for this mission focused on physical geography, and you learned that the components of our planet can be studied either independently or as they relate to one another. By applying this newly acquired knowledge, you are now a skilled observer, recognizing in detail many planetary features—continents, oceans, forests, deserts, cloud patterns, and storms. You can also see evidence of civilization. Using cameras, specialized imaging systems, and other scientific equipment, you can measure, monitor, map, and analyze characteristics of solid Earth, its atmosphere, its water bodies, and its environmental regions (Fig. 1.1). As an astronaut in orbit on the ultimate field trip, you are in an ideal position to understand and appreciate the Earth system—the physical geography of our planet.

The Earth System

A **system** may be thought of as a set of parts or components that are interrelated. These components, termed **variables,** are studied or grouped together as a system because the variables interact with one another as a functioning unit. Earth certainly fits this definition, because many continuously changing variables combine to make our home, the **Earth system,** function the way that it does. Virtually no part of a system oper-

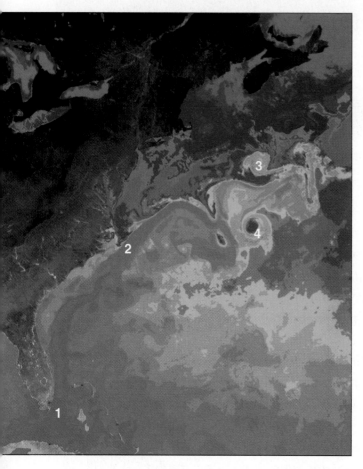

Figure 1.1 Thermal infrared image showing ocean temperatures and the Gulf Stream, a warm ocean current off the east coast of North America. Warm sea surface temperatures appear as reds and oranges (24 to 28°C), with the yellows, greens, and blues representing decreasing surface temperatures. The coldest temperatures appear as dark blue (2 to 9°C). Note that the Gulf Stream appears like a warm river as it moves from the tip of Florida (1) up the east coast. As it moves offshore near Cape Hatteras (2), it begins to form looping meander patterns. Some of these meanders pinch off to form warm-core eddies (3) and cold-core eddies (4). • **In which general direction is the Gulf Stream flowing?**

ates in isolation from another. A change in one part of the Earth system affects other parts, and the impact of these changes can be significant enough to appear in worldwide patterns, clearly demonstrating the interconnections among these variables. For example, the location and position of mountains influence the global distribution of rainfall, and variation in rainfall affects the density, type, and variety of vegetation. Plants, moisture, and the underlying rock are factors that affect the kind of soil that forms in an area. The characteristics of vegetation and soils influence the runoff of water from the land, leading to completion of the circle, since the amount of runoff is a major factor in stream erosion, which eventually can reduce the height of mountains. Many cycles such as this operate to change our planet daily, but the environment of Earth is complex, and these cycles and processes operate at widely varying rates.

We do know that the Earth system is dynamic (ever-changing), and that we can directly observe some of these changes—the four seasons, the ocean tides, earthquakes, floods, volcanic eruptions. Other aspects of our planet may take years, or even more than a lifetime, to accumulate enough total change to be directly noticed by humans. Such long-term changes in our planet are often difficult to understand or predict with certainty, so they must be carefully and scientifically studied in order to determine what is really happening and what the consequences might be. Changes of this type include shifts in world climates, drought cycles, the spread of deserts, worldwide rise or fall in sea level, erosion of coastlines, and major changes in river systems. Yet understanding changes in our planet is critical to human existence. We are, after all, a part of the Earth system; changes in the system may be naturally caused or human-induced, or result from a combination of these factors. To understand our planet, therefore, we must learn about its components and the processes that operate to change or regulate the Earth system. Such knowledge is in the best interest of not only humankind but also Earth, as a habitat for all living things.

Major Divisions

Four major divisions of our planet comprise the Earth system. The **atmosphere** is the gaseous blanket of air that envelops, shields, and insulates Earth. The movements and processes of the atmosphere create the changing conditions that we know as weather and climate. The solid Earth—landforms, rocks, soils, and minerals—

makes up the **lithosphere.** The waters of the Earth system—oceans, lakes, rivers, and glaciers—constitute the **hydrosphere.** The fourth major division, the **biosphere,** is composed of all living things: people, other animals, and plants. It is the nature of these four major subsystems and the interactions among them that create and nurture the conditions necessary for life on Earth (Fig. 1.2).

For example, the hydrosphere serves as the water supply for all life, including humans, and provides a home environment for many types

Figure 1.2 Environmental Systems: Earth's Major Subsystems. The study of Earth as a system is central to understanding changes in our planet's environments and adjusting to or dealing with these changes as they occur. A system as large and complex as Earth consists of many interconnected subsystems that lend themselves more readily to study through the use of computers, theoretical models, and mathematical computation. The Environmental Systems series of diagrams found throughout your textbook will focus on examples of such subsystems. Each example involves one or more of the four overlapping major Earth subsystems illustrated in this diagram: the atmosphere, hydrosphere, lithosphere, and biosphere. • How do each of these systems overlap? **For example, how does the atmosphere overlap with the hydrosphere, or the biosphere?**

of aquatic plants and animals. The hydrosphere directly affects the lithosphere as the moving water in streams, waves, and currents shapes landforms. It also influences the atmosphere through evaporation, condensation, and the effects of ocean temperatures on climate. The impact or intensity of interactions among Earth's subsystems is not identical everywhere on the planet, and it is this variation that leads to the patterns of environmental diversity that intrigued our orbiting astronaut.

Many other examples of overlap exist among the four divisions. Soil can be examined as part of either the biosphere, the hydrosphere, or the lithosphere. The water stored in plants and animals is part of both the biosphere and the hydrosphere, and the water in clouds is a component of the atmosphere as well as the hydrosphere. The fact that we cannot draw sharp boundaries between these divisions underscores the *interrelatedness* of the various parts of the Earth system. However, like a machine, a computer, or the human body, planet Earth is a system that functions well only when all of its parts (and its subsystems) work together harmoniously.

A Life-Support System

Certainly the most important attribute of Earth is that it is a **life-support system.** Like the space vehicle that supports the astronaut in orbit, the Earth system provides the necessary environmental constituents and conditions to permit life as we know it to exist. If a critical part of a life-support system is significantly changed or fails to operate properly, living organisms may no longer be able to survive. For instance, if all the oxygen in a spacecraft is used up, the crew inside will die. If there is no way to keep the spacecraft within the proper temperature range, its occupants may burn or freeze. If food supplies run out, the astronauts will starve. Similarly, on Earth, natural processes must provide an adequate supply of oxygen, the sun must interact with the atmosphere, oceans, and land to maintain tolerable temperatures, and photosynthesis or other continuous cycles of creation must provide new food supplies for living things.

Earth, then, is made up of a set of interrelated components that are vital and necessary for the existence of all living creatures. About 30 years ago, Buckminster Fuller, a distinguished scientist, philosopher, and inventor, coined the notion of Spaceship Earth—the idea that our planet is a life-support system, transporting us through space. Fuller also thought that knowing how Earth works is important, but that humans are only

slowly learning the processes involved, although this knowledge may be required for human survival. He compared this information to an operating manual, like the owner's manual for an automobile.

One of the most interesting things to me about our spaceship is that it is a mechanical vehicle, just as is an automobile. If you own an automobile, you realize that you must put oil and gas into it, and you must put water in the radiator and take care of the car as a whole. You know that you are going to have to keep the machine in good order or it's going to be in trouble and fail to function.

We have not been seeing our Spaceship Earth as an integrally-designed machine which to be persistently successful must be comprehended and serviced in total. . . . there is one outstandingly important fact regarding Spaceship Earth, and that is that no instruction book came with it.

Operating Manual for Spaceship Earth
R. Buckminster Fuller

Today, as we pass from one century to another, we realize that critical parts of our life-support system, the **natural resources,** can be abused, wasted, or exhausted, creating a potential threat to the functioning of planet Earth as a human life-support system. One such abuse is **pollution,** an undesirable or unhealthy concentration of a contaminant in an environment as a result of human activities (Fig. 1.3). We are aware that some of Earth's resources, such as air and water, can be polluted to the point where they become unusable or even lethal to some life-forms. By polluting the oceans, we may be killing off important fish species, allowing less desirable species to increase in number. Acid rain, caused by the release into the atmosphere of pollutants from industries and power plants, is damaging forests and killing fish in freshwater lakes. Air pollution has become a serious environmental problem for urban centers throughout the world (Fig. 1.4).

Because pollution is associated with human activity, it is not surprising that it represents a significant problem in places that have high population densities. What some people do not realize, however, is that pollutants are often transported by winds and waterways hundreds or perhaps thousands of kilometers from their source. Lead from automobile exhausts has been found even in the snow of Antarctica, as has the insecticide DDT. Pollution is a worldwide prob-

Figure 1.3 *Toxic chemicals, such as the ones discovered in this solid-waste dump, pose a serious health hazard and threaten local water supplies.* • **What pollutants form the major threat to the air and water supply in your community?**

lem that does not stop at political boundaries (Fig. 1.5).

Another concern is that humans may be rapidly depleting critical natural resources, especially those needed for fuel. Many natural resources on our planet are nonrenewable, meaning that nature will not replace them once they are exhausted. Coal and oil are nonrenewable resources. Although we probably have enough coal to last several hundred years, we have often been warned about future shortages in our petroleum supplies. When nonrenewable resources such as these mineral fuels are gone, the alternative resources may be less desirable or more expensive.

We are learning that, much like living on a spaceship, there are limits to the amount of suitable living space on Earth, and we must use it wisely. In our search for livable space, we occasionally construct buildings in locations that are not environmentally safe, or that are already overcrowded. Also, we sometimes plant crops in areas that are ill-suited to agriculture while at the same time prime farmland is being paved over for other uses.

Our space exploration programs are helping us learn more and more about the world in which we live. With the use of spacecraft-mounted cameras and sophisticated imaging systems, we can observe, and monitor over time, the changes on Earth's surface that are the result of human activity. All citizens of Earth must understand the impact of their actions on the complex Earth system. It is to the science of physical geography that we can turn to learn the consequences of these actions.

(a)

(b)

Figure 1.4 *(a) On a clear day it is not difficult to understand why Los Angeles, California, was named "The City of Angels." (b) But far too often this great metropolis justifies its designation as "Smog Capital" of the United States.* • **Would pollution affect your decision if you were to choose whether to live in a small town or a major city?**

(a)

(b)

Figure 1.5 *Pollution knows no political borders. Air pollution from distant coal-fired power plants, blown by winds over the southwestern United States, not only is a potential hazard, but it also has other impacts on the natural environment. A clear day in the Grand Canyon (**a**). Decreasing air quality has begun to affect the scenic beauty of Grand Canyon National Park, causing the air to be hazy and less transparent (**b**). As these photographs illustrate, both the visibility and clarity of the scenic landscape are being increasingly obscured by air pollution. Similar problems have been noted in other U.S. national parks.* • **How might air pollution affect the experience of visitors in their visit to a national park? Could it also affect the natural environment in the park?**

Models and Systems

Models

Physical geographers work to describe, understand, and explain the often complex features of planet Earth. To support these efforts they, like other scientists, develop representations of the real world called **models.** *A model is a useful simplification of a more complex reality that permits prediction, and each model is designed with a specific purpose in mind.* As examples, maps and globes are models—representations of Earth that provide us with useful information required to meet specific needs. But models are greatly simplified versions of what they depict, conveying the most important information about a feature or process without too much (and often confusing) detail. Models are essential to understanding and predicting the way that nature operates and they vary greatly in their levels of complexity. Today many models are generated by computers, because computers can handle great amounts of data and perform the mathematical calculations that are often necessary to construct and display certain types of models.

There are many kinds of models: **physical models** (three-dimensional models, such as globes or a representation of a mountain), **pictorial/graphic models** (photographs, maps, images, graphs, diagrams, drawings), and **mathematical or statistical models** (which are used to predict possibilities, such as the future flooding of major rivers, or to simulate changes in weather conditions due to a rise in atmospheric temperatures). Words, language, and the definitions of terms or ideas can also serve as models. Another important type of model is a **conceptual model,** or the mind imagery that we use for understanding our surroundings and experiences (Fig. 1.6).

Imagine for a minute (perhaps with your eyes closed) the image that the word "mountain" (or waterfall, cloud, tornado, beach, forest, desert) generates in your mind. Can you describe this feature and its characteristics in detail? Most likely what you "see" (conceptualize) in your mind is sketchy rather than detailed, but enough information is there to convey a mental idea of a mountain. This image is a conceptual model.

For geographers, a particularly important type of conceptual model is the **mental map,** which we use to think about places, travel routes, and the distribution of features in space. Psychologists have shown in many studies that such maps are very efficient media for conveying a great amount of **spatial** information that the brain can recognize, store, and access in an economical fashion. (Spatial [pronounced *spay-shul*] refers to the arrangement and location of features in space.) Try to think of other varieties of conceptual models that represent our

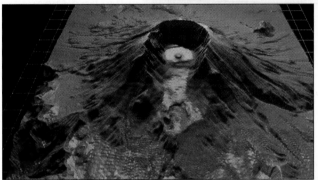

Figure 1.6 Examples of Models Used to Study Earth. *Geographers use many kinds of models to understand the Earth system, its subsystems, and how they operate, by simplifying the presentation of a complex reality. Models help us to focus attention on major features or processes, without unnecessary and distracting detail.* **(top)** *Globes are models, simplified presentations of Earth, that demonstrate many terrestrial characteristics—planetary shape, configuration of landmasses and oceans, their distributions, and spatial relationships. This globe shows sea ice over the north pole, glaciers, lakes, rivers, and shaded relief.* **(center)** *Digital landscape models. A computer-generated terrain model shows Mount St. Helens in Washington state, using digital elevation data (height measurements of the land surface taken at regular intervals). This model was rendered in shaded relief to produce a realistic looking landscape, and was produced by an undergraduate student in Oregon, using free, Internet-accessible data and software. The model to the right, of Mammoth, California in the Sierra Nevada, was produced by NASA and combines digital elevation data with a radar image.* **(bottom)** *This working model of the Kissimmee River in Florida was constructed to investigate ways to restore the Kissimmee River, Florida. Proposed modifications could be analyzed on this model before work was done on the actual river (see the discussion in Figure 1.12). Similar models exist of the Mississippi River and San Francisco Bay.*

planet's environment. How could we even begin to understand our world without conceptual models?

Systems Theory

But what if you try to think about Earth in its entirety, or to understand how even a *part* of the Earth system works? There are just too many factors to envision without resorting to excessive oversimplification. Our planet is too complex to permit a single model to represent its entire workings or to explain all of its environmental components and how they affect one another. Yet it is often said that in order to be environmentally responsible citizens of Earth, we should "think globally, but act locally." In order to begin to comprehend Earth as a whole, or to understand most of its environmental components, physical geographers use a powerful strategy for analysis called **systems theory.** Systems theory suggests that the way to understand how anything works is to use the following strategy:

1. **Clearly define the system that you are studying.**
   What are the boundaries (limits) of the system?

2. **Break the defined system down into its component parts (variables). The variables in a system are either matter or energy.**
   What are the important parts and processes that are involved in this system?

3. **Attempt to understand how these variables are related to (or affect, react with, or impact) one another.**
   How do the parts interact with each other in order to make the system work, and what will happen in the system if a part changes?

With environmental systems, an important question that we continually try to answer is how much change can a system tolerate without becoming drastically or irreversibly altered, particularly if the change is for the worse.

The systems approach is a beneficial tool for studying any level of environmental condition on Earth, from global to microscopic. Systems can be divided into **subsystems,** or units that demonstrate strong internal connections. For example, the Earth system consists of the atmosphere, hydrosphere, lithosphere, and biosphere, each a subsystem of the whole. The human body is a system that is composed of many subsystems (for example, the respiratory system, circulatory system, and digestive system). Subsystems can also be divided into subsystems, and so on.

Geographers often divide the Earth system into smaller subsystems in order to focus their attention on a particular part of the whole. Examples of subsystems examined by physical geographers include the atmospheric water cycle, climatic systems, storm systems, stream systems, the systematic heating of the atmosphere, and ecosystems. A great advantage of systems analysis is that it can be applied to environments at virtually any scale.

How Systems Work

Basically, the world "works" by the movement (or transfer) of matter and energy and the processes atttending these transfers. For example, as shown in Figure 1.7, **sunlight** (*energy*) **warms** (*process*) **a body of water** (*matter*) **and the water evaporates** (*process*) **into the atmosphere. Later, the water condenses** (*process*) **back into a liquid and the rain** (*matter*) **falls** (*process*) **on the land and runs off** (*process*) **downslope back to the sea.** The processes mentioned here each involve the movement of both matter and energy.

In a systems model, geographers can trace the movement of energy or matter into the system (**inputs**) and out of the system (**outputs**), as well as the interactions between energy and matter (**feedback**) within the system (Fig. 1.8). The subsystem involved in the heating of the atmosphere is an example of an **open system,** *because energy and matter are not confined to the boundaries of the system, but instead are constantly entering and leaving.*

A **closed system** *is one in which no substantial amount of matter crosses its boundaries, although energy can go in and out of a closed system.* Planet Earth, or the Earth system as a whole, is essentially a closed system, much like the spacecraft that provides the home for our orbiting astronaut. This shared characteristic is another reason why our planet has been referred to as Spaceship Earth: Except for meteorites that reach Earth's surface, the escape of gas molecules or spacecraft from the atmosphere, and a few Moon rocks brought back by astronauts, the Earth system is essentially closed to the movement of material. The subsystem that involves the movement of water from the ocean to the atmosphere to the land and back to the ocean again is another good example of a closed system. Water may exist in the system in all three of its states—as liquid, gas, or solid ice—and may be transformed from one state to another many times, but there is no gain or loss of water (no output of matter) in the system.

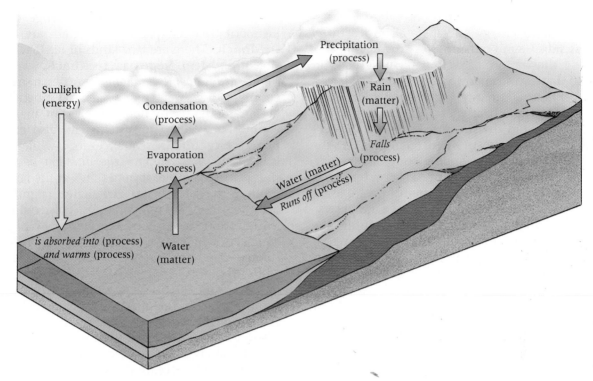

Figure 1.7 Simplified Example of Environmental Interactions—Energy, Matter, Process. *Being aware of energy and matter, and the interactive processes that link them, is an important part of understanding how environmental systems operate.* • **Can you think of another environmental system, and break it down into its components of energy, matter, and process?**

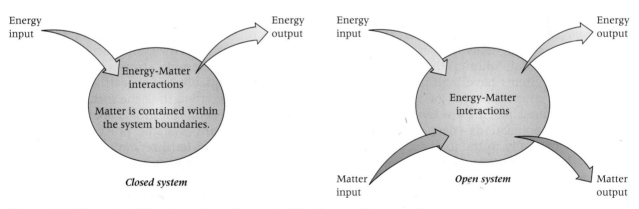

Figure 1.8 Diagram of Closed vs. Open Systems. *This diagram illustrates the difference between closed systems that allow only the passage of energy in and out of the system, and open systems that involve the inputs and outputs of both energy and matter. Earth is a good approximation of a closed system: solar energy enters the Earth system, and that energy is mainly dissipated (output) to space as heat. Energy interacts with matter in the Earth system, but external inputs of matter are virtually nil, mainly consisting of meteorites and moon rock samples. Except for outgoing space vehicles, equipment, or space "junk," virtually no matter is output from the Earth system. Because Earth is a closed system, humans and other living things on the planet have limits to their available natural resources.*

Although the Earth system as a whole operates (virtually) as a closed system, most subsystems on the planet are best described as being open systems, involving the flows (exchanges) of matter and energy into and out of the subsystems. • **The environment of a lake provides a good example of an open system. Can you outline some of the matter/energy inputs and outputs involved in such a system?**

Most Earth *subsystems,* however, are *open systems,* as both energy and material move freely across subsystem boundaries. A stream is an excellent illustration of an open subsystem, in which matter and energy in the form of soil, rock fragments, solar energy, and precipitation enter the stream while heat energy dissipates into the atmosphere and the stream bed. Water and sediments leave the stream where it empties into the ocean or some other standing body of water.

When we describe Earth as a system or a complex set of interrelated systems, we are using models to help us organize what we are observing. Models also assist us in explaining the processes involved in changing, maintaining, or regulating our planet's life-support systems. Throughout the chapters that follow we will use the concept of the Earth system as a model, as well as many kinds of other models, to help us simplify a complex planet so that we can focus on the most important elements in the physical environment.

Equilibrium in Earth Systems

The parts, or *variables,* of a system have a tendency to reach a balance with one another and with the factors that influence the system from outside its boundaries. If the amount of inputs entering the system is balanced by its outputs, the system is said to have reached a state of **equilibrium.** Regarding environmental systems, we often hear this called the **"balance of nature,"** and most natural systems have a tendency toward stability (equilibrium). What this means is that natural systems have built-in mechanisms that tend to counterbalance, or accommodate, change without changing the system dramatically. Animal populations—deer, for example—will adjust naturally to the food supply of their habitats. If the vegetation on which they browse is sparse in one particular year from drought, fire, overpopulation, or human impacts, deer may starve, reducing the population. The smaller deer population may mean that the vegetation can recover, and in the next season the deer will increase in numbers. Most systems are continually shifting slightly one way or another as a reaction to external conditions. This change within a range of tolerance is called **dynamic equilibrium;** that is, a balance exists, but maintaining it requires adjustment of the parts to changing conditions, much the way tightrope walkers sway back and forth and move their hands up and down to keep their balance. Dynamic equilibrium means that the *balance is not static.*

The interactions that cause change or adjustment between parts of a system are called **feedback.** There are two kinds of feedback possible in a system. **Negative feedback,** *where one change tends to offset another,* creates a natural counteracting effect that is generally beneficial because it tends to help the system maintain equilibrium. Earth subsystems can also exhibit **positive feedback** sequences for a while—that is, *changes that reinforce the direction of initial change.* For example, several times in the last two million years, Earth has experienced significant decreases in global temperatures. This cooling of the atmospheric system led to the growth of great ice sheets, which covered large portions of Earth's surface. The massive ice sheets increased the amount of solar energy that was reflected back to space from Earth's surface, thus increasing the cooling trend in the atmosphere and the further growth of the ice sheets. The result over a considerable period of time was positive feedback. But ultimately the climate got so cold that evaporation from the oceans decreased, cutting off the supply of moisture to storms that fed snow to the glaciers. The reduction of moisture is an example of what is called a **threshold,** a condition that causes a system to change dramatically, and in this case bring the positive feedback to a halt. The decrease in snowfall caused the glaciers to shrink and the climate to warm, thus beginning another cycle.

Thresholds are trigger mechanisms, or conditions that, if met or exceeded, can cause a fundamental change in the system and the way that it behaves. For example, stresses may build along an earthquake fault, but an earthquake will not occur until the strength of the rocks to resist that stress is exceeded. Thresholds are also common regulators of any systems process. As another example, putting fertilizer on a plant will help it to grow larger and faster. But will more and more fertilizer continue this positive feedback relationship forever? Too much fertilizer may actually poison the plant and cause it to die. Either exceeding or not meeting certain critical conditions (thresholds) changes the system dramatically.

To further illustrate how feedback works, let us consider a simplified example, a hypothetical scenario of what might happen if human-caused damage to the atmosphere's ozone layer continues unimpeded by human counteraction (Fig. 1.9). Figure 1.10 shows what is called a **feedback loop,** or a circular set of feedback operations that can be repeated as a cycle. Generally in nature, the overall result of a feedback loop is negative feed-

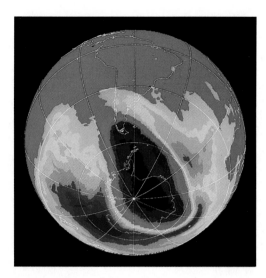

Figure 1.9 Satellite Image of the Ozone "Hole" Over Antarctica. *Concern about changes in the ozone layer led to a cooperative agreement between the United States and the former Soviet Union to launch a satellite with a special imaging apparatus especially designed to send back images of ozone levels in the upper atmosphere. The system, called TOMS (Total Ozone Mapping Spectrometer), beams data back to Earth, from which this kind of image is constructed through computer analysis and display. The ozone hole is shown in shades of purple, and its pattern changes both in size and position in response to wind patterns that circle the South Polar region, as well as other meteorological factors.* • **Is it important for scientists to study and monitor changes in the ozone layer? Why or why not?**

back, because the sequence of changes serves to counteract the direction of change in the initial element. The example is intended to show you how to think about Earth processes as a *system*. After you follow the steps (feedback relationships) in Figure 1.10, try to think of some other examples of feedback operations in natural systems.

Let us look at our example of a feedback loop and examine how the factors are related. First we must start with some facts.

1. We know that the ozone layer in the upper atmosphere protects us by blocking harmful ultraviolet (UV) radiation from space, radiation that could otherwise cause harmful skin cancers and cell mutations.
2. We also know that chlorofluorocarbons (CFCs), chemicals widely used in air conditioners (as freon), can migrate to the upper atmosphere and cause chemical reactions that destroy ozone.

(Knowing these facts, keep in mind that the following systems example is simplified, and like all models, is based upon assumptions which may or may not be scientifically verified. In fact, efforts have been undertaken in the last 20 years or so to minimize the use of CFCs in the United States. Today, new automobiles and trucks are sold with air conditioners that use an "ozone-friendly," non-CFC unit to cool the vehicles' interiors.)

The diagram in Figure 1.10 shows six of the most important factors related to ozone layer damage by CFCs. Each of these factors is linked by an interaction to the next variable in the loop. Systems analysis allows us to see how these processes will affect the variables and helps us to answer "What if?" questions. For example, if CFCs continue to erode the ozone layer, what would happen?

Follow Figure 1.10 starting with the human use of CFCs in the upper left and trace the feedback links outlined below.

1. If the amount of CFCs used by humans *increases,* the amount of CFCs in the atmosphere will also *increase.* (*An increase leads to an increase in the next factor, so this is a direct [positive] relationship.*)
2. *Increasing* the CFCs in the atmosphere will lead to a *decrease* of ozone in the ozone layer. (*Here an increase leads to a decrease in the next factor, so this is an inverse [negative] relationship between atmospheric CFCs and ozone.*)
3. *Decreasing* the ozone in the upper atmosphere will *decrease* the amount of harmful ultraviolet (UV) radiation that is blocked by the ozone layer. (*Here a decrease leads to a decrease, and this is a direct relationship because the decreasing effect is reinforced.*)
4. *Decreasing* the blocking of harmful UV radiation will cause an *increased* amount of harmful ultraviolet radiation at Earth's surface. (*A decrease leads to an increase, so this is an inverse relationship.*)
5. *Increasing* the level of UV radiation at Earth's surface will cause an *increased* amount of skin cancer in humans, which can be fatal. (*An increase leads to an increase, so this is a direct relationship.*)
6. *Increasing* skin cancer in humans could cause a decrease in the release of CFCs into the atmosphere, producing negative feedback relative to the initial variable in the feedback loop.

Finally, there remains an important question: What *is* likely to happen to the human use of

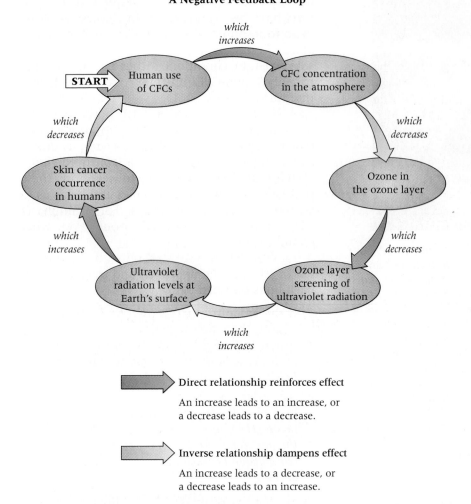

**Nature's Controlling Mechanism—
A Negative Feedback Loop**

Direct relationship reinforces effect

An increase leads to an increase, or
a decrease leads to a decrease.

Inverse relationship dampens effect

An increase leads to a decrease, or
a decrease leads to an increase.

Figure 1.10 Nature's Controlling Mechanism—A Negative Feedback Loop. This diagram of a feedback loop illustrates how negative feedback acts to moderate change and add stability to a system. Relationships between two interacting variables (one link to the next in the loop) can be either direct or inverse. A direct relationship means that either an increase or decrease in the first variable will lead to the same effect on the next. For example, a decrease in ozone leads to a decrease in ozone layer screening of ultraviolet radiation. An inverse relationship means that the change in the first variable will result in an opposite change in the next. For example, an increase in CFCs leads to a decrease in ozone in the ozone layer. After completing one pass through the negative feedback loop, an important shift will occur in how the first variable is affected and this will reverse all of the subsequent changes in the next cycle (although variables will maintain the same relationship to each other, either direct or inverse). You can follow a second pass through the feedback loop (reversing the increase/decrease interactions) to understand how this works. Now, decreasing human use of CFCs leads to a decrease in CFC concentration in the atmosphere, which scientists believe should lead to an increase in ozone. Again, note that the changes have reversed their effect, but the relationship between variables remains the same (as shown by black or gray arrows). This feedback loop shows that human decision making can play a role in environmental systems. The last link in the loop between skin cancer and human use of CFCs would likely be a result of people acting on their own to reduce the problem. • What might be the potential (and extreme) alternative resulting from a lack of corrective action by humans?

CFCs if the occurrence of skin cancer continues to increase? Will humans act to correct the problem, or will they be likely to do nothing? What would be the potential consequences in either case? Ironically, *negative* feedback loop operations are good because they regulate a system through a strong tendency toward balance. As we have seen in Figure 1.10, feedback loops in nature normally do not operate for extended periods on positive feedback because environmental limiting factors act to return the process to a state of equilibrium.

There are both advantages and disadvantages in viewing Earth as a vast system, made up of countless subsystems. As we have mentioned, systems are models, and like all models they are not the same as reality. They are products of the human mind and are only a way of looking at the real world. The examination of various Earth subsystems helps us to understand the natural processes involved in the development of the atmosphere, lithosphere, hydrosphere, and biosphere as we know them today. Models may even help us to simulate past events or predict future change. But we must be careful not to confuse simplified models with the complexities of the real world.

Humans and the Environment

Environment—Ecology—Ecosystem

One of the great advantages of considering subsystems when studying physical geography is that they serve to illustrate the relationships among the various elements within a system. As scientists, geographers are keenly interested in relationships, and they pay special attention to the relationships between humans and the physical environment. In no way can the human–environment relationship be better illustrated than through the examination of human impact on ecosystems. In the last half of the twentieth century, people have become increasingly conscious of their environment. We talk about the environment and ecology and worry about ecological damage caused by human activity. Newspapers and news magazines often devote entire sections to discussions of environmental issues. But what are we really talking about when we use words like *environment, ecology,* or *ecosystem*?

In the broadest sense, our **environment** can be defined as our surroundings; it is made up of all the physical, social, and cultural aspects of our world that affect our growth, our health, and

our way of living (Fig. 1.11). Humans also share environments with plants and animals. We can speak of a plant's environment and include in our discussion the soil in which the plant grows, the amount of sunlight and rainfall it receives, the gases that surround it, the range of air temperatures, and the nature of the plants that grow nearby, which also serve to block winds, sunlight, and rain.

Just as humans interact with their environment, so do other animals and plants. The study of relationships between organisms, whether animal or plant, and their environments is a science known as **ecology.** Ecological relationships are complex but naturally balanced "webs of life." Disrupting the natural ecology of a community of organisms may have negative results (although this is not always so). For example, filling in or polluting coastal marshlands may disrupt the natural ecology of such areas. As a result, fish spawning grounds may be destroyed and the food supply of some marine animals and migratory birds could be greatly depleted. The end product is the destruction of valuable plant and animal life.

The word **ecosystem** is a contraction of **ecological system.** An ecosystem is a community of organisms and the relationships of those organisms to their environment. An ecosystem is dynamic in that its various parts are always in flux. For instance, plants grow, rain falls, animals eat, and soil matures—all changing the environment of a particular ecosystem. Since each member of the ecosystem belongs to the environment of every other part of that system, a change in one alters the environment for all the others. As those components react to the alteration, they in turn continue to transform the environment for the others. A change in the weather, from sunshine to rain, affects plants, soils, and animals. Heavy rain may carry away soils and plant nutrients so that plants may not be able to grow as well and animals then may not be able to eat as much. On the other hand, the addition of moisture to the soil may help some plants grow, increasing the amount of shade beneath them and thus keeping other plants from growing.

The ecosystem concept (like other systems models) can be applied on almost any scale, in a wide variety of geographic locations, and under all environmental conditions in which life is possible. Hence, your backyard, a farm pond, a grass-covered field, a marsh, a forest, or a portion of a desert can be viewed as an ecosystem. Ecosystems are found wherever there is an exchange of ma-

Figure 1.11 *The physical and cultural attributes of a site combine to form a unique environment. The humans who occupy the site, such as these boat dwellers of Aberdeen, Hong Kong, both are influenced by the environment and are an integral part of it.* • **What can you learn about human–environment relationships by studying this photograph?**

terials among living organisms and where there are functional relationships between the organisms and their natural surroundings. Ecosystems are open systems, as both energy and material move across their boundaries. Although some ecosystems, such as a small lake or a desert oasis, have clear-cut boundaries, the limits of many others are not as precisely defined. Often the change from one ecosystem to another is obscure and transitional, occurring slowly over distance.

Ever since human beings first walked Earth, they have affected each ecosystem they have inhabited, and in modern times, humans' ability to alter the landscape has been increasing. For example, a century ago the interconnected Kissimmee River–Lake Okeechobee–Everglades ecosystem constituted one of the most productive and stable wetland regions on Earth. But sawgrass marsh and slow-moving water stood in the way of urban and agricultural development. Intricate systems of ditches and canals were built, and, since 1900, half of the original 4 million acres of the Everglades has disappeared (Fig.

1.12). The Kissimmee River has been channelized into an arrow-straight ditch, and wetlands along the river have been drained. Levees have prevented water in Lake Okeechobee from contributing sheet flow to the Everglades, and highway construction has divided the region, further disrupting natural drainage patterns. Fires have been more frequent and destructive, and entire biotic communities have been eliminated by lowered water levels. During excessively wet periods, portions of the Everglades are deliberately flooded to prevent drainage canals from overflowing. As a result, animals drown and birds cannot rest and reproduce. South Florida's wading bird population has decreased by 95 percent in the last hundred years. Without the natural purifying effects of wetland systems, water quality in South Florida has deteriorated, and with lower water levels, saltwater encroachment is a serious problem in coastal areas.

Today, backed by regional, state, and federal agencies, scientists are struggling to restore South Florida's ailing ecosystems. There are extensive

(text continues on page 18)

(a)

(b)

(c)

Figure 1.12 The mixed tree and grass vegetation characteristics of the Florida Everglades are shown in *(a)*. Large areas of this valuable ecosystem have been lost to farmland, industry, and housing developments. Drainage ditches and highways have altered the remainder at the expense of plants, animals, and human water supplies.

As a natural stream channel, the Kissimmee River meandered (flowed in broad, sweeping bends) on its flood plain for a 100-mile stretch between Lake Kissimmee downstream to Lake Okeechobee *(b)*. In the 1960s and early 1970s, the river was "channelized" (artificially straightened), disrupting the previously existing ecosystem. The artificially straightened channel is shown in *(c)*. Today, as part of an ongoing project to restore this riparian (river bank) habitat, the Kissimmee is reestablishing its flood plain, associated wetland environments, and its meandering channel. These changes are intended to restore, as much as is possible, the natural environment of the Kissimmee River and its flood-plain habitat. • **What factors should be considered prior to any attempts to return rivers and riparian habitats to their original condition?**

15

HUMAN–ENVIRONMENT RELATIONSHIPS

Some geographers consider the relationships or interactions between humans and the environments they occupy to be the central theme of geography. Whether or not this is the case, much of human geography throughout time has been a product of the adjustments various cultures have made to, and the modifications they have imposed on, their natural surroundings. The lower the level and the more primitive the skills and technology of the culture, the greater the adjustments that people have made to the environment. The more sophisticated the culture's technology, the greater the amount of environmental modification. Thus, human–environment interaction always was, and always will be, a two-way relationship, with the environment influencing human behavior and humans impacting on the environment.

Throughout much of history little thought has been given to the human–environment relationship, but two negative aspects of that relationship have recently gained serious attention. **Certain environmental processes, with little or no warning, can become dangerous to human life and property, and certain human activity threatens to cause major, and possibly irrevocable, damage to earth environments.** This environmental essay will serve to introduce both the nature of environmental hazards and the issues involving negative human impact on the environment.

Environmental Hazards

Catastrophes involving various aspects of the physical environment were merely unfortunate footnotes to human history until modern methods of communication such as radio, film, and television showed them

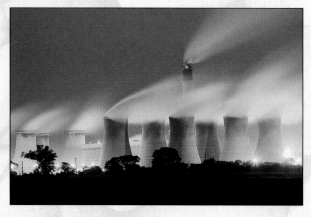

The Drax coal-fired power station near Selby, North Yorkshire, England; graphic evidence of atmospheric pollution.

to be real and ever-present dangers. In the past they were accepted as inevitable and unavoidable. Today, however, natural disasters and the uncertain fate of those living in hazard zones have become major concerns of both natural and social scientists.

Although all four of the major earth subsystems inevitably become involved, the environment becomes a hazard to humans and other life forms because of relatively uncommon and extraordinary events associated most directly with the atmosphere and the lithosphere. Phenomena such as excessive rainfall, drought, and violent storms (such as tornados and hurricanes) all stem from natural continuing processes that function within the atmosphere in an ordinary and understandable fashion. These processes are discussed in detail in Chapters 3 to 8, which follow in the text. Similarly, earthquakes, volcanic eruptions, mud slides, flooding rivers, and eroding shorelines are associated with normal processes that constantly alter the structure, shape, and appearance of the lithosphere. These processes are discussed in detail in Chapters 14 to 22 of the text.

The routine processes of the atmospheric and lithospheric subsystems become a problem and spawn environmental hazards for two reasons. First, on occasion and often unpredictably, they operate in excessive or violent fashion. Summer showers become torrential rains that repeatedly occur for days or even weeks. Ordinary tropical storms gain momentum as they travel over warm ocean waters and they reach coastlines as full-blown hurricanes. Molten rock and associated gases from deep beneath the earth move slowly toward the surface and suddenly trigger massive eruptions that literally blow apart volcanic mountains. Each of these examples of Earth systems operating in sudden or excessive fashion is a noteworthy environmental event, but it does not become an environmental hazard unless people or their property is affected. Thus, the second reason environmental hazards exist is because people live where catastrophic environmental events occur. The greater the number of people and the greater the value of the property involved, the greater the catastrophe.

Why do people live where environmental hazards pose a major threat? Why do they build their barns and farm houses on the lowlands bordering the Mississippi River under the constant threat of flood? Why do they return within weeks to villages destroyed by volcanic eruption on the slopes of Mt. Etna in Sicily? Why build million-dollar homes in the hills above Los Angeles despite the constant dangers from brush fire and mudflow? **There are many reasons why people continue to live in areas subject to natural hazards.** Some have no choice. The land they live on is their land by birthright. It was the land of their fathers and grandfathers. Especially in most densely populated, developing nations of the world, there is no other place to go, except to the overcrowded cities where even greater dangers await. Other people choose to live in hazardous areas because they believe the advantages outweigh the potential for natural disaster. They are attracted by the productivity of the farmland, the natural beauty of a building site, or the economic possibilities associated with a particular location. In addition, there are few populated areas of the world that are not associated with one or another environmental hazard. Forested regions are subject to fire; earthquake, landslide, and volcanic activities plague mountain regions; violent storms threaten interior plains; and many coastal regions experience periodic hurricanes or typhoons (the Pacific Ocean equivalent).

Destruction of property by a tornado that struck Andover, Kansas.

Environmental Issues

Officials in national governments and international organizations throughout the world have come to realize that certain human alterations of the environment are producing serious negative consequences for people today and far into the future. Issues such as global warming, acid precipitation, deforestation and the extinction of biological species in tropical areas, damage to the ozone layer of the atmosphere, and desertification have risen to the top of agendas when world leaders meet and international conferences are held. They are recurring subjects of magazine and newspaper articles, books, and television programs.

Much of the damage to the environment is a result of atmospheric pollution associated with industrialization in the wealthy, developed nations of the world. But as population pressures mount and the less wealthy nations struggle to industrialize, human activity is exacting an increasing toll on the soils, forests, air, and water of the developing world as well. Issues of environmental deterioration are worldwide, and solutions must involve all of the world's nations to be successful. As citizens of the wealthiest nation in the world, Americans should respond to each of the following questions. What are the major threats to the environment caused by human activity? What are the causes of these threats? Are the threats real and well documented? What steps must be taken to resolve the environmental issues, and what will be the impacts on both humans and their environments if these steps are taken? What can I personally do to become involved? The pages of text and the environmental essays that are included in this book should help you to answer most of these questions.

CRITICAL THINKING ▼

(1) Are there environmental hazards associated with the area in which you spent your childhood? If so, why did your family choose to live there? **(2)** If you were on a planning commission that had to choose between shutting down a power plant because it could not afford the cost of preventing atmospheric pollution or losing an industry because it would have inadequate power to continue production, what would you want to know to make your decision? **(3)** Will it be possible to end environmental deterioration within your lifetime? If not, what will be the result? If so, what must be done to resolve the issues?

plans to allow the Kissimmee River to meander again across its former flood plain, to return agricultural land to sawgrass marsh, and to restore historic water-flow patterns through the Everglades. The problems of South Florida should serve as a useful lesson. Alterations of the natural environment should not be undertaken without serious consideration of all the consequences.

The Human–Environment Equation

One lesson is obvious as we examine the varied physical environments that are produced by Earth's many subsystems. Despite the wealth of resources available in the form of air, water, soil, minerals, vegetation, and animal life on Earth, the capacity of our planet to support humans in growing numbers may have an ultimate limit, a *threshold*. And there are dangerous signs that such a limit may someday be reached. Human numbers on Earth have passed the 5 billion mark, and United Nations estimates indicate the total will reach 8 to 9 billion by 2025 if current growth rates continue. Today, over half of the world's people live in countries in which they must tolerate substandard living conditions and insufficient food to eat. A major problem today is the distribution of food supplies, but ultimately the size of the human population should not exceed the environmental resources to sustain them and each of those generations of humans who will succeed them.

Although our current objective is to study physical geography, we should not ignore the information shown in Figure 1.13, the World Map of Population Density. The map shows the distribution of people over the land areas of Earth and illustrates an important aspect of the human–environment equation. World population distributions are highly irregular; people have chosen to live, and have multiplied rapidly, in some places and not in others. One reason for this uneven distribution is the differing capacities of Earth's varied environments to support humans in large numbers. Figure 1.13 also shows where most of the problems that threaten Earth's environments originate. The spatial information presented in this map and the world distributional maps in the chapters that follow will also help us identify those locations where large numbers of people live in *hazard zones* under the constant threat of property damage or death from natural phenomena such as earthquakes, hurricanes, floods, fires, or volcanic eruptions.

The relationships between humans and the environments in which they live will be emphasized throughout this book. In fact, geographers consider human–environmental relationships to be a major theme of geographic study. They are also keenly aware that, as in most relationships, the nature or behavior of each of the parties in the relationship may have direct effects on the other (see "The Environment: Human–Environment Relationships"). However, when considering the human–environment equation and the sustaining of humans at acceptable living standards for generations to come, it is most important to note that environments do not change their nature to accommodate humans. Humans should make greater attempts to alter their behavior to accommodate the limitations and potentials of Earth environments. It has been said that humans are not passengers on Spaceship Earth; rather, they are the crew. This means we have the responsibility to maintain our own habitat. Poised at the interface between Earth and human existence, geography has much to offer in helping us understand the factors involved in meeting this responsibility.

The Study of Geography

Geography is a word that comes from two Greek roots. *Geo-* refers to Earth, and *-graphy* means picture or writing. Geography examines, describes, and explains Earth—its variation from place to place, and how places change over time. Geography is often called the **spatial science** because it includes recognizing, analyzing, and explaining the variations, similarities, or differences in phenomena located (or distributed) on Earth's surface.

In 1994, a set of national standards for geography education in the schools was published by the major professional geography organizations in the United States in a book entitled *Geography for Life*. The question "What is geography?" is answered in the book as follows:

Where is something? Why is it there? How did it get there? How does it interact with other things?

Geography is not a collection of arcane information. Rather it is the study of spatial aspects of human existence. People everywhere need to know about the nature of their world and their place in it. Geography has much more to do with asking questions and solving problems than it does with rote memorization of facts.

So what exactly is geography? It is an integrative discipline that brings together the physical and human dimensions of the world in the study of people, places, and environments. Its subject matter is the Earth's surface and the processes that shape it, the relationships between people and environments, and the connections between people and places.

Geography for Life
Geography Education Standards Project, 1994

In order to answer questions like the ones posed in *Geography for Life*, geographers gather, organize, and analyze geographic information. They also apply many skills, techniques, and tools to the task of answering these geographic—or *spatial*—questions that have infinite variations in the nature of the phenomena being studied. So it is not surprising that there are many divisions of the discipline of geography, yet the unifying factor among them is their focus on understanding and explaining spatial locations, distributions, and relationships. Each of these three topics is representative of geography as a spatial science—a study of where things are located on the surface of the Earth, how they are distributed, and how things in one place on Earth may be related to one another and to things in other places. Geographers study places, attempting to identify and explain the characteristics that two or more locations may have in common as well as why places vary in their geographical characteristics. Geographers are also interested in how to divide and synthesize areas into meaningful divisions called **regions.** Geographers study the processes that influenced the landscapes of the Earth in the past, how they continue to impact them today, how the landscape may change in the future, and the significance or impact of these changes.

The breadth of the subject may be seen in the Essential Elements from *Geography for Life* (Table 1.1). Even though that document was aimed at geography education in K–12 schools, most of the same concepts apply at the higher academic level of colleges and universities. In its concern with the natural environment, geography is very much a physical science, yet because geography also examines humanity's relationships with Earth, it is a social science as well.

Cultural or **human geography** is the study of human activities and of the results of those activities. Human geographers are concerned with such subjects as population distributions, cultural patterns, cities and urbanization,

natural resource utilization, industrial location, and transportation networks (Fig. 1.14). When a geographic study concentrates primarily on the physical and human features of a specific region, such as Canada, the Great Plains, Europe, or the Middle East, we call this **regional geography.**

Physical Geography

Physical geography encompasses the study of the processes and features that make up Earth, including human environment where it interfaces with the Earth system. In fact, physical geographers are concerned with all aspects of the Earth system and can be considered generalists because they are trained to view Earth in its entirety, functioning as a unit. However, after a thorough and broad education in basic physical geography, most physical geographers focus their expertise on advanced study in one or two specialties. For example, *meteorologists* and *climatologists* consider the atmospheric components that affect Earth's surface and that together influence weather and climate. Meteorologists are interested in the atmospheric processes that affect daily weather, and they use current weather data to forecast future weather conditions. Climatologists are interested in the averages and extremes of long-term weather data, the classification of climates into meaningful regions, monitoring and understanding climatic change and climatic hazards, and the long-range impacts of atmospheric conditions on human activity and the environment. The nature, development, and modification of landforms is a specialty called *geomorphology,* a major subfield of physical geography. Geomorphologists are interested in understanding and explaining variation in landforms, the processes that produce physical landscapes, and the nature and geometry of the land surface. The factors involved in landform development are as varied as the environments on Earth and include, for example, gravity, running water in streams, stresses in the Earth's crust, the flow of ice in glaciers, volcanic activity, and the erosion or deposition of Earth materials. *Biogeographers* examine natural and human-modified environments and the ecological processes that influence their nature and distribution, including vegetation change over time. They also study the ranges of vegetation and animal species in order to discover the environmental factors that limit or facilitate their distributions. Many *soil scientists* are geographers, and these individuals are often involved in the

(text continues on page 23)

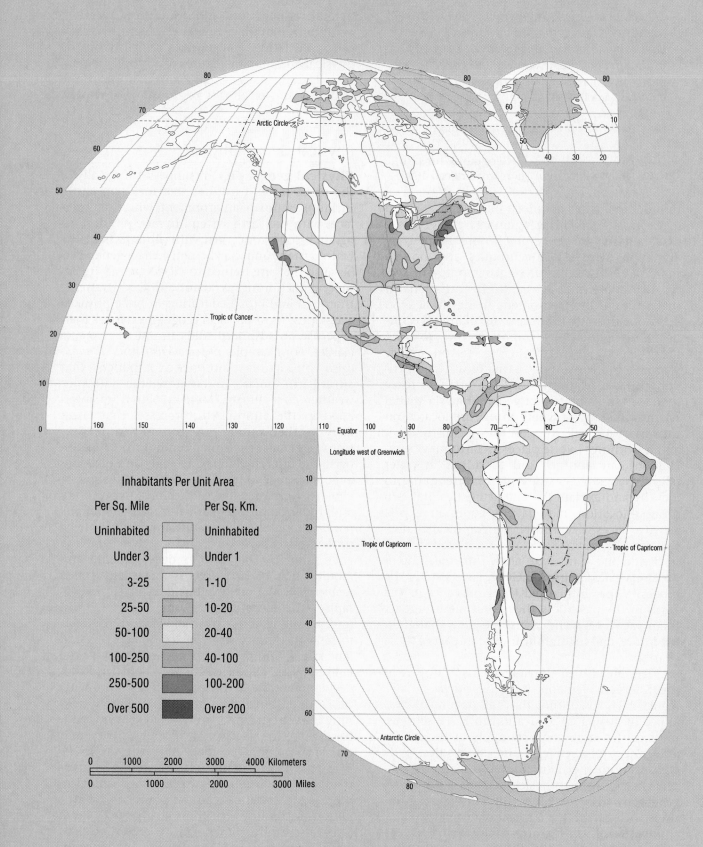

Figure 1.13 World Map of Population Density.

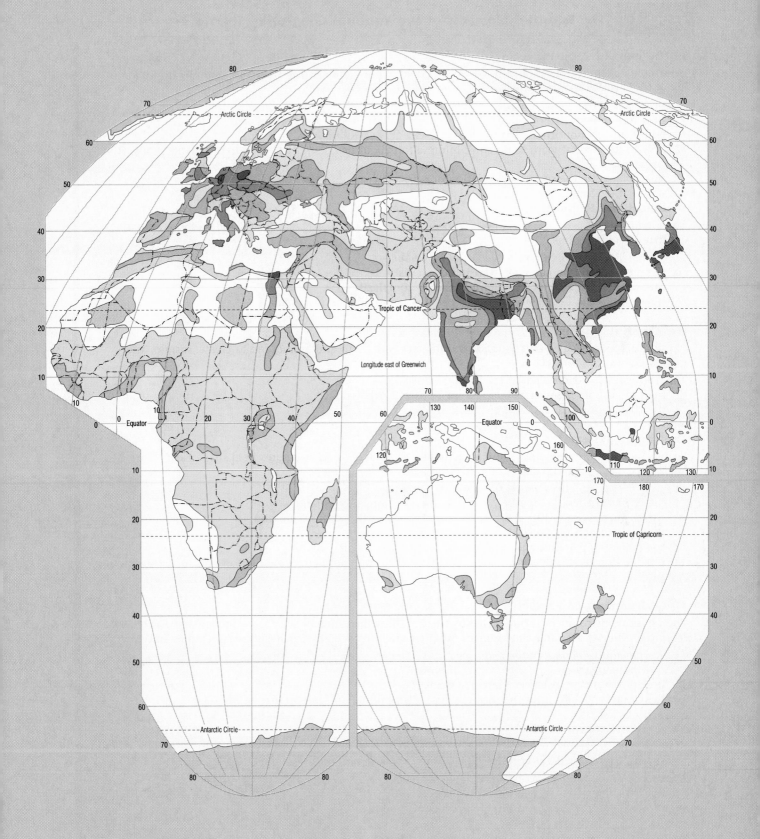

Arctic Circle

Arctic Circle

Tropic of Cancer

Longitude east of Greenwich

Equator

Equator

Tropic of Capricorn

Antarctic Circle

Antarctic Circle

A Western Paragraphic Projection
developed at Western Illinois University

TABLE 1.1 **The Essential Elements of the National Geography Standards**

I. The World in Spatial Terms

Geography studies the relationships between people, places, and environments by mapping information about them into a spatial context.

The geographically informed person knows and understands
1. how to use maps and other geographic representations, tools, and technologies to acquire, process, and report information from a spatial perspective
2. how to use mental maps to organize information about people, places, and environments in a spatial context
3. how to analyze the spatial organization of people, places, and environments on Earth's surface

II. Places and Regions

The identities and lives of individuals and peoples are rooted in particular places and in those human constructs called regions.

The geographically informed person knows and understands
4. the physical and human characteristics of places
5. that people create regions to interpret Earth's complexity
6. how culture and experience influence people's perceptions of places and regions

III. Physical Systems

Physical processes shape Earth's surface and interact with plant and animal life to create, sustain, and modify ecosystems.

The geographically informed person knows and understands
7. the physical processes that shape the patterns of Earth's surface
8. the characteristics and spatial distribution of ecosystems

IV. Human Systems

People are central to geography; human activities, settlements, and structures help shape Earth's surface, and humans compete for control of Earth's surface.

The geographically informed person knows and understands
9. the characteristics, distribution, and migration of human populations
10. the characteristics, distribution, and complexity of Earth's cultural mosaics
11. the patterns and networks of economic interdependence
12. the processes, patterns, and functions of human settlement
13. how the forces of cooperation and conflict among people influence the division and control of Earth's surface

V. Environment and Society

The physical environment is influenced by the ways in which human societies value and use Earth's natural resources, while at the same time human activities are influenced by Earth's physical features and processes.

The geographically informed person knows and understands
14. how humans modify the physical environment
15. how physical systems affect human systems
16. the changes that occur in the meaning, use, distribution, and importance of resources

VI. Uses of Geography

Knowledge of geography enables people to develop an understanding of the relationships between people, places, and environments over time—that is, of Earth as it was, is, and might be.

The geographically informed person knows and understands
17. how to apply geography to interpret the past
18. how to apply geography to interpret the present and plan for the future

Source: *Geography for Life: National Geography Standards 1994.* National Geographic Research and Exploration, Geography Education Standards Project, Washington, D.C.

Figure 1.14 *Chicago, an urban center in the midwestern United States that serves as a focal point for industry, commerce, financial institutions, interstate highways, air traffic, and millions of housing units, represents fertile ground for study by human geographers.*
• **What particular aspects and problems of urban areas are the subject of study by human geographers?**

mapping and analysis of soil types, determining the suitability of a soil for certain uses, such as agriculture, and working to conserve soil as a natural resource. Finally, because of the great importance of water to human existence, geographers are widely involved in the study of water bodies on Earth and their processes, movements, impacts, quality, and other characteristics. They may serve as *hydrologists, oceanographers,* and *glaciologists.* Many geographers involved with water studies also function as *water resource managers* to ensure that lakes, watersheds, springs, and groundwater sources are suitable for human use, provide an adequate supply of water, and are as free of pollution as possible.

Technology, Tools, and Methods

The technology being applied by physical geographers and other scientists in their efforts to learn more about the Earth system is rapidly changing. The abilities of computer systems to capture, process, model, and display spatial data—all functions that today are possible on a desktop computer—were only a dream 15 to 20 years ago. Computer access to the Internet provides instant connections to information sources on virtually any topic. The amounts of data, information, and imagery available for studying the environments of Earth have exploded. Graphic displays of these sorts of data and information are becoming more vivid and striking as a result of better methods of data processing and visual representation. Increased computer power allows the presentation of clearer images, three-dimensional images, and even animated images of Earth features, changes, and processes.

Continuous satellite imaging of Earth's environments has been ongoing for more than 25 years, giving us a better perspective on changes in the Earth system. It is certainly possible to

SANDRA VILLALOBOS

Sandra Villalobos at the "Weather" desk, KCEN–TV, Waco.

Sandra is a meteorologist for KCEN-TV in Waco/Temple, Texas. She provides local on-air weather updates and forecasts for both Texas Today and NBC's Today Show. But Sandra's on-air duties account for only a portion of her daily regimen. Much of her day is spent gathering the most accurate weather data for broadcast to Lone Star State residents.

High School Education: *JM Hanks High School, El Paso, Texas, 1987*

University Education: *Bachelor of Science, Physical and Applied Geography, Southwest Texas State University, San Marcos, Texas, 1997*

Early Interests: *Rock collecting, stargazing*

Prior Work Experience: *Intern, Texas Natural Resources Information Systems; Intern, KTBC-TV*

"I arrive at the television station by 4:00 a.m. Then I begin to collect data in order to prepare my forecast. I use several computer programs including Kavouras, which offers the Triton I-7, RADAC, and Metpac PC. I also use True Vision in which I use the ATVISTA Tips.

Once my information is ready, I create slides (the visuals) for my show. These slides consist of a day-part forecast, regional surface map, currents, jet stream, 7-day planner, and more. Texas Today is 1 hour and 35 minutes long. I normally have 7 twenty-second short weathercasts and 6 two-and-a-half-minute-long weathercasts.

Later, I do the local weather for two hours during the Today Show. In this two-hour period, I have 4 thirty-second weathercasts and 4 one-minute weathercasts.

The drought of 1998 (reported to be the worst of the decade) was very devastating to the people and economy of Texas. Many places experienced triple-digit temperatures for nearly a month, with little or no rain. If I can inform people of some inclement weather they need to be aware of, and be correct, then that feeling of accomplishment is what makes me happy.

My interest in science and the fact that weather is constantly changing drew me to meteorology. I guess ever since I was a little girl, I have been a scientist at heart."

track change in a single place over that time period, or to compare the differences between two different places at one point in time. Using various energy sources to produce an image from space, we are now able to see, measure, monitor, and map processes and the effects of these processes that otherwise would be invisible to the naked eye. Satellite technology can be used to determine the precise location of a positioning receiver on the Earth's surface, a capability that has many useful applications for geography and mapping.

Today, mapmaking and the analysis of maps are each becoming digital, computer-assisted operations. Making direct observations in the field and conducting fieldwork to gather data remain important to most physical geographers, but they also must keep up with new technologies in order to make use of them. Many of these technologies also directly support and facilitate traditional methods of fieldwork. In fact, the production of new technology outstrips our ability to understand or apply the full range of potential uses for these new tools. This situation of a technology-application lag creates many opportunities for geographers in the applied workplace. Many geographers are gainfully employed in positions that apply technology to the prob-

Figure 1.15 *Colorado National Monument near Grand Junction, Colorado. Natural environments, with their component rock structures, landforms, soils, vegetation, moisture conditions, and atmospheric elements, are the subjects of study in physical geography.* • **What characteristics of the physical geography near Grand Junction, Colorado, are suggested by this photograph? (Hint: remember the four major subsystems of Earth.)**

lems of understanding our planet and its environments, and their numbers are certain to increase in the future.

Themes in Physical Geography

This textbook provides a broad introduction to meteorological, biological, climatological, hydrological, and geomorphological aspects of geography (Fig. 1.15). Although the course you are taking is a useful beginning, the complexity of our planet is such that no one can fully understand everything about the Earth system, even after a lifetime of dedicated study. In fact, often the more you learn about a subject, the more unanswered questions begin to emerge in your mind. In learning about our planet, asking informed questions is really the beginning of the road to understanding. Even though physical geographers may be a varied lot with many divergent interests related to the Earth system, they share many common interests as well, particularly the effort to understand and explain the spatial variation displayed by our planet's environments.

As a part of this common ground, it is useful to consider and review some of the types of concerns physical geographers have about the Earth and some of the problems they seek to understand and solve. Remember the quote from *Geography for Life* suggesting that geography is about asking questions and solving problems? How do physical geographers think about the Earth, and what kinds of questions do they pose in performing their studies? What are some of the basic themes that physical geographers consider? Examples of these themes and related questions follow and can be found throughout this textbook. You will encounter them often during the course that you are taking.

Location.   As a spatial science, geography often begins with location. The location of a feature is usually expressed by one of two methods: **absolute location,** expressed by a coordinate system (or address), and **relative location,** which identifies where something is in relation to something else, usually a fairly well-known location. For example, the wreck of the Titanic lies in 3810 meters (12,500 ft) of water at latitude 41°44' North and longitude 49°57' West. A global address like this is an absolute location. Another way to report its location, though, would be to state that it is 530 kilometers (350 mi) southeast of Newfoundland, Canada. This is an example of relative location (relative to Newfoundland). Further locational questions include the following: *Where is a certain type of Earth feature found, and where is it not found? What methods can we use to locate a feature on Earth? How can we describe its location? What is the most likely or least likely location for a certain Earth feature? Does its location seem to fit with its surrounding environment?*

Characteristics of Places.   Physical geographers are interested in the environmental features that combine to make a place unique, and they are also interested in shared characteristics between places. For example, what physical geographic features make the Rocky Mountains appear as they do? This, of course, would require an assessment of features representing all four divisions of the Earth system. As another example, how are the Andes Mountains different from the Rockies, and what characteristics are common to both of these mountain ranges? Other examples might include: How does an Australian desert compare to the Sonoran desert of the southwestern United States? Or, how do the grasslands of the Great Plains of the United States compare to the grasslands of Argentina? *What are the environmental conditions at a particular site? How do places on Earth vary in their environments and why? In what ways are places unique and in what ways do they share similar characteristics with other places?*

Spatial Distributions and Spatial Patterns.   When studying how features are arranged in space, geographers are usually interested in two spatial fac-

tors. **Spatial distribution** means the extent of the area or areas where a feature exists. For example, where on Earth do we find the tropical rain forests? What is the distribution of rainfall in the United States on a particular day? **Spatial pattern** refers to the arrangement of features in space—are they regular or random, clustered or widely spaced? We have seen that population distribution on Earth can be either dense or sparse. The spatial pattern of earthquakes may be linear on a map, suggesting that a fault is located there, because faults display similar linear patterns. *Where are certain features abundant and where are they rare? How are particular factors or elements of physical geography arranged in space, and what spatial patterns exist, if any? What processes are responsible for these patterns? If a pattern exists, what does it signify?*

Spatial Interaction.   Few processes on Earth operate in isolation; this means that areas on our planet are interconnected, or linked to conditions elsewhere on Earth. A condition or occurrence in one place generally has an impact on other places. Unfortunately, the exact nature of these links is often difficult to establish with certainty except after years of intense study. A direct causal relationship is often only suspected, not known or proven. It is much easier to observe that changes seem to be *associated* with each other without knowing which event causes the other. For example, the presence of abnormally warm ocean waters off the west coast of South America, a condition called El Niño, seems to occur at times of unusual weather in other parts of the world. There is also a concern that clearing of the tropical rain forest may have an impact on world climates. Interconnections are one reason for considering the Earth system as a whole. *What are the relationships among places and features on Earth? How do they affect each other? What important interconnections link two or more of the four major divisions of the Earth system? If two events occur at the same time, why is it so difficult to prove which one caused the other to happen?*

Ever-Changing Earth.   As we have learned, Earth is a continuously changing, dynamic system. Weather and weather systems change from day to day and over the seasons, and from year to year. A new Hawaiian island is forming beneath the waters of the Pacific Ocean. Storms, earthquakes, landslides, and stream processes modify the landscape. Today, vegetation and wildlife are continuing to become reestablished in an area devastated by the volcanic eruption of Washington State's Mount St. Helens in 1980. World climates have changed throughout Earth's history, with attendant shifts in ecosystems. Change is always occurring, but not all of this change can be directly monitored. *How are Earth features (or systems) changing? What processes contribute to the change and what is the nature and impact of this change? What is the rate of change? Does change occur in a cycle? Can humans witness this change as it is taking place or is a long-term study required to recognize the change? Do all places on Earth experience the same levels of change, or is there spatial variation? Within one place on Earth, are all environmental components experiencing the same rates of change?*

Human Interactions with the Environment. Finally among our themes, we return to human existence and the ever-increasing ability that we have to affect our environment and the way that the Earth system operates. Humans are affecting environments on all scales, from global to local. As examples, cities can directly affect their own climate and weather, global climates may be changed by human activity, large artificial reservoirs provide new human-created environments and control river flow, and the vegetation communities of most places on Earth are influenced or overrun by introduced species. However, as we have noted, geographers are not simply concerned with how humans affect the environment; they are also interested in the ways that the environment affects the way we live. Where natural hazards exist, the operation of natural Earth processes (or those affected by human alteration) can adversely affect human settlements. *What pressures are humans exerting, or have they exerted, or will they exert on an environment? What challenges does the environment present to humans? How can the environment pose a hazard to human life? How can humans act to minimize the risk from natural hazards? Do interactions with the environment vary geographically or spatially? What part do humans play in this variation, and what is a direct result of environmental conditions?*

As the authors of this textbook, we have selected these six themes as representative of the broad range of emphases in the study of physical geography. Other geographers might favor other themes or state these same themes in another way. But the questions that accompany the themes will be much the same for all geographers as they apply their knowledge and expertise in a search for solutions to spatial-related problems.

Learning what questions to ask is the first step toward finding answers to these problems, and it is a major objective of your physical geography course.

Physical Geography and You

Many aspects of the physical environment affect our everyday lives. The principles, processes, and themes of physical geography provide keys that help us to be environmentally aware—that help us assess each environmental situation, analyze the factors involved, and then make an informed choice among possible courses of action. For example, should you plant a new lawn before or after the spring rains? What sort of environmental impacts might be expected from a proposed shopping center? What are the environmental advantages and disadvantages of a particular home site? What potential impacts of natural hazards—flooding, landslides, earthquakes, hurricanes, tornadoes—should you be aware of where you live? What can you do to minimize potential damage to your household from a natural hazard? What can you do to assure that you and your family are as prepared as possible for the kind of natural hazard that might affect your home?

It is apparent, then, that the study of physical geography, and the understanding of our natural environment that it provides, are valuable to all of us. Perhaps you have wondered, however—What do those people who call themselves physical geographers do in the workplace? What kinds of jobs do they hold? Physical geography sounds interesting and exciting, but can I make a living at it?

By applying their knowledge, skills, and techniques to real-world problems, physical geographers make major contributions to human well-being, environmental stewardship, and the economic development of society. Physical geographers emphasize the Earth system, but they do not ignore the effect of people on that system or the impact that our environment may have on people and the way they live. A knowledge of physical geography can help us analyze and solve environmental problems, such as whether we should continue to build nuclear power plants, allow offshore oil development, or drain coastal marshlands. Each of these questions may generate a different answer depending on the environmental conditions and geography of the location in question. Intelligence efforts by the

U.S. Department of Defense must predict the effects that weather and terrain may have on military or naval operations. Industries must evaluate how the development of a proposed plant site may alter the surrounding environment.

Applied physical geography takes many forms, and the *Career Vision Series* in this book will introduce you to physical geographers in the workplace. The geographers presented in these brief biographies each have a geography background, and they share a common experience with you—each began their geographic education as an undergraduate student in an introductory geography course like the one that you are now taking.

Finally, knowledge of physical geography can not only provide opportunities for personal *enrichment* and possible *employment,* it can also be a source of perpetual *enjoyment.* Geography is a visual science, and it is really more than just a subject. Geography is a way of looking at the world and of observing its features. It involves asking questions about the nature of those features and appreciating their beauty and complexity. It encourages you to seek explanations, gather information, and use geographic skills, tools, and knowledge to solve problems. Even if you forget many of the facts discussed in this book, you will have been shown new ways to consider, to see, and to evaluate the world around you. Just as you see a painting differently after an art course, so, too, will you see sunsets, waves, storms, deserts, rivers, forests, prairies, and mountains with an "educated eye." You should retain a knowledge of geography for life. You will see greater variety in the landscape, not because there is any more there, but because, like the astronaut in orbit, you will have been trained to observe Earth differently and with deeper understanding.

Define and Recall

system	model	closed system	spatial science
variable	physical model	equilibrium	human geography
Earth system	pictorial/graphic model	dynamic equilibrium	regional geography
atmosphere	mathematical/statistical	negative feedback	physical geography
lithosphere	model	positive feedback	absolute location
hydrosphere	conceptual model	threshold	relative location
biosphere	mental map	feedback loop	spatial distribution
life-support system	spatial		spatial pattern
natural resource	systems theory	environment	applied physical
pollution	subsystem	ecology	geography
	inputs and outputs	ecosystem	
	open system		

Discuss and Review

1. Why are geographers interested in systems? What is meant by the Earth system?

2. List the four major divisions of the Earth system. Give examples of how the divisions interact with one another.

3. How do open and closed systems differ? How does feedback affect the dynamic equilibrium of a system?

4. How does negative feedback operate to maintain a tendency toward balance in a system? What is a threshold in a system?

5. Give reasons why the study of ecosystems may illustrate the close relationships between humans and the environment.

6. What is meant by the *human–environment equation?* Why is the equation rapidly falling further out of balance?

7. How does geography fit into both the physical and the social sciences?

8. What career fields are attractive to applied physical geographers? What technologies, tools, and methods are used by physical geographers in studying aspects of the Earth system?

Consider and Respond

1. How have various kinds of pollution affected your life? List some sources of pollution in your city or town.

2. Give an example of an ecosystem in your local area that has been affected by human activity. In your opinion, was the change good or bad? What values are you using in making such a judgment?

CHAPTER

2

REPRESENTATIONS OF EARTH

CHAPTER PREVIEW

▶ Maps and globes are especially important to geographers, but they are everybody's business.

How are they my business? Who else would be especially interested? Why are geographers major users of maps?

▶ The globe is the only representation of the entire Earth with little or no distortion.

Why? What does this imply about maps?

▶ The geographic (Earth) grid is an arbitrary coordinate system designed to allow an accurate description of Earth location (position).

Why is it arbitrary? How was it developed? How is it used? In what ways is it associated with Earth navigation and standard time?

▶ The Public Lands Survey System has had a significant effect on human affairs in much of the United States.

In what ways? Why was the system developed? Can you think of a better system?

▶ Each map or map projection has its own specific properties that determine the purposes for which it should be used.

What does this mean? What does this suggest in relation to the development of a "perfect" map?

▶ There are certain characteristics (essentials) that should be present on most maps if they are to be of maximum use.

What are they? Can you make use of them all?

▶ Contour maps (usually United States Geological Survey topographic maps), computer maps, geographic information systems, global positioning systems, and the products of remote sensing are of special interest to the physical geographer.

Why? What are they? How are they used?

▲ This scene shows rugged mountains in southeast Tibet, and was taken from an orbiting Space Shuttle using an imaging radar system. In the lower right corner is the Lhasa River valley. Mountains in this area reach elevations of about 5800 meters (19,000 ft) and the valley floors are at about 4300 meters (14,000 ft) above sea level. The rugged topography has resulted from erosion in this part of the Tibetan Plateau. North is toward the upper left. The image is 49.8 kilometers by 33.6 kilometers (30.9 mi by 20.8 mi) and is centered at 30.2 degrees north latitude, 92.3 degrees east longitude.

The principle of location is essential to geographers in their attempts to describe and analyze different aspects of the Earth system. Although many of the principles used in dealing with locational problems have been known for hundreds of years, the methods that are being applied to these tasks are ever-changing. Today the tools that are used by geographers and other scientists to find, describe, record, and display locations are becoming more precise and accurate, and are being supported by computers and space-age technologies.

Location on Earth

Perhaps as soon as people began to communicate with language they also began to develop a language of location. They probably used landscape features as directional cues, saying, "Go along the river until it forks, then follow the widest branch until you come to the campsite." Today we use streets, traffic lights, and city blocks to find our way in built-up areas. In ancient times, however, humans used rivers, trees, hills, and other natural landmarks to find their way and to locate points of interest.

When ancient peoples began to sail the open seas, they recognized new needs for finding directions and ways to describe location. Long before the first compass was developed they discovered that positions of the sun and stars—rising, setting, or circling in the heavens—could provide accurate directions. Observing the stars and the sun and their relationships to Earth are still basic skills in **navigation,** the science of location and wayfinding. Navigation has also been called the process of getting *from where you are to where you want to go.*

No one knows when or where the first maps were made; their origin is lost in antiquity. Early humans certainly drew locational diagrams on rock surfaces or in the soil with sticks. Permanent maps were created by ancient cultures in many parts of the world, such as China, Egypt, the Pacific Islands, Greece, Mexico, and Peru. Early maps were constructed of sticks, or drawn on clay tablets, stone slabs, metal plates, papyrus, linen, or silk. These ancestors of modern maps were fundamental to the beginnings of geography, helping humans to communicate their spatial thinking (Fig. 2.1).

Maps and globes are visual representations of all or part of Earth, the Moon, other planets, and their moons. These representations convey a great deal of spatial information in symbolic form, and this standard "language" of location, using graphic symbols, must be understood to be fully appreciated.

Cartography is the science of mapmaking. Physical geographers who specialize in cartography supervise the development of maps and globes. They also ensure that the information and data on maps are accurate and presented in an effective manner. Most cartographers would agree that a map is primarily *a tool for communicating spatial information.* In recent years, cartography has been revolutionized through the application of computer technology.

The changes in map data collection and display that have occurred in the 20th century are comparable to the change from pedestrian to astronaut. Information that used to be collected little by little from ground observations can now be collected instantly by satellites hurtling through space, and recorded data can be flashed back to Earth at the speed of light. . . .

Cartographers can now gather spatial data and make maps faster then ever before— within hours—and the accuracy of these

Figure 2.1 When was the first map made by humans? Cave paintings by Cro-Magnon people in France depict the animals they were hunting, sometime between 17,000 and 35,000 years ago. Although this view shows detail of stags crossing a river, experts suggest that some of the artwork may represent a rudimentary map. The paintings include lines that apparently represent migration routes and other marks appear to represent locational information. If so, this is the earliest known example of humans recording their spatial knowledge.

*maps is excellent. Moreover, **digital mapping** enables mapmakers to experiment with a map's basic characteristics (for example, scale or projections), to combine and manipulate map data, to transmit entire maps electronically, and to produce unique maps on demand.*

USGS, *Exploring Maps,* page 1

Computer-assisted cartography and data-gathering methods, using measurements and images from aircraft or satellites, have contributed to a map "explosion." Maps are found everywhere. We can all think of many fields, such as navigation, political science, community planning, surveying, history, meteorology, and geology, in which representations of Earth, including globes or maps, are vital. In our everyday lives, through education, travel, television, recreation, and reading, we all have had experience with maps. Examine each map in one of your daily newspapers. How do they contribute to your understanding of the news? How many are there? How many would that equal in a year (365 daily papers)?

Globes and Great Circles

The world globe is a nearly perfect model of our planet. It shows the spherical shape of Earth and accurately displays spatial relationships between landforms and water bodies, comparative distances between locations, relative sizes and shapes of Earth's features, and accurate compass directions. A world globe is an extremely useful tool because it enables us to *see,* rather than merely imagine, many geographic aspects of our planet. For example, learning about the seasons or understanding how the length of daylight hours in one place changes annually is made a great deal easier. Because a globe has the same shape as our planet, it can represent almost without distortion the various features and relationships associated with Earth (Fig. 2.2).

Globes also have several limitations. A world globe would not help us find our way on a hiking or backpacking trail. It would be bulky and awkward to carry, and our location would appear as a tiny pinpoint, and with little, if any, local information. We would instead need a map that clearly showed elevations, trails, and rivers—a map that could be folded and put in a pocket or pack. However, despite the limitations presented

(a)

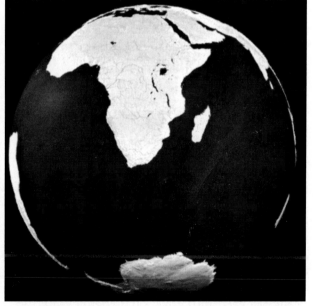

(b)

Figure 2.2 *Earth as photographed from space by Apollo astronauts, showing most of Africa, the surrounding oceans, storm systems in the Southern Hemisphere, and the relative thinness of the atmosphere **(a)**. A manufactured globe represents the same general viewpoint **(b)**.*

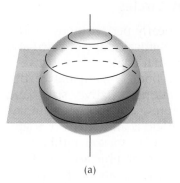
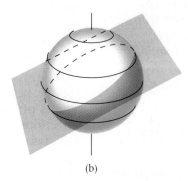
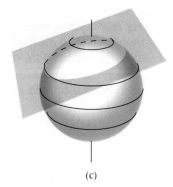

(a)   (b)   (c)

Figure 2.3 *Planes slicing through a globe's center (**a**) at the equator and (**b**) obliquely. In either case the globe is divided into equal halves, and the line where the plane intersects the surface of the globe is a great circle with the same circumference as the globe itself. In (**c**) the plane slices the globe into unequal parts. The line of intersection of such a plane with the globe is a small circle.*

by globes, if we want to view a representation of the entire Earth, a world globe provides the most accurate representation. Also, because maps attempt to depict a spherical Earth, or a part of Earth, on a flat surface, being familiar with the characteristics of a globe helps us to understand maps and how they are constructed.

An imaginary circle drawn in any direction around Earth's surface and whose plane passes through the center of Earth is called a **great circle** (Figs. 2.3a, b). It is "great" because it is the largest circle that can be drawn around Earth through two particular points. Great circles have several useful characteristics: (1) Every great circle divides Earth into equal halves called **hemispheres;** (2) every great circle is a circumference of Earth; and, perhaps most importantly, (3) great circles mark the shortest travel routes between locations on Earth's surface. Circles whose planes do not pass through the center of Earth are called **small circles** (Fig. 2.3c), and any circle on Earth's surface that does not divide the planet into equal halves is a small circle.

The shortest route between two places can be located by finding the great circle that connects them. Put a rubber band (or string) around a globe to help visualize this process. Connect any two cities, such as Moscow and New York, San Francisco and Tokyo, New Orleans and Paris, or Kansas City and Singapore, by stretching the rubber band around the globe so that it touches both cities and divides the globe in half. The rubber band then forms the shortest route between the two cities. These are called **great circle routes,** which navigators chart for aircraft and ships since traveling the shortest distance saves

time and fuel. The farther away two points are on Earth, the larger the distance savings will be by following a great circle route that connects these two places. The *arc of a great circle* is the shortest distance between two points *on any spherical surface.* Another important example of a great circle is the *circle of illumination,* which divides Earth into light and dark halves—a day hemisphere and a night hemisphere.

Latitude and Longitude

Imagine you are traveling across the United States and you want to visit the Football Hall of Fame in Canton, Ohio. Pulling the Ohio road map out of the glove compartment, you look up Canton in the map index, and find "G-6." Turning the map over, you look down the side of the map until you find 6, and then along the top until you reach G. The letter G and the number 6 meet in a box marked on the map. Scanning the area within box G-6, you locate Canton (Fig. 2.4).

What you have used is a coordinate system of intersecting lines, which on the map make up a system of *grid cells.* It is more difficult to describe location on a sphere like Earth than it is on a flat surface such as the road map. Imagine a smooth sphere like a playground ball. Let us say that there is a leak in the ball. You know where it is, and you want to describe its location in a note to the person who's going to fix it. However, soon it is apparent that there are no fixed reference points on a smooth ball. You cannot describe the location of the leak without some reference point.

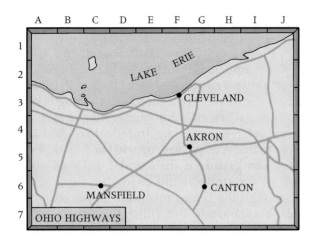

Figure 2.4 *Using one kind of rectangular coordinate system to locate a position accurately. This map employs an alphanumeric location system, similar to that used on many road maps.* • **What are the rectangular coordinates of Mansfield, and what is at location F-3?**

Measurement of Latitude. Unlike a ball, Earth has two fixed reference points. As Earth spins uniformly on an imaginary **axis,** the ends of that axis are points called the **North Pole** and the **South Pole.** The poles can be used to locate a great circle, the **equator,** that lies exactly between them. Now, using the equator, we can measure the angle north or south of it; this angular distance is called **latitude** and is measured in degrees.

Let us locate the latitude of Los Angeles. Imagine a line that goes from the center of Earth to Los Angeles and another line that goes from Earth's center to a point on the equator directly south of Los Angeles. These two lines form an angle, and the arc of this angle is the latitudinal distance (in degrees) that Los Angeles lies north of the equator (as indicated in Figure 2.5a). The angle made by the two imaginary lines is just over 34°—so the latitude of Los Angeles is about 34°N (north of the equator).

The equator is designated as 0° latitude. As we go farther north or south of the equator the angles and their arcs increase until we reach the North or South Pole and the maximum latitudes of 90° north or 90° south.

Because Earth's circumference is approximately 40,000 kilometers (25,000 mi) and there are 360 degrees in a circle, we can use division (40,000 km/360°) to find that a degree of latitude is equal to about 111 kilometers (69 mi). One degree of latitude covers a relatively large distance, so degrees are further divided into **minutes (')** and **seconds (")** of arc. Actually, Los Angeles is located at 34°03'N (34 degrees, 3 minutes north).

There are 60 minutes of arc in a degree. We can get even more precise: 1 minute is equal to 60 seconds of arc. So we could have a place at latitude 23°34'12"S, which we would read as 23 degrees, 34 minutes, 12 seconds south latitude. A minute of latitude equals 1.85 kilometers (1.15 mi) and a second is about 31 meters (102 ft).

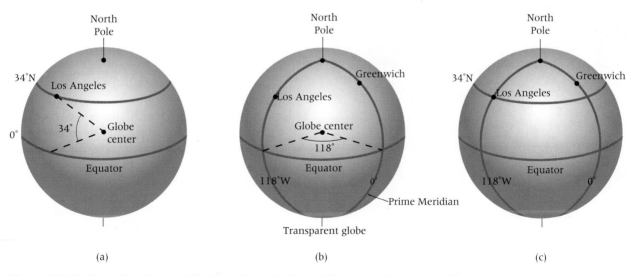

(a) (b) (c)

Figure 2.5 *Finding a location by latitude and longitude.* *(a) The geometric basis for the latitude of Los Angeles, California. Latitude is the angular distance in degrees either north or south of the equator. (b) The geometric basis for the longitude of Los Angeles. Longitude is the angular distance in degrees either east or west of the prime meridian, which passes through Greenwich, England. (c) The location of Los Angeles is 34°N, 118°W.* • **What is the latitude of the North Pole?**

A **sextant** can be used to visually determine latitude for celestial navigation (Fig. 2.6). This instrument measures the angle between the horizon and a celestial body, such as the noonday sun or the North Star (Polaris). However, the *latitude* of a location is only half of its global address. Los Angeles is located approximately 34° north of the equator, but an infinite number of points exist on the same line of latitude.

Measurement of Longitude. To fully describe Los Angeles's location, we must show *where* along the line of 34°N latitude it lies. However, to describe an east or west position, we must have a starting line, just as we used the equator as our latitudinal reference. Actually, any half of a great circle, drawn from pole to pole, could serve this function. The global position of this east-west reference line is arbitrary and, in the past, was a mat-

Figure 2.6 *Finding latitude for celestial navigation. A traditional way to determine latitude is by using a sextant to measure the angle between the horizon and a celestial body. Today most air and sea navigation (as well as certain land travel) is supported by a satellite-assisted technology called the Global Positioning System (GPS).* • **With high-tech location systems like GPS available, why might understanding how to use a sextant still be important?**

ter of national pride. Many countries used their own zero line, which passed through the country's capital, a situation that caused some confusion because longitudes were often dependent upon a map's nation of origin. This difficulty was finally resolved by international agreement in 1884, when the longitude line that passes through Greenwich, England (near London), was accepted as the **prime meridian,** or 0° longitude. **Longitude** is the angular distance east or west of the prime meridian.

Longitude is also measured in degrees, minutes, and seconds. Returning to our example of Los Angeles, let us imagine that same line drawn from the center of Earth to the point where the north-south half-circle passing through Los Angeles crosses the equator. This time, a second imaginary line will go from the center of Earth to a point where the prime meridian crosses the equator. Figure 2.5b shows that these two lines drawn from Earth's center define an angle the arc of which is the angular distance that Los Angeles lies west of the prime meridian. Los Angeles has a longitude of 118° west of Greenwich. Figure 2.5c provides the global address of Los Angeles by combining latitude and longitude.

As we go farther east or west from 0° longitude at the prime meridian, our longitude increases. Traveling eastward from the prime meridian, we will eventually be halfway around the world from Greenwich in the middle of the Pacific Ocean at 180°E. This line is also 180°W. Thus, longitude is measured in degrees up to a maximum of 180° east or west of the prime meridian.

The Geographic Grid

Any point on Earth's surface can be located by its latitude—north or south of the equator measured in degrees—and its longitude—east or west of the prime meridian measured in degrees. The geographic grid also forms a basis for determining locations on, and mapping of, the moon, the planets, and other celestial bodies.

Parallels and Meridians. The geographic grid consists of lines that run east and west around the globe and lines that run north and south from pole to pole (Fig. 2.7). East-west lines mark the latitude north or south of the equator. Latitude lines circle the globe, are evenly spaced, and are parallel to the equator and each other. Hence, they are known as **parallels.** The only parallel that is a great circle is the equator; all other lines of latitude are small circles. One degree of lati-

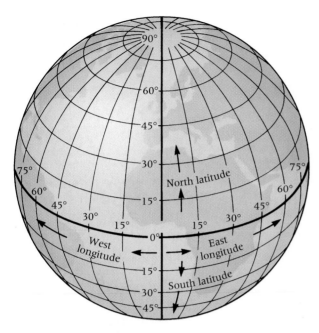

Figure 2.7 *A globelike representation of Earth, showing the global grid with parallels and meridians at 15° intervals.* • How do parallels and meridians differ?

tude equals about 111 kilometers (69 mi) anywhere on Earth.

Lines of longitude run north and south and converge at the poles and measure longitudinal distances east or west of the prime meridian. Each **meridian** of longitude, when joined with its mate on the opposite side of Earth, forms a great circle. Meridians *at any given latitude* are evenly spaced around the globe, although meridians get closer together as they move *poleward* from the equator. At the equator, meridians separated by 1° of longitude are about 111 kilometers (69 mi) apart, but at 60°N or 60°S latitude, they are only half that distance apart, about 56 kilometers (35 mi).

Longitude and Time

Time Zones

The relationship between longitude and time was used to establish the time zones that we know today. Until about a hundred years ago, each town or area used what was known as *local time.* That is, **solar noon** was determined by finding the precise moment in a day when a vertical stake cast its shortest shadow. This meant that the sun had reached its highest angle in the sky for that day at that location—noon, and all local clocks

were set to that time. Because of Earth's rotation, noon in a town to the east took place a little earlier, and towns to the west experienced noon a little later.

Using local time was not a great problem until the late 1800s, with the development of railroads and advances in communication, such as the telegraph. Increasing long-distance travel and technological changes made the use of local time by each community impractical. In 1884, the International Meridian Conference was held in Washington, D.C., to set up standardized time zones and to establish Greenwich as the location for the prime meridian and as 0° longitude. Earth was divided into 24 zones, one for each hour of the day. Ideally each time zone spans 15° of longitude (360° divided by 24 hours). The prime meridian was made the *central meridian* of its time zone, and the time when solar noon occurred at the prime meridian was established as noon for all places between $7\frac{1}{2}°$ E and $7\frac{1}{2}°$ W of that meridian. The same pattern was followed around Earth. Every line of longitude evenly divisible by fifteen degrees is the **central meridian** for a time zone of $7\frac{1}{2}°$ of longitude on either side. However, as shown in Figure 2.8, the time zones boundaries do not follow meridians exactly. Having a time zone boundary divide cities, towns, or states would be inconvenient, so jogs in the lines have been established to avoid most of these problems. In the United States, time zones often follow state boundaries. Imagine, for instance, the confusion that would result if Chicago was in one time zone and its suburbs were in another.

The time of day at the prime meridian, Greenwich Mean Time (sometimes called *GMT, Universal Time, UTC,* or *Zulu Time*) is used for a worldwide reference, and times to the east or west can be easily determined by comparing them to GMT. Time zones west of the prime meridian are said to be on *slow time* and places to the east are on *fast time.* Thus, a place 90° east of the prime meridian would be 6 hours fast (later), while time in the Pacific Time Zone of the United States and Canada, whose central meridian is 120°W, is said to be 8 hours slow (earlier).

For navigation, longitude can be determined with a *chronometer,* an extremely accurate clock. Two chronometers are used, one set on Greenwich time, and the other on local time. The number of hours between them, fast or slow, determines longitude (1 hour equals 15° of longitude). Until the advent of electronic navigation by ground- and satellite-based systems, the sextant and chronometer were the navigator's basic tools for determining location.

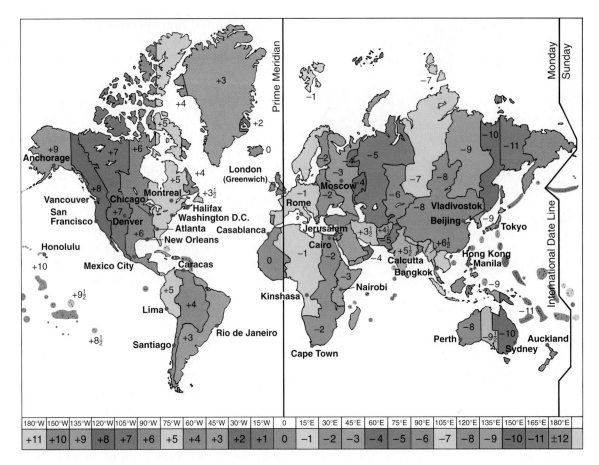

Figure 2.8 *World time zones reflect the fact that Earth turns through 15° of longitude each hour. Thus, time zones are approximately 15° wide. Political boundaries usually prevent the time zones from following a meridian perfectly.* • **What is the time difference between the time zone where you live and Greenwich, England?**

International Date Line

On the opposite side of Earth from the Greenwich meridian is the **International Date Line.** It is an imaginary line that follows the 180th meridian, except for jogs to separate Alaska and Siberia, and to skirt some Pacific island groups (Fig. 2.9).

At the International Date Line, we turn our calendar back a full day if we are traveling east and a full day forward if we are traveling west. Thus, if we are traveling east from Tokyo to San Francisco and it is 4:30 P.M. Monday just before we cross the International Date Line, it will be 4:30 P.M. Sunday on the other side. Or, if we are traveling west from Alaska to Siberia, and it is 10:00 A.M. Wednesday just before we reach the International Date Line, it will be 10:00 A.M. Thursday once we cross it. As a way of remembering this relationship, many globes have Monday and Sunday (or **M | S**) labeled in that order on the opposite sides of the International Date Line. To find the correct day, you just substitute the current day for Monday or Sunday, and use the same relationship, for example, *Wednesday | Tuesday.*

Although the International Date Line was not established officially until the 1880s, the need for such a line on Earth to adjust the day was inadvertently discovered by Magellan's crew, who circumnavigated Earth in the years 1519–21. Sailing westward from Spain around the world, the crew noticed when they returned that one day had apparently been missed in the ship's log. What had actually happened is that, in going around the world in a westward direction, the crew had experienced one less sunset and one less sunrise than had occurred in Spain during their absence.

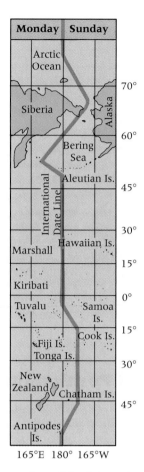

Figure 2.9 *A map view of the International Date Line. The new day officially begins at the International Date Line (IDL) and then sweeps westward around the Earth to disappear when it again reaches the IDL. Thus, west of the line is always a day later than east of the line. Maps and globes often have either "Monday|Sunday" or "M|S" shown on opposite sides of the line to indicate the direction of the day change.* • **Why does the International Date Line deviate from the 180° meridian in some places?**

The Public Lands Survey System

The global grid of parallels and meridians can be applied to *any* representation of Earth. Longitude and latitude, however, are used to locate the ***points*** *where these lines intersect.* Another system is used in the United States to define and locate land ***areas.*** This is the **U.S. Public Lands Survey System,** or the **Township and Range System.** Thought to have been suggested by Thomas Jefferson, it was proposed as a method for parceling public lands west of Pennsylvania for sale.

Using the Township and Range System, land was divided into parcels based on selected north-south lines called **principal meridians** and east-west lines called **base lines** (Fig. 2.10). The base lines were surveyed along parallels of latitude. The north-south meridians, though perpendicular to the base lines, had to be adjusted (jogged) along their length to counteract Earth's curvature. If adjustments were not made, the north-south lines would tend to converge and land parcels defined by this system would be smaller in northern regions of the United States.

The Township and Range System forms a pattern of nearly square parcels called *townships* laid out in horizontal *tiers* north and south of the base lines and in vertical *columns* ranging east and west of the principal meridians. A **township** is a square plot 6 miles on a side (36 sq mi or 93 sq km). As illustrated in Figure 2.11, townships are first labeled by their position north or south of a base line, so a township in the third tier south of a base line will be called *Township 3 South,* which is abbreviated T3S.

However, just as providing only the latitude of Los Angeles gave an insufficient locational description, labeling a township only by its tier position is also insufficient. We must also name a township according to its ***range,*** or its location east or west of the principal meridian for the survey area. Thus, if Township 3 South is in the second range east of the principal meridian, its full location can be given as T3S, R2E (Range 2 East).

The Public Lands Survey System further divides townships into 36 **sections** of *1 square mile or 640 acres* (2.6 sq km, or 259 ha). Sections are designated by numbers from 1 to 36, beginning in the northeasternmost section and going back and forth across the township, ending in the southeast corner with 36. Each section is divided into four quarters (quarter-sections), named by their location in the section—northeast, northwest, southeast, and southwest, each with 160 acres (65 ha). Quarter-sections are further subdivided into four quarters (quarter-quarter sections) of 40 acres each (16.25 ha), sometimes known as *forties.* These quarter-quarter sections, or forty-acre plots, are named after their position in the quarter: the northeast, northwest, southeast, and southwest forties. Thus, we can describe the location of the forty acres that are shaded in Figure 2.10 as being in the SW $\frac{1}{4}$ of the NE $\frac{1}{4}$ of Sec. 14, T3S, R2E, which we can find if we locate the principal meridian and the base line for the survey. The order is consistent from small division to

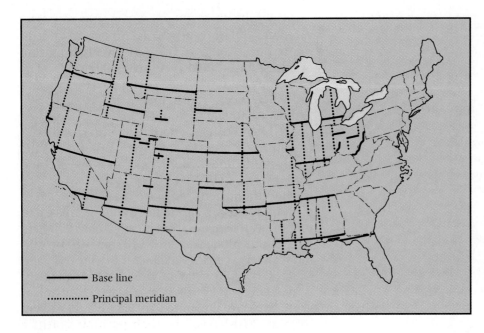

Figure 2.10 Principal base lines and meridians of the United States Public Lands Survey System (Township and Range System). • Why wasn't the Township and Range System applied throughout the eastern United States?

larger, and township location is always listed before range (T3S, R2E).

The Public Lands Survey System has had an enormous impact on the landscape of the United States; it is what gives most of our Midwest and West its regular checkerboard appearance from the air or from space (Fig. 2.12). The land was surveyed, staked, fenced, and farmed or ranched in parcels formed by the divisions of the Public Lands Survey System. Road maps in states that use this land survey system reflect the use of this grid, with many roads following angular boundaries surveyed using this government system.

The Global Positioning System

The newest way to determine our geographic location employs the Global Positioning System (GPS). The GPS uses a uniform network of satellites in orbits 11,000 miles above Earth. This high-tech locational system was originally created for military applications but today has a host of civilian uses—from surveying to navigation.

The GPS relies on the ability to measure exact distances from our location on Earth to several orbiting satellites. Radio signals transmitted by the satellites are detected with a ground re-

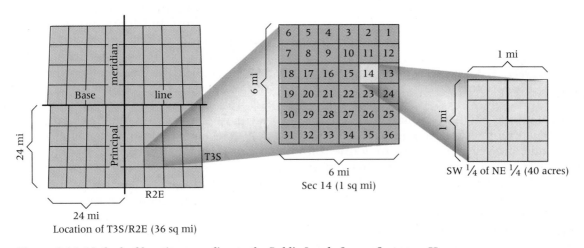

Figure 2.11 Method of location according to the Public Lands Survey System. • How would you describe the extreme southeastern 40 acres of section 20 in the middle diagram?

Figure 2.12 *Rectangular field patterns resulting from the Public Lands Survey System in the western United States. Note the slight jog in the field pattern to the right of the farm buildings in the center of the photo.* • **How do you know this photo was not taken in the midwestern United States?**

ceiver. The GPS receivers vary in size, but hand-held units are now available (Fig. 2.13). The distance from receiver to satellite is calculated by measuring the time it takes for the radio signals, traveling at the speed of light, to arrive. The GPS receivers perform these calculations and give a locational readout in longitude and latitude.

The Global Positioning System is based on the principle of triangulation. Assume, through the process just discussed, that we determine that our distance from the GPS satellite is 20,000 kilometers. Thus, we are now certain that our location is somewhere in space on an imaginary sphere with a radius of 20,000 kilometers centered on that satellite (Fig. 2.14a). This is an enormous area. However, if we had simultaneously

determined we were 15,000 kilometers from a second satellite, our locational possibilities would be greatly reduced. Our location now involves the intersection of two spheres, one with a radius of 20,000 kilometers, the other with a radius of 15,000 kilometers. The *only* common surface between two intersecting spheres is a circle formed where the two spheres intersect (Fig. 2.14b). A simultaneous measurement from a third satellite, perhaps 18,000 kilometers away, results in the intersection of a third sphere with the first two. The original *sphere* of possible locations, which was reduced to a *circle* of possible locations by the addition of a second satellite, is brought down to just *two possible points* by the addition of a third satellite (Figure 2.14c). A fourth satellite allows

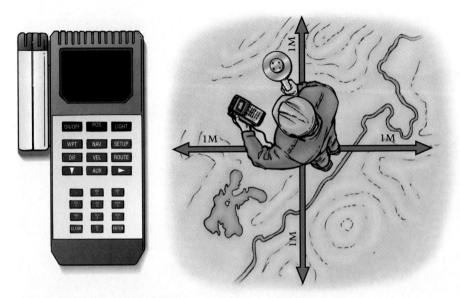

Figure 2.13 *Global Positioning System (GPS). A handheld GPS receiver provides a readout of its latitudinal and longitudinal position based on signals from a satellite network. A handheld unit provides an accuracy that is acceptable for many uses. Precise surveying, however, requires a second receiver (connected to the antenna with the small dome on top) to provide reference data from a point of known location. Using this two-receiver method, a locational accuracy within one meter is fairly common, but better accuracy can also be obtained using special techniques and equipment. • **Other than for surveying, what other uses can you think of for a small unit like this that displays your longitude and latitude as you move from place to place?***

us to determine which of the two points is our location.

The accuracy of the GPS depends on several factors. One is the number of satellites used, as a minimum of three satellites are needed to determine your position. Using signals from additional satellites, however, will give a more precise location. Also, satellite angles high above the horizon and low levels of interference between you and the satellites (open prairie vs. dense forest) are other factors that yield more accurate measurements. Accuracy to within several meters can be obtained, and locational coordinates within small fractions of a meter are possible.

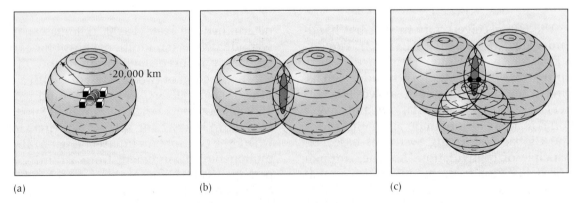

(a)   (b)   (c)

Figure 2.14 *The Global Positioning System uses signals from a network of satellites to determine a location. (a) If you are 20,000 kilometers from a known satellite you are somewhere on the sphere depicted. (b) Referencing another satellite at the same time narrows your location to somewhere on the circle of their intersection. (c) Adding yet a third satellite narrows your location to one of two points.*

Maps and Map Projections

Maps—representations of Earth or parts of Earth on flat paper—have many advantages. They can be reproduced easily and inexpensively, can depict the entire Earth or show a small area in great detail, and compared to a globe are easier to handle and transport. Maps also have limitations that vary with the type of map, but most of these limitations can be avoided by having an understanding of maps and cartography. There are many kinds of maps, and the key is knowing what kind of map is best for a particular job. No map fits all possible applications.

Advantages of Maps

It has been said that if a picture is worth a thousand words, then a map is worth a million. As visual representations of Earth, maps supply an enormous amount of information that would take pages and pages to describe—probably less successfully—in words. Because they are graphic representations and use symbolic language, maps show spatial relationships with great efficiency. Imagine trying to verbally explain all the information that a map of your city, county, state, or campus provides: sizes, areas, distances, directions, street patterns, railroads, bus routes, hospitals, schools, libraries, museums, highway routes, business districts, residential areas, population centers, and so forth. Maps can show true courses for navigation and true shapes of Earth features. They can be used to measure areas or distances, and they can show the best route from one place to another.

Geographers can produce maps to show almost any relationship in the environment. As the interest of physical geographers is essentially spatial—the relationships among environmental subsystems and their changes through time and space—geographers can use maps to develop a visual or a mathematical description of these relationships. For many reasons, then—whether it is presented on paper, on a computer screen, or in the mind—the map is the geographer's most important tool.

In addition to being vital tools to the professional geographer, maps are useful to people in every walk of life. They are valuable to tourists, political scientists, historians, geologists, pilots, soldiers, sailors, hikers, and even burglars (police departments also use maps to plot the spatial distribution of certain crimes). The Centers for Disease Control use maps to monitor the distributions of illness outbreaks worldwide. The potential applications of maps are practically infinite.

Limitations of Maps

Despite the positive aspects of maps, they can never depict the Earth with complete accuracy. Unlike globes, every map distorts the Earth in some way. Mathematically, it is impossible to make a spherical Earth flat (two-dimensional) and accurately maintain all of its properties. This process has been likened to trying to flatten out an eggshell. Distortion is an unavoidable problem of representing a globe on a flat surface.

Using a globe, we can directly compare the size, shape, and area of all the features of Earth, and we can also measure distance, direction, shortest routes, and true courses, all with great accuracy. But because of the inherent distortion of maps, we can never compare or measure all of these properties on one map with complete confidence. There is no such thing as a perfect map, at least not the perfect map for all uses.

We must accept the distortions of Earth that are represented on maps. We should realize, though, that when a map depicts only a small area, the distortion will be so slight as to be negligible. If we use a map of a state park for a day hike, the distortion will be too small to affect us, and we can have confidence in the map information. It is when we attempt to show large regions, or the entire Earth, that Earth's curvature has to be considered, and the map distortion becomes apparent and pronounced.

Obviously, maps are useful despite their limitations and distortions. We must, however, know what properties a map depicts accurately, what features it distorts, and for what purpose a map is best suited. Given this information, we can make accurate comparisons and measurements on maps.

The global grid has four important properties: (1) parallels of latitude are always parallel; (2) parallels are evenly spaced; (3) meridians of longitude converge at the poles; and (4) meridians and parallels always cross at right angles. Even though there are thousands of ways to transfer a spherical grid onto a flat surface to make a **map projection,** no map projection will maintain all four of these properties. Checking the grid system of a map for these four properties will help us discover areas of greatest and least distortion on a particular map projection.

Properties of Map Projections

Shape. *A flat map cannot depict large areas of the globe without distortion of either shape or area.* But small areas—regions, lakes, islands, bays—*can* be depicted in their true shape, with little apparent distortion. Maps that maintain true shape of areas are said to be **conformal.**

In order to preserve the shape of small areas, we would expect that the parallels and meridians, transferred from a globe to a flat surface conformal map, would maintain their global relationships. On all conformal maps, meridians and parallels always cross at right angles, just as they do on the globe.

Area. Mapmakers are able to create a world map that has true area for the whole Earth. That is, areas shown on the map have the same proportions to each other as they have in reality. Thus, if we cover any two parts of the map with, say, a nickel, no matter where the nickel is, at each location it will cover equivalent areas on Earth. Maps drawn with this property are called **equal-area** maps. Whenever a size comparison is being made between two or more areas, it is important to use an equal-area map. The property of equal area is also useful when examining spatial distributions. As long as the map has equal area and a symbol represents the same quantity throughout the map, we can get a good idea of the distribution of any feature—for example, people, churches, cornfields, hog farms, or volcanoes.

The only way to show equal area on a flat map, however, is to pull areas out of shape. It is impossible to show both equal area and true shape on a flat map. Most of us are familiar with the **Mercator** world projection, commonly used in schools and textbooks, although less so in recent years (Fig. 2.15). In the Mercator projection, areas have true *shape,* so it is conformal and cannot have the property of *equal area.* Areas are greatly out of proportion. For many years, the Mercator projection was widely used in schools, unfortunately leading generations of students to incorrectly believe that Greenland is as large as South America. (Mercator's projection shows Greenland as about the size of South America [see again Fig. 2.8], but actually, South America is about eight times larger!) A nickel placed on Greenland in this conformal map would cover an area of land much smaller than that covered by the same coin placed on South America.

Distortion is a major challenge in mapmaking. Because it is not possible to show both true shape and equal area correctly, cartographers must decide if they want to show one at the expense of the other. Another solution is to compromise and create a map that does a fairly good job of showing both properties but is not really correct for either.

Distance. Just as no flat map can show true shape for all of Earth or for large areas, neither can it maintain constant scale of distance over all Earth's surface. The scale on a map depicting a large area cannot be applied equally everywhere on that map. On a map of a small area, distance distortions will be insignificant, and the accuracy will usually be sufficient for most purposes.

It is possible for a map to have the property of **equidistance** in specific instances. That is, on a world map, the equator may have equidistance (a constant scale) along its length. Or all meridians, but not the parallels, may have equidistance. On another map all straight lines drawn from the center may have equidistance, but the scale will not be constant unless lines are drawn from the center.

Direction. Because the compass directions on Earth curve around the sphere, not all flat maps can show true compass directions. Thus, a given map may be able to show true north, south, east, and west, but the directions between those points may not be accurate in terms of the angle between them. So, if we are sailing toward an island, its location may be shown correctly according to its longitude and latitude, but the direction in which we must sail to get there may not be accurate, and we may pass right by it.

Maps that show true directions are called **azimuthal** map projections. These are drawn with a center or focus, and all straight lines drawn from that center are true compass directions (Fig. 2.16).

Map Projections

We have been discussing *map projections* without fully explaining what is specifically meant by this term. A map projection is a representation of the curved surface of a globe on a flat surface. As we have seen, some of the properties of a globe must be compromised in order to flatten it out into a map.

All map projections must maintain one aspect of the globe—the property of location. Every place shown on a map must be in the same location with respect to latitude and longitude as it is on the globe. No matter how much the

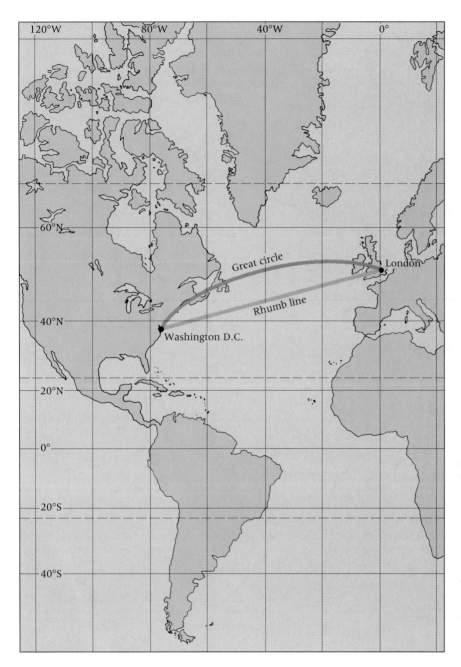

Figure 2.15 *The Mercator projection, often misused for general-purpose maps, was designed for the purpose of navigation. Its useful property is that lines of constant compass heading, called rhumb lines, are straight lines. The Mercator is developed from a cylindrical projection.* • **Compare the sizes of Greenland and South America on this map to their proportional sizes on a globe. Is the distortion great or small?**

arrangement of the global grid is changed by projecting parallels and meridians onto a flat surface, we must be sure that all places still have their proper location, that all bays, cities, lakes, and mountains are still at their accurate latitude and longitude.

Although maps are not actually made in this fashion, certain projections can be demonstrated by putting a light inside a transparent globe so that the grid lines are projected onto a plane or surface **(planar projection),** a cylinder **(cylindrical projection),** or a cone **(conic projection),** which can be flattened (Fig. 2.17). Map

projections today are developed by computers using mathematical computations involving the geographic grid.

The Mercator Projection. As previously mentioned, one of the best-known conformal projections is the Mercator, named for Gerhardus Mercator, who devised this map in 1569. The **Mercator projection,** shown in Figure 2.15, is actually a mathematically adjusted cylindrical projection. Imagine a transparent globe with a light inside along its equator that casts the shadows of the grid onto a sheet of paper wrapped

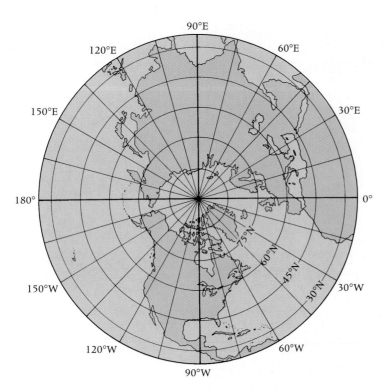

Figure 2.16 *Azimuthal map centered on the North Pole. Although this is the conventional orientation of such a map, it could be centered anywhere on Earth. It shows true direction to all other points. Azimuthal maps can show only one Earth hemisphere at one time.*

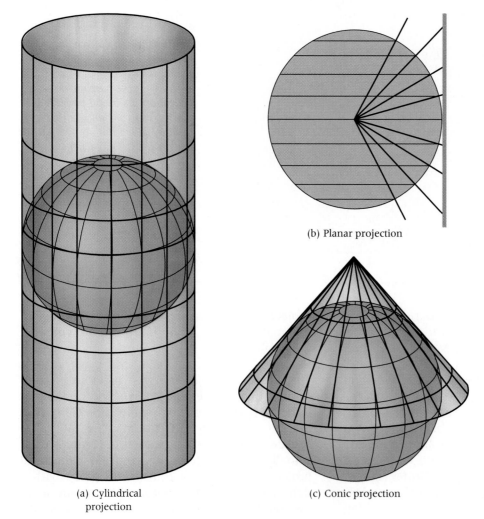

(b) Planar projection

(a) Cylindrical projection

(c) Conic projection

Figure 2.17 *The theory behind the development of the (a) cylindrical, (b) planar, and (c) conic projections. Although projections are not actually produced in this manner, they can be demonstrated by projecting light from a transparent globe.* • **Why do we use different map projections?**

around the globe to form a cylinder. If the paper touches the globe everywhere at the equator, then the equator on the flat map will be in scale. The rest of the grid, however, will not be to scale (Fig. 2.17a). Meridians will appear as parallel lines instead of converging at the poles. Obviously, there is enormous east-west distortion for the polar regions since the distance between meridians is stretched to the same width as that at the equator. The spacing of parallels on a Mercator projection is not even (equal) as it is on the globe, but it is proportional to the increased spacing of meridians as they approach the poles. The resulting grid consists of rectangles that increase in size toward the poles. Small areas on this map keep their true shape, but obviously this projection does not have equal area.

The Mercator projection has another important property that no other world projection has. A straight line drawn anywhere on a Mercator projection is a true compass heading. Such a line, *of constant direction,* is called a **rhumb line,** and it has great value to navigators (Fig. 2.15). Although the *distance* along a rhumb line drawn from place to place may not be accurate or to scale, the compass heading will always be accurate no matter where it is drawn on the map.

The Gnomonic Projection.   This is one of the oldest map projections. Gnomonic projections are planar projections, made by projecting the grid lines onto a plane (surface) (Fig. 2.17b). If we put a flat sheet of paper, which is a plane, tangent to (touching) the globe at the equator, the grid will be projected with great distortion. Parallels and meridians will be unevenly spaced, and the shapes and areas of land and water bodies will also be badly distorted.

Despite its distortion, the gnomonic projection has a valuable characteristic. In Figure 2.18, we can see that all the parallels appear curved except the equator, which appears as a straight line. In addition, *all* meridians appear as straight lines. One thing we have learned about the equator and the meridians is that they all follow great circles. Thus, if the equator and meridians appear as straight lines on a gnomonic projection, then all great circles must appear as straight lines on maps using this projection. In fact, gnomonic projections are the only maps that display all arcs of great circles as straight lines. This characteristic also makes gnomonic projections important to navigation. Navigators draw a straight line between their location and where they want to go, and this line will be a great circle route—the

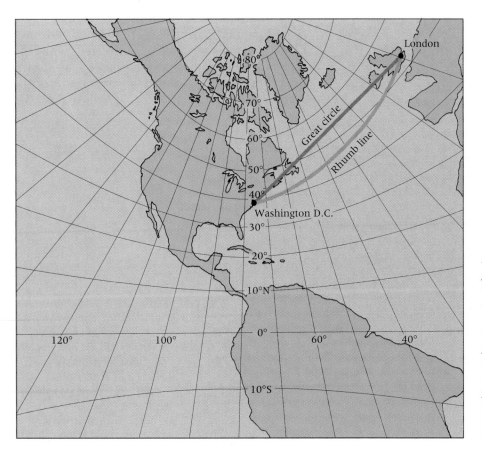

Figure 2.18 The gnomonic projection produces extreme distortion of distances, shapes, and areas. Yet it is extremely valuable for navigation, for it is the only projection that shows all great circles as straight lines. It is developed from a planar projection. • Compare this figure with Figure 2.15. How do the projections in the two figures differ?

shortest route between the two places. This projection has great utility for long-distance communications and radar, including some weather radar.

An interesting relationship exists between gnomonic and Mercator projections. Great circles on the Mercator projection appear as curved lines, while rhumb lines appear straight (Fig. 2.15). On the gnomonic projection the situation is reversed—great circles appear as straight lines, while rhumb lines are curves (Fig. 2.18).

Conic Projections.   Conic projections are used for making maps of midlatitude areas, such as the United States (other than Alaska and Hawaii), because they show regions in these latitudes with minimal distortion. In a simple conic projection, a cone is fitted over the globe with its pointed top centered over a pole. The sides of the cone are tangent to Earth at a predetermined parallel, known as the *standard parallel* (Fig. 2.17c). The parallels of a simple conic projection, like the lines of latitude they represent, are concentric circles that become smaller the closer they are to the pole. The meridians on a conic projection appear as straight lines radiating from the pole. However, the distortions are slight, although they increase with distance from the standard parallel. Two standard parallels are used to make a **Lambert conformal conic projection** of the United States (Fig. 2.19), and the result is a highly accurate and useful map. The distortion on this map is only 1 percent from the east coast to the west coast and from the Canadian to the Mexican borders.

Distortion of land masses can also be reduced by using an **interrupted projection** to produce several smaller segments (Fig. 2.20a). Each major continental area is based on its own central meridian, so no part of the projection suffers from extreme shape distortion. If our interest was centered on the world oceans, the projection could be interrupted in the continental areas to minimize distortion of the ocean basins. Some world maps are **compromise projections** that are neither conformal nor equal area, but an effort is made to balance and minimize the distortion to produce an "accurate looking" map, particularly in the areas where most of the world's population is located (Fig. 2.20b).

Map Essentials

We have examined only a small sample of the methods for projecting the global grid onto a flat surface. In addition, grid coordinates are only one element of a useful map. Each map is designed with a specific need in mind and it will include important information to meet that need. Among the essential items are title, date, legend, scale, and direction. In addition, cartographers must select the best way to depict the spatial phenomena that the map is designed to show.

Title, Date, and Legend.   Every map should have a title that tells what area is depicted, what the map is about, and what relationships it shows. For example, a map drawn to provide information for hiking and camping in Yellowstone Na-

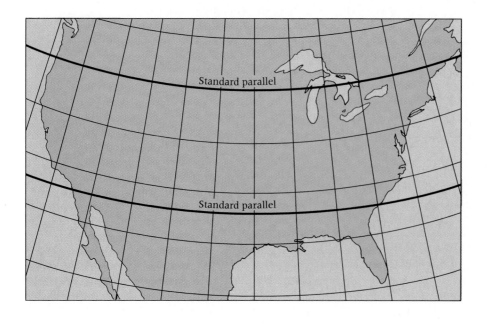

Figure 2.19 *The Lambert conformal conic projection is a mathematical projection having two standard parallels. This projection is used when angles and shapes of midlatitude areas are to be kept as accurate as possible.*
• **What is meant by a mathematical projection?**

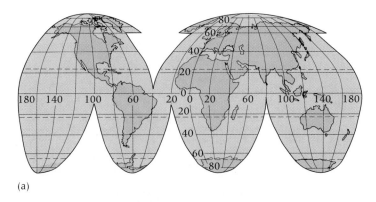

(a)

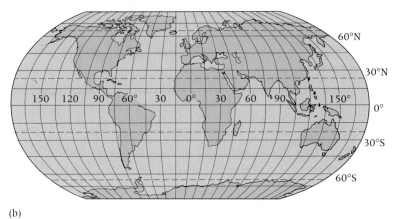

(b)

Figure 2.20 Distortion by projections can be reduced by interruption (**a**)—that is, by having a central meridian for each segment of the map. The Robinson projection (**b**) is considered a compromise projection because it departs from being equal-area to better depict the shape of the continents.
• Compare the distortion of these maps with the Mercator projection (Fig. 2.15). What is a disadvantage of (a) in terms of usage?

tional Park should have a title like this: "Yellowstone National Park: Trails and Camping Areas."

Each map should indicate its publication date or the date to which its information applies. For instance, a map of the population distribution in the United States should tell when the census was taken, to let us know whether the information or data are current, outdated, or whether the map is intended to show historical data.

A map should also have a **legend**—a key to symbols employed on the map. For example, if one dot represents 1000 people, or the symbol of a pine tree represents a roadside park, the legend should explain this information. If color shading is used on the map to represent elevations, different climatic regions, or other factors, then a key to the color coding should be provided. Map symbols can be designed to represent virtually any feature (Fig. 2.21).

Scale. Maps depict Earth features in a smaller size than they actually are. In order to measure size or distance, we need to know the map **scale** that we are using (Fig. 2.22). That is, we need to know the relationship or ratio between an actual distance on Earth and the same distance as it ap-

pears on the map. Nearly every map will indicate to what scale the map is drawn and to what parts of the map that scale is applicable. Scale representations enable us to accurately and precisely measure distances, determine areas, and compare sizes.

Because making a flat map involves stretching some places more than others, the scale given can never be totally accurate all over the map. However, if the map is of a small area, such as a city or a state, the distortion will be so slight that the scale can be accepted as applicable everywhere on the map. On world globes, however, one scale is applicable everywhere on its surface.

A **verbal scale** is a statement on the map that says something like "1 inch to 1 mile" (*1 inch on the map represents 1 mile on the ground*), or "1 inch to 1000 miles" (*1 inch represents 1000 miles*). A verbal scale, however, is no longer applicable if the original map has been reduced or enlarged. Nor can a verbal scale be used by people who do not understand the units of measure (such as miles or kilometers) stated on the map.

The **representative fraction (RF)** scale is presented as a ratio that is free of units of measurement. Thus, it can be used with any unit of measurement—feet, inches, meters, centimeters—

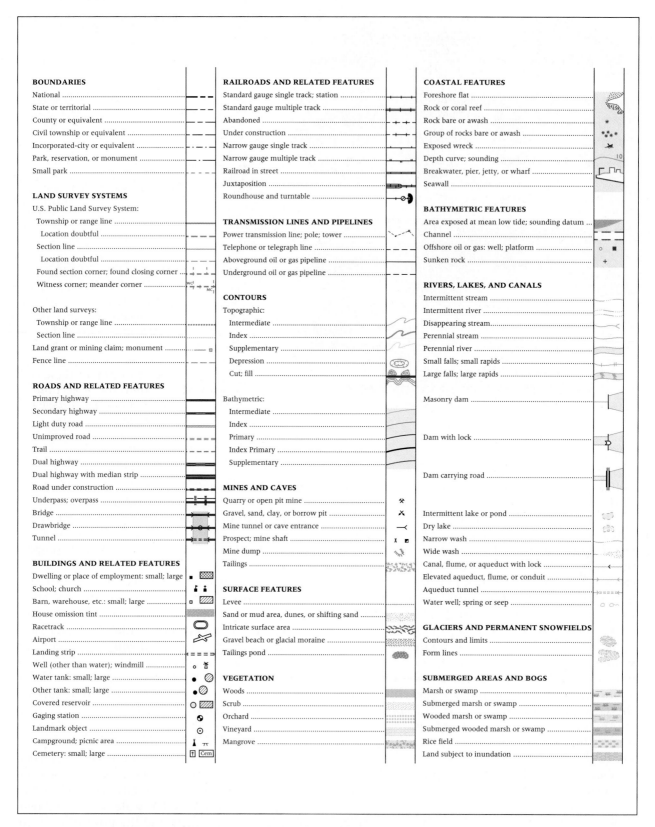

BOUNDARIES
National
State or territorial
County or equivalent
Civil township or equivalent
Incorporated-city or equivalent
Park, reservation, or monument
Small park

LAND SURVEY SYSTEMS
U.S. Public Land Survey System:
Township or range line
Location doubtful
Section line
Location doubtful
Found section corner; found closing corner
Witness corner; meander corner

Other land surveys:
Township or range line
Section line
Land grant or mining claim; monument
Fence line

ROADS AND RELATED FEATURES
Primary highway
Secondary highway
Light duty road
Unimproved road
Trail
Dual highway
Dual highway with median strip
Road under construction
Underpass; overpass
Bridge
Drawbridge
Tunnel

BUILDINGS AND RELATED FEATURES
Dwelling or place of employment: small; large
School; church
Barn, warehouse, etc.: small; large
House omission tint
Racetrack
Airport
Landing strip
Well (other than water); windmill
Water tank: small; large
Other tank: small; large
Covered reservoir
Gaging station
Landmark object
Campground; picnic area
Cemetery: small; large

RAILROADS AND RELATED FEATURES
Standard gauge single track; station
Standard gauge multiple track
Abandoned
Under construction
Narrow gauge single track
Narrow gauge multiple track
Railroad in street
Juxtaposition
Roundhouse and turntable

TRANSMISSION LINES AND PIPELINES
Power transmission line; pole; tower
Telephone or telegraph line
Aboveground oil or gas pipeline
Underground oil or gas pipeline

CONTOURS
Topographic:
Intermediate
Index
Supplementary
Depression
Cut; fill

Bathymetric:
Intermediate
Index
Primary
Index Primary
Supplementary

MINES AND CAVES
Quarry or open pit mine
Gravel, sand, clay, or borrow pit
Mine tunnel or cave entrance
Prospect; mine shaft
Mine dump
Tailings

SURFACE FEATURES
Levee
Sand or mud area, dunes, or shifting sand
Intricate surface area
Gravel beach or glacial moraine
Tailings pond

VEGETATION
Woods
Scrub
Orchard
Vineyard
Mangrove

COASTAL FEATURES
Foreshore flat
Rock or coral reef
Rock bare or awash
Group of rocks bare or awash
Exposed wreck
Depth curve; sounding
Breakwater, pier, jetty, or wharf
Seawall

BATHYMETRIC FEATURES
Area exposed at mean low tide; sounding datum
Channel
Offshore oil or gas: well; platform
Sunken rock

RIVERS, LAKES, AND CANALS
Intermittent stream
Intermittent river
Disappearing stream
Perennial stream
Perennial river
Small falls; small rapids
Large falls; large rapids

Masonry dam

Dam with lock

Dam carrying road

Intermittent lake or pond
Dry lake
Narrow wash
Wide wash
Canal, flume, or aqueduct with lock
Elevated aqueduct, flume, or conduit
Aqueduct tunnel
Water well; spring or seep

GLACIERS AND PERMANENT SNOWFIELDS
Contours and limits
Form lines

SUBMERGED AREAS AND BOGS
Marsh or swamp
Submerged marsh or swamp
Wooded marsh or swamp
Submerged wooded marsh or swamp
Rice field
Land subject to inundation

Figure 2.21 Standard USGS topographic map symbols. • **In general how are different types of roads indicated on these maps?**

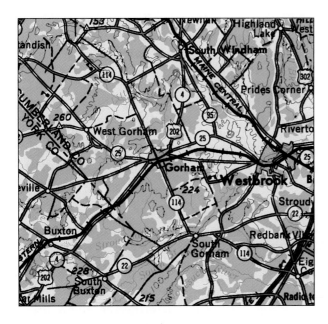

Top: 1:250,000 scale
1 inch = nearly 4 miles.
Area shown,
107 square miles.

Center: 1:62,500 scale
1 inch = nearly 1 mile.
Area shown,
6 3/4 square miles.

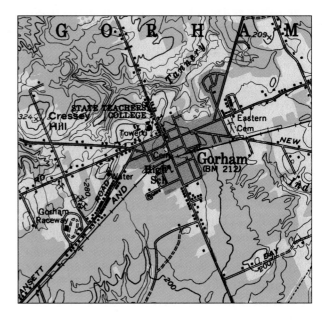

Bottom: 1:24,000 scale,
1 inch = 2000 feet.
Area shown,
1 square mile.

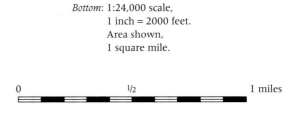

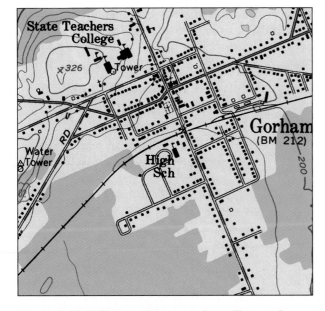

Figure 2.22 *Differences in map scale as illustrated on common U.S. Geological Survey
topographic maps.* • **Is the map at the top a small- or large-scale map?**

as long as the same unit is used on both sides of the ratio. As an example, a map may have an RF scale of 1:63,360, which can also be expressed 1/63,360. This means that 1 inch on the map represents 63,360 inches on the ground. It can also mean that 1 cm on the map represents 63,360 cm on the ground. In the case of our example using inches, knowing that 1 inch on the map represents 63,360 inches on the ground may be difficult to conceptualize unless we realize that 63,360 inches is the same as 1 mile. Thus, the representative fraction 1:63,360 means the map has the same scale as a map with a verbal scale of 1 inch to 1 mile.

A third kind of scale, a **graphic** or **bar scale,** is useful for directly measuring distances. A **graphic scale** is a graduated line marked with specific map distances proportional to distances on Earth. To use a graphic scale, take a straight edge, such as the edge of a piece of paper, and on it mark the distance between any two points on the map. Then use the graphic scale to measure the equivalent distance on Earth's surface. Graphic scales have two major advantages: (1) they are applicable even if the map is reduced or enlarged, because the graphic scale will also change proportionally in size; and (2) no mathematical calculations, or unit conversions, are necessary to determine true distances on the map, because the graphic scale can be used like a ruler to make direct measurements.

Maps are often described as of small, medium, or large scale. **Small-scale** maps show large areas (to do this the areas must be shown in a small size), include little detail, and have large denominators in their representative fractions. **Large-scale** maps show small areas of Earth's surface in greater detail and have smaller denominators in their representative fractions. To help avoid confusion, remember that 1/2 is a *larger* fraction than 1/100. Maps with representative fractions larger than 1:25,000 are large-scale. Medium-scale maps have representative fractions between 1:25,000 and 1:250,000. Small-scale maps have representative fractions less than 1:250,000. This classification follows the guidelines of the **United States Geological Survey,** publisher of many maps for the public and the federal government.

Direction. Direction can be shown on a map by the orientation of the global grid, because parallels of latitude are east-west lines and meridians of longitude run directly north-south. In addition to, or perhaps instead of, these lines, many maps have an arrow pointing to north (north on the map) to provide an orientation to any compass direction. A north arrow may indicate either **true north** or **magnetic north,** or two north arrows may be given, one for true north and one for magnetic north.

Our planet has a magnetic field that emanates from its core, making the planet act like a giant bar magnet. Earth has a magnetic north pole and a magnetic south pole, each with opposite charges, and a magnetic force field that radiates in a distinctive pattern into space. Although the magnetic poles shift position slightly from time to time, they are located in the Arctic and Antarctic regions. The north magnetic pole is located in northern Canada at 79°N, 103°W, near Ellef Ringnes Island, and the south magnetic pole is at 65°S, 139°E, in Commonwealth Bay, Antarctica.

A magnetized compass needle aligns itself with Earth's magnetic field, and the north-seeking end of the needle points toward the magnetic north pole. If we know the **magnetic declination,** the angular difference between magnetic north and true geographic north, we can then adjust our direction accordingly (Fig. 2.23). Thus, if our compass points north and we know that the magnetic declination for our location is 20°E, we can adjust our course knowing that our compass is pointing 20° east of true north. To do this we should turn 20° west from the direction indicated by our compass in order to face true north.

Magnetic declination varies from place to place over Earth's surface and also changes position through time. For this reason, maps of magnetic declination are revised periodically, and using a recent map is very important.

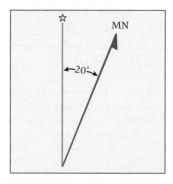

Figure 2.23 Map symbol showing true north, symbolized with a star representing Polaris, the North Star, and magnetic north, symbolized by an arrow. The example indicates 20° east magnetic declination. • **In what circumstances would we need to know the magnetic declination of our location?**

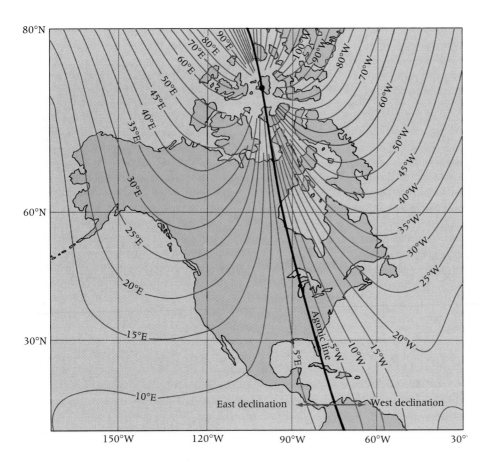

Figure 2.24 *Isogonic map of North America, showing the magnetic declination that must be added (west declination) or subtracted (east declination) from a compass reading to determine true directions.* • **What is the magnetic declination of your hometown to the nearest degree?**

A map of magnetic declination is an **isogonic map** (Fig. 2.24). On such a map, **isogonic lines** connect locations that have the same magnetic declination. It is possible to be in a position where magnetic north and true north are aligned. On an isogonic map, the line that connects these points is called the **agonic line** (line of zero magnetic declination).

Location. Having examined globes and maps and learned some of their characteristics and symbols, we now have the ability to comprehend and communicate the language of location. We can report a location on a map or globe by latitude and longitude. We can also show location by our compass direction from a known position to our position, and by the distance from that known position to the point where we are. Compass directions can be given by using either the *azimuth* system or the *bearing* system. In the **azimuth** system, direction is given in degrees of a full circle (360°) with respect to north. That is, if we imagine a circle of 360° with north at 0° (and at 360°) and read the degrees in a clockwise manner, we can describe a direction by its number of degrees away from north. For instance, straight

east would have an azimuth of 90°, and due south would be 180°. The **bearing** system divides compass directions into four quadrants of 90° (N, E, S, W), each numbered by directions in degrees away from either north or south. Using this system, an azimuth of 20° would be North, 20°East (20° east of due north), and an azimuth of 210° would be South, 30° West (30° west of due south). Both azimuths and bearings are used for giving precise directional instructions for mapping, surveying, military, and navigational purposes.

Differentiation and Distribution on Maps

A limitless variety of information may be shown on a map. Maps that are designed to focus attention on only one feature (or a few related ones) are called **thematic** maps. Examples include maps of climate, vegetation, soils, earthquake epicenters, tornadoes, or any other singular theme. To fulfill a particular purpose, mapmakers choose methods of cartographic representation that best display and communicate the desired information. For example, a common type of map used by geographers is one that shows the outline or shape of areas or *regions* which exhibit a common characteristic within their boundaries. Adjacent *areas* are differentiated from one another by different colors, patterns of lines, or depths of shading. Maps that use this kind of representation are used to show the distribution of soil, climate, and vegetation types (see the world maps throughout the book). They are also used to show different political divisions. A map of the United States that shows the separate states in different colors (at least no neighboring states would be the same color) is a good example.

Another type of map in everyday use is one that shows how certain kinds of features are located or arranged on Earth's surface. Maps of this type reveal patterns of such diverse phenomena as streams, highways, railroads, buildings, types of land use, and deposits of natural resources like coal, petroleum, and iron ore.

A particularly valuable map to the geographer is one that shows the *distribution* of amounts or quantities of something across Earth's surface. Numerous techniques and symbols have been developed to prepare this type of map. A familiar example is the **dot map,** in which each dot represents a specific quantity of a particular feature. Sometimes squares, cubes, or circles of different sizes are used instead of dots to represent quantities. The *pattern* of the symbols shows the distribution. Dot maps are often employed to show population density, but they can also show the density of many other features, such as earthquakes, hog farms, oil wells, automobiles, or registered voters (Fig. 2.25).

Distributions can also be shown using **isolines,** lines on a map that connect all points with the same numerical value. We have already encountered one example of an isoline—the isogonic line, which connects points with the same magnetic declination (see Fig. 2.24). Isolines that we will be using later on include **isotherms,** lines connecting points of equal temperature; **isobars,** lines connecting points of equal barometric pressure; **isobaths** (also called bathymetric contours), lines connecting points with equal depths of water; and **isohyets,** lines connecting points having equal amounts of precipitation.

Topographic Maps

Contour lines, isolines connecting points on a map that are at the same elevation above mean sea level (or sometimes below, as in Death Valley, California), are of special interest to physical geographers. For example, if we walk around a

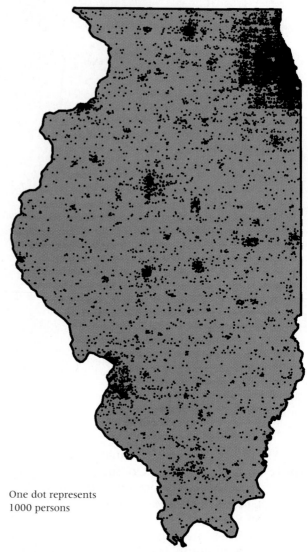

**Population Density
Illinois, 1990**

One dot represents
1000 persons

SOURCE: U.S. Bureau of the Census, 1990
Map produced by The Illinois Institute for Rural Affairs

Figure 2.25 *This map of population density in Illinois (1990) illustrates the use of a dot map to show spatial distribution.* • **How would you describe the population distribution of Illinois?**

hill along the 1200-foot contour line shown on the map, we would be 1200 feet above sea level at all times and we would walk on a level line, not uphill or downhill, maintaining a constant elevation.

Contour lines are important because they are an excellent method for depicting the land surface configuration on a map (see Map Interpretation: Contour Maps). The arrangement, spacing, and changing shapes of the contours give a map reader an accurate idea of what the topog-

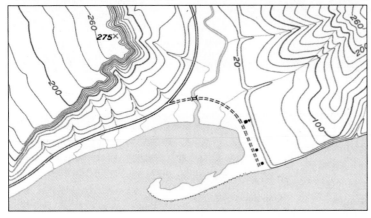

Figure 2.26 A bird's-eye view of a river valley and surrounding hills, shown on a shaded-relief diagram. Note that the river flows into a bay partly enclosed by a hooked sand spit. The hill on the right has a smoothly eroded form, whereas the one on the left rises to a cliff at the edge of an inclined tableland; (bottom) the same features represented on a contour map.
• **If you had only a contour map, could you visualize the terrain shown in the shaded-relief diagram?**

raphy (the "lay of the land") of the map area looks like (Fig. 2.26).

Figure 2.27 illustrates how contour lines show the geometric character of the land surface. The bottom portion of the diagram is a simple contour map of an asymmetrical volcanic island. Note that the difference in elevation between adjacent contour lines is a constant interval of

20 feet. A constant difference in elevation between contour lines is found on all contour maps and is called the **contour interval.**

What kind of terrain would we cover if we were to walk across this volcanic island from point A to point B? We start from point A at sea level but immediately begin to climb. We soon cross the 20-foot contour line, then the 40-foot,

(text continues on page 58)

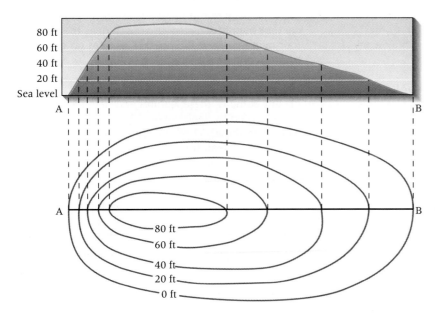

Figure 2.27 Topographic contour lines connect points of equal elevation in relation to mean sea level. The top portion of the figure shows the topographic profile (side view) of an island. Horizontal lines mark 20-foot intervals of elevation above sea level. The lower portion of the figure shows how these contour lines look from directly overhead. • **Study this figure and Figure 2.26. What is the relationship between the spacing of contour lines and steepness of slope?**

CONTOUR MAPS

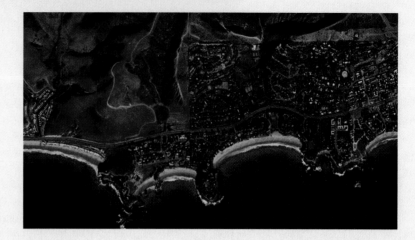

Aerial photograph of Laguna Beach, California. ▶

The Map

A contour map is the most widely used method of depicting elevational variations within an area. An individual contour line connects points of equal elevation above some reference datum—usually sea level. By drawing numerous contour lines on a map and by noting their spacing and configuration, a great deal of information about the topography of the land surface depicted on the map can be ascertained. For example, the location of mountains and valleys, the steepness of slopes, and the direction and flow of rivers within the area can all be determined by studying a contour map.

The interval at which the contour lines are drawn depends on the scale of the map and the relief of the area. For example, most maps drawn at a scale of 1:62,500 use a contour interval of 20 feet. However, in mountainous areas a larger interval may be needed to keep the contour lines from becoming too crowded. On the other hand, a flatter area may require a smaller contour interval to pick up the subtle relief of the area.

There are several rules to keep in mind while interpreting a contour map:

Closely spaced contours indicate a steep slope, while widely spaced contours indicate a gentle slope.

Evenly spaced contours indicate a uniform slope.

Closed contour lines represent a hill or depression.

Contour lines never cross but may converge along a vertical cliff.

A contour line will bend upstream when it crosses a valley.

Interpreting the Map

1. What is the contour interval of the map?

2. The scale of the contour map is 1:24,000. Consequently, one inch on the map represents how many feet on the Earth's surface?

3. What is the highest elevation on the map? Where is it located?

4. What is the lowest elevation on the map? Where is it located?

5. Note the mountain ridge between Boat and Emerald Canyons. Is it steeper on its east or west side? How can you tell?

6. In what direction does the stream in Boat Canyon flow? How can you tell?

7. The aerial photograph above depicts a portion of the contour map on the left. What portion of the contour map is depicted? How well does the contour map represent the physical features on the aerial photograph?

8. The contour map also depicts some cultural features. Identify some of these features and describe the symbols used to depict them. The contour map is older than the aerial photograph. Can you identify some cultural features on the aerial photograph not depicted on the contour map?

Laguna Beach, California ▶
Scale 1:24,000
Contour Interval = 20 feet
U.S. Geological Survey

the 60-foot, and, near the top of our island, the 80-foot level. After walking over a relatively broad summit that is above 80 feet but not as high as 100 feet (or we would cross another contour line), we once again cross the 80-foot contour line, which means we must be starting down. During our descent, we cross each lower level in turn until we arrive back at sea level (point B).

In the top portion of Figure 2.27 a **profile** (side view) helps us to visualize the ground we covered in our walk. We can see why the trip up the mountain was so much more difficult than the trip down. The closely spaced contour lines near point A represent a steeper slope than the more widely spaced contour lines near point B. Actually, we have discovered something that is true of all isoline maps: The closer together the lines are on the map, the steeper the gradient (the greater the rate of change per unit of horizontal distance).

We should understand when studying a contour map that the slope of Earth's surface almost always changes gradually. Thus, the land between two contour lines nearly always slopes gradually toward the line with the lower value; it is unlikely that the land drops off in steps downslope as the contour lines might suggest.

Most contour maps show other Earth features in addition to elevations (see Fig. 2.21). For instance, a contour map may show water bodies such as streams, lakes, rivers, and oceans, or cultural features such as towns, cities, bridges, and railroads. When a contour map provides such a variety of surface features in an area, it is called a **topographic map.** The U.S. Geological Survey produces topographic maps of the United States at several different scales (Fig. 2.22). Some of these maps—1:24,000, 1:62,500, 1:63,360 (1 inch to 1 mile), and 1:250,000—use English units for their contour intervals. The most recent maps are produced at scales of 1:25,000 and 1:100,000 and use metric units. Contour maps that show undersea topography are called **bathymetric charts.** In the United States they are produced by the National Ocean Service.

Modern Mapping Technology

Computer Mapping

Technology has revolutionized cartography from a manual process, slow and sometimes tedious, to a fully automated process employing computers to process data and computer-assisted equipment to draw the final map. For most mapping projects, computer systems are faster, more precise, more efficient, and less expensive than the manual cartographic techniques that they are replacing. However, an understanding of basic cartographic principles is still required to make a good map, because the computer mapping system will only draw what an operator instructs it to draw.

Map data can be digitally stored until a map is ready to be produced or revised. Hundreds of millions of bits of mapping data, such as elevations, depths, temperatures, or populations, can be stored in a digital database. A typical USGS topographic map has over 100 million bits of information to be stored and thousands of bits of data to be plotted on the finished map. A database for a map may include information on coastlines, political boundaries, city locations, river systems, map projections, and coordinate systems. Systems are now being developed in which data can be entered immediately into a data bank while the field survey is being conducted. The data may even be recorded by voice recognition systems, called in by the field surveyors and directly entered into the map data system.

An important advantage of computer mapping is that maps can be revised without painstakingly redrawing them. Existing map data and new survey data can be displayed on computer screens and corrected, changed, and improved until the cartographer decides that the final map is ready to be produced in hard copy by a computer printer or plotter. The cartographer may *tile* together separate maps to cover a large area or may *zoom* in on a particular area of the map to give a more detailed, larger-scale view. In addition to scale changes, computer map revision also allows easy metric conversions and rapid changes in projections, contour intervals, symbols, colors, and direction of view (orientation).

Computer map revision is essential for rapidly changing phenomena, such as weather systems, air pollution, ocean currents, volcanic eruptions, and forest fires. Weather maps in particular must be constantly revised, and computer mapping allows rapidly changing atmospheric conditions to be monitored and plotted continually.

Of particular interest to physical geographers, geologists, and engineers are **digital terrain models,** three-dimensional (3-D) views of a particular area (Fig. 2.28). A 3-D model is particularly useful for making calculations of the volume of rock material, oil, or water resources. Digital terrain models may be designed to show vertical exaggeration to enhance the relief of an area or to show a cross-sectional view of a par-

Figure 2.28 Digital Terrain Model. Digital terrain models are computer-generated, three-dimensional views of a land surface. Many different kinds of terrain displays can be generated from digital elevation data, and the models can be rotated on a computer screen to be viewed from any angle or direction.

ticular location. Many types of terrain displays and maps can be produced from digital elevation data. Some of these are color-scaled contour maps (where areas between assigned contours are a certain color), conventional contour maps, shaded relief maps, and three-dimensional relief maps.

Though computer-assisted cartography may be a much different process than the mapmaking of the past, it retains the basic purpose of all mapping: to communicate geographic and spatial knowledge to the user in visual form.

Geographic Information Systems

A tremendously versatile and relatively recent innovation in computerized handling of spatial data and maps is called a **geographic information system (GIS).** A GIS is a computer-based technology that assists the user in the entry, analysis, manipulation, and display of geographic information derived from map layers. In essence, a GIS blends computer cartography and database management as users model and analyze spatial and nonspatial data (Fig. 2.29). It is especially useful to geographers, who often solve problems that require large amounts of spatial data from many different sources. According to a corporation that develops GIS systems, this technology—software, hardware, service, and applications—represents a multi-billion-dollar industry employing hundreds of thousands of people around the world. The Geographic Infor-

mation Systems (GIS) are also widely used by governmental agencies, and many geographers are employed in career fields that apply GIS technology. GIS can not only analyze, manipulate, and display various layers, as in computer cartography, but it can also make the scale and map projection of the layers compatible, thus allowing several or all of the layers to be integrated into new, more meaningful composite maps.

What a GIS Does and How It Works.   Imagine you are in a giant map library with thousands of maps, all of the same area but each showing a different aspect of the same place: one map shows roads, another highways, another trails, another rivers (or soils, or vegetation, or slopes, or rainfall, and on almost to infinity). Further, each map is at a different scale, and there are many different projections (some that do not even preserve shape or direction). These factors will make comparing these different maps a difficult process. There are also digital terrain models and satellite images that you would like to compare directly to the maps. Being interested in geography, you want to be able to put some of this information together since few aspects of the environment are the result of only one factor or exist in spatial isolation. You have a problem, and to solve that problem, you need some way of making all of these representations of a part of Earth directly comparable. What you need in order to accomplish this is a GIS, and the knowledge of how to use this system.

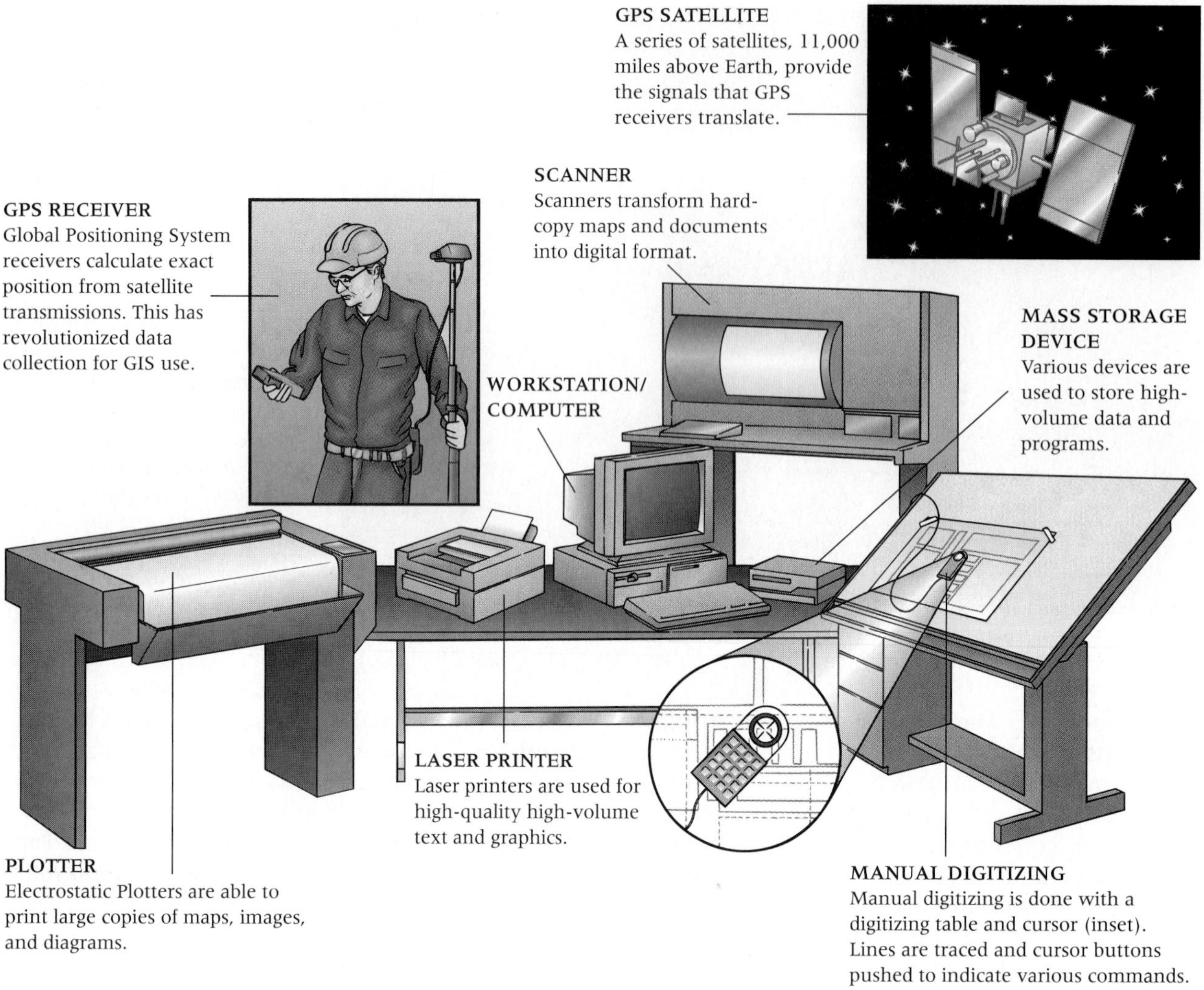

GPS SATELLITE
A series of satellites, 11,000 miles above Earth, provide the signals that GPS receivers translate.

SCANNER
Scanners transform hard-copy maps and documents into digital format.

GPS RECEIVER
Global Positioning System receivers calculate exact position from satellite transmissions. This has revolutionized data collection for GIS use.

WORKSTATION/ COMPUTER

MASS STORAGE DEVICE
Various devices are used to store high-volume data and programs.

LASER PRINTER
Laser printers are used for high-quality high-volume text and graphics.

PLOTTER
Electrostatic Plotters are able to print large copies of maps, images, and diagrams.

MANUAL DIGITIZING
Manual digitizing is done with a digitizing table and cursor (inset). Lines are traced and cursor buttons pushed to indicate various commands.

Figure 2.29 *The components of a Geographic Information System. A GIS consists of several subsystems: (1) input systems which provide spatial data into the GIS through the use of scanners that optically convert paper images into a digital format, manual digitizing that is generally used to update maps, and the Global Positioning System, which can feed locational coordinates directly from the field to computer storage in digital format; (2) data storage systems, such as CD-ROMs, computer floppy disks, and hard drives; (3) computer hardware and software systems for data management, access, analysis, and for updating the database (computer hardware and software also facilitate temporary display on the computer screen); and (4) output systems like printers and plotters (which have the ability to generate large hard-copy maps and images). In addition to these four basic subsystems of a GIS, access to many kinds of spatial data (maps, photographs, satellite imagery, digital elevation data) is important. However, it is often said that people are the most important part of a Geographic Information System. •* **How can computer technology help us to store, retrieve, and update maps and other spatial information?**

Data and Attribute Entry. The first step in solving this problem is to enter the map and image data into a computer system. There are three general ways to do this: (1) use a device called a **digitizer** to enter the map information into the computer; (2) use a computer scanner to scan in each map; and (3) directly enter in digital data from a floppy disk, CD-ROM, or tape. Each map is entered and stored as a separate data file that represents an individual *thematic* map—*separate layers of information* (Fig. 2.30). Another step, **geocoding** of the spatial information, is done

Environmental   **Map Layer**

Geology

Hazard areas

Existing land use

Noise contours

Flood plain

Soils

Vegetation

Surficial hydrology

EIR study areas

Planning study
index reference

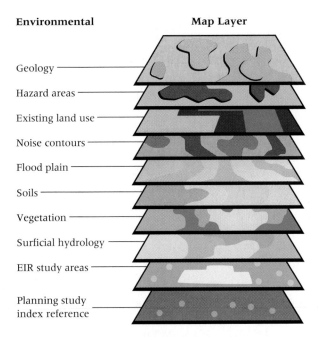

*Figure 2.30 Geographic Information Systems (GIS) allow geographers to find solutions to complex problems, by storing different information and data files for a place as individual "layers." GIS technology is widely used in environmental studies where several different variables need to be assessed and compared spatially to solve a problem. In the example shown here, information for the variables listed will be gathered and entered into a database. This information can then be analyzed and mapped separately or overlaid to provide more detailed information. • **Can you think of applications, other than environmental, for Geographic Information Systems?***

either simultaneously or afterward. Geocoding means to locate all spatial information in terms of its location in relation to some grid coordinates, such as latitude and longitude, to give them a locational address. Further, a list of **attributes** (information) about each mapped feature (or grid cell; see below) can be stored and easily accessed.

Registration and Display. Each map layer is registered to the same set of coordinates, as is the geocoded information. The computer will now display any layer (or set of layers) that you wish, stretched to fit any map projection that you specify, at a scale that you specify. The maps, images, and data sets can now be directly compared at the same size and on the same configuration of grid coordinates.

Now it is possible to also combine the geocoded, registered data and maps in any set of thematic layers that you wish. If you want to see

the location of homes on a river flood plain, the computer will overlay these two thematic layers for you, giving you an instant map of what you want just from entering the commands to retrieve separate homes and flood-plain layers. If you want to see earthquake faults, artificially filled areas, and the locations of fire and police stations, that will require four layers, but this is no problem for a GIS to display.

GIS in the Workplace. As another example of a GIS application, suppose you are a geographer working for the Natural Resources Conservation Service. Your current problem is to control erosion along the banks of a newly constructed reservoir. Although it would be nice to initiate erosion control measures throughout the lakefront, limited resources will not permit this broad action. You know that erosion is a function of many variables: soil types, slopes, vegetation characteristics, and others. Through a GIS you could enter data for each of these variables into the computer as a separate thematic layer. These variables could be analyzed individually; however, by integrating information from individual layers (soils, slope, vegetation, etc.), you could identify the locations *most* susceptible to erosion. Your resources and personnel could then be directed toward controlling erosion in those target areas.

The capacity of a GIS to integrate and analyze a wide variety of geographic information, from census data to landform characteristics, makes it useful to both human and physical geographers. Chris Friel, while working for the State of Illinois GIS Division, used GIS to analyze the relationship between the locations where retirees were settling in Illinois and their impact on local communities. Chris is now a GIS specialist with the Florida Marine Research Institute. One of his projects involves assessing the impacts of rapid growth and urban development on the delicate ecosystems of the surrounding waters in the Florida Keys. With many applications in geography and other disciplines, GIS is now, and will continue to be, an important tool for spatial analysis.

Remote Sensing of the Environment

Beginning in the early 1900s, cartographers made extensive use of **aerial photographs** taken from airplanes. In the 1960s, new technologies for obtaining and creating images from aircraft or satellites were declassified from military ap-

CHRISTOPHER A. FRIEL

University Education: *Bachelor of Science, Geography and Land Use Planning, Northern Michigan University, 1987. Master of Science, Geography, Western Illinois University, 1990*

Early Interests: *Outdoor activities, environmental concerns*

Prior Work Experience: *Urban forestry, cabin construction, landscaping, survival jobs during post–high school travel in Alaska*

Chris Friel is a research administrator and geographic information system (GIS) specialist for the Florida Marine Research Institute (FMRI) in St. Petersburg, Florida. At Northern Michigan University, he took his life-long love of the out- *doors and applied it to his geographic studies and interest in environmental issues. He now manages both a marine resources GIS and a cartographic laboratory, pulling together and displaying critical data on Florida coastal regions such as the Florida Keys, where urban development, diving, and boating now threaten coral reefs and other natural resources. Here, Chris discusses the challenges he faces in his work and the opportunities the GIS field offers.*

"Over the last 20 years," Chris says, "the state of Florida has experienced phenomenal urban growth and degradation of its natural resources. The Keys are a classic example of that. That area is the most ecologically rich and fragile part of Florida, which is also a popular tourist spot.

The types of questions that are being asked down there are really involved, and the stakes are high. Land developers have a lot of money invested there. Management policies being discussed range from 'do nothing'—which would let the Keys die—to the most extreme suggestion, which is to literally go in there, push everyone out, and tear the buildings down. I'm convinced neither one of those extremes will happen. Scientists and managers want to find a middle ground that balances human desire to enjoy the Keys with the need to protect its vital natural resources.

Answering even the simple questions about what is going on in the Keys requires the manipulation of vast quantities of data. We routinely work with more than 40 databases of environmental information. That's where GIS is so useful. It's what I call 'The Great Integrator' because it combines so many different types of information. A GIS is a computerized database and spatial analysis tool that can bring together different types of geographic information to give you a unique perspective on an environmental problem.

I believe GIS is a field that is exploding. Early GIS was used in labs by PhDs in white coats. Now, it is being used by a wide variety of individuals and organizations using standard, affordable computers. The potential uses of GIS are as broad as the discipline of geography itself, and the bottom line is that as GIS creates new ways of looking at the world, there will be more and more job opportunities for those who study geography."

plications and made generally available. A geographer, Evelyn Pruitt, coined the term *remote sensing* to refer to the gathering and interpretation of all types of aerial and space imagery. **Remote sensing** is the collection of information about objects or environments from a distance. A **remote sensor** is the device that collects, stores, and retrieves this information without being in physical contact with the object of interest. Many remote sensing systems can generate images of objects and scenes that would otherwise be invisible to the narrow range of human vision. In other words, images made by remote sensors that "see" types of energy that are nearly or totally in-visible can be converted into images that humans can see and interpret.

Remote sensing is commonly divided into *photographic and nonphotographic techniques.* **Photographic** remote sensing uses cameras and lenses to record a picture on film. **Nonphotographic** techniques produce an image, usually a **digital image,** that is derived by computers from a stream of numbers electronically beamed back from the sensor (Figure 2.31). One way to recognize nonphotographic imagery is to look for scan lines produced when the data were assembled into an image. Most images returned from space are digital because recovering film from

Figure 2.31 Digital Image. This scene, part of Chesapeake Bay, is not a photograph. Beamed back from an orbiting satellite as digital data, computer processing produced this image—the digital equivalent of a color-infrared photo. • **Using color information from Figure 2.33a, what features can you recognize on this image?**

space is difficult, and digital-image data can be easily broadcast back to Earth continuously from space. Digital imagery also offers the advantage

of computer-assisted data processing, image enhancement, and interpretation.

Aerial Photography

One common remote sensor, the camera, records data photographically from reflected light energy. Aerial photographs can provide us with a new perspective on our environment (Fig. 2.32), and most are **vertical** photographs (looking straight down), although some photos are **oblique** (taken at an acute angle to Earth's surface). Air photo interpreters use aerial photographs to examine and describe relationships among objects on Earth's surface. A device called a *stereoscope* allows the overlapping of pairs of aerial photographs taken from different positions to show Earth's surface three-dimensionally. Advanced stereographic equipment, along with the use of computers, has greatly facilitated the production of contour maps.

When astronauts first used cameras on their space missions, many of their first Earth photos were disappointing. One of the reasons was that atmospheric scattering of ultraviolet, violet, and blue light causes haze, and normal photographs taken from high altitudes often appear murky and hazy. **Near-infrared** (NIR) film cuts though

Figure 2.32 An oblique aerial photograph in natural color showing farmland, countryside, and forest.

haze better than normal film does, giving a clearer, sharper picture. It is sensitive to part of visible light as well as to longer-wavelength, near-infrared energy—basically light just beyond red, wavelengths (expressed as colors) that are too long for our eyes to see.

Color NIR film is sometimes called **false color** film, because strong NIR reflectance shows up as red on the photo. This color was chosen so that a person can instantly tell a color NIR photograph of a vegetated area from a normal color photo. On NIR, healthy grasses, trees, and most plants will show up as bright red, rather than green (Fig. 2.33). Originally NIR film was used for military purposes as "camouflage detection" film, because healthy vegetation shows up bright red on color NIR, and fake or dead vegetation (camouflage) has a different, easily detectable appearance. *A widely-held, but incorrect, notion of infrared film is that it images heat.* No photographic processes clearly record heat emissions on film (see "TIR Scanning" below).

Nonphotographic Remote Sensing

Nonphotographic remote sensors may utilize ultraviolet, visible, near-infrared, thermal infrared, and microwave parts of the electromagnetic spectrum. Two of these, *thermal infrared scanning* and *imaging radar,* are employed in environmental remote sensing. Nonphotographic sensors are either **passive,** which sense natural radiation, or **active,** which transmit their own energy signal.

Thermal Infrared (TIR) Scanning.   The process that creates true images of heat radiation, called **thermal infrared scanning,** images energy that we sense as heat. The way heat is sensed to make an image (remember, it is not a photograph) is complicated, but basically the image is scanned line by line with a mirror that focuses heat energy on a supercold sensor that records heat variations. The heat data are beamed back, recorded on magnetic tape, and converted into a visual image. If this system sounds mysterious, it is—but really, a supercooled sensor is sensitive to heat variation just as a light-sensitive film records a photographic image. TIR sensors are used on aircraft or spacecraft (see again Fig. 1.1).

Thermal-infrared images have scan lines like a television picture, and they *record contrasts in temperature whether they are cold or hot.* How well an object will show up on a thermal image will depend on how different the temperature of the object is from its surroundings. Hot objects show up in light tones, and cool objects will be dark, but often the gray tones are colorized by computer to emphasize heat differences. Original thermal images are in black and white, but they are often colorized later by computer enhancement.

Thermal scanning is expensive and tends to be used for specific purposes. Some civilian applications include: finding leaks in steam pipes or other pipelines, detecting thermal pollution, finding volcanic hot spots and geothermal sites, finding leaks in insulation of buildings, and locating forest fires through dense smoke.

Weather satellites also use thermal scanners to image weather patterns for forecasting. You have all seen these on television, where the meteorologist says, "Let's see what the weather satel-

(a) Color-infrared photo—near Burlington, VT.

(b) Color photo—same area.

Figure 2.33 A comparison of a "false color," near-infrared photograph (a) to the same view in normal color (b). Note that, in general, environmental details can be seen better and are more easily recognized on the near-infrared photograph.
• **Remembering that red represents healthy growing vegetation and clear or deep water is dark blue, what kinds of details can you recognize more easily on the CIR photo?**

**Thunderstorm Tour
25 June 1997
2145 UTC**

GOES Project
NASA-GSFC

Figure 2.34 Thermal infrared scanner images such as this one show true temperature differences with gray tones, darker meaning warmer. In a sense, they are like a "map" of heat (or cold) patterns. This scene shows the southeastern part of North America as beamed back from one of a series of U.S. weather satellites called GOES (Geostationary Operational Environmental Satellite). The whiter the cloud tops, the colder (and higher) they are. Note the roughly circular thunderstorm cells and other storm patterns. • **What other details can you see in this heat image?—the day is June 25.**

lite shows." Clouds show up as black on the original thermal image because they are colder than their background, Earth below. Since we don't like to see black clouds, the image is reversed, like a photo negative, so that the clouds will appear white (Fig. 2.34).

Radar. *RA*dio *D*etection *A*nd *R*anging is an *active* sensor that transmits microwave energy and reads the reflected signals to produce an image. It was used at first for military purposes in detecting aircraft and ships. *Radar imaging operates day or night and sees through clouds.* These are great advantages for mapping hostile regions as well as for gathering information about places that are frequently cloudy, such as tropical rainforest areas.

Today specific types of weather radar designed to monitor and track thunderstorms, hurricanes, and tornadoes (Fig. 2.35a) are used all over the world. Weather radar systems produce maplike images of precipitation patterns because radar penetrates clouds (day or night) but bounces off of raindrops and other precipitation, producing a signal on the radar screen. Precipitation patterns are typically the kind of weather radar image that we see on television. The latest systems include **Doppler radar,** a system that can determine the precipitation pattern, direction of movement, and speed of a storm system (much like police radar measures vehicle speed).

Imaging radar includes several specialized radar systems that sense the ground (topography, rock, water, ice, sand dunes, and so forth) by bouncing microwaves off the surface and converting the reflections into a landscape image. **Side-Looking Airborne Radar (SLAR)** was designed to be airplane-mounted to image areas located to the side of the aircraft (without flying over the region). Other imaging radar systems are mounted on satellites and the space shuttle (Shuttle Imaging Radar, or SIR). Imaging radar generally does not "see" trees (depending on the system), and will make an image of the *landscape* rather than a *crown of trees.* So SLAR is excellent for terrain and hydrologic mapping (Fig. 2.35b). SLAR is probably most often used to map remote, inhospitable, inaccessible, cloudy, or heavily forested regions.

Multispectral Imagery and Analysis

Perhaps the most important development for gaining information about our environment is the ability to use remote sensing from Earth-orbiting space satellites. The term multispectral really has two related meanings; first, **multispectral analysis** means using and comparing more than one type of image of the same place whether taken from space or not (for example, radar and thermal images, or near-infrared and normal photos). **Multispectral scanning,**

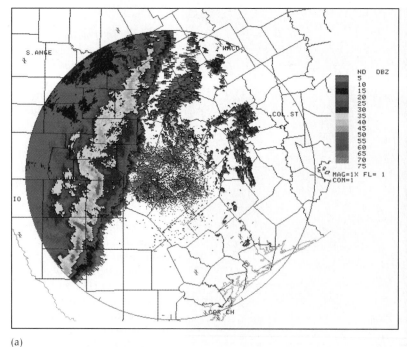

(a)

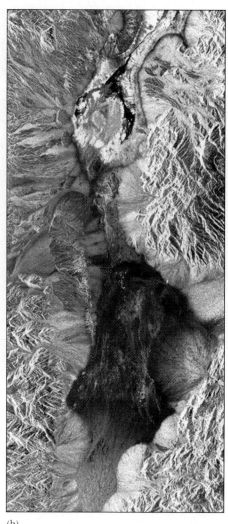

(b)

Figure 2.35 *Weather Radar and Imaging Radar. NEXRAD (next generation weather radar) image (a) of thunderstorms associated with a storm front. NEXRAD uses radar reflections to provide detail of the storm (circle is 124 nautical miles). Blue tones show cloud cover, with no precipitation (except in the very center where blue indicates stray reflections, called ground clutter). Other colors show rainfall intensity: green—light rain, yellow—moderate, and orange-red—heavy rain. The storm is moving to the east-southeast. (b) Imaging radar "bounces" a signal off a land surface and monitors variations in the reflected signals to produce a picture of a landscape. Radar images are originally in black and white, but can be "colorized" to enhance detail. Radar reflections are affected by many factors, particularly the materials reflecting the radar, as well as steepness and orientation of the terrain. This radar image shows part of Death Valley National Park, California, a desert basin flanked by mountains.* • **How are weather and imaging radar scenes different in terms of what they record about the environment?**

however, refers to devices widely used on satellites to simultaneously scan the same area at a variety of wavelengths. Combinations of visible and near-infrared bands are most common, but some satellites also have thermal-infrared capability.

Landsat. Landsat, a series of satellites initially launched in 1972, was the first space program to provide continuous monitoring and imaging of Earth's surface. There have been several Landsat satellites, and configurations vary. Landsat scans a 100-nautical-mile (185 km) swath in several wavelength bands (visible light, near-IR, and thermal IR), and transmits digital image data to receiving stations. The receiving stations record

and store the data for later conversion into digital images of 100 × 100 nautical miles (185 km).

From 705 kilometers (438 mi) high in a near polar orbit, the satellite keeps in the same orbit and relies on the rotation of Earth below to get nearly full coverage of the planet. Landsat's orbit is also sun-synchronized, which means that it always overflies and images a place at the same time of day. Every 16 days Landsat can image the same location, and a great benefit of this system is its excellent sequential coverage, important for environmental monitoring.

Each Landsat image consists of millions of **pixels,** which is short for picture element. A pixel is the smallest area resolved by an image scanning system. We can think of a Landsat image as a mosaic, or a grid, similar to graph paper. Each cell (pixel) of the grid has a location address and a value inside the box that represents the brightness of the ground area that the pixel represents. These grid cell values are translated into an image by computer.

Many types of images can be generated from digital, multispectral data, but the most familiar is the **color composite** (Fig. 2.36). A color composite is an image that was digitally created by blending three images of the same location but using pixel data from three different wavelengths. The usual Landsat color composite was designed to resemble a false-color near-infrared photograph, with the same color assignment as color near-infrared photos. The current Landsat imaging system, **Thematic Mapper (TM),** collects information at a ground resolution of 30 meters by 20 meters.

On a standard Landsat color composite, red is healthy vegetation, barren areas show up as white or brown, clear, deep water bodies are dark blue, and muddy water appears light blue. Clouds and snow are bright white. Urbanized areas are

Figure 2.36 *Landsat Thematic Mapper color composite showing Salt Lake City, Utah; the western front of the Rocky Mountains; the Great Salt Lake; and the deserts west of the city. North is at the top of this image. A railroad causeway isolates the north end from the rest of the lake, and the difference in tone following the causeway results from differences in the salinity of the water on each side.* • **What kinds of features can you identify on this false color image?**

blue-gray. Although the colors are visually important, the greatest benefit of multispectral imagery is that computers can identify, classify, and map (in a first approximation) these kinds of areas automatically.

Multispectral Remote Sensing Applications. The future of remote sensing will involve an increased emphasis on computer-automated methods of analyzing imagery, and in using digital imagery to capture thematic layers to be integrated into a GIS. Computer analysis permits rapid image processing, although our ability to collect data outstrips our ability to use, apply, and understand the information contained in the images that have already accumulated. This information explosion is complicated by the fact that many geographic and environmental problems are best understood by studying images of several types simultaneously, using **multispectral remote sensing.** Each part of the spectrum may yield different information about an aspect of the environment, and again, geographic information systems will play a strong role in this kind of analysis.

Weather satellites have been operating since the first TIROS was launched by the United States in 1960. Today two NOAA (National Oceanic and Atmospheric Administration) weather satellites observe weather patterns from a polar orbit 854 kilometers (530 mi) above Earth. They produce visible and thermal IR coverage of weather and cloud patterns of entire Earth. The GOES (Geostationary Operational Environmental Satellite) satellites watch major storm systems. These satellites use a **geostationary orbit.** They are positioned over the equator and orbit at a speed that keeps them in synch with Earth rotation so that they are always over the same Earth position. A GOES orbital height is approximately 36,000 kilometers (22,300 mi) from Earth. GOES-East is centered on a longitude over the U.S. east coast, while GOES-West is centered on a longitude over the west coast and watches for Pacific storm patterns. GOES images are seen on our television weather shows and are reproduced in daily newspapers (see again Fig. 2.34). In addition to NOAA and GOES satellites, the U.S. Air Force, the European Space Agency, Russia, and Japan also have weather satellites in operation.

Weather satellites have saved many lives and millions of dollars by increasing our storm and weather forecasting capabilities. An excellent example of this occurred when Hurricane Andrew struck the south coast of Florida in August 1992. Although this was the most devastating hurricane in U.S. history in terms of property damage ($25–30 billion), the advance warning provided by weather satellites allowed people to evacuate, saving hundreds of lives.

Sonar (*SO*und *NA*vigation and *R*anging) allows us to probe the ocean depths and to map undersea topography. The use of seismic waves (from natural and artificially generated earthquakes) has given us knowledge of the interior of the planet and helps us find resources such as gas and oil hidden in solid Earth. Each remote sensor allows us to "see" our planet better and to gather data to map and plot its complex ever-changing systems.

The great advantage of remote sensing techniques is that they provide rapid and worldwide coverage of environmental conditions for geographers, cartographers, and other scientists. Satellite coverage is especially important in the less accessible parts of Earth's surface such as ice caps,

mountains, tropical forests, and deserts. Important resources such as mineral ores, water, and potential energy sources can be found more rapidly and with greater accuracy. Through continuous monitoring of Earth's surface, global, regional, and even local changes can be detected and mapped. Geographic information systems offer the ability to match and combine thematic data layers of any sort, including digitized maps, scanned photographs, digital terrain models, and multispectral images, instantly accessing any combination of these that we need to solve complex spatial problems related to the environment.

Though maps remain a major tool of the geographer and other Earth scientists, the ability to gather and present data has greatly changed in the last few decades. The use of remote sensing, GPS, GIS, computer-assisted data handling, and computer mapping has revolutionized both cartography and the field of geography.

Define and Recall

cartography	township	graphic (bar) scale	near-infrared
digital mapping	section	magnetic declination	photography
great circle	Global Positioning	azimuth	thermal infrared
small circle	System (GPS)	bearing	scanning
latitude		thematic map	radar
prime meridian	conformal map	isoline	multispectral analysis
longitude	projection	contour line	Landsat
parallel	equal-area map	contour interval	color composite
meridian	projection	profile	Thematic Mapper
	Mercator projection		geostationary orbit
solar noon	rhumb line	digital terrain model	sonar
International Date Line	gnomonic projection	Geographic Information	
	map legend	System (GIS)	
Public Lands Survey	scale	geocoding	
System	verbal scale		
principal meridian	representative fraction	remote sensing	
base line	(RF)	digital image	

Discuss and Review

1. Why is the concept of a great circle so useful to navigators?

2. What great circle has been chosen as the zero point for latitude?

3. What is the latitude and longitude of your city?

4. Approximately how accurate in meters could you be if you tried to locate a building in your city to the nearest second of latitude and longitude? Using a GPS?

5. What time zone are you in? What is the time difference between Greenwich time and your time zone?

6. If it is 2 A.M. Tuesday in New York (EST), what time and day is it in California (PST)? What time is it in London (GMT)?

7. If you fly across the Pacific Ocean from the United States to Japan, how will the International Date Line affect you?

8. How has the use of the Public Lands Survey System affected the landscape of the United States? Has your local area been affected by its use? How?

9. Why can't maps give an accurate representation of Earth's surface? What is the difference between a conformal map and an equal-area map?

10. How does the Mercator projection differ from the gnomonic projection? What advantages does the Mercator offer the navigator?

11. What is the difference between RF and a verbal map scale?

12. Why is the date of publication of a map important? List several reasons, including one related to Earth's magnetic field.

13. Why is the scale of a map so important? What does a small-scale map show in comparison with a large-scale map?

14. Why are contour maps so important to physical geographers? What specific information can be obtained from a contour map that cannot be obtained from other maps?

15. What does the concept of "thematic map layers" mean in a geographic information system? What are attributes?

16. How have computers revolutionized cartography and the handling of spatial data? What specific advantages did computers offer to the map-making process?

17. What is the difference between a photograph and a digital image?

18. Describe the difference between imagery from a weather satellite and from Landsat Thematic Mapper. How do the monitoring purposes of these satellites affect the different imagery produced by each?

19. What does a weather radar image show in order to help us understand weather patterns?

Consider and Respond

1. Use an atlas and globe to determine what cities are located at the following grid coordinates: 40°N, 75°W; 34°S, 151°E; and 41°N, 112°W.

2. What are the grid coordinates of the following cities: Portland, Oregon; Rio de Janeiro; your hometown?

3. You are located at 10°S latitude, 10°E longitude; you travel 30° north and 30° east. What are your new geographic coordinates?

4. You are located at 40°N latitude, 90°W longitude. You travel due north 40°, then due east 60°. What are your new geographic coordinates?

5. Using an atlas and Figure 2.24, identify a city in North America with each of the following magnetic declinations: 20°East; 20°West; 25°East; 0°.

6. Select one place within the United States that you would most like to visit for a vacation. If you could take along only one map and you had your choice between a state highway map, a USGS topographic map, or an official map of a city or town, which would you choose and why would you choose it?

7. If you were an applied geographer and wanted to use a Geographic Information System to build a database of spatial information about the environment of a park (pick a state or national park near you, for example), what are the seven most important "layers" of mapped information that you would want to have? What combinations of two or more layers would be particularly important to your purpose?

8. GIS has been called a "power tool" for spatial (geographic) analysis. Why are people so important to its effective application?

CHAPTER
3

EARTH AS A PLANET

CHAPTER PREVIEW

▶ The universe is made up of billions of galaxies, each containing so many stars that the mathematical possibility of other planets containing life as we know it seems unlimited.

What is a galaxy and what is its relationship to the universe? How is distance involved in the determination of whether life beyond our solar system does exist in the universe?

▶ Earth is one of nine planets that, together with the sun, comprise the major components of our solar system.

In what ways is the sun of most importance to life on Earth? What are the chief characteristics of the other planets? What are the other components of our solar system?

▶ Although Earth is often considered spherical, it is actually an oblate spheroid.

What is an oblate spheroid and what causes Earth to deviate from true sphericity? How do we know that Earth is nearly spherical?

▶ The regular movements of Earth, termed *rotation* and *revolution,* are the fundamental elements of Earth-sun relationships, which initially control the dynamics of our atmosphere and the phenomena related to it.

Why is this one of the most important understandings in physical geography? What other understandings follow from this concept?

▶ The relationship of Earth's axis to the plane of Earth's orbit is the key to an explanation of seasons on Earth.

What is the relationship? How does it operate in conjunction with Earth's revolution to produce seasons? How does it influence variations in the amounts of insolation reaching different portions of Earth's surface?

▲ Sunset along rocky coastline at Toleak Point in Olympic National Park, Washington.

Everyone has wondered about environmental changes that take place from time to time throughout the year and from place to place over Earth's surface. Perhaps when you were young you wondered why it got so much warmer in summer than in winter and why some days were long while those in other seasons were much shorter. These questions, and many like them, are probably as old as humanity, and the answers to them help provide us with an understanding of the physical geography of our world.

Physical geographers' concerns take them beyond planet Earth to a consideration of the sun and Earth's position in the solar system. Geographers examine the relationship between the sun and Earth to explain such earthly phenomena as the alternating periods of light and dark that we know as day and night. Special relationships between Earth and sun also help to explain the seasonal variations in climate that we know as spring, summer, fall, and winter. Although the universe and solar system are not strictly within the province of physical geography, an acquaintance with each can be of help in an examination of Earth as an environment for life as we know it.

The Solar System and Beyond

If you look at the sky on a clear night, all of the stars that you see are part of a single collection of stars called the Milky Way Galaxy. A **galaxy** (Fig. 3.1) is an enormous island in the universe— an almost incomprehensible cluster of stars, dust, and gases. Our sun is one of hundreds of billions of stars that compose the Milky Way Galaxy. In turn, the observable universe appears to contain billions of other galaxies.

Because distances within the universe are so vast, it is convenient to use the distance light travels in one year, termed a **light-year,** as a unit of measure. A light-year is equal to 6 trillion miles. Light travels at the amazing speed of 298,000 kilometers per second (186,000 mi per second). Thus in one second light could travel seven times around the circumference of Earth. While that may seem like a great distance, the *closest* star to Earth, other than the sun, is 4.3 light-years away, and the *closest* galaxy to our galaxy is 75,000 light-years away.

Figure 3.1 *Photograph of the Milky Way Galaxy. This photograph was taken from Laguna Mountains, California, looking toward the center of our galaxy.*

The Sun and the Solar System

The sun is the center of our solar system (Table 3.1). A **solar system** can be defined as all the heavenly bodies associated with a particular star due to the star's dominant mass and gravitational attraction (Fig. 3.2). The major heavenly bodies in our sun's system consist of nine major **planets** (a celestial body that revolves around a star and reflects the star's light rather than producing its own) with their 61 **satellites** (like Earth's moon, these bodies orbit the planets), numerous **asteroids** (a very small planet, usually with a diameter of less than 500 miles), **comets** (made up of a head that is a collection of solid fragments and a tail, sometimes millions of miles long, comprised of gases; a comet usually follows an orbit around the sun), and **meteors** (small stonelike bodies that, when they enter Earth's atmosphere, burn and often appear as a streak of light, or "shooting star," crossing the sky. A meteor that survives the fall through the atmosphere and reaches Earth's surface is called a **meteorite**).

The intimate and life-producing relationship between Earth and sun is the result of the amount and distribution of radiant energy received by Earth from the sun. Such factors as our planet's size, its distance from the sun, its atmosphere, and the movement of Earth around the sun as well as its rotation on an axis all affect the amount of radiant energy that Earth receives. Though there are processes of our physical environment that result from Earth forces not related to the sun, these processes would have little relevance were it not for the life-giving, life-sustaining energy of the sun.

As far as we know with certainty, within our solar system, only on Earth has the energy from the sun been used to create life—to create something that grows, changes, and that can reproduce itself. Yet there remains a possibility of life, or at least the basic organic building blocks, on Mars and perhaps even on one or two of the moons of Saturn or Jupiter. What fascinates scientists and philosophers alike, however, is the likelihood that there are millions of planets in the universe like Earth on which forms of life have developed that might be even more sophisticated than humans, Earth's most complex and intelligent life-form.

Earth revolves around the sun at an average distance of 149 million kilometers (93 million mi). The sun's size and its distance from us challenge our comprehension. About 130 million Earths could fit inside the sun, and a plane, flying at 500 miles per hour, would take 21 years to reach the sun.

TABLE 3.1   Comparison of the Planets

Name	Distance from Sun	Revolution Period	Diameter	Mass	Density
	(AU)*	(yrs)	(km)	(10^{23} kg)	(g/cm^3)
Mercury	0.39	0.24	4,878	3.3	5.4
Venus	0.72	0.62	12,102	48.7	5.3
Earth	1.00	1.00	12,756	59.8	5.5
Mars	1.52	1.88	6,787	6.4	3.9
Jupiter	5.20	11.86	142,984	18,991	1.3
Saturn	9.54	29.46	120,536	5,686	0.7
Uranus	19.18	84.07	51,118	866	1.2
Neptune	30.06	164.82	49,660	1,030	1.6
Pluto	39.44	248.60	2,200	0.01	2.1

*An AU (or astronomical unit) is the distance from the Earth to the sun.

The sun, like all the other stars in the universe, is a self-luminous mass of gases that emits radiant energy. A slightly less than average-sized star, our sun is the only such self-luminous body in our solar system and is the source for almost all the light and heat for the surfaces of the various planets and other satellites within that system. It has an estimated surface temperature of between 5500°C and 6100°C (10,000°F and 11,000°F). The energy emitted by the sun comes from nuclear reactions that take place in its interior. There, under high pressure, hydrogen is

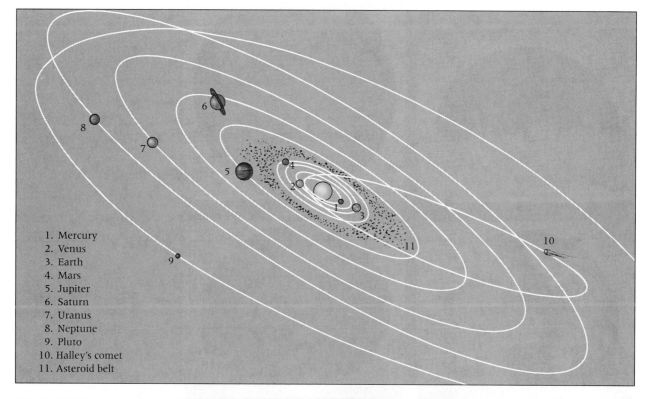

1. Mercury
2. Venus
3. Earth
4. Mars
5. Jupiter
6. Saturn
7. Uranus
8. Neptune
9. Pluto
10. Halley's comet
11. Asteroid belt

Figure 3.2 The solar system, showing the sun and planets in their proper order. The approximate size relationships between the individual planets is shown. However, the planetary orbits are much condensed, and the scale of the sun and planets is greatly exaggerated. The planets would be much too small to be visible at the scale of the orbits shown.

changed into helium through nuclear fusion in a process similar to that in a hydrogen bomb. This nuclear reaction releases tremendous amounts of energy that radiate out from the sun in all directions at the speed of light.

Earth intercepts only about 1/2,000,000,000 of the radiation given off by the sun, but even this much energy affects the biological and physical characteristics of Earth's surface. While some of the remainder of the sun's radiant energy is also intercepted by other bodies in the solar system, the vast proportion of it travels out through space unimpeded.

The Planets

The four planets closest to the sun (Mercury, Venus, Earth, and Mars) are called the **terrestrial planets** (Fig. 3.3). They are relatively small, warmed by their proximity to the sun, and composed of rock and metal. They all have solid surfaces that exhibit records of geological forces in the form of craters, mountains, and volcanoes.

The next four planets (Jupiter, Saturn, Uranus, and Neptune) are much larger and composed primarily of lighter ices, liquids, and gases. These planets are termed the **giant planets** (Fig. 3.4). While they have solid cores at their centers, they are more like huge balls of gas and liquid with no solid surface on which to walk. Finally, at the outer edge of the solar system is Pluto, which is neither a terrestrial nor a giant planet. It is more like one of the moons of the giant planets and may, at one time, have been just that.

The nine major planets that are known to revolve around the sun have several phenomena in common. From a point far out in space above the sun's "north pole," they would all appear to move around the sun in the same counterclockwise direction. Their orbits follow an elliptical, almost circular, path. All the planets also rotate, or spin, on their own axes. With the exception of Venus and Uranus, all rotate in the same direction. All but Pluto lie close to the same plane passing through the sun's equator, and all have an atmospheric layer of gases with the exception

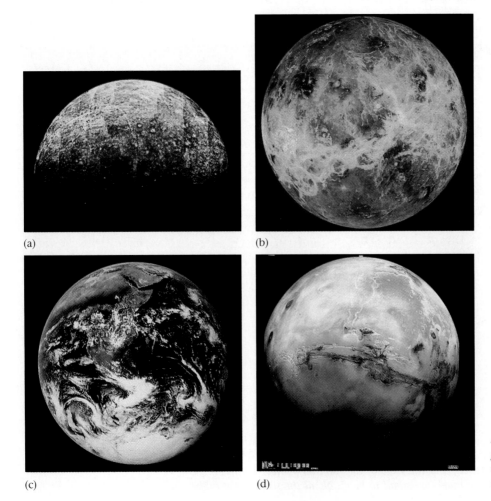

(a)   (b)   (c)   (d)

Figure 3.3 The terrestrial planets: (a) Mercury; (b) Venus; (c) Earth; and (d) Mars.

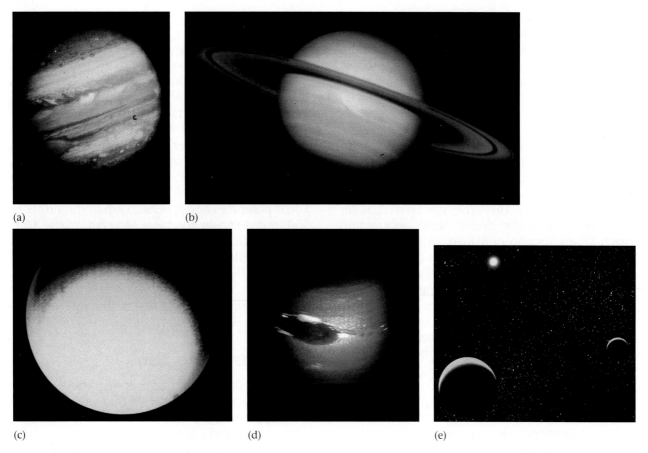

(a)   (b)

(c)   (d)   (e)

Figure 3.4 *The giant planets and Pluto:* *(a)* *Jupiter;* *(b)* *Saturn;* *(c)* *Uranus;* *(d)* *Neptune; and* *(e)* *Pluto.*

of Pluto and Mercury, which are not dense or heavy enough to hold gases.

Although the planets are similar in many ways, they differ in a number of vital characteristics (Table 3.1). Mercury, the smallest terrestrial planet, has a diameter less than half that of Earth, and its **mass** (the total amount of material in the planet) is only 5 percent of that of Earth's. This means that Mercury does not have sufficient mass and gravitational pull to hold onto a meaningful atmosphere. (**Gravitation** is the attractive force one body has for another. The greater mass or amount of matter a body has, the greater the gravitational pull it will exert on other bodies.)

Also, partly because of the very thin atmosphere around Mercury, surface temperatures on that planet reach over 425°C (800°F) on the side facing the sun. As we will see later, Earth's atmosphere protects our planet's surface from much of the sun's radiation, which can be dangerous to life, and keeps temperatures on Earth within a livable range.

Further, because Mercury rotates on its axis only once every 59 days, the same side of the planet is exposed to the sun for long periods, while the opposite side receives none of the sun's energy. This situation compounds the problem of Mercury's surface temperatures. The side facing the sun is much too hot and the opposite side is much too cold (−175°C, or −280°F) to support life.

Uranus, Neptune, and Pluto, the three planets furthest from the sun, are too distant to receive enough solar energy to have surface temperatures conducive to life as we know it. In addition, though Uranus and Neptune are large enough to have atmospheres, their atmospheres are made up primarily of hydrogen and helium, and they apparently have no free oxygen, a factor that seems to be necessary for the development of life like that on Earth.

Jupiter and Saturn are the two largest planets; yet, because significant parts of their masses are gaseous, both have very low densities. For instance, Saturn's density is less than that of water, while Earth's density is more than five times as great. The atmospheres of Jupiter and Saturn also have a high proportion of hydrogen and helium and, again, there is no free oxygen. And even though Jupiter and Saturn are closer to the sun than Uranus, Neptune, and Pluto, they still do not receive enough solar energy to produce livable surface temperatures; their temperatures are down around −95°C to −150°C (−200°F to −300°F). It is unlikely that life exists on the two largest planets, and scientists have begun to look instead at the moons of Saturn and Jupiter for that possibility. In 1997 the spacecraft Cassini was launched by the United States and should reach Titan, a moon of Saturn, in 2002.

Venus and Mars, in the orbits closest to that of Earth, are the planets most similar to our own. Venus, in fact, has been called Earth's twin because it is most like Earth in size, density, and mass. However, we cannot see the surface of Venus since it is covered by a permanent thick cloud layer. Through information gathered by the spacecraft Magellan in 1991–1993, we have a much better understanding of surface conditions on Venus. The surface consists primarily of lowland lava plains, much like the basaltic ocean basins of Earth, with two continents rising above the lowlands. There is no liquid or frozen water. The atmosphere of Venus is 96 percent carbon dioxide. This thick layer of carbon dioxide allows very little energy to escape from the planet. This results in a surface temperature for Venus of >450°C (850°F).

Mars is better known to us because thousands of pictures have been taken by numerous spacecraft. In addition, the Viking spacecrafts landed on the surface of Mars, providing exceptional images of surface features (Fig. 3.5). Like Venus, the atmosphere of Mars is 96 percent carbon dioxide. However, it is so thin that energy is not trapped like it is on Venus. Thus surface temperatures range from −125°C (−190°F) at the poles to 25°C (77°F) at the equator. There are seasonal polar ice caps of frozen carbon dioxide (dry ice), but no water has been detected on the Martian surface. However, evidence has been found that suggests that rain once fell and rivers once flowed on Mars. If life has existed on Mars, it is most likely that it existed at a time when water was available.

Earth and the Sun

The color views of Earth transmitted by the Apollo spacecrafts on their lunar missions gave us some unforgettable views of our planet. The cameras showed Earth as a sphere of blue oceans, green and brown landmasses, and swirls of white clouds. One astronaut has described Earth as it appears to someone who has traveled close to the moon:

Figure 3.5 This image of the Martian landscape was taken by Pathfinder (unmanned exploratory vehicle). The ramp that the Imager used to exit Pathfinder is in the foreground.

*E*arth looked so tiny in the heavens that there were times during the Apollo 8 mission when I had trouble finding it. If you can imagine yourself in a darkened room with only one clearly visible object, a small blue-green sphere about the size of a Christmas tree ornament, then you can begin to grasp what Earth looks like from space. I think that all of us subconsciously think that Earth is flat or at least almost infinite. Let me assure you that, rather than a massive giant, it should be thought of as the fragile Christmas tree ball which we should handle with considerable care.

This Island Earth
edited by Oran W. Nicks, NASA, SP-250, 1970

As we begin the study of our planet we should not lose sight of the image of Earth as an exceedingly isolated island in a seemingly endless sea. Scientists have always speculated that there could be a planet in another galaxy that has intelligent life. However, it was not until 1995 that scientists finally proved other planets exist. And the fact remains that we have not actually seen other planets but only proved their existence through mathematical modeling. Thus, we should learn as much as we can about our planet and treat it with exceptional care because, in all likelihood, Earth is the only home the human race will ever know.

Size and Shape of Earth

For most of their history humans have pondered the size and shape of the world in which they live. Though it was not until the 1960s that we were able to travel deep enough into space to see the shape of Earth, ancient Greeks as early as Pythagoras in 540 B.C. theorized that Earth was a sphere. However, it was not until about 200 B.C. that a philosopher named Eratosthenes made a fairly accurate estimate of the circumference of Earth. His estimate was within a few hundred miles of Earth's actual circumference. The accuracy of Eratosthenes' calculation is all the more amazing if we imagine ourselves trying to estimate the shape and size of our planet without the benefit of today's maps, globes, or navigational devices. What kind of description could we provide of Earth, its landforms and oceans, its shape and dimensions, if we knew nothing but what we can see around us? What things can you think of that might prove Earth is a sphere?

The apparent boundary line between the sky and Earth is called the **horizon.** If you hold a basketball to represent Earth in front of you with the sky as a background, you can see a similar boundary line, which is curved no matter which way you turn the ball. If you were far enough from Earth's surface, the horizon would similarly appear to be curved.

Many aspects of our world are related to the curvature of Earth. For example, the curvature affects the intensity and duration of solar radiation received at different locations on Earth. Differences in temperature from place to place and currents in the oceans and atmosphere are also related to Earth's near-sphericity. Further, we determine time by a system based on Earth as a sphere. We have devised a full system of direction and location by means of a grid based on the shape of the globe, and part of our navigation system is based on Earth's spherical shape.

For most purposes Earth can be considered a perfect sphere with an equatorial circumference of 39,840 kilometers (24,900 mi). However, due to forces associated with Earth rotation, the area near Earth's equator actually bulges out somewhat, and the two poles are accordingly flattened slightly. So instead of a perfect sphere, Earth is more properly an **oblate spheroid** or **ellipsoid of rotation.**

Earth's deviation from a perfect sphere is exceedingly minor. Nevertheless, these irregularities do affect navigation, mapping, and distance accuracies. People working in navigation, surveying, aeronautics, and cartography must include in their calculations the deviations of Earth's shape from true sphericity. Scientists have been able to confirm the extent of such deviations from measurements of variations in gravitational pull acting on satellites. They have found that the diameter of Earth at the equator is 12,758 kilometers (7927 mi), while from pole to pole it is 12,714 kilometers (7900 mi). On a globe with a diameter of 12 inches, this difference of 44 kilometers would be 4/100 of an inch, a deviation of about one-third of 1 percent and not noticeable to the naked eye.

Landforms also cause Earth to deviate from true sphericity. Mount Everest in the Himalayas is the highest point of land on Earth at 8847 meters (29,028 ft) above sea level. The lowest known point on Earth's surface is in the Challenger Deep, a part of the Mariana Trench in the Pacific Ocean southwest of Guam. This spot is 11,033 meters (36,198 ft) below sea level. The difference between these two points, 19,880 meters, or just over 12 miles, is insignificant and invisible when reduced in scale to a globe with a 12-inch diameter.

Movements of Earth

Earth has three basic movements: **galactic movement, rotation,** and **revolution.** The first of these is the movement of Earth with the sun and the rest of the solar system in an orbit around the center of the Milky Way Galaxy. This movement has limited effect upon the changing environments of Earth and is generally the concern of astronomers rather than of geographers. The other two movements of Earth, rotation on an axis and revolution around the sun, are of vital interest to the physical geographer. The consequences of these movements are the phenomena of day and night, variations in the length of day, and, to a major extent, the changing seasons.

Rotation.   Rotation refers to the turning of Earth on its own axis, an imaginary line extending from the North Pole to the South Pole. Earth rotates on its axis at a uniform rate, making one complete turn with respect to the sun in 24 hours.

Earth turns in an eastward direction (Fig. 3.6). The sun "rises" in the east and appears to move westward across the sky, but it is actually Earth, not the sun, that is moving, rotating toward the morning sun (that is, toward the east).

Earth, then, rotates in a direction opposite to the apparent movement of the sun, moon, and stars across the sky. If we look down on a globe from above the North Pole, the direction of rotation is counterclockwise. This eastward direction of rotation not only defines the movement of the zone of daylight on Earth's surface but also helps define the circulatory movements of the atmosphere and oceans.

The velocity of rotation at the Earth's surface varies with the distance of a given place from the *equator* (the imaginary circle around Earth halfway between the two poles). All points on the globe take 24 hours to make one complete rotation (360°). Thus the *angular velocity* for all locations on Earth's surface is the same—360° per 24 hours, or 15° per hour. However, the *linear velocity* depends on the distance (not the angle) covered in that 24 hours. The linear velocity at the poles is zero. You can see this by spinning a globe with a postage stamp affixed to the North Pole. The stamp rotates 360° but covers no distance and therefore has no linear velocity. Place the stamp anywhere between the North and South Poles and it will cover a measurable distance during one rotation of the globe. The greatest linear velocity is found at the equator, where the distance traveled by a point in 24 hours is largest. At Kampala, Uganda, near the equator, the velocity is about 460 meters (1500 ft) per second, or approximately 1660 kilometers (1038 mi) per hour (Fig. 3.7). In comparison, at St. Petersburg, Russia (60° north latitude), where the distance traveled during one complete rotation of Earth is about half that at the equator, Earth rotates about 830 kilometers per hour.

We are unaware of the speed of rotation, however, because (1) the angular velocity is constant for each place on Earth's surface, (2) the atmosphere rotates with Earth, and (3) there are no nearby objects, either stationary or moving at a different rate with respect to Earth, to which we can relate Earth's movement. Without such references we are unable to perceive the speed of rotation.

Rotation accounts for our alternating days and nights. This can be demonstrated by shining a light at a globe while rotating the globe slowly toward the east. You can see that half the sphere is always illuminated while the other half is not, and that new points are continually moving into the illuminated section of the globe while others are moving into the darkened sector. This corre-

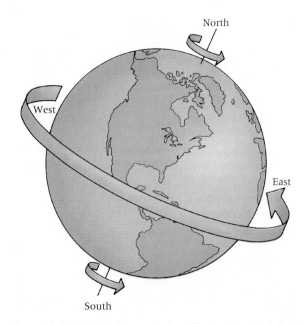

Figure 3.6   Earth spins around a tilted axis as it follows its orbit around the sun. Earth's rotation is from west to east, making the stationary sun appear to rise in the east and set in the west.

Figure 3.7 *The speed of rotation of Earth varies with the distance from the equator.*

sponds to Earth's rotation and the sun's energy striking Earth. While one half of Earth receives the light and energy of solar radiation, the other half is in darkness. As noted in Chapter 2, the great circle separating day from night is known as the **circle of illumination.**

Revolution. While Earth rotates on its axis, it also revolves around the sun in a slightly elliptical orbit at an average distance from the sun of about 150,000,000 kilometers (93,000,000 mi) (Fig. 3.8). On about January 3, Earth is closest to the sun and is said to be at **perihelion** (from Greek: *peri,* close to; *helios,* sun); its distance from the sun then is approximately 147,500,000 kilometers.

At around July 4, Earth is about 152,500,000 kilometers from the sun. It is then that Earth has reached its farthest point from the sun and is said to be at **aphelion** (Greek: *ap,* away; *helios,* sun).

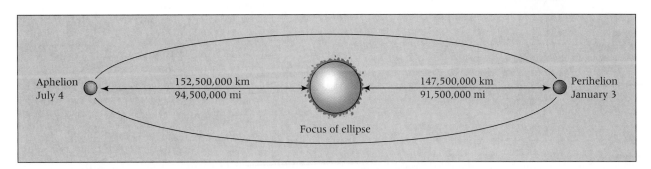

Figure 3.8 *Oblique view of the elliptical orbit of Earth around the sun. Earth is closest to the sun at perihelion and farthest away at aphelion. Note that in the Northern Hemisphere summer (July), Earth is farther from the sun than at any other time of year.*

Five million kilometers is insignificant in space, and these varying distances from Earth to the sun only minimally (±3.5 percent) affect the receipt of energy on Earth. Hence, they have no relationship to the seasons.

The period of time Earth takes to make one revolution around the sun determines the length of one year. Because Earth makes $365\frac{1}{4}$ rotations on its axis during the time it takes to complete one revolution of the sun, a year is said to have $365\frac{1}{4}$ days. Because of the difficulty of dealing with a fraction of a day, it was decided that a year would have 365 days, and that every fourth year, called *leap year,* an extra day would be added in February.

Plane of the Ecliptic, Inclination, and Parallelism. Earth in its orbit around the sun moves in a constant plane. This plane is called the **plane of the ecliptic.** Earth's axis is tilted at an angle of $23\frac{1}{2}°$ from the perpendicular to the plane of the ecliptic and thus makes a constant angle of inclination of $66\frac{1}{2}°$ with the plane (Fig 3.9).

In addition to a constant angle of inclination, Earth's axis maintains another characteristic called **parallelism.** As Earth revolves around the sun, Earth's axis remains parallel to its former positions. That is, at every position in Earth's orbit, the axis remains pointed toward the same spot in the sky. For the North Pole that spot is close to the star that we call the North Star, or Polaris. Thus, Earth's axis is fixed with respect to the stars outside our solar system but not with respect to the sun.

To get a better picture of what happens to Earth in its movement around the sun, pick up a globe and carry it around an imaginary sun. Keep certain facts in mind: (1) Earth in its orbit around the sun and the sun itself lie in the plane of the ecliptic, (2) Earth's axis is inclined with respect to that plane so as to make a constant angle of inclination with it of $66\frac{1}{2}°$, and (3) Earth's axis as Earth moves around the sun remains parallel to itself in all former positions. As you will see as you walk the globe around the sun under these conditions, Earth's axis changes position with respect to the sun as Earth revolves around the sun. Sometimes one pole of the axis is tilted toward the sun, sometimes away, and sometimes neither. These systematically changing positions of Earth's axis relative to the sun result in variations in intensity of the sun's energy from place to place and time to time. Understanding these facts of Earth's relationship with the sun leads us directly into an examination of the seasonal changes on Earth and a discussion of how the intensity of the sun's rays varies from place to place throughout the year.

The Seasons

Many people assume that the seasons must be caused by the changing distance between Earth and the sun during Earth's revolution around the sun. As noted earlier, the change in this distance is very small. Farther, for people in the Northern Hemisphere, Earth is actually closer to the sun in January and farthest away in July (see Fig. 3.8). This is exactly opposite of that hemisphere's seasonal variations. As we will see, seasons are caused by the $23\frac{1}{2}°$ tilt of Earth's axis from the perpendicular to the plane of the ecliptic (see Fig. 3.9) and the parallelism the axis maintains as Earth orbits the sun. About June 22 Earth is in a position in its orbit so that the northern tip of its axis is inclined toward the sun at an angle of $23\frac{1}{2}°$ from a line perpendicular to the plane of the ecliptic. This time in Earth's orbit is called the summer **solstice** (from Latin: *sol,* sun; *sistere,* to stand) in the Northern Hemisphere. We can best see what is happening if we refer to Figure 3.10, position A. On that diagram we can see that the Northern and Southern Hemispheres receive unequal amounts of light from the sun. That is, as we imagine rotating Earth under these conditions, a larger portion of the Northern Hemisphere than the Southern Hemisphere remains in

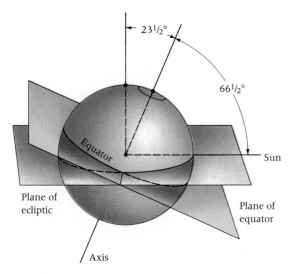

Figure 3.9 *The plane of the ecliptic is defined by the orbit of Earth around the sun. The $23\frac{1}{2}°$ inclination of Earth's rotational axis causes the plane of the equator to cut across the plane of the ecliptic.*

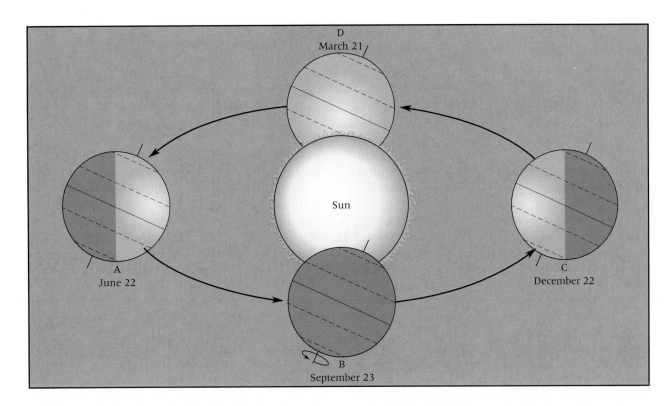

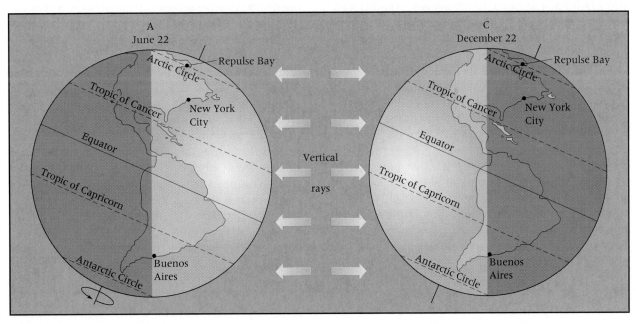

Figure 3.10 *Geometric relationships between Earth and the sun at the solstices. Note the differing day lengths at the summer and winter solstices in the Northern and Southern Hemispheres.*

daylight. Conversely, a larger portion of the Southern Hemisphere than the Northern Hemisphere remains in darkness. Thus, referring again to Figure 3.10, position A, a person living at Repulse Bay, Canada, north of the Arctic Circle, experiences a full 24 hours of daylight at the June solstice and can go hunting at 1:00 A.M. We can also see that someone living in New York City will experience a longer period of daylight than of darkness. And someone living in Buenos Aires,

Argentina, will have a longer period of darkness than daylight on that day. This day is called the winter solstice in the Southern Hemisphere.

Now let us imagine the movement of Earth from its position at the June solstice toward a position a quarter of a year later, in September. As Earth moves toward that new position, we can imagine the changes that will be taking place in our three cities. First, in Repulse Bay there will be an increasing amount of darkness through July, August, and September. In New York, sunset will be arriving earlier, although it will still be light enough to play softball after dinner. And in Buenos Aires the situation will be reversed; as Earth moves toward its position in September, we can see that the periods of daylight in the Southern Hemisphere will begin to get longer, the nights shorter.

Finally, on or about September 23, Earth will reach a position known as an **equinox** (Latin: *aequus,* equal; *nox,* night). On this date (the autumnal equinox in the Northern Hemisphere), day and night will be of equal length at all locations on Earth. Thus, on the equinox conditions are identical for both hemispheres. As you can see in Figure 3.11, position B, Earth's axis points neither toward nor away from the sun; the circle of illumination passes through both poles, and it cuts Earth in half along its axis.

Imagine again the revolution and rotation of Earth while moving from September 23 toward a new position another quarter of a year later in December. We can see that in Repulse Bay the nights will be getting longer and longer until, on the winter solstice, which occurs on or about December 22, this northern town will experience 24 hours of darkness (Fig. 3.10, position C). The only natural light at all in Repulse Bay will be a faint glow at noon refracted from the sun below the horizon. And in New York, too, the days will get shorter and the sun will set earlier, until by the time you do your Christmas shopping, it will be dark at 5:30, before the stores close. Again, we can see that in Buenos Aires the situation is reversed, and on December 22 that city will experience its summer solstice; conditions will be much as they were in New York City in June. It may be a sweltering day in Buenos Aires, and everyone will go to the beach for the Christmas holidays.

Moving from late December through another quarter of a year to late March, Repulse Bay will have longer and longer periods of daylight, as will New York, while in Buenos Aires the nights will be getting longer, though they still will not be as long as the days. Then on or about March 21, Earth will again be in an equinox position (the vernal equinox in the Northern Hemisphere) similar to the one in September (Fig. 3.11, position D). Again, days and nights will be equal all over Earth. In other words, it will be 12 hours from sunrise to sunset and from sunset to sunrise. Finally, moving through another quarter of the year toward the June solstice where we began, Repulse Bay and New York City are both experiencing longer periods of daylight than darkness, and the sun is setting earlier and earlier in Buenos Aires, until on or about June 22 Repulse Bay and New York City will have their longest day of the year and Buenos Aires its shortest. Further, we can see that on June 22, a point on the Antarctic Circle in the Southern Hemisphere will experience a winter solstice similar to that which Repulse Bay had on December 22, with no daylight in 24 hours except what will appear at noon as a glow of twilight in the sky.

Lines Related to Earth Revolution

Looking at the diagrams of Earth in its various positions as it revolves around the sun, we can see that the angle of inclination is important. Because Earth's axis on June 22 is tilted $23\frac{1}{2}°$ toward the sun with respect to a line drawn perpendicular to the plane of the ecliptic, the sun's rays can reach that far ($23\frac{1}{2}°$) beyond the North Pole. The **Arctic Circle,** an imaginary line drawn around Earth $23\frac{1}{2}°$ from the North Pole (or $66\frac{1}{2}°$ north of the equator), marks this limit. We can see from the diagram that all points on or north of the Arctic Circle will experience no darkness on the June solstice, and, further, that all points south of the Arctic Circle will have some darkness on that day. The **Antarctic Circle** in the Southern Hemisphere ($23\frac{1}{2}°$ north of the South Pole, or $66\frac{1}{2}°$ south of the equator) marks a similar limit.

Furthermore, it can be seen from the diagrams that the sun's **vertical,** or **direct, rays** (rays that strike Earth's surface at right angles) also shift position in relation to the poles and the equator as Earth revolves around the sun. At the time of the June solstice, the sun's rays are vertical, or directly overhead, at noon at all points located $23\frac{1}{2}°$ *north* of the equator. This imaginary line around Earth marks the northernmost position at which the solar rays will ever be directly overhead during a full revolution of our planet around the sun. The imaginary line marking this limit is called the **Tropic of Cancer.** Six months later, at the time of the December solstice, the solar rays are vertical and the noon sun is directly

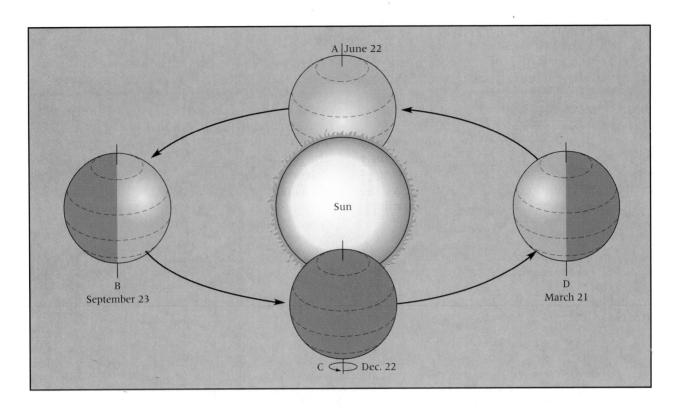

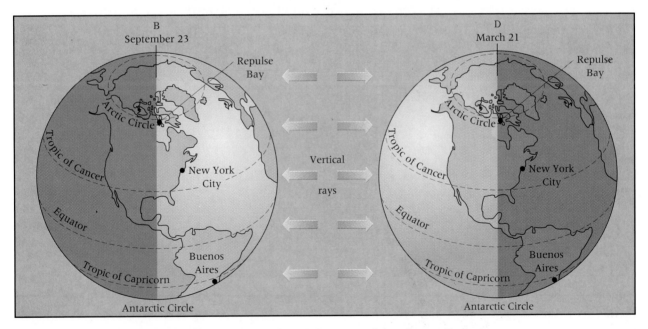

Figure 3.11 *Geometric relationships between Earth and the sun at the equinoxes. Day length is 12 hours everywhere, since the circle of illumination crosses the equator at right angles and cuts through both poles.* • **If earth were not inclined on its axis, would there still be latitudinal temperature variations? Would there be seasons?**

overhead at all points $23\frac{1}{2}°$ *south* of the equator. The imaginary line marking this limit is known as the **Tropic of Capricorn.** At the times of the March and September equinoxes, the vertical solar rays will strike directly only at the equator; the noon sun is directly overhead at all points on that line.

Note also that on any day of the year the sun's rays will strike Earth at a 90° angle at only one position either on or between the two tropics. All other positions that day will receive the sun's rays at an angle of less than 90° (or will receive no sunlight).

Analemma

The latitude at which the noon sun is directly overhead is also known as the sun's **declination.** Thus, if the sun appears directly overhead at 18° south latitude, the sun's declination is 18°S. A figure called an **analemma,** which is often drawn on globes as a big-bottomed "figure 8" in the middle of the Pacific Ocean, shows the declination of the sun throughout the year. A modified analemma is presented in Figure 3.12. Thus, if you would like to know where the sun will be directly overhead on March 12, you can look on

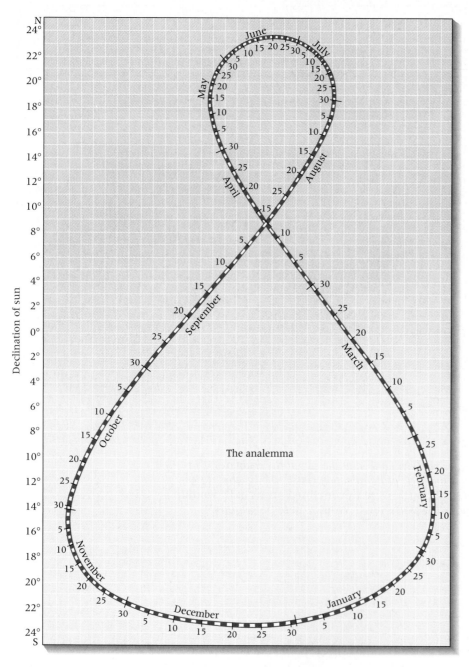

Figure 3.12 *An analemma shows the location of the vertical noon sun throughout the year.*

the analemma and see that it will be at 4°S. The analemma actually charts the passage of the direct rays of the sun over the 47 degrees of latitude they cover during a year.

Insolation and the Seasons

Solar radiation received by the Earth system is known as **insolation** (for *in*coming *sol*ar radi*ation*), and is the main source of energy on our planet. The seasonal variations in temperature that we experience are due primarily to fluctuations in insolation.

What causes these variations in insolation and thus in the seasons? It is true that Earth's atmosphere affects the amount of insolation received. Heavy cloud cover, for instance, will keep more solar radiation from reaching Earth's surface than will a clear blue sky. However, cloud cover is irregular and unpredictable, and it affects total insolation to only a minor degree over long periods of time.

The real answer to the question of what causes variations in insolation can be found by once again studying Figures 3.10 and 3.11, which illustrate the inclination and parallelism of Earth's axis as it revolves around the sun. Two major phenomena vary regularly for a given position on Earth as our planet rotates on its axis and revolves around the sun: the duration of daylight and the angle of the solar rays. The amount of daylight controls the duration of solar radiation, and the angle of the sun's rays directly affects the intensity of the solar radiation received. Together, the intensity and the duration of radiation are the major factors that affect the amount of insolation available at any location on Earth's surface (Table 3.2).

This situation is like an oven in which a roast is being cooked. The roast will cook faster and

TABLE 3.2	Radiation Intensity for Certain Solar Angles Expressed as a Percentage of the Maximum Possible (perpendicular beam)						
Solar Angle (degrees above the horizon)	0.0	10.0	20.0	23.5	26.5	30.0	40.0
% of Maximum	00.0	17.4	34.2	39.9	44.6	50.0	64.3
Solar Angle	50.0	60.0	63.5	66.5	70.0	80.0	90.0
% of Maximum	76.6	86.6	89.5	91.7	94.0	98.5	100.0

TABLE 3.3	Duration of Daylight for Certain Latitudes		
	Length of Day (N. Hemisphere) (read down)		
Latitude (in degrees)	March 21–Sept. 21	June 21	Dec. 21
0.0	12 hr.	12 hr. 0 min.	12 hr. 0 min.
10.0	12 hr.	12 hr. 35 min.	11 hr. 25 min.
20.0	12 hr.	13 hr. 12 min.	10 hr. 48 min.
23.5	12 hr.	13 hr. 35 min.	10 hr. 41 min.
30.0	12 hr.	13 hr. 56 min.	10 hr. 4 min.
40.0	12 hr.	14 hr. 52 min.	9 hr. 8 min.
50.0	12 hr.	16 hr. 18 min.	7 hr. 42 min.
60.0	12 hr.	18 hr. 27 min.	5 hr. 33 min.
66.5	12 hr.	24 hr.	0 hr.
70.0	12 hr.	24 hr.	0 hr.
80.0	12 hr.	24 hr.	0 hr.
90.0	12 hr.	24 hr.	0 hr.
Latitude	Equinox	Dec. 21	June 21
	Length of Day (S. Hemisphere) (read up)		

get browner if (1) the temperature is turned up and/or (2) someone leaves the oven on longer than usual. Likewise, a spot on Earth will receive more insolation if (1) the sun shines more directly, or (2) the sun shines longer, or (3) both. One reason that locations along the Tropics of Cancer and Capricorn are so hot during their summer solstice is that the sun's rays are intense and the day is long (there are many hours of daylight).

The intensity of solar radiation received at any one time varies from place to place because Earth presents a spherical surface to insolation.

Therefore, only a portion of Earth's surface can receive radiation at right angles, while the rest is struck at varying oblique angles (Fig. 3.13a). As we can see from Figure 3.13b, solar energy that strikes Earth at a vertical angle covers less area than an equal amount striking Earth at an oblique angle. And we can see that the closer to a right angle that the sun's rays strike Earth, the smaller will be the area covered. Since the amount of energy is the same no matter what kind of angle the rays make with the surface, it follows that the smaller the area that is struck,

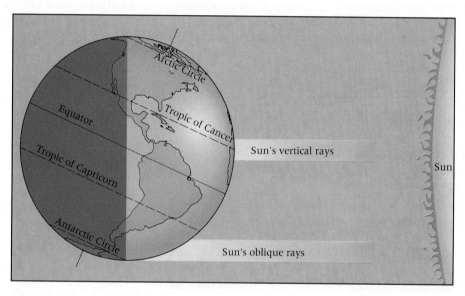

Figure 3.13a *The angle at which the sun's rays strike Earth determines the amount of solar energy received per unit of surface area. This amount in turn controls the seasons. The diagram represents the June condition, in which solar radiation strikes the surface perpendicularly only in the Northern Hemisphere, creating summer conditions there. In the Southern Hemisphere, oblique rays are spread over large areas, producing less receipt of energy per unit of area and making this the winter hemisphere.* • **How would this figure differ in December?**

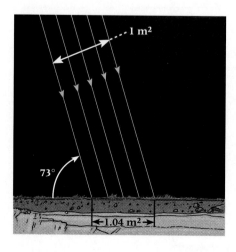
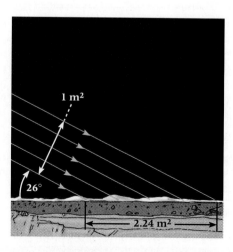

Figure 3.13b *The sun's rays in summer and winter. In summer the sun appears high in the sky, and its rays hit Earth more directly, spreading out less. In winter the sun is low in the sky, and its rays spread out over a much wider area, becoming less effective at heating the ground.*

the greater will be the intensity per unit area. Conversely, the more oblique the angle at which the sun's rays strike Earth, the greater the area over which those rays will be spread, and so less energy will strike Earth per unit area. In addition, the atmosphere limits to some extent the amount of insolation that reaches Earth's surface, and oblique rays must pass through a greater thickness of atmosphere than vertical rays.

The duration of solar energy is related to the length of daylight received at a particular point on Earth, since no insolation is received at night (Table 3.3). Obviously, the longer the period of daylight, the greater the amount of solar radiation that will be received at that location. And, as we have seen, periods of daylight vary in length through the seasons of the year as well as from place to place on Earth's surface.

Variations of Insolation with Latitude

Neglecting for the moment the influence of the atmosphere on variations in insolation during a 24-hour period, a place will receive its greatest insolation at solar noon when the sun has reached its zenith or highest point in the sky for that day. At any location, no insolation will be received during the hours of darkness. The amount of energy received after daybreak in-

creases as Earth rotates toward the time of solar noon. The amount of insolation then decreases until the next period of darkness begins.

We also know that the amount of daily insolation received at any one location on Earth varies with the seasons. There are three distinct patterns in the distribution of the seasonal receipt of solar energy in each hemisphere. These patterns serve as the basis for recognizing six latitudinal zones, or bands, of insolation and temperature that circle Earth (Fig. 3.14).

If we look first at the Northern Hemisphere, we may take the Tropic of Cancer and the Arctic Circle as the dividing lines for three of these distinctive zones. The area between the equator and the Tropic of Cancer can be called the north **tropical zone.** Here, insolation is always high but is greatest at the two times during the year that the sun is directly overhead at noon. These dates vary according to latitude. The north **midlatitude zone** is the wide band between the Tropic of Cancer and the Arctic Circle. In this belt, insolation is greatest on the June solstice, when the sun reaches its highest noon altitude and the period of daylight is long. Insolation is least at the December solstice when the sun is low in the sky and the period of daylight is short. The north **polar zone,** or **Arctic zone,** extends from the Arctic Circle to the pole. In this region, as in the midlatitude zone, insolation is greatest at the June solstice, but it ceases during the period that the sun's rays are blocked entirely by the tilt of Earth's axis. This period lasts for six months at the North Pole but is as short as one day directly on the Arctic Circle.

Similarly, there are the south tropical zone, the south midlatitude zone, and the south polar or **Antarctic zone,** all separated by the Tropic of Capricorn and the Antarctic Circle in the Southern Hemisphere. These areas get their greatest amounts of insolation at opposite times of the year from the northern zones.

Despite various patterns in the amount of insolation received in these zones, there are generalizations that we can make. For example, total annual insolation at the top of the atmosphere at a particular latitude remains constant from year to year. Furthermore, annual insolation tends to decrease from lower latitudes to higher latitudes. And the closer to the poles a place is located, the greater will be its seasonal variations caused by fluctuations in insolation.

The amount of insolation received by Earth is an important concept in understanding atmospheric dynamics and the distribution of climate and vegetation. Such climatic elements as

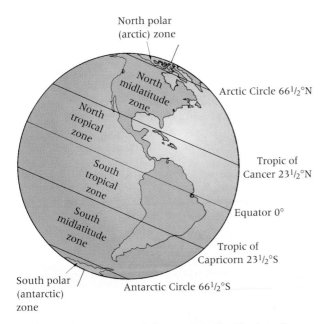

North polar (arctic) zone

North midlatitude zone

North tropical zone

South tropical zone

South midlatitude zone

South polar (antarctic) zone

Arctic Circle 66$\frac{1}{2}$°N

Tropic of Cancer 23$\frac{1}{2}$°N

Equator 0°

Tropic of Capricorn 23$\frac{1}{2}$°S

Antarctic Circle 66$\frac{1}{2}$°S

Figure 3.14 *The line of the equator, the Tropics of Cancer and Capricorn, and the Arctic and Antarctic Circles define six latitudinal zones that have distinctive insolation characteristics.* • **Which zone(s) would have the least annual variation in insolation? Why?**

temperature, precipitation, and winds are controlled in part by the amount of insolation received by Earth. People depend on certain levels of insolation for physical comfort, and plant life is especially sensitive to the amount of available insolation. You may have noticed plants that have wilted in too much sunlight or that have grown brown in a dark corner away from a window. Over a longer period of time, deciduous plants have an annual cycle of budding, flower-

ing, leafing, and losing their leaves. This cycle is apparently determined by the fluctuations of increasing and decreasing solar radiation that mark the changing seasons. Even animals respond to seasonal changes: some animals hibernate, many North American birds fly south toward warmer weather as winter approaches, and many animals breed at such a time that their offspring will be born in the spring, when warm weather is approaching.

Define and Recall

galaxy	terrestrial planet	rotation	solstice
light-year	giant planet	revolution	equinox
solar system		circle of illumination	Arctic Circle
planet	mass	perihelion	Antarctic Circle
satellite	gravitation	aphelion	vertical ray
asteroid		plane of the ecliptic	Tropic of Cancer
comet	horizon	angle of inclination	Tropic of Capricorn
meteor	oblate spheroid	parallelism	declination
meteorite	galactic movement		analemma
			insolation

Discuss and Review

1. What is a solar system? What bodies constitute our solar system?

2. How is the energy emitted from the sun produced?

3. Name the terrestrial planets. What do they have in common? Name the giant planets. What do they have in common?

4. Give a unique characteristic for each of the nine planets.

5. What unique physical characteristics of Earth are directly related to the planet's size, shape, and distance from the sun?

6. In the twentieth century we tend to accept as a given that the world is round. What are some of the ways to prove that it really is round?

7. Is Earth an absolutely perfect sphere? How are Earth's imperfections important in the study of geography?

8. Describe briefly how Earth's rotation and revolution affect life on Earth.

9. If the sun is closest to Earth on January 3, why isn't winter in the Northern Hemisphere warmer than winter in the Southern Hemisphere?

10. Identify the two major factors that cause regular variation in insolation throughout the year. How do they combine to cause the seasons?

Consider and Respond

1. Do you think we have discovered all the galaxies in the universe? Why?

2. Given what you know of the sun's relation to life on Earth, explain why the solstices and equinoxes have been so important to cultures all over the world. What are some of the major festivals associated with these times of the year?

3. Use the analemma presented in Figure 3.12 to determine the latitude where the noon sun will be directly overhead on:
   (a) February 12
   (b) July 30
   (c) November 2
   (d) December 30

4. Use the discussion of solar angle, presented on page 86 and depicted in Figures 3.13a and 3.13b to explain why we can look directly at the sun at sunrise and sunset but not at the noon hour.

5. Imagine you are at the equator on March 21. The noon sun would be directly overhead. However, for every degree of latitude that you travel to the north or south the noon solar angle would decrease by the same amount. For example, if you travel to 40°N latitude, the solar angle would be 50°. Can you explain this relationship? Can you develop a formula or set of instructions to generalize this relationship? What would be the solar angle at 40°N on June 22? on December 22?

6. Describe in your own words the relationship between insolation and latitude.

CHAPTER
4

THE ATMOSPHERE, ATMOSPHERIC HEATING, AND TEMPERATURE

CHAPTER PREVIEW

▶ Our planet's atmosphere is essential to life as we know it here on Earth.

How is this true? What is the significance of this statement for humans? How should this fact affect human behavior?

▶ Atmospheric elements are affected by atmospheric controls to produce weather and climate.

How do the elements differ from the controls? How does weather differ from climate?

▶ The sun is the original and ultimate source of the energy that drives the various components of the Earth system.

How does the sun's energy reach Earth? How does this energy affect the Earth system?

▶ Earth has an energy budget with a multiplicity of inputs and outputs (exchanges) that ultimately remains in balance despite recurring deficits and surpluses from time to time and place to place.

How is the budget concept useful to an understanding of atmospheric heating and cooling? How can we tell that the budget remains in balance?

▶ Water plays the single most important role in the exchanges of energy that fuel atmospheric dynamics.

What characteristics of water are responsible for its importance in energy exchange? In what ways is water involved in the heating of the atmosphere?

▶ As a direct result of differences in insolation and the mechanics of atmospheric heating, air temperature varies over time and both horizontally and vertically through space.

What are the most obvious variations? Why do they occur? How do temperatures stay within the ranges suitable for life if there are such great differences in the amounts of insolation received?

▲ Brilliant sky at sunset on the Gulf Coast of Florida.

All living things on our planet need water and oxygen to survive. Plants need carbon dioxide as well. Most living things we know cannot survive extreme temperatures, nor can they live long if exposed to large doses of the sun's ultraviolet radiation. It is the atmosphere, the layer of air that surrounds Earth, that supplies most of the oxygen and carbon dioxide and that helps maintain a constant level of water and radiation in the Earth system.

Although actually a thin film of air, the atmosphere serves as an insulator, maintaining the livable temperatures we find on Earth. Without the atmosphere, Earth would experience temperature extremes of as much as 260°C (500°F) between day and night. The atmosphere also serves as a shield, blocking out much of the sun's ultraviolet radiation and protecting us from showers of meteors. At other times the atmosphere is described as an ocean of air surrounding Earth. This description reminds us of the currents and circulation of the atmosphere—its dynamics—which create the changing conditions on Earth that we know as weather.

For contrast, we can look at our moon—a celestial body with virtually no atmosphere—in order to see the importance of our own atmosphere. Most obviously, a person standing on the moon without a space suit would not have any oxygen to breathe. Also, astronauts have recorded temperatures of up to 204°C (400°F) on the hot, sunlit side of the moon. On the dark side, temperatures approaching −121°C (−250°F) would kill an unprotected human.

The next thing our astronaut on the moon might notice is the "unearthly" silence. On Earth we hear sounds because sound waves are moved by the vibrating molecules of the atmosphere. Since the moon has no atmosphere and no molecules to vibrate and carry the sound waves, the lunar visitor is not able to hear any sounds. Also, because there is no atmosphere, the astronaut cannot fly aircraft or helicopters, and it would be fatal to try to use a parachute. In addition, lack of atmosphere means no protection from the bombardment of meteors that fly through space and collide with the moon; nearing Earth, most meteors burn up before reaching the surface because of friction within the atmosphere. And without an atmosphere for protection, a visitor to the moon might also be burned by the ultraviolet rays of the sun. On Earth we are protected to a large degree from ultraviolet radiation because the ozone layer of the upper atmosphere absorbs the major portion of this harmful radiation.

We can see that, in contrast to the stark lifelessness of the moon, Earth presents a hospitable environment for life almost solely because of its atmosphere. All living things are adapted to its presence. For example, many plants reproduce by pollen carried by winds. Birds can fly only because of the air, and the water cycle of Earth is maintained through the atmosphere, as is the "heat budget." The atmosphere diffuses sunlight as well, giving us our blue skies and the fantastic reds, pinks, oranges, and purples of sunrise and sunset. Without this diffusion the sky would appear black, as it does from the moon.

Further, the atmosphere provides a means by which the systems of Earth attempt to reach equilibrium. Changes in weather are ultimately the result of the atmospheric effects that equalize temperature and pressure differences on Earth's surface by transferring heat and moisture through Earth's atmospheric and oceanic circulation systems.

Characteristics of the Atmosphere

The atmosphere extends as far as 9600 kilometers (6000 mi) above Earth's surface. Its density decreases rapidly with altitude, and, in fact, 97 percent of the air is concentrated in the first 29 kilometers (18 mi) or so. Since air has mass, the atmosphere exerts pressure on Earth's surface. At sea level this pressure is about 1034 grams per square centimeter (14.7 lb per sq in), but the higher the elevation, the lower the atmospheric pressure. In Chapter 5 we will examine the relationship between atmospheric pressure and elevation in more detail.

Composition of the Atmosphere

The atmosphere is composed of numerous gases (Table 4.1). Most of these gases remain in the same proportions regardless of the density of the atmosphere. About 78 percent of the atmosphere's volume is made up of nitrogen and nearly 21 percent consists of oxygen. Argon comprises most of the remaining 1 percent. The percentage of carbon dioxide in the atmosphere varies but is about 0.03 percent by volume. There are traces of other gases as well: ozone, hydrogen, neon, xenon, helium, methane, nitrous oxide, and krypton.

Nitrogen, Oxygen, and Carbon Dioxide.   Nitrogen makes up the largest proportion of air. It is an important element supporting plant growth. In addition, some of the other gases in the atmosphere are vital to the development and mainte-

TABLE 4.1	Composition of Earth's Atmosphere	
Name of Gas	Chemical Symbol	Percent of Mass
Nitrogen	N	78.09
Oxygen	O	20.95
Argon	Ar	0.93
Carbon dioxide	CO_2	0.03 (variable)
Ozone	O_3	Trace (variable)
Water vapor	H_2O	Trace (variable)
Hydrogen	H	Trace
Inert gases: Neon, krypton, helium	Ne, Kr, He	Traces

nance of life on Earth. One of the most important of the atmospheric gases is, of course, oxygen, which all animals, including humans, use to oxidize (burn) the food they eat. Oxidation, which is technically the chemical combination of oxygen with other materials to create new products, occurs in situations outside animal life as well. Rapid oxidation takes place, for instance, when we burn fossil fuels or wood and thus release tremendous amounts of heat energy. The decay of certain rocks or organic debris and the development of rust are examples of slow oxidation and are processes that depend upon the existence of oxygen in the atmosphere.

Carbon dioxide is also an important atmospheric gas since it absorbs heat from Earth. The atmosphere then emits about half of that absorbed heat energy back to Earth. This process helps maintain the warmth of Earth and is a factor in Earth's heat energy budget. However, human activity—largely the burning of fossil fuels—has greatly increased the amount of carbon dioxide in the atmosphere. There is concern that this will upset our current temperature structure.

Carbon dioxide is also involved in the *system* known as the carbon cycle. Plants, through a process known as **photosynthesis,** use carbon dioxide and water to make carbohydrates (sugars and starches), in which amounts of energy, derived originally from the sun, are stored (Fig. 4.1). Oxygen is given off as a by-product. Animals then use the oxygen to oxidize the carbohydrates, releasing the stored energy. A by-product of this process in animals is the release of carbon dioxide, which completes the cycle when it is in turn used by plants in photosynthesis.

Ozone.   Another vital gas in Earth's atmosphere is ozone. The ozone molecule (O_3) is a cousin of the oxygen molecule (O_2); it is made up of three atoms of oxygen while regular oxygen is made up of only two. Ozone is formed in the upper atmosphere when an oxygen molecule is split into two atoms by short-wave solar radiation and the free unstable atoms join two other oxygen molecules to form two molecules of ozone consisting of three oxygen atoms each.

Ozone is important to climate because it is capable of absorbing large amounts of the sun's ultraviolet radiation. Without the ozone of the upper atmosphere, the ultraviolet radiation reaching Earth would severely burn human skin, increase the incidence of skin cancer, destroy certain microscopic forms of life, and damage plants. There is, therefore, increasing concern that human activity, especially the addition of chlorofluorocarbons (CFCs) to the atmosphere, may permanently damage this fragile ozone layer.

The small proportion of ultraviolet radiation that the ozone layer allows to reach Earth does serve useful purposes. For instance, it is important in the production of certain vitamins, and it helps the growth of some beneficial viruses and bacteria. It also has a function in the process of photosynthesis. Least important, ultraviolet radiation produces painful sunburns or sensible suntans depending on individual skin tolerance and exposure time.

Water Vapor, Liquids, and Solids.   Water vapor is always mixed in some proportion with the dry air of the lower part of the atmosphere, although it varies from 0.02 percent by volume in a cold, dry climate to a high of nearly 5 percent in the humid tropics. The percentage of water vapor in the air will be discussed later under the broad topic of humidity, but it is important to note here that the variations in this percentage over time and place are an important consideration in the examination and comparison of climates.

Water vapor also absorbs heat in the lower atmosphere and so prevents its rapid escape from Earth. Thus, like carbon dioxide, water vapor plays a large role in the insulating action of the atmosphere. In addition to gaseous water vapor, liquid water also exists in the atmosphere as rain

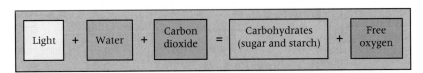

Figure 4.1 *The equation of photosynthesis shows how solar energy (light) is used by plants to manufacture sugars and starches from atmospheric carbon dioxide and water, liberating oxygen in the process. The stored food energy is then eaten by animals, which also breathe the oxygen released by photosynthesis.*

and as fine droplets in clouds, mist, and fog. Solid water is found in the atmosphere in the form of ice crystals, snow, and hail. Suspended in the atmosphere are many other solids, such as dust, soil particles, pollen, microscopic animals, bacteria, smoke particles, seeds, spores, and salts from ocean spray, all of which can play an important role in absorption of energy and in the formation of raindrops.

Vertical Layering of the Atmosphere

Though people function primarily in the lowest levels of the atmosphere, there are times, such as when we fly in aircraft or climb a mountain, when we leave our normal altitude. The thinness of the atmosphere at these higher altitudes may affect us if we are not accustomed to it. Visitors to Inca ruins in the Andes or high-altitude Himalayan climbers may experience "altitude sickness," and even skiers in the Rockies near "mile-high" Denver may need time to adjust. The air at these levels is much *thinner* than most of us are used to; there is more empty space between air molecules, and thus there is less oxygen and other gases in each breath of air inhaled.

The atmosphere can be divided into several layers according to differences in temperature and rates of temperature change (Fig. 4.2). The first of these layers, lying closest to Earth's surface, is the **troposphere** (from Greek: *tropo,* turn—the turning or mixing zone), which extends about 10 to 16 kilometers (6 to 10 mi) above Earth. Its thickness, which tends to vary seasonally, is least at the poles and greatest at the equator. It is within the troposphere that people live and work, plants grow, and virtually all Earth's weather takes place.

The troposphere has two distinct characteristics that differentiate it from other layers of the atmosphere. One is that the water vapor and dust particles of the atmosphere are concentrated in this one layer; they are rarely found in the atmospheric layers above the troposphere. The other characteristic of this layer is that temperature normally decreases with increased altitude.

The altitude at which the temperature ceases to fall with increased altitude is called the **tropopause,** which separates the troposphere from the **stratosphere**—the second layer of the atmosphere. The temperature of the lower part of the stratosphere remains fairly constant (about

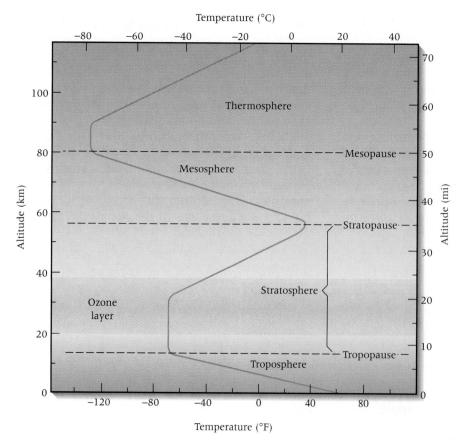

Figure 4.2 *Vertical temperature changes in Earth's atmosphere are the basis for its subdivision into the troposphere, stratosphere, mesosphere, and thermosphere.*
• **At what altitude is our atmosphere the coldest?**

−57°C, or −70°F) to an altitude of about 32 kilometers (20 mi). It is in the stratosphere that we find the concentration of ozone that does so much to protect life on Earth from the sun's ultraviolet radiation. Because of this ozone layer, however, temperatures increase in the upper parts of the stratosphere as the ozone absorbs ultraviolet radiation. Temperatures at the **stratopause,** which is about 56 kilometers (35 mi) above Earth, are about the same as temperatures found on Earth's surface, although little of that heat can be conducted because the air is so thin.

Above the stratopause are the **mesosphere,** in which temperatures tend to drop with increased altitude, and the **thermosphere,** where temperatures increase until they approach 1100°C (2000°F) at noon. However, the air is so thin at this altitude that there is practically a vacuum and little heat can be conducted.

The thermosphere was once called the ionosphere because of the ionization of molecules and atoms that occurs in this layer, mostly as a result of ultraviolet rays but also because of X rays and gamma radiation. (Ionization refers to the process whereby atoms are changed to ions through the removal or addition of electrons, giving them an electrical charge.) The thermosphere merges gradually into the exosphere, the zone where Earth's atmosphere gives way to interplanetary space.

Weather and Climate

Weather refers to the condition of atmospheric elements at a given time and for a specific area. That area could be as large as the New York metropolitan area or a spot as small and specific as a weather observation station. Although the high-level atmospheric zones are important in such fields as space research, remote sensing, and telecommunications, it is the lowest layer, the troposphere, that is of the greatest interest to physical geographers and weather forecasters who survey the changing conditions of the atmosphere in the field of study known as **meteorology.**

Many observations of the weather of a place over a period of years provide us with a description of its climate. **Climate** describes an area's average weather, but it also includes those common deviations from the norm or average that are likely to occur, as well as extreme situations, which can be very significant. Thus we could describe the climate of the southeastern United States in terms of average temperatures and precipitation through a year, but we would also have to include mention of the likelihood of hurricanes during certain periods of the year. **Climatology** is the study of the varieties of climates, both past and present, found on our planet and their distribution over its surface.

Weather and climate are of prime interest to the physical geographer because they affect and are interrelated with other parts of the Earth system. The changing conditions of atmospheric elements such as temperature, rainfall, wind, and so forth, affect soils and vegetation, erode landforms, and cause flooding of towns and farms.

Elements of Weather and Climate

There are five basic elements of the atmosphere that serve as the "ingredients" of weather and climate. They are (1) solar energy (or insolation), (2) temperature, (3) pressure, (4) wind, and (5) precipitation (and moisture). We must examine these **elements** in order to understand and categorize weather and climate. Thus, a weather forecast will generally include the probable temperature range, the present temperature, a description of the cloud cover, the chance of precipitation, the speed and direction of the winds, and air pressure.

In Chapter 3 we noted that the amount of solar energy received at one place on Earth's surface varies during a day and throughout the year. The amount of insolation a place receives is the most important weather element since the other four elements are in part dependent upon the intensity and duration of solar energy.

The temperature of the atmosphere at a given place on or near the surface of Earth is largely a function of the insolation received at that location. It is also influenced by many other factors, such as land and water distribution and altitude. Unless there is some form of precipitation occurring, the temperature of the air may be the first element of weather we describe when someone asks us what it is like outside.

However, if it is raining, or the fog is in, or it is snowing, we will probably notice and mention that condition first. We are less aware of the amount of water vapor or moisture in the air (except in very arid or humid areas). However, moisture in the air is a vital weather element in the atmosphere, and its variations play an important role in the likelihood of precipitation.

We are probably least aware of variations in air pressure, although the fluctuations in air pressure are basic to the development of winds and storms. However, there are some people who say they can feel a change in the weather "in their bones" because they have arthritis and can probably sense the movement of fluids under pressure in their joints.

We all know that weather varies. Since it is the momentary state of the atmosphere at a given location, it varies from time to time and from place to place. There are even variations in the amount that weather varies. In some places or at some times of year, the weather changes almost daily—from rain to sunshine to clouds to rain to snow. And in other places there may be weeks of uninterrupted sunshine, blue skies, and moderate temperatures and then weeks of persistent rain. There are a few places where there are only minor differences in the weather throughout the year. The language of the original people of Hawaii is said to have no word for weather because conditions there varied so little.

Controls of Weather and Climate

Variations in the elements of weather and climate over Earth's surface are caused by several **controls.** The major controls are (1) latitude, (2) land and water distribution, (3) ocean currents, (4) altitude, (5) landform barriers, and (6) human activity.

Latitude. Latitude is the most important control of weather and climate. Recall that, because of the inclination and parallelism of Earth's axis as it revolves around the sun, there are distinct patterns in the latitudinal distribution of the seasonal and annual receipt of solar energy over Earth's surface. This has a direct effect on tem-

peratures. In general, annual insolation tends to decrease from lower latitudes to higher latitudes (see again Fig. 3.14). Table 4.2 shows the average annual temperatures for several locations in the Northern Hemisphere. We can see that, responding to insolation (with one exception), a poleward decrease in temperature is true for these locations. The exception is near the equator itself. Due to the heavy cloud cover in the equatorial regions, average annual temperatures there tend to be lower than at places slightly to the north or south, where skies are clearer.

Another very simple way to see this general trend of decreasing temperatures as we move toward the poles is to think about the kinds of clothes we would take along for one month—say, Jaunary—if we were to visit Ciudad Bolivar, Venezuela; Raleigh, North Carolina; or Point Barrow, Alaska.

Land and Water Distribution. Not only do the oceans and seas of Earth serve as storehouses of water for the whole system, but they also store tremendous amounts of energy. Their widespread distribution makes them an important atmospheric control that does much to modify the atmospheric elements.

All things heat at different rates. On Earth's surface, bodies of water heat and cool more slowly than do land surfaces. This is because water is *transparent,* and solar energy passes through the surface into the layers below, while in more opaque materials like soil and rock, the energy is concentrated on the surface. Thus, a given unit of heat energy will spread through a greater volume of water than land. In addition, *evaporation*

TABLE 4.2	Typical Temperatures in the Northern Hemisphere		
		Average Annual Temperature	
Location	Latitude	(°C)	(°F)
Libreville, Gabon	0°23′N	26.5	80
Ciudad Bolivar, Venezuela	8°19′N	27.5	82
Bombay, India	18°58′N	26.5	80
Amoy, China	24°26′N	22.0	72
Raleigh, North Carolina	35°50′N	18.0	66
Bordeaux, France	44°50′N	12.5	55
Goose Bay, Labrador, Canada	53°19′N	−1.0	31
Markova, Russia	64°45′N	−9.0	15
Point Barrow, Alaska	71°18′N	−12.0	10
Mould Bay, NWT, Canada	76°17′N	−17.5	0

from the ocean's surface is constantly occurring. Evaporation is a cooling process which further lowers the ocean's surface temperature. Also, since liquid water flows and *mixes,* it is able to transfer heat to even deeper layers within its mass. Finally, the *specific heat* of water is greater than that of land, which means that water must absorb more heat energy than land in order to be raised the same number of degrees in temperature. In fact, the specific heat of water is over four times that of earth or rock.

For these same reasons water cools off more slowly than does land. The result is that as summer changes to winter, the land cools more rapidly than bodies of water, and as winter becomes summer, the land heats more rapidly. Since the air gets much of its heat from the surface with which it is in contact and which it overlies, the differential heating of land and water surfaces sets up inequalities and variations in the temperature of the atmosphere above these two surfaces.

The mean temperature in Seattle, Washington, in July is 18°C (64°F), while the mean temperature during the same month in Minneapolis, Minnesota, is 21°C (70°F). Since the two cities are at similar latitudes their annual pattern and receipt of solar energy are also similar. Therefore, their different temperatures in July must be related to a control other than latitude. Much of this difference in temperature can be attributed to the fact that Seattle is near the Pacific coast, while Minneapolis is in the heart of a large continent and far from the moderating influence of an ocean. Consequently, Seattle stays cooler than Minneapolis in the summer because the surrounding water warms up slowly, keeping the air relatively cool. Minneapolis, on the other hand, is in the center of a large landmass that warms very quickly and in turn warms the layer of air above it. In the winter, the opposite is true. Seattle is warmed by the water while Minneapolis is not. The mean temperature in January is 4.5°C (40°F) in Seattle and −15.5°C (4°F) in Minneapolis.

Not only do water and land heat and cool at different rates, but so do various land surface materials. Soil, forest, grass, and rock surfaces all heat differentially and set up variations in the overlying temperatures of the air; this can in turn affect the other climatic elements.

Ocean Currents. Surface ocean currents are large movements of water pushed by the winds. They may flow from a place of warm temperatures to one of cooler temperatures or vice versa. These movements result, as we saw in Chapter 1, from the attempt of Earth systems to reach a balance—in this instance, a balance of temperature and density.

The rotation of Earth affects the movements of the winds, which in turn affect the movement of the ocean currents. In general the currents move in a clockwise direction in the Northern Hemisphere and in a counterclockwise direction in the Southern Hemisphere (Fig. 4.3).

Since the temperature of the ocean greatly affects the temperature of the air above it, an ocean current that moves warm equatorial water toward the poles or cold polar water toward the equator can significantly modify the air temperatures of those locations into which it flows. If the currents pass close to land and are accompanied by onshore winds, they can have a significant impact upon the coastal climate.

The Gulf Stream, with its extension, the North Atlantic Drift, is an example of an ocean current that moves warm water northward, keeping the coasts of Great Britain and Norway ice-free in wintertime and moderating the climates of nearby land areas (Fig. 4.4). We can see the effects of the Gulf Stream if we compare the winter conditions of the British Isles with those of Labrador in northeastern Canada. Though both are at the same latitude, the climate of the British Isles is moderated by the effects of the Gulf Stream (North Atlantic Drift). For example, the average temperature in Glasgow, Scotland, in January is 4°C (39°F), while during the same month it is −21.5°C (−7°F) in Nain, Labrador.

The California Current is a current off the west coast of the United States that helps moderate the climate of that coast as it brings cold water south into relatively warm areas. As the current swings southwest from the coast of central California, cold bottom water is brought to the surface, causing further chilling of the air above. San Francisco's cool summers (July average: 14°C, or 58°F) reflect the effect of this current.

Altitude as a Control. As we have seen, temperatures within the troposphere decrease with increasing altitude. In Southern California you can find snow for skiing if you go to an altitude of 2400 to 3000 meters (8000 to 10,000 ft). Mount Kenya, 5199 meters (17,058 ft) high and located at the equator, is cold enough to have glaciers. Anyone who has hiked upward 500, 1000, or 1500 meters in midsummer has experienced a decline in temperature with increasing altitude. Even if it is hot on the valley floor, you may need a sweater once you climb a few thousand meters. The city of Quito, Ecuador, only 1° south of the

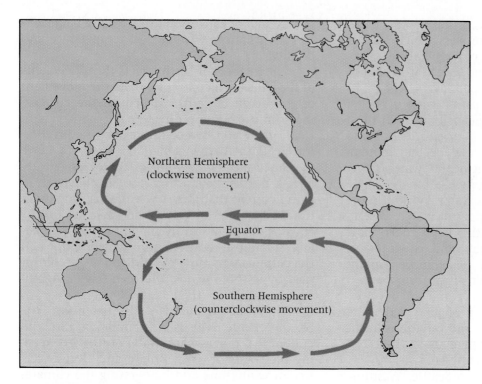

Figure 4.3 A highly simplified map of currents in the Pacific Ocean to show their basic rotary pattern. The major currents move clockwise in the Northern Hemisphere and counterclockwise in the Southern Hemisphere. A similar pattern exists in the Atlantic. • **What path would a hurricane, forming off of western Africa, take as it approached the United States?**

equator, has an average temperature of only 13°C (55°F) because it is located at an altitude of about 2900 meters (9500 ft).

Change in altitude has a direct bearing on another atmospheric element, air pressure, which, like temperature, decreases with increasing altitude. As a result, the air pressure on top of a 4000-meter (13,000-ft) mountain will be less than that in the plain far below, and it will affect many everyday things. Water at the higher altitude boils at 85°C (185°F) instead of at 100°C

(212°F), making it nearly impossible to make a good cup of tea. Automobile carburetors do not work effectively, and people traveling rapidly up or down the mountain have a popping sensation in their ears because of the change in pressure.

Landform Barriers. Landform barriers, especially large mountain ranges, can block movements of air from one place to another and thus affect the weather and climate of an area. For example, the Himalayas keep cold, winter, Asiatic air out of In-

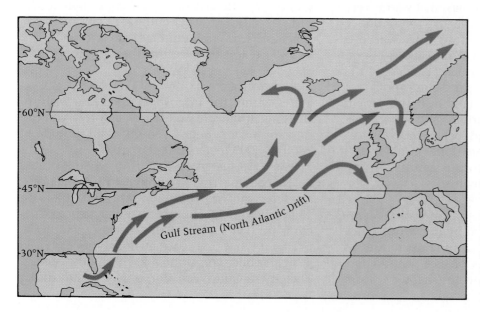

Figure 4.4 The Gulf Stream (the North Atlantic Drift farther eastward) is a warm current that greatly moderates the climate of northern Europe. • **Use this figure and the information gained in Figure 4.3 to discuss the route sailing ships would follow from the United States to England and back.**

dia, giving the Indian subcontinent a year-round tropical climate.

If the prevailing winds are from the west, and if they tend to bring rain and moisture with them, then a mountain range that runs north-south will generally have a wet climate on its west-facing, windward slope and a dry one on its east-facing, leeward (sheltered) slope. Although mountain ranges that run north–south, like the Rockies, Cascades, or Sierra Nevada in North America, block the movement of moisture-carrying air from the western oceans to the interior of the continent, thus helping create and maintain desert areas on their eastern sides, they do little to block the movement of cold polar air flowing toward the equator. Therefore, because they are not protected by an east–west mountain range to the north, areas in the southern United States can be subjected to unusual cold spells from the invasion of polar air.

Human Activities. Human beings, too, may be considered "controls" of weather and climate. Such activities as building cities, burning fossil fuels, destroying forests, draining swamps, or creating large reservoirs can significantly affect local climatic patterns and, possibly, world climatic patterns as well. In addition, people have tried to modify weather since almost the beginning of time. Though we have had only slight success, our potential for influencing weather and climate is considerable.

Solar Energy and Atmospheric Dynamics

As we noted in Chapter 3, our sun is the major source of energy, either directly or indirectly, for the entire Earth system. Earth does receive very small proportions of energy from other stars and from the interior of Earth itself (volcanoes and geysers provide certain amounts of heat energy); however, when compared with the amount received from the sun, these other sources are insignificant.

Energy from the Sun

Energy is emitted by the sun in the form of **electromagnetic energy,** which travels at the speed of light in a spectrum of waves of varying lengths (Fig. 4.5). It takes slightly more than eight minutes for these waves to reach Earth. About 41 percent of this spectrum of waves is in the form of visible light rays, but much of the sun's energy cannot be seen by the human eye. About half of the sun's radiant energy is in waves that are longer than visible light rays, and these include some *infrared waves*. While these cannot be seen, they can sometimes be sensed by the human skin. The remaining 9 percent of solar energy is made up of X rays, gamma rays, and *ultraviolet* rays, all of which are shorter in length than those of visible light. These also cannot be seen but can affect other tissues of the human body (thus the danger in absorbing too many X rays). Collectively, visible light, ultraviolet rays, X rays, and gamma rays are known as **short-wave radiation.** We have learned to harness some of these energy waves for communications (radio, microwave transmission, television), health (X rays), and use in the field of remote sensing (photography, radar, infrared imagery).

Energy is radiated into space by the sun at a steady rate. At its outer edge, Earth's atmosphere intercepts an amount of energy equivalent to 1.97 calories per square centimeter per minute. A **calorie** is the amount of *energy* required to raise the temperature of 1 gram of water 1°C. This

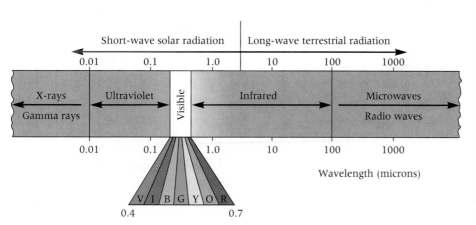

Figure 4.5 Radiation from the sun travels toward Earth in a wide spectrum of wavelengths, which are measured in micrometers (1 micrometer equals one ten-thousandth of a centimeter). Visible light occurs at wavelengths of approximately 0.4 to 0.7 micrometers. Solar radiation is short-wave radiation (less than 4.0 micrometers), whereas terrestrial (Earth) radiation is of long wavelengths (more than 4.0 micrometers).

can also be expressed in units of *power*—in this case, 1367 watts per square meter. The rate of a planet's receipt of solar energy is known as the **solar constant** and has been measured with great precision outside Earth's atmosphere by orbiting satellites. The atmosphere affects the amount of solar radiation received on the surface of Earth because some energy is absorbed by clouds, some is reflected, and some is refracted. If we could remove the atmosphere from Earth, we would find that the solar energy striking the surface at a particular location for a particular time would be a constant value determined by the latitude of the location (see Fig. 3.13b).

Of course, the measured value of the solar constant varies with distance from the sun as the same amount of energy radiates out into larger and larger areas. Because of this, if we measured the solar constant for the planet Mercury, it would be much higher than that for Earth. When Earth is closest to the sun in its orbit, its solar constant is slightly higher than the yearly average, and when it is farthest away, the solar constant is slightly lower than average. However, this difference does not have a significant effect on Earth's temperatures. When Earth is at aphelion in July and the solar constant is lowest because of the distance from the sun, the Northern Hemisphere is in the midst of a summer with temperatures that are not significantly different from those in the Southern Hemisphere six months later. The solar constant also varies slightly with changes in activity on the sun; during intense sunspot or sun storm activity, for example, the solar constant will be slightly higher than usual. However, these variations are not even as great as those caused by Earth's elliptical orbit.

The Role of Water

As it penetrates our atmosphere, some of the incoming solar radiation is involved in energy exchanges, as water in the Earth system is altered from one state to another. Water is the only material that can exist in all three states of matter—as a solid, as a liquid, and as a gas—within the normal temperature range of Earth's surface. In the atmosphere, water exists as a clear, odorless gas called **water vapor.** It is also a liquid in the atmosphere, as well as in the oceans and other water bodies of Earth, in vegetation and animals, and underground. Water is a solid in snow and ice in the atmosphere as well as on and under the surface of the colder parts of Earth.

Not only is water stored in all three states of matter, but it can change from one state to another, as illustrated in Figure 4.6. In doing so it is involved in the heat energy exchange of the Earth system. The molecules of a gas move faster than do those of a liquid. Thus, during the process of **condensation,** when water vapor changes to water, its molecules slow down and some of their energy is released (590 calories per gram). The molecules of a solid move even more slowly than those of a liquid, so during the process of **freezing,** when water changes to ice, additional energy is released (80 calories per gram). When the process is reversed, heat must be added. Thus, **melting** requires the addition of 80 calories per gram, and **evaporation** requires the addition of 590 calories per gram. This added energy is stored as **latent** (or hidden) **heat.**

Some of these energy exchanges can be easily demonstrated. For example, if you hold an ice

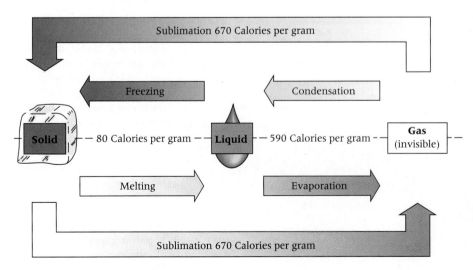

Figure 4.6 The three physical states of water and energy exchanges between them.

cube in your hand, your hand feels cold because it is giving off the heat needed to melt the ice. We are cooled by perspiration evaporating from our skin, since heat must be absorbed both from our skin and from the remaining perspiration, thereby lowering the temperature of both.

Scientists have become increasingly aware of the importance of the energy exchange between the atmosphere and the hydrosphere, especially at the ocean's surface. Three fourths of Earth's surface is covered by water. Because water heats up more slowly than land, it retains its heat longer. Thus, the oceans act as huge reservoirs of heat energy to power the atmosphere, which directly influences our weather and climate on land.

Effects of the Atmosphere on Solar Radiation

In addition to its involvement in these latent energy exchanges, the sun's energy, as it passes through Earth's atmosphere, loses over half its intensity through various processes. In fact, the amount of insolation actually received at a particular location depends not only on the latitude, the time of day, and the time of year (all of which are related to the angle at which the sun's rays strike Earth), but also on the transparency of the atmosphere (or the amount of cloud cover, moisture, carbon dioxide, and solid particles in the air).

When the sun's energy passes through the atmosphere, several things happen to it (the following figures represent approximate averages for entire Earth; at any one location or time they may differ): (1) Twenty-six percent of the energy is *reflected* directly back to space by clouds and the ground, (2) 8 percent is *scattered* by minute atmospheric particles and returned to space as diffuse radiation, (3) 20 percent reaches Earth's surface as diffuse radiation after being scattered, (4) 27 percent reaches Earth's surface as direct radiation, and (5) 19 percent is *absorbed* by the ozone layer and by water vapor in the clouds of the atmosphere (Fig. 4.7). In other words, on a worldwide average, 47 percent of the incoming solar radiation eventually reaches the surface, 19 percent is retained in the atmosphere, and 34 percent is returned to space. Since Earth's energy budget is in equilibrium, the 47 percent received at the surface is ultimately returned to the atmosphere by processes that we will now examine.

Heating the Atmosphere

Since the 19 percent of direct solar radiation that is retained by the atmosphere is "locked up" in the clouds and the ozone layer and thus is not available to heat the troposphere, some other source must be found to explain the creation of atmospheric warmth. The explanation lies in the 47 percent of incoming solar energy reaching Earth's surface (on both land and bodies of water) and in the transfer of heat energy from Earth back to the atmosphere through such physical processes as (1) radiation, (2) conduction, (3) convection (along with the related phenomenon, advection), and (4) the latent heat of condensation.

Methods of Heat Energy Transfer

Radiation. The process by which electromagnetic energy is transferred from the sun to Earth is called **radiation.** We should be aware that *all* objects emit electromagnetic radiation. The characteristics of that radiation depend on the temperature of the radiating body. In general we can state that the warmer the object, the *more* energy it will emit and the shorter the wavelengths of peak emission. Since the sun's absolute temperature is 20 times that of Earth's, we can predict that the sun will emit more energy, and at shorter wavelengths, than Earth. This is borne out by the facts: The solar energy output per square meter is approximately 160,000 times that of Earth! Further, the majority of solar energy is emitted at wavelengths shorter than 4.0 micrometers, whereas most of Earth's energy is radiated at wavelengths longer than 4.0 micrometers (see Fig. 4.5). Thus, short-wave radiation from the sun reaches Earth and heats its surface, which, being cooler than the sun, gives off energy in the form of long waves. It is this **long-wave radiation** from Earth's surface that heats the lower layers of the atmosphere.

Conduction. The means by which heat is transferred from one part of a body to another or between two touching objects is called **conduction.** Heat flows from the warmer to the cooler (part of a) body in order to equalize temperature. Conduction actually occurs as heat is passed from one molecule to another in chainlike fashion. It is conduction that makes the bottom of your soup bowl too hot to touch. Conduction also occurs when a spoon left in your coffee gets hot.

Atmospheric conduction occurs at the interface of (zone of contact between) the atmosphere and Earth's surface. However, it is actually

(Continued on page 104)

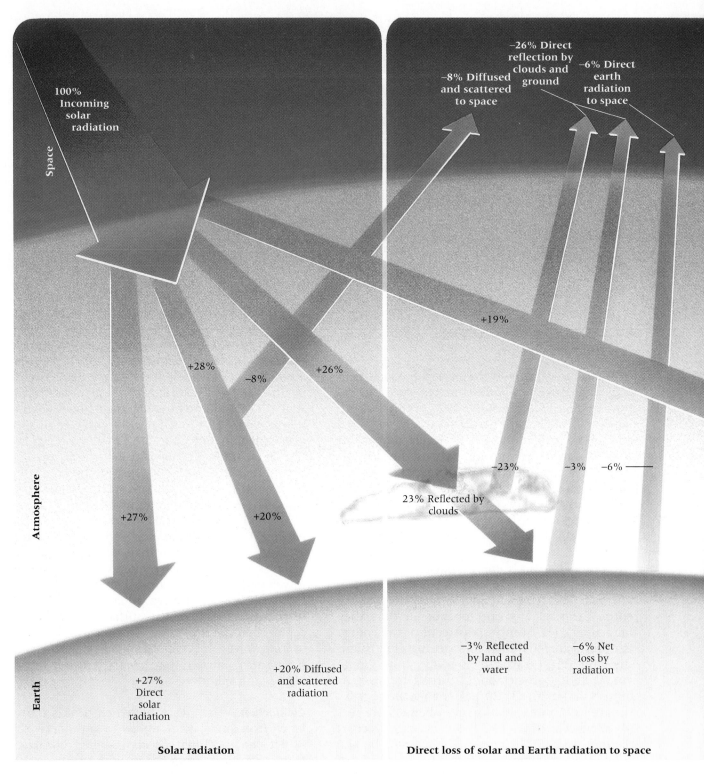

Figure 4.7 Environmental Systems: Earth's Radiation Budget *If you live outside the tropics you are certainly aware that temperatures can vary greatly from day to day, month to month, or year to year. However, from one year to the next, Earth's overall average temperature varies very little. This fact indicates that a long-term global balance, or equilibrium, must exist between the energy received and emitted by the Earth system.*

The source of all energy for the Earth system is the sun. As the energy from the sun enters our atmosphere and flows toward Earth's surface, several things happen to it. Some of the solar energy is reflected back to space by clouds and Earth's surface. Additional solar energy is scattered by dust, water, and other small particles in the atmosphere or

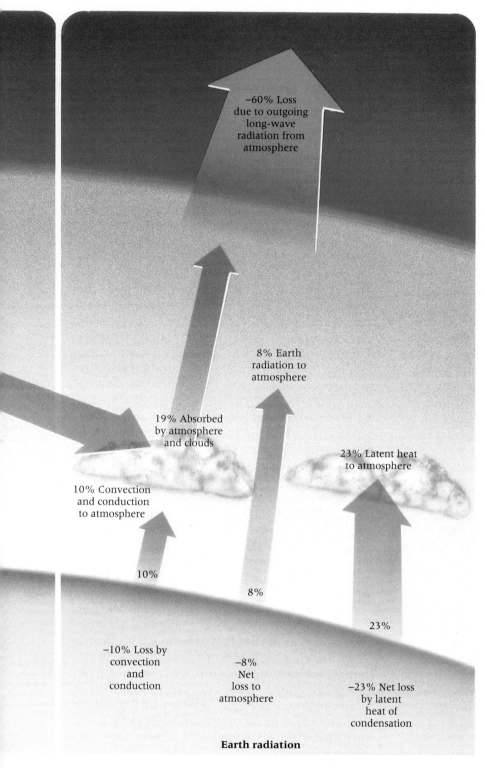

−60% Loss
due to outgoing
long-wave
radiation from
atmosphere

8% Earth
radiation to
atmosphere

19% Absorbed
by atmosphere
and clouds

23% Latent heat
to atmosphere

10% Convection
and conduction
to atmosphere

10%

8%

23%

−10% Loss by
convection
and
conduction

−8%
Net
loss to
atmosphere

−23% Net loss
by latent
heat of
condensation

Earth radiation

absorbed by the atmosphere and clouds. Note in the diagram that only 47 percent of the incoming solar energy reaches and is absorbed by Earth's surface.

The energy absorbed by Earth's surface heats the atmosphere through various processes such as conduction, convection, or latent heat of condensation. The atmosphere readily absorbs Earth-emitted energy. Eventually, all the energy gained by the atmosphere is lost to space. However, the radiation budget is a dynamic one. In other words, alterations to one element affect the other elements. As a result, there is growing concern that one of the elements, human activity, will cause the atmosphere to absorb more Earth-emitted energy, thus raising global temperatures.

a minor method of heat transfer in terms of warming the atmosphere because it affects only the layers of air closest to Earth's surface. This is because air is a poor conductor of heat (unlike certain metals). In fact, air is just the opposite of a good conductor; it is a good insulator. This property of air is why a layer of air is sometimes put between two panes of glass to help keep heat indoors. The same principle is used in a thermos bottle. The air sandwiched between layers of glass keeps the contents of the bottle warm or cold, as the case may be. Air is also used as a layer of insulation in sleeping bags and ski parkas. In fact, if air were a good conductor of heat our kitchens would become unbearable every time we turned on the stove or oven.

Convection.   In the atmosphere, as pockets of air near the surface are heated, they expand in volume, become less dense than the surrounding air, and therefore rise. This vertical transfer of heat through the atmosphere is called **convection,** and it is the same type of process by which heated water circulates in a pan on the stove. The water in the center near the bottom is heated first, becoming lighter and less dense as it is heated. As this water tends to rise, colder, denser water flows down to replace it. As this new water is warmed, it too flows up, while additional colder water moves downward.

The currents set into motion by the heating of a fluid (liquid or gas) make up a convectional system. Such systems account for much of the vertical transfer of heat within the atmosphere and the oceans and are a major cause of clouds and precipitation.

Advection.   **Advection** is the term applied to *horizontal* heat transfer. There are two major advection agents within the Earth-atmosphere system: winds and ocean currents. Both agents help to horizontally transfer energy between the equatorial and polar regions, thus maintaining the energy balance in the Earth-atmosphere system.

Latent Heat of Condensation.   As we have seen, when water evaporates, a significant amount of energy is stored in the water vapor as latent or potential heat (see Fig. 4.6). This water vapor is then transported by advection or convection to new locations, where condensation takes place and the stored energy is released. This process plays a major role in the transfer of energy within the Earth system: The heat required for evaporation helps cool the atmosphere while the **latent**

heat of condensation helps warm the atmosphere and, in addition, is a source of energy for storms.

The Heat Energy Budget

The Budget at Earth's Surface.   Now that we know the various means of heat transfer, we are in a position to examine what happens to the 47 percent of solar energy that reaches Earth's surface (look again at Fig. 4.7). Approximately 14 percent of this energy is emitted in the form of long-wave radiation. This 14 percent includes a net loss of 6 percent (of the total originally received by the atmosphere) directly to outer space while 8 percent is captured by the atmosphere. In addition, there is a net transfer back to the atmosphere (by conduction and convection) of 10 of the 47 percent that reached Earth. The remaining 23 percent returns to the atmosphere through the release of latent heat of condensation. Thus, the 47 percent of the sun's original insolation that reached Earth's surface is all returned to other segments of the system. There has been no long-term gain or loss. Therefore, at Earth's surface, the heat energy budget is in balance.

Examination of the heat energy budget of Earth's surface helps us to understand the *open energy system* that is involved in the heating of the atmosphere. The *input* in the system is that of the incoming short-wave solar radiation that reaches Earth's surface; this is balanced by the *output* of long-wave terrestrial radiation back to the atmosphere and to space. Since these are in balance, we can say that the overall temperature of Earth's surface is in a state of *dynamic equilibrium.*

Of course, it should be noted that the percentages mentioned earlier represent an oversimplification in that they refer to *net* losses that occur over a long period of time. In the shorter term, heat may be passed from Earth to the atmosphere and then back to Earth in a chain of cycles before it is finally released into space. In fact, it is the transfer of heat and energy back and forth between Earth and atmosphere that produces unusually high atmospheric temperatures over short periods.

We are all familiar with what happens to the inside of a car on a sunny day if all the windows are left closed. Short-wave radiation from the sun is able to penetrate the glass windows (Fig. 4.8). When insolation strikes the interior of the car it is absorbed and heats up the exposed surfaces. Energy, emitted from the surfaces as long-wave

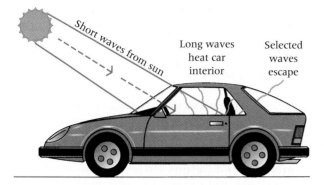

Figure 4.8 *Heat buildup in a closed car results from penetration of short-wave radiation through the car window and inability of long-wave radiation to return outward through the glass.*

radiation, cannot escape through the glass as freely. The result is that the interior of the vehicle gets hotter and hotter throughout the day. In extreme cases, windows in some cars have cracked due to differential expansion, or, what is more serious, temperatures have become so great in automobiles with closed windows that pets or babies inside have died.

A similar phenomenon also occurs in the atmosphere. Like glass, carbon dioxide and water vapor can impede the escape of long-wave radiation by absorbing it and then radiating it back to Earth. This is termed the **greenhouse effect** and is the primary reason for the moderate temperatures observed on Earth.

The Budget in the Atmosphere. At one time or another, about 60 percent of the solar energy intercepted by the Earth system is temporarily retained by the atmosphere. This includes 19 percent of *direct solar radiation* absorbed by the clouds and the ozone layer, 8 percent emitted by *long-wave radiation* from Earth's surface, 10 percent transferred from the surface by *conduction*

and *convection,* and 23 percent released by the *latent heat of condensation.* Some of this energy is recycled back to the surface for short periods of time, but eventually all of it is lost into outer space after being replaced by other solar energy. Hence, just as was the case at Earth's surface, the heat energy budget in the atmosphere is in balance over long periods of time—a dynamically stable system.

Many scientists believe that an imbalance in the heat energy budget, with possible negative effects, is developing. Since the Industrial Revolution, human beings have been adding more and more carbon dioxide to the atmosphere through their burning of fossil (carbon) fuels. Since carbon dioxide absorbs the long-wave radiation from Earth's surface, restricting its escape to space, such heat retention could increase the greenhouse effect. The rising temperatures would also have significant effects on other Earth features, such as the extent of the polar ice caps and world sea level.

Variations in the Heat Energy Budget

Remember that the figures we have seen for the heat energy budget are averages for the *whole* Earth over many years. For any *particular* location, the heat energy budget is most likely not balanced. Some places have a surplus of incoming solar energy over outgoing energy loss in their budget, while others have a deficit. The main causes of these variations are differences in latitude and seasonal fluctuations.

As we have previously noted, the amount of insolation received is directly related to latitude (Fig. 4.9). In the tropical zones where insolation is high throughout the year, more solar energy is received at Earth's surface and in the atmosphere than can be emitted back into space. In the Arctic and Antarctic zones, on the other hand, there is so little insolation during the winter, when Earth is still emitting long-wave radiation, that

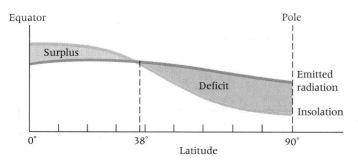

Figure 4.9 *Latitudinal variation in the energy budget. Low latitudes receive more insolation than they lose by reradiation and have an energy surplus. High latitudes receive less energy than they lose outward and therefore have an energy deficit.* • **How do you think the surplus energy in the low latitudes is transferred to higher latitudes?**

there is a large deficit for the year. Locations in the midlatitude zones have lower deficits or surpluses, but only at about latitude 38° is the budget balanced. If it were not for the heat transfers within the atmosphere and the oceans, the tropical zones would get hotter and hotter, and the polar zones would get colder and colder.

At any location, the heat energy budget varies throughout the year according to the seasons, with a tendency toward a surplus in the summer or high-sun season, and a tendency toward a deficit six months later. Seasonal differences may be small near the equator, but they are great in the midlatitude and polar zones.

Air Temperature

Temperature and Heat

Although heat and temperature are highly related, they are not the same. **Heat** is a form of energy—the total kinetic energy of all the atoms that make up a substance. All substances are made up of atoms that are constantly in motion (vibrating) and therefore possess kinetic energy—the energy of motion. On the other hand, **temperature** is the average kinetic energy of the individual atoms of a substance. When something is heated, its atoms vibrate faster and its temperature increases. It is important to remember that the amount of heat energy depends on the mass of the substance under discussion, while the temperature refers to the energy of individual molecules. Thus a burning match has a high temperature but minimal heat energy while the oceans have moderate temperatures but high heat energy content.

Scales

Three different scales are used in measuring temperature. The one with which Americans are most familiar is the **Fahrenheit** scale, devised in 1714 and included in the English system of measurements. By this scale the temperature at which water boils at sea level is 212°F, while the temperature at which water freezes is 32°F.

The **Celsius** scale (also called the **centigrade** scale) was devised in 1742 by Anders Celsius, a Swedish astronomer. It is part of the metric system. The temperature at which water freezes at sea level by this scale was arbitrarily set at 0°C, while the temperature at which water boils was identified as 100°C.

The Celsius scale is used nearly everywhere but in the United States, though even in the United States the Celsius scale is the one used by the majority of the scientific community. By this time you have undoubtedly noted that throughout this book comparable figures in both the Celsius (centigrade) and Fahrenheit scales are given side by side for all important temperatures. Similarly, whenever important figures for distance, area, weight, or speed are given, we use the metric system followed by the English system.

Figure 4.10 can help you compare the Fahrenheit and Celsius systems as you encounter temperature figures outside this book. In addition, the following formulas can be used for conversion from Fahrenheit to Celsius or vice versa:

$$°C = \tfrac{5}{9}(°F - 32) \qquad \text{or} \qquad °F = \tfrac{9}{5}°C + 32$$

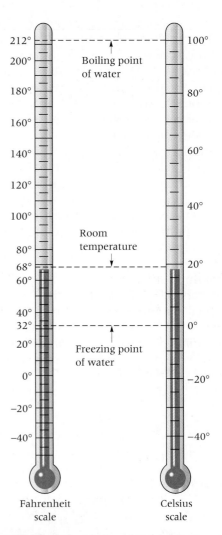

Figure 4.10 *The Fahrenheit and Celsius temperature scales. The scales are aligned to permit direct conversion of readings from one to the other.* • **Can you find the temperature that is the same on the Fahrenheit and Celsius scales?**

The third temperature scale, used primarily by scientists, is the **Kelvin** scale. It is based upon the fact that the temperature of a gas is related to the molecular movement within the gas. As the temperature of a gas is reduced, the molecular motion within the gas is reduced. There is a temperature at which all molecular motion stops and no further cooling is possible. This temperature is approximately −273°C and is termed *absolute zero*. The Kelvin scale uses absolute zero as its starting point. Thus 0°K equals −273°C. Conversion of Celsius to Kelvin is expressed by the following formula:

$$°K = °C + 273$$

Short-Term Variations in Temperature

Local changes in atmospheric temperature can have a number of causes. These are related to the mechanics of the receipt and dissipation of energy from the sun and to various properties of Earth's surface and the atmosphere.

The Daily Effects of Insolation.   As we noted earlier, at any particular location the amount of insolation varies both throughout the year (annually) and throughout the day (diurnally). Annual fluctuations are associated with the sun's changing declination and hence with the seasons. Diurnal changes are related to the rotation of Earth about its axis. Each day, insolation receipt begins at sunrise, reaches its maximum at noon (local solar time), and returns to zero at sunset.

Although insolation is greatest at noon, you may have noticed that temperatures usually do not reach their maximum until two or three o'clock in the afternoon (Fig. 4.11). This is be-cause the insolation received by Earth from shortly after sunrise until the afternoon hours exceeds the energy being lost through Earth radiation. Hence, during that period, as Earth and atmosphere continue to gain energy, temperatures normally show a gradual increase. Sometime around 3 P.M., when outgoing Earth radiation begins to exceed insolation, temperatures start to fall. The daily lag of Earth radiation and temperature behind insolation is accounted for by the time it takes for Earth's surface to be heated to its maximum and for this energy to be radiated to the atmosphere.

Insolation receipt ends with sunset, but on into the night energy that has been stored in Earth's surface layer during the day continues to be lost and there is a decreasing ability to heat the atmosphere. The lowest temperatures occur just after dawn, when the maximum amount of energy has been emitted and before replenishment from the sun can occur. Thus, if we disregard other factors for the moment, we can see that there is a predictable hourly change in temperature called the **daily march of temperature.** There is a gentle decline from midafternoon until dawn and a rapid increase in the eight hours or so from dawn until the next maximum is reached.

Cloud Cover.   The extent of cloud cover is another factor that affects the temperature of Earth's surface and the atmosphere (Fig. 4.12). Weather satellites have shown that at any time about 50 percent of Earth is covered by clouds. This is important because a heavy cloud cover can reduce the amount of insolation a place receives, thereby causing daytime temperatures to be lower than when the sky is clear. On the other

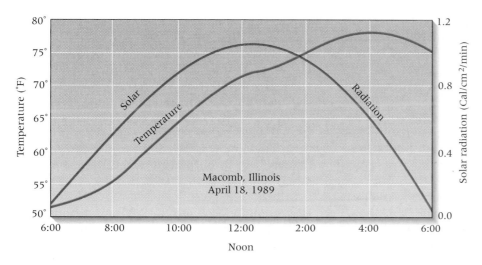

Figure 4.11 Diurnal changes in insolation and temperature for Macomb, Illinois (April 18, 1989). • **Why does temperature rise even after solar energy declines?**

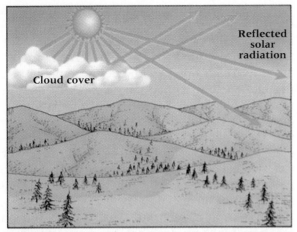

(a) Day

(b) Night

Figure 4.12 *The effect of cloud cover on temperatures. (a) By intercepting insolation, clouds produce lower air temperatures during the day. (b) By trapping long-wave radiation from Earth, clouds increase air temperatures at night. The overall effect is a great reduction in the diurnal temperature range.* • **Desert regions have large diurnal variations in temperature. Why is this so?**

hand, we also have the greenhouse effect, in which clouds, which are composed in large part of water droplets, are capable of absorbing heat energy radiating from Earth, thereby keeping temperatures near Earth's surface warmer than they would otherwise be, especially at night. The general effect of cloud cover, then, is to moderate temperature by lowering the potential maximum and raising the potential minimum temperatures.

Differential Heating of Land and Water. Earlier we saw that bodies of water heat and cool more slowly than the land. The air above Earth's surface is heated or cooled in part by what is be-

neath it. Therefore, temperatures over bodies of water or on land subjected to ocean winds (**maritime** locations) tend to be more moderate than those of land-bound places at the same latitude. Thus, the greater the **continentality** of a location (the distance removed from a large body of water), the less its temperature pattern will be modified.

Reflection. The capacity of a surface to reflect the sun's energy is called its **albedo;** a surface with a high albedo has a high percentage of reflection. The more solar energy that is reflected back into space by an Earth surface, the less that is available for heating the atmosphere. Temperatures will be higher at a given location if its surface has a low albedo rather than a high albedo.

As you may know from experience, snow and ice are good reflectors; they have an albedo of 90 to 95 percent. This is one reason why glaciers on high mountains do not melt away in the summer or why there may still be snow on the ground on a warm day in the spring: solar energy is reflected away. A forest, on the other hand, has an albedo of only 10 to 12 percent, which is good for the trees because they need solar energy for photosynthesis. The albedo of cloud cover varies according to the thickness of the clouds, and it can vary from 40 percent to 80 percent. The high albedo of many clouds is why much solar radiation is reflected directly back into space by the atmosphere. Cities have an albedo of only about 10 percent. This is one reason why hot summer days can be so miserable in the city, yet in the surrounding countryside they can be several degrees cooler.

The albedo of water varies greatly, depending on the depth of the water body and the angle of the sun's rays. If the angle of the sun's rays is high, smooth water will reflect little; in fact, if the sun is vertical over a calm ocean, the albedo will be only about 2 percent. Yet a low-angle sun, such as just before sunset, causes an albedo of over 90 percent from the same ocean surface. Likewise, a snow surface in winter, when solar angles are lower, can reflect up to 95 percent of the energy striking it, and skiers must constantly be aware of the danger of severe burns from reflected solar radiation. In a similar fashion, the high albedo of sand causes the sides of sunbathers' legs to burn faster when they lie on the beach.

Horizontal Air Movement. We have already seen that advection is the major mode of horizontal transfer of heat and energy over Earth's surface.

Any movement of air due to the wind, whether on a large or small scale, can have a significant short-term effect on the temperatures of a given location. Thus, wind blowing from an ocean to land will generally bring cooler temperatures in summer and warmer temperatures in winter. Large quantities of air moving from polar regions into the midlatitudes can cause sharp drops in temperature, while air moving poleward will usually bring warmer temperatures.

Vertical Distribution of Temperature

Normal Lapse Rates.   We have learned that Earth's atmosphere is primarily heated from the ground up as a result of long-wave terrestrial radiation, conduction, and convection. Thus, temperatures in the troposphere are usually highest at ground level and decrease with increasing altitude. For every 1000 meters of altitude, the temperature decreases an average of 6.5°C (3.6°F/ 1000 ft). This rate, in the free air, is known as the **normal** or **environmental lapse rate.**

The lapse rate at a particular place can vary for a variety of reasons (Fig. 4.13). Low lapse rates can exist if denser and colder air is drained into a valley from a higher elevation or if advectional winds bring air in from a cooler region at the same altitude. In each case, the surface is cooled so that its temperature is closer to that at higher elevations directly above it. On the other hand, if the surface is heated strongly by the sun's rays on a hot summer afternoon, the air near Earth will be disproportionately warm, and the lapse rate will be steep. Fluctuations in lapse rates due to abnormal temperature conditions at various altitudes can play an important role in the weather a place may have on a given day.

Inversions.   Under certain circumstances the normal observed *decrease* of temperature with increased altitude may be reversed; temperature may actually *increase* for several hundred meters. This is called a **temperature inversion.**

Some inversions take place 1000 or 2000 meters above the surface of Earth where a layer of warmer air interrupts the normal decrease in temperature with altitude (Fig. 4.14). Such inversions tend to stabilize the air, causing less turbulence and discouraging both precipitation and the development of storms. Upper air inversions may occur when air settles slowly from the upper atmosphere. Such air is compressed as it sinks and rises in temperature, becoming more stable and less buoyant. Inversions caused by descending air are common at about 30° to 35° north and south latitudes.

An upper-air inversion common to the coastal area of California results when cool marine air blowing in from the Pacific Ocean moves under stable, warmer, and lighter air aloft created by subsidence and compression. Such an inversion layer tends to maintain itself; that is, the cold underlying air is heavier and cannot rise through the warmer air above. Not only does the cold air resist rising or moving, but pollutants, such as smoke, dust particles, and automobile exhaust, which are created at Earth's surface, also fail to rise and spread out. They therefore accumulate, filling the lower atmosphere. This situation is particularly acute in the Los Angeles area, which is a basin surrounded by higher mountainous areas (Fig. 4.15). Cooler air blows into the

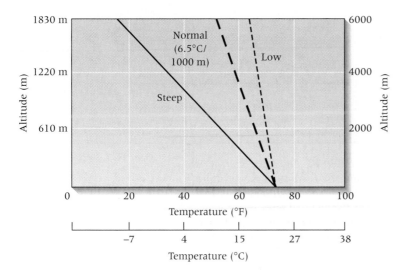

Figure 4.13 *Steep, normal, and low atmospheric lapse rates.*

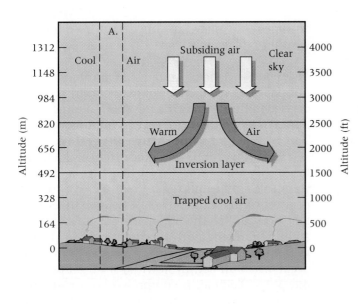
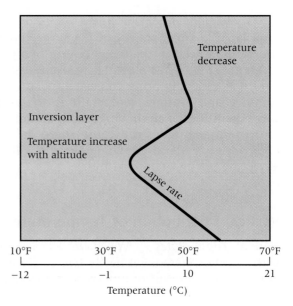

Figure 4.14 (Left) Temperature inversion caused by subsidence of air. (Right) Lapse rate associated with the column of air (A) in left-hand drawing.

basin from the ocean and then cannot escape either horizontally, because of the landform barriers, or vertically, because of the inversion.

Some of the most noticeable temperature inversions are those that occur near the surface when Earth cools off the lowest layer of air through conduction and radiation (Fig. 4.16). In this situation the coldest air is nearest the surface

and the temperature rises with altitude. Inversions near the surface most often occur on clear nights in midlatitudes and are encouraged by snow cover and the recent advection of cool, dry air into an area. Such conditions produce extremely rapid cooling of Earth's surface at night as it loses the day's insolation through radiation. Then the layers of the atmosphere that

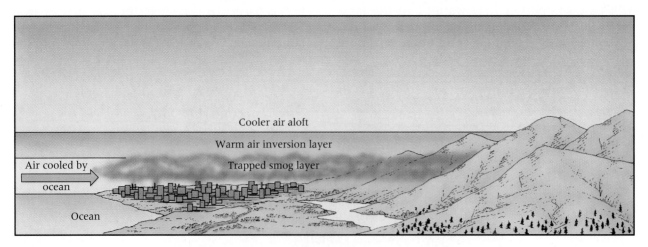

Figure 4.15 Conditions producing smog-trapping inversion in the Los Angeles area. Air moving onshore is cooled by the ocean surface and cannot rise because of the resulting temperature inversion. The air pools against the surrounding mountains and absorbs pollutants from the heavily urbanized area below. Often the polluted layer grows high enough to spill out through mountain passes, bringing smog to the interior desert area.
• **Can you think of a local example of pollution caused by inversion?**

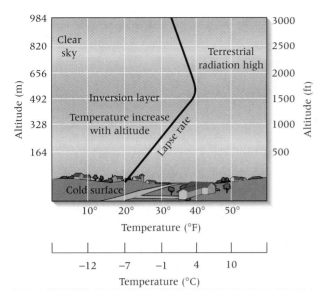

Figure 4.16 *Temperature inversion caused by the rapid cooling of the air above the cold surface of Earth at night.* • **What is the significance of an inversion?**

Figure 4.17 *Fans (propellers) are used to protect Washington apple orchards from frost.*

are closest to Earth are cooled by radiation and conduction more than those at higher altitudes. Calm air conditions near the surface help produce, and partially result from, these temperature inversions.

Surface Inversions and Frost.   Frost often occurs as the result of a surface inversion. Especially where Earth's surface is hilly, dense cold surface air will tend to flow down the sides of the hills and accumulate in the lower valleys. This air drainage causes colder air to build up in the valleys. Temperatures will decrease there, sometimes resulting in a killing frost, while temperatures on the hillsides remain above freezing.

Farmers use a variety of methods to prevent such frosts from destroying their crops. For example, fruit trees in California that can be destroyed by a frost during the growing season are often planted on the warmer hillsides instead of in the valleys. Farmers may also put blankets of straw, cloth, or some other poor conductor over their plants. These take the place of the missing water vapor in the clear atmosphere, preventing the escape of Earth's radiation to outer space and thereby keeping the plants warmer.

Fans are sometimes used to stir up the air in an effort to mix the layers and disturb the inversion (Fig. 4.17). Another device used to prevent frost is huge orchard heaters that heat the air, disturbing the temperature layers. Smudge

pots, an older method of preventing frost, have declined in favor because they are major air polluters; the smoke they pour into the air provides an insulation blanket much like the straw or blankets just mentioned, preventing the escape of terrestrial radiation.

Temperature Distribution at the Earth's Surface

Isotherms (from Greek: *isos*, equal; *therm*, heat) are defined as lines that connect places of equal temperature. When constructing isothermal maps showing temperature distribution over Earth's surface, elevation has to be accounted for by adjusting temperature readings to what they would be at sea level. This adjustment means adding 6.5°C for every 1000 meters of elevation (the normal lapse rate). The rate of temperature change on an isothermal map is called the **temperature gradient.** Closely spaced isotherms indicate a steep temperature gradient (or rapid temperature change over distance), and widely

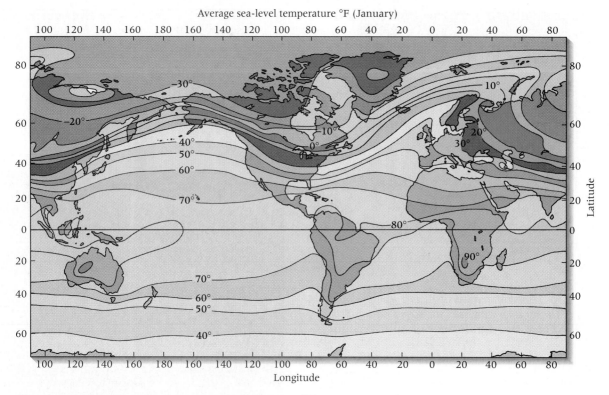

Figure 4.18 *Average sea-level temperatures in January (°F).*

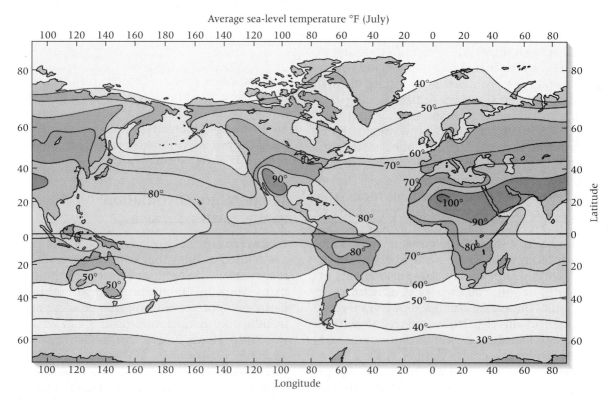

Figure 4.19 *Average sea-level temperatures in July (°F).* • **Observe the temperature gradients between the equator and northern Canada in July and January. Which is greater? Why?**

spaced lines indicate a weak one (or slight temperature change over distance).

Figures 4.18 and 4.19 show the horizontal distribution of temperatures for Earth at two critical times, during January and July, when the seasonal extremes of high and low temperatures are most obvious in the Northern and Southern Hemispheres. The easiest feature to recognize on both maps is the general orientation of the isotherms; they run nearly east-west around Earth, as do the parallels of latitude.

A more detailed study of Figures 4.18 and 4.19 and a comparison of the two maps reveal some additional important features. The highest temperatures in January are in the Southern Hemisphere; in July they are in the Northern Hemisphere. Look up the latitudes of Lisbon, Portugal, and Melbourne, Australia, in an atlas. Now note on the July map that Lisbon in the Northern Hemisphere is nearly on the 70°F isotherm, while at Melbourne in the Southern Hemisphere the average July temperature is less than 50°F, even though the two cities are approximately the same distance from the equator. The temperature differences between the two hemispheres are again a product of insolation, this time changing as the sun shifts north and south across the equator between its positions at the two solstices.

Note that the greatest deviation from the east-west trend of isotherms occurs where the isotherms leave large landmasses to cross the oceans. As the isotherms leave the land, they usually bend rather sharply toward the pole in the hemisphere experiencing winter and toward the equator in the summer hemisphere. This behavior of the isotherms is a direct reaction to the differential heating and cooling of land and water. The continents are hotter than the oceans in the summer and colder in the winter.

Other interesting features on the January and July maps can be mentioned briefly. Note that the isotherms poleward of 40° latitude are much more regular in their east-west orientation in the Southern than in the Northern Hemisphere. This is because in the Southern Hemisphere (often called the "water hemisphere") there is little land south of 40° latitude to produce land and water contrasts. Note also that the temperature gradients are much steeper in winter than in summer in both hemispheres. The reason for this can be understood when you recall that the tropical zones have high temperatures throughout the year, whereas the polar zones have large seasonal differences. Hence, the difference in temperature between tropical and polar zones is much greater in winter than in summer. As a final point, observe the especially sharp swing of the isotherms off the coasts of eastern North America, southwestern South America, and southwestern Africa in January, and off southern California in July. In these locations the normal bending of the isotherms due to land-water differences is augmented by the presence of warm or cool ocean currents.

Annual March of Temperature

Isothermal maps are commonly plotted for January and July because there is a lag of about 30 to 40 days from the solstices, when the amount of insolation is at a minimum or maximum (depending on the hemisphere), to the time of minimum or maximum temperature. This **annual lag of temperature** behind insolation is similar to the daily lag of temperature explained previously. It is a result of the changing relationship between incoming insolation and outgoing Earth radiation. Temperatures continue to rise for a month or more after the summer solstice because insolation continues to exceed radiation. Temperatures continue to fall after the winter solstice until the increase in insolation finally matches Earth's radiation. In short, the lag exists because it takes time for Earth to heat or cool and for those temperature changes to be transferred to the atmosphere.

The annual changes of temperature for a location may be plotted in a graph (Fig. 4.20). The mean temperature for each month in a place such as Peoria, Illinois, is recorded and a line drawn connecting the 12 temperatures. The mean monthly temperature is the average of the daily mean temperature recorded at a weather station during a month. The daily mean temperature is the average of the maximum and minimum temperatures for a 24-hour period.

Such a temperature graph, depicting the **annual march of temperature,** is able to show both the decrease in solar radiation, as reflected by a decrease in temperature, from midsummer to midwinter, and the increase in temperature from midwinter to midsummer caused by the increase in solar radiation.

It is these seasonal fluctuations that impose annual rhythms on our agricultural activities, our recreational pursuits, our clothing styles, and our heating bills. Human activities are constantly influenced by temperature changes, which reflect the input-output patterns of Earth's energy systems.

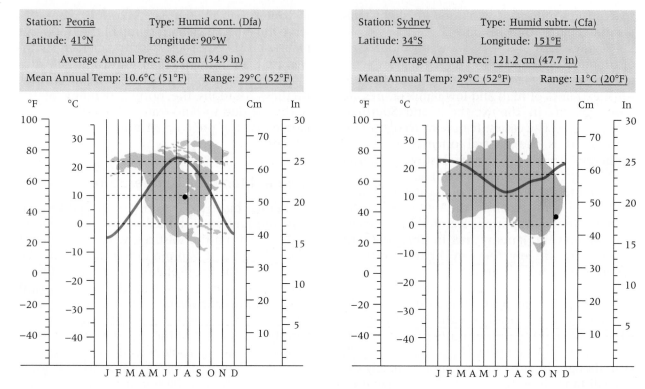

Figure 4.20 *The annual march of temperature at Peoria, Illinois, and Sydney, Australia.*
• **Why do these two locations have opposite temperature curves?**

Define and Recall

photosynthesis
troposphere
stratosphere

weather
climate
element (weather and
   climate)
control (weather and
   climate)

solar constant
radiation
conduction
convection
advection
latent heat of
   condensation
greenhouse effect

heat
temperature
daily (annual) march of
   temperature
maritime
continentality
albedo

normal (environmental)
   lapse rate
temperature inversion
isotherm
temperature gradient
annual (daily) lag of
   temperature

Discuss and Review

1. Why is it useful to think of the atmosphere as a thin film of air? As an ocean of air?

2. Name the gases of which the atmosphere is composed and give the percentage of the total that each supplies.

3. What function does ozone play in the support of life on Earth? Where and how is ozone formed?

4. How is the atmosphere subdivided? In what level do you live? Have you been in any of the other levels?

5. What is the difference between meteorology and climatology?

6. What are the basic characteristics that we call *elements* of weather and climate?

7. What factors cause variation in the elements of weather and climate?

8. Would you expect an area like Seattle to have a milder or a harsher winter than Grand Forks, North Dakota? Why?

9. What is the solar constant? What would happen if there were a significant change in the solar constant?

10. Discuss the role of water in energy exchange. What characteristics of water make it so important?

11. How does Earth's atmosphere affect incoming solar radiation (insolation)? By what processes is insolation prevented from reaching the Earth's surface? What percentages are involved in a generalized situation? What percentages reach the surface and by what processes?

12. How is the atmosphere heated from the Earth's surface? What processes and percentages are involved in a generalized situation?

13. What is meant by Earth's energy budget? List and define the important energy exchanges that keep it in balance.

14. What is the temperature in Fahrenheit degrees today in your area? In Celsius degrees?

15. At what time of day does insolation reach its maximum? Its minimum? Compare this to the daily temperature maximum and minimum.

16. How is albedo a factor in your selection of outdoor clothes on a hot, sunny day? On a cold, clear, winter day?

17. What is a temperature inversion? Give several reasons why temperature inversions occur.

18. Why do citrus growers use wind machines and heaters? Describe any techniques you are familiar with to prevent frost damage to plants in your area.

19. Describe the behavior of the isotherms in Figures 4.18 and 4.19. What factors cause the greatest deviation from an east-west trend? What factors cause the greatest differences between the January and July maps?

Consider and Respond

1. Convert the following temperatures to Fahrenheit: 20°C, 30°C, 15°C.

2. Convert the following temperatures to Celsius: 60°F, 15°F, 90°F.

3. Convert the temperatures in (1) and (2) to Kelvin.

4. Refer to Figure 4.7. List the major means by which the atmosphere gains heat; loses heat.

5. What are the major weather and climate controls that operate in your area?

6. The normal lapse rate is 6.5°C/1000 meters. If the surface temperature is 25°C (77°F), what is the air temperature at 10,000 meters (32,800 ft) above the Earth's surface? Convert your answer to degrees Fahrenheit.

7. Refer to Figures 4.18 and 4.19. What location on the Earth's surface exhibits the greatest annual range of temperature? Why?

CHAPTER

5

ATMOSPHERIC PRESSURE AND WIND

CHAPTER PREVIEW

▶ Latitudinal differences in temperature (as a result of differential receipt of insolation) provide a partial explanation for latitudinal differences in pressure.

What is the relation between temperature and pressure? Why is this only a partial explanation?

▶ There are seasonal (January and July) variations in world patterns of pressure and winds that correspond to similar seasonal changes in temperature and insolation.

What variations occur? What are the root causes of all these variations? What might be the effects on the physical environment and on humans if these seasonal shifts did not exist?

▶ The fact that land heats and cools more rapidly than water is of significance not only to world patterns of temperature but also to world patterns of pressure, winds, and precipitation.

How can you explain this fact? What effect does this fact have on world patterns? In what ways can it affect human beings?

▶ The horizontal transfer of air (wind) is a direct result of atmospheric movement to adjust for pressure inequalities.

How do these inequalities develop? How does wind adjust for high and low pressure?

▶ Because of the Coriolis effect, free-moving objects (including winds) are deflected to the right in the Northern Hemisphere and to the left in the Southern Hemisphere.

Why is this so and how does this affect atmospheric circulation patterns near Earth's surface in the Northern and Southern Hemispheres?

▶ Convergent and divergent wind systems (cyclones and anticyclones) in the Northern Hemisphere are characterized by air systems spiraling in opposite directions from one another and in opposite directions from their counterparts in the Southern Hemisphere.

In what directions (clockwise or counterclockwise) do these systems spiral in the Northern and Southern Hemispheres?

▲ Strong winds associated with a storm over Miami Beach, Florida.

An individual air molecule weighs almost nothing; however, the atmosphere as a whole has considerable weight and exerts an average pressure of 1034 grams per square centimeter (14.7 lb/sq in.) on Earth's surface. The reason why people are not crushed by this atmospheric pressure is that we have air inside us—in our blood, tissues, and cells—exerting an equal outward pressure that balances the inward pressure of the atmosphere. Atmospheric pressure is important because variation in pressure within the Earth-atmosphere system causes our atmospheric circulation and thus plays a major role in determining our weather and climate.

In 1643, Evangelista Torricelli, a student of Galileo, performed an experiment that was the basis for the invention of the **mercury barometer,** an instrument that measures atmospheric pressure. Torricelli took a tube filled with mercury and inverted it in an open pan of mercury. The mercury inside the tube fell until it was at a height of about 76 centimeters (29.92 in.) above the mercury in the pan, leaving a vacuum bubble at the closed end of the tube (Fig. 5.1). At this point, the pressure exerted by the atmosphere on the open pan of mercury was equal to the pressure from the mercury in the tube. Torricelli observed that as the air pressure increased, it pushed the mercury up into the tube, increasing the height of the mercury until the pressure exerted by the mercury would equal the pressure of the air. On the other hand, as the air pressure decreased, the mercury column fell.

In the strictest sense, a mercury barometer does not actually measure the pressure exerted by the atmosphere on Earth's surface, but instead measures the *response* to that pressure. That is, when the atmosphere exerts a specific pressure, the mercury will respond by rising to a specific height. Meteorologists usually prefer to work with actual pressure units. The unit most often used is the millibar (mb). Standard sea-level pressure of 1013.2 mb will cause the mercury to rise 76 centimeters (29.92 in.).

Variations in Atmospheric Pressure

Vertical Variations in Pressure

Imagine a pileup of football players during a game. The player on the bottom gets squeezed

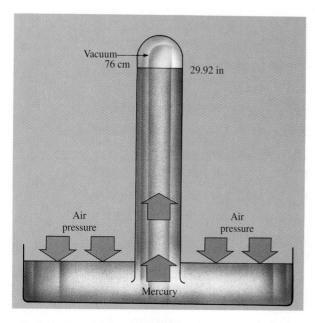

Figure 5.1 *A simple mercury barometer. Standard sea-level pressure of 1013.2 millibars will cause the mercury to rise 76 centimeters (29.92 in.) in the tube.* • **When air pressure increases, what happens to the mercury in the tube?**

more than a player near the top because he has the weight of all the others on top of him. Similarly, air pressure decreases with elevation, for the higher we go, the more diffused the air molecules become. The increased intermolecular space results in less pressure. In fact, at the top of Mount Everest (elevation: 8708 meters, or 29,028 ft), the air pressure is only about one-third the pressure at sea level.

Humans are usually not sensitive to small, everyday variations in air pressure. However, when we climb or fly to altitudes significantly above sea level, we become aware of the effects of air pressure on our system. When jet aircraft fly at 10,000 meters (33,000 ft), they have to be pressurized and nearly airtight so that a near-sea-level pressure can be maintained. Even then, the pressurization does not work perfectly, so our ears may pop as they adjust to a rapid change in pressure when ascending or descending. Hiking or skiing at heights that are a few thousand meters in elevation will affect us if we are used to the air pressure at sea level. The reduced air pressure and resulting rarefied atmosphere mean less oxygen is contained in each breath of air. Thus, we sometimes find that we get out of breath far more easily at high elevations until our bodies adjust to the reduced air pressure.

Changes in air pressure are not solely related to altitude. At Earth's surface small but important variations in pressure are related to the intensity of radiation, the general movement of global circulation, and local humidity and precipitation. Consequently, a change in air pressure at a given locality often indicates a change in the weather.

Horizontal Variations in Pressure

The causes of horizontal variation in air pressure are grouped into two types: thermal (determined by temperature) and dynamic (related to motion).

We will look at the more simple thermal type first. In Chapter 4 we saw that Earth is heated unevenly because of unequal distribution of insolation, differential heating of land and water surfaces, and different albedos of surfaces. One of the basic laws of gases is that the pressure and density of a given gas vary inversely with temperature. Thus, during the day, as Earth heats the air above it, the air expands in volume and decreases in density. Such air has a tendency to rise as its density decreases. When the warmed air rises, there is less air near the surface, with a consequent decrease in surface pressure. The equator is an area where such low pressure occurs.

In an area with cold air, there is an increase in density and a decrease in volume. This causes the air to sink and pressure to increase. The poles are an area where such high pressure occurs. Thus, the constant low pressure in the equatorial zone and the high pressure at the poles are thermally induced.

From this we might expect a gradual increase in pressure from the equator to the poles to accompany the gradual decrease in average annual temperature. However, actual readings taken at Earth's surface indicate that pressure does not increase in a regular fashion latitudinally poleward from the equator. Instead, there are regions of high pressure in the subtropics and regions of low pressure in the subpolar regions. The dynamic causes of these zones, or *belts,* of high and low pressure are more complex than the thermal causes.

These dynamic causes are related to the rotation of Earth and the broad patterns of circulation. As air rises steadily at the equator, it moves toward the poles. Earth's rotation, however, causes the poleward-flowing air to drift to the east. In fact, by the time it is over the subtropical regions, the air is flowing from west to east. This bending of the flow as it moves poleward impedes movement and causes the air to pile up over the subtropics, which results in increased pressure at Earth's surface there.

With high pressure over the polar and subtropical regions, dynamically induced areas of low pressure are created between them, in the subpolar region. As a result, air flows from the highs to the lows, where it rises. Thus, both the subtropical and subpolar pressure regions are dynamically induced.

Idealized World Pressure Belts

Using what we have just learned about pressure on Earth's surface, we can construct a theoretical model of the pressure belts of the world (Fig. 5.2). Later, we will see how real conditions depart from

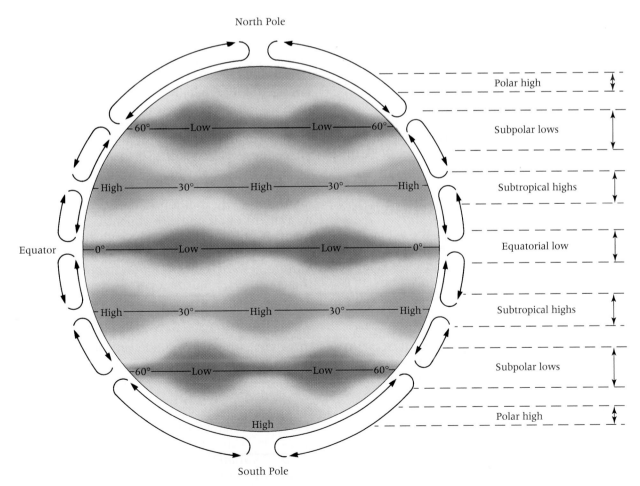

Figure 5.2 *Idealized world pressure belts. Note the arrows on the perimeter of the globe that illustrate the cross-sectional flow associated with the surface pressure belts.*

our model and examine why these differences occur.

Centered approximately over the equator in our model is a belt of low pressure, or a **trough.** Because this is the region on Earth of greatest annual heating, we can conclude that the low pressure of this area, the **equatorial low,** is determined primarily by thermal factors which cause the air to rise.

North and south of the equatorial low and centered on the so-called "horse latitudes," which are about 30°N and 30°S, are cells of relatively high pressure. These are the **subtropical highs,** which are the result of dynamic factors related to the sinking of convectional cells initiated at the equatorial low.

Poleward of the subtropical highs in both the Northern and Southern Hemispheres are large belts of low pressure that extend through the upper middle latitudes. Pressure decreases through these **subpolar lows** until about 65° latitude. Again, dynamic factors play a role in the existence of subpolar lows.

In the polar regions are high-pressure systems called the **polar highs.** These are determined by the cold temperatures and consequent sinking of the dense air in those regions.

This system of pressure belts that we have just developed is a generalized picture. Just as temperatures change from month to month, day to day, and hour to hour, so do pressures vary from time to time at any one place. Our model disguises these changes, but it does give an idea of broad pressure patterns on the surface of Earth.

Mapping Pressure Distribution

We will turn to maps that show the actual distribution of atmospheric pressure on Earth's surface in order to see how well our idealized model of world pressure belts represents the "real world." We have noted that atmospheric pressure varies vertically with altitude as well as horizontally with temperature and other factors. Although we should be aware of the decrease in pressure that takes place as we move vertically through the atmosphere, it is the variations from place to place and time to time over Earth's surface that are basic to an understanding of weather and climate. Therefore, when we map air pressure, we reduce all pressures to what they would be at sea level, just as we changed temperature to sea level in order to eliminate altitude as a factor. This is especially important for atmospheric pressure because the variations due to altitude are

far greater than those due to atmospheric dynamics and would tend to mask the more meteorologically important regional differences.

Figures 5.3 and 5.4, which show the average sea-level pressure patterns for January and July, are isobaric maps. **Isobars** (from Greek *isos,* equal; *baros,* weight) are lines drawn on these maps to connect places of equal pressure. When the isobars appear close together, they portray a significant difference in pressure between places, hence a strong **pressure gradient.** When the isobars are far apart, a weak pressure gradient is depicted.

The atmosphere tends to form cells of high and low pressure. Depicted on a map, these cells are outlined by concentric isobars that form a closed system around centers of low pressure or high pressure. The cells of low pressure are commonly referred to as **lows,** or **cyclones;** the cells of high pressure are called **highs,** or **anticyclones.**

The General Pattern of Atmospheric Pressure

As our idealized model suggests, the atmosphere tends to form belts of high and low pressure along east-west axes in areas where there are no large bodies of land. These belts are latitudinally arranged and generally maintain their bandlike pattern. However, where there are continental landmasses, belts of pressure are broken and tend to form cellular pressure systems. The landmasses affect the development of belts of atmospheric pressure in several ways. Most influential is the effect of the differential heating of land and water surfaces. In addition, landmasses affect the movement of air and consequently the development of pressure systems through friction with their surfaces. Landform barriers such as mountain ranges also block the movement of air and thereby affect atmospheric pressure.

Seasonal Variations in the Pattern

In general, the atmospheric pressure belts shift northward in July and southward in January, following the changing position of the sun's direct rays as they migrate between the Tropics of Cancer and Capricorn. Thus, there are thermally induced seasonal variations in the pressure patterns, as seen in Figures 5.3 and 5.4. These seasonal variations tend to be small in low latitudes, where there is little temperature variation, and large in high latitudes, where there is an increasing contrast in length of daylight and angle

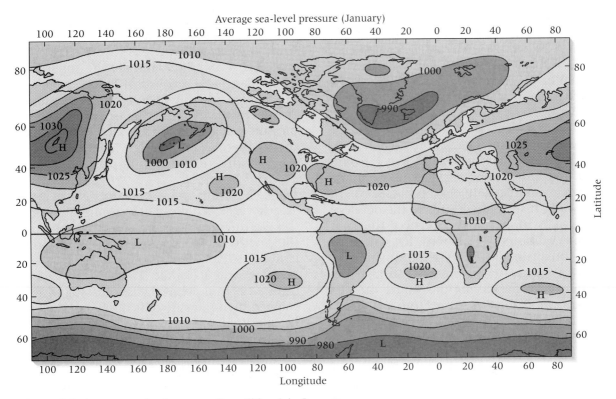

Figure 5.3 *Average sea-level pressure (in millibars) in January.*

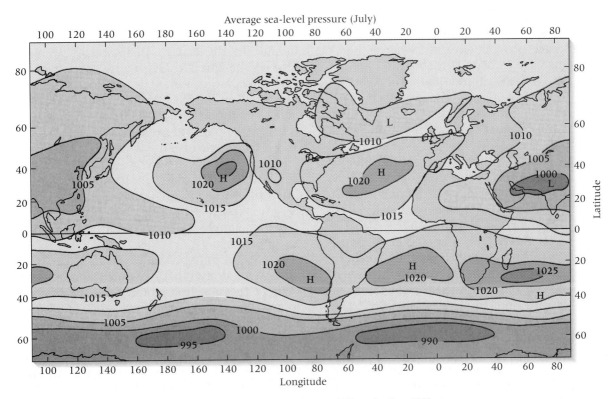

Figure 5.4 *Average sea-level pressure (in millibars) in July.* • **What is the difference between the January and July average sea-level pressures at your location? Why do they vary?**

of the sun's rays. Furthermore, landmasses tend to alter the general pattern of seasonal variation. This is an especially important factor in the Northern Hemisphere, where land accounts for 40 percent of the total Earth surface, as opposed to less than 20 percent in the Southern Hemisphere.

January. Because land cools more quickly than the oceans, its temperatures will be lower in winter than the surrounding seas. Figure 5.3 shows that in the middle latitudes of the Northern Hemisphere, this variation leads to the development of cells of high pressure over the land areas. In contrast, the subpolar lows develop over the oceans because they are comparatively warmer. Over eastern Asia there is a strongly developed anticyclone during the winter months that is known as the **Siberian High.** Its equivalent in North America, known as the **Canadian High,** is not nearly so well developed.

In addition to the Canadian High and the Siberian High, two low-pressure centers develop: one in the North Atlantic, called the **Icelandic Low,** and the other in the North Pacific, called the **Aleutian Low.** The air in them has relatively lower pressure than either the subtropical or the polar high systems. Consequently, air moves toward these low-pressure areas from both north and south. Such low-pressure regions are associated with cloudy, unstable weather and are a major source of winter storms, whereas high-pressure areas are associated with clear, blue-sky days, calm, starry nights, and cold, stable weather. Therefore, during the winter months cloudy weather tends to be associated with the two oceanic lows and clear weather with the continental highs.

We can also see that the polar high in the Northern Hemisphere is well developed. This development is primarily due to thermal factors, since it is now the coldest time of the year. The subpolar lows have developed into the Aleutian and Icelandic cells described earlier. At the same time, the subtropical highs of the Northern Hemisphere appear slightly south of their average annual position because of the migration of the sun toward the Tropic of Capricorn. The equatorial trough also appears centered south of its average annual position over the geographic equator.

In January in the Southern Hemisphere, the subtropical belt of high pressure appears as three cells centered over the oceans because the belt of high pressure has been interrupted by the continental landmasses where temperatures are much higher and pressure tends to be lower than over the oceans. Because there is virtually no land between 45°S and 70°S latitude, the subpolar low circles Earth as a real belt of low pressure and is not divided into cells by interrupting landmasses. Seasonally there is little change in this belt of low pressure other than that in January (summer in the Southern Hemisphere), when it lies a few degrees north of its July position.

July. The anticyclone over the North Pole is greatly weakened during the summer months in the Northern Hemisphere primarily because of the lengthy (24 hours) heating of the oceans and landmasses in that region (see Fig. 5.4). The Aleutian and Icelandic Lows nearly disappear from the oceans, while the landmasses, which developed high-pressure cells during the cold winter months, have extensive low-pressure cells slightly to the south during the summer. In Asia, a low-pressure system develops, but it is divided into two separate cells by the Himalayas (see Fig. 6.8). The low-pressure cell over northwest India is so strong that it combines with the equatorial trough, which has moved north of its position six months earlier. The subtropical highs of the Northern Hemisphere are more highly developed over the oceans than over the landmasses. In addition, they migrate northward and are highly influential factors in the climate of landmasses nearby. In the Pacific, this subtropical high is termed the **Pacific High;** this system of pressure plays an important role in moderating the temperatures of the west coast of the United States. In the Atlantic Ocean the corresponding cell of high pressure is known as the **Bermuda High** to Americans and as the **Azores High** to Europeans. As we have already mentioned, the equatorial trough of low pressure moves north in July, following the migration of the sun's vertical rays, and the subtropical highs of the Southern Hemisphere lie slightly north of their January locations.

In examining pressure systems at Earth's surface, we have seen that there are essentially seven belts of pressure (two polar highs, two subpolar lows, two subtropical highs, and one tropical low), which are broken into cells of pressure in some places, primarily because of the influence of certain large landmasses. We have also seen that these belts and cells vary in size, intensity, and location with the seasons and with the migration of the sun's vertical rays over Earth's surface.

Wind

Wind is the horizontal movement of air in response to differences in pressure. Winds are the means by which the atmosphere attempts to balance the uneven distribution of pressure over Earth's surface. The movements of the wind also play a major role in correcting the imbalances in radiational heating and cooling that occur over Earth's surface. On average, locations below 38° latitude receive more radiant energy than they lose, while locations poleward of 38° lose more than they gain. Our global wind system transports energy poleward to help maintain an energy balance. The global wind system also gives rise to the ocean currents, which are another significant factor in equalizing the energy imbalance. Thus, without winds and their associated ocean currents, the equatorial regions would get hotter and hotter and the polar regions colder and colder.

Besides serving a vital function in the advectional transport of heat energy over Earth's surface, winds also transport water vapor from the air above bodies of water, where it has evaporated, to land surfaces, where it condenses and releases latent energy. This allows greater precipitation over land surfaces than could otherwise occur. In addition, winds exert influence on the rate of evaporation itself. Furthermore, as we become more aware and concerned about the effect that the burning of fossil fuels has on our atmosphere, we look for alternate energy sources.

Natural sources, such as water, solar energy, and wind become increasingly attractive alternatives to fossil fuels. They are clean, abundant, and renewable. As we enter the next century they will likely become important sources of energy for our society (see The Environment: Alternative Energy Sources).

Pressure Gradients and Winds

Winds vary widely in speed, intensity, duration, and direction. Much of their strength depends upon the size or strength of the *pressure gradient* to which they are responding. As we noted previously, pressure gradient is the term applied to the rate of change of atmospheric pressure between two points (at the same elevation). The greater this change—that is, the steeper the pressure gradient—the greater will be the wind response (Fig. 5.5). Winds tend to flow down a pressure gradient from high pressure to low pressure, just as water flows down a slope from a high point to a low one. The steeper the pressure gradients involved, the faster and stronger the winds. Yet wind does not flow directly from high to low, as we might expect, for there are other factors involved that affect the direction of wind.

The Coriolis Effect and Wind

Two factors, both related to our Earth's rotation, greatly influence wind direction. First, our fixed grid system of latitude and longitude is

(Text continues on page 126)

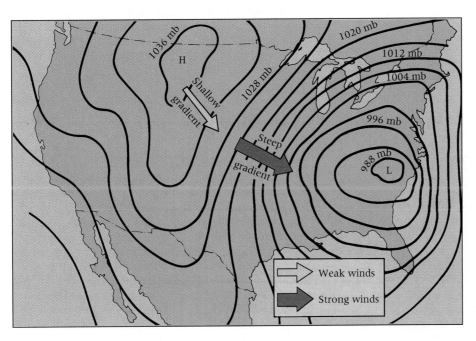

Figure 5.5 The relationship of wind to the pressure gradient: The steeper the pressure gradient, the stronger the resulting wind. • Where else on this figure (other than the area indicated) would winds be strong?

ALTERNATIVE ENERGY SOURCES

Ninety percent of the energy consumed in the United States is derived from the burning of fossil fuels. As these conventional energy sources dwindle and as we become more concerned with their role in escalating the greenhouse effect, alternative energy sources become more viable as new methods to heat and cool our homes and provide power to our factories. Nuclear power is a pollution-free alternative energy source. However, the possibility of the accidental release of radioactive material into the atmosphere through a nuclear power plant accident like that at Chernobyl (Ukraine) in 1986 is a major concern. It is estimated that as many as 4 million people *could* develop health problems directly related to the radiation emitted from the explosion of that nuclear reactor. Even though the probability of a catastrophic nuclear accident is remote, the environmental problem of how to dispose of the spent radioactive fuel is still another stumbling block to its use.

Natural alternative energy sources, such as wind, water, and solar energy, are attractive, since they are not only clean and safe sources of energy, but also renewable—indeed, infinite—resources. Hydropower, power derived from the energy of moving water, is used in those regions where the damming of streams permits its use. Although the cost of generating electricity by this method is reasonable, the high cost of transmitting electrical power over long distances hinders widespread distribution. In the near future, the two other energy sources mentioned—wind and solar energy—might offer more acceptable solutions to our growing energy crisis.

Wind Power

Wind power is an inexhaustible source of clean energy. Windmills were used to pump water and to grind grain before the widespread availability of inexpensive electricity. Although wind power, unlike solar power, is not limited to daytime use, it does have some problems. First, it is quite costly; energy production by windmills costs more than twice that of conventional fuels. Second, a steady wind source is needed. Because wind is so variable, a wind power system must be able to store the energy generated for use during calm periods. This usually necessitates the use of expensive storage batteries.

Rather than use individual wind turbines, it is much more economical to use a cluster of wind turbines, called *wind farms*. Wind farms consist of 50 or more wind generators, each producing at least 1 megawatt of electricity.

Most wind power systems can extract 30 to 40 percent of the wind's energy, with some experimental models reaching up to 60 percent efficiency. A typical wind power system requires that wind speeds must be at least 20 kilometers (12 mi) per hour 40 percent of the time to operate economically. **Since the power generated is proportional to the cube of the wind speed, a doubling of the wind speed increases energy production eight times.** Thus, we want persistent strong winds. However, if the winds are too strong or gusty, damage to the expensive turbines could result. Possible sites within the United States that hold the greatest potential for wind power are the western Great Plains from Wyoming to Oklahoma, the New England coast, the Pacific Northwest, and coastal California. In fact, California is the leading producer of wind-derived energy in the United States.

Solar Energy

Earth receives, in just two weeks' time, an amount of solar energy that would equal our entire known global supply of fossil fuels. Although the potential use of solar radiation for power production is enormous, various problems inhibit its use on a larger scale. Primarily, the intermittent nature of solar radiation (no radiation at night, cloud cover) coupled with the high cost of converting solar energy to electricity precludes anything but local application at this time. An economical means is needed of first collecting the energy, then storing it for use during nonsunshine periods, and finally converting it to electricity. Until such technology is available, solar energy will not be a major source of power production.

Wind farm close to the North Sea in the Weiringermeer Polder of the Netherlands. Why is a coastal location an advantage when choosing a site for a wind farm?

Small-scale use of solar energy, where cost is not a factor, is already with us. For many centuries, fishermen have used the power of the sun to dry fish. In the space program and in sea buoys, road signs, and off-shore oil rigs, solar power lights and heats small-scale operations. On the domestic scene, the use of solar space heating and hot water heating is a reality. One approach, the use of **passive systems,** employs good architectural design and directional siting to warm interiors in winter and prevent overheating in summer.

Native American adobe structures of the Southwest were well adapted to the desert sun, with their thick walls, small windows, and south-facing exposures set in overhangs on canyon walls to shelter inhabitants from the near-vertical summer rays, thus making maximum use of solar heating and cooling. **Active systems** include flat-plate and collector panels that heat water to 70°C (158°F). The water is then circulated and/or stored for domestic heating arrangements. Obviously, initial installation costs are expensive, but estimates suggest that after about a year and a half, there are savings of 70 percent on the cost of heating water.

Large-scale energy operations are mainly in the experimental stage and are too costly at present. Two types of solar technology are being actively developed in various parts of the world, especially in the United States, France, Israel, and Australia. Photovoltaic cells that convert sunlight directly into electrical power (not unlike a camera's light meter) are already in use in Arizona and California.

The other major type of solar technology involves solar thermal towers, where racks of tracking mirrors (a heliostat field) follow the sun and focus its heat on a steam boiler perched on a high tower. Temperatures in the boiler may be raised to over 500°C (900°F). In principle no different from our youthful experiments with a magnifying glass to set fire to paper, this device already is operative at experimental sites in France and elsewhere. In California, Solar One in the Mojave Desert is the world's largest solar thermal electric power plant of this kind. However, development is slow because additional solar thermal towers must be consid-

One of three solar energy complexes in the Mojave Desert of California that together produce 90 percent of the world's *grid-connected* solar energy.

ered in terms of current economic feasibility. Only as the costs of other energy sources rise and those of solar devices fall significantly will these become fully competitive.

Whether from Earth-based solar energy stations, from insolation-collecting satellites beaming electricity to Earth via microwave transmission, or from some other as yet undeveloped technology, the sun will become increasingly important as humanity enters the twenty-first century.

CRITICAL THINKING ▼

(1) How is electricity produced in your hometown? If production does not make use of alternative energy sources, is there a potential to employ one in the future? **(2)** If a nuclear reactor exploded in Mexico City, what geographic area would most likely face the possibility of contamination? Why? **(3)** What types of practices might people employ to reduce the consumption of energy in their homes?

constantly rotating. Thus, our frame of reference for tracking the path of any free-moving object—whether it is an air mass or a missile—is constantly changing its position. Second, the speed of rotation of Earth's surface increases as we move equatorward and decreases as we move toward the poles (see Fig. 3.7). Thus, to use our previous example, someone in St. Petersburg (60° north latitude), where the distance around a parallel of latitude is about half that at the equator, moves at about 840 kilometers per hour (525 mph) as Earth rotates, while someone in Kampala, Uganda, near the equator, moves at about 1680 kilometers per hour (1050 mph).

Because of these Earth rotation factors, anything moving horizontally appears to be deflected to the right of the direction in which it is traveling in the Northern Hemisphere and to the left in the Southern Hemisphere.

This apparent deflection is termed the **Coriolis effect.** The degree of deflection, or curvature, is a function of the speed of the object in motion and the latitudinal location of the object. The higher the latitude, the greater will be the Coriolis effect. In fact, not only does the Coriolis effect decrease at lower latitudes, but it does not exist at the equator. Also, the faster the object is moving, the greater will be the apparent deflection, and the greater the distance something must travel, the greater will be the Coriolis effect.

A few diagrammatic examples may help you better understand the Coriolis effect. In Figure 5.6 a disk rotating counterclockwise illustrates the apparent deflection of objects caused by our Earth's rotation. Assume the center of the disk is the North Pole and the outer edge is the equator.

First, let us consider east or west movement over Earth's surface (Fig. 5.6a). You are located at Point A, and you launch a rocket with just enough fuel and thrust to hit Point B. However, while your rocket was airborne, Earth (our disk) rotated counterclockwise so that Points A and B are now located at Points A' and B', respectively. Since we are really unaware that Earth is constantly rotating, our mind assumes the rocket will follow path A'–B'. The rocket launched from Point A, however, will follow the initial path and land at Point B. Therefore, from our present position at Point A', it appears the rocket followed a path A'–B, which represents an apparent deflection to the right of the expected path A'–B'.

Movement north or south (Fig. 5.6b) is slightly more complicated, since we must take into account not only the rotation of Earth but

also the variation in rotational speed with latitude. As in the previous example, you are at Point A launching a rocket toward Point B. While the rocket is airborne, Earth rotates, so that Points A and B are now located at Points A' and B', respectively.

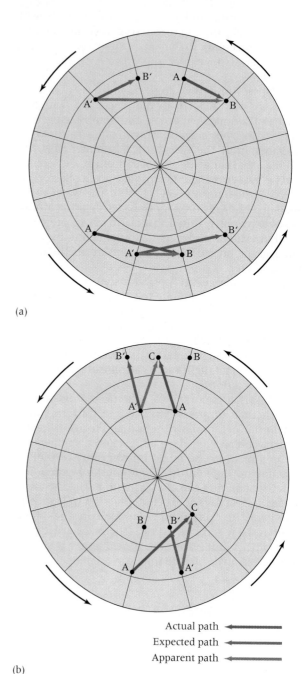

(a)

(b)

Actual path ⟵
Expected path ⟵
Apparent path ⟵

Figure 5.6 *Schematic illustration of the apparent deflection (Coriolis effect) of an object caused by Earth's rotation when (a) the object is moving eastward or westward in the Northern Hemisphere, and (b) the object is moving northward or southward in the Northern Hemisphere.*

In north or south movement two velocities are operating on the rocket. First, you have the north or south component supplied by the rocket's engine and calculated to take the rocket from Point A to Point B. Second, the rocket will also have an eastward velocity component associated with Earth's rotational speed at the point of launch. It is the composite of these two velocities that determines where the rocket will land. Thus, a rocket launched poleward will actually overshoot its target and land at Point C. This occurs because its eastward velocity component (the equivalent of distance A–A′) is greater than the eastward velocity of the target, Point B. This results in an apparent deflection to the right of the expected path.

A rocket launched equatorward will undershoot the target because its Earth-induced eastward velocity component is less than that of its target. This again results in an apparent deflection to the right of the expected path.

As we have said, anything that moves horizontally over Earth's surface exhibits the Coriolis effect. Thus, both the atmosphere and the oceans are deflected in their movements. Winds in the Northern Hemisphere moving across a gradient from high to low pressure are apparently deflected to the right of the direction in which they originally blow (and to the left in the Southern Hemisphere). In addition, when considering winds at Earth's surface, we must take into account another force. This force, **friction,** interacts with the pressure gradient and the Coriolis effect.

Friction and Wind

Above Earth's surface, frictional drag is of little consequence to wind development. At this level the wind starts down the pressure gradient and turns 90 degrees in response to the Coriolis effect. At this point, the pressure gradient is balanced by the Coriolis effect, and the wind, termed a **geostrophic wind,** flows parallel to the isobars (Fig. 5.7a).

However, at or near Earth's surface (up to about 1000 meters above the surface) frictional drag is important because it reduces the wind speed. A reduced wind speed in turn reduces the Coriolis effect, while the pressure gradient is not affected. With the pressure gradient and Coriolis effect no longer in balance, the resultant surface wind does not flow between the isobars like its upper-level counterpart. Instead, a surface wind flows obliquely across the isobars toward the low-pressure area (Fig. 5.7b).

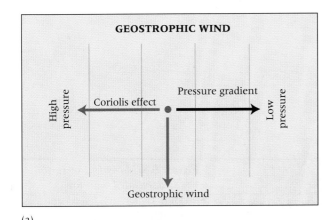

(a)

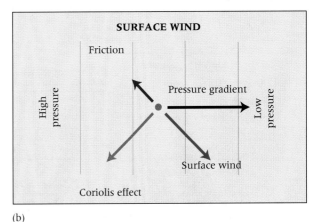

(b)

Figure 5.7 *The Northern Hemisphere examples illustrate that (a) in a geostrophic wind, as a parcel of air starts to flow down the pressure gradient, the Coriolis effect causes it to veer to the right until the pressure gradient and Coriolis effect reach an equilibrium and the wind flows between isobars; (b) in a surface wind, this equilibrium is upset by friction, which reduces the wind speed. Since the Coriolis effect is a function of wind speed, it also is reduced. With the Coriolis effect reduced, the pressure gradient dominates and the wind now flows across isobars in the direction of low pressure.* • **If the amount of friction increased, would the surface wind be closer to the pressure gradient, closer to the Coriolis effect, or unchanged?**

Cyclones and Anticyclones

Imagine a high-pressure cell (anticyclone) in the Northern Hemisphere in which the air is moving from the center in all directions down pressure gradients. As it moves, the air will be deflected to the right, no matter which direction it was originally going. Therefore, the wind moving out of an anticyclone in the Northern Hemisphere will move from the center of high pressure in a clockwise spiral (Fig. 5.8).

Air tends to move down pressure gradients from all directions toward the center of a low-pressure area (cyclone). However, since the air is apparently deflected to the right in the Northern Hemisphere, the winds move into the cyclone in a counterclockwise spiral. Because all objects including air and water are apparently deflected to the left in the Southern Hemisphere, spirals there are reversed. Thus, in the Southern Hemisphere, winds moving away from an anticyclone do so in a counterclockwise spiral and winds moving into a cyclone move in a clockwise spiral.

Convergent and Divergent Circulation

As we have just seen, winds blow toward the center of a cyclone and can be said to *converge* toward it. Hence, a cyclone is a closed system of isobars whose center serves as the focus for **convergent** wind circulation (Fig. 5.9). The winds of an anticyclone blow away from the center of high

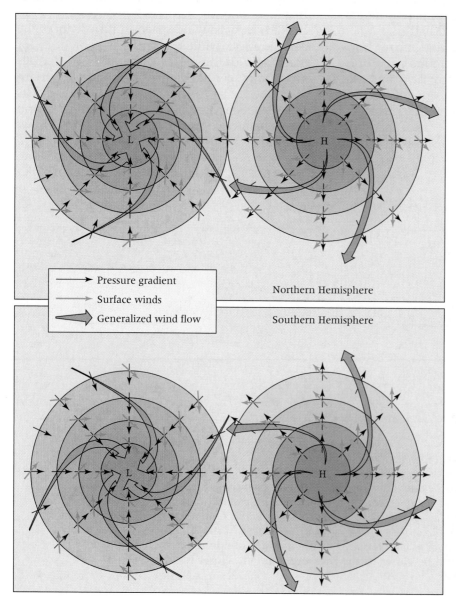

Pressure gradient
Surface winds
Generalized wind flow

Northern Hemisphere

Southern Hemisphere

Figure 5.8 *Movement of surface winds associated with low-pressure centers (cyclones) and high-pressure centers (anticyclones) in the Northern and Southern Hemispheres. Note that the surface winds are to the right of the pressure gradient in the Northern Hemisphere and to the left of the pressure gradient in the Southern Hemisphere.* • **What do you think might happen to the diverging air of an anticyclone if there is a cyclone nearby?**

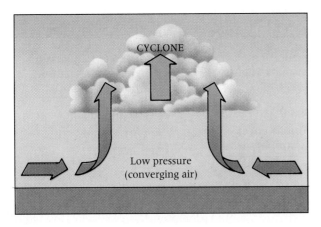
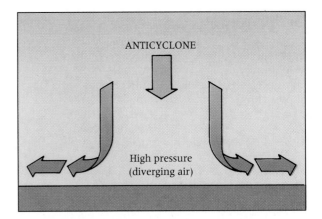

Figure 5.9 Winds converge in cyclones (low-pressure centers) and diverge from anticyclones (high-pressure centers).

pressure and are said to be *diverging.* In the case of an anticyclone, the center of the system serves as the source for **divergent** wind circulation.

Cyclonic circulation is said to be converging whether it is in a counterclockwise spiral, as in the Northern Hemisphere, or in a clockwise spiral, as in the Southern Hemisphere. Anticyclonic circulation is said to be diverging whether it is doing so in the counterclockwise outward spiral of the Southern Hemisphere or the clockwise outward spiral of the Northern Hemisphere.

Wind Terminology

Winds are named after their source. That is, a wind that comes out of the northeast is called a *northeast wind.* One coming from the south, even though going toward the north, is called a *south wind.*

Windward refers to the direction from which the wind blows. The side of something that faces the direction from which the wind is coming is called the *windward* side. Thus, a windward shore is the one that receives onshore winds from the ocean and a windward slope is the side of a mountain against which the wind blows (Fig. 5.10). **Leeward,** on the other hand, means the direction toward which the wind is blowing. Thus, a leeward shore would have offshore winds, since it faces the way winds are blowing. When the winds are coming out of the west, the leeward slope of a mountain would be the east slope. We know that winds can blow from any direction, yet in some places winds may tend to blow more from one direction than any other. We speak of these as the **prevailing winds.**

As we have seen, winds result from variations in pressure. In this chapter we have detailed the global pattern and variation in atmospheric pressure. In Chapter 6 we will discuss the resulting worldwide movement of winds as a major component of global circulation patterns.

Figure 5.10 Illustration of the meaning of windward (facing into the wind) and leeward (facing away from the wind). • How might vegetation differ on the windward and leeward sides of an island?

Define and Recall

equatorial low
  (doldrums)
subtropical high
subpolar low
polar high

isobar

pressure gradient
cyclone
anticyclone
Siberian High
Canadian High
Icelandic Low
Aleutian Low

Pacific High
Bermuda High (Azores
  High)

wind
Coriolis effect
geostrophic wind

convergent circulation
divergent circulation
windward
leeward
prevailing wind

Discuss and Review

1. What is atmospheric pressure at sea level? How do you suppose Earth's gravity is related to atmospheric pressure?

2. Horizontal variations in air pressure are caused by thermal or dynamic factors. How do these two factors differ?

3. How does incoming insolation affect pressure in the atmosphere? Give an example of an area where incoming insolation would create a pressure system. Would high or low pressure occur?

4. What is the difference between a cell and a belt of pressure?

5. What kind of pressure (high or low) would you expect to find in the center of an anticyclone? Describe and diagram the wind pattern of an anticyclone in the Northern and Southern Hemispheres.

6. Explain how water and land surfaces affect the pressure overhead during summer and winter. How does this relate to the afternoon sea breeze?

7. How do landmasses affect the development of belts of atmospheric pressure over Earth's surface?

8. What are the major obstacles to widespread use of solar energy as a major source of power production?

9. What kinds of winds are characteristic of the area in which you live? What causes these winds? How is plant life affected by them? How are your outdoor activities affected?

10. Explain how surface friction causes the surface winds to flow *across* isobars rather than *parallel to* the isobars, as in the case of a geostrophic wind.

11. What is the circulation pattern around a center of low pressure (cyclone) in the Northern Hemisphere? In the Southern Hemisphere? Draw diagrams to illustrate these circulation patterns.

Consider and Respond

1. Atmospheric pressure decreases at the rate of 0.036 mb per foot as one ascends through the lower portion of the atmosphere.
   (a) The Sears Tower in Chicago, Illinois, is one of the world's tallest building at 1450 ft. If the street-level pressure is 1020.4 mb, what is the pressure at the top of the Sears Tower?
   (b) If the difference in atmospheric pressure between the top and ground floor of an office building is 13.5 mb, how tall is the building?
   (c) A single story of a building is 12 ft. You enter an elevator on the top floor of the building and wish to descend five floors. The elevator has no floor markings—only a barometer! If the initial reading was 1003.2, at what pressure reading would you get off?

2. Look at the January (Fig. 5.3) and July (Fig. 5.4) maps of average sea-level pressure. Answer the following questions:
   (a) Why is the subtropical high-pressure belt more continuous (linear, not cellular) in the Southern Hemisphere than in the Northern Hemisphere in July?
   (b) During July, what area of the United States exhibits the lowest average pressure? Why?

3. The amount of power that can be generated by wind is determined by the equation

$$P = \frac{1}{2} D \times S^3$$

where P is the power in watts, D is the density,

and S is the wind speed in meters per second (m/s). Since $D = 1.293$ kg/m^3, we can rewrite the equation as

$$P = 0.6 \times S^3$$

(a) How much power (in watts) is generated by the following wind speeds: 2 m/s; 6 m/s; 10 m/s; 12 m/s?
(b) Since wind power increases significantly with increased wind speed, very windy locations are ideal locations for "wind farms." Cities A and B both have average wind speeds of 6 m/s. However, City A tends to have very consistent winds, while City B tends to have half of its winds at 2 m/s and the other half at 10 m/s. Which site would be the best location for a wind generation plant?

4. Refer to the mercury barometer in Figure 5.1.
(a) Convert the following pressure valves to millimeters: 1020 mb; 1008 mb; 998 mb.
(b) Convert the following pressure values to millibars: 750 mm; 800 mm; 775 mm.

CHAPTER

6

CIRCULATION PATTERNS

CHAPTER PREVIEW

▶ Global circulation systems are essential to the maintenance of Earth's energy balance and are a major key to the understanding of weather and climate.

Why is the circulation of energy over Earth so important to the preservation of equilibrium in the Earth system? Why are physical geographers especially interested in global circulation systems?

▶ Planetary wind systems in association with global pressure patterns play a major role in global circulation.

What are the primary sources of the planetary winds? What are the six major planetary wind belts or zones, and what are their chief characteristics? Why do the wind belts migrate with the seasons?

▶ Ocean currents constitute a system of global circulation that can exert significant influence on the atmospheric conditions of nearby land areas.

What is the driving force behind the ocean current circulation patterns? How do ocean currents affect atmospheric conditions of land areas?

▶ El Niños can have a devastating impact on our global weather.

What is an El Niño? How does it influence global weather? Why have scientists recently become better able to predict the onset of an El Niño?

▶ In addition to our global wind system, smaller-scale wind systems exert significant influence over the lives of those humans who experience them.

Why is the monsoon circulation system so well developed in Eurasia? How is the circulation of the monsoon system similar to land- and sea-breeze circulation? How do the two circulation systems differ?

▶ Upper air winds and atmospheric circulation play a major role in controlling surface weather and climatic conditions.

What is upper air circulation like? How does it affect surface conditions? How are human beings affected?

▲ Circulation pattern of Hurricane Elena.

O ur study of the atmospheric elements that combine to produce weather and climate has to this point focused on the fundamental influence of solar energy on the global distributional patterns of temperature and pressure. The unequal receipt of insolation by latitude over Earth's surface has produced temperature patterns that vary from the equator to the poles, and these temperature differences are one of the major causes of the development of patterns of higher and lower pressure that also vary with latitude. Now, in this chapter, we will examine patterns of another kind—patterns of movement or, more properly, circulation, where both energy and matter travel cyclically through Earth subsystems.

Geographers are particularly interested in circulation patterns because they illustrate spatial interaction, one of geography's major themes introduced in Chapter 1. Patterns of movement between one place and another reveal that the two places have a relationship and prompt geographers to seek both the nature and effect of that relationship. It is also important to understand the causes of the spatial interaction taking place, and, as we examine the circulation patterns featured in this chapter, you should make a special effort once again to trace each pattern back to the fundamental influence of solar energy.

We will begin with the review of Earth's major wind systems, followed by the study of the ocean flow (currents) that mirrors atmospheric circulation patterns. The transfer of energy between latitudes through both the atmosphere and the ocean is essential to the maintenance of heat and energy balances over Earth's surface. After the examination of global atmospheric and oceanic circulation patterns, the chapter will conclude with a review of selected regional and local surface patterns as well as circulation in the upper atmosphere.

Global Surface Wind Systems

The planetary or global wind system that is a response to the global pressure patterns examined in Chapter 5 also plays a role in the maintenance of those same pressures. This wind system, which is the major means of transport for energy and moisture through Earth's atmosphere, can be examined in an idealized state. To do so, however, we must ignore the influences of landmasses and seasonal variations in solar energy. By assuming, for the sake of discussion, that Earth has a homogeneous surface and that there are no seasonal variations in the amount of solar energy received at different latitudes, we can examine a theoretical model of the atmosphere's planetary circulation. Such an understanding will help to explain specific features of climate like the rain and snow of the Sierra Nevada and Cascade mountains and the existence of arid regions farther to the east. It will also account for the movement of great surface currents in our oceans that are driven by this atmospheric engine.

The Idealized Model of Atmospheric Circulation

Since winds are related to pressure, various types of winds are associated with different kinds of pressure cells. Therefore, a system of global winds can be demonstrated using the model of pressures that we previously developed (see Fig. 5.2).

The characteristics of convergence and divergence are very important to our understanding of global wind patterns. We know that be-

cause of the pressure gradient, surface winds blow away from high-pressure cells and toward low-pressure cells. In other words, surface air diverges from zones of high pressure and converges on areas of low pressure.

Knowing that surface winds originate in areas of high pressure, and taking into account the global system of pressure cells, we can develop our model of the wind systems of the world (Fig. 6.1). This model takes into account differential heating, Earth rotation, and atmospheric dynamics.

Note that the winds do not blow in a straight north-south line. The variation is due, of course, to the Coriolis effect, which causes an apparent deflection to the right in the Northern Hemisphere and to the left in the Southern Hemisphere.

Our idealized model of global atmospheric circulation includes six wind belts or zones in addition to the seven pressure zones we have previously identified. Two wind belts, one in each hemisphere, are located where winds move out of the polar highs and move down the pressure gradients toward the subpolar lows. As these winds are deflected to the right in the Northern Hemisphere and to the left in the Southern, they become the **polar easterlies.**

The remaining four wind belts are closely associated with the divergent winds of the subtropical highs. In each hemisphere winds flow out of the poleward portions of these highs toward the subpolar lows. Because of their general movement from the west, the winds of the upper middle latitudes are labeled the **westerlies.** The winds blowing from the highs toward the equator have been called the **trade winds.** Because of the Coriolis effect, they are the **northeast trades** in the Northern Hemisphere and the **southeast trades** south of the equator.

Our model does not conform exactly to actual conditions. First, as we know, the vertical rays of the sun do not stay precisely over the equator but move north and south during the year as far as the Tropics of Cancer and Capricorn. Therefore the pressure systems, and consequently the winds, must move in order to adjust to the change in the position of the sun. Then, as we have already discovered, the existence of the continents, especially in the Northern Hemisphere, causes longitudinal pressure differentials that affect the zones of high and low pressure. It should prove interesting to compare our model of the planetary wind and pressure systems with conditions as they actually exist on Earth's surface.

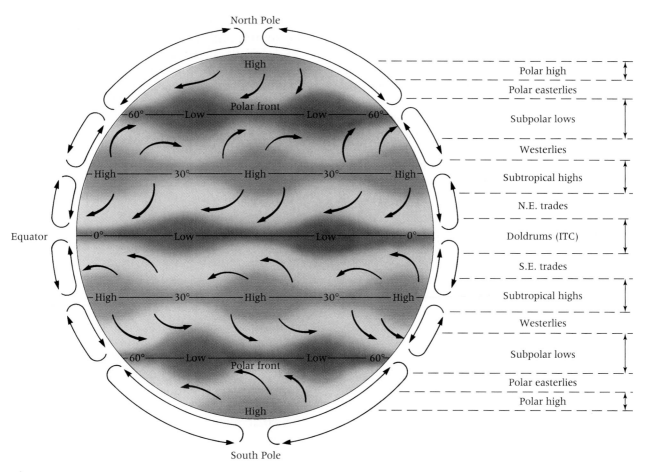

Figure 6.1 *Idealized model of Earth's pressure and wind systems.*

Conditions Within Latitudinal Zones

Trade Winds.   A good place to begin our examination of wind and associated weather patterns
as they actually occur is in the vicinity of the subtropical highs. On Earth's surface, the trade
winds, which blow out of the subtropical highs
toward the equatorial trough in both the Northern and Southern Hemispheres, can be identified
between latitudes 5° and 25°. Because of the Coriolis effect, the northern trades move away from
the subtropical high in a clockwise direction out
of the northeast. In the Southern Hemisphere the
trades diverge out of the subtropical high toward
the tropical low from the southeast, as their
movement is counterclockwise. Because the
trades tend to blow out of the east, they are also
known as the **tropical easterlies.**

The trade winds tend to be constant, steady
winds, consistent in their direction. This is most
true when they cross the eastern sides of the
oceans (near the eastern portion of the subtropical high). The area of the trades varies somewhat

during the solar year, moving north and south a
few degrees of latitude with the sun. Near their
source in the subtropical highs, the weather of
the trades is clear and dry, but after crossing large
expanses of ocean the trades have a high potential for stormy weather.

Early Spanish sea captains depended upon
the northeast trade winds to drive their galleons
to destinations in Latin America in search of gold,
spices, and new lands. Going eastward toward
home, navigators usually tried to plot a course
using the westerlies to the north. The trade winds
are one of the reasons that the Hawaiian Islands
are so popular with tourists; the steady winds
help to keep temperatures pleasant, even though
Hawaii is located south of the Tropic of Cancer.

Doldrums.   Where the trade winds converge in
the equatorial trough (or tropical low) lies a zone
of calms and weak winds of no prevailing direction. Here the air, which is very moist and heated
by the sun, tends to expand and rise, maintaining the low pressure of the area. This zone, which

is roughly between 5°N and 5°S, is generally known as the "doldrums." It is also called the **Intertropical Convergence** zone (ITC) and the "equatorial belt of variable winds and calms." Because of the converging moist air and high potential for rainfall in the doldrums, this region coincides with the world's latitudinal belt of heaviest precipitation and most persistent cloud cover.

Old sailing ships often remained becalmed in the doldrums for days at a time. A description of a ship becalmed in the doldrums appears in *The Rime of the Ancient Mariner* by Samuel Taylor Coleridge (lines 103–108). The ship is sailing northward from the tropical southeasterlies (trades) when it gets to the doldrums.

The fair breeze blew, the white foam flew,
The furrow followed free;
We were the first that ever burst
Into that silent sea.

Down dropt the breeze, the sails dropt down,
'Twas sad as sad could be;
And we did speak only to break
The silence of the sea!

All in a hot and copper sky,
The bloody Sun, at noon,
Right up above the mast did stand,
No bigger than the Moon.

Day after day, day after day,
We stuck, nor breath nor motion;
As idle as a painted ship
Upon a painted ocean.

Subtropical Highs. The areas of subtropical high pressure, generally located between latitudes 25° and 35°N and S, and from which winds blow equatorward as do the trades, are often called the subtropical belts of variable winds, or the "horse latitudes." This name comes from the occasional need by the Spanish conquistadors to throw their horses overboard in order to conserve drinking water when their ships were becalmed in these latitudes. The subtropical highs are areas, like the doldrums, in which there are no strong prevailing winds. However, unlike the doldrums, which are characterized by convergence, rising air, and heavy rainfall, the subtropical highs are areas of sinking and settling air from higher altitudes, which tend to build up the atmospheric pressure. Weather conditions are typically clear, sunny, and rainless, especially over the eastern portions of the oceans where the high-pressure cells are strongest.

Westerlies. The winds that flow poleward out of the subtropical high-pressure cells in the Northern Hemisphere are deflected to the right and thus blow from the southwest. Those in the Southern Hemisphere are deflected to the left and blow out of the northwest. Thus, these winds have correctly been labeled the *westerlies*. They tend to be less consistent in direction than the trades, but they usually are stronger winds and are often associated with stormy weather. The westerlies occur between about 35° and 65°N and S latitudes. In the Southern Hemisphere, where there is less land than in the Northern Hemisphere to affect the development of winds, the westerlies attain their greatest consistency and strength. Most of the United States, except Florida, Hawaii, and Alaska, is under the influence of the westerlies.

Polar Winds. Accurate observations of pressure and wind are very sparse in the two polar regions; therefore, we must rely on remotely sensed information. Our best estimate is that pressures are consistently high throughout the year at the poles and that prevailing easterly winds blow from the polar regions to the subpolar low-pressure systems.

Polar Front. Despite our limited knowledge of the wind systems of the polar regions, we do know that the winds can be highly variable, blowing at times with great speed and intensity. When the cold air flowing out of the polar regions and the warmer air moving in the path of the westerlies meet, they do so like two warring armies: one does not absorb the other. Instead, the denser, heavier cold air pushes the warm air out of the way, forcing it to rise rapidly. The line along which these two great wind systems battle is appropriately known as the **polar front.** The weather that results from the meeting of the cold polar air and the warmer air from the subtropics can be very stormy. In fact, most of the storms that move slowly through the middle latitudes in the path of the prevailing westerlies are born at the polar front.

The Effects of Seasonal Migration

Just as insolation, temperature, and pressure systems migrate north and south as Earth revolves around the sun, Earth's wind systems also migrate with the seasons. During the summer months in the Northern Hemisphere, maximum insolation is received north of the equator. This condition causes the pressure belts to move north as well, and the wind belts of both hemispheres shift accordingly. Six months later, when maximum heating is taking place south of the equator, the various wind systems have migrated south in response to the migration of the pressure systems. Thus, seasonal variation in wind and pressure conditions is one important way in which actual atmospheric circulation differs from our idealized model.

The seasonal migration will most affect those regions near the boundary zone between two wind or pressure systems. During the winter months such a region will be subject to the moods of one system. Then, as summer approaches, that system will migrate poleward and the next equatorward system will move in to influence the region. Two such zones in each hemisphere have a major effect on climate. The first lies between latitudes 5° and 15°, where the wet equatorial low of the high-sun season (summer) alternates with the dry subtropical high and trade winds of the low-sun season (winter). The second occurs between 30° and 40°, where the subtropical high dominates in summer but is replaced by the wetter westerlies in winter.

California is an example of a region located within a zone of transition between two wind or pressure systems (Fig. 6.2). During the winter this region is under the influence of the westerlies blowing out of the Pacific High. These winds, turbulent and full of moisture from the ocean, bring winter rains and storms to "sunny" California. As summer approaches, however, the subtropical high, and its associated westerlies, moves north. As California comes under the influence of the calm and steady high-pressure system, it experiences again the climate for which it is famous: day after day of warm, clear, blue, cloudless skies. This alternation of moist winters and dry summers is typical of the western sides of all landmasses between 30° and 40° latitude.

Longitudinal Differences in Winds

We have seen that there are sizeable latitudinal differences in pressure and winds. In addition, there are significant longitudinal variations, especially in the zone of the subtropical highs.

As we have previously mentioned, the subtropical high-pressure cells, which are generally

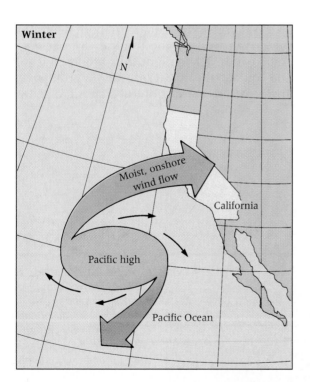
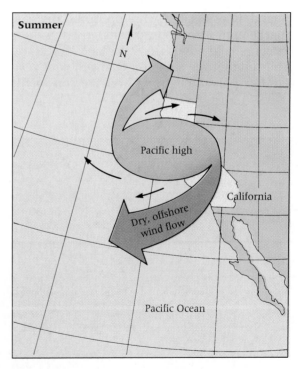

Figure 6.2 Winter and summer positions of the Pacific anticyclone in relation to California. In the winter, the anticyclone lies well to the south and feeds the westerlies that bring the cyclonic storms and rain from the North Pacific to California. The influence of the anticyclone dominates during the summer. The high pressure blocks out cyclonic storms and produces warm, sunny, and dry conditions. • **In what ways would the seasonal migration of the Pacific anticyclone affect agriculture in California?**

centered over the oceans, are much stronger on their eastern sides than on their western sides. Thus, over the eastern portions of the oceans (west coasts of the continents) in the subtropics, subsidence and divergence are especially noticeable. The above-surface temperature inversions so typical of anticyclonic circulation are close to the surface, and the air is calm and clear. The air moving equatorward from this portion of the high produces the classic picture of the steady trade winds with clear, dry weather.

Over the western portions of the oceans (eastern sides of the continents), conditions are markedly different. In its passage over the ocean, the diverging air is gradually warmed and moistened; the above-surface inversion occurs at higher elevations, and turbulent, stormy weather conditions are likely to develop. As indicated in Figure 6.3, wind movement in the western portions of the anticyclones may actually be poleward and directed toward landmasses. Hence, the trade winds in these areas are especially weak or nonexistent much of the year.

As we have pointed out in discussing Figures 5.3 and 5.4, there are great land-sea contrasts in temperature and pressure throughout the year farther toward the poles, especially in the Northern Hemisphere. In the cold continental waters, the land is associated with pressures that are higher than those over the oceans, and thus there are strong, cold winds from the land to the sea. In the summer the situation changes, with relatively low pressure existing over the continents because of higher temperatures. Wind directions are thus greatly affected, and the pattern is reversed so that winds flow from the sea to the land.

Ocean Currents

Like the planetary wind system, surface ocean currents play a significant role in helping to equalize the energy imbalance between the tropical and polar regions. In addition, surface ocean currents greatly influence the climate of coastal locations.

Earth's surface wind system is the primary control of the major surface currents and drifts. Other controls are the Coriolis effect and the size, shape, and depth of the sea or ocean basin. Other currents may be caused by differences in density due to variations in temperature and salinity, tides, and wave action.

The major surface currents move in broad circulatory patterns, called **gyres,** around the subtropical highs. Due to the Coriolis effect, the gyres flow clockwise in the Northern Hemisphere and counterclockwise in the Southern Hemisphere (Fig. 6.4). As a general rule, the surface currents do not cross the equator.

Waters near the equator in both hemispheres are driven west by the tropical easterlies or the trade winds. The current thus produced is called the Equatorial Current. At the western margin of the ocean, its warm tropical waters are deflected

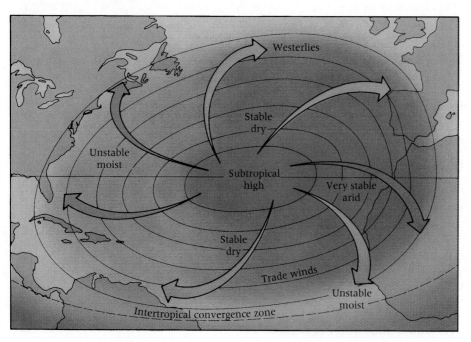

Figure 6.3 *Circulation pattern in a Northern Hemisphere subtropical anticyclone. Subsidence of air is strongest in the eastern part of the anticyclone, producing calm air and arid conditions over adjacent land areas. The southern margin of the anticyclone feeds the persistent northeast trade winds.* • **What wind system is fed by the northern margin?**

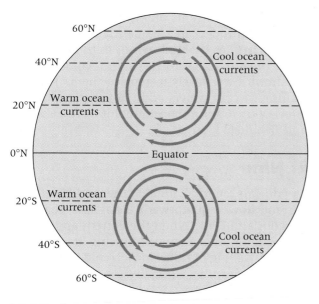

Figure 6.4 *The major ocean currents flow in broad gyres in opposite directions in each hemisphere.*

poleward along the coastline. As these warm waters move into higher latitudes, they move through waters cooler than themselves and are identified as *warm* currents (Fig. 6.5).

In the Northern Hemisphere, warm currents, such as the Gulf Stream and the Kuroshio Current, are deflected more and more to the right (or east) because of the Coriolis effect. At about 40°N, the westerlies begin to drive these warm waters eastward across the ocean, as in the North Atlantic Drift and the North Pacific Drift. Eventually, these currents run into the land at the eastern margin of the ocean, and most of the waters are deflected toward the equator. By this time, these waters have lost much of their warmth, and while moving equatorward into the subtropical latitudes, they are cooler than the adjacent waters. They have become *cool* currents. These waters complete the circulation pattern when they rejoin the westward-moving Equatorial Current.

On the eastern side of the North Atlantic, the North Atlantic Drift moves into the seas north of the British Isles and around Scandinavia, keeping

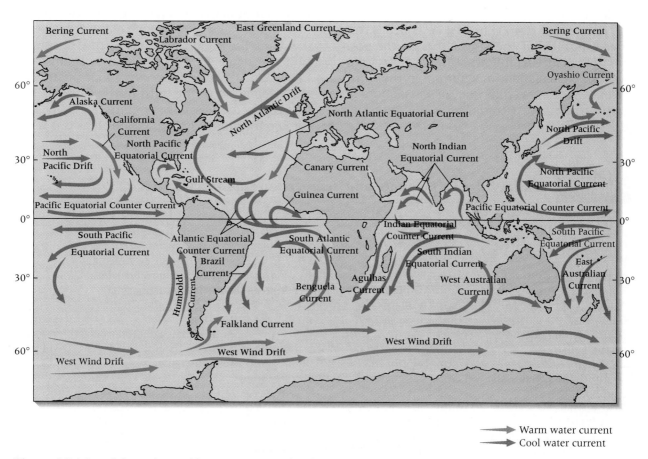

Figure 6.5 *Map of the major world ocean currents, showing warm and cool currents.*
• **How does this map of ocean currents help explain the mild winters in London, England?**

those areas warmer than their latitudes would suggest. Some Norwegian ports north of the Arctic Circle remain ice-free because of this warm water. Cold polar water—the Labrador and Oyashio Currents—flows southward into the Atlantic and Pacific oceans along their western margins.

The circulation in the Southern Hemisphere is comparable to that in the Northern except that it is counterclockwise. Also, because there is little land poleward of 40°S, the West Wind Drift (or Antarctic Circumpolar Drift) circles Earth as a cool current across all three major oceans almost without interruption. It is cooled by the influence of the Antarctic ice cap (Table 6.1).

In general, then, warm currents move poleward as they carry tropical waters into the cooler waters of higher latitudes, as in the case of the Gulf Stream or the Brazil Current. Cool currents deflect water equatorward, as in the California Current and the Humboldt Current. Warm currents tend to have a humidifying and warming effect on the east coasts of continents along which they flow, while cool currents tend to have a drying and cooling effect on the west coasts of the land masses.

The general circulation just described is consistent throughout the year, although the position of the currents follows seasonal shifts in atmospheric circulation. In addition, in the North Indian Ocean, the direction of circulation reverses seasonally according to the monsoon winds.

The cold currents along west coasts in subtropical latitudes are frequently reinforced by **upwelling.** As the trade winds in these latitudes drive the surface waters offshore, colder waters rise from lower levels to replace them. This upwelling of cold waters adds to the strength and effect of the California, Humboldt (Peru), Canary, and Benguela Currents.

El Niño

As you can see in Figure 6.5, the cold Humboldt current flows equatorward along the coasts of Ecuador and Peru. When the current approaches the equator, the westward-flowing trade winds cause upwelling of nutrient-rich cold water along the coast. Fishing, especially for anchovies, is a major local industry.

Every year during the months of December and January, a weak, warm countercurrent replaces the normally cold coastal waters. Without the upwelling of nutrients from below to feed the fish, fishing comes to a standstill. Fishermen in this region have known of the phenomenon for hundreds of years. In fact, this is the time of year they traditionally set aside to tend to their equipment and await the return of cold water. The residents of the region have given this phenomenon the name **El Niño,** which is Spanish for "The Child," because it occurs about the time of the celebration of the birth of the Christ Child.

While the warm-water current usually lasts for two months or less, there are occasions when

TABLE 6.1   Primary Ocean Currents and Temperature Characteristics

Pacific Ocean		Atlantic Ocean		Indian Ocean	
Oyashio	Cool	East Greenland	Cool	North Indian monsoon	
Bering	Cool	Labrador	Cool	currents (reverse seasonally	
North Pacific Drift	Warm	North Atlantic Drift	Warm	with the monsoon winds)	
Kuroshio (Japan)	Warm	Gulf Stream	Warm	North Indian Equatorial	Warm
Alaska	Warm	Canary	Cool	Indian Equatorial Counter	
California	Cool	Guinea	Warm	Current	Warm
North Pacific Equatorial	Warm	North Atlantic		South Indian Equatorial	Warm
Pacific Equatorial Counter		Equatorial	Warm	Agulhas (Mozambique)	Warm
Current	Warm	North Atlantic		West Australian	Cool
South Pacific Equatorial	Warm	Equatorial Counter			
Humboldt (Peru)	Cool	Current	Warm		
East Australian	Warm	South Atlantic			
West Wind Drift	Cool	Equatorial	Warm		
(Antarctic Circumpolar		Brazil	Warm		
Drift; also present in		Benguela	Cool		
South Atlantic and		Falkland	Cool		
South Indian Oceans)					

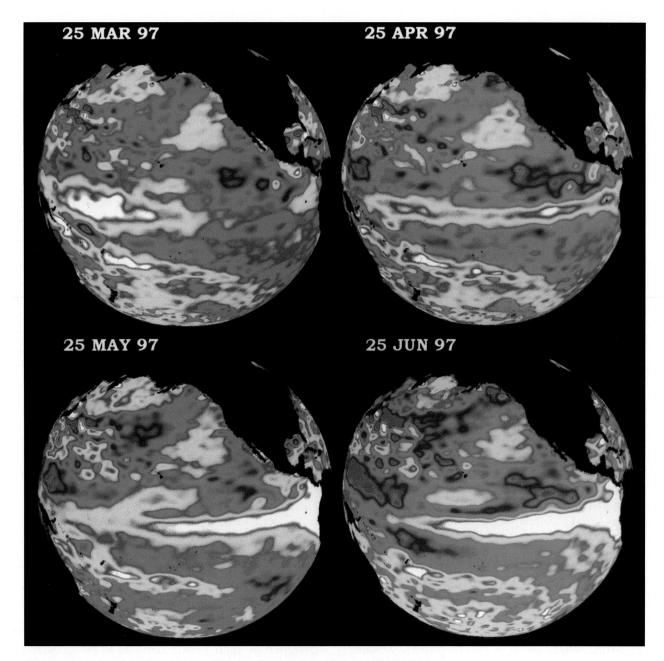

Figure 6.6 *Colored satellite images showing development of the 1997 El Niño event. The areas colored red and white depict warmer water. The top-left image taken on March 25 shows the warm water confined to the western Pacific. The bottom-right image, taken on June 25, illustrates a well-developed El Niño with warm water reaching South America.*

the disruption to the normal flow lasts for many months. In these situations, water temperatures are raised not just along the coast, but for thousands of kilometers offshore (Fig. 6.6). Over the past decade, "El Niño" has come to be used to describe these exceptionally strong episodes and not the annual event. During the past fifty years, ten El Niños have been observed. Not only do the El Niños affect the temperature of the equa-

torial Pacific, but the strongest of them impact worldwide weather.

El Niño and the Southern Oscillation

To completely understand the processes that interact to produce an El Niño requires that we study conditions all across the Pacific, not just in the waters off South America. Over fifty years ago

Sir Gilbert Walker, a British scientist, discovered a connection between surface pressure readings at weather stations on the eastern and western sides of the Pacific. He noted that a rise in pressure in the eastern Pacific is usually accompanied by a fall in pressure in the western Pacific and vice versa. He called this seesaw pattern the **Southern Oscillation.** This link between El Niño and the Southern Oscillation is so great that they are often referred to jointly as ENSO (*El Niño/Southern Oscillation*).

During a typical year, the eastern Pacific has a higher pressure than the western Pacific. This east-to-west pressure gradient enhances the trade winds over the equatorial Pacific waters. This results in a warm surface current that moves from east to west at the equator. The western Pacific develops a thick, warm layer of water while the eastern Pacific has the cold Humboldt current enhanced by upwelling.

Then, for unknown reasons, the Southern Oscillation swings in the opposite direction, dramatically changing the usual conditions described above, with pressure increasing in the western Pacific and decreasing in the eastern Pacific. This change in the pressure gradient causes the trade winds to weaken or, in some cases, to reverse. This causes the warm water in the western Pacific to flow eastward, increasing sea-surface temperatures in the central and eastern Pacific. This eastward shift signals the beginning of El Niño.

El Niño and Global Weather

Cold ocean waters impede cloud formation. Thus, clouds tend to develop over the warm waters of the western Pacific but not over the cold waters of the eastern Pacific. However, during an El Niño, when warm water migrates eastward, clouds develop over the entire equatorial water of the Pacific (Fig. 6.7). These clouds can build to heights of 18,000 meters (59,000 ft). Clouds of this magnitude can disrupt the high-altitude wind flow above the equator. As we will see later in this chapter, a change in the upper-air wind flow in one portion of the atmosphere will trigger wind flow changes in other portions of the atmosphere. Alterations in the upper-air winds result in alterations to surface weather.

Scientists try to document as many past El Niño events as possible by piecing together bits of historical evidence, such as sea-surface temperature records, daily observations of atmospheric pressure and rainfall, fisheries' records from South America, and the writings of Spanish colonists living along the coasts of Peru and Ecuador dating back to the fifteenth century. Additional evidence comes from the growth patterns of coral and trees in the region.

Based on this historical evidence, we know that El Niños have occurred as far back as records go. One disturbing fact is that they are occurring more often. Records indicate that during the sixteenth century, an El Niño occurred on average every six years. Evidence gathered over the past few decades indicate El Niños are now occurring on average every 2.2 years. Even more alarming is the fact that they appear to be getting stronger. The record-setting El Niño of 1982–83 was recently surpassed by the one in 1997–98.

The 1997–98 El Niño brought copious and damaging rainfall to the southern United States, from California to Florida. Snowstorms in the northeast portion of the United States were more frequent and stronger than in most years. The warm El Niño winters fueled Hurricane Linda, which devastated the western coast of Mexico. Linda was the strongest hurricane ever recorded in the eastern Pacific.

In recent years scientists have become better able to monitor and forecast El Niño events. An elaborate network of weather buoys plus satellite observations provide an enormous amount of data that can be studied by computer analysis to help predict the formation and strength of El Niño events. Our improved observation skills have led to the discovery of the **Atlantic Cycle**—a strengthening of the subtropical high and subpolar low over the Atlantic Ocean. During an Atlantic Cycle, increased precipitation and temperatures occur in the southeastern United States while cold, dry conditions occur over the northeastern United States.

Will scientists ever be able to predict the occurrence of such phenomena as El Niño or the Atlantic Cycle? No one can answer that question, but as our technology improves, our forecasting ability will also increase. We have made tremendous progress: In the last few decades we have come to recognize the close association between the atmosphere and hydrosphere as well as to understand the complex relationship between these Earth systems.

Subglobal Surface Wind Systems

As we have seen, winds develop whenever differential heating causes differences in pressure. The global wind system is a response to the con-

Equatorial Cloud Development over
Pacific Ocean in El Niño Year

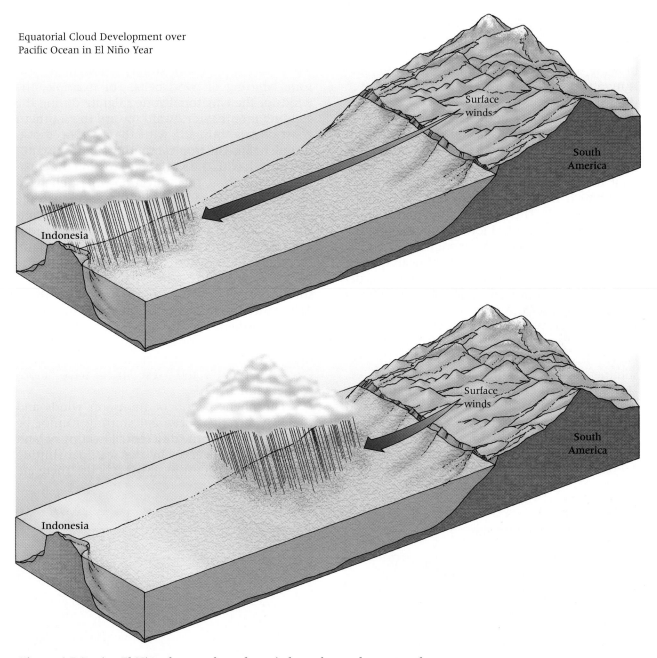

Figure 6.7 *During El Niño the easterly surface winds weaken and retreat to the eastern Pacific, allowing the central Pacific to warm and the rain area to migrate eastward.*

stant temperature imbalance between tropical and polar regions. On a smaller, or subglobal, scale, additional wind systems develop. Monsoon winds are continental in size and develop in response to the *seasonal* variations in temperature between large landmasses and adjacent oceans. On an even smaller scale are local winds which develop in response to *diurnal* variations in heating.

Monsoon Winds

The term *monsoon* comes from the Arabic word *mausim*, meaning season. This word has been used by Arab seamen for many centuries to describe seasonal changes in wind direction across the Arabian Sea between Arabia and India. As a meteorological term, **monsoon** refers to the directional shifting of winds from one season to the next. Usually, the monsoon occurs when a humid wind blowing from the ocean toward the land in the summer shifts to a dry, cooler wind blowing seaward off the land in the winter, and it involves a full 180-degree direction change in the wind.

The monsoon is most characteristic of southern Asia, although it occurs on other continents as well. As the large landmass of Asia cools more quickly than the surrounding oceans, the continent develops a strong center of high pressure from which there must be an outflow of air in winter (Fig. 6.8). This outflow blows across much land toward the tropical low before reaching the oceans. It brings cold, dry air south.

In summer the Asian continent heats quickly and develops a large low-pressure center. This development is reinforced by a poleward shift of the Intertropical Convergence to a position over southern Asia. Warm, moist air from the oceans is attracted into this low. Though full of water vapor, this air does not in itself cause the wet summers with which the monsoon is associated. However, any turbulence or landform barrier that makes this moist air rise and, as a result, cool off will bring about precipitation. This precipitation is particularly noticeable in the foothills of the Himalayas, the western Ghats of India, and the Annamese Highlands of Vietnam. This is the time

of year when the rice crop is planted in many parts of Asia.

In the lower latitudes, a monsoonal shift in winds can come about simply through the seasonal shifting of wind belts. For example, the winds of the equatorial zone migrate during the summer months northward toward the southern coast of Asia, bringing with them warm, moist, turbulent air. The tropical southeasterlies also migrate north with the sun, some crossing the equator. As they do so, their direction will be changed by the Coriolis effect, and they will become southwesterly trades in the Northern Hemisphere. They also bring warm, moist air (from their travels over the ocean) to the southern (and especially southeastern) coast of India. In the winter months the equatorial winds and the southern trades migrate south, leaving southern Asia under the influence of the dry, calm trades of the Northern Hemisphere. Asia and northern Australia are true monsoon areas, with a full 180-degree wind shift. Other regions, like the southern United States and West Africa, have "monsoonal tendencies."

The phenomenon of monsoon winds and their characteristic seasonal shifting cannot be fully explained by the differential heating of land and water, however, or by the seasonal shifting of tropical and subtropical wind belts. There are

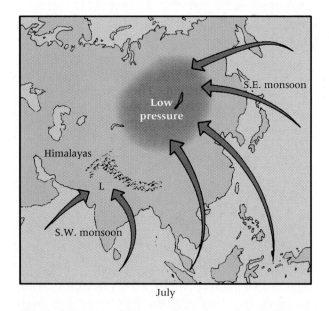

July

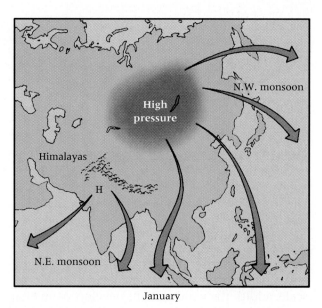

January

Figure 6.8 *Seasonal changes in surface wind direction that create the Asiatic monsoon system. The "burst" of the "wet monsoon," or the sudden onshore flow of tropical humid air in July, is apparently triggered by changes in the upper air circulation, resulting in heavy precipitation. The offshore flow of dry continental air in winter creates the "dry monsoon" and drought conditions in southern Asia.* • **How do the seasonal changes of wind direction in Asia differ from those of the southern United States?**

aspects of the monsoon system—for example, its "burst" or sudden transition between dry and wet in southern Asia—that must have other causes. Meteorologists looking for a more complete explanation of the monsoon are examining the role played by the jet stream (described in a later section) and other wind movements of the upper atmosphere.

Local Winds

Earlier in this chapter, we discussed the major circulation patterns of Earth's atmosphere, a knowledge of which is vital to understanding the climatic regions of Earth and the fundamental climatic differences between those regions. Yet we are all aware that there are winds that affect weather on a far smaller scale. These local winds are often a response to local landform configurations and add further complexity to the problem of understanding the dynamics of weather.

Land Breeze–Sea Breeze. The **land breeze–sea breeze cycle** is a diurnal (daily) one in which the differential heating of land and water again plays a role (Fig. 6.9). During the day, when the land—and consequently the air above it—is heated more quickly and to a higher temperature than the nearby ocean (or large lake or sea), the air above the land expands and rises. This process creates a local area of low pressure, and the rising air tends to be replaced by the denser, cooler air from over the ocean. Thus, a sea breeze of cool, moist air blows in over the land during the

day. This sea breeze helps explain why seashores are so popular in summer, since cooling winds help alleviate the heat. These winds can mean a 5° to 9°C (9°–16°F) reduction in temperature along the coast as well as a lesser influence on land perhaps as far from the sea as 15 to 50 kilometers (9–30 mi). During hot summer days, such winds cool cities like Chicago, Milwaukee, and Los Angeles. At night, the land and the air above it cool more quickly and to a cooler temperature than the nearby water body and the air above it. Consequently, the pressure builds up over the land and air flows out toward the lower pressure over the water, creating a land breeze.

For thousands of years, fishermen in sailboats have left their coasts at dawn, when there is still a land breeze, and have returned with the sea breeze of the late afternoon.

Mountain Breeze–Valley Breeze. Under the calming influence of a high-pressure system, there is a daily **mountain breeze–valley breeze cycle** (Fig. 6.10) that is somewhat similar in mechanism to the land breeze–sea breeze cycle just discussed. During the day, when the valleys and slopes of mountains are heated by the sun, the air expands and rises up the sides of the mountains. This warm daytime breeze is the valley breeze, named for its place of origin. Clouds, which can often be seen hiding mountain peaks, are actually the visible evidence of condensation in the warm air rising from the valleys. At night, when the valley and slopes are cooled because Earth is giving off more radiation than it is

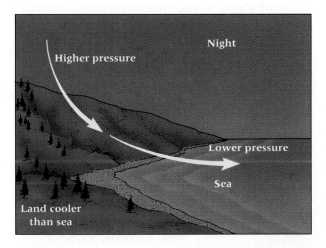
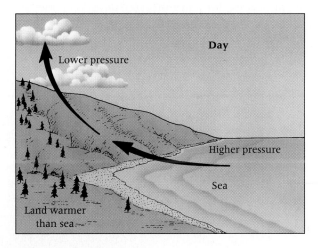

Figure 6.9 Land and sea breezes. This day-to-night reversal of winds is a consequence of the different rates of heating and cooling of land and water areas. The land becomes warmer than the sea during the day and colder than the sea at night; the air flows from the cooler to the warmer area. • What is the impact on daytime coastal temperatures of the land and sea breeze?

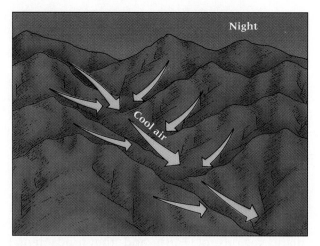

Figure 6.10 *Mountain and valley breezes. This daily reversal of winds results from heating of mountain slopes during the day and their cooling at night. Warm air is drawn up slopes during the day, and cold air drains down the slopes at night.* • **Do you think fog formation in valleys is more likely during the day or at night? Why?**

receiving, the air cools and sinks once again into the valley as a cool mountain breeze.

Drainage Winds. Also known as **katabatic winds, drainage winds** are local to mountainous regions and can occur only under calm, clear conditions. Cold, dense air will accumulate in a high valley, plateau, or snowfield within a mountainous area. Because the cold air is very dense, it tends to flow downward, escaping through passes and pouring out onto the land below. Drainage winds can be extremely cold and strong, especially when they result from cold air accumulating over ice caps such as Greenland and Antarctica. These winds are known by many local names; for example, in Yugoslavia they are

called the *bora*, in France the *mistral*, and in Alaska the *Taku*.

Foehn Winds. A fourth type of local wind is also known by several names in different parts of the world—for example, **Chinook** in the Rocky Mountain area and **foehn** (pronounced "fern") in the Alps. Foehn-type winds occur when air originating elsewhere must pass over a mountain range. As these winds flow down the leeward slope after crossing the mountains, the air is compressed and heated at a greater rate than it was cooled when it ascended the windward slope (Fig. 6.11). Thus, the air enters the valley below as warm, dry winds. The rapid temperature rise brought about by such winds has been known to

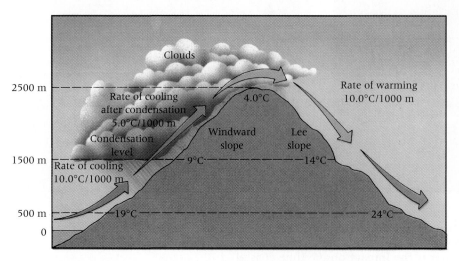

Figure 6.11 *Foehn winds result when air ascends a mountain range and undergoes condensation that dries it out and adds latent heat that slows its rate of cooling. As the dry air descends the leeward side of the range, it is compressed and heated at a greater rate than it was initially cooled. This produces the relatively warm, dry conditions with which foehn winds are associated.* • **The term Chinook, a type of foehn wind, means "snow eater." Can you offer an explanation for how this name came about?**

damage crops, increase forest-fire hazard, and set off avalanches. In addition, some research supports the notion that foehn-type winds affect people's personalities. The research indicates that suicides and crimes of passion increase when these winds are blowing.

An especially dry foehn-type wind is the **Santa Ana** of southern California. It forms when high pressure develops over the interior desert regions of southern California. The clockwise circulation of the high drives the air of the desert southwest over the mountains of eastern California, accentuating the dry conditions as the air moves down the western slopes.

There is no question that winds, both local and global, are effective elements of atmospheric dynamics. We all know that a hot, windy day is not nearly as unpleasant as a hot day without the wind. This difference exists because winds increase the rate of evaporation and thus the rate of removal of heat from our bodies, the air, animals, and plants. For this same reason, the wind on a cold day increases our discomfort.

Upper Air Winds

Thus far we have closely examined the wind patterns near Earth's surface. Of equal, or perhaps even greater, importance is the flow of air *above* Earth's surface—in particular, the flow of air at altitudes above 5000 meters (16,500 ft), in the upper troposphere. The formation, movement, and decay of surface cyclones and anticyclones depend to a great extent on the flow of air high above Earth's surface.

The circulation of the upper air winds is a far less complex phenomenon than surface wind circulation. In the upper troposphere, an average westerly flow, the *upper air westerlies,* is maintained poleward of about 15° to 20° latitude in both hemispheres. Because of the reduced frictional drag, the upper air westerlies move much more rapidly than their surface counterparts. Between 15° and 20°N and S latitudes are the *upper air easterlies,* which can be considered the upper air extension of the trade winds. The flow of the upper air winds became very apparent during World War II, when high-altitude bombers moving eastward were found to cover similar distances faster than those flying westward. American pilots had encountered the upper air westerlies, or perhaps even the **jet streams**— very strong air currents embedded within the upper air westerlies.

The upper air westerlies form as a response to the temperature difference between warm tropical air and cold polar air. The air in the equatorial latitudes is warmed, rises convectively to high altitudes, and then flows toward the polar regions. At first this seems to contradict our previous statement, relative to surface winds, that air flows from cold areas (high pressure) toward warm areas (low pressure). This apparent discrepancy disappears, however, if you recall that the pressure gradient, down which the flow takes place, must be assessed between two points *at the same elevation.* A column of cold air will exert a higher pressure at Earth's surface than a column of warm air. Consequently, the pressure gradient established at Earth's surface will result in a flow from the cooler air toward the warmer air. However, cold air is denser and more compact than warm air. Thus pressure decreases with height more rapidly in cold air than in warm air. As a result, at a specific height above Earth's surface, a lower pressure will be encountered in cold air than in warm air. This will result in a flow (pressure gradient) from the warmer air toward the colder air at that height. Figure 6.12 illustrates this concept.

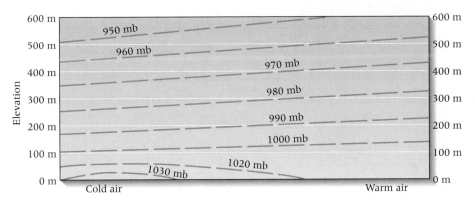

Figure 6.12 Variation of pressure surfaces with height. Note that the horizontal pressure gradient is from cold to warm air at the surface and in the opposite direction at higher elevations (such as 400 meters). • **In what direction would the winds flow at 300 meters?**

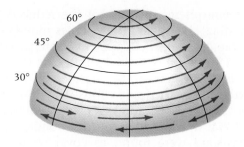

Figure 6.13 *The upper air westerlies form a broad circumpolar flow throughout most of the upper atmosphere.*

Returning to our real-world situation, as the upper air winds flow from the equator toward the poles (down the pressure gradient), they are turned eastward because of the Coriolis effect. The net result is a broad circumpolar flow of westerly winds throughout most of the upper atmosphere (Fig. 6.13). Since the upper air westerlies form in response to the thermal gradient between tropical and polar areas, it is not surprising that they are strongest in winter (the low-sun season), when the thermal contrast is greatest. On the other hand, during the summer (the high-sun season), when the contrast in temperature over the hemisphere is much reduced, the upper air westerlies move more slowly.

The temperature gradient between tropical and polar air, especially in winter, is not uniform but rather is concentrated where the warm tropical air meets cold polar air. This boundary, or front, with its stronger pressure gradient, marks the location of the **polar jet stream.** Ranging from 40 to 160 kilometers (25–100 mi) in width and up to 2 or 3 kilometers (1–2 mi) in depth, the polar jet stream can be thought of as a faster, internal ribbon of air within the upper air westerlies. While the polar jet stream flows over the midlatitudes, another westerly **subtropical jet stream** flows above the sinking air of the subtropical highs in the lower midlatitudes. Like the upper air westerlies, both jets are best developed in winter, when hemispherical temperatures exhibit their steepest gradient (Fig 6.14). During the summer, both jets weaken in intensity and the subtropical jet stream frequently disappears completely.

We can now go one step further and combine our knowledge of upper air circulation and surface circulation to yield a more realistic portrayal of the vertical circulation pattern of our atmosphere (Fig. 6.15). In general, the upper air westerlies and the associated polar jet stream flow in a fairly smooth pattern (Fig. 6.16a). At times, however, the upper air westerlies develop oscillations, termed *long* waves or **Rossby waves,** after the meteorologist who proposed and proved

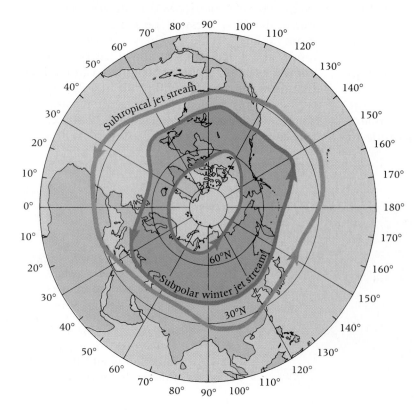

Figure 6.14 *Approximate location of the subtropical jet stream and area of activity of the polar jet stream (shaded) in the Northern Hemisphere winter.*

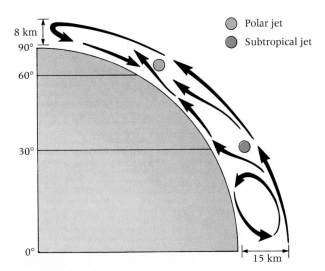

Figure 6.15 *A more realistic schematic cross section of the average circulation in the atmosphere.*

their existence (Fig. 6.16b). Rossby waves result in cold polar air pushing into the lower latitudes and forming *troughs* of low pressure, while warm tropical air moves into higher latitudes, forming *ridges* of high pressure. It is when the upper air circulation is in this configuration that surface weather is most influenced. In Chapter 8 we will examine this influence in more detail.

Eventually the upper air oscillations become so extreme that the "tongues" of displaced air are cut off, forming upper air cells of warm and cold air (Fig. 6.16c). This process helps to maintain a net poleward flow of energy from equatorial and tropical areas. The cells eventually dissipate and the pattern returns to normal (Fig. 6.16a). The complete cycle takes four to eight weeks. Although it is not completely clear why the upper atmosphere goes into these oscillating patterns, we are currently gaining additional insights. One

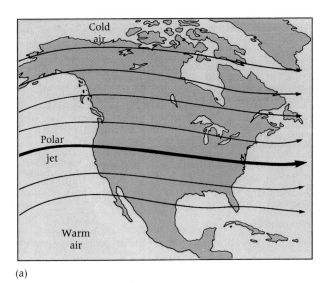

(a)

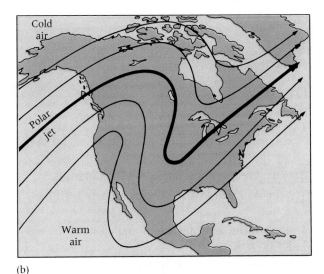

(b)

Figure 6.16 *Development and dissipation of Rossby waves in the upper air westerlies. (a) A fairly smooth flow prevails. (b) Rossby waves form, with a ridge of warm air extending into Canada and a trough of cold air extending down to Texas. (c) The trough and ridge are cut off and will soon dissipate. The flow will then return to a pattern similar to (a).* • **How are Rossby waves closely associated with the changeable weather of the central and eastern United States?**

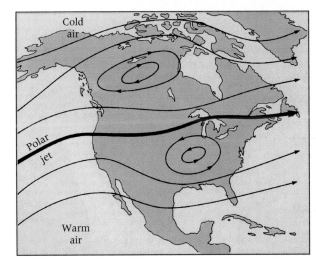

(c)

possible cause is variation in ocean surface temperatures. If the oceans in, say, the northern Pacific or near the equator become unusually cold or warm, this apparently triggers oscillations, which continue until the ocean surface temperature returns to normal. Other causes are also likely.

In addition to this influence on weather, jet streams are important for other reasons. They can carry pollutants, such as radioactive wastes or volcanic dust, over great distances and at relatively rapid rates. It was the polar jet stream that carried ash from the Mount St. Helens eruption eastward into Idaho and Montana. Nuclear fallout from the Chernobyl incident in the former Soviet Union could be monitored in succeeding days as it crossed the Pacific, and later the United States, in the jet streams. Pilots flying eastward, for example from America to Europe, take advantage of the jet stream, so the flying times in this direction may be significantly shorter than those in the reverse direction.

Define and Recall

polar easterlies	polar front	Southern Oscillation	katabatic wind
westerlies			foehn wind
northeast trades	gyre	monsoon	
southeast trades	upwelling	land breeze–sea breeze	polar jet stream
Intertropical		mountain breeze–valley	subtropical jet stream
Convergence (ITC)	El Niño	breeze	Rossby wave

Discuss and Review

1. Map the trade winds of the Atlantic Ocean and compare your map with one of trade routes in the nineteenth century or earlier.

2. What are the horse latitudes? The doldrums?

3. Why do Earth's wind systems migrate with the seasons?

4. Explain the general pattern and effect of the major oceanic currents.

5. How are the land breeze–sea breeze and monsoon circulations similar? How are they different?

6. What are monsoons? Have you ever experienced one? What causes them? Name some nations that are concerned with the arrival of the "wet monsoon."

7. What effect on valley farms could a strong drainage wind have?

8. What effect would foehn-type winds have on farming, forestry, and ski resorts?

9. El Niño was particularly damaging in 1997–98. What areas of the United States were damaged the worst? How does El Niño impact global weather?

10. Why are meteorologists concerned with upper air observations? What methods do they use to make these observations?

11. Describe the movements of the upper air. How have pilots applied their experience of the upper air to their flying patterns?

12. What is the relationship between the jet stream and upper air westerlies?

13. How can you apply knowledge of pressure and winds in your everyday life?

Consider and Respond

1. If you wished to *sail* from London to New York and back, what route would you take? Why?

2. Is the jet stream stronger in the summer or the winter? Why?

3. The amount of power that can be generated by wind is determined by the equation

$$P = \frac{1}{2} D \times S^3$$

where P is the power in watts, D is the density, and S is the wind speed in meters per second (m/s). Since $D = 1.293$ kg/m^3, we can rewrite the equation as

$$P = 0.6 \times S^3$$

(a) How much power (in watts) is generated by the following wind speeds: 2 m/s, 6 m/s, 10 m/s, 12 m/s?

(b) Since wind power increases significantly with increased wind speed, very windy areas are ideal locations for "wind farms." Cities A and B both have average wind speeds of 6 m/s. However, City A tends to have very consistent winds, while in City B, half of its winds tend to be at 2 m/s and the other half at 10 m/s. Which site would be the best location for a wind generation plant?

CHAPTER
7

MOISTURE, CONDENSATION, AND PRECIPITATION

CHAPTER PREVIEW

▶ Because the hydrologic cycle involves the circulation of water throughout all the major Earth spheres, it is fundamental to the nature and operation of the entire Earth system.

How does the hydrologic cycle involve all of Earth's spheres? In what ways does the hydrologic cycle affect the Earth system?

▶ There is a direct relationship between the temperature of the atmosphere and the amount of water vapor the atmosphere can hold.

What is this relationship? Why is it important?

▶ Although water may evaporate from all Earth surfaces and transpiration may add considerable moisture to the air, the oceans are the most important source of water vapor in the atmosphere.

Why is this so? What portion (latitudes) of the oceans would have the highest evaporation rates and why?

▶ There is only one way for significant condensation, cloud formation, and precipitation to occur: Air must be forced to rise so that sufficient adiabatic cooling will take place.

How are condensation, clouds, and precipitation related? Why does adiabatic cooling take place in rising air? Why might the rate of cooling differ in wet and dry air?

▶ Stability is associated with low (or weak) environmental lapse rates; instability is associated with high (or steep) environmental lapse rates.

What is stability? Instability? Why does the lapse rate affect the tendency of air to rise?

▶ The horizontal distribution of precipitation over Earth can be explained by either one or a combination of the following: frontal (cyclonic), orographic, or convectional precipitation.

Which is most important in your community? Which is most closely related to landform? How is this statement related to the fourth statement above?

▶ As a general rule, the less precipitation a region receives, the greater will be the variability of precipitation in that region from year to year.

What is the effect of this variability on human beings? Where do you think the effect would be greatest? In deserts or in semiarid grasslands?

▲ Atmospheric diamonds: Glaze on trees after an ice storm in Illinois.

Water is vital to all life on Earth. Although some living things can survive without air, nothing can survive without water. Water is necessary for photosynthesis, cell growth, protein formation, soil formation, and the absorption of nutrients by plants and animals.

Water affects Earth's surface in innumerable ways. Because of the structure of the water molecule, water is able to dissolve an enormous number of substances—so many, in fact, that it has been called the universal solvent. Because water acts as a solvent for so many substances, it is almost never found in a pure state. Even rain is filled with impurities picked up in the atmosphere. Indeed, without these impurities to condense around, neither clouds nor precipitation could occur. In addition, rainwater usually contains some dissolved carbon dioxide from the air. Therefore, rain is a very weak form of carbonic acid. We shall see later (Chapter 18) that this fact affects how water shapes certain landforms. The weak "material" acidity of rainwater should not be confused with the environmentally damaging *acid rain*, which is at least ten times more acidic.

Not only can water dissolve and transport many minerals, it also can transport solid particles in suspension. These characteristics make water a unique transportation system for Earth. Water supplies nutrients that would not otherwise be available to plants. Water carries minerals and nutrients down streams, through the soil, through the openings in subsurface rocks, and through plants and animals. It deposits solid matter on stream floodplains, in river deltas, and on the ocean floor.

The surface tension of water and the behavior of water molecules make possible **capillary action**—the ability of water to pull itself upward through small openings against gravity. Capillary action also permits transport of dissolved material in an upward direction. Capillary action moves water into the stems and leaves of plants—even to the topmost needles of the great California redwoods and the top leaves of the rainforest trees. Capillary action is also important in the movement of blood through our bodies. Without it, many of our cells could not receive the necessary nutrients carried by the blood.

Another important and highly unusual property of water is that it expands when it freezes. Most substances contract when cooled and expand when heated. Water follows these rules until it is cooled below 4°C (39°F); then it begins to expand. Ice is therefore less dense than water and consequently will float on water, as do ice floes and icebergs.

Finally, compared with solids, water is slow to heat and slow to cool. Therefore, as we saw in Chapter 4, large bodies of water on Earth act as reservoirs of heat during winter and have a cooling effect in the summer. This moderating effect on temperature can be seen in the vicinity of lakes as well as on seacoasts.

Earth's water, or the hydrosphere (from the Latin: *hydros,* water), is found in all three states: as a liquid in rivers, lakes, oceans, and rain; as a solid in the form of snow and ice; and as a gas in our atmosphere. Even the water temporarily stored in living things can be considered part of the hydrosphere. About 73 percent of Earth's surface is covered by water, with the largest proportion contained within the world's oceans (Fig. 7.1). In all, the total water content of the Earth system, whether liquid, solid, or vapor, is about 1.33 billion cubic kilometers (326 million cubic mi). Although water cycles in and out of the atmosphere, lithosphere, and biosphere, this total amount of water in the hydrosphere remains constant.

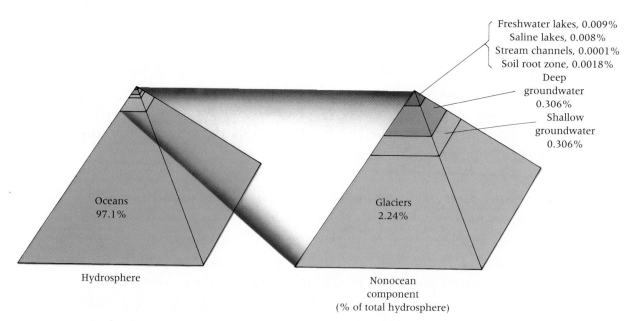

Figure 7.1 *This illustration emphasizes the fact that the vast majority of water in the hydrosphere is salt water, stored in the world's oceans. The bulk of the supply of fresh water is relatively unavailable because it is stored in polar ice caps.* • **How might global warming or cooling alter this figure?**

The Hydrologic Cycle

The circulation of water from one part of the general Earth system to another is known as the **hydrologic cycle.** The air contains water vapor that has entered the atmosphere through evaporation from Earth's surface. When water vapor condenses and falls as precipitation, several things may happen to it. First, it may go directly into a body of water, where it is immediately available for evaporation back into the atmosphere. Alternatively, it may fall onto the land, where it may run off the surface to form streams, ponds, or lakes. Or it may be absorbed into the ground, where it can either be contained by the soil or flow through open spaces, called *interstices,* that exist in loose sand, gravel, clay, and voids in solid rock. Ultimately, much of the water in or on Earth's surface reaches the oceans. Some water that reaches the surface as snow becomes a part of the large amounts of ice over Greenland and Antarctica as well as in high mountain glaciers in other parts of the world. Other water is used by plants and animals and temporarily becomes a part of living things. Thus, there are six storage areas for water in the hydrologic cycle: the atmosphere, the oceans, bodies of fresh water on the surface, plants and animals, open spaces beneath Earth's surface, and glacial ice.

Liquid water is returned to the atmosphere as a gas through evaporation. Water evaporates from all bodies of water on Earth, from plants and animals, and from soils; it can even evaporate from falling precipitation. Once the water is an atmospheric gas again, the cycle can be repeated.

The hydrologic cycle is basically one of condensation, precipitation, and evaporation. It is considered a *closed system* because there is no gain or loss of water from the system. Although it is a closed system, it is not static but exceedingly dynamic. The percentage of water associated with any one component of the system changes constantly from place to place as well as over time. For example, during the last ice age, evaporation and precipitation were greatly reduced. Also, some changes are human induced; the cutting down of a forest or the damming of a river will cause adjustments among the components.

The hydrologic cycle is one of the most important subsystems of the larger Earth system. It is linked to numerous other subsystems that rely on water as a transportation agent. For example, it plays a major role in the redistribution of en-ergy over Earth's surface. Figure 7.2 provides a schematic illustration of the circulation of water in the hydrologic cycle.

Water in the Atmosphere
The Water Budget and Its Relation to the Heat Budget

We are most familiar with, and most often take notice of, water in its liquid form, as it pours from a tap or as it exists in fine droplets within clouds or fog. Most commonly, however, water exists as a tasteless, odorless, transparent gas known as water vapor, which is mixed with the other gases of the atmosphere in varying proportions. Water vapor is found within approximately the first 5500 meters (18,000 ft) of the troposphere and makes up a small but highly variable percentage of the atmosphere by volume. Atmospheric water is the source of all condensation and precipitation. Through these processes, as well as through evaporation, water plays a significant role as Earth's temperature regulator and modifier. In addition, as we noted in Chapter 4, water vapor in the atmosphere absorbs and reflects a significant portion of both incoming solar energy and outgoing Earth radiation. By preventing great losses of heat from Earth's surface, water vapor helps to maintain the moderate range of temperature found on this planet.

As we noted previously, Earth's hydrosphere is a closed system (that is, water is neither received from outside the Earth system nor given off from it). Thus, an increase in water within one subsystem must be accounted for by a loss in another. Put another way, we say that the Earth system operates with a *water budget,* in which the total quantity of water remains the same and in which the deficits must balance the gains throughout the entire system.

We know that the atmosphere gives up a great deal of water, most obviously in clouds and through several forms of precipitation (rain, snow, hail, sleet) and condensation (fog, dew). If the quantity of water in the atmosphere remains at the same level through time, the atmosphere must be absorbing from other parts of the system an amount of water equal to that which it is giving up. During one minute, as many as one billion tons of water are given up by the atmosphere through some form of precipitation or condensation, while another billion tons are evaporated and absorbed as water vapor by the atmosphere.

(Continued on page 158)

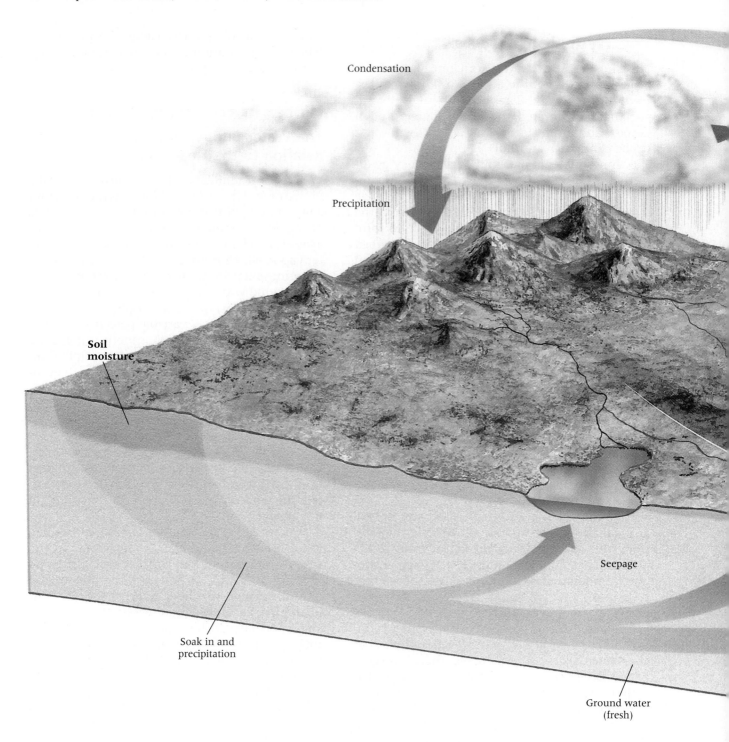

Figure 7.2 Environmental Systems: The Hydrologic System—The Water Cycle *The hydrologic system is concerned with the circulation of water from one part of the Earth system to another. The subsystem of the hydrologic system illustrated in this diagram is referred to as the hydrologic cycle. Largely through condensation, precipitation, and evaporation, water is cycled endlessly between the atmosphere, the soil, subsurface storage, lakes and streams, plants and animals, glacial ice, and the principal reservoir—the oceans.*

The radiation budget of the Earth system depicted in Figure 3.11 is an example of an open system, since energy flows into and out of that subsystem. The hydrologic cycle is a closed system; that is, water may appear in the system in all three of its major states—as liquid, gas, or solid ice—and may be transferred from one state to another, but there is no gain or loss of water by the system.

Study of the hydrologic system helps scientists understand how changes in one subsystem can greatly

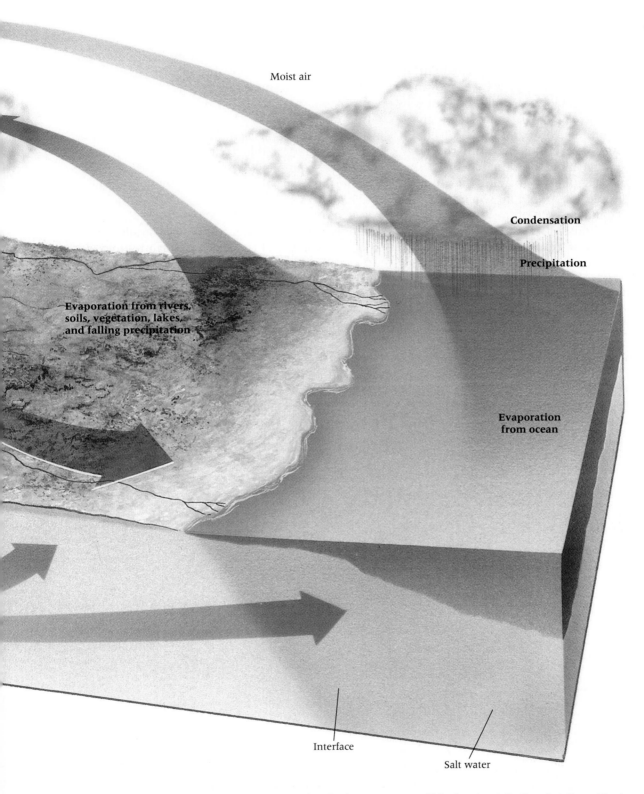

Moist air

Condensation

Precipitation

Evaporation from rivers, soils, vegetation, lakes, and falling precipitation

Evaporation from ocean

Interface

Salt water

impact other subsystems. For example, it is theorized that changes in the radiation budget of the Earth system can produce changes in the hydrologic cycle that, in turn, can explain changes in glacial subsystems and bring about continental glaciation. If the energy available to heat Earth's atmosphere were reduced, the colder temperatures would affect the hydrologic cycle. A colder atmosphere would store less water vapor. Thus, there would be less precipitation, but the colder temperatures would result in more snowfall. With time, glacial ice would increase and sea level would drop. The high albedo of ice would further reduce the energy available to heat our atmosphere, which would allow glacial ice to increase and eventually result in continental ice sheets covering large portions of Earth.

If we look again at our discussion of the heat energy budget in Chapter 4, we can see that a part of that budget is the latent heat of condensation. Of course, this energy is originally derived from the sun. The sun's energy is used in evaporation and is then stored in the molecules of water vapor, to be released only during condensation. Although the heat transfers involved in evaporation and condensation within the total heat energy budget are proportionately small, the actual energy is significant. Imagine the amount of energy released every minute when a billion tons of water condense out of the atmosphere. It is this vast storehouse of energy, the latent heat of condensation, that provides the major source of power for Earth's storms: hurricanes, tornadoes, and thunderstorms.

There are limits to the amount of water vapor that can be held by any parcel of air. The most important determinant of the amount of water vapor that can be held by the air is temperature. The warmer air is, the greater the quantity of water vapor it can hold. Therefore, we can make a generalization that air in the polar regions can hold far less water vapor (approximately 0.2 percent by volume) than the hot air of the tropics and equatorial regions of Earth, where the air can contain as much as 5 percent by volume.

Saturation and Dew Point

When air of a given temperature holds all of the water vapor that it possibly can, it is said to be in a state of **saturation** and has reached its **capacity.** If a constant temperature is maintained in a quantity of air, there will come a point, as more water vapor is added, when the air will be saturated and unable to hold any more water vapor. For example, when you take a shower, the air in the room becomes increasingly humid until a point is reached at which the air cannot contain more water. Then, excess water vapor condenses onto the colder mirrors and walls.

We know that the capacity of air to hold water vapor varies with temperature. In fact, as we can see in Figure 7.3, this capacity of air to contain moisture increases with rising temperatures. Some examples will help illustrate the relationship between temperature and water vapor capacity. If we assume that a parcel of air at 30°C is saturated, then it will contain 30 grams of water vapor in each cubic meter of air (30 g/m^3). Now suppose we increase the temperature of the air to 40°C *without* increasing the water vapor content. The parcel is no longer saturated, since air at 40°C can hold more than 30 g/m^3 of water vapor (actually, 50 g/m^3). Conversely, if we de-

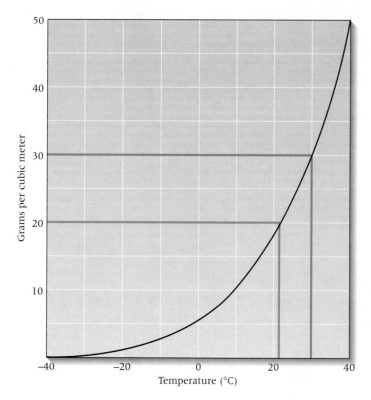

Figure 7.3 The graph shows the maximum amount of water vapor that can be contained in a cubic meter of air over a wide range of temperatures. • **Compare the change in capacity when the air temperature is raised from 0° to 10°C versus a change from 20° to 30°C. What does this indicate about the relationship between temperature and capacity?**

crease the temperature of saturated air from 30°C (which contains 30 g/m³ of water vapor) to 20°C (which has a water vapor capacity of only 17 g/m³), some 13 grams of the water vapor will be forced to condense out of the air because of the reduced capacity.

It is also evident that if an unsaturated parcel of air is cooled, it will eventually reach a temperature where the air will become saturated. This critical temperature is known as the **dew point.** For example, we know that if a parcel of air at 30°C contains 20 g/m³ of water vapor, it is not saturated, since it can hold 30 g/m³. However, if we cool that parcel of air to 21°C, it would become saturated, since the capacity of air at 21°C is 20 g/m³. Thus, that parcel of air at 30°C has a dew-point temperature of 21°C. It is the cooling of air to below its dew-point temperature that brings about the condensation that must precede precipitation.

Because the capacity of air to hold water vapor increases with rising temperatures, air in the equatorial regions has a higher dew point than does air in polar regions. Thus, because the atmosphere can hold more water in these equatorial regions, there is greater potential for large quantities of precipitation than in polar regions. Likewise, in the middle latitudes, summer months, because of their higher temperatures, have more potential for large-scale precipitation than do winter months.

Humidity

The amount of water vapor in the air at any one time and place is called **humidity.** There are three common ways to express the humidity content of the air, and each method provides information that contributes to our discussion of weather and climate.

Absolute and Specific Humidity. **Absolute humidity** is the measure of the mass of water vapor that exists within a given *volume* of air. It is expressed either in the metric system as the number of grams per cubic meter (g/m³), or in the English system as grains per cubic foot (gr/ft³). **Specific humidity** is the mass of water vapor (given in grams) per mass of air (given in kilograms). Obviously, both are measures of the actual amount of water vapor in the air. Since most water vapor gets into the air through the evaporation of water from Earth's surface, it stands to reason that absolute and specific hu-

midity will decrease with *vertical* distance from Earth.

We have also learned that air is compressed as it sinks and expands as it rises. Thus, a given parcel of air changes its volume as it moves vertically, although its weight remains the same and there may be no change in the amount of water vapor in that quantity of air. We can see, then, that absolute humidity, although it measures the amount of water vapor, can vary simply as a result of the vertical movement of a parcel of air. Specific humidity, on the other hand, changes *only* as the quantity of the water vapor changes. For this reason, when assessing the changes of water vapor content in large masses of air, which often have vertical movement, specific humidity is the preferred measurement.

Relative Humidity. **Relative humidity,** which is commonly given on television and radio weather reports, is probably the best-known means of describing the content of water vapor in the atmosphere. It is simply the ratio between the amount of water vapor in air of a given temperature and the maximum amount of vapor that the air could hold at that temperature; it expresses how close the air is to saturation.

If the temperature and absolute humidity of an air parcel are known, its relative humidity can be determined by using Figure 7.3. For instance, if we know that a parcel of air has a temperature of 30°C and an absolute humidity of 20 g/m³, we can look at the graph and determine that if it were saturated, its absolute humidity would be 30 g/m³. Then, to determine relative humidity, all we do is divide 20 grams (actual content) by 30 grams (potential content) and multiply by 100 (to get an answer in percentage):

$$(20 \text{ grams} \div 30 \text{ grams}) \times 100 = 67\%$$

The relative humidity in this case is 67 percent. In other words, the air is holding only two-thirds of the water vapor it could contain at 30°C ; it is only at 67 percent of its capacity.

There are two important factors in the *horizontal* distribution and variation of relative humidity. One of these is the availability of moisture. For example, air above bodies of water is apt to contain more moisture than similar air over land surfaces because there is simply more water available for evaporation. Conversely, the air overlying a region like the central Sahara Desert is usually very dry because it is far from the oceans and there is little water to be evaporated. The second factor in the horizontal variation of

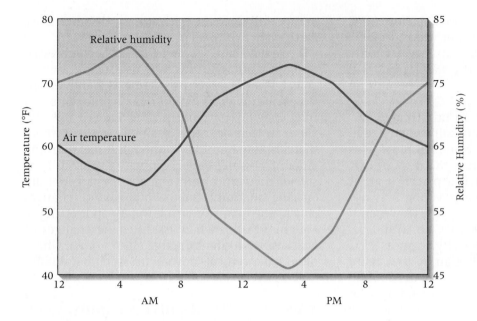

Figure 7.4 This graph illustrates the relationship between air temperature and relative humidity throughout the average day in May in Washington, DC.

relative humidity is temperature. In regions of higher temperature, relative humidity for the same amount of water vapor will be lower than it would be in a cooler region.

At any one point in the atmosphere, relative humidity varies if the amount of water vapor increases as a result of evaporation *or* if the temperature increases or decreases. Thus, although the quantity of water vapor may not change through a day, the relative humidity will vary with the daily temperature cycle. As air temperature increases from shortly after sunrise to its maximum in midafternoon, the relative humidity decreases as the air becomes capable of holding greater and greater quantities of water vapor. Then, as the air becomes cooler, decreasing toward its minimum temperature right after sunrise, the relative humidity increases (Fig. 7.4.).

Relative humidity affects our comfort through its relationship to the rate of evaporation. Perspiration evaporates into the air, leaving behind a salty residue, which you can taste if you lick your lips after sweating a great deal. Evaporation is a cooling process, since the heat used to change the perspiration to water vapor (and which becomes locked in the water vapor as latent heat) is subtracted from your skin. This is the reason that, on a hot August day when the temperature approaches 35°C (95°F), you will be far more uncomfortable in Atlanta, Georgia, where the relative humidity is 90 percent, than in Tucson, Arizona, where it may be only 15 percent at the same temperature. At 15 percent, your perspiration will be evaporated at a faster rate than at a higher relative humidity, and you will benefit from the resultant cooling effects. When the relative humidity is 90 percent, the air is nearly saturated and far less evaporation can take place.

Sources of Atmospheric Moisture

In our earlier discussion of the hydrologic cycle, we saw that the atmosphere receives water vapor through the process of evaporation. Water evaporates into the atmosphere from many different places, most important of which are the surfaces of Earth's bodies of water. Water also evaporates from wet ground surfaces and soils, from droplets of moisture on vegetation, from city pavements and other man-made surfaces, and even from falling precipitation.

Vegetation provides another source of water vapor. Plants give up water in a complex process known as **transpiration,** which can be a significant source of atmospheric moisture. A mature oak tree, for instance, can give off 400 liters (105 gal) of water per day, and a cornfield may add 11,000 to 15,000 liters (2900–4000 gal) of water to the atmosphere per day for each acre under cultivation. In some parts of the world—notably tropical rainforests of heavy, lush vegetation—transpiration accounts for a significant amount of atmospheric humidity. Together, evaporation and transpiration, or **evapotranspiration,** account for virtually all the water vapor in the air.

Rate of Evaporation

The rate of evaporation is affected by several factors. First, it is affected by the amount of accessible water. Thus, as Table 7.1 shows, the rate of evapotranspiration tends to be greater over the oceans than over the continents. The only time this generalization is not true is in equatorial regions between 0° and 10°N and S, where the vegetation is so lush on the land that transpiration provides a large amount of water for the air.

Second, temperature also affects the rate of evaporation. As the temperature of the air increases, its capacity to contain moisture also increases, providing room for additional water in the atmosphere. Also, as air temperature increases, so does the temperature of the water at the evaporation source. Such increases in temperature mean that more energy is available to the water molecules for their escape from a liquid state to a gaseous one. Consequently, more molecules can make the transition.

Mentioned previously but deserving more attention here is a third factor affecting the rate of evaporation: the degree to which the air is saturated with water vapor. The drier the air and the lower the relative humidity, the greater the rate of evaporation can be. Some of us have had direct experience with this principle. Compare the length of time it takes your bathing suit to dry on a hot, humid day with how long it takes on a day when the air is dry.

Wind, too, affects the rate of evaporation. If there is no wind, the air that overlies a water surface will approach saturation as more and more molecules of liquid water change to water vapor. Thus, evaporation will cease once saturation is reached. However, if there is a wind, it will blow the saturated or nearly saturated air away from the evaporating surface, replacing it with air of a lower humidity. This allows evaporation to continue as long as the wind keeps blowing saturated air away and bringing in drier air. Anyone who has gone swimming on a windy day has experienced the chilling effects of rapid evaporation.

Potential Evapotranspiration

So far, we have discussed actual evaporation and transpiration (evapotranspiration). However, geographers and meteorologists are also concerned with **potential evapotranspiration** (Fig. 7.5). This term refers to the idealized conditions in an area under which there would be sufficient moisture for all possible evapotranspiration to occur. Various formulas have been derived for estimating the potential evapotranspiration at a location because it is difficult to measure directly. These formulas commonly employ temperature, latitude, vegetation, and soil character (permeability, water retention ability) as factors that could affect the potential evapotranspiration.

In places where precipitation exceeds potential evapotranspiration, there is a surplus of water for storage in the ground and in bodies of water, and water can even be exported to other places if canal construction is feasible. When potential evapotranspiration exceeds precipitation, as it does during the dry summer months in California, then there is no water available for storage, and in fact the water stored during previous rainy months evaporates quickly into the warm, dry air (Fig. 7.6). Soil becomes dry and vegetation turns brown as any available water is soaked up by the atmosphere. For this reason, fires

TABLE 7.1	**Distribution of Actual Mean Evapotranspiration**					
Zone	**Latitude**					
	60°–50°	*50°–40°*	*40°–30°*	*30°–20°*	*20°–10°*	*10°–0°*
	Northern Hemisphere					
Continents	36.6 cm(14.2 in.)	33.0(13.0)	38.0(15.0)	50.0(19.7)	79.0(31.1)	115.0(45.3)
Oceans	40.0(15.7)	70.0(27.6)	96.0(37.8)	115.0(45.3)	120.0(47.2)	100.0(39.4)
Mean	38.0(15.0)	51.0(20.1)	71.0(28.0)	91.0(35.8)	109.0(42.9)	103.0(40.6)
	Southern Hemisphere					
Continents	20.0 cm(7.9 in.)	NA	51.0(20.1)	41.0(16.1)	90.0(35.4)	122.0(48.0)
Oceans	23.0(9.1)	58.0(22.8)	89.0(35.0)	112.0(44.1)	119.0(47.2)	114.0(44.9)
Mean	22.5(8.8)	NA	NA	99.0(39.0)	113.0(44.5)	116.0(45.7)

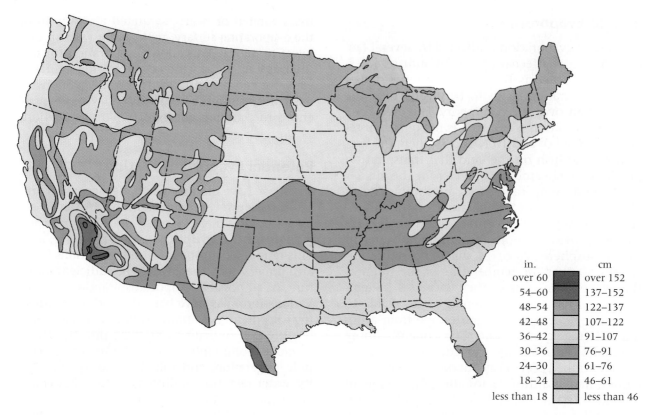

Figure 7.5 *Potential evapotranspiration for the contiguous 48 states* • **Why is potential evapotranspiration so high over the desert southwest?**

become a potential hazard during the late summer months in California.

Knowledge of potential evapotranspiration is used by irrigation engineers to learn how much water will be lost through evaporation so that they can determine whether the water that is left is enough to justify a canal. Farmers, by assessing the daily or weekly relationship between potential evapotranspiration and precipitation, can determine when and how much to irrigate their crops.

Condensation

Condensation is the process by which a gas is changed to a liquid. In our present discussion of atmospheric moisture and precipitation, condensation refers to the change of water vapor to liquid water.

Condensation occurs when air saturated with water vapor is cooled. Viewed in another way, we can say that if we lower the temperature of air until it has a relative humidity of 100 percent (the air has reached the dew point), condensation will occur with additional cooling.

It follows, then, that condensation is dependent upon (1) the relative humidity of the air and (2) the degree of cooling. In the arid air of Death Valley, California, a huge amount of cooling must take place before the dew point is reached. In contrast, on a humid summer afternoon in Mississippi, a minimal amount of cooling will bring on condensation.

This is the principle behind the formation of droplets of water on the side of a can of cold cola on a warm afternoon: The temperature of the air is lowered when it comes in contact with the cold can. Consequently, the air's capacity to hold water vapor is diminished. If air touching the can is cooled sufficiently, its relative humidity will reach 100 percent. Any cooling beyond that point will result in condensation in the form of water droplets on the can.

Condensation Nuclei

For condensation to occur, one other factor is necessary: the presence of **condensation nuclei.** These are minute particles in the atmosphere that provide a surface upon which condensation can take place. Condensation nuclei are most often sea-salt particles in the air from the evaporation of seawater. They also can be particles of dust, smoke, dirt, pollen, or volcanic material. More and more commonly, they are chemical particles that are the by-products of industrialization. The condensation that takes place on such chemical nuclei is often corrosive and dangerous to human health; when it is, we know it as smog.

Theoretically, if all such particles were removed from a volume of air, we could cool that air below its dew point without condensation occurring. Conversely, if there is a superabundance of such particles, condensation may take place at relative humidities just below 100 percent. For example, ocean fogs, which are an accumulation of condensation droplets formed on sea-salt particles, can form when the relative humidity is as low as 92 percent.

In nature, condensation appears in a number of forms. Fog, clouds, frost, and dew are all the results of condensation of water vapor in the atmosphere. The type of condensation produced depends upon a number of factors, including the cooling process itself. The cooling that produces condensation in one form or another can occur

as a result of radiation cooling, through advection, through convection, or through a combination of these processes.

Fogs

Fogs and clouds appear when water vapor condenses on nuclei and a large number of these droplets form a mass. Not being transparent to light in the way that water vapor is, these masses of condensed water droplets appear to us as fog or clouds, in any of a number of shapes and forms, and usually in shades of white or gray.

In the water budget and the hydrologic cycle, fog is a minor form of condensation. Yet, in certain areas of the world, it has important climatic effects. The "drip factor" helps to sustain vegetation along desert coastlines where fog occurs. Fog also plays havoc with our modern transportation systems. Navigation on the seas is made more difficult by fog, and air travel can be greatly impeded. In fact, fog sometimes causes major airports to shut down until visibility improves. Highway travel is also greatly hampered by heavy fogs, which can lead to huge, chain-reaction pile-ups of cars.

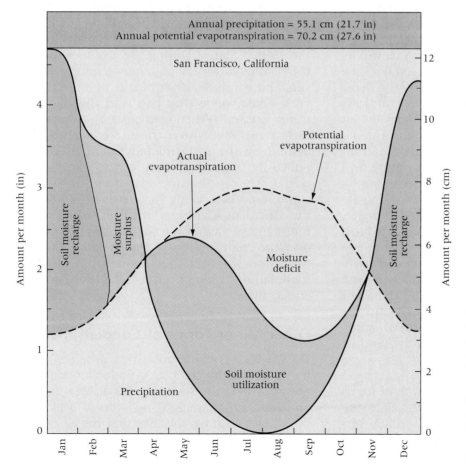

Figure 7.6 This is an example of the Thornthwaite water budget system, which "keeps score" of the balance between water input by precipitation and water loss to evaporation and transpiration, permitting month-by-month estimates of both runoff and soil moisture.
• **When would irrigation be necessary at this site?**

Radiation Fog.   Radiation cooling can lead to **surface-inversion or radiation fog.** This kind of fog is likely to occur on a cold, clear, calm winter night, usually in the middle latitudes. These conditions allow for maximum outgoing radiation from the ground with no incoming radiation. The ground gets colder and colder during the night as it gives up more and more of the heat that it has received during the day. As time passes, the air directly above is cooled by conduction through contact with the ground. Since the cold ground can cool only the lower few meters of the atmosphere, an inversion is created in which cold air at the surface is overlain by warmer air above. If this cold layer of air at the surface is cooled to a temperature below its dew point, then condensation will occur, usually in the form of a low-lying fog. However, wind strong enough to disturb this inversion layer can prevent the formation of the fog by not allowing the air to stay at the surface long enough to become cooled below its dew point.

The chances of a surface-inversion fog occurring are increased by certain types of land formations. In valleys and depressions, cold air accumulates through air drainage. During a cold night, this air can be cooled below its dew point, and a fog forms like a lake in the valley. It is common in mountainous areas to see an early morning radiation fog on the valley floor while snow-capped mountain tops shine against a clear blue sky. Radiation fogs have a diurnal cycle. They form during the night and are usually the densest right after sunrise, when temperatures are lowest. They then "burn off" during the day, when the heat from the sun slowly penetrates the fog and warms Earth. Earth in turn warms the air directly above it, increasing its temperature and consequently its capacity to hold water vapor. This greater capacity allows the fog to evaporate into the air. As Earth's heat penetrates to higher and higher layers of air, the fog continues to burn off—from the ground up!

Radiation fog often forms even more densely in industrial areas, where the high concentration of chemical particles in the air provides abundant condensation nuclei. Such a fog is usually thicker and denser than conventional "natural" radiation fogs and less easily dissipated by wind or sun.

Advection Fog.   Another common type of fog is **advection fog,** which occurs through the movement of warm, moist air over a colder surface, either land or water. When the warm air is cooled below its dew point through heat loss by radiation and conduction from the colder surface below, condensation occurs in the form of fog. Advection fog is usually less localized than radiation fog. It is also less likely to have a diurnal cycle, though if not too thick, it can be burned off early in the day to return again in the afternoon or early evening. More common, however, is the persistent advection fog that spreads itself over a large area for days at a time. Advection fog is a major reason why ski resorts are forced to close. Warm, moist air moving over the cold snow causes the dense fog.

Advection fog forms over land during the winter months in middle latitudes. It forms, for example, in the United States when warm, moist air from the Gulf of Mexico flows northward over the cold, frozen, and sometimes snow-covered upper Mississippi Valley.

During the summer months, advection fog may form over large lakes or over the oceans. Formation over lakes occurs when warm continental air flows over a colder water surface, such as when a warm air mass passes over the cool surface of Lake Michigan. An advection fog also can be formed when a warm air mass moves over a cold ocean current and the air is cooled sufficiently to bring about condensation. This variety of advection fog is known as a *sea fog*. Such a situation accounts for fogs along the West Coast of the United States. During the summer months the Pacific subtropical high moves north with the sun, and winds flow out of the high toward the West Coast where they pass over the cold California Current. When condensation occurs, fogs form that flow in over the shore, pushed from behind by the eastward movement of air and pulled by the low pressure of the warmer land (Fig. 7.7). Advection fogs also occur in New England, especially along the coasts of Maine and the Canadian Maritime Provinces, when warm, moist air from above the Gulf Stream flows north over the colder waters of the Labrador Current. Advection fog over the Grand Banks off Newfoundland has long been a hazard for cod fishermen there.

Other Minor Forms of Condensation

Dew, which is made up of tiny droplets of water, is formed by the condensation of water vapor at or near the surface of Earth. Dew collects on surfaces that are good radiators of heat (your car, blades of grass, the bike that stayed outside last night). These good radiators give up large

Figure 7.7 *An advection sea fog caused by warm, moist air passing over colder water.*
• **What unique problems might coastal residents face as a result of sea fogs?**

amounts of heat during the night hours when there is no incoming solar radiation. When the air comes in contact with these cold surfaces, it cools, and if cooled sufficiently, droplets of water will form as beads on the surface. When the temperature of the air is below 0°C (32°F), **white frost** forms. It is important to note that frost is not frozen dew but instead represents a sublimation process—water vapor changing directly to the frozen state.

Sometimes under very still conditions with low air pressure, though air temperatures may be below 0°C (32°F), the liquid droplets that make up clouds or fogs are not frozen into solid particles. When such *supercooled* water droplets come in contact with a surface, like the edge of an airplane wing or a tree branch or a window, ice crystals are created on that surface in a formation known as **rime.**

Clouds

Clouds are the most common form of condensation and are important for several reasons. First, they are the source of all precipitation. **Precipitation** is made up of condensed water particles, either liquid or solid, which fall to Earth. Obviously, not all clouds result in precipitation, but we cannot have precipitation without the formation of a cloud first. Also, clouds serve an important function in the heat energy budget. We have already noted that clouds absorb some of the incoming solar energy. They also reflect some of that energy back to space and scatter and diffuse other parts of the incoming energy before it strikes Earth as diffuse radiation. In addition, clouds absorb some of Earth's radiation so that it is not lost to space, and then reradiate it back to the surface. Finally, clouds are a beautiful and ever-changing aspect of our environment. The colors of the sky and the variations in the shapes and hues of clouds have provided us all with a beautiful backdrop to the natural scenery here on Earth.

Adiabatic Heating and Cooling. The cooling process that leads to cloud formation is quite different from that associated with the other condensation forms we have already examined. The cooling process that produces fog, frost, and dew is either radiation or advection. On the other hand, clouds usually develop from a cooling process that results when a parcel of air on Earth's surface is lifted into the atmosphere.

The rising parcel of air will expand as it encounters decreasing atmospheric pressure with height. This expansion allows the air molecules to spread out, which causes the parcel's temperature to decrease. This is known as **adiabatic cooling** and occurs at the constant lapse rate of approximately 10°C/1000 meters (5.6°F/1000 ft). By the same token, air descending through the atmosphere is compressed by the increasing pressure and undergoes **adiabatic heating** of the same magnitude.

However, the rising and cooling parcel of air will eventually reach its dew-point temperature at which water vapor begins to condense out, forming cloud droplets. From this point on, the adiabatic cooling of the rising parcel will decrease as latent energy released by the condensation process is added to the air. To differentiate between these two adiabatic cooling rates, we refer to the precondensation rate (10°C/100 m) as the **dry adiabatic lapse rate** and the lesser, after-condensation rate, as the **wet adiabatic lapse rate.** Although the latter averages 5°C/1000 meters (3.2°F/1000 ft), it varies according to the amount of water vapor that condenses out of the air.

A rising air parcel will cool at one of these two adiabatic rates. Which rate is in operation depends upon whether condensation is (wet adiabatic rate) or is not (dry adiabatic rate) occurring. On the other hand, the warming temperatures of descending air allow it to hold greater quantities of water vapor. Consequently, condensation will not occur, so the heat of condensation will not affect the rate of rise in temperature. Thus, the temperature of air that is descending and being compressed always *increases* at the dry adiabatic rate.

It is important to note that adiabatic temperature changes are the result of volumetric change and do not involve the addition or subtraction of heat from external sources.

It is also extremely important to differentiate between the *environmental lapse rate* and *adiabatic lapse rates*. In Chapter 4 we found that, in general, the temperature of our atmosphere decreases with increasing height above Earth's surface—called the *normal lapse rate* or environmental lapse rate. Although it averages 6.5°C/1000 meters (3.6°F/1000 ft), this rate is quite variable and must be measured through the use of meteorological instruments sent aloft. Whereas the environmental lapse rate reflects nothing more than the vertical temperature structure of the atmosphere, the adiabatic lapse rates are concerned with temperature changes as a parcel of air moves through the atmospheric layers (Fig. 7.8).

Stability and Instability.   Although adiabatic cooling results in the development of clouds, the various forms of clouds are related to differing degrees of vertical air movement. Some clouds are associated with rapidly rising, buoyant air, while other forms result when air resists vertical movement.

An air parcel will rise of its own accord as long as it is warmer than the surrounding atmospheric air. When it reaches a layer of the atmosphere that is the same temperature as itself, it will stop rising. Thus, an air parcel warmer than the surrounding atmospheric air will rise and is said to be **unstable.** On the other hand, an air parcel that is colder than the surrounding atmospheric air will resist any upward movement and is said to be **stable.**

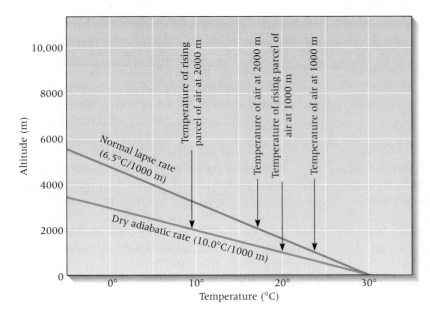

Figure 7.8 *Comparison of the dry adiabatic lapse rate and the normal lapse rate. The normal lapse rate is the average vertical change in temperature. Air displaced upward will cool (at the dry adiabatic rate) because of expansion.*
• **In this example, what is the temperature of the layer of air at 4000 meters?**

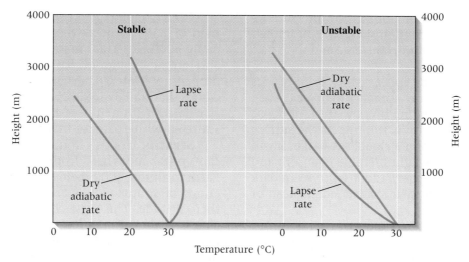

Figure 7.9 *Relationship between lapse rates and air-mass stability. When air is forced to rise, it cools adiabatically. Whether it continues to rise or resist vertical motion depends on whether adiabatic cooling is less rapid or more rapid than the prevailing vertical temperature lapse rate. If the adiabatic cooling rate exceeds the lapse rate, the lifted air will be colder than its surroundings and will tend to sink when the lifting force is removed. If the adiabatic cooling rate is less than the lapse rate, the lifted air will be warmer than its surroundings and will therefore be buoyant, continuing to rise even after the original lifting force is removed.* • **In this example, what would be the temperature of the air if it rose to 2000 meters?**

Determining the stability or instability of an air parcel involves nothing more than asking the question, If an air parcel were lifted to a specific elevation (cooling at an adiabatic lapse rate), would it be warmer, colder, or the same temperature as the atmospheric air (determined by the environmental lapse rate at that time) at that same elevation?

If the air parcel is warmer than the atmospheric air at the selected elevation, then the parcel would be unstable and continue to rise, since warmer air is less dense and, therefore, buoyant. Thus, under conditions of **instability** the environmental lapse rate must be *greater than* the adiabatic lapse rate in operation. For example, if the environmental lapse rate is 12°C/1000 meters and the ground temperature is 30°C, then the atmospheric air temperature at 2000 meters would be 6°C. On the other hand, an air parcel (assuming no condensation occurs) lifted to 2000 meters would have a temperature of 10°C. Since the air parcel is warmer than the atmospheric air around it, it is unstable and will continue to rise (Fig. 7.9). However, let us assume that it is another day and all the conditions are the same except that measurements indicate the environmental lapse rate on this day is 2°C/1000 meters. Consequently, although our air parcel if lifted to

2000 meters would still have a temperature of 10°C, the temperature of the atmosphere at 2000 meters would now be 26°C. Thus, the air parcel would be colder and would sink back toward Earth as a result of its greater density (see Fig. 7.9). As you can see, under conditions of **stability,** the environmental lapse rate is *less than* the adiabatic lapse rate in operation. If an air parcel, upon being lifted to a specific elevation, has the same temperature as the atmospheric air surrounding it, it is neither stable nor unstable. Instead, it is considered **neutral** and will neither rise nor sink but will remain at that elevation.

Whether an air parcel will be stable or unstable is related to the amount of cooling and heating of air at Earth's surface. With chilling of the air through radiation and conduction on a cool, clear night, air near the surface will be relatively close in temperature to that aloft, and the environmental lapse rate will be low, thus enhancing stability. With the rapid heating of the surface on a hot summer day, there will be a very steep environmental lapse rate because the air near the surface is so much warmer than that above, and instability will be enhanced.

Pressure zones can also be related to atmospheric stability. In areas of high pressure, stability is maintained by the slow subsiding of relatively cool air from aloft. In low-pressure regions, on the other hand, instability is promoted by the tendency for air to converge and then rise.

Cloud Forms.   Clouds appear white or in shades of gray, even deep gray approaching black. They differ in color depending on how thick or dense they are and if the sun is shining on the surface that we see. The thicker a cloud, the darker it will appear, for the more light it is able to absorb and thus block from our view. Clouds also seem dark when we are seeing their shaded side instead of their sunlit side.

There are three basic forms of clouds: cirrus, stratus, and cumulus. Classification systems categorize these cloud formations into many subtypes; however, most such subtypes are overlaps of the three basic forms. Figure 7.10 illustrates the appearance and the general heights of common clouds; Figure 7.11 provides a pictorial summary of the major cloud types.

Cirrus clouds (from Latin: *cirrus,* a lock or wisp of hair) form at very high altitudes, normally 6000 to 10,000 meters (19,800–36,300 ft), and are made up of ice crystals rather than droplets of water. They are thin, wispy, white clouds that trail like feathers across the sky. When associated with fair weather, cirrus clouds are scattered white patches in a clear blue sky.

Stratus clouds (from Latin: *stratus,* layer) can appear anywhere from near the surface of Earth to almost 6000 meters (19,800 ft). The variations of stratus clouds are based in part upon their altitude. The basic characteristic of stratus clouds is their horizontal sheetlike appearance, lying in layers with fairly uniform thickness. The horizontal configuration indicates they form in stable atmospheric conditions, which inhibit vertical development.

Often stratus clouds cover the entire sky with a gray cloud layer. It is stratus clouds that make up the dull, gray, overcast sky common to win-

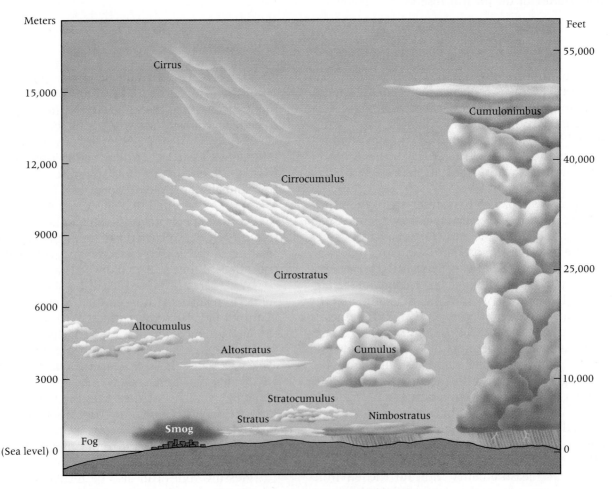

Figure 7.10 Clouds are named based upon their form and height. • **Observe this figure and Figure 7.11; what cloud type is present in your area today?**

ter days in much of the midwestern and eastern United States. The stratus cloud formation may overlie an area for days and any precipitation will be steady and persistent.

Cumulus, or **cumuliform,** clouds (from Latin: *cumulus,* heap or pile) develop vertically rather than forming the more horizontal structures of the cirrus and stratus forms. Cumulus are massive piles of clouds, usually with a flat base, which can be anywhere from 500 to 12,000 meters (1650–39,600 ft) above sea level. From this base, they pile up into great rounded structures, often with tops like cauliflowers. The cumulus cloud is the visible evidence of an unstable atmosphere; its base is the point where condensation has begun in a column of air as it moves upward.

Another term used in describing clouds is **nimbus,** meaning precipitation (rain is falling). Thus there is the nimbostratus cloud, which may bring a long-lasting drizzle, and the cumulonimbus or the thunderhead. This latter cloud has a flat top, called an anvil head, as well as a flat base, and it becomes darker as condensation within it increases and blocks the sun. The cumulonimbus is the source of the gusty winds, torrential rain, and lightning common on hot afternoons in humid regions.

Precipitation

Condensed droplets within cloud formations stay in the air and do not fall to Earth because of their general buoyancy and the upward movement of the air. These droplets of condensation are so minute that they are kept floating in the cloud formation, since their mass and the consequent pull of gravity are insufficient to overcome the buoyant effects of air and the vertical currents, or updrafts, within the clouds.

Precipitation occurs when the droplets of water, ice, or frozen water vapor coalesce and develop masses too great to be held above Earth. They then fall to Earth as rain, snow, hail, or sleet. The form that precipitation takes largely depends upon the method of formation and the temperature during formation.

Forms of Precipitation

Rain, consisting of droplets of liquid water, is by far the most common form of precipitation. In most cases raindrops develop through the grouping of water droplets around ice crystals formed within cumulus clouds that reach into altitudes with subfreezing temperatures. However, rain is also produced by clouds that do not reach such high altitudes. To explain the development of raindrops within these clouds, meterologists suggest the fusion of small droplets into bigger drops, which in beginning their fall attract even more droplets. It has been estimated that one raindrop is due to the coalescence of about one million cloud droplets. Raindrops vary in size but are generally about 2.5 to 6 millimeters (0.1–0.25 in.) in diameter. As we all know, rain can come in many ways: as a brief afternoon shower, a steady three-day rainfall, or the deluge of a tropical rainstorm. When the temperature of an air mass is only slightly below the dew point, the raindrops may be very small (about 0.5 millimeters or less in diameter) and close together. The result is a fine mist or haze called **drizzle.** Drizzle is so light that it is greatly affected by the direction of air currents and the variability of winds. Consequently, drizzle seldom falls vertically.

Snow is the second most common form of precipitation. When water vapor is frozen directly into a solid without first passing through a stage as liquid water, it forms minute ice crystals around certain types of nuclei, and these crystals are usually in six-sided, symmetric shapes. Combinations of these ice-crystal shapes make up the intricate and delicate patterns of snowflakes.

Sleet is frozen rain, formed when rain, in falling to Earth, passes through a cold layer of air and freezes. The result is the creation of solid particles of clear ice. In English-speaking countries outside the United States, sleet refers not to this phenomenon of frozen rain but rather to a mixture of rain and snow.

Hail is a less common form of precipitation than the three just described. It occurs most often during summer months and is the result of certain phenomena that are especially peculiar to the cumulonimbus cloud form. Hail appears as rounded lumps of ice, called **hailstones,** which can vary in size from 5 millimeters (0.2 in.) in diameter up to the size of a baseball. The world record is a hailstone 30 centimeters (12 in.) in diameter that fell in Australia. Dropping from the sky, hailstones can be highly destructive to crops and other vegetation, as well as to cars and buildings. Hailstones have even been known to kill animals and humans. Children think it is strange that they must leave a pool or lake where they are swimming simply because hailstones are falling, but this is a sensible precaution because the atmospheric conditions that produce hailstones also produce thunder and lightning. In

Cirrocumulus

Cirrostratus

Altostratus

Altocumulus

Fog

Cumulonimbus

***Figure 7.11** Cloud types.*

fact, these phenomena often occur in conjunction with one another.

Hail forms when ice crystals are lifted by strong updrafts in a cumulonimbus cloud. Then, as these ice crystals fall through the cloud, su-

percooled water droplets attach themselves and are frozen in a layer. Sometimes these pellets are lifted up into the cold layer of air and then dropped again and again. The resulting hailstone, made up of concentric layers of ice, has a frosty,

Sun through stratus

Cirrus

Cumulus

Nimbostratus

Stratocumulus

opaque appearance when it finally breaks out of the strong updrafts of the cloud formation and falls to Earth.

On occasion, a raindrop can form and have a temperature below 0°C (32°F). These super-cooled droplets will freeze the instant they fall onto a cold surface. The resulting icy covering on trees, plants, and wires is known as **glaze.** People usually call the rain and its blanket of ice an "ice storm." Because of the weight of ice, glazing can break off large branches of trees, bringing down telephone and power lines. It can also make roads practically impassable. A small counterbalance against the negative effects of glazing is the beauty of the natural landscape after an ice storm. Against the background of a clear blue sky, sunlight catches on the ice, reflecting and making diamonds out of the most ordinary weeds and tree branches.

Conditions Causing Precipitation

Adiabatic cooling is the only process that can lower the temperature of a large mass of air to its dew point in order to produce sufficient condensation for precipitation, and adiabatic cooling can take place only in an air mass that rises and expands.

There are three major ways in which a parcel of air may be forced to rise, and each of these produces its own characteristic types of precipitation (Fig. 7.12). There is **convectional precipitation,** which results from the displacement of warm air upward in a convectional system. **Cyclonic** or **frontal precipitation** takes place when a warm air mass rises after encountering a colder and denser air mass. In **orographic precipitation,** an air mass encounters a land barrier, usually a mountain, and must rise above it in order to pass.

Convectional Precipitation. The simple explanation of convection is that when air is heated near the surface, it expands, becomes lighter, and rises. It is displaced by the cooler, denser air around it to complete the convection cycle. The important factor in convection for our discussion of precipitation is that the heated air rises and thus fulfills the one essential criterion for significant condensation and ultimately precipitation.

To enlarge our understanding of convectional precipitation, let us apply what we have learned about instability and stability. Figure 7.13 illustrates two cases where air rises due to convection. But case (a) is quite different from case (b). In both, the lapse rate in the free atmosphere is the same, and it is especially high during the first few thousand meters, but slows after that (as on a hot summer day). In case (a) the air parcel is not very humid, and thus the dry adiabatic rate applies throughout its ascent. By the time the air reaches 3000 meters (9900 ft), its temperature and density are the same as the surrounding atmospheric air. At this point convectional lifting ceases.

In case (b) we have introduced the latent heat of condensation. As in case (a), the as yet unsaturated rising column of air cools at the dry

adiabatic rate of 10°C/1000 meters (5.6°F/1000 ft) for the first 1000 meters (3300 ft). However, since the air parcel is humid, the dew point is soon reached in the rising air column, condensation takes place, and cumulus clouds begin to form. As condensation occurs, the heat locked up in the water vapor is released and heats the moving parcel of air, retarding the adiabatic rate of cooling so that the rising air is now cooling at the wet adiabatic rate (5°C/1000 meters). Hence, the temperature of the rising air parcel remains warmer than the atmospheric air it is passing through, and the air parcel will continue to rise and rise and rise. It is obvious that in case (b), which incorporates the latent heat of condensation, we have massive condensation, towering

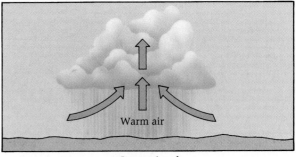
Convectional

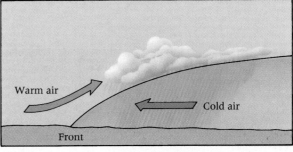
Cyclonic (frontal)

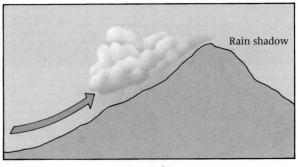
Orographic

Figure 7.12 The principal causes of precipitation are upward movement of moist air resulting from convection, frontal activity, and orographic lifting. • **What do all three diagrams have in common?**

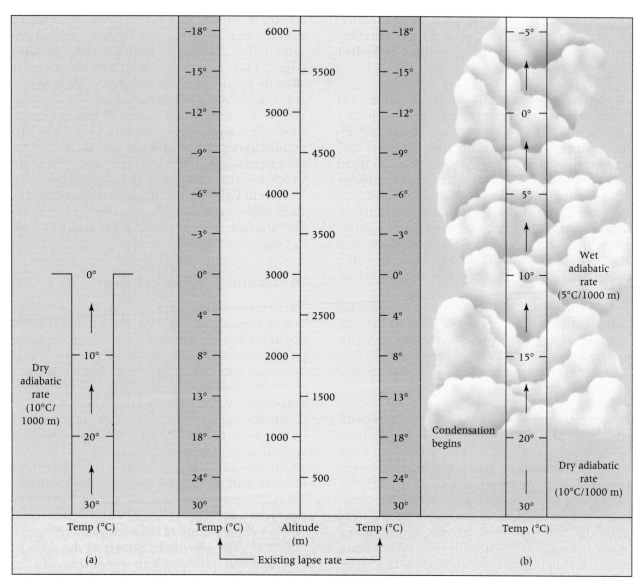

Figure 7.13 *Effect of humidity on air-mass stability. In **(a)**, warm, dry air rises and cools at the dry adiabatic rate, soon becoming the same temperature as the surrounding air, at which point convectional uplift terminates. Since the rising dry air did not cool to its dew-point temperature by the time convectional lifting ended, no cloud formed. In **(b)**, rising warm, moist air soon cools to its dew-point temperature. The upward moving air subsequently cools at an assumed average wet adiabatic rate, which keeps the air warmer than the surrounding atmosphere so that the uplift continues. Only when all moisture is removed by condensation will the air cool rapidly enough at the dry adiabatic rate to become stable.* • **What would be necessary for the cloud in (b) to stop its upward growth at 4500 meters?**

cumulus clouds, and potentially heavy precipitation.

Convectional precipitation is most common in the humid equatorial and tropical areas that receive much of the sun's energy and during summers in middle latitudes. Though differential heating of land surfaces plays an important role in convectional precipitation, it is not the sole factor.

Other factors, such as surface topography and atmospheric dynamics associated with the upper air winds, may provide the initial upward push for air that is potentially unstable. Once condensation begins in a convectional column, additional energy is available from the latent heat of condensation for further lifting.

It is such convectional lifting that can result in the heavy precipitation, thunder, lightning, and tornadoes of summer afternoon thunderstorms. When the convectional currents are strong in the characteristic cumulus clouds, hail can result.

Frontal Precipitation. The zones of contact between relatively warm and relatively cold bodies of air are known as **fronts.** When two large bodies of air with masses of different densities and temperatures meet, the warmer one is lifted above the colder. When this happens, the major criterion for large-scale condensation and precipitation is once again met. Frontal precipitation thus occurs as the moisture-laden warm air rises above the front caused by contact with the cold air. Continuous frontal precipitation caused the devastating floods of the Mississippi River in 1993.

For fronts to be fully understood, we must examine what causes unlike bodies of air to come together, and what happens when they do. This will be discussed in Chapter 8, where we will take a more detailed look at frontal disturbances and precipitation.

Orographic Precipitation. As was the case with convectional rainfall, there is a simple definition of orographic rainfall and a somewhat more complex explanation. When land barriers—such as mountain ranges, hilly regions, or even the escarpments (steep edges) of plateaus or tablelands—lie in the path of prevailing winds, large portions of the atmosphere are forced to rise above these barriers. This fills the one main criterion for significant precipitation—that large masses of air are cooled by ascent and expansion until large-scale condensation takes place. The resultant precipitation is termed orographic (from Greek: *oros,* mountains). As long as the air parcel rising up the mountainside remains stable (cooling at a greater rate than the environmental lapse rate), any resulting cloud cover and precipitation will be stratiform. However, the story is not really complete until we realize that the situation can be complicated by the same circumstances described in case (b) of Figure 7.13. A potentially unstable air parcel may need only the initial lift provided by the orographic barrier to set it in motion. In this case it will continue to rise of its own accord (no longer forced) as it seeks air of its own temperature and density. The land barrier only serves to provide the initial thrust; it has performed its function as a lifting mechanism.

Because the air has deposited most of its moisture on the windward side of the mountain, there will normally be a great deal less precipitation on the leeward side, since on this side the air will be much drier and the dew point consequently much lower. The leeward side of the mountain is thus said to be in the **rain shadow** (Fig. 7.14a). Just as being in the shade, or in shadow, means that you are not receiving any direct sun, so being in the rain shadow means that you do not receive much rain. If you live near a mountain range, you can see the effects of orographic precipitation and the rain shadow in the pattern of vegetation (Figs. 7.14b and 7.14c). The windward side of the mountains (say, the Sierra Nevada in California) will be heavily forested and thick with vegetation. The opposite slopes in the rain shadow will usually be drier and the cover of vegetation sparser.

Distribution of Precipitation

There are different ways to describe the precipitation a region receives. We can look at its average annual precipitation to get an overall picture of the amount of moisture it gets during a year. We could also look at its number of raindays (in a rainday, 0.25 millimeter [0.01 in.] or more of rain is received during 24 hours). If we were to divide the number of raindays in a year or month by the total number of days in that period, we would have a percentage that would be the probability of rain. Such a measure is important to farmers and to ski or summer resort owners whose incomes may depend on precipitation or the lack of it.

We can also look at the average monthly precipitation. This provides a picture of the seasonal variations in precipitation (Fig. 7.15). For instance, in describing the climate of the west coast of California, we would not be giving the full story were we just to give the average annual precipitation, for this figure would not show the distinct wet and dry seasons that characterize this region.

Horizontal Distribution of Precipitation. Figure 7.16 shows average annual precipitation for the world's continents. We can see that there is great variability in the distribution of precipitation over Earth's surface. Although there is a zonal distribution of precipitation related to latitude, this distribution is obviously not the only factor involved in the amount of precipitation an area receives.

The likelihood and amount of precipitation are based on two factors. First, precipitation depends on the degree of lifting that occurs in air of a particular region. This lifting, as we have already seen, may be due to the convergence of different air masses, to differential heating of Earth's surface, to the lifting that results when an air

(Continued on page 178)

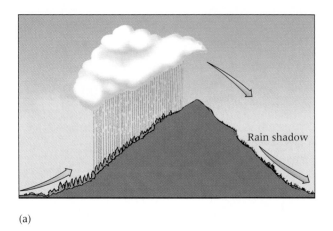

(a)

(b)

(c)

Figure 7.14 *(a) and (b) Orographic uplift over the windward (western) slope of the Sierras produces condensation, cloud formation, precipitation, and the resulting dense stands of forest. (c) Semiarid or rain-shadow conditions occur on the leeward (eastern) slope of the Sierras.* • **Can you identify a mountain range in Eurasia in which the leeward side of that range is in the rain shadow?**

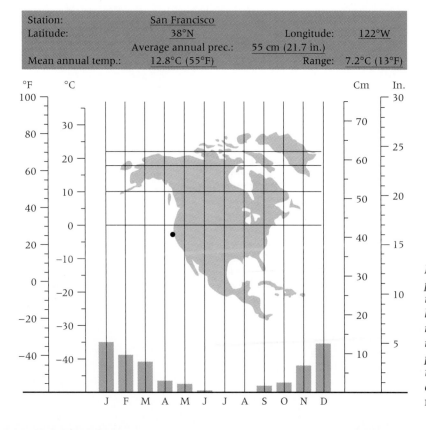

Station:	San Francisco		
Latitude:	38°N	Longitude:	122°W
	Average annual prec.:	55 cm (21.7 in.)	
Mean annual temp.:	12.8°C (55°F)	Range:	7.2°C (13°F)

Figure 7.15 *Average monthly precipitation in San Francisco, California, is represented by colored bars along the bottom of the graph. Such a graph of monthly precipitation figures gives a much more accurate picture than the annual precipitation total, which does not tell us that nearly all the precipitation occurs in only half of the year.* • **How would this rainfall pattern affect agriculture?**

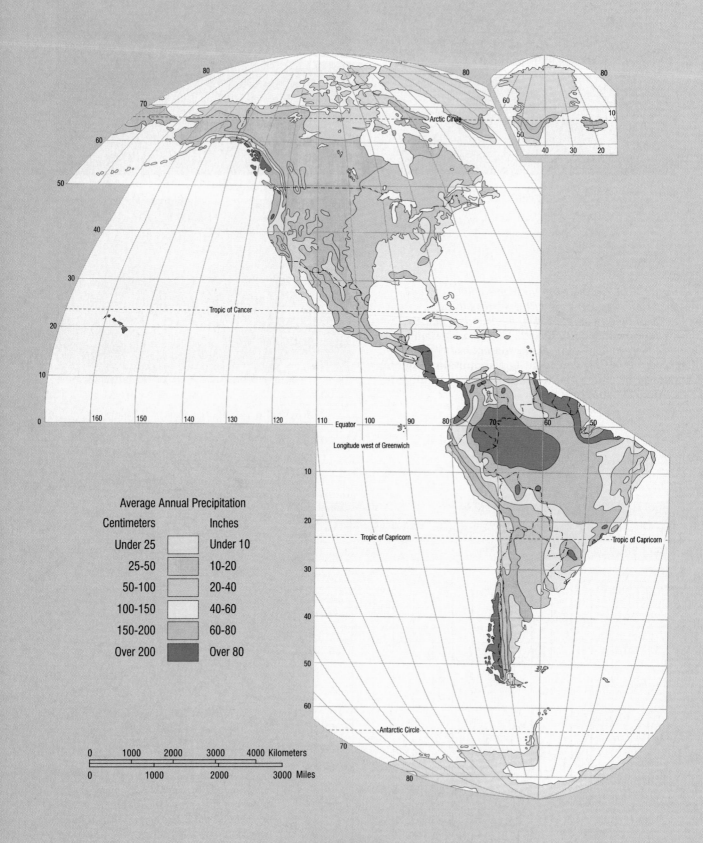

Figure 7.16 *World Map of Average Annual Precipitation.* • In general, where on Earth's surface does the heaviest rainfall occur? Why?

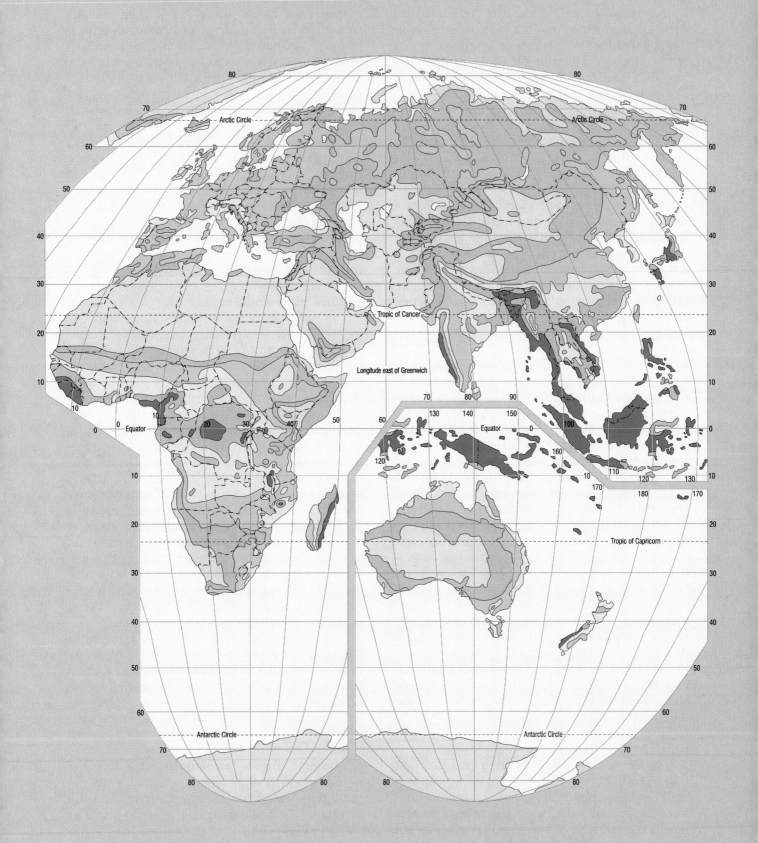

A Western Paragraphic Projection
developed at Western Illinois University

mass encounters a rise in Earth's surface, or to a combination of these factors. Also, precipitation depends on the internal characteristics of the air itself, which include its degree of instability, its temperature, and its humidity.

Since higher temperatures, as we have seen, allow air masses to hold greater amounts of water vapor, and since, conversely, cold air masses can hold less water vapor, we can expect a general decrease of precipitation from the equator to the poles that is related to the unequal zonal distribution of incoming solar energy discussed in Chapter 3.

However, if we look again at Figure 7.16, we see that there is a great deal of variability in average annual precipitation beyond the general pattern of a decrease with increased latitude. In the following discussion, we will examine some of these variations and give the reasons for them. We will be applying what we have already learned about temperature, pressure systems, wind belts, and precipitation.

Distribution Within Latitudinal Zones.   The equatorial zone is generally an area of high precipitation (over 200 centimeters, or 79 inches, annually), largely due to the zone's high temperatures, high humidity, and the instability of its air. High temperatures and instability lead to a general pattern of rising air, which in turn allows for precipitation. This tendency is strongly reinforced by the convergence of the trades as they move toward the equator from opposite hemispheres. In fact, the Intertropical Convergence Zone is one of the two great zones where air masses converge. (The other is the polar front within the westerlies zone.)

In general, the air of the trade wind zones is stable compared with the instability of the equatorial zone. Under the control of these steady winds, there is little in the way of atmospheric disturbances to lead to convergent or convectional lifting. However, since the trade winds are basically easterlies, when they move onshore along east coasts or high islands they bring moisture from the oceans with them. Thus, within the trade wind belt, continental east coasts tend to be wetter than continental west coasts.

In fact, where the air of the equatorial and trade wind regions—with its high temperatures and vast amounts of moisture—moves onshore from the ocean and meets a landform barrier, record rainfalls can be measured. The windward slope of Mount Waialeale on Kauai, Hawaii, at approximately 22°N latitude, holds the world's record for greatest average annual rainfall—1160 centimeters (460 in.).

Moving poleward from the trade wind belts, we enter the zones of subtropical high pressure where the air is subsiding. As it sinks lower, it is warmed adiabatically, increasing its moisture-holding capacity and consequently reducing the amount of precipitation in this area. In fact, if we look at Figure 7.17, which shows average annual precipitation on a latitudinal basis, we can see a dip in precipitation level corresponding to the latitude of the subtropical high-pressure belts and cells. These areas of subtropical high pressure are, in fact, where we find most of the great deserts of the world: in northern and southern Africa, Arabia, North America, and Australia. The exceptions to this subtropical aridity occur along the eastern sides of the landmasses, where, as we have already noted, the subtropical high-pressure cells are weak and wind direction is often onshore. This exception is especially true of regions affected by the monsoons.

In the zones of the westerlies, from about 35° to 65°N and S latitude, precipitation occurs largely as a result of the meeting of cold, dry polar air masses and warm, humid subtropical air masses along the polar front. Thus, there is much cyclonic or frontal precipitation in this zone.

Naturally, the continental interiors of the middle latitudes are drier than the coasts because they are farther away from the oceans. Furthermore, where air in the prevailing westerlies is forced to rise, as it is when it crosses the Cascades and Sierra Nevada of the Pacific Northwest and California, especially during the winter months, there is heavy orographic precipitation. Thus, in the middle latitudes, continental west coasts tend to be wet and precipitation decreases with movement eastward toward continental interiors. Along eastern coasts within the westerlies, precipitation usually increases once again because of proximity to humid air from the oceans.

In the United States, the interior lowlands are not as dry as we might expect within the prevailing westerlies. This is because of the great amount of frontal activity resulting from the conflicting northward and southward movements of polar and subtropical air. If there were a high east-west mountain range extending from central Texas to northern Florida, the lowlands of the continental United States north of that range would be much drier than they actually are because they would be cut off from moist air originating in the Gulf of Mexico.

Also characteristic of the belt of the westerlies are desert areas that occur in the rain shadows of prevailing winds that are forced to rise over mountain ranges. This effect is in part responsible for the development and maintenance

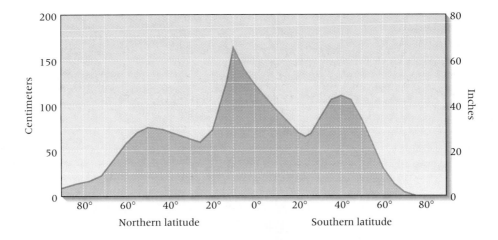

Figure 7.17 *Latitudinal distribution of average annual precipitation. This figure illustrates the four distinctive precipitation zones: high precipitation caused by convergence of air in the tropics and in the middle latitudes along the polar front, and low precipitation caused by subsidence and divergence of air in the subtropical and polar regions.* • **Compare this figure with Figure 5.2. What is the relationship between world rainfall patterns and world pressure distribution?**

of California's Death Valley as well as the desert zone of eastern California and Nevada in the United States, the mountain-ringed deserts of eastern Asia, and Argentina's Patagonian Desert, which is in the rain shadow of the Andes. Note in Figure 7.17 that there is greater precipitation in the middle latitudes of the Southern Hemisphere than there is in the Northern. This occurs largely because there is a lot more ocean and less landmass in the Southern Hemisphere westerlies than in the corresponding zone of the Northern Hemisphere.

Moving poleward, we find that temperatures decrease along with the moisture-holding capacity of the air. The low temperatures also lead to low evaporation rates and, in addition, the air in the polar regions shows a general pattern of subsidence that yields areas of high pressure. This settling of the air in the polar regions is the opposite of the lifting needed for precipitation. All these factors combine to cause low precipitation values in the polar zones.

Variability of Precipitation

The rainfall depicted in Figure 7.16 is an annual average. It should be remembered, however, that for many parts of the world, there are significant variations in precipitation, both within any one year and between years. For example, areas like the Mediterranean region, California, Chile, South Africa, and Australia, which are on the west sides of the continents and roughly between 30° and 40° latitude, get much more rain in the winter than in the summer. And there are areas between 10° and 20° latitude that get much more of their precipitation in the summer (high-sun season) than in the winter (low-sun season).

Rainfall totals can change markedly from one year to the next, and, tragically for many of the world's people, the drier a place is on the average, the greater will be the statistical variability in its precipitation (compare Fig. 7.18 with Fig. 7.16). To make matters worse for people in dry areas, a year with a particularly high amount of rainfall may be balanced with several years of below-average precipitation. This situation has occurred recently in West Africa's Sahel, the Russian steppe, and the American Great Plains.

Thus, there are years of drought and years of flood, each bringing its own kind of disaster upon the land. Farmers, resort owners, construction workers, and others whose economic well-being depends in one way or another on precipitation or the lack of it can determine only a probability of rainfall on an annual, monthly, or even a seasonal basis.

Meteorologists are not able to predict rainfall with 100 percent accuracy. This inability is due to the many factors involved in causing precipitation (temperature, available moisture, atmospheric disturbances, landform barriers, frontal activity, air-mass movement, upper air winds, and differential surface heating, among others). In addition, the interaction of these factors in the development of precipitation is very complex and not completely understood.

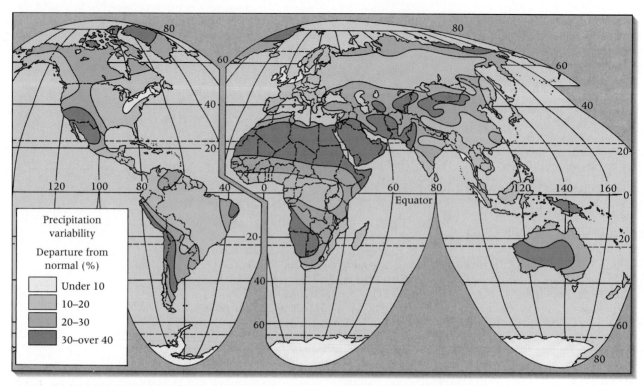

Figure 7.18 *Precipitation variability. The greatest variability in year-to-year precipitation totals is in the dry regions, accentuating the critical problem of moisture supply in those parts of the world.*

Define and Recall

capillary action	radiation fog	instability	glaze
hydrologic cycle	advection fog	stability	convectional
saturation	dew	cirrus	precipitation
capacity	white frost	stratus	cyclonic (frontal)
dew point	rime	cumulus	precipitation
absolute humidity	precipitation	nimbus	orographic precipitation
specific humidity	adiabatic cooling (and		rain shadow
relative humidity	heating)	drizzle	
	dry adiabatic lapse rate	sleet	
evapotranspiration	wet adiabatic lapse rate	hail	

Discuss and Review

1. How is the hydrologic cycle related to Earth's water budget?

2. Is the saturation or dew point in your area generally higher or lower than that in Siberia? Honduras?

3. What is the difference between absolute and specific humidity? What is relative humidity?

4. Follow the weather report in your area for a week by watching TV, listening to the radio, or reading the newspapers. What figures were given for temperature and humidity? What did each day feel like in terms of your own comfort with the weather?

5. Why is the inside of a greenhouse generally more humid than an ordinary room?

6. Why is wearing wet clothes sometimes bad for your health? Under what conditions might wearing wet clothes be good for your health?

7. Imagine that you are deciding when, in your daily schedule, to water the garden. What time of day would be best for conserving water? Why?

8. What factors must be taken into consideration when calculating potential evapotranspiration? How is evapotranspiration related to the water budget of a region?

9. What factors affect the formation of ground-inversion fogs?

10. What kinds of fogs occur in the area where you live?

11. What causes adiabatic cooling? Differentiate between the environmental lapse rate and adiabatic lapse rates.

12. How and why does the wet adiabatic lapse rate differ from the dry adiabatic lapse rate?

13. How is atmospheric stability related to the adiabatic lapse rates?

14. Describe the most recent cloud formations in your area. With what kinds of weather (temperature, humidity, precipitation) were the cloud types associated?

15. What atmospheric conditions are necessary for precipitation to occur?

16. Find out how many inches of precipitation have fallen in your area this year. Is that average or unusually high or low?

17. Compare and contrast convectional, orographic, and frontal precipitation.

18. How is rainfall variability related to total annual rainfall? How might this relationship be considered a double problem for people?

Consider and Respond

1. Refer to Figure 7.3.
   (a) What is the water vapor capacity of air at 0°C? 20°C? 30°C?
   (b) If a parcel of air at 30°C has an absolute humidity (actual water vapor content) of 20.5 g/m^3, what is the parcel's relative humidity?
   (c) If the relative humidity of a parcel of air is 33 percent and the air temperature is 15°C, what is the absolute humidity of the air in g/m^3?
   (d) A major concern of northern climate residents is the low relative humidity within their homes during the winter. Low relative humidity is not healthy, and it has an adverse effect on the homes' furnishings. The problem results when cold air, which can hold little water vapor, is brought indoors and is heated up. The following example will illustrate the problem. Assume the air outside is 5°C and has a relative humidity of 60 percent. What is the actual water vapor content of this air? If it is brought indoors (through the doors, windows, and cracks in the home) and heated to 20°C with no increase in water vapor content, what is the new relative humidity?

2. Recall that as a parcel of air rises it expands and cools. The rate of cooling is 10°C/1000 meters and is termed the dry adiabatic lapse rate. (A descending parcel of air will always warm at this rate.) In addition, the dew-point temperature decreases 2°C/1000 meters within a rising parcel of air. At the height at which the dew-point temperature is reached, condensation begins and, as dis-

cussed in the textbook, the wet adiabatic lapse rate of 5°C/1000 meters becomes operational. When the wet adiabatic lapse rate is in operation, the dew-point temperature will be the same as the air temperature. When a parcel descends through the atmosphere, its dew-point temperature increases 2°C/1000 meters. The height at which condensation begins is termed the lifting condensation level (LCL) and can be determined by the following formula:

$$\text{LCL (meters)} = \frac{T - T_d}{8.0} \times 1000$$

where T = temperature of the parcel

T_d = dew-point temperature of the parcel

8.0 = difference between the rate of change of T and T_d

For example, a parcel of air with a temperature of 31°C and a dew-point temperature of 18°C would have an LCL of 1625 meters: $(31 - 18)/8.0 \times 1000 = 1.625 \times 1000 = 1625$ meters.
   (a) A parcel of air has a temperature of 25°C and a dew-point temperature of 14°C. What is the height of the LCL? If the parcel were to rise to 4000 meters, what would be its temperature?
   (b) A parcel of air at 6000 meters has a temperature of −5°C and a dew-point temperature of −10°C. If it descended to 2000 meters, what would be its temperature and its dew-point temperature?

CHAPTER
8

AIR MASSES AND
ATMOSPHERIC DISTURBANCES

CHAPTER PREVIEW

▶ The movement of relatively large portions of the atmosphere (air masses) is responsible for the transportation of widely varied conditions of temperature and humidity to regions far from their original sources.

How is this important to the operation of the Earth system? What is the significance of air mass movement to human beings?

▶ The meeting of two unlike air masses occurs along a sloping surface of discontinuity called a front.

How do air masses differ? What kind of air masses meet along the polar front? Why are fronts important to an explanation of middle-latitude weather?

▶ The major explanation for the variable and nearly unpredictable weather of the middle latitudes may be found in the irregular migration of relatively short-lived low pressure systems (cyclones) in the path of the prevailing westerlies.

Why do cyclones play such a significant role? In which direction would you normally look to obtain some forewarning of future weather? What are the human consequences of variable and unpredictable weather?

▶ Because meteorology is an inexact science and there is much yet to be learned about the behavior of air masses, fronts, and pressure systems, we should anticipate that weather forecasting will long be a hazardous art.

What questions about the weather remain unanswered? How accurate is weather prediction? How successful are humans at altering the weather?

▶ The latent heat of condensation is the major source of energy behind the violent atmospheric disturbances of the middle latitudes, such as tornadoes, hurricanes, and thunderstorms.

How might this information be useful? What have we done in the United States to protect against violent storms?

▶ Although the associated atmospheric disturbances are weak and the daily changes are subtle, the weather of tropical regions is not as monotonous as once was thought.

Why has knowledge of tropical weather been so limited in the past? What gave early observers the impression of monotonous weather?

▲ Lightning illuminates the night sky over the Midwestern plains.

In the previous four chapters, we have looked at the elements of the atmosphere and investigated some of the controls that act upon those elements, causing them to vary from place to place and through time. However, even more is involved in the examination of weather. We have not yet looked at storms (atmospheric disturbances)—their types and characteristics, their origin, and their development. Storms are an important part of the weather story. They help illustrate the interactions among the weather elements. Further, they represent a major means of energy exchange within the atmosphere.

Air Masses

Before we begin to study atmospheric disturbances, we should understand the nature and significance of air masses. An **air mass** is a large body of air, sometimes subcontinental in size, that may move over Earth's surface as a distinct entity. An air mass is relatively homogeneous in temperature and humidity. That is, at approximately the same altitude within the air mass, the temperature and humidity will be similar. Of course, since an air mass may extend over 20 or 30 degrees of latitude, we can expect some slight modifications due to variations in insolation, which are significant over that distance, and to changes caused by contact with differing land and ocean surfaces. As a result of this temperature and moisture uniformity, the density of air will be much the same throughout any one level within an air mass.

The similar characteristics of temperature and humidity within an air mass are determined by the nature of its **source region**—that is, the place where the air mass originates. Only a few areas on Earth make good source regions. In order for the air mass to have similar characteristics throughout, the source region must have a nearly homogeneous surface. In addition, the air mass must have sufficient time to acquire the characteristics of the source region. Hence, gently settling, slowly diverging air will accompany a source region, while converging, rising air will not.

Six terms are used to describe air masses, and each reflects a property of a source region. They are: (1) equatorial (very warm; symbolized by capital *E*); (2) tropical (warm, *T*); (3) polar (cold, *P*); (4) arctic (very cold, *A*); (5) continental (dry, *c*); and (6) maritime (wet, *m*). These six terms can be combined to give us the classification of air masses first described in 1928 and still used today: Maritime Equatorial (*mE*), Maritime Tropical (*mT*), Continental Tropical (*cT*), Continental Polar (*cP*), Continental Arctic (*cA*), and Maritime Polar (*mP*). These six types are described more fully in Table 8.1. From now on, we will use the symbols rather than the full names as we discuss each type of air mass.

Stability of Air Masses

As a result of the general circulation patterns within the atmosphere, air masses do not remain stationary over their source regions indefinitely. When an air mass begins to move over Earth's surface along a path known as a **trajectory,** it retains its distinct and homogeneous characteristics to a large extent. However, modification does occur as the air mass gains or loses some of its thermal energy and moisture content to the surface below. Although this modification is generally slight, the gain or loss of thermal energy can make an air mass more stable or unstable.

Because it has a bearing on its stability, an air mass is further classified by whether it is warmer or colder than the surface over which it is in contact. If an air mass is colder than the surface over which it passes, then the surface will heat the air mass from below. This will, in turn, increase the environmental lapse rate, enhancing the prospect of instability. To describe such a situation, the letter *k* (from German: *kalt,* cold) is added to the other letters that symbolize the air mass. For example, an *mT* air mass originating over the Gulf of Mexico in summer that moves onshore over the warm land would be denoted *mTk.* Such an air mass is often unstable and can produce copious convective precipitation. On the other hand, this same *mT* air mass moving onshore during the winter would now be warmer than the land surface. Consequently, the air mass would be cooled from below, decreasing its environmental lapse rate, which enhances the prospect of stability. We describe this situation with the letter *w* (from German: *warm,* warm), and the air mass would be denoted *mTw.* In this case stratiform, not convective, precipitation is most likely.

North American Air Masses

Because most of us are familiar with the weather in at least one region of the United States, we will concentrate in this chapter on the air masses of North America and their effects on weather. What we learn will be applicable to the rest of the world, and, as we examine climatic regions in some of the following chapters, we will be able to understand that weather everywhere is most often affected by the movements of air masses. Especially in middle-latitude regions, the majority of atmospheric disturbances result from the confrontations of different air masses.

Five types of air masses (*cA, cP, mP, cT,* and *mT*) influence the weather of North America, some more than others. Air masses assume characteristics of their source regions (Fig. 8.1). Consequently, as the source regions change with the seasons, primarily because of changing insolation, the air masses also will vary.

TABLE 8.1   Types of Air Masses

	Source Region	Usual Characteristics at Source	Accompanying Weather
Maritime Equatorial (*mE*)	Equatorial oceans	Ascending air, very high moisture content	High temperature and humidity, heavy rainfall; never reaches the United States
Maritime Tropical (*mT*)	Tropical and subtropical oceans	Subsiding air; fairly stable, but some instability on western side of oceans; warm and humid	High temperatures and humidity, cumulus clouds, convectional rain in summer; mild temperatures, overcast skies, fog, drizzle, and occasional snowfall in winter; heavy precipitation along *mT/cP* fronts in all seasons
Continental Tropical (*cT*)	Deserts and dry plateaus of subtropical latitudes	Subsiding air aloft; generally stable, but some local instability at surface; hot and very dry	High temperatures, low humidity, clear skies, rare precipitation
Maritime Polar (*mP*)	Oceans between 40° and 60° latitude	Ascending air and general instability, especially in winter; mild and moist	Mild temperatures, high humidity; overcast skies and frequent fogs and precipitation, especially during winter; clear skies and fair weather common in summer; heavy orographic precipitation, including snow, in mountainous areas
Continental Polar (*cP*)	Plains and plateaus of subpolar and polar latitudes	Subsiding and stable air, especially in winter; cold and dry	Cool (summer) to very cold (winter) temperatures, low humidity; clear skies except along fronts; heavy precipitation, including winter snow, along *cP/mT* fronts
Continental Arctic (*cA*)	Arctic Ocean, Greenland, and Antarctica	Subsiding very stable air; very cold and very dry	Seldom reaches United States, but when it does, bitter cold, subzero temperatures, clear skies, often calm conditions

Continental Arctic Air Masses. The frigid, *frozen* surface of the Arctic Ocean serves as the source region for this air mass. It is extremely cold, very dry, and very stable. Even during the winter, when this air mass is best developed, it seldom travels far enough south to affect the United States. However, on those few occasions when it does extend down into the midwestern and eastern United States, its impact is awesome. Below-zero, record-setting cold temperatures often result. If the *cA* air mass remains in the Midwest for an extended period, vegetation—not accustomed to the extreme cold—can be severely damaged or killed.

Continental Polar Air Masses. At its source in north-central North America, a continental polar air mass is cold, dry, and stable since it is warmer than the surface beneath it; the weather of a *cP* air mass is cold, crisp, clear, even sparkling. Because there are no east-west landform barriers in North America, *cP* air can migrate south across Canada and the United States. A tongue of *cP* air can sometimes reach as far south as the Gulf of Mexico or Florida. When winter *cP* air extends into the United States, its temperature and humidity are raised only slightly, and the movement of such an air mass into the Midwest and South brings with it a cold wave characterized by

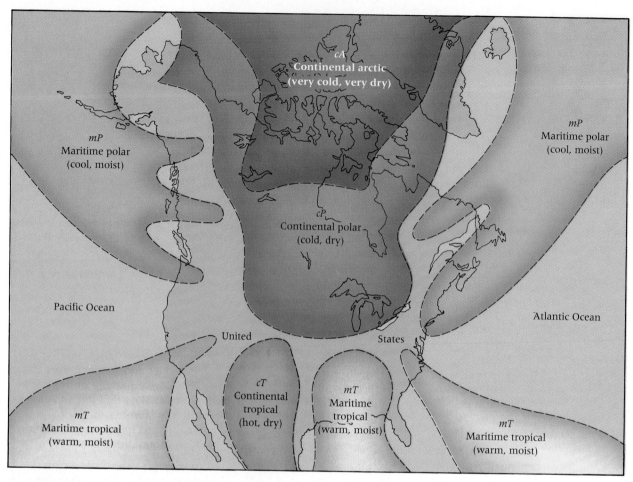

Figure 8.1 *Source regions of North American air masses. Air-mass movements import the temperature and moisture characteristics of these source regions into far distant areas.* • **Use Table 8.1 and this figure to determine which air masses affect your location. Are these seasonal variations?**

colder-than-average temperatures and clear, dry air, causing freezing temperatures as far south as Florida and Texas.

Because of the general westerly direction of atmospheric circulation in the middle latitudes, a *cP* air mass is rarely able to break through the great western mountain ranges to the west coast of the United States. When such an air mass does reach the California, Washington, and Oregon coasts, however, it brings with it unusual freezing temperatures that do great damage to agriculture.

Maritime Polar Air Masses. During winter months, because the oceans tend to be warmer than the land, an *mP* air mass tends to be warmer than its cousin on land, the *cP* air mass. Much *mP* air is originally cold, dry, *cP* air that has moved to a position over the ocean. There it is heated by the warmer water and collects mois-

ture. Thus, at its source, *mP* air is cold (although not nearly so cold as *cP* air) and damp, with a tendency toward instability. The northern Pacific Ocean serves as the source region for maritime polar air masses, which, because of the general westerly circulation of the atmosphere in the middle latitudes, affect the weather of the United States. As such an air mass moves south and east, it gains moisture and warmth until it reaches land. When this *mP* air meets any lifting agent (such as a mass of colder, denser air or coastal mountain ranges), the result is usually very cloudy weather with a great deal of precipitation. An *mP* air mass is still the source of many midwestern snowstorms even after crossing the western mountains.

Generally, an *mP* air mass that develops over the northern Atlantic Ocean does not affect the weather of the United States since such an air mass tends to flow eastward toward Europe. However, on some occasions there may be a reversal of wind direction accompanying the migration of a low-pressure system, and New England can be made miserable by the cool, damp winds, rain, and snow of a "northeaster."

Maritime Tropical Air Masses. The Gulf of Mexico and subtropical Atlantic Ocean serve as a source region for maritime tropical air masses that have a great influence on the weather of the United States. During winter the waters are warm and the air above is warm and moist. As the warm, moist air moves northward up the Mississippi Lowlands, it travels over increasingly cooler land surfaces. The lower layers of air are chilled, and dense advection fog often results. And when it reaches the *cP* air migrating southward from Canada, the warm air is forced to rise, and significant precipitation can occur.

The longer days and more intense insolation of summer months modify a Gulf and Atlantic air mass at its source region by increasing its temperature and moisture content. However, during summer the land is warmer than the nearby waters, and as the *mT* air mass moves onto the land, the instability of the air mass is increased. This air mass is the source of great thunderstorms and convective precipitation on hot, humid days, and it is also responsible for much of the hot, humid weather of the southeastern and eastern United States.

Other *mT* air masses form over the Pacific Ocean in the subtropical latitudes. These air masses tend to be slightly cooler than those that form over the Gulf and the Atlantic, partly because of their passage over the cooler California Current. A Pacific *mT* air mass is also more stable because of the strong subsidence associated with the eastern portion of the Pacific subtropical high. This air mass contributes to the dry summers of southern California and occasionally brings moisture in winter as it rises over the mountains of the Pacific Coast.

Continental Tropical Air Masses. There is a fifth type of air mass, but it is the least important to the weather of the United States. This is the continental tropical air mass that develops over large, homogeneous land surfaces in the subtropics. The Sahara Desert of North Africa is a prime example of a source region for this type of air mass. The weather typical of the *cT* air mass is usually very hot and dry, with clear skies and major heating from the sun during daytime.

In North America, there is little land in the correct latitudes to serve as a source region for a *cT* air mass of any significant proportion. A small *cT* air mass can form over the deserts of the southwestern United States and northwestern Mexico in the summer. In its source region, a *cT* air mass provides hot, dry, clear weather. When it moves eastward, however, it is usually greatly modified as it comes in contact with larger and stronger air masses of different temperatures, humidities, and densities.

Fronts

We have seen that air masses migrate with the general circulation of the atmosphere. Over the United States, which is influenced primarily by the westerlies, there is a general eastward flow of the air masses. In addition, air masses tend to diverge from areas of high pressure and converge toward areas of low pressure. This tendency means that the tropical and polar air masses, formed within systems of divergence, tend to flow toward the areas of continental low pressure within the United States. An important feature of an air mass is that it maintains the primary characteristics first imparted to it by its source region, although some slight modification may occur during its migration.

Air masses differ primarily in their temperature and in their moisture content, which in turn affect the air masses' density and atmospheric pressure. Air masses with different properties do not mix easily but instead come in contact along sloping boundaries called *fronts.* Although usually depicted on maps as a one-dimensional boundary line separating two different air masses, a front is actually a three-dimensional surface with length, width, and height. To emphasize this concept, a front is sometimes referred to as a **surface of discontinuity.** Because the surface of discontinuity is a zone that can cover an area from 2 or 3 kilometers (1–2 mi) wide to as wide as 150 kilometers (90 mi), it is more accurate to speak of a frontal zone rather than a frontal line.

The sloping surface of a front is created as the warmer and lighter of the two contrasting air masses is lifted or rises above the cooler and denser air mass. Such rising is known as **frontal lifting** and is a major source of precipitation in middle-latitude countries like the United States, where contrasting air masses are most likely to converge.

The steepness of the frontal surface is governed primarily by the degree of difference between the two converging air masses. When there is a sharp difference between the two air masses, as when an *mT* air mass of high temperature and moisture content meets a *cP* air mass with its cold, dry characteristics, the slope of the frontal surface will be steep. With a steep slope, there will be greater frontal lifting. Provided other conditions (for example, temperature and moisture content) are equal, a steep slope, with its greater frontal lifting, will produce heavier precipitation than will a gentler slope.

Fronts are differentiated by determining whether the colder air mass is moving on the warmer one or vice versa. The weather that occurs along a front is also dependent on which air mass is the "aggressor."

Cold Front

A **cold front** occurs when a cold air mass actively moves upon a warmer air mass and pushes it upward. The colder air, denser and heavier than the warm air it is displacing, stays at the surface while forcing the warmer air to rise. As we can see in Figure 8.2, a cold front usually results in a relatively steep slope in which the warm air may rise 1 meter in the vertical for every 40 to 80 meters of horizontal distance. If the warm air mass is unstable and has a high moisture content, heavy precipitation can result, sometimes in the form of violent thunderstorms. In any case, cold fronts are usually associated with strong weather disturbances or sharp changes in temperature, air pressure, and wind.

Warm Front

When a warmer air mass is the aggressor and invades a region occupied by a colder air mass, a **warm front** results. In a warm front, the warmer air, as it slowly pushes against the cold air, also rises over the colder, denser air mass, which again stays in contact with Earth's surface. The slope of the surface of discontinuity that results is usually far gentler than that occurring in a cold front. In fact, the warm air may rise only 1 meter in the vertical for every 100 or even 200 meters of horizontal distance. Thus, the frontal lifting that develops will not be as great as that occurring along a cold front. The result is that the warmer weather associated with the passage of a warm front *tends* to be less violent and the changes less abrupt than those associated with cold fronts.

If we look at Figure 8.3, we can see why the advancing warm front affects the weather of areas ahead of the actual surface location of the frontal zone. Changes in the weather resulting from fronts that have not yet reached us can be indicated by the cloud forms that precede them.

Stationary and Occluded Fronts

When two air masses have converged and formed a frontal boundary, but neither then moves, we have a situation known as a **stationary front.** Locations under the influence of a stationary front are apt to experience clouds, drizzle, and rain for several days. In fact, a stationary front and its accompanying weather will remain until the contrasts between the two air masses are reduced or the circulation of the atmosphere finally causes one of the air masses to move.

An **occluded front** occurs when a faster-moving cold front overtakes a warm front, pushing all of the warm air aloft. This frontal situa-

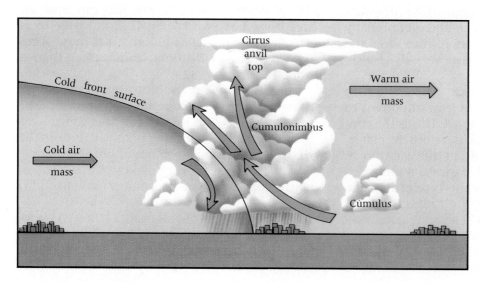

Figure 8.2 Cross section of a cold front. Cold fronts generally move rapidly, with a blunt forward edge that drives adjacent warmer air upward, producing violent precipitation from the warmer air.

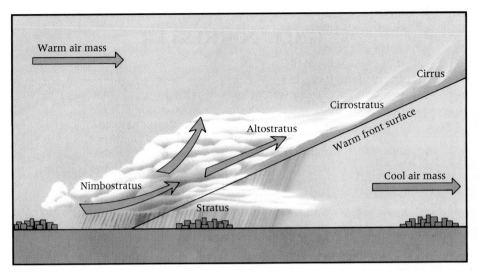

Figure 8.3 *Cross section of a warm front. Warm fronts advance more slowly than cold fronts and replace (rather than displace) cool air by sliding upward over it. The gentle rise of the warm air produces stratus clouds and gentle drizzles. •* Compare Figures 8.2 and 8.3. How are they different? How are they similar?

tion usually occurs in the latter stages of a midlatitude cyclonic storm, which will be discussed next. (Map symbols of the four frontal types are shown in Fig. 8.4).

Atmospheric Disturbances

Imbedded within the wind belts of the general atmospheric circulation (see Chapter 6) are secondary circulations. These are made up of storms and other atmospheric disturbances. We use the term **atmospheric disturbance** because it is more general than **storm** and includes variations

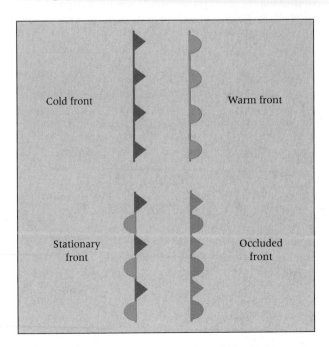

Figure 8.4 *The four major frontal symbols used on weather maps.*

in the secondary circulation of the atmosphere that cannot correctly be classified as storms.

Partly because our primary interest is in the weather of North America, we will concentrate on an examination of **extratropical disturbances,** as the atmospheric disturbances of the middle latitudes are sometimes called.

During World War I, Norwegian meteorologists Vilhelm and Jacob Bjerknes (a father and son) put forth the *polar front theory,* which provided insight into the development, movement, and dissipation of middle-latitude storms. They recognized that the middle latitudes are an area of convergence where unlike air masses, such as cold polar air and warm subtropical air, commonly meet, forming the *polar front.* While the polar front may be a continuous boundary circling the entire globe, it is most often fragmented into several individual line segments. Furthermore, the polar front tends to move north and south with the seasons and is apt to be stronger in winter than in summer.

Middle-latitude storms develop at the front and then travel along it. These migrating storms, with their juxtaposed cold, dry polar air and warm, humid tropical air, can cause significant variation in the day-to-day weather of the locations over which they pass. It is not unusual in some parts of the United States for people to go to bed at the end of a beautiful warm day in early spring and wake up to falling snow the next morning. Such changeability is common for middle-latitude weather, especially during certain times of the year when the weather changes from a period of cold, clear, dry days to a period of snow, only to be followed by one or two more moderate but humid days.

GREG FORRESTER

Greg Forrester

High School Education: *South Beloit High School, Illinois, 1982*

University Education: *Bachelor of Science, Geography, Western Illinois University, 1986, University of Nebraska, Lincoln*

Early Interests: *Biking, travel, weather phenomena, map-reading*

Prior Work Experiences: *Graduate Teaching Assistant*

Greg Forrester is a meteorologist with the National Weather Service in Glasgow, Montana, where his responsibilities include forecasting and observing. Prior to that, Greg worked at the NWS in Marquette, Michigan, keeping track of weather patterns over parts of Michigan and Lake Superior. Although he was a geography major, Greg took classes to learn about the use of radar and other meteorological tools—remote sensing, dynamic meteorology, synoptic charts (or weather maps)—as well as an extra physics course. Here, Greg talks about his work at the NWS, how meteorology has changed, and how technology drives that change.

I am currently assigned to watch the weather over eastern Montana. This territory includes about 40,000 square miles of plains and rolling hills. The Missouri River flows east through this area. While the average annual precipitation is only 11 inches, most of it comes in summer thunderstorms, which can occasionally produce flash flooding.

Temperatures can change very rapidly in eastern Montana, especially in the winter, when it can go from 30 degrees above to 30 degrees below in only a few hours. High winds frequently accompany the frontal passage producing the temperature change. Cattle ranching is a major way of life out here, and an accurate forecast is vital to protecting livestock.

Based on my forecasts, a boater may decide whether or not to go fishing on a particular day. My daily routine also includes answering questions from the public. People call me and ask, 'What is the weather going to be like in Billings?' The information that I provide might allow them to avoid hazardous driving conditions.

Simple meteorological tools such as the wind vane and the rain gauge have been used since ancient times. By the late eighteenth century, scientists had invented more advanced instruments, including the anemometer (wind speed and direction), barometer (air pressure), hygrometer (humidity), and thermometer (temperature). When the NWS was established in the 1870s, meteorologists often relied on folklore and simple visual clues to foretell certain types of weather. Today we also use satellites, radar, and airplanes to collect weather data. At the NWS, we rely on conventional as well as Doppler radar systems to provide forecasts for two weather-radio stations.

Developing a forecast begins with good weather observations. I also use computer models generated by the National Forecast Center in Washington, D.C. These models attempt to predict atmospheric conditions for the next 48- to 72-hour period. Computer models also predict the location of certain high pressure/low pressure conditions, or 'ridge and trough' locations. Combined with wind direction, this information makes it possible for me to predict when specific weather conditions might come our way.

The NWS is currently modernizing its forecast systems to include Doppler radar. Doppler radar relies on the Doppler effect, which is a change you can observe in a sound wave or electromagnetic radiation as you move toward or away from the source of the wave. Conventional radar determines the location of a storm essentially by bouncing radio waves off an object. Doppler radar uses sound waves to determine how fast that storm is moving toward or away from a given region.

Students who are interested in meteorology should take extra computer courses and learn about radar and other technologies. It is also important to recognize that a career in weather service may require personal sacrifices—overtime due to developing weather emergencies as well as rotating shifts. For me, long hours and rotating shifts are small sacrifices because I continue to be thrilled by weather. I learn something new every day, and my job never bores me."

Cyclones and Anticyclones

Nature, Size, and Appearance on Maps. We have previously distinguished cyclones and anticyclones according to differences in pressure and wind direction. Also, when studying maps of world pressure distribution, we identified large areas of semipermanent cyclonic and anticyclonic circulation in Earth's atmosphere (the subtropical high, for example). Now, when examining middle-latitude atmospheric disturbances, we will use the terms *cyclone* and *anticyclone* to describe the moving cells of low and high pressure that drift with varying regularity in the path of the prevailing westerly winds. As systems of higher pressure, the anticyclones are usually characterized by clear skies, gentle winds, and a general lack of precipitation. As centers for converging, rising air, the cyclones are the true storms of the middle latitudes, with associated fronts of various types.

As we know from experience, no middle-latitude cyclonic storm is ever exactly like any other. The storms vary in their intensity, in the number of hours or days they last, in the speed with which they pass, in the strength of their winds, in the amount and type of cloud cover, in the quantity and kind of precipitation they deposit, and in the surface area they affect.

Because of the seemingly endless variety of cyclones, we will be describing *model* cyclones in the following discussions. Not every storm will act in the way we describe, but there are generalizations that will be helpful in understanding middle-latitude storms.

Because a cyclone has a low-pressure center, winds tend to converge toward that center in an attempt to equalize pressure. If we visualize this air moving in toward the center of the low-pressure system, we can see that the air that is already at the center must be displaced upward. The lifting that occurs in a cyclone results in clouds and precipitation, and the warm air rises like a corkscrew.

Anticyclones are high-pressure systems in which atmospheric pressure decreases toward the outer limits of the system. Visualizing an anticyclone, or high, we can see that air in the center of the system must be subsiding, in turn displacing surface air outward, away from the center of the system. Hence, an anticyclone has diverging winds. In addition, an anticyclone tends to be a fair-weather system, since the subsiding air in its center increases in temperature and stability, reducing the opportunity for condensation.

We should note here that the pressures we are referring to in these two systems are relative. What is important is that in a cyclone, pressure *decreases* toward the center, and in an anticyclone, pressure *increases* toward the center. Furthermore, the intensities of the winds involved in these systems depend on the steepness of the pressure gradients involved. Thus, if there is a steep pressure gradient in a cyclone, with the pressure much lower at the center than at the outer portions of the system, the winds will converge toward the center with considerable velocity.

The situation is easier to visualize if we imagine these pressure systems as landforms. A cyclone is shaped like a basin (Fig. 8.5). If we are filling the basin with water, we know that the water will flow in faster the steeper the sides and the deeper the depression. If we visualize an anticyclone as a hill or mountain, then we can also see that just as water flowing down the sides of such landforms will flow faster with increased height and steepness, so will the air blowing out of an area of very high pressure move rapidly.

On a surface weather map, cyclones and anticyclones are depicted by concentric isobars of increasing pressure toward the center of a high and of decreasing pressure toward the center of a low. Usually a high will cover a larger area than a low. Both pressure systems are capable of covering and affecting extensive areas. There are times when nearly the entire midwestern United States is under the influence of the same system. The average diameter of an anticyclone is about 1500 kilometers (900 mi), while that of a cyclone is about 1000 kilometers (600 mi).

General Movement. The cyclones and anticyclones of the middle latitudes are steered, or guided, along a path reflecting the configuration and speed of the upper air westerlies. The upper air flow can be quite variable with wild oscillations. However, a general west-to-east pattern does prevail. Consequently, people in most of the eastern United States look at the weather occurring to the west to see what they might expect in the next few days. Most storms that develop in the Great Plains or Far West move across the United States during a period of a few days at an average speed of about 36 kilometers per hour (23 mph), and then travel on into the North Atlantic before occluding.

Although neither cyclones nor anticyclones develop in exactly the same places at the same times each year, they do tend to develop in certain areas or regions more frequently than in

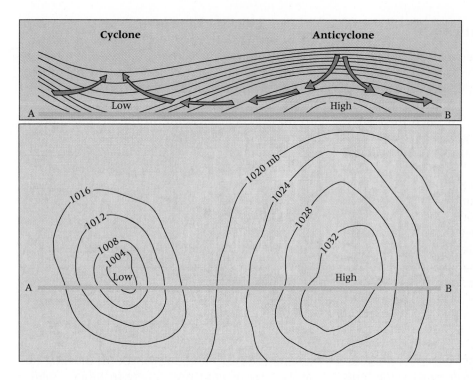

Figure 8.5 *Close spacing of isobars around a cyclone or anticyclone indicates a steep pressure gradient that will produce strong winds. Wide spacing of isobars indicates a weaker system.* • Where would the strongest winds be in this figure? Where would the weakest winds be?

others, and they do follow the same general paths, which are known as **storm tracks** (Fig. 8.6). These storm tracks vary with the seasons. In addition, because the temperature variations between the air masses are stronger during the winter months, the atmospheric disturbances that develop in the middle latitudes during those months are greater in number and intensity.

Cyclones. Now let us look more closely at cyclones—their origin, development, and characteristics. Warm and cold air masses meet at the

polar front, where most cyclones develop. These two contrasting air masses do not merge but may move in opposite directions along the frontal zone. Although there may be some slight uplift of the warmer air along the edge of the denser, colder air, the uplift will not be significant. There may be some cloudiness and precipitation along such a frontal zone, though not of storm caliber.

For reasons not completely understood, but certainly related to the wind flow in the upper troposphere, a wavelike kink may develop along the polar front. This is the initial step in the for-

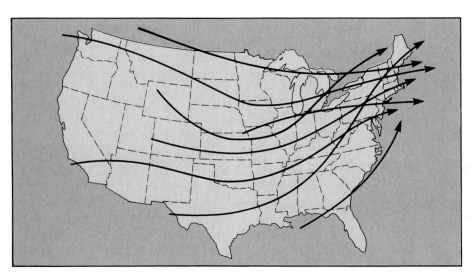

Figure 8.6 *Common storm tracks for the United States. Virtually all cyclonic storms move from west to east in the prevailing westerlies and swing northeastward across the Atlantic Coast. Storm tracks originating in the Gulf of Mexico represent tropical hurricanes.* • What storm tracks influence your location?

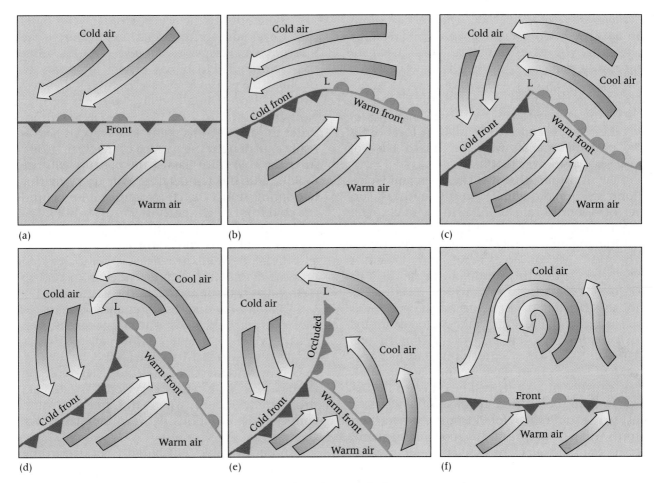

Figure 8.7 *Stages in the development of a midlatitude cyclone. Each view represents the development somewhat eastward of the preceding view as the cyclone travels along its storm track. Note the occlusion in (e).* • In (c), where would you expect rain to develop? Why?

mation of a fully grown wave cyclone (Fig. 8.7). At this bend in the polar front, we now have warm air pushing poleward—a warm front—and cold air pushing equatorward—a cold front—with a center of low pressure at the location where the two fronts are joined.

As the contrasting air masses jockey for position, the clouds and precipitation that exist along the fronts are greatly intensified and the area affected by the storm is much greater. Along the warm front, precipitation will be more widespread but less intense than along the cold front. One factor that can vary the kind of precipitation occurring at the warm front is the stability of the warm air mass. If it is stable, then its upgliding over the cold air mass may cause only a fine drizzle—or a gentle, powdery snow if the temperatures are low enough. On the other hand, if the warm air mass is moist and unstable, the

upgliding may set off heavier precipitation. As you can see by referring again to Figure 8.3, the precipitation that falls at the warm front may *appear* to be coming from the colder air, and the weather may feel cold and damp, but the precipitation is actually coming from the overriding warmer air mass above, though it must fall through the colder air mass to reach Earth's surface.

Because the cold front moves faster, it will eventually overtake the warm front. This produces the situation we previously identified as an occluded front. Because additional warm, moist air will not be lifted after occlusion, condensation and the release of latent energy will diminish and the storm will soon die. Occlusions are usually accompanied by rain and are the major process by which midlatitude wave cyclones are destroyed.

Cyclones and Local Weather. Different portions of a wave cyclone exhibit different weather. Therefore, the weather that a location experiences at a *particular* time depends upon which portion of the wave cyclone is over the location. Also, since the entire cyclonic system tends to travel from west to east, a specific *sequence* of weather can be expected at a given location as the cyclonic system, with its "mixed bag" of weather, passes over that location.

Let us assume that it is late spring and a cyclonic storm has originated in the southeast corner of Nebraska and is following a track (refer again to Fig. 8.6) across northern Illinois, northern Indiana, northern Ohio, through Pennsylvania, and finally out over the Atlantic Ocean. A bird's-eye view of this storm, at one point in its journey, is presented in Figure 8.8a. A cross-sectional view *north* of the center of the cyclone is presented in Figure 8.8b, while a cross-sectional view *south* of the center of the cyclone is presented in Figure 8.8c. As the storm continues eastward, at 9 to 13 meters per second (20–30 mph), the sequence of weather will be different for Detroit, where the warm and cold fronts will pass just to the south, than for Pittsburgh, where both fronts will pass overhead. To illustrate this point, let us examine, element by element, the variation in weather that will occur in Pittsburgh, with reference, where appropriate, to the differences that occur in Detroit as the cyclonic system moves east.

As we have previously noted, atmospheric temperature and pressure are closely related. As temperature increases, air expands and pressure decreases. Therefore, these two elements will be discussed together. Since a cyclonic storm is composed of two dissimilar air masses, there are usually important temperature contrasts. The sector of warm, humid *mT* air between the two fronts of the cyclone is usually considerably warmer than the cold *cP* air surrounding it. The temperature contrast is accentuated in the winter, when the source region for *cP* air is the cold cell of high pressure normally found in Canada at that time of year. During the summer the contrast between these air masses is greatly reduced.

As a consequence of the temperature difference, the atmospheric pressure in the warm sector is considerably lower than the atmospheric pressure in the cold air behind the cold front. Far in advance of the warm front the pressure is also high, but as the warm front (see Fig. 8.3) approaches, increasingly more cold air is replaced by overriding warm air, thus steadily reducing the surface pressure.

Therefore, as the warm front of this late-spring cyclonic storm approaches Pittsburgh, the pressure will decrease. After the warm front passes through Pittsburgh (where the temperature may have been 8°C [46°F] or more), the pressure will stop falling and the temperature may easily rise to 18° to 20°C (64°–69°F) as *mT* air invades the area. At this point, Indianapolis has already experienced the passage of the warm front. After the cold front passes, the pressure will rise rapidly and the temperature will drop. In this late-spring storm the *cP* air temperature behind the cold front might be 2° to 5°C (35°–40°F). Detroit, which is to the north of the center of the cyclone, will miss the warm air sector entirely and therefore will experience a slight increase in pressure and a temperature change from cool to cold as the cyclone moves to the east.

Changes in wind direction are one signal of the approach and passing of a cyclonic storm. Since a cyclone is a center of low pressure, winds usually flow toward its center. Also, winds are caused by differences in pressure. Therefore, the winds associated with a cyclonic storm are stronger in winter when the pressure (and temperature) variations between air masses are greatest.

In our example, Pittsburgh is located to the south and east of the center of low pressure and ahead of the warm front, and it is experiencing winds from the southeast. As the entire cyclonic system moves east, the winds in Pittsburgh will shift to the south-southwest after the warm front passes. Indianapolis is currently in this position. After the cold front passes, the winds in Pittsburgh will be out of the north-northwest. St. Louis has already experienced the passage of the cold front and currently has winds from the northwest. The changing direction of wind, clockwise around the compass from east to southeast to south to southwest to west and northwest, is called a **veering wind shift** and indicates that the center of a low has passed to the north of your position. On the other hand, Detroit, which is also experiencing winds from the southeast, will undergo a completely different sequence of directional wind changes as the cyclonic storm moves eastward. Detroit's winds will shift to the northeast as the center of the storm passes to the south. Chicago has just undergone this shift. Finally, after the storm has passed, the winds will blow from the northwest. Des Moines, to the west of the storm, currently has northwest winds. Such a change of wind direction, from east to northeast to north to northwest, is called a **backing wind shift,** as the wind "backs" counterclockwise around the compass. A backing wind shift indicates that you were north of the cyclone's center.

The type and intensity of precipitation and cloud cover also vary as a cyclonic disturbance moves through a location. In Pittsburgh, the first sign of the approaching warm front will be high cirrus clouds. As the warm front continues to approach, the clouds will thicken and lower. When the warm front is within 150 to 300 kilometers (90–180 mi) of Pittsburgh, light rain and drizzle will begin and stratus clouds will blanket the sky.

After the warm front has passed, precipitation will stop and the skies will clear. However, if the warm, moist *mT* air is unstable, convective showers may result.

As the cold front passes, warm air in its path will be forced to move aloft rapidly. This may mean that there will be a cold, hard rain, but the band of precipitation normally will not be very wide because of the steep angle of the surface of discontinuity along a cold front. In our example, the cold front and the band of precipitation have just passed St. Louis.

Thus, Pittsburgh can expect three zones of precipitation as the cyclonic system passes over its location: a broad area of cold showers and drizzle in front of the warm front; a zone within the moist, subtropical air from the south where there can be scattered convectional showers; and a narrow band of hard rainfall associated with the cold front (Fig. 8.8c). However, locations to the north of the center of the cyclonic storm, such as Detroit, will usually experience a single, broad band of light rains resulting from the lifting of warm air above cold air from the north (Fig. 8.8b). In winter, the precipitation is likely to be snow, especially in locations just to the northwest of the center of the storm, where the humid *mT* air overlies extremely cold northern air.

As you can see, different portions of a wave cyclone are accompanied by different weather. If we know where the cyclone will pass relative to our location, we can make a fairly accurate forecast of what our weather will be like as the storm moves east along its track (see Map Interpretation: Weather Maps).

Cyclones and the Upper-Air Flow. The upper-air wind flow greatly influences our surface weather. We have already discussed the role of these upper-air winds in the steering of surface storm systems. Another less obvious influence of the upper-air flow is related to the undulating, wavelike flow so often exhibited by the upper air. As the air moves its way through these waves it un-

dergoes divergence or convergence because of the rather complex dynamics associated with curved flow. This upper-air convergence and divergence greatly influence the surface storms below.

The region between a ridge and the next downwind trough (A–B in Fig. 8.9) is an area of upper-level convergence. In our atmosphere an action taken in one part of the atmosphere is compensated for by an opposite reaction somewhere else. In this case the upper-air convergence is compensated for by *divergence at the surface,* which will inhibit the formation of a midlatitude storm or cause an existing storm to weaken or even dissipate. On the other hand, the region between a trough and the next downwind ridge (B–C in Fig. 8.9) is an area of upper-level divergence, which in turn is compensated for by surface convergence. Convergence at the surface will enhance the prospects of storm development or strengthen an already existing storm.

In addition to storm development or dissipation, upper-air flow (see Fig. 8.9) will have an impact on temperatures as well. If we assume that our "average" upper-air flow is from west to east, then any deviation from that pattern will cause either colder air from the north or warmer air from the south to be advected into an area. Thus, after the atmosphere has been in a wavelike pattern for a few days, the areas in the vicinity of a trough (B in Fig. 8.9) will be colder than normal as polar air from higher latitudes is brought into that area. Just the opposite occurs at locations near a ridge (C in Fig. 8.9). In this case, warmer air from more southerly latitudes than would be the case with west-to-east flow is advected into the area near the ridge.

Weather Forecasting. Weather forecasting, at least in principle, is fairly straightforward. Meteorological observations are made, collected, and mapped to depict the current state of the atmosphere. From this information, the probable movement, as well as any anticipated growth or decay, of the current weather systems is projected for a specific amount of time into the future.

When a forecast is wrong—which we all know occurs—it is usually because either limited or erroneous information has been collected and processed in the first place, or, more likely, because errors have been made in anticipating the path or growth of the storm systems. Little errors will compound themselves over time. For example, a few degrees' shift in a storm's path may result in an error of a few miles in the projected location of that storm in a 2-hour forecast, but this same few-degree error may result in a

(Continued on page 200)

(a)

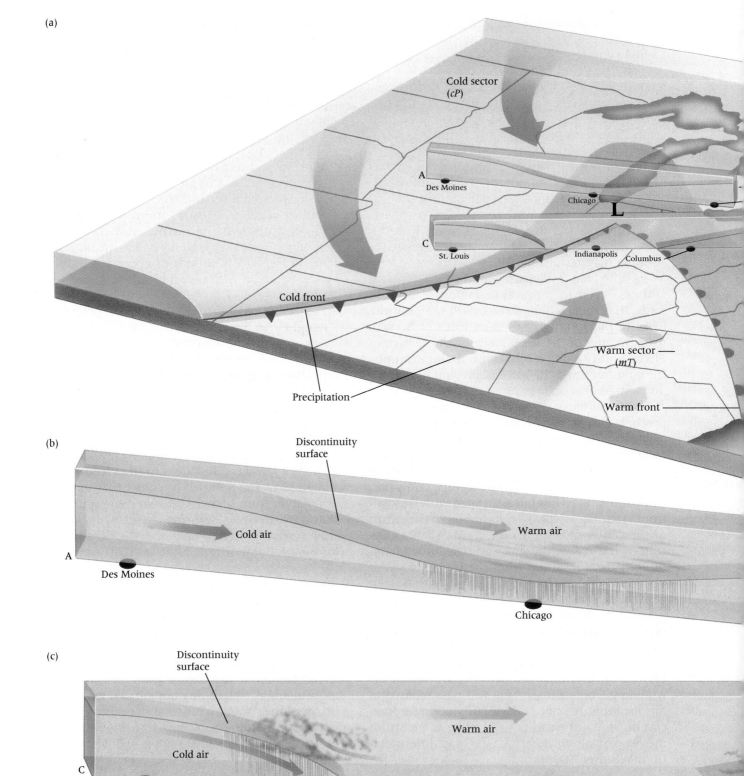

(b)

(c)

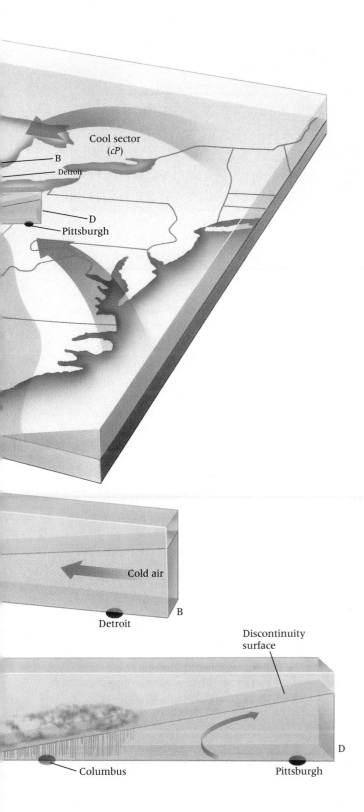

Figure 8.8 Environmental Systems: Midlatitude Cyclonic Systems *This diagram models a midlatitude cyclone positioned over the Midwest on its movement to the east:* **(a)** *A bird's-eye view of the cyclonic system.* **(b)** *A cross section along line AB to the north of the center of low pressure.* **(c)** *A cross section along line CD to the south of the center of low pressure.*

Midlatitude cyclones are dynamic systems created when air masses with different temperatures and humidity characteristics converge along the polar front. The contrasting air masses try to achieve equilibrium; as this happens, fronts, or boundaries, form between the warm and cold air. Along the fronts, warm, moist air is forced to rise. The rising warm air cools adiabatically, leading to condensation and often precipitation. Condensation supplies energy to the storm system in the form of latent heat. Since the water vapor that is undergoing condensation in this model was initially evaporated largely from the Gulf of Mexico, this midlatitude storm plays a significant role in the transfer of energy from lower latitudes to higher latitudes.

In most midlatitude cyclonic systems, the cold front moves faster than the warm front and eventually overtakes it. At that time the warm and cold air become stratified, with the warm air aloft and the cold air near the surface—much like cross section 8.8b. Since the air is now more or less in equilibrium and no additional warm tropical air is rising, condensation and the release of latent energy diminishes and the storm system soon dies.

WEATHER MAPS

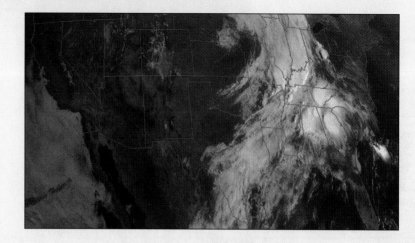

A weather satellite image (GOES) of the conditions on April 15, 1994. ▶

The Map

There are a wide variety of weather maps. Some depict surface conditions, while others depict conditions at various levels of the atmosphere. Also, some maps display current conditions, whereas others are historical or prognostical in nature.

One of the most widely used weather maps is a *surface weather map* (opposite page, top). The purpose of these maps is to portray the weather conditions, over a large area at a given moment in time, that are important to weather depiction and forecasting. Simultaneous observations of atmospheric pressure, temperature, humidity, state of the weather, precipitation measurements, cloud cover, visibility, and wind velocity are taken at locations across North America. In fact, observations are taken at locations around the world for use in the construction of other maps. This information is then electronically relayed to the National Meteorological Center

near Washington, D.C., where the surface observations are analyzed and mapped. The map is drawn on a Lambert conformal conic projection with standard parallels at 30° and 60°N latitude.

The surface map itself provides a wealth of information. The weather information observed at the individual locations is displayed using internationally agreed upon symbols (opposite page, bottom right). Meteorologists at the Center then use the individual pieces of information to develop and depict the larger picture over the entire area. For example, isobars are drawn to reveal the location of cyclones (L) and anticyclones (H). Temperatures and wind directions allow frontal boundaries to be indicated. Areas of precipitation are shaded in. The end result is a wealth of surface weather information that can then be used in forecasting future changes in weather patterns.

Interpreting the Map

1. What is the interval (in millibars) between adjacent isobars?

2. Based upon the isobaric pattern, are surface winds stronger over Iowa or Georgia? Explain.

3. Study the map of upper-air wind flow. What path will the center of low pressure likely follow in the next 24 hours?

4. The circulation around a low-pressure area is inward and counterclockwise. Identify the wind direction at several stations to confirm this circulation.

5. What air mass is influencing Atlanta, Georgia; New York, New York; Wichita, Kansas?

6. Does the surface map accurately depict the cloud cover indicated by the satellite image at the top of this page?

7. At a location where precipitation is not occurring, how could you use the station information to forecast the *type* of precipitation that might occur in the near future?

8. Describe the current weather in Indiana. In a *relative* sense, how do you expect the weather will change in the next 24 to 36 hours? Specifically, discuss changes in wind speed and direction, atmospheric pressure, cloud cover, precipitation, and temperature.

9. Does the location of the fronts and areas of precipitation agree with the idealized relationship depicted in Figure 8.8?

(Top) Surface weather map depicting conditions ▶ at 7 A.M. EST on April 15, 1994. (Lower left) The upper-air wind flow at 500 millibars (approximately 18,000 ft) for the same time period. (Lower right) Station model illustrating the type of information reported at each weather station on the surface map.

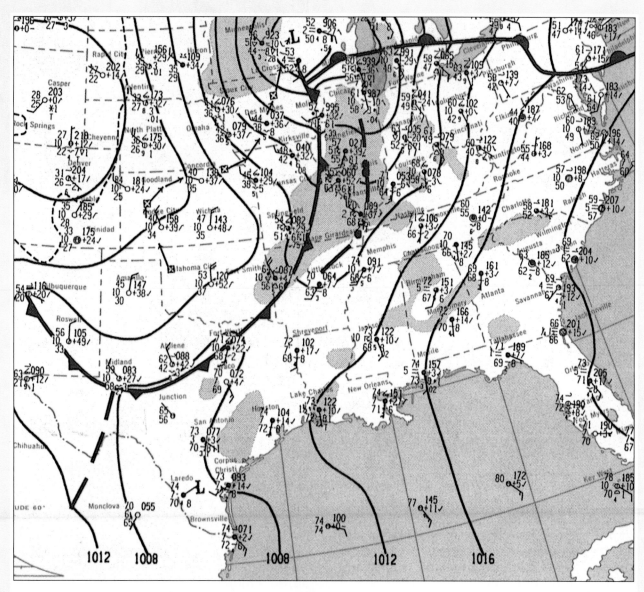

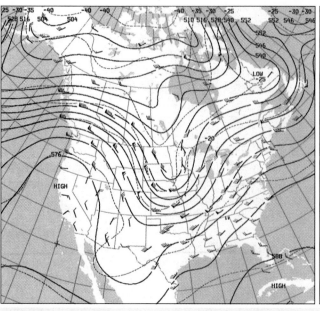

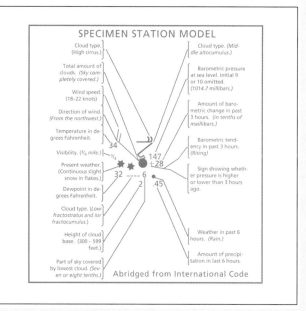

SPECIMEN STATION MODEL

Cloud type. (High cirrus.)

Cloud type. (Middle altocumulus.)

Total amount of clouds. (Sky completely covered.)

Barometric pressure at sea level. Initial 9 or 10 omitted. (1014.7 millibars.)

Wind speed. (18–22 knots)

Amount of barometric change in past 3 hours. (in tenths of millibars.)

Direction of wind. (From the northwest.)

Temperature in degrees Fahrenheit.

Barometric tendency in past 3 hours. (Rising)

Visibility. (³/₄ mile.)

Present weather. (Continuous slight snow in flakes.)

Sign showing whether pressure is higher or lower than 3 hours ago.

Dewpoint in degrees Fahrenheit.

Cloud type. (Low fractostratus and /or fractocumulus.)

Height of cloud base. (300 – 599 feet.)

Weather in past 6 hours. (Rain.)

Part of sky covered by lowest cloud. (Seven or eight tenths.)

Amount of precipitation in last 6 hours.

Abridged from International Code

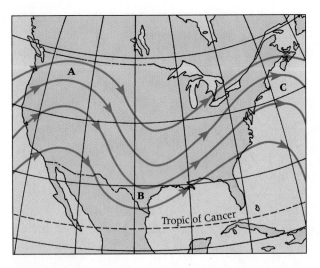

Figure 8.9 *The upper-air wind pattern, such as that depicted above, can have a significant influence on temperatures and precipitation at Earth's surface.*
• **Where would you expect storms to develop?**

Figure 8.10 *Satellite image of U.S. west coast weather pattern low-pressure system.*

projected locational error of hundreds of miles in a 48-hour forecast. Consequently, the further into the future one tries to forecast, the greater the chance of error.

Although forecasts are not perfect, they are much better today than in the past. Much of this improvement can be attributed to our current sophisticated technology and equipment. Increased knowledge and surveillance of the upper atmosphere have improved the accuracy of weather prediction. Weather satellites have helped tremendously by providing meteorologists with a better understanding of weather and weather systems. They have been of particular value to forecasters on the West Coast of the United States (Fig. 8.10). Before the advent of weather satellites, these forecasters had to rely on information relayed from ships, leaving enormous areas of the Pacific unobserved. Thus, forecasters were often caught off-guard by unexpected weather events.

In addition, high-speed computers allow rapid processing and mapping of observed weather conditions. Computers also allow the processing of numerical forecasts, which are based on the solution of physical equations that govern our atmosphere. Numerical forecasts and long-term forecasts based on statistical relationships would not be possible without computers. In fact, computers now play such an important role in forecasting that some of the world's largest and fastest computers are used to forecast the weather.

Though forecasters now possess a great deal of knowledge and a variety of highly sophisticated devices that were previously unavailable, none of these devices are foolproof. Understanding some of the problems the weather forecaster faces may make us more understanding when a forecast fails. No one can promise a sunny day. Nor can anyone say that it will definitely rain tomorrow, for no one knows the future. The weather forecaster combines science and art, fact and interpretation, data and intuition, to come up with some probabilities about future weather conditions.

Anticyclones. Just as cyclones are centers of low pressure that are typified by the convergence of air, so anticyclones are cells of high pressure in which air diverges. The subsidence of air in the center of an anticyclone encourages stability, as the air is warmed adiabatically while sinking toward the surface. Consequently, the air is able to hold additional moisture as its capacity increases with increasing temperatures. The weather resulting from the influence of an anticyclone is often clear, with no rainfall. There are, however, certain conditions under which there can be some precipitation within a high-pressure system. When such a system passes near or crosses a large body of water, the resulting evaporation can cause variations in humidity significant enough to result in precipitation.

There are two sources for the relatively high pressures that are associated with anticyclones in the midlatitudes of North America. Some anticyclones move into the middle latitudes from northern Canada and the Arctic Ocean, in what are called outbreaks of cold polar air. These outbreaks can be quite extensive, covering much of the midwestern and eastern United States. The temperatures in an anticyclone that has developed in a *cA* air mass can be markedly lower than those expected for any given time of year. They may be far below freezing in the winter and are often associated with the first frost in the fall and the last frost in the spring.

Other anticyclones are generated in zones of high pressure in the subtropics. When they move across the United States toward the north and northeast, they bring waves of hot, clear weather in summer and unseasonably warm days in the winter months.

Thunderstorms

Thunderstorms are common local storms of the middle and lower latitudes. Very simply, a **thunderstorm** is a convectional storm accompanied by thunder and lightning. **Lightning** is an intense discharge of electricity. For lightning to occur, positive and negative electrical charges must be generated within a cloud. It is believed that the intense friction of the air on moving ice particles within a cumulonimbus cloud generates these charges. A clustering of positive charges tends to occur in the upper portion of the cloud, with negative charges clustering in the lower portion. When the potential difference between these charges becomes large enough to overcome the natural insulating effect of the air, a lightning flash, or discharge, takes place. These discharges, which often involve over a million volts, can occur within the cloud, between two clouds, or from cloud to ground. The air immediately around the discharge is momentarily heated to temperatures in excess of 25,000°C (45,000°F)! The heated air expands explosively, creating the shock wave we call **thunder.**

Thunderstorms usually cover a small area of a few miles, although there may be a series of related thunderstorms covering a larger region. The intensity of a thunderstorm depends on the degree of instability of the air and the amount of water vapor it holds. A thunderstorm will die out when most of its water vapor has condensed, for there will no longer be energy available for continued vertical movement. In fact, most thunderstorms last less than an hour.

As an intense form of convectional precipitation, thunderstorms result from the uplift of moist air. As is the case for other convectional precipitation, the trigger actions causing that uplift can be thermal convection, orographic lifting, or frontal lifting.

Thunderstorms are most common in lower latitudes during the warmer months of the year and during the warmer hours of the day. It is apparent, then, that the amount of solar heating affects the development of thunderstorms. This is true because the intense heating of the surface steepens the environmental lapse rate, which in turn leads to increased instability of the air, allowing for greater moisture-holding capacity and adding to the buoyancy of the air.

Orographic thunderstorms occur when air is forced to rise over land barriers, providing the necessary initial trigger action leading to the development of convectional cells. Thunderstorms of such orographic origin play a large role in the tremendous precipitation of the monsoons of South and Southeast Asia. In North America, they occur over the mountains in the West (the Rockies and the Sierra Nevada), especially during summer afternoons, when heating of south-facing slopes increases the air's instability. For this reason, pilots of small planes try to avoid flying in the mountains during the afternoon in summer for fear of getting caught in the turbulence of a thunderstorm.

Thunderstorms are often associated with cold fronts, where a cooler air mass forces a warmer air mass to rise. This action can bring about the strong, vertical updrafts necessary for convectional precipitation. In fact, a cold front may be immediately preceded by a line of thunderstorms, part of a **squall line** resulting from such frontal lifting.

As we mentioned in the discussion of precipitation types in Chapter 7, hail can be a product of thunderstorms when the vertical updrafts of the convection cells are sufficiently intense to carry water droplets repeatedly into a freezing layer of air. Fortunately, since thunderstorms are primarily associated with warm-weather areas, only a very small percentage around the world produce hail. In fact, hail seldom occurs in thunderstorms in the lower latitudes. In the United States, there is little hail along the Gulf of Mexico, where thunderstorms themselves are most common.

Tornadoes

Tornadoes are the most violent storms on the face of Earth (Fig. 8.11). They can occur almost anywhere but are far more common in the interior of North America than anywhere else in the world (Fig. 8.12). In fact, Oklahoma and Kansas lie in the path of so many "twisters" that together they are sometimes referred to as "tornado alley."

Figure 8.11 *The aftermath of a tornado that devastated Stillwater, New York, in June 1998.*

Systematic government documentation of tornado activity, such as that depicted in Figure 8.12a, began in 1875. Accounts of tornadoes occurring prior to 1875 must be tracked down through other sources. These accounts, while often unverifiable and vague, do offer interesting and informative insights into our forebears' perceptions of tornadoes. The accounts below describe a tornado that killed several people as it swept across several counties in western Illinois on May 21, 1859.

It was "a violent storm or hurricane [which] did immense damage to houses, barns, fences, and also caused some destruction of life." It was described as having a "frightful, . . . balloon or funnel shape, and appeared . . . peculiarly bright and luminous, not at all black or dark in any of its parts, except its base or bottom."

A vivid account of what surely must be related to the output of static electricity associated with a tornado is given in this account of the same tornado as it swept across Morgan county: "Mr. Cowell was plowing his field. . . . He saw the frightful cloud approaching . . . and at once attempted to drive his horses and plow to the house. . . . The horses suddenly took fright . . . their manes and tails and all their hair 'stood right out straight' as he expressed it, and . . . the iron in the harness . . . and plow, in his language 'seemed all covered with fire.' He felt a violent pulling of his own hair which left 'his head sore for some days' and the hair itself rigid and inflexible."

In addition, although unconfirmed by others, Mr. Cowell was one of the few individuals to have a tornado pass directly over him and live to tell about it. He described the light in the center of the tornado as being "so brilliant that he could not endure it with his eyes open, and for the most part kept them shut. . . . Yet [inside the tornado] there was no wind, no thunder and no noise whatever. . . ." Another interesting feature of this same tornado can be attributed to the low pressure of the vortex: "When the terrific whirl struck . . . [it] stripped all of the feathers off from the hens and turkeys, as perfectly clean as if picked for the table. Some, though badly plucked, and made entirely blind, still lived." Such a bizarre occurrence probably resulted when the hollow quills of the feathers expanded so suddenly—as the low pressure vortex moved over the area—that the birds' feathers "exploded."

Transactions of the Illinois Natural History Society
Phillips Bros., 1861

A **tornado** is actually a small, intense cyclonic storm of very low pressure, violent updrafts, and converging winds of enormous contrast. Fortunately, they are small and short-lived. Even in tornado alley, a tornado is likely to strike a given locale only once in 250 years.

Although only 1 percent of all thunderstorms produce a tornado, 80 percent of all tornadoes are associated with thunderstorms and midlatitude cyclones. The remaining 20 percent of tornadoes are spawned by hurricanes that make landfall. Approximately 700 to 1000 tornadoes occur each year in the United States, most of them from March to July in the late afternoon or early evening in the central part of the country.

Because of their small size and limited life span, tornadoes are extremely difficult to detect and forecast. However, relatively new radar technology, **Doppler radar,** improves tornado detection and forecasting significantly. Doppler radar has more power concentrated in a narrower beam than previous radar units. This allows meteorologists to assess storms in much greater detail (Fig. 8.13). Even more important is a Doppler radar's ability to measure the speed of the wind flow toward or away from the radar site, as well as any turbulence or sheer within the wind flow.

When the energy emitted by radar strikes precipitation, a small portion is scattered back to the radar. Depending on whether the precipitation is moving toward or away from the radar site, the wavelength of the returned energy is either compressed or elongated. The faster the movement, the greater the change. Previous

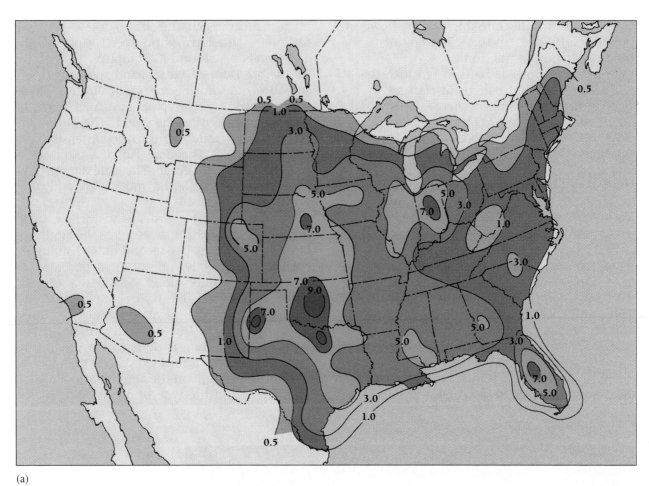

(a)

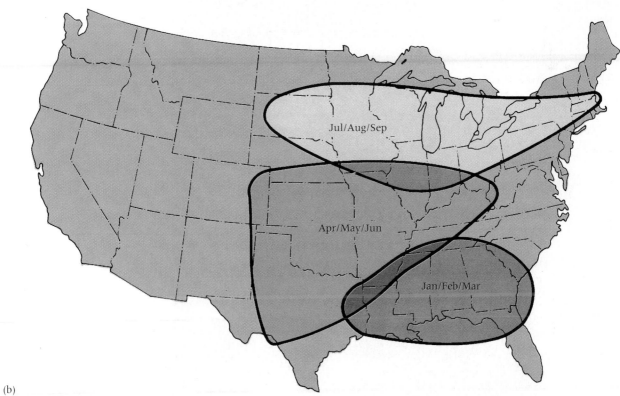

(b)

Figure 8.12 **(a)** *Average number of tornadoes per 26,000 square kilometers/10,000 square miles.* **(b)** *Seasonal march of peak tornado activity.*

radars could not measure this change; however, Doppler radar does and uses it to estimate the wind circulation of the storm.

In fact, Doppler radar is so sensitive that it can detect the wind pattern in *clear* air by detecting the back-scattered energy from clouds, pollution, insects, and so forth—something previous radars could not achieve. This allows meteorologists to see the formation of a tornado, thus increasing warning time to the public. Also, around airports, clear air turbulence (CAT), a major factor in airline accidents, can now be detected. The United States government, through the NEXRAD program (NEXt-generation weather RADar), has installed 151 Doppler radar sites across the country.

Doppler radar observations indicate that most tornadoes (62 percent) are fairly *weak* and have wind speeds of 45 meters per second (100 mph) or less. About 35 percent of tornadoes can be classified as *strong,* with wind speeds reaching 90 meters per second (200 mph). Nearly 70 percent of all tornado fatalities result from *violent* tornadoes. Although very rare (only 2 percent of all tornadoes reach this stage), these may last for hours and have wind speeds approaching 135 meters per second (300 mph).

Before Doppler radar, wind speeds within a tornado could not be measured directly. Therefore, tornado intensity was estimated from the damage produced by the storm. The most commonly used scale of tornado intensity was developed by T. Theodore Fujita, a professor at the University of Chicago. The scale is termed the Fujita Intensity Scale, or, more commonly, the F-scale (Table 8.2).

A tornado first appears as a swirling, twisting funnel cloud that moves across the landscape at 10 to 15 meters per second (22–32 mph). Its narrow end may be only 100 meters (330 ft) across. The funnel cloud becomes a tornado when its narrow end is in contact with the ground, where the greatest damage is done. Above the ground, the end can swirl and twist, but little or nothing is done to the ground below. The color of a tornado can be milky white to black depending on the amount and direction of sunlight and the type of debris being picked

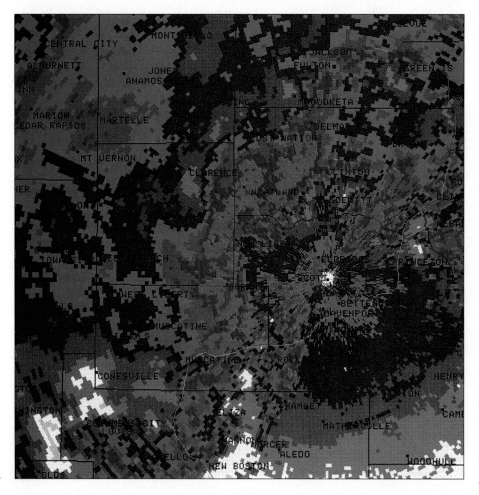

Figure 8.13 *Doppler radar image illustrating rainfall rates. The dark red areas are receiving rainfall at a rate of approximately 2.0 inches per hour.*

TABLE 8.2 Fujita Intensity Scale

Wind Speed		Expected Damage
(KPH)	(MPH)	
<116	<72	Light Damage Damage to chimneys and billboards; broken branches; shallow-rooted trees pushed over.
116–180	72–112	Moderate Damage The lower limit is near the beginning of hurricane wind speed. Surfaces peeled off roofs; mobile homes pushed off foundations or overturned; moving autos pushed off the road.
181–253	113–157	Considerable Damage Roofs torn off houses; mobile homes demolished; boxcars pushed over, large trees snapped or uprooted; light-object missiles generated.
254–332	158–206	Severe Damage Roofs and some walls torn off well-constructed houses; trains overturned; most trees in forest uprooted; heavy cars lifted off ground and thrown.
333–419	207–260	Devastating Damage Well-constructed houses leveled; structures with weak foundations blown some distance; cars thrown and large missiles generated.
>419	>260	Incredible Damage Strong frame houses lifted off foundations and carried considerable distance to disintegrate; automobile-sized missiles fly through the air farther than 100 m; trees debarked; incredible phenomena occur.

up by the storm as it travels across the land. Although most tornado damage is caused by the violent winds, most tornado injuries and deaths result from flying debris. The small size and short duration of a tornado greatly limit the number of deaths caused by tornadoes. In fact, more people die from lightning strikes each year than from tornadoes.

Weak Tropical Disturbances

Until World War II, the weather of the tropical regions was described as hot and humid, generally fair, but basically pretty monotonous. The only tropical disturbance given any attention was the tropical cyclone (also called a *hurricane* or *typhoon*), a spectacular, though relatively uncommon, storm that affects only islands, coastal lands, and ships at sea.

Even a few decades ago, an aura of mystery remained about the weather of this region. One reason for this lack of information was that the few weather stations located in the tropical areas were widely scattered and often poorly equipped. As a result, it was difficult to understand completely the passing weather disturbances in the tropics.

Largely through satellite technology and computer analysis, it is now known that there are a variety of weak atmospheric disturbances that affect the weather and relieve the monotony, although it is likely that the full number of these disturbances has not yet been recognized. The primary impact of these weak tropical disturbances on the weather of the tropical region is not on the temperature but rather on the cloud cover and on the amount of precipitation. Temperatures in the tropics are largely unaffected during the passage of a tropical storm, except that as the cloud cover is increased, temperature extremes are consequently reduced.

Easterly Wave. The **easterly wave** is the best known of the weak tropical disturbances. It shows up in Figure 8.14 as a trough-shaped, weak, low-pressure cell that is generally aligned on an approximate north-south axis. Traveling slowly in the trade wind belt from east to west, it is preceded by fair, dry weather and followed by cloudy, showery weather. This occurs because air tends to converge into the low from its rear, or the east, causing lifting and convectional showers. The resulting divergence and subsidence to the west account for the fair weather. Meteorologists believe that this type of disturbance can, on occasion, develop into a tropical hurricane.

Polar Outbreak. Occasionally, there may be an outbreak of polar air that follows a low into the subtropics and tropics. Such an outbreak would, of course, be preceded by the squalls and clouds and rain of a cold front. Following, however, would be a period of cool, clear, fair weather as the modified polar air influences are felt. On rare occasions near the equator in the Brazilian Amazon, such an Antarctic outburst, known locally as a *friagem,* can bring freezing temperatures and widespread damage to vegetation. Farther to the south, near São Paulo, the coffee crop can be ruined, causing coffee prices in North America to rise.

Hurricanes

Hurricanes are severe tropical cyclones which receive a great deal of attention from scientists and lay people alike, primarily because of their great destructive powers (Fig. 8.15). Abundant, even torrential, rains and winds often exceeding 45 meters per second (100 mph) characterize hurricanes. Though hurricanes develop over the oceans, their paths at times do take them over islands and coastal lands. The results can be devastating destruction of property and loss of life. It is not just the rains and winds that cause damage, for accompanying the hurricane are unusually high seas, called **storm surges,** which can flood entire coastal communities.

A **hurricane** is a circular, cyclonic system with wind speeds in excess of 33 meters per second (74 mph). It has a diameter of 160 to 640 kilometers (100–400 mi). Extending upward to heights of 12 to 14 kilometers (40,000–45,000 ft), the hurricane is a towering column of spiraling air (Fig. 8.16). At its base, air is sucked in by the very low pressure at the center and then spirals inward. Once within the hurricane structure, air rises rapidly to the top and spirals outward. This rapid upward movement of moisture-laden air produces enormous amounts of rain. Furthermore, the release of latent energy provides the power to drive the storm.

At the center of the hurricane is the eye of the storm, an area of calm, clear, usually warm

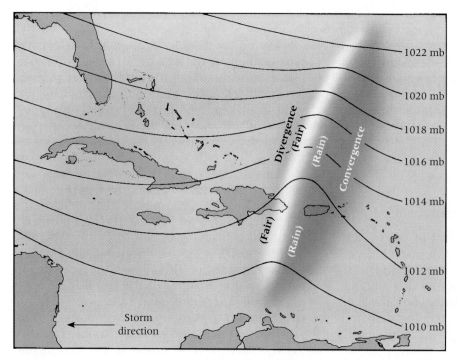

Figure 8.14 *A typical easterly wave in the tropics. Note that the isobars (and resulting winds) do not close in a circle but merely make a poleward "kink," indicating a low-pressure trough rather than a closed cell. The resulting weather is a consequence of convergence of air coming into the trough, producing rains, and divergence of air coming out of the trough, producing clear skies.* • **Why do easterly waves move toward the west?**

Labels on figure: 1022 mb, 1020 mb, 1018 mb, 1016 mb, 1014 mb, 1012 mb, 1010 mb, Divergence (Fair), (Rain), Convergence, (Fair), (Rain), Storm direction

Figure 8.15 An aerial view of damage caused by Hurricane Fran to the coastal area of North Carolina. The damage was most likely caused by the storm surge that accompanied the hurricane.

and humid but rainless air. Sailors traveling through the eye have been surprised to see birds flying there. Unable to leave the eye because of the strong winds surrounding it, these birds will often alight on the passing ship as a resting spot.

Hurricanes have very strong pressure gradients because of the extreme low pressure at their centers. The strong pressure gradients in turn cause the powerful winds of the hurricane. In contrast to the midlatitude cyclone, a hurricane is formed from a single air mass and does not have the different temperature sectors like a frontal system. Rather, a hurricane has a fairly

even, circular temperature distribution, which we might have been led to expect from its circular winds.

The Saffir-Simpson hurricane scale provides a means for classifying hurricane intensity, and potential damage, by assigning a number from 1 to 5 based on a combination of central pressure, wind speed, and the height of the storm surge (Table 8.3).

Although a great deal of time, effort, and money has been spent on studying the development, growth, and paths of hurricanes, much is still not known. For example, it is not yet possible to predict the path of a hurricane, even though it can be tracked with radar and studied by planes and weather satellites. In addition, meteorologists can list factors favorable for the development of a hurricane but cannot say that in a certain situation a hurricane will definitely develop and travel along a particular path.

Among the factors leading to hurricane development are a warm ocean surface of about 25°C (77°F) and warm, moist overlying air. These factors are probably the reasons why hurricanes occur most often in the late summer and early fall, when air masses have maximum humidity and ocean surface temperatures are highest. Also,

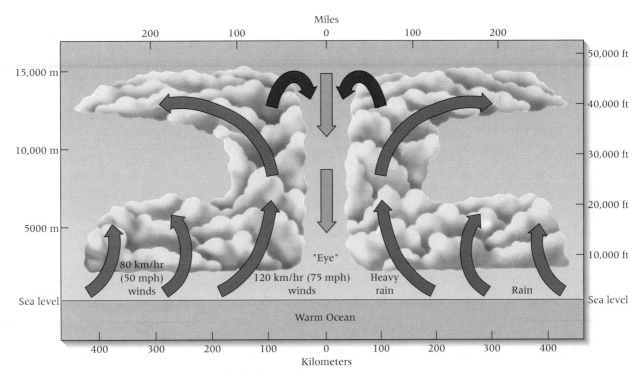

*Figure 8.16 Circulation patterns within a hurricane, showing inflow of air in the spiraling arms of the cyclonic system, rising air in the towering circular wall cloud, and outflow in the upper atmosphere. Subsidence of air in the storm's center produces the distinctive calm, cloudless "eye" of the hurricane. • **Why is this so?***

TABLE 8.3   Saffir-Simpson Hurricane Scale

Scale Number	Central Pressure	Wind Speed		Storm Surge		Damage
(Category)	(Millibars)	(KPH)	(MPH)	(Meters)	(Feet)	
1	≥980	119–153	74–95	1.2–1.5	4–5	Minimal
2	965–979	154–177	96–110	1.6–2.4	6–8	Moderate
3	945–964	178–209	111–130	2.5–3.6	9–12	Extensive
4	920–944	210–250	131–155	3.7–5.4	13–18	Extreme
5	<920	>250	>155	>5.4	>18	Catastrophic

the Coriolis effect must be sufficient to support the rapid spiraling of the hurricane. Because of this, hurricanes neither develop nor survive in the equatorial zone from about 8°S to 8°N, for there the Coriolis effect is far too weak. Hurricanes begin as weak tropical disturbances such as the easterly wave and, in fact, will not develop without the impetus of such a disturbance. It is further speculated that some sort of turbulence in the upper air may play a part in a hurricane's initial development.

Names are assigned to storms once they reach tropical storm status (winds greater than 17 meters per second/38 mph). Each year the names are selected from a different list of alternating female and male names—one list for the North Atlantic and one for the North Pacific. If a hurricane is especially destructive, its name is retired and never used again.

Hurricanes do not last long over land, for their source of moisture (and consequently their source of energy) is cut off. Also, friction with the land surface produces a drag on the whole system. North Atlantic hurricanes that move first toward the west with the trades and then north and northeast as they intrude into the westerlies become polar cyclones if they remain over the ocean and eventually die out. Over land, they will also become simple cyclonic storms, but even when they have lost some of their power, hurricanes can still do great damage.

Hurricanes occur over subtropical and tropical oceans and seas (the South Atlantic is an exception, though it is not known why). In Australia and the South Pacific they are called *tropical cyclones.* Near the Philippines, they are known as *bagyos,* but in most of East Asia they are called **typhoons.** In the Bay of Bengal they are referred to as *cyclones.*

Two hurricanes caused extensive damage in the United States during 1992. Hurricane Iniki struck the Hawaiian Islands, doing extensive damage to the island of Kauai. On August 24, 1992, Hurricane Andrew crashed into the populated areas of southern Florida with wind speeds of 150 miles per hour and then proceeded across the Gulf of Mexico to Louisiana. Andrew was the costliest natural disaster in U.S. history, causing over $20 billion in damage.

Some people have suggested that we seek ways to control these destructive storms. On the other hand, hurricanes are a major source of rainfall and an important means of transferring energy within Earth's system away from the tropics. Eliminating them might cause unwanted and unforeseen climate changes.

Define and Recall

air mass	Continental Tropical (cT)	stationary front	backing wind shift
source region		occluded front	squall line
Continental Arctic (cA)	Maritime Equatorial (mE)		tornado
Continental Polar (cP)		extratropical disturbance	easterly wave
Maritime Polar (mP)	cold front	polar front	storm surge
Maritime Tropical (mT)	warm front	storm track	hurricane
		veering wind shift	typhoon

Discuss and Review

1. What is an air mass?

2. Do all areas on Earth produce air masses? Why or why not?

3. What letter symbols are used to identify air masses? How are these combined? What air masses influence the weather of North America? Where and at what time of the year are they most effective?

4. Use Table 8.1 and Figure 8.1 to find out what kinds of air masses are most likely to affect your local area. How do they affect weather in your area?

5. What forces modify the behavior of air masses? What kinds of weather may be produced when an air mass begins to move?

6. Why do you suppose air masses can be classified by whether they develop over water or over land?

7. Why does *mP* air affect the United States? Are there any deviations from this tendency?

8. What kind of air mass forms over the southwestern United States in summer? Have you ever experienced weather in such an air mass? What was it like? What kind of weather might you expect to experience if such an air mass met an *mP* air mass?

9. What is a front? How does it occur?

10. In a meeting of two contrasting air masses, how can the aggressor be determined?

11. Compare warm and cold fronts. How do they differ in duration and precipitation characteristics?

12. What kind of weather often results from a stationary front? What kinds of forces, do you suppose, tend to break up stationary fronts?

13. How does the westerly circulation of winds affect air masses in your area? What kinds of weather result?

14. What are the major differences between middle-latitude cyclones and anticyclones?

15. Can you draw a diagram of a mature (fully developed) midlatitude cyclone that includes the center of the low with several isobars, the warm front, the cold front, wind direction arrows, appropriate labeling of warm and cold air masses, and zones of precipitation?

16. If a wind changes to a clockwise direction, what is the shift called? Where does it locate you in relation to the center of a low-pressure system? Explain why this happens.

17. How does the configuration of the upper-air wind patterns play a role in the surface weather conditions?

18. How can a knowledge of cyclones be used to help forecast weather changes? For how long in advance? What changes might occur to spoil your forecast?

19. Describe the sequence of weather events over a 48-hour period in St. Louis, Missouri, if a typical low-pressure system (cyclone) passes 300 kilometers (180 mi) north of that location in the spring.

20. List three major causes of thunderstorms. How might the storms that develop from each of these causes differ?

21. Have you ever experienced a tornado or hurricane? Describe your feelings during it and the events surrounding it. How do the news media prepare us for such natural disasters?

Consider and Respond

1. Collect a three-day series of weather maps from your local newspapers. Based upon the migration of high- and low-pressure systems during that period, discuss the likely pattern of the upper-air winds.

2. List the ideal conditions for the development of a hurricane.

3. Look at Figure 8.8a. Assume you are driving from Point A to Point B. Describe the changes in weather (temperature, wind speed and direction, barometric pressure, precipitation, and cloud cover) you would encounter on your trip. Do the same analysis for a trip from Point C to Point D.

4. The location of the polar front changes with the season. Why? In what way is Figure 8.12b related to the seasonal migration of the polar front?

5. Redraw Figure 8.7 so that it depicts a Southern Hemisphere example.

CHAPTER

9

CLIMATE AND CLIMATE CHANGE

CHAPTER PREVIEW

▶ Because atmospheric elements vary so greatly from place to place over Earth's landmasses, scientists have classified climates by combining those elements with similar statistics to identify a manageable number of groups or types.

What two atmospheric elements are most often used when classifying climates, and why are they the elements selected? Why is there a need for more than one system of climate classification?

▶ On a global or macro scale, the Köppen system of climate classification is the most widely used.

What are some of the advantages and disadvantages of the Köppen system? What are some advantages of the Thornthwaite system, and how does it differ from the Köppen system? At what other scales can climate be studied?

▶ A knowledge of past climates is critical to an understanding of the nature of present-day climates and to the attempted prediction of future climates.

What important climatic changes have occurred over the past 2.4 million years? How has the Holocene, or most recent climatic epoch, differed from much of the rest of the Pleistocene? What recent evidence has been most important to the deciphering of climatic changes during the Pleistocene?

▶ Scientists today believe that dramatic changes in climate can occur over surprisingly short periods of time.

On what evidence are climatologists basing this belief? What type of feedback system is probably responsible for such rapid climatic change? What major climatic changes have occurred over the past 18,000 years?

▶ Because there are a number of plausible causes of climatic change, it is difficult for scientists to determine which cause is responsible for a specific climatic event.

What are some of the major causes? Which causes seem to best explain some of the most important climatic events of the past? In which of the causes is human activity most likely involved?

▶ Despite the application of the best scientific methods and the use of the most modern technology, there remains great difficulty in accurately predicting future climate.

Why is this so? What methods and technology are most commonly used? What climate changes are most likely in the immediate future? In the more distant future?

▲ Gravel deposits in an open valley where ice once stood on the slopes of Mt. Rainier: clear evidence of climate change.

In Chapter 4 weather and climate were defined briefly. In this chapter we will begin the study of climate in much greater detail. Unlike weather, which describes the state of the atmosphere over short periods of time, climatic analysis relies heavily on averages and statistical probabilities involving data accumulated for the atmospheric elements over periods of many years. Climatic descriptions include such things as average annual precipitation and temperature; patterns of change in wind direction, temperature, and precipitation throughout the year; anticipated number of days of sunshine or cloud cover; and ranges between the statistical highs and lows that may be expected of each atmospheric element.

In this chapter we first will introduce the characteristics and classification of modern climates. Since climate can be defined at different scales, from a single hillside to a region as large as the Sahara stretching across much of northern Africa, two systems of describing and classifying climates will be discussed. The first of those systems has been widely adopted by physical geographers and other scientists and, in a modified version, it will be the basis for the worldwide regional study of present-day climates included in Chapters 10 and 11. The second system is introduced because it is one of the most effective available to scientists for classification and study of climates on a local scale.

Although the first part of this chapter will deal with modern climate classification, the remainder of the chapter will focus on climatic change. Climates in the past were not the same as they are today and there is every reason to believe that future climates will be different as well.

It is now also recognized that humans can alter Earth's climate. The most direct human impact on climate results from a change in the composition and relative abundance of atmospheric gases. The concentrations of gases can change either directly, through the release of enormous quantities of carbon dioxide (CO_2) and other "greenhouse" gases, or indirectly, from such actions as paving over fields or forests, thus reducing the number of plants that absorb CO_2 from the atmosphere.

For decades scientists have realized that Earth has experienced major climate shifts during the past 30,000 years. But it was believed that the shifts were gradual, and could not be detected by humans during a lifetime or two. However, recent research into the record of the deep sea sediments and polar ice caps reveals that the climate has shifted repeatedly between extremes over some exceedingly short intervals. Moreover, the research has revealed that climate during the most recent 10,000 years has been extraordinarily stable compared to similar intervals in the past.

In order to predict future climates, it is critical that the details of past climate changes be examined, including both the magnitude and rates of prehistoric climate change. During the past 18,000 years, Earth has experienced both an ice age and a lengthy period that was warmer than today. These fluctuations serve as indicators of the natural variability of climate in the absence of significant human impacts. Using knowledge of present and past climates, as well as models of how and why climate changes, we will conclude this chapter with predictions of future climate trends.

Classifying Climates

Knowledge that climate varies from region to region dates to ancient times. The Greeks classified the known world into Torrid, Temperate, and Frigid zones based on average temperatures. It was also recognized that these zones varied systematically with latitude, and that the flora and fauna reflected these changes as well. With the further exploration of the world by Europeans, naturalists noticed that the distribution of climates could be explained using factors such as sun angles, prevailing winds, elevation, and proximity to large water bodies.

The two weather variables used most often as indicators of climate are temperature and precipitation. To classify climates accurately, climatologists need many years of records of these two atmospheric elements. The invention of an instrument to reliably measure temperature—the thermometer—dates only to Galileo in the early seventeenth century. European settlement of and sporadic collection of temperature and precipitation data from distant colonies began in the 1700s but was not routine until the mid–nineteenth century. This was soon followed in the early twentieth century by some of the first attempts to classify global climates using actual temperature and precipitation data.

As we have seen in earlier chapters, temperature and precipitation vary greatly over Earth. But climatologists can reduce the infinite number of worldwide variations in atmospheric elements to a comprehensible number of groups or varieties by combining elements with *similar* statistics (Fig. 9.1). That is, they can classify climates strictly on the basis of atmospheric elements, ignoring the causes of those variations (such as air-mass analysis). Such a classification, based on statistical parameters or physical characteristics, is called an **empirical classification.** On the other hand, if climatologists base their classification on the causes or the *genesis* of climatic variation, they would have a **genetic classification.**

Ordering the vast wealth of available climatic data into descriptions of major climatic groups, on either an empirical or a genetic basis, enables geographers to concentrate on the larger-scale causes of climatic differentiation. In addition, they are also able to examine exceptions to the general relationships, the causes of which are often one or more of the minor atmospheric controls. And, finally, differentiating climates helps to explain the distribution of other climate-related phenomena of importance to humans.

Despite its value, climate classification is not without its problems. One reason is that climate is a generalization about observed facts based on the averages and probabilities of weather; therefore, it does not describe a real weather situation; instead, it presents a composite weather picture.

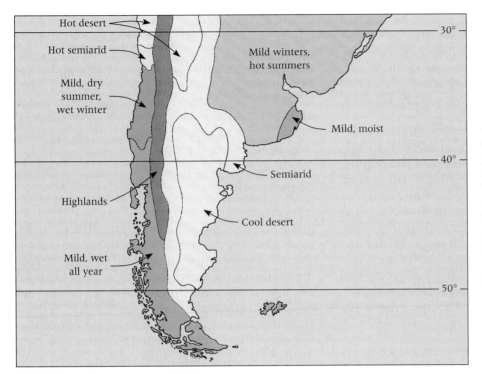

Hot desert

Hot semiarid

Mild, dry summer, wet winter

Mild winters, hot summers

Mild, moist

Semiarid

Highlands

Cool desert

Mild, wet all year

30°

40°

50°

Figure 9.1 *This map shows the diversity of climates possible in a relatively small area, including portions of Chile, Argentina, Uruguay, and Brazil. The climates range from dry to wet and from hot to cold, with many possible combinations of temperature and moisture characteristics.* • **What can you suggest as the causes for the major changes in climate as you follow the 40° south latitude line from west to east across South America?**

In addition, it is impossible within such a generalization to include the many variations that actually exist. On a global scale, generalizations, simplifications, and compromises are made to distinguish between climatic types and regions.

The Köppen System

The most widely used climate classification is based on temperature and precipitation patterns and is referred to as the **Köppen system,** named for the German botanist and climatologist who developed it. Dr. Wladimir Köppen recognized that major vegetation associations reflect the local climate. Hence, his climatic regions were formulated to coincide with well-defined vegetation regions, and each climatic region can be described by the natural vegetation most often found there. Evidence of the strong influence of Köppen's system is seen by the wide usage of his climatic terminology, even in nonscientific literature (for example, steppe climate, tundra climate, rainforest climate).

Advantages and Limitations of Köppen's System. Not only are temperature and precipitation two of the easiest weather elements to measure, they are also measured more often and in more parts of the world than any other variables. By using temperature and precipitation statistics to define his boundaries, Köppen was able to develop pre-cise definitions for each climatic region, eliminating the imprecision that can develop in verbal and sometimes in genetic classifications.

Moreover, temperature and precipitation are the most important and effective weather elements. Variations caused by the atmospheric controls will show up most obviously in temperature and precipitation statistics. On the other hand, temperature and precipitation are the weather elements that most directly affect humans, other animals, vegetation, soils, and the form of the landscape. Because he tied his classification to these two elements, Köppen's system is closely related to the visible aspects of our environment.

Köppen's climatic boundaries were designed to define the vegetation regions included in a classification of world vegetation produced earlier by the Swiss botanist Alphonse de Candolle. Thus, Köppen's climatic boundaries reflect "vegetation lines." For example, the 10°C (50°F) monthly isotherm is used by Köppen because it has relevance to the timberline, the line beyond which it is too cold for trees to thrive. Trees need a certain amount of warmth to grow, and unless at least one month of the year has an average temperature exceeding 10°C, trees will ordinarily be unable to survive. For this reason, Köppen defined the treeless polar climates as including those areas where the mean temperature of the warmest month is below 10°C. Clearly, if climates are divided according to associated vegetation types, and if the division is based on the atmospheric elements of temperature and precipitation, the result will be a visible association of vegetation with climatic types. The relationship with the visible world in Köppen's climate classification system is one of its most appealing features to geographers.

There are, of course, limitations to Köppen's system. For example, Köppen considered only average monthly temperature and precipitation in making his climatic differentiations. These two climatic elements permit estimates of precipitation effectiveness but do not measure it with enough precision to permit comparison from one locality to another. In addition, for the purposes of generalization and simplification, Köppen ignored winds, cloud cover, intensity of precipitation, humidity, and daily temperature extremes—much, in fact, of what makes local weather and climate distinctive.

Simplified Köppen Classification. The Köppen system, as modified by later climatologists, divides the world into six major climate categories. The first four are based on the annual range of temperatures: humid tropical climates *(A)*, humid mesothermal (mild winter) climates *(C)*, humid microthermal (severe winter) climates *(D)*, and polar climates. Another category, the arid and semiarid climates *(BW and BS)*, identifies regions that are characteristically dry based on both temperature and precipitation values. Since plants need more moisture to survive as the temperature increases, the arid and semiarid climates include regions where the temperatures range from cold to very hot. The final category, highland climates *(H)*, identifies mountainous regions where vegetation and climate vary rapidly as a result of changes in elevation and exposure.

Within each of the first five major climatic categories there are individual climatic types and subtypes differentiated from one another by specific parameters of temperature and precipitation. Some are characterized by seasonal rainfall in either summer or winter whereas others exhibit evenly distributed monthly precipitation. Other climatic types are differentiated by aspects of their annual temperature ranges that mostly reflect the degree of continentality associated with the climate (for example, hotter summers or colder winters inland; milder summers or winters in coastal locations).

Köppen gave names to each of his individual climatic types, and, as indicated when we introduced his six major climate categories, he also used combinations of letters to symbolize them. The letter symbols provide an international shorthand describing climatic types that are often difficult to characterize in words. In the interest of generalization, we will use only the names when identifying climates in the discussion of climatic regions included later in Chapters 10 and 11. However, a detailed presentation of Köppen's letter symbols, along with the definitions used to classify climates in the system, may be found in the Appendix.

Climographs

It is possible to summarize the nature of the climate at any point on Earth by graphic means, as in Figure 9.2. Given information on mean monthly temperature and rainfall, we can express the nature of the changes in these two elements throughout the year simply by plotting their values as points above or below (in the case of temperature) a zero line. To make the pattern of the monthly temperature changes clearer, we can connect the monthly values with a continuous line, producing an annual temperature curve. To avoid confusion, monthly precipitation amounts are usually shown as bars reaching to various heights above the line of zero precipitation. Such a display of a location's or station's climate is called a **climograph.** To read it, one must relate the temperature curve to the values given along the left margin of the graph, while the precipitation amounts are read on the right margin. Other information may also be displayed, depending upon the type of climograph used. Figure 9.2 represents the type we will use in Chapters 10 and 11. This climograph can be used to determine the Köppen classification of the station as well as to show its specific temperature and rainfall regimes.

The Thornthwaite System

For a farmer interested in growing a specific crop in a particular area, the Köppen system is inadequate. Although the Köppen system correctly identifies the major vegetation type of the region and the annual range of both temperature and precipitation, it does not provide a farmer with information concerning the amounts and timing of annual soil moisture surpluses or deficits. From a farmer's perspective, it is much more important to know that moisture will be available in the growing season, whether it comes directly in the form of precipitation or from the soil.

The **Thornthwaite system** was developed by an American climatologist, Dr. C. Warren Thornthwaite, to examine moisture availability at the subregional scale (Fig. 9.3). It is the system

(Continued on page 216)

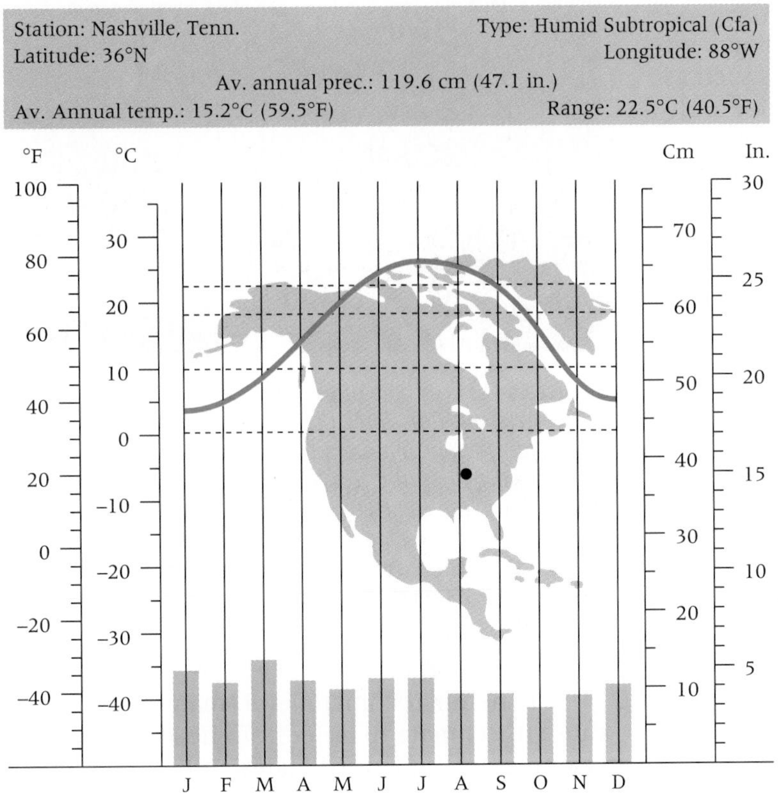

Station: Nashville, Tenn.
Latitude: 36°N

Type: Humid Subtropical (Cfa)
Longitude: 88°W

Av. annual prec.: 119.6 cm (47.1 in.)

Av. Annual temp.: 15.2°C (59.5°F)

Range: 22.5°C (40.5°F)

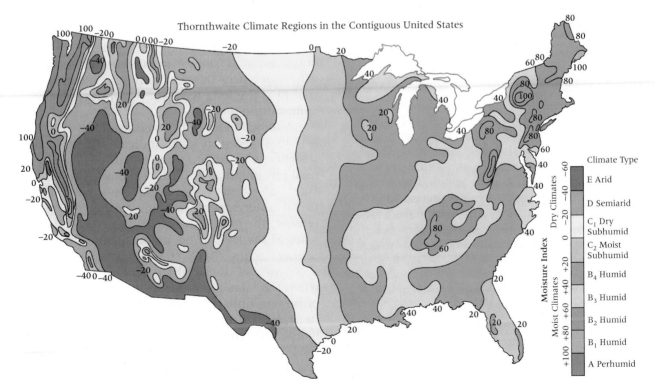

Figure 9.2 A standard climograph showing average monthly temperature (curve) and rainfall (bars). The horizontal index lines at 0°C (32°F), 10°C (50°C), 18°C (64.4°F), and 22°C (71.6°F) are the Köppen temperature parameters by which the station is classified. • **What specific information can you read from the graph that identifies Nashville as a specific climate type (humid subtropical) in the Köppen classification?**

Figure 9.3 Thornthwaite climate regions in the contiguous United States may be based on the relationship between precipitation (P) and potential evaporation (PE). The moisture index (MI) for a region is determined by the simple equation: $MI = 100 \times ((P - PE) \div PE)$. Where precipitation exceeds potential evaporation the index is positive and where potential evaporation exceeds precipitation the index is negative. • **What are the moisture index and Thornthwaite climatic type for coastal southern California?**

preferred by many soil scientists and others examining climates on a local scale. Development of detailed climate classification systems such as the Thornthwaite system became possible only after temperature and precipitation data were widely collected at numerous locations beginning in the latter half of the nineteenth century.

The Thornthwaite system focuses on the concept of **Potential Evapotranspiration** (PE), which approximates the water use of plants with an unlimited water supply. Potential evapotranspiration increases with increasing temperature, winds and, daylight length. However, it decreases with increasing humidity. **Actual Evapotranspiration** (AE), which reflects actual water use by plants, can be supplied during the dry season by soil moisture if the soil is saturated, the climate is relatively cool, and/or the daylengths are short. Thus, measurements of AE relative to PE and available soil moisture are the determining factors for most vegetation and crop growth.

The Thornthwaite system recognizes three climate zones based on PE values: low-latitude climates (PE greater than 130 cm [51 in.]); mid-latitude climates (PE less than 130 cm but greater than 52.5 cm [20.5 in.]); and high-latitude climates (PE less than 52.5 cm). Climate zones may be subdivided based on how long, and by how much, AE is below PE. Moist climates have either surpluses or a minor deficit (less than 15 cm [6 in.]). Dry climates have an annual deficit greater than 15 cm. Because of extreme moisture seasons, two classes do not fit into the normal classification system: the Tropical Wet and Dry regions and the Mediterranean climate regions.

Thornthwaite's original equations for PE were based on analyses of data collected in the midwest and eastern United States. The method was subsequently used with less success in other parts of the world. Over the past few decades many attempts have been made to improve the accuracy of the Thornthwaite system for regions outside the United States.

Scale and Climate

Climate can be measured at different scales (macro, meso, or micro). The climate of a large (macro) region, such as the Sahara, may be described correctly as hot and dry. Climate can also be described at mesoscale levels; for example, the climate of coastal southern California is sunny and warm, with dry summers and wet winters. Finally, climate can be described at local scales, such as on the slopes of a single hill. This is termed a **microclimate.**

Figure 9.4 *This photograph facing west near Desert View, Arizona, illustrates the significance of slope aspect. The south-facing slopes on the right of the photograph are receiving the full effect of the desert sun's energy. These slopes are clearly hotter and drier than the more shaded north-facing slopes.* • **Why do the differing angles that the sun's rays strike the two opposite slopes affect temperatures?**

At the microclimate level, many factors will cause the climate to differ from nearby areas. For example, in the United States and other regions north of the Tropic of Cancer, south-facing slopes tend to be warmer and dryer than the north-facing slopes because they receive more sunlight (Fig. 9.4). This variable is referred to as *slope aspect,* the direction a mountain slope faces in respect to the sun's rays. Microclimatic differences such as slope aspect can cause significant differences in vegetation and soil moisture that over the long term affect soil thickness and composition.

Microclimates can also be affected by human activities. Recent research indicates that the construction of a large reservoir leads to greater annual precipitation immediately downwind of the lake. This occurs because the lake supplies additional water vapor to passing storms, which intensifies the rainfall or snows immediately downwind of the lake. These microclimatic effects are similar to the "lake effect" snows that occur downwind of the Great Lakes in the early winter when the lakes are not frozen. Another example of human impact on microclimates is the urban **heat island** effect, which leads to changes in temperature (urban centers tend to be warmer than their outlying rural areas), rainfall, wind speeds, and many other phenomena (Table 9.1).

Past Climates

To try to predict future climates it is critical to understand the magnitude and frequency of previous climate changes. Knowledge that Earth ex-

TABLE 9.1	Effects of Urbanization on Climatic Elements*	
Elements	Comparison with Rural Environment
Pollutants	
Solid particles	10 times more
Gases	5–25 times more
Cloud cover	5–10 percent greater
Fog, winter	100 percent more
Fog, summer	30 percent more
Precipitation	5–10 percent more
Snowfall	5 percent less
Rain days with less than 5 mm	10 percent more
Relative humidity, winter	2 percent less
Relative humidity, summer	8 percent less
Radiation	15–20 percent less
Ultraviolet radiation, winter	30 percent less
Ultraviolet radiation, summer	5 percent less
Duration of sunshine	5–15 percent less
Annual mean temperature	0.5°–1.0°C higher
Heating degree days	10 percent fewer
Annual mean windspeed	20–30 percent less
Calms	5–20 percent more

* From World Meteorological Organization, 1970.

perienced major climate changes in the past is not new. In 1837, Louis Agassiz, a European naturalist, proposed that Earth had experienced major periods of *glaciation,* periods known as ice ages, when large areas of the continents were covered by sheets of ice. He presented evidence, controversial at the time, that glaciers (flowing ice) once had covered most of England, northern Europe and Asia, as well as the foothill regions of the Alps. Agassiz' ideas were not readily accepted in Europe, in part because the continental glaciers covering Greenland and Antarctica had yet to be discovered. Agassiz arrived in the United States in 1846, and similar evidence of widespread glaciation was soon recognized throughout North America.

The Pleistocene Epoch

The nomenclature that defines Earth's geologic history was developed and defined before knowledge of ice ages was thorough. The **Pleistocene** epoch has become synonymous with the time of the most recent ice age. Unfortunately, the Pleistocene was originally defined on the basis of the first appearance of cold-water marine species in a sequence of Italian sedimentary rocks. Subsequent radiometric dating of the sediments reveals that this occurred only about 1.6 million years ago in the Mediterranean Sea. Today, it is recog-

nized that major glacial advances occurred as early as 2.4 million years ago, several hundred thousand years before the Pleistocene "officially" begins.

Until the 1960s, it was widely believed that Earth had experienced four major glacial advances followed by warmer interglacial periods (glacial cycles) during the Pleistocene. In the foothills of the Alps near the Danube River, evidence of four glacial advances was discovered. These glacial epochs were termed the Gümz (oldest), Mindel, Riss, and Würm. Likewise, in the United States, evidence of four glacial periods was recognized. Based on the southward limit of the glacial advances, these were termed the Nebraskan (oldest), Kansan, Illinoian, and Wisconsinan glaciations.

A major problem with studying the advance and retreat of glaciers on land is that each subsequent advance of the glaciers tends to destroy, bury, or greatly disrupt the sedimentary evidence of previous glaciers. The evidence of the fourfold record of glacial advances was largely recognized on the basis of minor glacial deposits lying beyond the limit of the two most recent glaciations. Evidence of "average" glacial advances that were subsequently overridden by more recent glaciers was rarely recognized.

Before the advent of radiometric dating techniques following World War II, the timing of the glacial advances in both Europe and the United States was only crudely known. For example, estimates of the age of the last deglaciation were based on the rates at which Niagara Falls and St. Anthony's Falls had eroded headward after the areas were first exposed when the glaciers retreated. Calculations ranging from 8000 to 30,000 years ago—surprisingly recent—were produced. In Sweden, the age of deglaciation was based on the number of sedimentary couplets, or **varves,** that are present in some newly exposed lake basins. A varve is a pairing of organic-rich summer sediments and organic-poor winter sediments, and each couplet represents one year of time. This technique suggested that the region had been deglaciated only about 13,000 years ago. Many scientists were skeptical about such studies because it was difficult to believe that several thousand feet of ice had covered large areas of Europe and North America so recently.

Modern Research

Two major advances in scientific knowledge about climate change occurred in the 1950s. First, radiometric techniques, such as radiocarbon dating, that measured the absolute ages of landforms produced by the glaciers began to be widely used.

Radiocarbon dating of Pleistocene lake deposits in the western United States and glacial debris in the midwest conclusively showed that the last ice advance peaked a mere 18,000 years ago (Fig. 9.5). Given that the Pyramids of Egypt were built about 4000 years ago, it was mind-boggling for scientists to realize that major glaciers dominated landscapes as far south as the Ohio and Missouri Rivers only 18,000 years ago, covering modern-day city sites such as Boston, New York, Indianapolis, and Des Moines.

The second major discovery was that evidence of detailed climate changes have been recorded in the sediments on the ocean floors. Unlike the continental record, the deep-sea sedimentary record has not been disrupted by subsequent glacial advances. Rather, the slow, continuous sediment record provides a complete history of climate changes during the past several million years. The most important discovery of the deep-sea record is that Earth has experienced numerous major glacial advances during the Pleistocene, not just the four that had been identified previously. Today, the names of only two of the glacials, the Illinoian and Wisconsinan, have been retained.

Since the deep-sea sedimentary record is so important to climate change studies, it is impor-

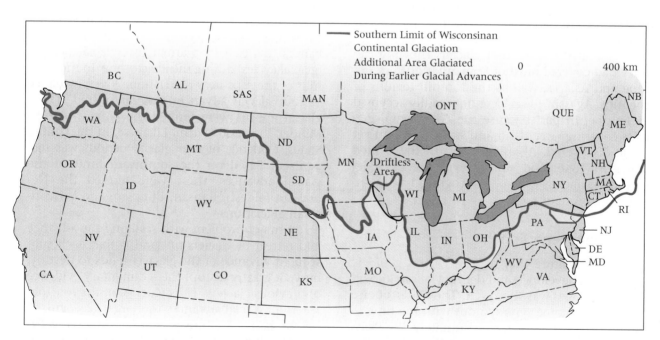

Figure 9.5 *This map identifies the extensive areas of Canada and the northern United States that were covered by moving sheets of ice as recently as 18,000 years ago. It also shows that ice covered large highland regions in the western United States and extended even further south during the earlier glacial advances.* • **Can you suggest a relationship between glaciation and the location of the Missouri and Ohio Rivers?**

tant to understand how the record is deciphered. First, the deep-sea ooze contains the microscopic record of innumerable surface-dwelling marine animals, collectively known as **foraminifera,** who have died and sunk to the sea floor. Different species thrive in different surface water temperatures, and so the stratigraphic record of foraminifera fossils (tests) produces a detailed history of water temperature fluctuations.

Because the foraminifera tests are composed of calcium carbonate ($CaCO_3$), the tests also record the oxygen composition of the sea water in which they lived. One common measurement technique for determining oxygen composition is known as **oxygen-isotope analysis.** An isotope of a given element has a different number of neutrons, and thus a slightly different mass (weight), than the other atoms of the same element. Two of the most common isotopes of oxygen are ^{16}O and the "heavier" ^{18}O. Modern sea water has a fixed ratio of the two oxygen isotopes. When water evaporates from the ocean, slightly more ^{16}O relative to ^{18}O evaporates because water containing the lighter-weight oxygen evaporates more readily. During an ice age, the evaporated water is stored on the continents in the form of glaciers rather than returned to the oceans. Not only does the storage of water on land result in a sea level drop of more than 100 meters (325 ft), but the ratio of ^{16}O to ^{18}O in the ocean changes slightly to reflect the ^{16}O-enriched water being stored in the glaciers. Analysis of the oxygen ratios recorded in the foraminifera tests reveals a record of sea level changes that reflect the growth and decline of ice sheets. The combined record of cold and warm foraminifera species and the associated oxygen-isotope ratios has produced a detailed reconstruction of climate changes.

Because analyses of the deep-sea sedimentary record produced a very complex record of climate changes, a new system was required to describe the major shifts in global climates. The record of oxygen-isotope changes was divided into intervals, termed "stages," of cold and warm climates. These shifts were numbered back in time, beginning with the number one for the current warm interval. Thus, all warm intervals have odd-numbered oxygen-isotope stages (OIS), and all glacial intervals have even-numbered oxygen-isotope stages. A review of the oxygen-isotope record indicates that the last glacial advance about 18,000 years ago was only one of many major glacial advances during the past 2.4 million years (Fig. 9.6).

Today, climatologists are aware that the present climate is but a short interval of relative stability in a time of major climate shifts. Moreover, the modern climatic epoch, known as the **Holocene** or **Recent,** is a time of extraordinarily stable, warm temperatures compared to most of the last 2.4 million years. Based on the deep-sea record, it appears that global climates tend to rest at one of two extremes: a very cold interval characterized by major glaciers and lower sea levels, and shorter intervals between the glacial advances marked by unusually warm temperatures and high sea levels. With the realization that global climates have changed dramatically numerous times, two obvious questions arise: how

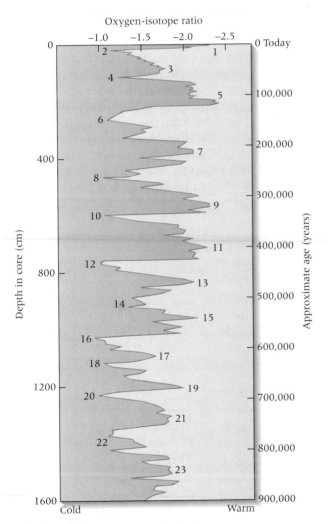

Figure 9.6 *As this graph indicates, the analysis of oxygen-isotope ratios in fossils that have accumulated in deep-sea sediments over the past 900,000 years reveals frequent changes between periods of colder and warmer climates.* • **How does this recent reconstruction of past climates differ from the widely accepted theories of fifty years ago?**

quickly does global climate change from one extreme to the other, and what causes global climate to change so often?

Rates of Climate Change

Radiocarbon dating of glacial features confirmed that full-glacial conditions existed about 18,000 years ago. In the United States, glaciers covered most areas north of the Missouri and Ohio Rivers. In the west, freshwater lakes more than 500 feet deep covered much of Utah and Nevada. The same scientific methods that showed the extent of glacial features also revealed that the United States was mostly glacier-free, and the western lake basins dry, by about 9000 years ago. Abundant evidence was even found that the climate about 7000 years ago (a time known as the **Altithermal**) was hotter than today.

For glaciers several thousand feet thick to melt completely, and for deep lakes to evaporate, a substantial increase in insolation is required over a few thousand years. Where did so much extra energy come from?

To answer questions about such rapid rates of climate change requires a more detailed record of climate than the deep-sea sediments can provide. This is because the deep-sea sedimentary record is extraordinarily slow—a few centimeters of ooze accumulates in a thousand years. Rapid shifts in climate during periods of a few hundred years are not recorded clearly in the seafloor sediments. This problem was solved by coring the thick glaciers covering Antarctica and Greenland. Glacial ice records yearly amounts of snowfall and is much more likely to provide short-term evidence of climate changes. Analyses of oxygen isotopes in the glacial ice of Antarctica and, most recently, Greenland, have revealed a detailed record of climate changes during the past 250,000 years (Fig. 9.7).

A surprising discovery of the ice sheet analyses is the speed at which climate changes. Rather than gradually changing from glacial to interglacial conditions over thousands of years, the ice record indicates that the shifts can occur in a few years or decades. Thus, whatever is most responsible for major climate changes must develop rapidly. This probably requires a *positive feedback system*, which means, as explained in Chapter 1, that a change in one variable will cause changes in other variables that magnify the amount of original change. For example: Most glaciers have high albedos, reflecting significant amounts of sunlight back to space. However, if the ice sheets begin to decline for whatever reason, low-albedo land begins to replace the high-

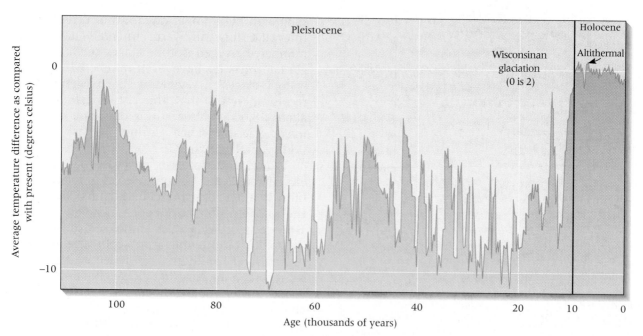

Figure 9.7 *Analyses of oxygen-isotope ratios in ice cores taken from the glacial ice of Antarctica and Greenland provide evidence of surprising shifts of climate over short periods of time.* • **Has the general trend of temperatures on Earth been warmer or colder during the Holocene?**

albedo ice, increasing the amount of energy available to melt the ice. Thus, the more ice that melts, the more energy that is available to melt the ice further, magnifying the initial glacial retreat.

In contrast, a *negative feedback system,* where changes in one of the variables induce the system to remain stable, will also affect the likelihood or rate of climate change. For example, since increasing global temperatures cause evaporation rates to increase, the more water that evaporates from the ocean surface, the more clouds that will form. The more clouds that exist, the more insolation is reflected back to space, cooling Earth's surface. (A counter argument to this effect is that clouds also operate as a greenhouse blanket, trapping heat in the lower atmosphere.) Thus, for climate changes to occur rapidly, negative feedback cycles such as this one must be overwhelmed by positive feedback cycles.

Causes of Climate Change

Although theories about the causes of climate change are numerous, they can be organized into five broad categories. They are (1) astronomical variations in Earth's orbit; (2) changes in the output of solar energy and/or the transparency of outer space; (3) changes in atmospheric gases or dust, causing changes in incoming or outgoing radiation; (4) changes in oceanic circulation; and (5) changes in landmasses that affect albedo and oceanic circulation.

Orbital Variations

Astronomers have detected slow changes in Earth's orbit that affect the distance between the sun and Earth as well as the deviation of Earth's orbit from the plane of the ecliptic. These orbital cycles produce regular changes in the amount of solar energy that reaches Earth. The longest is known as the **eccentricity cycle,** which is a 100,000-year variation in the shape of Earth's orbit around the sun. In simple terms, Earth's orbit changes from a slight ellipse to a more circular orbit, and then back. In addition, the planet slightly changes its position relative to the plane of the ecliptic. A second cycle is termed the **obliquity cycle,** and represents a 41,000-year variation in the tilt of Earth's axis from a maximum 24.5 degrees to a minimum of 22.1 degrees, and back. The greater Earth's tilt, the greater the seasonality at mid- and high latitudes. Finally, a **precession cycle** has been recognized that has

a major periodicity of 23,000 years and a secondary periodicity of 19,000 years. The precession cycle determines the time of year that perihelion occurs. Today, Earth is closest to the sun on January 3 and, as a result, receives about 3.5 percent greater insolation than the average in January. Because of the precession cycle, this date advances one calendar day each 71 years, and thereby alters the timing of maximum insolation receipt: the Northern Hemisphere is closest to the sun on January 3, so that the winter season in the Northern Hemisphere is somewhat warmer than average today. When aphelion occurs on January 3 in about 11,500 years, the Northern Hemisphere winters should be somewhat colder.

Because these cycles operate collectively, the combined effect of the three cycles must be calculated (Fig. 9.8). The first person to examine all three of these cycles in detail was a mathematician, Milutin Milankovitch. In 1924, prior to the invention of calculators, Milankovitch completed the complex mathematical calculations and published graphs showing how these subtle changes in Earth's orbit would affect insolation at three different latitudes (Fig. 9.9). Computers permit the values to be calculated for all latitudes (Fig. 9.10). Milankovitch's calculations indicated that there should be numerous glacial cycles during a million-year interval. But because many scientists believed that there was field evidence for only four major glaciations, and because some early radiocarbon evidence appeared contradictory, Milankovitch's ideas were largely dismissed until the recovery and interpretation of the deep-sea record in the 1950s. By the 1980s, most paleoclimate scientists were convinced that there is an unusually good correlation between the deep-sea record and Milankovitch's predictions. This suggests that the primary driving force behind glacial cycles is regular orbital variations, and it indicates that long-term climate cycles are entirely predictable! Unfortunately for humans, the Milankovitch theory indicates that the warm Holocene interglacial will soon end and that Earth is destined to experience full glacial conditions (glacial ice as far south as the Ohio and Missouri Rivers) in about 23,000 years.

Changes in Solar Output and Outer Space

Astronomers have also detected regular cycles in the energy output of the sun. The most famous of these cycles is the approximately 11-year sunspot cycle, with the greatest solar output correlating with the maximum number of sunspots.

(Continued on page 224)

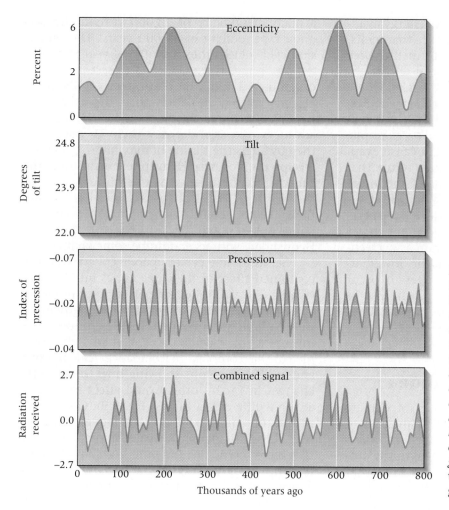

Figure 9.8 *The bottom line graph in this figure is the most significant because it demonstrates the combined effect that eccentricity, tilt, and precession cycles have had on the amounts of solar radiation received by Earth during the past 800,000 years. Because there is close correlation between this graph and climate change as indicated by the deep-sea record, many scientists believe that variations in Earth's orbit exert a major influence on glacial cycles.* • **Which two of the top three graphs exhibit the greatest similarity?**

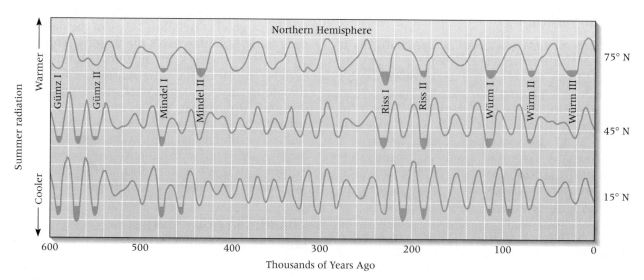

Figure 9.9 *These graphs, produced by Milutin Milankovitch, show how variations in Earth's orbit would have affected insolation at three different latitudes over the past 600,000 years.*

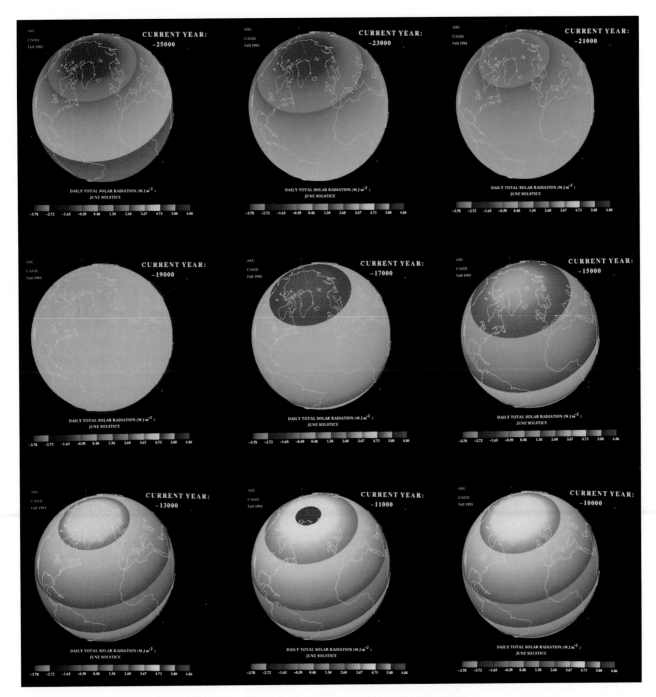

Figure 9.10 *The nine computer images in this figure are based on the Milankovitch theory that links changes in receipt of insolation on Earth to changes in Earth's orbital characteristics. The blue end of the spectrum indicates values below those of the present time; the red end of the spectrum indicates values above the present time.* • **What effect should these changes in receipt of insolation have had on global climates?**

These changes have been confirmed recently with satellite measurements of solar fluxes. Solar maxima occurred in 1979 and 1991, and a solar minimum occurred in 1986. Another minimum should occur in the very near future.

Climatic effects associated with the 11-year sunspot cycle are less certain. Because deviations from average weather are so difficult to recognize, and detailed meteorological records are so short, nothing definite can be concluded based on direct measurements of climate variables. However, some indirect records of climate (proxy records), such as tree-ring records, show rather good evidence of an 11-year cycle.

Other cycles related to variations in solar output are more speculative. Cycles of 2 to 2.2, 5.5, 20 to 23, 50, and 100 years have been reported in the scientific literature.

Analysis of the deep-sea sedimentary record indicates that the 41,000-year obliquity cycle is the most important cycle in the climate record from 2,400,000 to 900,000 years ago. But for the past 900,000 years, the 100,000-year eccentricity cycle has dominated the record. One recent theory attributes the increased impact of the eccentricity cycle to orbital changes that cause Earth to travel through "dustier" Earth-sun pathways slightly off of the plane of the ecliptic. While most studies of climate changes presume that the transparency of outer space to sunlight remains constant over geologic time, it is possible, however, that dust in outer space is not evenly distributed, therefore affecting the amount of sunlight reaching Earth over time. A difficult task currently facing climate scientists is to determine what kind of evidence might be available to test such a theory.

Changes in Earth's Atmosphere

Many theories attribute climate changes to variations in atmospheric dust levels. The primary villain is volcanic activity, which pumps enormous quantities of ash into the stratosphere, where strong winds spread it around the world. The dust reduces the amount of insolation reaching Earth's surface for periods of one to three years (Fig. 9.11).

Volcanic Activity. The climatic cooling effect of volcanic activity is unquestioned: all of the coldest years of record over the past 200 years have occurred in the year following major eruptions. Following the massive eruption of Tambora in 1815, the summer of 1816 was known as "the year without a summer." Killing frosts in July ruined crops in New England and Europe, resulting in famines. Several decades later, following the massive eruption of Krakatoa in 1883, temperatures decreased significantly during 1884 (Fig. 9.12). Although no twentieth-century eruptions have approached the magnitude of these two, the 1991 eruption of Mt. Pinatubo produced

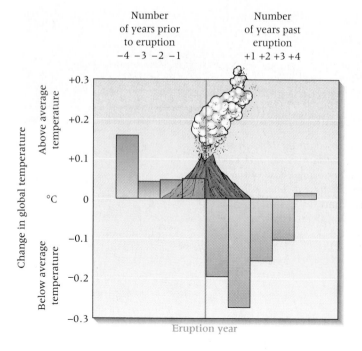
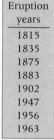

Number of years prior to eruption
−4 −3 −2 −1

Number of years past eruption
+1 +2 +3 +4

Change in global temperature

Above average temperature

Below average temperature

°C

+0.3
+0.2
+0.1
0
−0.1
−0.2
−0.3

Eruption year

Eruption years
1815
1835
1875
1883
1902
1947
1956
1963

Figure 9.11 *Examination of global temperatures within four years before and after major volcanic eruptions provides compelling evidence that volcanic activity can have a direct effect upon the amounts of insolation reaching Earth's surface.* • **At what period after an eruption year does the effect seem the greatest?**

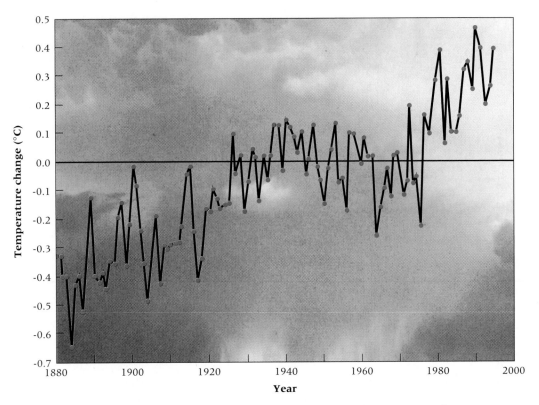

Figure 9.12 *This graph shows the gradual warming trend in global temperatures since 1880. It also documents the sharp reversal of the trend and the cooling of temperatures after the eruptions of Krakatoa in 1883, and Mt. Pinatubo in 1991.*

a substantial respite of cool conditions in an otherwise continuous series of record warm years (see again Fig. 9.12).

Volcanic activity is an important variable in global temperatures. In early climate models attempting to replicate global temperature variations, volcanic activity was found to be so important that it was given twice the weight of any other variable. Interestingly, volcanic ashes that accumulated about 2.6 million years ago are ten times thicker than average on the floor of the North Pacific Ocean. The modern series of major global climate changes began soon thereafter, suggesting that the massive eruptions could have set in motion a series of feedback cycles that continue today.

Atmospheric Gases. Another phenomenon closely correlated with average global temperatures is the composition of atmospheric gases. For many years scientists have known that CO_2 acts as a "greenhouse gas." Although the analogy to a greenhouse is problematic (because gases are not solids and do not physically restrict air move-

ment), there is no question that CO_2 transmits incoming short-wave radiation and absorbs outgoing long-wave radiation, similar to the effect of the glass panes in a greenhouse or in your automobile on a sunny day (refer again to Chapter 4). Thus, as the atmospheric content of greenhouse gases rises, so will the amount of heat trapped in the lower atmosphere. Although the amount of CO_2 in the atmosphere is about .0036 percent, its effect on climate is considerable. It has been estimated that the average temperature of Earth would be below freezing, rather than a balmy 16°C (61°F), if CO_2 was not present in the atmosphere.

A consequence of the combustion of fossil fuels is that CO_2 is released in huge quantities. Detailed measurements of atmospheric CO_2 content since 1958 reveal a cyclical and ever-increasing atmospheric CO_2 level (Fig. 9.13). The cyclicity is related to the dominance of plant growth in the Northern Hemisphere summer, which temporarily lowers the global CO_2 level.

Trapped in the glacial ice of Antarctica and Greenland are air bubbles containing minor

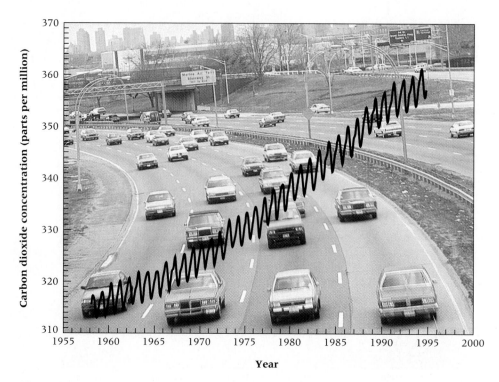

Figure 9.13 *The steady increase since 1958 in the concentration of carbon dioxide in the atmosphere is demonstrated by this graph. Although motor vehicle exhaust contributes major quantities of CO_2 to the atmosphere worldwide, other sources of CO_2 such as heavy industry, power production, and burning of wood for heat, cooking, and clearing of land are just as important.* • **What causes the annual fluctuations of CO_2 levels that are indicated on the graph?**

samples of the atmosphere that existed at the time the ice formed. One of the important discoveries of the ice core projects is that prehistoric atmospheric CO_2 levels increased during interglacials and decreased during major glacial advances. It is still not known why the two are so closely correlated. One explanation involves the oceans, since they serve as the primary short-term storage site for CO_2. Unlike many gases, the solubility of CO_2 increases as water temperature decreases. This means that cold polar waters absorb CO_2 from the atmosphere. In turn, CO_2 is released into the atmosphere as the water warms in tropical regions. Global cooling leads to colder ocean water, which increases the amount of CO_2 it can hold. The decreasing atmospheric CO_2 levels then cause temperatures to drop and glaciers to advance, resulting in a positive feedback cycle of global cooling. Operating in the reverse, warming ocean water releases CO_2, which should produce a positive feedback cycle of global warming by enhancing the greenhouse effect.

The fact that average global temperatures and CO_2 levels are so closely correlated suggests

that Earth will experience record warmth as the atmospheric level of CO_2 increases. The present level of approximately 360 parts per million of CO_2 is already higher than at any time in the past million years.

Carbon dioxide is not the only greenhouse gas. Molecule for molecule, methane (CH_4) is more than 20 times more effective than CO_2 as a greenhouse gas but is considered less important because the atmospheric concentrations and the lengths of time the molecules of gas remain in the atmosphere (residence times) are much smaller. Widespread publicity about the need to reduce the increase in methane levels has led to proposals to monitor (and perhaps control) dump emissions and termite mounds, both of which produce substantial quantities of CH_4. A much more important source of atmospheric methane may be from the tundra regions or the deep sea, where vast quantities of methyl hydrates could be released if oceanic temperatures rise. If warming tundra or ocean water indeed release large amounts of methane, the resulting positive feedback cycle of warming could be enormous.

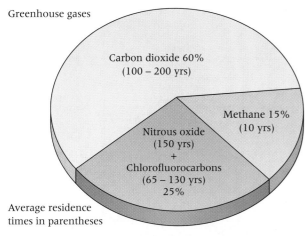

Greenhouse gases

Carbon dioxide 60%
(100 – 200 yrs)

Methane 15%
(10 yrs)

Nitrous oxide
(150 yrs)
+
Chlorofluorocarbons
(65 – 130 yrs)
25%

Average residence
times in parentheses

Figure 9.14 *Gases, other than carbon dioxide, released to the atmosphere by human activity contribute approximately 40% to the greenhouse effect. The figures in parentheses indicate the average number of years that the different gases remain in the atmosphere and contribute to temperature change.*

Other greenhouse gases are CFCs (chlorofluorocarbons) and N_2O (nitrous oxide). The relative greenhouse contribution of common greenhouse gases and their average residence times in the atmosphere are presented in Figure 9.14.

Changes in the Ocean

Oceans cover 70 percent of Earth's surface. Because of their enormous volume, high heat capacity, and thermal inertia, oceans are the single largest buffer against changes in Earth's climate. Whenever changes occur in oceanic temperatures, chemistry, or circulation, significant changes in global climate are certain to follow.

Surface oceanic currents are driven mostly by winds. However, there is a much slower circulation deep below the surface that moves large volumes of water between the oceans. A major driving force of the deep circulation appears to be differences in water buoyancy caused by differences in salinity. Where surface evaporation is rapid, the rising salinity content causes the seawater density to increase, inducing subsidence. On the other hand, when major influxes of freshwater flow from adjacent continents, or concentrations of melting icebergs, called *Heinrich Events,* flood into the North Atlantic, the salinity is reduced, thereby increasing the buoyancy of the water. When the surface water is buoyant, deep-water circulation slows. In many cases the freshwater influx is immediately followed by a major flow of warm surface waters into the North Atlantic, causing an abrupt warming of the Northern Hemisphere.

In modern times, short-term changes in Pacific circulation are primarily responsible for El Niño events. Because the onset of El Niño/Southern Oscillation (ENSO) climatic events is both rapid and global in extent, it is widely believed that changes in oceanic circulation may be responsible for similar rapid climate changes during the Pleistocene.

Changes in Landmasses

The final category of climate change theories invokes changes in Earth's surface to explain lengthy periods of cold climates. Ice ages, some with more than 125 glacial advances, occurred during past periods of Earth's history (Fig. 9.15). To explain some of the previous glacial periods, scientists have proposed several factors that might be responsible. For example, one characteristic that all of these glacial periods have in common with the Pleistocene is the presence of a continent in polar latitudes. Polar continents permit glaciers to accumulate on land, which results in lowered sea levels and consequent global effects. For example, when sea level was lowered between 40 and 50 meters (130 and 160 ft) some 10 million years ago (see Fig. 9.15), the Mediterranean Sea was cut off from the Atlantic Ocean. As a result, evaporation of seawater in the Mediterranean basin caused the precipitation of large quantities of salt. When the link between the Mediterranean Sea and the global ocean was reestablished, the average salinity of the oceans was reduced by 6 percent. As oceanic salinity decreases, seawater is more likely to freeze, enhancing global cooling. In short, the presence of a polar continent may set feedback mechanisms in motion that could eventually lead to a series of glacial cycles.

Another geologic factor sometimes invoked as a cause of climate change is the formation of a landmass that restricts oceanic or atmospheric circulation, or the disappearance of such a barrier because of landmass movements. For example, eruptions of the Panama isthmus volcanoes would have restricted ocean currents flowing between the Atlantic and Pacific, thereby closing a pathway of significant energy transfers. Another example is the uplift of the Himalayas, altering atmospheric flows, monsoonal effects, and weathering rates in Asia. Both of these events, and several other significant changes, immediately

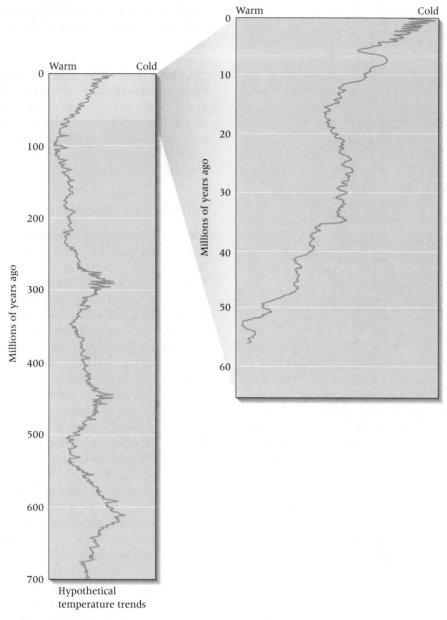

From Sheckleton & Kennett, 1975.

Figure 9.15 *These graphs show hypothetical temperature trends and periods of glaciation as suggested by evidence in the rocks and fossils that have accumulated over 700 million years of Earth history.* • **Why would the simulation of climates of the last few million years be considered more reliable than that of previous periods of Earth history?**

predate the onset of the modern series of glaciations. Which ones caused climate changes and which changes are simply coincidences is yet to be determined.

A final group of theories involve changes in albedo, caused either by major snow accumula-tions on high-latitude landmasses or by large oceanic ice sheets drifting into tropical regions. The increased reflection of sunlight starts a pos-itive feedback cycle of cooling that may end when the polar oceans freeze, shutting off the primary moisture source for the polar ice sheets.

Future Climates

With so many variables potentially responsible for climate change, reliably predicting future climate is a very difficult proposition at best. The primary difficulty in climate prediction is posed by natural variability. Figure 9.16 displays the frequency and magnitudes of climate changes that have occurred naturally over the last 150,000 years. Although the Holocene has been the most stable interval of the whole period, a detailed examination of the Holocene record reveals a wide range of climates. For example, a long interval of climates hotter than today's occurred during the

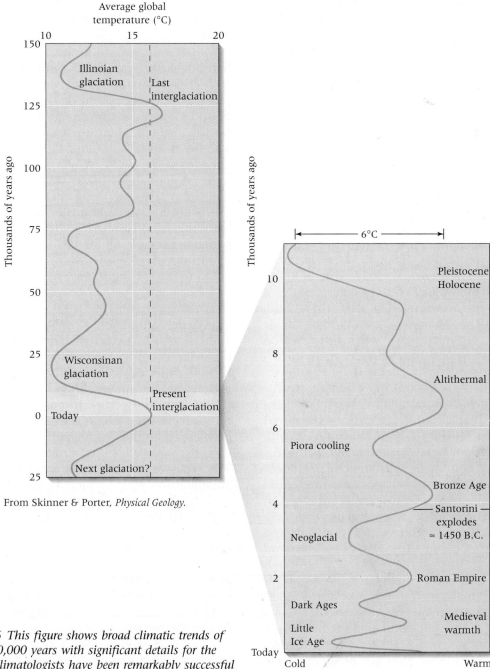

From Skinner & Porter, *Physical Geology.*

Figure 9.16 *This figure shows broad climatic trends of the past 150,000 years with significant details for the Holocene. Climatologists have been remarkably successful in dating recent climatic change but predicting future climates remains extremely difficult.* • **Why is this so?**

After Imbrie & Imbrie, *Ice Ages: Solving the Mystery,* Enslow Publishers, Short Hills, NJ, p. 179.

Altithermal. This interval was characterized by grasslands in the Sahara and severe droughts on the Great Plains. Other warm intervals occurred during the Bronze Age, the second half of the Roman Empire, and medieval times. An unusually cold interval began with the eruption of Santorini (the site of a civilization that was the basis for the Atlantis myth) in the Aegean Sea. Other cold periods occurred during the Dark Ages and again beginning about 1197 A.D. in the North Atlantic and 1309 A.D. in continental Europe. This latter episode has been termed the **Little Ice Age.** The Little Ice Age had major impacts on civilizations—from the isolation of the Greenland settlements established during the medieval warm period to the abandonment of the Colorado Plateau region by the Anasazi cultures. An important point to remember is that, with the exception of the cold interval that began with the eruption of Santorini, climate scientists do not know what variables changed to cause each of these major climate fluctuations.

Attempts to reliably predict future climates are complicated further by the many feedback cycles that operate. Simply increasing the amount of heat that is trapped by atmospheric gases in the lower atmosphere may or may not result in long-term warming. First, negative feedback processes such as increased cloud formation and increased plant uptake of CO_2 may operate to counteract the warming. On the other hand, warming of the oceans and tundra may release additional greenhouse gases, setting into motion some significant positive feedback cycles. Which feedback mechanisms will dominate is not certain, and so all predictions must be tentative.

There have been numerous attempts to simulate the variables affecting climate. **General Circulation Models** (GCMs) are complex computer simulations based on the relationships between variables discussed throughout this book: sun angles, temperature, evaporation rates, land versus water effects, energy transfers, and so on (Fig. 9.17). The models often simulate Earth's climate by dividing the surface into grid cells and estimating how all of the variables will respond when a change is made in one of them. Although the complexity and usefulness of GCMs is increasing rapidly, so far none of the models has accurately predicted most previous climate changes, and so their ability to predict future climate changes is still suspect. Nevertheless, GCMs appear to do a good job predicting how conditions will change in specific regions as the globe warms or cools, and they have added new insights into how some climatic variables interact.

Based on the record of climate changes during the past, only one thing can be concluded about future climates: They will be variable. Looking into the distant future, the Milankovitch cycles indicate that another glacial cycle is probably on the way. The most rapid cooling should occur between 3000 and 7000 years from now. The end result will be glacial ice covering most of Canada and northern Europe, and the United States can expect glacial ice as far south as Long Island, Indianapolis, and Des Moines about 25,000 years from now.

In the near term, however, global warming is most likely. The rise of atmospheric CO_2, the widespread destruction of vegetation, and the most likely feedback cycles that will result are bound to increase the average global temperature for the foreseeable future. An average increase of 1°C (or nearly 2°F) would be equivalent to the change that has occurred since the end of the Little Ice Age in about 1850. A 2°C warming would be greater than anything that has happened in the Holocene, including the Altithermal. A 3°C warming would exceed anything that has happened in the past million years. Current estimates and the most reliable GCMs predict a 1°C to 3.5°C (2° to 6°F) warming *in the twenty-first century.*

Detailed studies about conditions during the Altithermal, as well as GCMs, permit climatologists to determine how global warming will likely affect Earth. It is clear that not all areas will be impacted equally. One of the most important effects is expected to be a more vigorous hydrologic cycle, fueled largely by increases in evaporation from the ocean. Intense rainfalls will be more likely everywhere, as will droughts in regions such as the Great Plains (for a discussion of the recent effects of drought in Africa, see The Environment: Drought and Desertification). Temperatures will rise most in the polar regions, during nighttime rather than daytime, and during winter months. As a result of the warming, sea levels will rise mostly because of the thermal expansion of the oceans. By 2100, sea levels should be between 15 and 95 centimeters (0.5 and 3.1 ft) higher than today. In addition, the ranges of tropical diseases will expand toward higher latitudes, tree lines will rise, and many alpine glaciers will disappear.

However, some greenhouse effects will be beneficial to humans. Growing seasons in the high latitudes will increase in length. Because of the increase in atmospheric CO_2, some crops

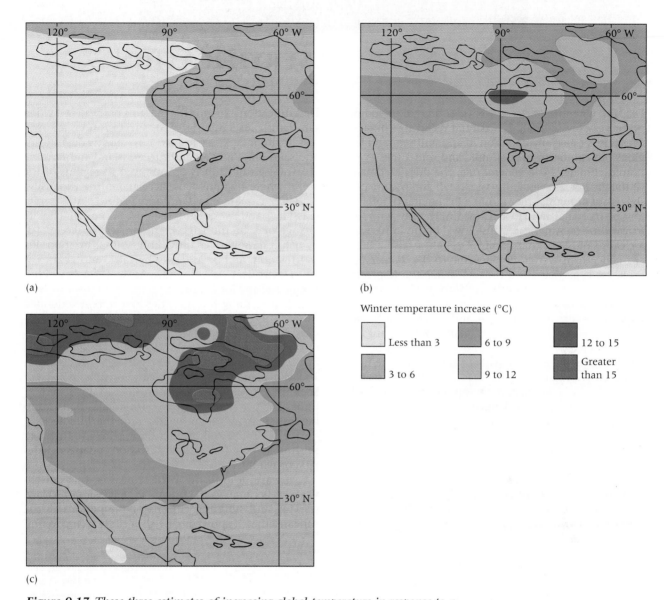

Figure 9.17 *These three estimates of increasing global temperature in response to a doubling of atmospheric carbon dioxide were produced as a result of different General Circulation Models (GCMs). All three estimates agree that global warming will occur.*
• **However, why do you think they vary so widely from one another?**
(a) National Center for Atmospheric Research, (b) NASA/Goddard Institute for Space Studies, (c) United Kingdom Meteorological Office.

such as wheat, rice, and soybeans should grow larger faster. In the United States, a 1°C increase in average temperature should decrease heating bills by about 11 percent.

Many scientists believe that the warming is already occurring, since the five hottest years of the twentieth century have occurred during the 1990s and each subsequent year usually sets a new record. Average annual global temperatures have already risen between 0.3° and 0.6°C (0.5 and 1.1°F), and sea level has risen between 10 and 25 centimeters (4 and 10 in.) during the past 100 years. Given the long residence times of many greenhouse gases (see again Fig. 9.14) and the thermal inertia of the oceans, some warming is inevitable and will likely continue for hundreds of years no matter what actions are taken to reduce greenhouse gas emissions. On the other

(Continued on page 234)

DROUGHT AND DESERTIFICATION

Drought, when there is a longer than normal period of no or very low rainfall, is a natural recurring climatic event. It is especially common in arid and semiarid regions but can occur in subhumid and sometimes even humid climates. As a general rule, the drier a climate is, the greater the variability of rainfall and the greater the risk of drought. Thus, farming in climates that have low or marginal rainfall is exceedingly risky without irrigation. The American "Dust Bowl" of the 1930s was a clear example of that hazard. Drought is also a hazard concern in wetter climates due to the dangers of fire. This is particularly a concern in subhumid places with brushfire potential such as California, but has also been a cause of forest fire danger in most of the mountainous western United States and eastern Australia.

The causes of drought are many and not always fully understood. Two periodic events seem to be related to drought in some parts of the world. The first is the 22-year sunspot cycle. Tree-ring research seems to indicate a relationship of this cycle to drought episodes in the western United States. The second is El Niño, which also appears to be related, as severe drought occurs in Australia and even southern Africa during such an event.

Desertification is the natural process of desert expansion caused by climate change, but accelerated by human activities. **It is an even more serious situation than drought because it involves long-term environmental and human consequences.**

Desertification expands the margins of the desert when rare rains cause gully erosion, sheet erosion, and soil loss. It also increases wind erosion, causing dust storms and sand dune movement into grassland and farming areas. Desertification is pronounced in regions of the world where humans have accelerated the expansion of desert climate and landform features into former grassland and woodland regions. Although climate change may be the trigger, the process is accelerated by deforestation, overcultivation, soil salinization due to irrigation, and overgrazing by cattle and goats.

Desertification is not new. Archeological evidence from Israel and Jordan indicates that as far back as 4000 B.C., early farming communities may have deteriorated the soil and deforested the hills, causing desertification. Recent research into ancient environmental catastrophes has shown a similar pattern of denudation of the hilly landscape of Greece as early as 3000 B.C. Today, evidence of desertification is visible in areas of Spain that exhibit deep gully erosion, in northwestern India as the Thar Desert expands into Rajasthan's farming areas, and throughout much of the Middle East, northern China, and Africa. Along with the threat to the human population, desertification endangers habitats for wildlife.

It was not until the 1970s, however, that desertification became well known, as television revealed starving and suffering citizens of the nations of the African Sahel. It showed bone-thin cattle trying to find

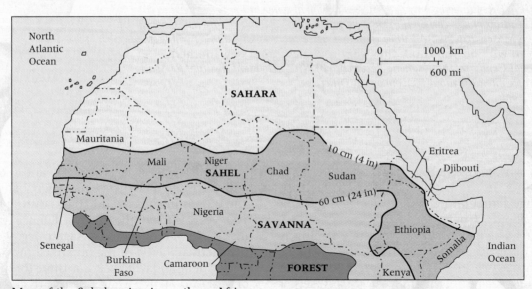

Map of the Sahel region in northern Africa.

Desertification in the Sahel region in the west African country of Mali.

Line-up at food station in Somalia.

a blade of grass in a dusty landscape. The TV also revealed villages being invaded by sand dunes. The *Sahel* is the semiarid zone bordering the southern margin of the Sahara. It extends across northern Africa from Mauritania on the Atlantic coast to Somalia on the Indian Ocean. A United Nations conference soon popularized the term *desertification,* and most people associated it with the ongoing suffering in the African Sahel.

What does the future hold for millions of people who are at risk of starvation in the Sahel? In 1994, the UN Food and Agricultural Organization (UNFAO) estimated that there were "exceptional food emergencies" in 15 African nations that include portions of the region. In many cases only massive food and medical aid relief could avert grave human tragedy. Added to the growing desert problems are periodic locust swarms and continuing ethnic warfare in many of these nations. During the driest years, 1983–1984, over a million people died and millions more became environmental refugees in neighboring nations. **Past UN peacekeeping efforts to help deliver food to starving Somalis serve as a realistic example of the serious nature of the physical, economic, and political problems this region faces.** As the UN forces left in 1995, great numbers of Somalis remained at risk of famine.

Fortunately, for many in the Sahel the foreign food aid has helped avert starvation, and in some years the "good rains" do return. Unfortunately, due to a characteristic of desertification, the rains do not come as often as in the past. The rapidly growing numbers of people, with marginal food output and decreasing cattle herds, wonder how they would cope if a "big dry" returns as it did in the 1980s.

The United Nations Environment Program (UNEP) estimates that the cost of successfully fighting worldwide desertification would be between $10 and 20 billion annually for 20 years. In 1994, 87 nations signed the Desertification Convention in Paris. It will become an international treaty when ratified by 50 nations. This treaty would budget funds to help protect the fertility of lands that are at greatest risk of desertification. It is hoped that the nations of the world will be able to muster future successful antidesertification programs. Only a major international effort can deal with a natural hazard that causes such large-scale environmental deterioration and human suffering.

CRITICAL THINKING ▼

(1) What methods could be used to hold back the desert? **(2)** Do you think that desertification occurs in areas of the United States? **(3)** What changes in human activity may help to fight desertification?

hand, the lesson of this chapter is that short-term climate trends, and some longer-term climate trends, will never be reliably predictable. Major volcanic eruptions, changing oceanic circulation, or human impacts (such as nuclear wars) can significantly disrupt climate trends at any time.

Moreover, because sudden, major shifts in the climate system that were not caused by human activity are abundant in the record of the last 2.4 million years, predicting climate will always be a risky endeavor.

Define and Recall

empirical classification
genetic classification
Köppen system
climograph
Thornthwaite system
Potential
  Evapotranspiration (PE)

Actual
  Evapotranspiration
  (AE)
microclimate
heat island

Pleistocene
varve
oxygen-isotope analysis
Holocene (Recent)

Altithermal
eccentricity cycle

obliquity cycle
precession cycle
Little Ice Age
General Circulation
  Model (GCM)

Discuss and Review

1. Why is it important to study the nature and possible causes of past climates when attempting to predict future climate change?

2. What is the difference between genetic and empirical classification? Is a classification based on statistics empirical or genetic?

3. Why are temperature and precipitation the two atmospheric elements most widely used as the sources of statistics for climatic classification? How are these two elements used in the Köppen system to identify six major climate categories?

4. What are the advantages and disadvantages of the Köppen system for geographers? Why are the Köppen climatic boundaries often referred to as "vegetation lines"?

5. What is a climograph? What is its function?

6. How does the Thornthwaite system of climate classification differ from the Köppen system? What are the advantages of the Thornthwaite system?

7. What examples of microclimates can be found in your local area? How did you decide that these are examples of microclimates?

8. Why is the occurrence, frequency, and dating of glacial advances and retreats so important to the study of past climates? How has modern research changed earlier theories of glacial coverage and associated climate change during the Pleistocene?

9. In what way has the Holocene been unusual compared to climatic conditions during most of the rest of the Pleistocene?

10. How have scientists been able to document the rapid shifts of climates that have occurred during the latter part of the Pleistocene?

11. What are the major possible causes of global climatic change? What contribution did the mathematician Milankovitch make to theories regarding glaciation?

12. What is the evidence that volcanic activity can affect global temperatures? How does this occur?

13. What effects can changes in the amounts of CO_2 and other "greenhouse" gases in the atmosphere have on global temperatures? How can past changes in amounts of CO_2 be determined?

14. How might changes in Earth's oceans and landmasses affect global climates?

15. What is the primary difficulty for any climatologist who attempts to predict future climates? How effective has the use of computer simulations been in the prediction of future climate change?

16. What are the likely changes that will occur in Earth's major subsystems if global warming continues for the near term as many scientists believe?

Consider and Respond

1. Study the definitions for the individual Köppen climatic types described in the Appendix. Why do you think Köppen and later climatologists who modified the system selected the particular temperature and precipitation parameters that separate the individual types from one another?

2. Examine the climograph in Figure 9.2. During what month does Nashville experience the greatest precipitation? What major change would immediately identify this graph as representing a Southern Hemisphere location? What do you think the four horizontal dashed lines represent?

3. Review Figures 9.6, 9.7, 9.15, and 9.16. State in your own words the general conclusions you would draw from a study of all four figures.

4. After reading and examining the content of Chapters 8 and 9 in your textbook, which do you believe is the most important to you now and in the future: the subject of weather or of climate? Defend your answer.

CHAPTER
10

LOW-LATITUDE AND ARID CLIMATIC REGIONS

CHAPTER PREVIEW

▶ Regardless of the chosen system of classification, the world's climates are distributed in patterns that occur on each landmass in similar positions and in similar latitudes.

What system of classification has been modified for the presentation of climatic types in this textbook? Why are the climatic types of this system so closely related to major vegetation zones?

▶ A knowledge of the controls of weather and climate is sufficient preparation to permit a reasonably accurate prediction of the broad distribution of climatic types over Earth's land surfaces.

What control has the greatest influence over climatic distribution? What other controls produce significant deviation from the pattern established by the major control?

▶ The geographer employs regions for much the same reasons scholars in other disciplines utilize arbitrary systems for the organization of information—to create an orderly presentation of diverse phenomena.

How does a geographer identify and define a region? What type of phenomena can be organized into regions?

▶ Although the humid tropical climates are all characterized by high temperatures throughout the year, they exhibit significant differences from one another based on either the amounts or the distribution of the precipitation they receive.

What temperature parameter do these climates have in common? How do they differ from one another on the basis of precipitation? What controlling factors explain these differences?

▶ With the exception of a few unusual circumstances, the tropical rainforest climatic regions are among the least populated areas of the world despite being coincident with the belt of heaviest rainfall, insolation, and vegetative growth.

Why are there so few people in these regions? What are the exceptions? In what ways are these regions valuable to humankind?

▶ Although rainfall is seasonal in both tropical savanna and tropical monsoon climatic regions, the differences in total rainfall between the two climates cause major dissimilarities in their appearance, resource characteristics, and human use.

What are the chief dissimilarities? How is rainfall responsible? How are humans affected?

▶ A knowledge of the location of the world's deserts is similar to an understanding of the distribution of the world's steppe regions as well.

What is the association between deserts and steppes? How do the regions differ? In what ways are they similar?

▲ Typical vegetation of the tropical rainforest climatic region along the trade wind coast of Maui, Hawaii.

The odds are overwhelming that within your lifetime, if you have not already done so, you will travel extensively either within or beyond the borders of the United States to destinations far from where you are living today. You may be moving to a new home or place of employment. You may be traveling on business or simply for pleasure. But whatever the reason, it is likely that you will ask the question almost every other traveler asks: "I wonder what the weather will be like?"

Realistically, as you have learned from reading previous chapters, the question should probably be, "I wonder what the climate is like?"

The constant variability associated with weather in many areas of the world makes it difficult to predict, but the long-term ranges and averages upon which climate is based allow geographers to provide the traveler with a general idea of the atmospheric conditions that should be experienced at specific locations throughout the world during different times of the year.

Of course, there are many reasons other than travel why a knowledge of climate and its variation over Earth's surface is a valuable asset. An understanding of climate in other areas of the world helps us to better understand the adaptations to atmospheric conditions that have been made by the people who live there. We can better appreciate some of their economic activities and certain aspects of their cultures. In addition, the climate of any place on Earth has a dominant effect on native vegetation and animal life. It influences the rate and manner by which rock is destroyed and soil is formed. It is a contributing factor in the way landforms are reduced and physical landscapes are sculptured. In short, knowledge of climates provides endless clues not only to atmospheric conditions but to numerous other aspects of the physical environment as well.

In this chapter and the next, you will be provided with a broad descriptive survey of world climates: their locations, distributions, general characteristics, associated features, and pertinent related human activities. The information contained in the two chapters can serve as a valuable knowledge base as you prepare for the future. Throughout the chapters we will use a modified and simplified version of the Köppen climate classification that was introduced in Chapter 9. (As previously mentioned, a detailed presentation of the Köppen system is included in the Appendix.) It is interesting to note that each of the climates discussed in the following chapters can be found within the United States, so, even if you never travel beyond your nation's borders, you still will find the discussions of all climates valuable preparation as you move about your own country.

The Distribution of Climatic Types

Within five of the six major climate categories of the Köppen classification there are sufficient differences in the ranges, total amounts, and seasonality of either temperature or precipitation to produce the thirteen distinctive climatic types listed in Table 10.1. The tropical and arid climatic types will be discussed in some detail in the latter portion of this chapter, while the various mesothermal, microthermal, and polar climates will be presented in Chapter 11, along with a brief coverage of undifferentiated highlands. But before we begin a study of individual climates, it would be worthwhile to review what we already have learned about the atmospheric controls that produce the climates (Fig. 10.1). Applying our knowledge of these controls, we should be able to predict, with a reasonable amount of accuracy, the climatic patterns that exist throughout Earth's land areas. For example, suppose a large, unexplored continent divided the vast Pacific Ocean, stretching almost from pole to pole. Based upon our knowledge of the atmospheric controls, and assuming that this continent does not have a mountainous landscape, it would not be difficult to construct a map of its climatic patterns (see Fig. 10.2). Throughout the following section, frequent reference should be made to this figure and to Figure 10.3, which includes photographs related to each climatic type.

Tropical Climates

Near the equator of our hypothetical continent we would find high temperatures year-round, since the noon sun would never be far from the zenith. Humid climates of this type with no winter season are Köppen's tropical climates. As his boundary for tropical climates, Köppen chose 18°C (64.4°F) for the average temperature of the coldest month because it closely coincides with the geographic limit of certain tropical palms.

Table 10.1 and Figure 10.3 indicate that there are three humid tropical climates, reflecting major differences in the amount and distribution of rainfall within the tropical regions. Tropical climates extend poleward to 30° latitude or higher in the continent's interior but to lower latitudes near the coasts, because of the moderating influence of the oceans on coastal temperatures. Note also that the humid tropical climates extend further poleward along the east coast than along the west coast because cool currents flow along tropical west coasts while warm currents flow along the tropical east coasts.

Regions near the equator are influenced by the Intertropical Convergence Zone (ITC). But the convergent and rising air of the ITC, which brings rain to the tropics, is not anchored in one place; it instead follows the overhead sun, migrating with the seasons. Within 5° to 10° of the

(Continued on page 241)

Latitude

Tropical (*A*) climate. Island of Jamaica.

Mesothermal, mild winter (*C*) climate. Summer in southern Spain.

Microthermal, severe winter (*D*) climate. Winter in Illinois.

Polar (*E*) climate. Glaciers in the Alaska Mountain Range.

Atmospheric Pressure

Altitude

Arid (*B*) climate. Mojave Desert, southern California.

Highland (*H*) climate. Teton Range, near Jackson Lake, Wyoming.

Figure 10.1 *Photographs representing Köppen's six major climate categories.* • **What advantages or disadvantages for human use does each photograph suggest?**

TABLE 10.1   Simplified Köppen Climate Classes

Climates	Climograph Abbreviation
Humid Tropical Climates (*A*)	
Tropical Rainforest Climate	Tropical Rf.
Tropical Monsoon Climate	Tropical Mon.
Tropical Savanna Climate	Tropical Sav.
Arid Climates (*B*)	
Steppe Climate	Low-lat. and Mid.-lat. Steppe
Desert Climate	Low-lat. and Mid.-lat. Desert
Humid Mesothermal (Mild Winter)	
Climates (*C*)	
Mediterranean Climate	Medit.
Humid Subtropical Climate	Humid Subt.
Marine West Coast Climate	Marine W.C.
Humid Microthermal (Severe Winter)	
Climates (*D*)	
Humid Continental, Hot Summer Climate	Humid Cont. H.S.
Humid Continental, Mild Summer Climate	Humid Cont. M.S.
Subarctic Climate	Subarctic
Polar Climates (*E*)	
Tundra Climate	Tundra
Ice-cap Climate	Ice-cap
Highland Climates (*H*)	

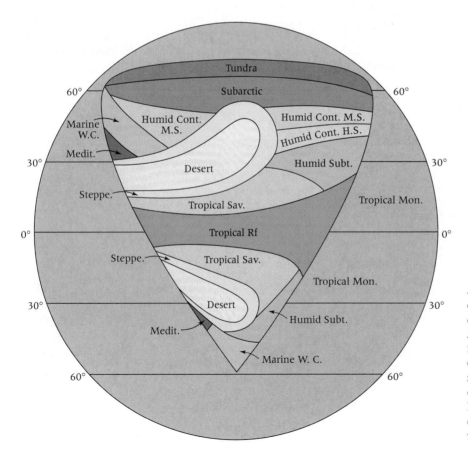

Figure 10.2 A generalized continent showing the characteristic locations of the Köppen climatic regions. • **How might a mountain range extending from north to south along the west coasts of the northern middle latitudes affect the distribution of climates in this model?**

equator, rainfall occurs year-round since the ITC moves through twice a year and is never far away. Likewise, tropical coasts facing the easterly trade winds rarely experience drought since there is a steady onshore flow of humid air. But poleward of this zone, precipitation becomes seasonal. When the ITC is over the region (during the high-sun period or summer) there is adequate rainfall. However, during the low-sun period (winter) the subtropical highs and trade winds invade the area, bringing clear, dry weather.

Thus, on our hypothetical continent, we find the tropical rainforest climate in the equatorial region flanked both north and south by the dry-winter tropical savanna climate. Finally, along coasts facing the strong, moisture-laden inflow of air associated with the summer monsoon, we find the tropical monsoon climate. This climate owes its existence to the seasonal reversal of atmospheric pressure and winds over both land and water, described in Chapter 6.

Polar Climates

Just as the tropical climates lack winters (cold periods), the polar climates—at least statistically—lack summers. Polar climates were defined by Köppen as areas in which no month has an average temperature exceeding 10°C (50°F). Poleward of this temperature boundary, trees cannot survive. On our hypothetical continent, the 10°C isotherm for the warmest month more or less coincides with the Arctic Circle, poleward of which the sun does not rise above the horizon in midwinter and in summer strikes at a low angle, spending its energy over a large area. Again, the effects of ocean currents prevent this isotherm from being a straight line. It is deflected northward on the west side of our hypothetical continent, where tropical waters move poleward, and dips to the south on the east side, where cold arctic water leaks into the general oceanic circulation.

The polar climates are subdivided into tundra and ice-cap climates. Tundra climate occurs where at least one month averages above 0°C (32°F), while in the ice-cap climate no month has an average temperature above 0°C.

Mesothermal and Microthermal Climates

Except where arid climates intervene, the lands between the tropical and polar climates are occupied by the transitional midlatitude mesothermal and microthermal climates. As they are neither tropical nor polar, the mild and severe winter climates must have at least one month averaging below 18°C (64.4°F) and one month averaging above 10°C (50°F). Although both midlatitude climatic categories have distinct temperature seasons, the microthermal climates have severe winters with at least one month averaging below freezing. Once again, vegetation reflects the climatic differences. In the severe winter climates all but needle-leaf trees defoliate naturally during the winter as soil water is temporarily frozen and unavailable. Much of the natural vegetation of the mild winter climates retains its foliage throughout the year since liquid water is always present in the soil. The line separating mild from severe winters usually lies in the vicinity of the fortieth parallel.

A number of important internal differences within the mesothermal and microthermal climatic groups produce individual climatic types based on precipitation patterns or seasonal temperature contrasts. On our hypothetical continent the Mediterranean, or dry summer, mesothermal climate appears along west coasts between 30° and 40° latitude, where the dry subtropical highs of summer alternate with the humid conditions of the westerly wind belt in winter. On the east coasts, in generally the same latitudes, the humid subtropical climate is found. The contrast between coasts should be expected, since the subsidence, dry weather, and temperature inversions of the subtropical highs during the summer months are only in evidence along the west coasts of continents. In contrast, continental east coasts receive both cyclonic and convective precipitation year-round.

The distinction between the humid subtropical and marine west coast climates illustrates a second important criterion for the internal subdivisions of midlatitude climates: seasonal contrasts. Both mesothermal climates have year-round precipitation, but humid subtropical summer temperatures are much higher than those in the marine west coast climate. Because the humid subtropical climate is located along continental east coasts it lies directly in the path of maritime tropical air masses moving onshore from source regions over high-temperature tropical oceans, and summers are hot. In contrast, the mild summers of the marine west coast climate, located poleward of the Mediterranean climate along continental west coasts, often extending beyond 60° latitude, are a direct result of mild, moist conditions carried onshore by the westerly

(Continued on page 244)

Tropical rainforest. Windward coast of Jamaica.

Desert. Sonoran Desert of Arizona.

Tropical monsoon. Himalayan foothills, West Bengal, India.

Steppe. Sand Hills of Nebraska.

Tropical savanna. East African highlands.

Mediterranean. Village in southern Spain.

Figure 10.3 *Examples of climates in the Köppen classification system.*

Humid subtropical. Grapefruit grove in central Florida.

Humid continental, mild summer. Fall season in New Hampshire.

Marine west coast. North Sea coast of Scotland.

Subarctic. Southern edge of the Canadian shield.

Humid continental, hot summer. Corn Belt farm in Missouri.

Tundra. Ice meets ocean in southern Greenland.

winds in these latitudes. This is but one example of how seasonal temperatures are used to distinguish between climatic types in the Köppen system.

Another example is found among microthermal climates which usually receive year-round precipitation associated with storms traveling along the polar front. Internal subdivision into climatic types is based on summers that become shorter and cooler and winters that become longer and more severe with increasing latitude and continentality. On our generalized continent the microthermal climates are found exclusively in the Northern Hemisphere. As in the real world, there is no land in the Southern Hemisphere latitudes that would normally be occupied by the humid microthermal climates. In the Northern Hemisphere these climates progress poleward through the humid continental, hot-summer climate, to the humid continental, mild-summer climate, and finally, to the subarctic climate.

Arid Climates

Climatic differentiation thus far has been based on temperature boundaries and the seasonality of precipitation. However, our ideal continent will contain a large area in which these variations are less important than the fact that the environment is dominated by year-round moisture deficiency. This dry area will penetrate deep into the continent, interrupting the neat zonation of climates that would otherwise exist. This is an area of Köppen's arid climates.

The definition of climatic aridity is that precipitation received is less than potential evaporation. Aridity does not depend solely on the amount of precipitation received. Evaporation rates and temperature must also be taken into account. In a low-latitude climate with relatively high temperatures, the evaporation rate is greater than in a higher, colder-latitude climate. As a result, more rain must fall in the lower latitudes to produce the same effects (especially noticeable on vegetation) that smaller amounts of precipitation produce in areas with lower temperatures and consequently lower evaporation rates.

On our hypothetical continent the arid climates are concentrated in a zone about 15°N and S latitude to about 30°N and S along the western coasts, expanding much further poleward over the heart of the landmass. This eastward expansion is a consequence of remoteness from the oceanic moisture supply. The main factor that causes the deficiency of precipitation is air subsidence on the east sides of the subtropical high-pressure systems. The correspondence between the arid climates and the belt of subtropical high pressure is quite unmistakable.

Ocean currents reinforce the effect of the subtropical anticyclones in suppressing precipitation along west coasts in the subtropical zone. On the eastern sides of the oceanic anticyclones, cold water from the polar seas moves equatorward along the continental west coast. Air moving onshore from the west is cooled from below by these currents, increasing its stability. This allows the zone of deficient precipitation to be accentuated on the coast itself. In fact, coastal deserts receive the lowest precipitation amounts of any region on Earth.

In desert climates the annual amount of precipitation is less than half the annual potential evaporation. Bordering the deserts are steppe climates, or semiarid climates that are transitional between the extreme aridity of the deserts and the moisture surplus of the humid climates. The definition of the steppe climate is an area where annual precipitation is less than potential evaporation but more than half the potential evaporation.

Highland Climates

We have been speaking of our hypothetical continent as if it were a uniform surface with no significant relief features. Of course, the pattern of climates and extent of aridity would be affected by irregularities in configuration, such as the presence of deep gulfs, interior seas, or significant highlands. The climatic patterns of Europe and North America are quite different because of such variations.

Highlands can channel air-mass movements and create abrupt climatic divides. Their own microclimates form an intricate pattern related to elevation, cloud cover, and exposure. One significant effect of highlands aligned at right angles to the prevailing wind direction is the creation of arid regions extending tens to hundreds of kilometers leeward.

Climatic Regions

Because each of our modified Köppen climatic types are defined by specific parameters for monthly averages of temperature and precipitation, it is possible to draw boundaries between the types on a world map. The areas within these boundaries are examples of one type of world region. A **region,** as the term is used by geographers, refers to an area that has recognizably similar internal characteristics that are distinct from those of other areas. A region may be described on any basis that unifies it and differentiates it from others.

As we examine the climatic regions of the world in the pages that follow, you should make frequent reference to the map of world climatic regions (Fig. 10.4). It shows the patterns of Earth's climates as they are distributed over each of Earth's continents. It is not surprising that these patterns closely resemble the distribution of climates on our hypothetical continent as we applied real-world climatic controls to our model. However, a word of caution is in order. On a map of climatic regions, distinct lines separate one region from another. Obviously the lines do not mark points where there are abrupt changes in temperature or precipitation conditions. Rather, the lines signify **zones of transition** between different climatic regions. Furthermore, these zones or boundaries between regions are based on monthly and annual averages and may shift as temperature and moisture statistics change over the years.

The actual transition from one climatic region to another is gradual except in cases where the change is brought about by an unusual climatic control such as a mountain barrier. It would be more accurate to depict climatic regions and their zones of transition on a map by showing one color fading into another. Always keep in mind as we describe Earth's climates that it is the core areas of the regions that best exhibit the characteristics that distinguish one climate from another.

Now, let us look more closely at Figure 10.4. One thing that is immediately noticeable is the change in climate with latitude. This is especially apparent in North America. Looking at the East Coast of the United States and Canada, we see that the southern tip of Florida has a tropical savanna climate. This gives way almost immediately to the humid subtropical climate of the southern states. As we move poleward from Virginia toward Maryland and southern New Jersey, the humid subtropical climate changes to a humid continental, hot-summer climate. Further north, Vermont, New Hampshire, and Maine have a humid continental climate with mild summers. Moving still further north, we see that most of Canada has a subarctic climate until we reach a latitude of 55° to 60°N, or 65° to 70°N in the western part of the country. Poleward of these latitudes the climate is described as polar tundra.

It is also interesting to discover locations that have similar climates. For instance, how many areas in the world have climates similar to the one with which you are most familiar? The Mediterranean climate is found in many parts of the world: California, southern Europe, the Middle East, central Chile, and small sections of the southwest coasts of Africa and Australia. These places are all in similar latitudes (between about 30° and 40°) and receive winds from oceans to the west. The marine west coast climate often is found poleward of the Mediterranean type: in Washington and Oregon, southern Chile, and much of northwest Europe, at latitudes between 40° and 60°. This climate is found as well in New Zealand, on the southeast coast of Australia, and in southeast Africa. The subarctic category of the humid microthermal climatic category is found over the vast continental areas of both northern Canada and northern Eurasia. Follow the equator around the world, and you will spend most of your time in a tropical rainforest climate except for excursions into the mountains and savanna of eastern Africa.

A first look at the world climate map may have given you the impression that climatic regions are scattered about haphazardly. A closer inspection, however, will show that in many areas there is a latitudinal progression of climate. Furthermore, we have seen that similar climates usually appear in similar latitudes and/or in similar locations with respect to landmasses or topography. These climatic patterns emphasize the close relationship between climate, the weather elements, and the climatic controls. As we predicted when we examined the hypothetical continent, there is an order to Earth's atmospheric conditions and so also to its climatic regions.

Carrying this idea even further, let us compare the world maps of temperature (page 112), precipitation (pages 176–177), and pressure and winds (page 121) with Figure 10.4, which shows world climates. To a large degree, the superposition of the maps showing the individual atmospheric elements or the combination of them has produced the map of climatic regions. We know, of course, that ultimately the order and pattern of the climatic regions are based *first* on the patterns produced by Earth-sun relationships. The exceptions to those patterns are the result of the uneven distribution and irregular sizes and shapes of land and water bodies, of the Coriolis

(Continued on page 248)

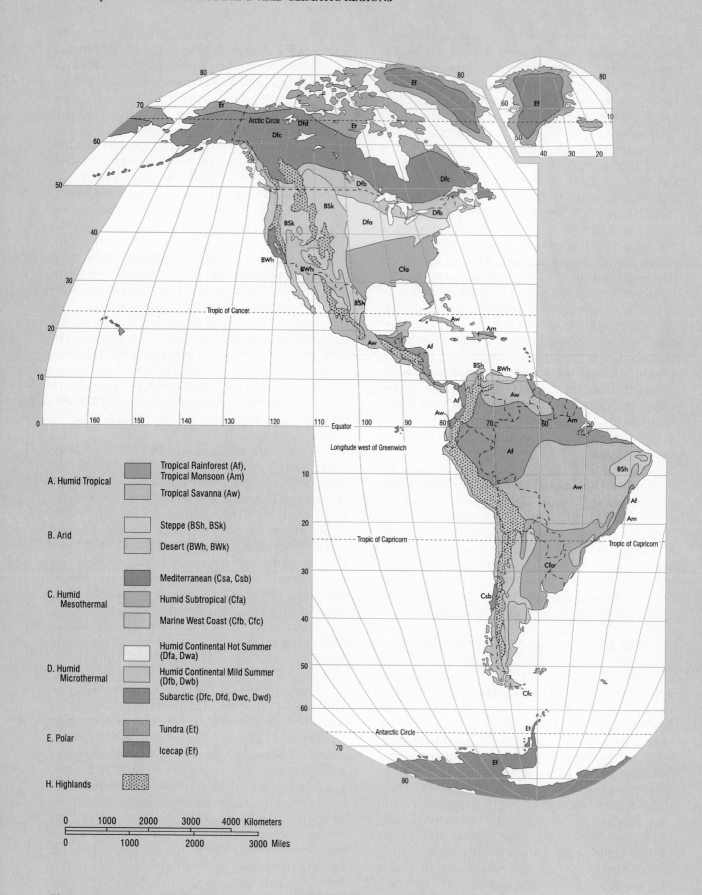

Figure 10.4 World Map of Climates in the Modified Köppen Classification System.

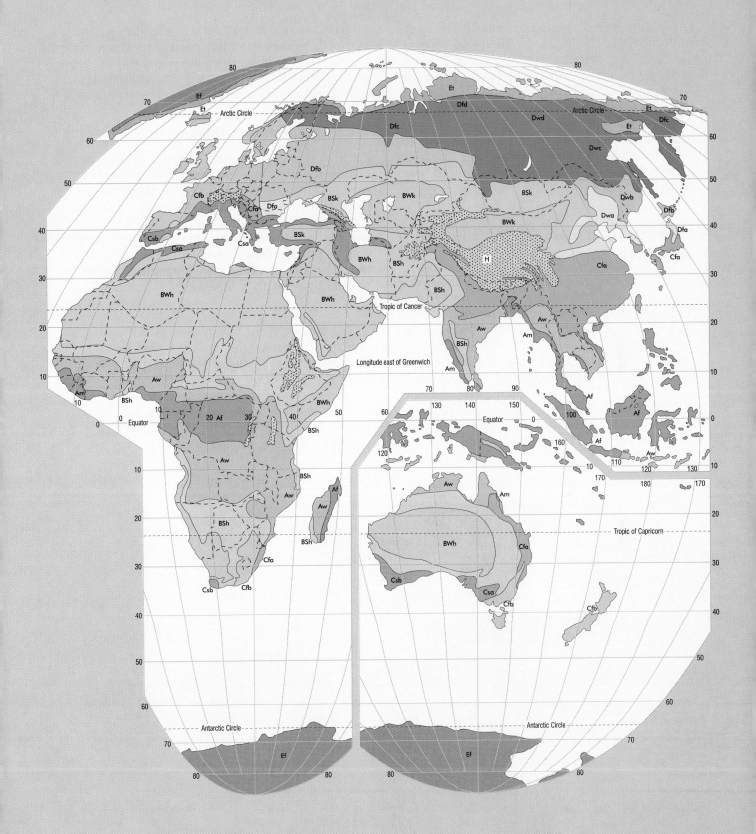

A Western Paragraphic Projection
developed at Western Illinois University

effect on wind and ocean currents, and of interruption of the continents by mountain barriers.

A striking variation in these global climatic patterns becomes apparent when we compare the Northern and Southern Hemispheres. Because the Southern Hemisphere lacks the large landmasses of the Northern Hemisphere, no climates in the higher latitudes (in land regions) can be classified as humid microthermal, and only one small peninsula of Antarctica can be said to have a tundra climate.

Humid Tropical Climatic Regions

We have already learned a good deal about the climatic regions of the humid tropics through our study of a hypothetical continent and our examination of the maps throughout this book. It now remains for us to identify the major characteristics of each humid tropical climatic type in turn, along with its associated world regions.

A glance at Figure 10.5 and a careful reading of Table 10.2 reviews the locations and provides a preview of the significant facts associated with the humid tropical climates. The table also re-

minds us that although each of the three humid tropical climates has high average temperatures throughout the year, there are major differences among them based on the amount and distribution of precipitation.

Tropical Rainforest Climate

Tropical rainforest regions probably come most readily to mind when someone says the word *tropical*. Hot and wet throughout the year, the tropical rainforest climate has been the stage for many stories of both fact and fiction. One cannot easily forget the classic struggle with the elements portrayed by Bogart and Hepburn in the film *The African Queen*. Even scientists are impressed by what they observe.

The heat increased rapidly towards two o'clock (92° and 93°F), by which time every voice of bird or mammal was hushed; only in the trees was heard at intervals the harsh whirr of a cicada. The leaves, which were so moist and fresh in the early morning, now become lax and drooping; the flowers shed their petals. . . . On most days, a heavy shower

(Continued on page 250)

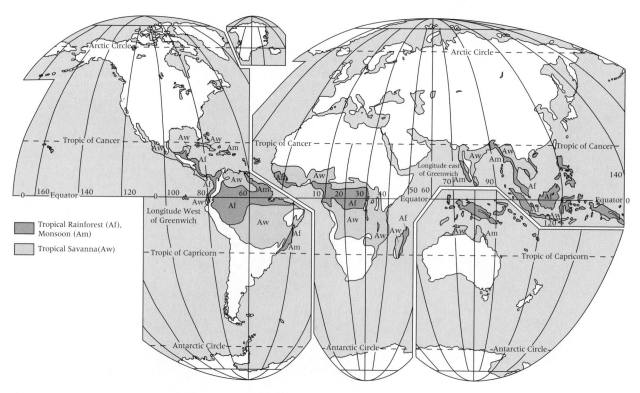

Humid Tropical Climates

Figure 10.5 Index Map of Humid Tropical Climates

TABLE 10.2   The Humid Tropical Climates

Name and Description	Controlling Factors	Geographic Distribution	Distinguishing Characteristics	Related Features
Tropical Rainforest				
Coolest month above 18°C (64.4°F); driest month with at least 6 cm (2.4 in.) of precipitation	High year-round insolation and precipitation of doldrums (ITC); rising air along trade wind coasts	Amazon R. Basin, Congo R. Basin, east coast of Central America, east coast of Brazil, east coast of Madagascar, Malaysia, Indonesia, Philippines	Constant high temperatures; equal length of day and night; lowest (2°–3°C/3°–5°F) annual temperature ranges; evenly distributed heavy precipitation; high amount of cloud cover and humidity	Tropical rainforest vegetation (selva); jungle where light penetrates; tropical iron-rich soils; climbing and flying animals, reptiles, and insects; slash-and-burn agriculture
Tropical Monsoon				
Coolest month above 18°C (64.4°F); one or more months with less than 6 cm (2.4 in.) of precipitation; excessively wet during rainy season	Summer onshore and winter offshore air movement related to shifting ITC and changing pressure conditions over large landmasses; also transitional between rainforest and savanna	Coastal areas of southwest India, Sri Lanka, Bangladesh, Myanmar, southwestern Africa, Guyana, Surinam, French Guiana, northeast and southeast Brazil	Heavy high-sun rainfall (especially with orographic lifting), short low-sun drought; 2°–6°C (3°–10°F) annual temperature range, highest temperature just prior to rainy season	Forest vegetation with fewer species than tropical rainforest; grading to jungle and thorn forest in drier margins; iron-rich soils; rainforest animals with larger leaf-eaters and carnivores near savannas; paddy rice agriculture
Tropical Savanna				
Coolest month above 18°C (64.4°F); wet during high-sun season, dry during lower-sun season	Alternation between high-sun doldrums (ITC) and low-sun subtropical highs and trades caused by shifting winds and pressure belts	Northern and eastern India, interior Myanmar and Indo-Chinese Peninsula; northern Australia; borderlands of Congo R., south central Africa; llanos of Venezuela, campos of Brazil; western Central America, south Florida, and Caribbean Islands	Distinct high-sun wet and low-sun dry seasons; rainfall averaging 75–150 cm (30–60 in.); highest temperature ranges for humid tropical climates	Grasslands with scattered, drought-resistant trees, scrub, and thorn bushes; poor soils for farming; grazing more common; large herbivores, carnivores, and scavengers

*would fall some time in the afternoon, pro-
ducing a most welcome coolness. The approach of the rainclouds was after a uniform
fashion very interesting to observe. First, the
cool sea-breeze, which commenced to blow
about ten o'clock, and which had increased in
force with the increasing power of the sun,
would flag and finally die away. The heat
and electric tension of the atmosphere would
then become almost insupportable. Languor
and uneasiness would seize on everyone; even
the denizens of the forest betraying it by their
motions. White clouds would appear in the
east and gather into cumuli, with an increasing blackness along their lower portions. The
whole eastern horizon would become almost
suddenly black, and this would spread upwards, the sun at length becoming obscured.
Then the rush of a mighty wind is heard
through the forest, swaying the tree-tops; a
vivid flash of lightning bursts forth, then a
crash of thunder, and down streams the del-
uging rain. Such storms soon cease, leaving
bluish-black motionless clouds in the sky until
night. Meantime all nature is refreshed; but
heaps of flower-petals and fallen leaves are
seen under the trees. Towards evening life re-
vives again, and the ringing uproar is resumed
from bush and tree. The following morning
the sun again rises in a cloudless sky, and the
cycle is completed; spring, summer and autumn, as it were, in one tropical day. The
days are more or less like this throughout the
year in this country. . . . With the day and
night always of equal length, the atmospheric
disturbances of each day neutralizing themselves before each succeeding morn; with the
sun in its course proceeding midway across
the sky, and the daily temperature the same
within two or three degrees throughout the
year—how grand in its perfect equilibrium
and simplicity is the march of Nature under
the equator.*

The Naturalist on the River Amazon
H. W. Bates

This account of a day in a tropical rainforest
brings its climate vividly to life. We feel the high
temperatures, oppressive humidity, and the frequent heavy rains for which it is known.

Constant Heat and Humidity. Most weather stations in the tropical rainforest climatic regions
record average monthly temperatures of 25°C

(77°F) or more (Fig. 10.6). Because these regions
are usually located within 5° or 10° of the equator, the sun's noon rays are always close to being
directly overhead. Days and nights are of almost
equal length, and the amount of insolation received remains nearly constant throughout the
year. Consequently, there are no appreciable temperature variations that can be linked to the sun
and therefore considered seasonal.

The **annual range** of temperature, or the
difference between the average temperatures of
the warmest and coolest months of the year, reflects the consistently high angle of the sun's
rays. As indicated in Figure 10.6, the annual range
is seldom more than 2° or 3°C (4° or 5°F). In fact,
at Ocean Island in the central Pacific the annual
range is 0°C because of the additional moderating influence of a large water body on the already
nearly uniform pattern of insolation.

One of the most interesting features of the
tropical rainforest climate is that the **daily** (or
diurnal) **temperature ranges,** or the differences between the highest and lowest temperatures during the day, are usually far greater than
the annual range. Highs of 30° to 35°C and lows
of 20° to 24°C produce daily ranges of 10° to 15°C
(18°–27°F). However, the drop in temperature at
night is small comfort. The high humidity causes
even the cooler evenings to seem oppressive.

The climographs of Figure 10.6 illustrate that
significant variations in precipitation can occur
even within rainforest regions. Although most
rainforest locations receive more than 200 centimeters (80 in.) a year of precipitation and the
average is in the neighborhood of 250 centimeters (100 in.), there are some that record an annual precipitation of over 500 centimeters. Ocean
locations, near the greatest source of moisture,
tend to receive the most rain. As a group, climate
stations in the humid tropics experience much
higher annual totals than typical humid midlatitude stations. Compare, for example, the 365
centimeters in Akassa with the average 112 centimeters received annually in Portland, Oregon,
or the 61 centimeters received in London.

We should recall that the heavy precipitation
of the tropical rainforest climate is associated
with the warm, humid air of the doldrums and
the unstable conditions along the ITC. Both convection and convergence serve as trigger mechanisms, causing the moist air to rise and resulting
in the heavy rains that are characteristic of this
climate. There is heavy cloud cover during the
warmer, daylight hours when convection is at
its peak, although the nights and early morn-
ings can be quite clear. Variations in rainfall can

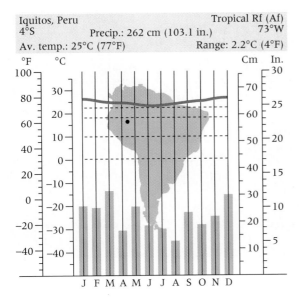

Iquitos, Peru   Tropical Rf (Af)
4°S   Precip.: 262 cm (103.1 in.)   73°W
Av. temp.: 25°C (77°F)   Range: 2.2°C (4°F)

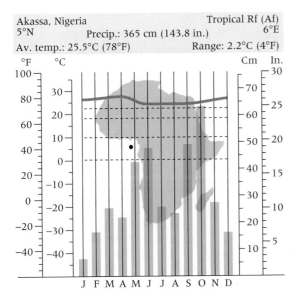

Akassa, Nigeria   Tropical Rf (Af)
5°N   Precip.: 365 cm (143.8 in.)   6°E
Av. temp.: 25.5°C (78°F)   Range: 2.2°C (4°F)

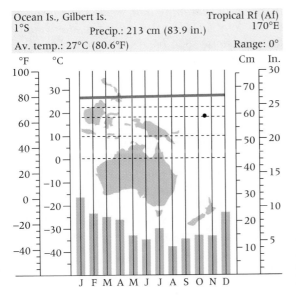

Ocean Is., Gilbert Is.   Tropical Rf (Af)
1°S   Precip.: 213 cm (83.9 in.)   170°E
Av. temp.: 27°C (80.6°F)   Range: 0°

usually be traced to the ITC and its weak low-pressure systems. Many tropical rainforest locations (Akassa, for example) exhibit two maximum precipitation periods during the year, one during each appearance of the ITC as it follows the migration of the sun's direct rays. In addition, although no season can be called dry, during some months it may rain on only 15 or 20 days.

A Delicate Balance. The most common vegetation of tropical rainforest climatic regions is multistoried, broad-leaf evergreen forest made up of many species whose tops form a thick, almost continuous cover that blocks out much of the sun's light (Fig. 10.7). This type of rainforest is called **selva.** Within the selva there is usually little undergrowth on the forest floor because sunlight cannot penetrate.

The relationship between the soils beneath the selva and the vegetation the soils support is so close that there exists a nearly perfect ecological balance between the two, threatened only by people's efforts to earn a living from the soil. The trees of the selva supply the tropical soils with the nutrients the trees need for growth. As leaves, flowers, and branches fall to the ground, or as roots die, the numerous soil animals and bacteria act on them, transforming the organic matter into usable nutrients. However, if the trees are removed, there is no replenishment of these nutrients and no natural counterbalance to the processes by which large amounts of percolating soil moisture remove the soluble minerals from the soil. The intense activities of microorganisms, worms, termites, ants, and other insects cause rapid deterioration of the remaining organic debris and soon all that remains is an infertile mixture of insoluble manganese, aluminum, and iron compounds.

In recent years there has been large-scale harvesting of the tropical rainforests as land has been cleared for agriculture, especially in the Amazon River basin. Such deforestation can have a significant and, unfortunately, permanent impact upon the delicate balance that exists among Earth's systems (see The Environment: Destruction of the Amazon Rainforest).

Figure 10.6 Climographs for tropical rainforest climate stations. • Why is it difficult without looking at the climograph key to determine whether each station is located in the Northern or Southern Hemisphere?

Figure 10.7 *The multistoried broad-leaf evergreen forest of the Amazon Basin near Iquitos, Peru.* • **Why might this photograph taken along the river bank give an incorrect impression of vegetation at ground level in the forest?**

Because environmental conditions vary from place to place within climatic regions, the typical rainforest situation we have just described does not apply everywhere in the tropical rainforest climate. Some regions are covered by the true jungle, a term often misused when describing the rainforest. **Jungle** is a dense tangle of vines and smaller trees that develops where direct sunlight does reach the ground, as in clearings and along streams. Other regions have soils that remain fertile or have bedrock that is chemically basic and provides the soils above with a constant supply of soluble nutrients through the natural weathering processes. Examples of the former region are found along major river floodplains, and examples of the latter are the volcanic regions of Indonesia and the limestone areas of Malaysia and Vietnam. Only in such regions of continuous soil fertility can agriculture be intensive and continuous enough to support population centers in the tropical rainforest climate.

Human Activities. Throughout much of the tropical rainforest climate humans are far outnumbered by other forms of animal life. A comparison of Figures 1.8 and 10.4 suggests that most rainforest regions are among the least inhabited areas of the world. Though there are few large animals of any kind, a great variety of smaller tree-dwelling and aquatic species live in the rainforest. Birds, monkeys, bats, alligators, crocodiles, snakes, and amphibians such as frogs of many varieties abound. Animals that can fly or climb into the food-rich leaf canopy have become the dominant animals in this world of trees.

Most common of all, though, are the insects: mosquitoes, ants, termites, flies, beetles, grasshoppers, butterflies, and bees live everywhere in the forest. Because they can breed continuously in this climate without danger from cold or drought, insects thrive. The problem this poses for humans is best described by someone who has been there.

One of our chief troubles at night was from insects. In the daytime they did not trouble us much, for there are not nearly so many mosquitoes in the great jungle as in the rubber, or near cultivation in the plains; but at night they bit us severely. Worse even than the bite is the shrill humming that seems to be just beside your ear. However much you slap your face, the noise soon breaks out again. Far worse than the mosquitoes were the midges, whose wings made no noise, but whose bite was really a bite and itched like a nettle-sting. They were particularly bad in the early morning and often woke us up long before dawn. As a result of the bites we received in the night, our faces would be so enlarged and distorted that in the morning we were almost unrecognizable. Often our cheeks were so swollen that the eyes were closed and we could not see until we had bathed them in the cool water of a stream.

The Jungle Is Neutral
F. Spencer Chapman

252

In addition to the insects themselves, there are genuine health hazards for tropical rainforest inhabitants. Not only does the oppressive, sultry weather impose physical discomfort and pain, it also allows a variety of human parasites and disease-carrying insects to thrive. Malaria, yellow fever, and sleeping sickness are all insect-borne diseases of the tropics, frequently fatal, and uncommon in middle-latitude climates.

Whenever native populations have existed in the rainforest, subsistence hunting and gathering of fruits, berries, small animals, and fish have been important. Since the introduction of agriculture, land has been cleared and crops such as manioc, yams, beans, maize (corn), bananas, and sugar cane have been grown. It has been the practice to cut down the smaller trees, burn the resulting debris, and plant the desired crops (Fig. 10.8). With the forest gone, this kind of farming is only possible for a year or two, or perhaps even three, before the soil is completely exhausted of its small supply of nutrients and the surrounding area is depleted of game. At this point the native population moves to another area of forest to begin the practice over again. This kind of subsistence agriculture is variously known as **slash-and-burn, swidden,** or simply **shifting cultivation.** Its impact upon the close ecological balance between soil and forest is obvious in many rainforest regions. Sometimes the damage done to the system is irreparable, and only jungle, thornbush, or scrub vegetation will return to the cleared areas.

In terms of numbers of people supported, the most important agricultural use of the tropical rainforest climate is the wet-field (paddy) rice agriculture on the river floodplains of southeastern Asia. However, this is best developed in the *monsoon* variant of this climate. Commercial plantation agriculture is also significant. The principal plantation crops are rubber and cacao, both of which originally grew wild in the forests of the Amazon Basin but are now of greatest importance elsewhere—rubber in Malaysia and Indonesia and cacao in West Africa and the Caribbean area.

Tropical Monsoon Climate

We associate the monsoon most closely with the peninsula lands of Southeast Asia. Here the alternating circulation of air (from land in winter, from water in summer) is strongly related to the shifting of the ITC. During the summer the ITC

(Continued on page 256)

Figure 10.8 *Preparing an area for planting in Jamaica; an example of subsistence slash-and-burn (swidden) agriculture.* • **Would you expect shifting cultivation to be on the increase or decrease in tropical rainforests?**

DESTRUCTION OF THE AMAZON RAINFOREST

It has been estimated that prior to the beginnings of agriculture, *closed forests* with dense canopies and little undergrowth, combined with *open forests* or woodlands with open canopies and grasses between trees, covered more than 6 billion hectares (15 billion acres) of Earth's land surface. Calculations provided by United Nations sources indicate that by 1980 the area covered by forests had been reduced by almost one third, to approximately 4.3 billion hectares (10.5 billion acres). Such change, however, attracted little attention until recently and caused even less concern because it was associated with the clearing of land for timber and the raising of crops in middle-latitude nations in which industrialization was expanding rapidly. Until the twentieth century the luxuriant tropical forests remained virtually untouched except where replaced by plantation agriculture or altered by shifting cultivation. However, rapid destruction of tropical forests currently is under way, and this is nowhere more evident than in the Amazon rainforest of South America.

Tropical closed forests comprised nearly 50 percent of Earth's forested land in the 1980 survey by the United Nations, and more recent satellite data indicate that such forests still cover nearly 13 percent of Earth's total land surface. A total forested area of more than 2 billion hectares would seem adequate to meet human needs forever until one considers the current rate of deforestation. Between 1981 and 1990 nearly 170 million hectares (420 million acres) of virgin tropical forest were lost to ax, saw, and fire. **Each year an area of forest somewhat larger than a football field is currently being destroyed *every second*.** To add to the problem, the rate of deforestation has steadily increased. The rate of deforestation for all tropical forests doubled between 1980 and 1989, and in the Amazon rainforest, where more than half of rainforest resources are located, the rate nearly tripled. Simple mathematics indicates that even at the present rate of deforestation, the Amazon rainforest will virtually disappear in 150 years.

There are a number of reasons for the clearing of the forests that throughout human history have covered major portions of the Amazon River basin. Most of these forest resources lie in Brazil, with smaller areas located in Peru, Ecuador, and Columbia. These are all developing countries desperate for income to fund national growth and repay staggering debts owed to foreign banks. Tropical timber sales provide the prospect of short-term profit, even though in many instances as little as 15 percent of the timber is harvested before the land is cleared by burning. In addition, the Brazilian government in particular has viewed the Amazon rainforest as frontier land available for agricultural development and for resettlement of the poor from overcrowded urban areas. Government funds have long financed the building of roads into the forested interior and have supported the development of cattle ranches through long-term loans and tax credits sufficient to cover most investment costs.

The irony is that clearing tropical forests for timber sales and agricultural development is economically unsound. Over a few years the income from forest

Cattle grazing along the Rio Salimoes in Brazil in an area of former rainforest.

A section of Amazon rainforest cleared by slash-and-burn techniques for potential farming or grazing.

products such as nuts, fruits, resins, fibers, and medicinal supplies (as opposed to the one-time sale of timber) and the selective cutting of timber far outweighs the economic gain from one-time timber sales and agricultural production. It has been estimated that the 12 million hectares (32 million acres) of existing cattle ranches in the Amazon have already cost the Brazilian government more than $2 billion in income from forest products.

Decisions concerning the future of tropical rainforests rest with the governments of nations that control these valuable resources. Why do people in other nations wish to become involved? Why has the destruction of the Amazon rainforest, for example, become a serious international issue? Answers to these questions are found in the concerns of atmospheric scientists, biologists, and other environmentalists throughout the world. **When the tropical forests are removed, the hydrologic cycle and energy budgets in the previously forested areas are dramatically altered, often irreversibly.** In tropical rainforests the canopy shades the forest floor, thus helping to keep it cooler. In addition, the huge mass of vegetation provides a tremendous amount of water vapor to the atmosphere through transpiration. The water vapor condenses to form clouds, which in turn provide rainfall to nourish the forest. With the forests removed, transpiration is diminished, which leads to less cloud cover and less rainfall. With fewer clouds and no forest canopy, more solar energy reaches Earth's surface.

The unfortunate outcome is that areas that are deforested soon become hotter and drier, and any *ecosystem* in place is seriously damaged or destroyed. Once the rainforests are removed, the soils lose their source of plant nutrients and this precludes the growth of any significant crops or plants. In addition, the rainforest, which was in harmony with the soil, cannot reestab-

lish itself. Thus the multitude of flora and fauna species indigenous to the rainforest is lost forever. It is impossible to calculate the true cost of this reduction in biodiversity (the total number of different plant and animal species in the Earth system). The lost species may have held secrets to increased food production; a cure for AIDS, cancer, or other health problems; or a base for better insecticides that do not harm the environment. Similar services to humanity already have been provided by tropical forest species.

Tropical deforestation is also threatening the natural chemistry of the atmosphere. The tropical forests are a major source of the atmospheric oxygen so essential to all animal life. And deforestation encourages global warming by enhancing the greenhouse effect, since forests act as a major reservoir of carbon dioxide. It has been estimated that forest clearing since the mid-1800s has contributed over 130 billion tons of carbon to the atmosphere, an amount more than two-thirds that which has been added by the burning of coal, oil and natural gases combined.

What can be done? The reasons for tropical deforestation and the solutions to the problem may be economic, but the issues are extremely complex. **It is not sufficient for the rest of the world to point out to governments of tropical nations that their forests are a major key to human survival.** It is unacceptable for scientists and politicians from nations where barely one fourth of the original forests remain to insist that the citizens of the tropics cease cutting trees and establish forest plantations on deforested land. These are desired outcomes, but it is first the responsibility of all the world's people to help resolve the serious economic and social problems that have prevented most tropical nations from considering their forests as a sustainable resource.

CRITICAL THINKING ▼

(1) What should be the policy of the United States government in regard to the destruction of the Amazon rainforest? **(2)** What lessons concerning the management of United States forest resources might be learned from examining the results of tropical deforestation? **(3)** Should Earth resources be considered the concern of all the world's people or just the concern of the people who control them?

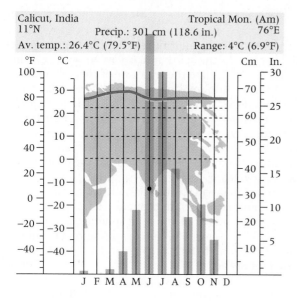

Calicut, India   Tropical Mon. (Am)
11°N   Precip.: 301 cm (118.6 in.)   76°E
Av. temp.: 26.4°C (79.5°F)   Range: 4°C (6.9°F)

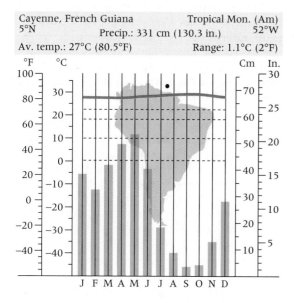

Cayenne, French Guiana   Tropical Mon. (Am)
5°N   Precip.: 331 cm (130.3 in.)   52°W
Av. temp.: 27°C (80.5°F)   Range: 1.1°C (2°F)

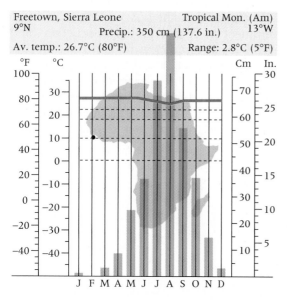

Freetown, Sierra Leone   Tropical Mon. (Am)
9°N   Precip.: 350 cm (137.6 in.)   13°W
Av. temp.: 26.7°C (80°F)   Range: 2.8°C (5°F)

moves suddenly north into the Indian subcontinent and adjoining lands to a latitude of 20° or 25°. This is due in part to the attracting force of the deep, low-pressure area of the Asian continent. But, as we have previously noted, the mechanism is complex and involves changes in the upper-air flow as well as in surface currents. Several months later the moisture-laden summer monsoon is replaced by an outflow of dry air from the massive Siberian high-pressure area that develops in the winter season over central Asia. By this time the ITC has shifted far to the south.

However, Figures 10.5 and 10.9 confirm the fact that there are climatic regions outside of Asia that fit the simplified Köppen classification of tropical monsoon. Although a modified version of the monsoonal wind shift occurs at Freetown, Sierra Leone, in Africa, the climate there might also be described as transitional between the constantly wet rainforest climate and the sharply seasonal wet and dry conditions of the tropical savanna. Cayenne, French Guiana, on the other hand, receives much of its rainfall during the low-sun (winter) season, when the trades move onshore along the northeast coast of South America, bringing abundant moisture from the nearby ocean.

Distinctions Between Rainforest and Monsoon. Whatever the factors that produce tropical monsoon climatic regions, there are strong similarities between these regions and those that are classified as tropical rainforest. In fact, although their core regions are distinctly different from one another, the two climates are often intermixed over transition areas. A major reason for the similarity between monsoon and rainforest climates is that a monsoon area has enough precipitation to allow continuous vegetative growth with no dormant period during the year. Rains are so abundant and intense and the dry season is so short that the soils usually do not dry out completely. As a result, this climate and its soils support a

Figure 10.9 *Climographs for tropical monsoon climate stations.* • **Why does the distribution of precipitation at Cayenne differ so much from that of the other two stations, although all three are in the Northern Hemisphere?**

plant cover much like that of the tropical rainforests.

However, there are clear distinctions between rainforest and monsoon climatic regions. The most important, of course, concern precipitation, including both distribution and amount. The monsoon climate has a short dry season while the rainforest does not. Perhaps even more interesting is the fact that the average rainfall in monsoon regions varies more widely from place to place. It usually totals between 150 and 400 centimeters (60 and 150 in.) and may be massive where the onshore monsoon winds are forced to rise over mountain barriers. Mahabaleshwar, altitude 1362 meters (4467 ft), on the windward side of India's Western Ghats, averages more than 630 centimeters (250 in.) of rain during the five months of the summer monsoon.

The annual march of temperature of the monsoon climate differs appreciably from the monotony of the rainforest climate. The heavy cloud cover of the rainy monsoon considerably reduces insolation and temperatures during that time of year. But during the period of clear skies just prior to the onslaught of the rains, higher temperatures are recorded. As a result, the annual temperature range in a monsoon climate is from 2° to 6°C (compared with 2° to 3°C in the tropical rainforest). This is illustrated by the temperature curves on the climographs of Figure 10.9.

Some additional distinctions between monsoon and rainforest regions can be found in vegetation and animal life. Toward the wetter margins, the tropical monsoon forest resembles the tropical rainforest, but fewer species are present and certain ones become dominant, as teak does in Myanmar. The seasonality of rainfall in the monsoon narrows the range of species that will prosper. Toward the drier margins of the climate, the trees grow farther apart and the monsoon forest often gives way to jungle or a dwarfed thorn forest. The composition of the animal kingdom here also changes. The climbing and flying species that dominate the forest are joined by larger, hoofed leaf-eaters and by carnivores such as the famous tigers of Bengal.

Effects of Seasonal Change.
The wet and dry seasonality of the monsoon regions has often been compared to the four temperature seasons of the middle latitudes. Just as some writers have sought to catch the essence of spring, summer, fall, and winter on paper, so have those whose lives have been intertwined with the monsoon sought to capture its unique qualities. Following is such a description by an Indian writer.

What the four seasons of the year mean to the European, the one season of the monsoon means to the Indian. It is preceded by desolation; it brings with it the hopes of spring; it has the fullness of summer and the fulfillment of autumn all in one.

Those who mean to experience it should come to India some time in March or April. The flowers are on their way out and the trees begin to lose their foliage. The afternoon breeze has occasional whiffs of hot air to warn one of the days to come. For the next three months the sky becomes a flat and colorless gray without a wisp of a cloud anywhere. People suffer great agony. Sweat comes out of every pore and the clothes stick to the body. Prickly heat erupts behind the neck and spreads over the body till it bristles like a porcupine and one is afraid to touch oneself. The thirst is unquenchable, no matter how much one drinks. The nights are spent shadowboxing in the dark trying to catch mosquitoes and slapping oneself in an attempt to squash those humming near one's ears. One scratches and curses when bitten, knowing that the mosquitoes are stroking their bloated bellies safely perched in the farthest corners of the nets, that they have gorged themselves on one's blood. When the cool breeze of the morning starts blowing, one dozes off and dreams of a paradise with ice cool streams running through lush green valleys. Just then the sun comes up strong and hot and smacks one in the face. Another day begins with its heat and its glare and its dust.

After living through all this for ninety days or more, one's mind becomes barren and bereft of hope. It is then that the monsoon makes its spectacular entry. Dense masses of dark clouds sweep across the heavens like a celestial army with black banners. The deep roll of thunder sounds like the beating of a billion drums. Crooked shafts of silver zigzag in lightning flashes against the black sky. Then comes the rain itself. First it falls in fat drops; the earth rises to meet them. She laps them up thirstily and is filled with fragrance. Then it comes in torrents which she receives with the supine gratitude of a woman being ravished by her lover. It impregnates her with life which bursts forth in abundance within a few hours. Where there was nothing, there is everything: green grass, snakes, centipedes, worms, and millions of insects.

I Shall Not Hear the Nightingale
Khushwant Singh

Figure 10.10 *Winnowing rice outside a small village in West Bengal, India. Rice is the staple in the diet of villagers throughout monsoon Asia.* • **In what countries other than India would you expect to find such ancient methods of separating the rice grain from the chaff?**

The seasonal precipitation of the tropical monsoon climate is also important for economic reasons, especially to the people of Southeast Asia and India. Most of the people living in those areas are farmers, and their major crop is rice, which is the staple food for millions of Asians (Fig. 10.10). Rice most often is an irrigated crop, so the monsoon rains are very important to its growth. Harvesting, on the other hand, must be done during the dry season (Fig. 10.11).

Each year an adequate food supply for much of South and Southeast Asia depends on the arrival and departure of the monsoon rains. The difference between famine and survival for many people in these regions is very much associated with the climate.

Tropical Savanna Climate

Because of its location well within the tropics (usually between latitudes 5° and 20° on either side of the equator), the tropical savanna climate has much in common with the tropical rainforest and monsoon. The sun's vertical rays at noon are never far from overhead, and temperatures remain constantly high. Days and nights are of nearly equal length throughout the year, as they are in other tropical regions.

However, as previously noted, its distinct seasonal precipitation pattern identifies the tropical savanna. As the latitudinal wind and pressure belts shift with the sun, savanna regions are under the influence of the rain-producing ITC (doldrums) for part of the year and the rain-suppressing subtropical highs for the other part.

In fact, the poleward limits of the savanna climate are approximately the poleward limits of migration of the ITC, while the equatorward limits of this climate are the equatorward limits of movement by the subtropical high-pressure systems.

As you can see in Table 10.2 and Figure 10.5, the greatest areas of savanna climate are found peripheral to the rainforest climates of Latin America and Africa. Lesser but still important savanna regions occur in India, peninsular Southeast Asia, and Australia. In some instances the climate extends poleward of the tropics as it does in the southernmost portion of Florida.

Transitional Features of the Savanna. Of particular interest to the geographer is the transitional nature of the tropical savanna. Often located between the humid rainforest climate on the one hand and the rain-deficient steppe climate on the other, the savanna experiences some of the characteristics of both. During the rainy, high-sun season, atmospheric conditions resemble those of the rainforest, while the low-sun season can be as dry as nearby arid lands are all year. Because of the gradational nature of the climate, precipitation patterns vary considerably (Fig. 10.12). Savanna locations close to the rainforest may have rain during every month, and their total annual precipitation may exceed 180 centimeters (70 in.). In contrast, the drier margins of the savanna have longer and more intensive periods of drought and lower annual rainfalls (less than 100 centimeters).

Other characteristics of the savanna help to demonstrate its transitional nature. The higher temperatures just prior to the arrival of the ITC produce annual temperature ranges 3° to 6°C (5°–8°F) wider than those of the rainforest but still not as wide as those of the steppe and desert. Although the typical savanna vegetation (known as **llanos** in Venezuela and **campos** in Brazil) is a mixture of grasslands and trees, scrub, and thorn bushes, there is considerable variation. Near the equatorward margins of these climates, grasses are higher, and trees, where they exist, grow fairly close together. Toward the drier, poleward margins, trees are more widely scattered and smaller, and the grasses are shorter. Soils, too, are affected by the climatic gradation, as the iron-rich reddish soils of the wetter sections are replaced by darker-colored wet and dry tropical soils in the drier regions.

Both vegetation and soils have made special adaptations to the alternating wet-dry seasons of the savanna. During the wet (high-sun) period

258

(Continued on page 260)

(a)

(b)

Figure 10.11 *Crop selection and agricultural production adjust to the wet (a) and dry (b) seasons throughout southeast Asia.* • Would it be beneficial to the people of southeast Asia if the traditional rice farming methods were replaced by mechanized rice agriculture as practiced in the United States?

Figure 10.12 *Climographs for tropical savanna climate stations.* • Consider the differences in climate, and human use of the environment, between Key West and Raipur. Which are the most important in the geography of the two places, the physical or the human factors?

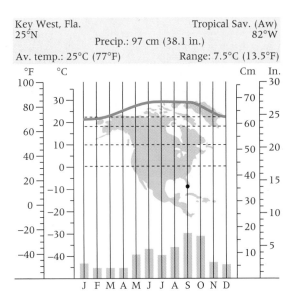

Key West, Fla.
25°N
Tropical Sav. (Aw)
82°W
Precip.: 97 cm (38.1 in.)
Av. temp.: 25°C (77°F)   Range: 7.5°C (13.5°F)

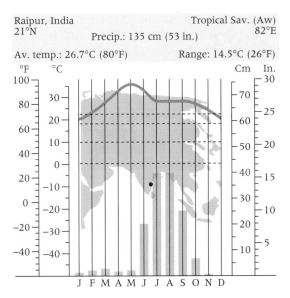

Raipur, India
21°N
Tropical Sav. (Aw)
82°E
Precip.: 135 cm (53 in.)
Av. temp.: 26.7°C (80°F)   Range: 14.5°C (26°F)

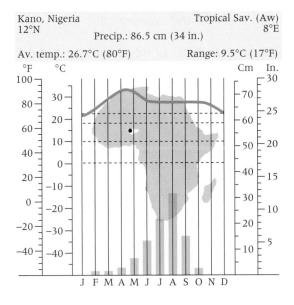

Kano, Nigeria
12°N
Tropical Sav. (Aw)
8°E
Precip.: 86.5 cm (34 in.)
Av. temp.: 26.7°C (80°F)   Range: 9.5°C (17°F)

Figure 10.13 Laterite cut for building stone and stacked along a village road in the state of Orissa, India. • **Why is building with brick or stone rather than wood so important in heavily populated, less developed nations such as India?**

the grasslands are green and the trees are covered with foliage. During the dry (low-sun) period the grass turns dry, brown, and lifeless, and most of the trees lose their leaves as an aid in reducing moisture loss through transpiration. The trees have deep roots that can reach down to water in the soil during the dry season. They are also fire-resistant, an advantage for survival in the savanna since the grasses may burn during the winter drought.

Laterite is often associated with the soils. This is a zone in the subsoil, as much as 6 meters (20 ft) thick, depleted of all but oxides and hydroxides of iron, aluminum, and manganese. When wet, this material may be cut into bricks that become rock-hard when dried out (Fig. 10.13). Laterite results from the concentration of these oxides as the subsoil zone alternates between wet and dry periods in response to the seasonal rainfall of the savanna climate.

Savanna Potential. Conditions within tropical savanna regions are not well suited to agriculture, although many of our domesticated grasses (grains) are presumed to have grown wild there. Rainfall is far more unpredictable than in the rainforest or even the monsoon climate. For example, Nairobi, Kenya, has an average rainfall of 86 centimeters. Yet from year to year the amount of rain received may vary from 50 to 150 centimeters. As in the steppe and desert, the drier the savanna station, the more unreliable the rainfall. However, the rains are essential for human and animal survival in this region. When they are late or deficient, as they have been in West Africa in recent years, severe drought and famine result. On the other hand, when the rains last

longer than usual or are excessive, they can cause major floods, often followed by outbreaks of disease.

Savanna soils (except in areas of recent stream deposits) also limit productivity. During the rains of the wet season they become gummy, while during the dry season they are hard and almost impenetrable. Consequently, people in the savannas have often found the soils better suited to grazing than to farming. The Masai cattle herders of East Africa are world-famous examples (Fig. 10.14). However, even animal husbandry has its problems. Many savanna regions make poor pasture lands at least part of the year, and large-scale commercial cattle raising must await the successful introduction of more nutritious subtropical grasses. In addition, soil erosion can be a problem in the savannas during the rainy season, especially in areas that have been over-grazed by both wild and domesticated animals.

The savannas of Africa have exhibited the greatest potential of the world's savanna regions. They have been veritable zoological gardens for the larger tropical animals. These grassland regions support many different herbivores (plant eaters), such as the elephant, rhinoceros, giraffe,

Figure 10.14 The grasslands of the East African savanna are well suited to the support of large numbers of grazing animals. The Masai cattle herders shown here in Kenya count their wealth by the numbers of animals they own. • **What problems are created for the Kenyan government as cattle herds grow?**

Figure 10.15 Elephants have always been a majestic sight in the savanna climate of East Africa. • Why are opportunities for photographs such as this in serious danger of disappearing completely?

zebra, and antelope (Fig. 10.15). The herbivores in turn are eaten by the carnivores (flesh eaters), such as the lion, leopard, cheetah, hyena, jackal, and wild dog. During the dry season the herbivores find grasses and water along stream banks and forest margins and at isolated water holes.

The carnivores follow the herbivores to the water, and a few human hunters still follow them both. However, the days of the great herds of game animals in Africa are numbered. Most are now crowded into overgrazed national parks, which are themselves under pressure from those who desire more range for their growing cattle herds. In a few decades the great numbers of wild animals may be reduced to a few zoo specimens.

Arid Climatic Regions

Arid climatic regions in the simplified Köppen system are widely distributed over Earth's surface. A brief study of Figure 10.16 confirms that they are found from the vicinity of the equator to more than 50°N and S latitude. There are two major concentrations of desert lands, and each illustrates one of the important causes of climatic aridity. The first is centered on the Tropics of Cancer and Capricorn ($23\frac{1}{2}°$ N and S latitudes) and spreads 10° to 15° poleward and equatorward from there. The second is located poleward of the first and occupies continental interiors, particularly in the Northern Hemisphere.

The concentration of deserts in the vicinity of the tropical sun lines is directly related to the subtropical high-pressure cells. Although the

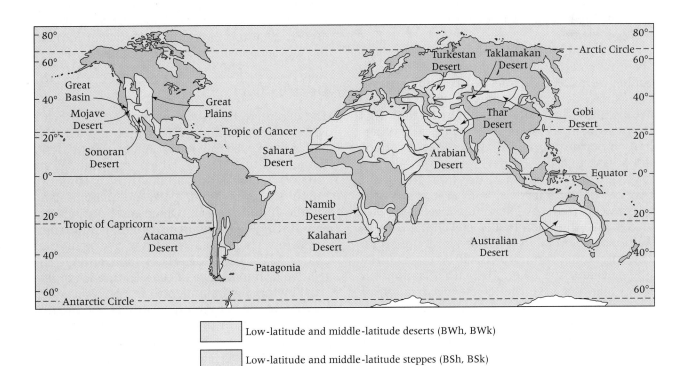

Low-latitude and middle-latitude deserts (BWh, BWk)

Low-latitude and middle-latitude steppes (BSh, BSk)

Figure 10.16 A map of the world's arid lands. • What does a comparison of this map with the Map of World Population Density (Fig. 1.8) suggest?

boundaries of the subtropical highs may migrate north and south with the sun, their influence remains constant in these latitudes. We have already learned that the subsidence and divergence of air associated with these cells is strongest along the eastern portions of the oceans (western portions of the continents). Hence, the clear weather and dry conditions of the subtropical high pressure extend inland from the western coasts of each landmass in the subtropics. The Atacama, Namib, and Kalahari Deserts and the desert of Baja California are restricted in their development by the small size of the landmass or by landform barriers to the interior. But the western portion of the Afro-Eurasian landmass comprises the greatest stretch of desert in the world and includes the Sahara, Arabian, and Thar Deserts. Similarly, the Australian Desert occupies most of the interior of the Australian continent.

The second concentration of deserts is located within continental interiors remote from moisture-carrying winds. Such arid lands include the vast cold-winter deserts of inner Asia and the Great Basin of interior drainage in the western United States. The dry conditions of the latter region extend northward into the Columbia Plateau and southward into the Colorado Plateau and are increased by the mountain barriers that restrict the movement of rain-bearing air masses from the west. Similar rain-shadow conditions help to explain the Patagonia Desert of Argentina and the arid lands in western China (Xinjiang/Sinkiang).

Both wind direction and ocean currents can accentuate aridity in coastal regions. When prevailing winds blow parallel to a coastline instead of onshore, desert conditions are likely to occur because little moisture is brought inland. This seems to be the case in eastern Africa and perhaps in northeastern Brazil. Where a cold current flows next to a coastal desert, foggy conditions may develop. Warm, moist air from the ocean may be cooled to its dew point as it passes over the cooler current. A temperature inversion is created, increasing stability and preventing the upward movement of air required for precipitation. The unique, fog-shrouded coastal deserts in Chile (the Atacama), southwest Africa (the Namib), and Baja California have the lowest precipitation of any regions on Earth.

The map of Figure 10.16 shows deserts of the world to be core areas of aridity, usually surrounded by the slightly moister steppe regions. Hence our explanations for the location of deserts hold true for the steppes as well. The steppe climates either are subhumid borderlands of the humid tropical, mesothermal, and microthermal climates, or are transitional between these climates and the deserts. As previously noted, we classify both steppe and desert on the basis of the relation between precipitation and potential evaporation. In the desert climate the amount of precipitation received is less than half the potential evaporation. In the steppe climate the precipitation is more than half but less than the total potential evaporation.

The criterion for determining whether a climate is desert, steppe, or humid is *precipitation effectiveness*. The amount of precipitation actually available for use by plants and animals is the effective precipitation. Precipitation effectiveness is related to temperature. At higher temperatures it takes more precipitation to have the same effect on vegetation and soils than at lower temperatures. The result is that areas with higher temperatures and therefore greater potential for evaporation can receive more precipitation than cooler regions and yet have a more arid climate.

Because of the temperature influence, precipitation effectiveness depends on the season in which an arid region's meager precipitation is concentrated. Obviously, precipitation received during the low-sun period will be more effective than that received when temperatures are higher, for less will be lost through evaporation. Graphs based on the concept of precipitation effectiveness are included in the Appendix and may be used to determine whether a particular location has a desert, steppe, or humid climate.

Desert Climates

The deserts of the world extend through such a wide range of latitudes that the simplified Köppen system recognizes two major subdivisions: the low-latitude deserts, where temperatures are relatively high year-round and frost is absent or infrequent even along poleward margins, and middle-latitude deserts, which have distinct seasons, including below-freezing temperatures during winter (Table 10.3). However, the significant characteristic of all deserts is their aridity. The relative unimportance of temperature is emphasized by the small number of occasions on which we distinguish between low-latitude and middle-latitude deserts in the discussion that follows.

Land of Extremes. By definition, deserts are associated with a minimum of precipitation, but they represent the extremes in other atmospheric

TABLE 10.3 **The Arid Climates**

Name and Description	Controlling Factors	Geographic Distribution	Distinguishing Characteristics	Related Features
Desert				
Precipitation less than half of potential evaporation; mean annual temperature above 18°C (64.4°F) (low-lat.), below (mid.-lat.)	Descending, diverging circulation of subtropical highs; continentality often linked with rain-shadow location	Coastal Chile and Peru, southern Argentina, southwest Africa, central Australia, Baja California and interior Mexico, North Africa, Arabia, Iran, Pakistan and western India (low-lat.); inner Asia, and western United States (mid.-lat.)	Aridity; low relative humidity; irregular and unreliable rainfall; highest percentage of sunshine; highest diurnal temperature range; highest daytime temperatures; windy conditions	Xerophytic vegetation; often barren, rocky, or sandy surface; desert soils; excessive salinity; usually small, nocturnal, or burrowing animals; nomadic herding
Steppe				
Precipitation more than half but less than potential evaporation; mean annual temperature above 18°C (64.4°F) (low-lat.), below (mid.-lat.)	Same as deserts; usually transitional between deserts and humid climates	Peripheral to deserts, especially in Argentina, northern and southern Africa, Australia, central and southwest Asia, and western United States	Semiarid conditions, annual rainfall distribution similar to nearest humid climate; temperatures vary with latitude, elevation, and continentality	Dry savanna (tropics) or short grass vegetation; highly fertile black and brown soils; grazing animals in vast herds, predators and smaller animals; ranching, dry farming

conditions as well. Because there are few clouds and little water vapor in the air, as much as 90 percent of insolation reaches Earth in desert regions. This is why the highest insolation and the highest temperatures are recorded in low-latitude desert areas and not in the more humid tropical climates that are closer to the equator. Again because of light cloud cover, there is little atmospheric effect and much of the energy received by Earth during the day is radiated back to the atmosphere at night. Consequently, night temperatures in the desert drop far below their daytime highs. The excessive heating and cooling give low-latitude deserts the greatest diurnal temperature ranges in the world, and middle-latitude deserts are not far behind. In the spring and fall

these ranges may be as great as 40°C (72°F) in a day. More common diurnal temperature ranges in deserts are 22° to 28°C (40°–50°F).

Daytime air temperatures, especially in low-latitude deserts, must be experienced to be appreciated, but the following quotation may convey the idea of desert heat. Note that at the height of summer in this example, the daytime heat lasts throughout much of the night.

The first summer in Kharga was rather a shock; though I had been in Egypt for over four years. . . . At 6 A.M. in the morning the mercury stood at 98°, and this being the cool of the day the house was shut right up until *evening, so that coming from the glare and heat at mid-day one got the impression—a totally false one—that the house was cool. At mid-day one saw the most ghastly sight I have ever seen and that was a bright patch of sunlight in the fireplace caused by the sun shining straight down the chimney.[*] A warm red glow from a fireplace in midwinter is one of the pleasantest things I know, but a staring yellow patch of sunlight where the glowing coals ought to be, lightening up the gloom of a darkened room that is pretending to be cool, has a most grisly effect.*

During the whole of the day the temperature remained at from 110° to 115° with a hot wind. At 6 P.M. the wind dropped and it seemed to get hotter till 11 P.M. when the wind started again, feeling quite as blistering and unpleasant as it had been at mid-day. It was quite impossible to sleep, and I used to walk about on the verandah throwing water on the mosquito-curtains in a vain attempt to bring down the temperature and do something to moisten the intense dryness that caused the tables and chests of drawers in the house to split with loud reports. At about 2 A.M. there was a slight but appreciable cooling off and one could usually get to sleep then till 5:30 A.M. when the heat of the newly-born sun awoke one to another day of hell.

Three Deserts
C. S. Jarvis

Because the sun's rays are so intense in the clear, dry desert air, temperatures in shade are much lower than those a few steps away in direct sunlight. (Keep in mind that all temperatures recorded for statistical purposes, including those in the desert, are shade temperatures.) Khartoum (in the Sahara of the Sudan) has an *average* annual temperature of 29.5°C (85°F), which is a *shade* temperature. Temperatures in the bright desert sun under cloudless skies at Khartoum are often 43°C (110°F) or more. Soil temperatures rise to close to 95°C (200°F) in midsummer in the Mojave Desert of southern California.

During low-sun or winter months, deserts experience colder temperatures than more humid areas at the same latitude, and in summer they experience hotter temperatures. Just as with the high diurnal ranges in deserts, these high annual temperature ranges can be attributed to the lack of moisture in the air.

Annual temperature ranges are usually greater in middle-latitude deserts, like the Gobi in Asia, than in low-latitude deserts because of the colder winters experienced at higher latitudes. Compare, for example, the climograph for Aswan in south central Egypt—at 24°N, a low-latitude desert location—with the climograph for Turtkul, Uzbekistan—at 41°N, a middle-latitude desert location (Fig. 10.17). The annual range for Aswan is 17°C; in Turtkul, it is 34°C.

Precipitation in the desert climate is irregular and unreliable, but when it comes, it may arrive in an enormous cloudburst, bringing more precipitation in a single rainfall than has been recorded in years. This happened in the extreme at the port of Walvis Bay, located on the coast of the Namib Desert, a cold-current coastal desert of southwest Africa. The equivalent of 10 years of rains was received in one night when a freak storm dumped 3.2 centimeters of rain.

Absolute humidity, or the actual amount of water vapor in the air, may be high in desert areas. However, the hot daytime temperatures increase the capacity of the air so that the relative humidity during daylight hours is quite low (10 to 30 percent). Desert nights are a different story.

Radiation of energy is rapid in the clear air, and as temperature drops, relative humidity increases and the formation of dew in the cool hours of early morning is common. Where measurements have been made, the amount of dew formed has sometimes considerably exceeded the annual rainfall for that location. It has been suggested, in fact, that dew may be of great importance to plant and animal life in the desert. Studies are now being carried out on the use of dew as a moisture source for certain crops, thereby minimizing the need for large-scale irrigation.

The convection currents set up by the intense heating of the land during the day help to make the desert a windy place. In addition, the sparseness of vegetation and the absence of topographic interruptions in some deserts allow winds to sweep across these arid lands unimpeded. Sand and dust are carried by the desert winds, lowering visibility and irritating eyes and throats.

Adaptations by Plants and Animals. Deserts tend to have sparse vegetation, and large tracts may be barren bedrock, sand, or gravel. The plants that do exist are **xerophytic,** or adapted to extreme drought. They may have thick bark,

---

*Kharga is at latitude 25°26′N, so the sun at the June solstice would be about 2° from the zenith at noon.

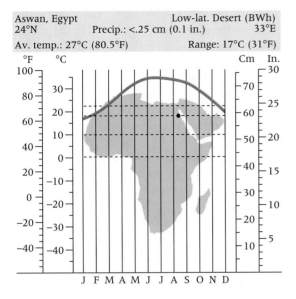

Aswan, Egypt — Low-lat. Desert (BWh)
24°N — Precip.: <.25 cm (0.1 in.) — 33°E
Av. temp.: 27°C (80.5°F) — Range: 17°C (31°F)

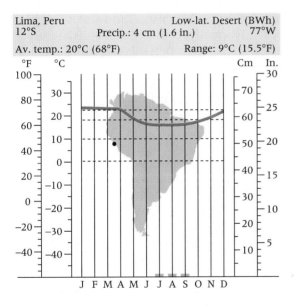

Lima, Peru — Low-lat. Desert (BWh)
12°S — Precip.: 4 cm (1.6 in.) — 77°W
Av. temp.: 20°C (68°F) — Range: 9°C (15.5°F)

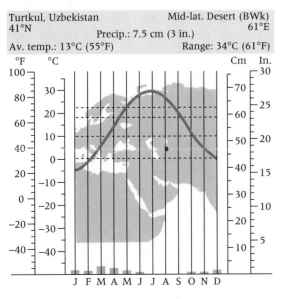

Turtkul, Uzbekistan — Mid-lat. Desert (BWk)
41°N — Precip.: 7.5 cm (3 in.) — 61°E
Av. temp.: 13°C (55°F) — Range: 34°C (61°F)

thorns, little foliage, and waxy leaves, all of which reduce loss of water by transpiration (Fig. 10.18). Additional characteristic adaptations are the storage of moisture in stem or leaf cells, as in the cactus (barrel cactus, prickly pear, saguaro). Other vegetation, like creosote bush, mesquite, and acacias, have deep root systems to reach water, while still others, such as the Joshua tree, spread their roots widely near the surface for their moisture supply. Many desert plants react to prevent moisture loss through the surface of their leaves in the process of transpiration. The ocotillo, for example, sheds its leaves in dry periods. Others, like the darning-needle cactus and the palo verde, have no leaves. The acacia tree folds its leaves, protecting and shading the inner surface of each leaf from the desert sun. Many desert plants, such as the cactus, desert dandelion, and desert primrose, produce brightly colored flowers, which encourage pollination and thus ensure survival of the species. Still others, like creosote and brittle bush, secrete poison, serving to keep out competitors. In addition to the xerophytic shrub forms, herbaceous annuals spring up periodically as a result of rains. During early spring and again in the fall, California's Mojave Desert may be carpeted with flowers and grass. The richness of the herbaceous cover varies from season to season and year to year as the rains vary in location and amount.

Survival is the key question of everyday life in the desert for animal, plant, and human, as we can see in the following excerpt about the Mojave Desert.

And there are true secrets in the desert. In the war of sun and dryness against living things, life has its secrets of survival. Life, no matter on what level, must be moist or it will disappear. I find most interesting the conspiracy of life in the desert to circumvent the death rays of the all-conquering sun. The beaten earth appears defeated and dead, but it only appears so. A vast and inventive organization of living matter survives by seeming to have lost. The gray and dusty sage wears oily armor to protect its inward small moistness. Some

Figure 10.17 *Climographs for desert climate stations.* • **If you consider the serious limitations of desert climates, how do you explain why some people choose to live in desert regions?**

Figure 10.18 This diverse association of vegetation is well adapted to survive the year-round drought conditions prevailing in desert climates • **Under what circumstances might significant numbers of people be attracted to regions with desert climates?**

plants engorge themselves with water in the rare rainfall and store it for future use. Animal life wears a hard, dry skin or an outer skeleton to defy the desiccation. And every living thing has developed techniques for finding or creating shade. Small reptiles and rodents burrow or slide below the surface or cling to the shaded side of an outcropping. Movement is slow to preserve energy, and it is a rare animal which can or will defy the sun for long. A rattlesnake will die in an hour of full sun. Some insects of bolder inventiveness have devised personal refrigeration systems. Those animals which must drink moisture get it at second hand—a rabbit from a leaf, a coyote from the blood of a rabbit.

One may look in vain for living creatures in the day time, but when the sun goes and the night gives consent, a world of creatures awakens and takes up its intricate pattern. Then the hunted come out and the hunters, and hunters of the hunters. The night awakes to buzzing and to cries and barks.

Travels with Charley
John Steinbeck

Even humans, the most adaptable of animals, find the desert environment a lasting challenge. For the most part people have been hunters and gatherers, nomadic herders, and subsistence farmers wherever there was a water supply from wells or *exotic streams* (streams that bring water from outside the region) like the Nile, Tigris, Euphrates, Indus, and Colorado Rivers. Desert people have learned to adjust their habits to the environment. For example, they wear loose clothing to protect themselves from the burning rays of the sun and to prevent moisture loss by evaporation from the skin. At night, when the temperatures drop, the clothing keeps them warm by insulating and minimizing the loss of body heat.

Given an adequate water supply that is carefully applied, desert regions can be agriculturally productive (Fig. 10.19). Soils are rich in plant nutrients, but they present numerous problems. These often include excessive salinity, the presence of a dense lime subsoil, low organic content, few microorganisms, and excessive permeability. Normally, only alluvial soils are used for agriculture, other types being thin and often rocky.

Great care must be used in irrigating desert lands. Excessive irrigation may cause water to evaporate at or near the surface, leaving a load of dissolved salts there. Vast agricultural areas in Iraq, Iran, Pakistan, and elsewhere have been destroyed and abandoned as a consequence of human-induced salinization. The solution to the problem is adequate drainage and the use of irrigation not only to provide water to the crops but also to flush the soil.

Of course, permanent agriculture has been established in desert regions all around the world, wherever river or well water is available. Every desert has its irrigated oases, although they vary considerably in size. Some produce mainly subsistence crops, but others have become significant producers of commercial crops for export.

Steppe Climates

Further study of Figure 10.16 and Table 10.3 provides a reminder that the distribution of the world's steppe lands is closely related to the location of the deserts. Both of the moisture-deficient climate types share the controlling factors of continentality, rain-shadow location, the subtropical high-pressure cells, or some combination of the three. The transitional nature of the steppes may make them seem like better-watered deserts at one time and like slightly subhumid versions of their humid climate neighbors at another. Herein lies the major problem of steppe regions. How and to what extent should these vari-

Figure 10.19 *Irrigated cotton, shown here in the Imperial Valley of California, owes its existence to water from the Colorado River.* • **What country might dispute this example of irrigated agriculture? Why?**

able and unpredictable climatic regimes be used by humans?

Similarities to Deserts. We are already aware that steppe regions are differentiated from deserts by their greater precipitation. For example, while most low-latitude desert locations receive fewer than 25 centimeters (10 in.) of rain annually, low-latitude steppe regions usually receive between 25 and 50 centimeters (10 and 20 in.). However, the similarities between deserts and steppes are often greater than the differences. In both climates the potential evaporation exceeds the precipitation. As in the deserts, precipitation in steppe regions is unpredictable and varies widely in total amount from year to year. Annual rainfall differs significantly from place to place within both desert and steppe regions, and vegetation varies accordingly.

To be more specific, both the general precipitation pattern and the nature of the vegetation of a steppe region are usually closely related to the more humid climate immediately adjacent

to it. That is, when the steppe is located between the desert and the tropical savanna, the steppe's rains come with the high-sun season. Next to a Mediterranean climate, the steppe receives primarily winter precipitation. Similarly, the short, shallow-rooted grasses most commonly associated with the steppe climate occur in the areas of transition from mesothermal and microthermal climates to the desert. But in the areas of transition from tropical savanna to the desert, the vegetation is the dry savanna type, including scrub tree and bush growth, which becomes more stunted and sparse toward the drier margins until the typical desert shrub vegetation type is dominant.

Due to similar conditions of atmospheric moisture and cloud cover, temperatures in the steppe regions also have much in common with those of the desert. Both low-latitude and middle-latitude varieties are identified by mean annual temperature (see Table 10.3 and Fig. 10.20). As in the desert, steppe temperatures vary throughout the climate type with latitude, distance from the

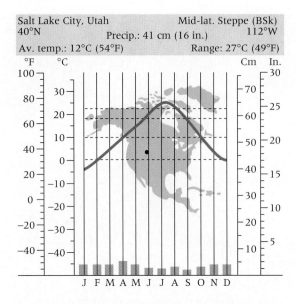

Salt Lake City, Utah
40°N
Mid-lat. Steppe (BSk)
112°W
Precip.: 41 cm (16 in.)
Av. temp.: 12°C (54°F)
Range: 27°C (49°F)

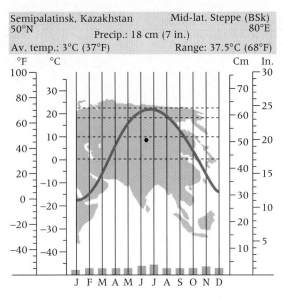

Semipalatinsk, Kazakhstan
50°N
Mid-lat. Steppe (BSk)
80°E
Precip.: 18 cm (7 in.)
Av. temp.: 3°C (37°F)
Range: 37.5°C (68°F)

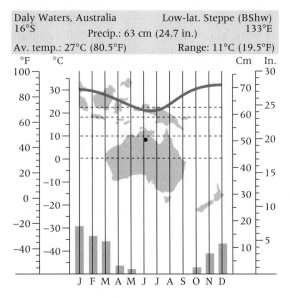

Daly Waters, Australia
16°S
Low-lat. Steppe (BShw)
133°E
Precip.: 63 cm (24.7 in.)
Av. temp.: 27°C (80.5°F)
Range: 11°C (19.5°F)

sea, and elevation. The climographs of Figure 10.20 demonstrate that although summer temperatures are high in all steppe regions, the differences in winter temperatures can produce annual ranges in middle-latitude steppes that are two or three times as great as those in low-latitude steppes.

A Dangerous Appeal. Although the surface cover is often incomplete, in more humid regions of the middle-latitude steppes the grasses have been excellent for pasture. In North America this was the realm of the bison (Fig. 10.21) and antelope, and in Asia it was the domain of wild horses. Steppe soils are usually high in organic matter and soluble minerals. Attributes such as these have attracted farmers and herdsmen alike to the rich grasslands but not without penalty to both humans and land.

The climate is dangerous for agriculture, and people take a sizeable risk when they attempt to farm. Although dry-farmed wheat and drought-resistant barley and sorghum can be successfully raised because both farming methods and crops are adapted to the environment, the use of techniques employed in more humid regions can lead to serious problems. Where farmers have attempted to grow crops by normal methods near the wetter margins of the steppe, they have suffered during the years of drought, which are inevitable and occur at unpredictable intervals. During dry cycles, crops fail year after year, and with the land stripped of its natural sod, the soil is exposed to wind erosion. Even using the grasses for grazing domesticated animals is not always the answer, for overgrazing can just as quickly create "Dust Bowl" conditions.

The difficulties in making steppe regions more productive point out again the sensitive ecological balance of Earth's systems. The natural rains in the steppe are usually sufficient to support a vegetation cover of short grasses that in turn can feed the roaming herbivores that graze on them. The herbivore population in turn is kept in check by the carnivores who prey on them. But when people enter the scene, sending out more animals to graze, plowing the land, or merely killing off the predators, the ecological balance is tipped and sometimes the results are disastrous.

Figure 10.20 *Climographs for steppe climate stations.* • **What are the chief differences between the climographs for Salt Lake City and Daly Waters? What are the causes of these differences?**

Figure 10.21 *It is hard to imagine the magnitude of the animal population on the North American steppes little more than a century ago. One is now fortunate to see a single remnant such as this one of the former herds of millions of bison and other grazing animals.* • **Where would a person go today to take a similar photograph?**

Define and Recall

tropical	region	selva	campos
polar	tropical rainforest	jungle	laterite
mesothermal	climate	shifting cultivation	
microthermal	annual temperature	tropical monsoon	desert climate
arid	range	climate	xerophytic
	daily (diurnal)	tropical savanna climate	steppe climate
	temperature range	llanos	

Discuss and Review

1. Where would you expect each of the 13 distinctive climatic types of the modified Köppen system used in your text to be located on a hypothetical continent? What reasons explain the location of each?

2. Describe the three tropical (*A*) climates. Why does each exist?

3. How do ocean currents affect tropical and polar climates?

4. How are the mesothermal (*C*) and the microthermal (*D*) climates similar? In what important respect do they differ?

5. In what two important ways do the *C* and *D* climates differ internally?

6. What is the major factor producing the dryness of arid climates? What factors produce the variations in highland climates?

7. What is a region as defined by geographers? Why is regionalization such a useful concept?

8. How many areas in the world have climates similar to the one with which you are most familiar? (Refer to Figure 10.4 for help in identifying them.)

9. What exceptions are there to the general rule that the pattern of climatic regions is produced by Earth-sun relationships? How do the Northern and Southern Hemispheres differ in climatic patterns, and why?

10. What are the controlling factors that explain the tropical rainforest climate? Give a brief verbal description of this climatic type.

11. What aspects of tropical rainforests are favorable to human use? Unfavorable? Describe the delicate balance between forest and soil. How might humans affect that balance?

12. What are the differences in climate, vegetation, and human use between the tropical rainforest and the monsoon climates?

13. Explain the seasonal precipitation pattern of the tropical savanna climate. State some of the transitional features of this climate. How have vegetation and soils adapted to the wet-dry season?

14. What conditions give rise to desert climates?

15. Describe the characteristics of desert soils. How can they be used for agriculture?

16. What kinds of adaptations have desert animals made to their environment?

17. How do steppes differ from deserts? Why might human use of steppe regions in some ways be more hazardous than use of deserts?

18. Do each of the following for the listed climates: tropical rainforest; tropical monsoon; tropical savanna; desert; steppe.
   (a) Identify the climate from a set of data or a climograph indicating average monthly temperature and precipitation for a representative station within a region typical of that climate.
   (b) Match the climatic type with a written statement that includes one or more of the following: the statistical parameters of the climate in the modified Köppen classification, the particular climatic controls (controlling factors) that produce the climate, the geographic distributions of the climate stated in terms of physical or political location, the unique climatic characteristics or combinations of characteristics that distinguish the climate from others, types of plants, animals, and soils associated with the climate, and the human utilization typical of the climate.
   (c) Distinguish between the important subtypes (if any exist) of each climate by identifying the characteristics which separate them from one another.

Consider and Respond

1. Based on the classification scheme presented in the Appendix, classify the following climatic stations from the data provided.

		J	F	M	A	M	J	J	A	S	O	N	D	Yr.
a.	Temp. (°C)	20	23	27	31	34	34	34	33	33	31	26	21	29
	Precip. (cm)	0.0	0.0	0.0	0.0	0.2	0.1	1.1	0.2	0.0	0.0	0.0	0.0	1.6
b.	Temp. (°C)	21	24	28	31	30	28	26	25	26	27	25	22	26
	Precip. (cm)	0.0	0.0	0.2	0.8	7.1	11.9	20.9	31.1	13.7	1.4	0.0	0.0	87.2
c.	Temp. (°C)	19	20	21	23	26	27	28	28	27	26	22	20	24
	Precip. (cm)	5.1	4.8	5.8	9.9	16.3	18.0	17.0	17.0	24.0	20.0	7.1	3.0	149.0
d.	Temp. (°C)	2	4	8	13	18	24	26	24	21	14	7	3	14
	Precip. (cm)	1.0	1.0	1.0	1.3	2.0	1.5	3.0	3.3	2.3	2.0	1.0	1.3	20.6
e.	Temp. (°C)	27	26	27	27	27	27	27	27	27	27	27	27	27
	Precip. (cm)	31.8	35.8	35.8	32.0	25.9	17.0	15.0	11.2	8.9	8.4	6.6	15.5	243.8
f.	Temp. (°C)	13	14	17	19	22	24	26	26	26	24	19	15	20
	Precip. (cm)	6.6	4.1	2.0	0.5	0.3	0	0	0	0.3	1.8	4.6	6.6	26.7

2. The six locations represented by the above data are listed below, although not in the order of presentation. Use an atlas and your knowledge of climates to match the climatic data with the following locations: Albuquerque, New Mexico; Belém, Brazil; Benghazi, Libya; Faya, Chad; Kano, Nigeria; Miami, Florida.

3. Benghazi, Libya, and Albuquerque, New Mexico, both exhibit steppe climates, yet the cause (or control) of their dry conditions is quite different. Describe the primary factor responsible for the dry conditions at each location.

4. Tropical *B* climates are located more poleward than *A* climates, yet their daytime high temperatures are often higher. Why?

5. Kano, Nigeria, has copious amounts of rainfall during the summer season. There are two sources, or causes, of this rainfall. What are they?

6. It has been said that there is no "climate" in mountain regions, just a great variety of microclimates. Based on your previous reading, what do you believe are the causes of such variety?

CHAPTER

11

MIDDLE-LATITUDE, POLAR, AND HIGHLAND CLIMATIC REGIONS

CHAPTER PREVIEW

▶ Perhaps the most distinctive feature of Mediterranean climatic regions is that they receive most of their precipitation during the season opposite that during which greatest plant growth normally occurs.

Why is this so? What plant adaptations have developed as a result? What human use patterns have developed in response to these climatic conditions?

▶ The locations of the Mediterranean and the humid subtropical climates on opposite sides of continents in the lower middle latitudes provide convincing evidence of the dominance of the subtropical highs as the climatic control of these latitudes.

In what ways are the subtropical highs responsible for these contrasting climates? How do the highs influence climate both poleward and equatorward of these latitudes?

▶ Some geographers refer to the marine west coast climate as the "temperate oceanic," or simply the "marine," climate.

Why do we call this climate the marine west coast climate? Why do all these names emphasize maritime terms? In what ways does the climate exemplify its names? What factors produce the climate?

▶ Although humid continental hot summer and mild summer regions have more in common than not, there are distinct differences in nat-

ural and cultivated vegetation between the two climates that may be used to distinguish them from one another, particularly in North America.

What are these differences and why do they exist? What does this statement suggest about scientific classification?

▶ The climatic and related physical characteristics that distinguish the subarctic climate and the two polar climates are accompanied by human utilization patterns significantly different from those found in the humid continental climates.

In what ways does the physical environment in each of these high-latitude climates affect human use patterns? In what ways are cultural factors more influential?

▶ Highlands are occupied by a complex pattern of widely varying microclimates far too intricate to be shown on anything but large-scale maps.

What factors are most important in explaining the existence of a particular highland microclimate? What other phenomena are most directly affected by the different microclimates? What are the major peculiarities of mountain climates?

▲ Fall in the humid continental, mild summer climatic region of New Hampshire.

As we have noted, the tropical climates exhibit the constant characteristic of heat, whereas polar climates exhibit the constant characteristic of cold. The arid climates have inadequate precipitation in common. But what is the constant characteristic of the mesothermal and microthermal middle-latitude climates that we are about to examine? If there is one, then it is the oxymoron "constant inconsistency." Each of the middle-latitude climates is dominated by the changing of the seasons and the variability of atmospheric conditions associated with migrating air masses or cyclonic activity along the polar front.

Middle-Latitude Climates

If change is constant in the mesothermal and microthermal climatic groups, then degree of change is what distinguishes one climatic type from another. In one or another of the middle-latitude climates, summers vary from hot to cool and winters from mild to extremely cold. Adequate monthly precipitation for some climates is year-round while others experience winter drought or, even more challenging to humans who live there, dry months during the normal summer growing season. But despite the changing atmospheric conditions of middle latitude climates, their regions are home to the majority of Earth's people, and they are major factors in some of Earth's most attractive, interesting, and productive physical environments.

Humid Mesothermal Climatic Regions

When we use the term *mesothermal* (from the Greek: *mesos,* middle) in describing climates, we are usually referring to the *moderate* temperatures that characterize such regimes. However, we could also be referring to their *middle* position between those climates that have high temperature throughout the year and those that experience severe cold. By definition, the mesothermal climates experience seasonality, with distinct summers and winters that distinguish them from the humid tropics. But their summers are long and their winters are mild, and this separates them climatically from the microthermal climates, which lie poleward.

Table 11.1 presents the three distinct mesothermal climates introduced in Chapter 10. In all three the annual precipitation exceeds the annual potential evaporation, but in the Mediterranean there is a lengthy period of precipitation deficit in the summer season that distinguishes this climate from the humid subtropical and the marine west coast climates. The latter two are further differentiated by the fact that the humid subtropical regions have hot summers, while the marine west coast regions experience mild summers.

Mediterranean Climate

The Mediterranean climate is one of the best arguments a geographer can use for organizing a study of the environment or developing an understanding of world regions on the basis of climatic classification. Such a climate appears with remarkable regularity in the vicinity of 30° to 40° latitude along the west coasts of each landmass (Fig. 11.1). The alternating controls of subtropical high pressure in summer and westerly wind movement in winter are so predictable that all Mediterranean lands have notably similar and easily recognized temperature and precipitation characteristics (Fig. 11.2). The special appearance, combinations, and climatic adaptations of Mediterranean vegetation not only are unusual but also are clearly distinguishable from those of other climates. Agricultural practices, crops, recreational activities, and architectural styles all exhibit strong similarities within Mediterranean lands.

Warm, Dry Summers; Mild, Moist Winters. The major characteristics of the Mediterranean climate are a dry summer, a mild, moist winter, and abundant sunshine (90 percent of possible sunshine in summer and as much as 50 to 60 percent even during the rainy winter season). Summers are warm throughout the climate, but there are enough differences between the monthly temperatures in coastal and interior locations to recognize two distinct subtypes. The moderate-summer subtype has the lower summer temperatures associated with a strong maritime influence. The hot-summer subtype is located further inland and reflects an increased continental influence. Because the inland version has higher summer and daytime temperatures and slightly cooler winter and nighttime temperatures than its coastal counterpart, it has greater annual and diurnal temperature ranges as well. Compare, for example, the annual range for Red Bluff, California, an inland station, with that of San Francisco, a coastal station about 240 kilometers (150 mi) further south (see Fig. 11.2).

Whichever the subtype, Mediterranean summers clearly show the influence of the subtropical highs. Weeks go by without a sign of rain, and evaporation rates are high. Effective precipitation is lower than actual precipitation, and the summer drought is as intense as that of the desert. Days are warm to hot, skies are blue and clear, and sunshine is abundant. The high percentage of insolation coupled with nearly vertical rays of the noon sun may drive daytime temperatures as high as 30° to 38°C (86°–100°F), except where moderated by a strong ocean breeze or coastal fog.

Fog is common throughout the year in coastal locations and is especially noticeable during the summer. As moist maritime air moves onshore, it passes over the cold ocean currents that

TABLE 11.1   The Mesothermal Climates

Name and Description	Controlling Factors	Geographic Distribution	Distinguishing Characteristics	Related Features
Mediterranean				
Warmest month above 10°C (50°F); coldest month between 18°C (64.4°F) and 0°C (32°F); summer drought; hot summers (inland); mild summers (coastal)	West coast location between 30° and 40° N and S latitudes; alternation between subtropical highs in summer and westerlies in winter	Central California; central Chile; Mediterranean Sea borderlands, Iranian highlands; Capetown area of South Africa; southern and southwestern Australia	Mild, moist winters, hot, dry summers inland with cooler, often foggy coasts; high percentage of sunshine; high summer diurnal temperature range; frost danger	Sclerophyllous vegetation; low, tough brush (chaparral); scrub woodlands; varied soils, erosion in Old World regions; winter-sown grains, olives, grapes, vegetables, citrus, irrigation
Humid Subtropical				
Warmest month above 10°C (50°F); coldest month between 18°C (64.4°F) and 0°C (32°F); hot summers; generally year-round precipitation, winter drought (Asia)	East coast location between 20° and 40° N and S latitudes; humid onshore (monsoonal) air movement in summer, cyclonic storms in winter	Southeastern United States; southeastern South America; coastal southeast South Africa and eastern Australia; eastern Asia from northern India through south China to southern Japan	High humidity; summers like humid tropics; frost with polar air masses in winter; precipitation 65–250 cm (25–100 in.), decreasing inland; monsoon influence in Asia	Mixed forests, some grasslands, pines in sandy areas; soils productive with regular fertilization; rice, wheat, corn, cotton, tobacco, sugarcane, citrus
Marine West Coast				
Warmest month above 10°C (50°F); coldest month between 18°C (64.4°F) and 0°C (32°F); year-round precipitation; mild to cool summers	West coast location under the year-round influence of the westerlies; warm ocean currents along some coasts	Coastal Oregon, Washington, British Columbia, and southern Alaska; southern Chile; interior South Africa; southeast Australia and New Zealand; northwest Europe	Mild winters, mild summers, low annual temperature range; heavy cloud cover, high humidity; frequent cyclonic storms, with prolonged rain, drizzle, or fog; 3- to 4-month frost period	Naturally forested, green year-round; soils require fertilization; root crops, deciduous fruits, winter wheat, rye, pasture and grazing animals; coastal fisheries

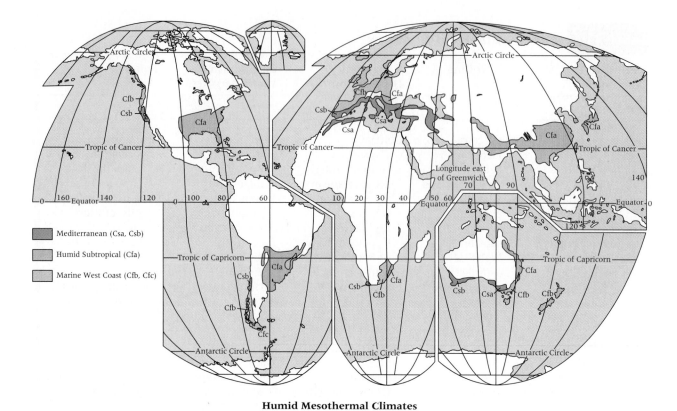

Humid Mesothermal Climates

Figure 11.1 Index Map of Humid Mesothermal Climates.

typically parallel west coasts in Mediterranean latitudes. The air is cooled, condensation takes place, and fog regularly creeps in during the late afternoon, remains through the night and "burns off" during the morning hours. As in the desert, radiation loss is rapid at night, and even in summer, nighttime temperatures are commonly only 10° to 15°C (50°–60°F). People accustomed to the humid summer nights of the eastern United States do not recognize, without experience, the need to bring a sweater for an evening at Disneyland.

Winter is the rainy season in the Mediterranean climate. The average annual rainfall in these regions is usually between 35 and 75 centimeters (15 and 30 in.), with 75 percent or more of the total rain falling during the winter months. The precipitation results primarily from the cyclonic storms and frontal systems common in the westerlies. Annual amounts increase with elevation and decrease with increased distance from the ocean. Only because the rain comes during the cooler months when evaporation rates are lower is there sufficient precipitation to make this a humid climate.

Despite the rain during the winter season, there are often many days of fine, mild weather.

Insolation is still usually above 50 percent, and the average temperature of the coldest month rarely falls below 4° to 10°C (40°–50°F). Frost is uncommon, and because of its rarity, many less hardy tropical varieties of fruits and vegetables are grown in these regions. Hence, when frost does occur, it can do great damage.

Special Adaptations. The summer drought, not frost, is the great challenge to vegetation in Mediterranean regions. The natural vegetation reflects the wet-dry seasonal pattern of the climate. During the rainy season the land is covered with lush, green grasses that turn golden and then brown under the summer drought. Only with winter and the return of the rains does the landscape become green again. Much of the natural vegetation is **sclerophyllous** (hard-leaved) and drought-resistant. Like xerophytes, these plants have tough surfaces, shiny, thick leaves that resist moisture loss, and deep roots to help combat aridity.

One of the most familiar plant communities is made up of many low, scrubby bushes that grow together in a thick tangle. In the western United States this is called **chaparral** (Fig. 11.3). A common low bush in California is the man-

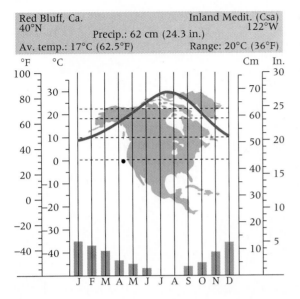

Red Bluff, Ca.   Inland Medit. (Csa)
40°N   122°W
                    Precip.: 62 cm (24.3 in.)
Av. temp.: 17°C (62.5°F)   Range: 20°C (36°F)

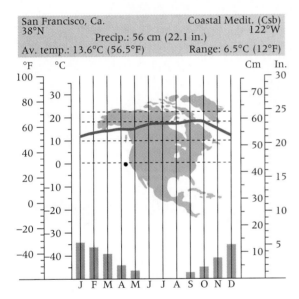

San Francisco, Ca.   Coastal Medit. (Csb)
38°N   122°W
                    Precip.: 56 cm (22.1 in.)
Av. temp.: 13.6°C (56.5°F)   Range: 6.5°C (12°F)

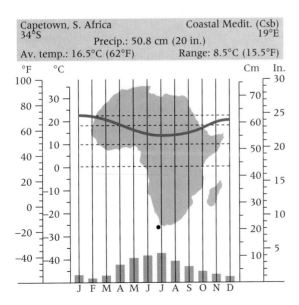

Capetown, S. Africa   Coastal Medit. (Csb)
34°S   19°E
                    Precip.: 50.8 cm (20 in.)
Av. temp.: 16.5°C (62°F)   Range: 8.5°C (15.5°F)

Figure 11.2 *Climographs for Mediterranean climate stations.*
• **In what ways is the climograph of Capetown clearly distinguishable from the two California climographs? What causes the differences?**

zanita, a tough, red-stemmed shrub with crooked limbs that interlock to form an almost impenetrable barrier. Most chaparral is less than 25 years old because of the frequent fires that occur in this dry brush. The fires help to perpetuate the chaparral because the associated heat is required to open seed pods and allow many chaparral species to reproduce. People have often removed chaparral as a preventive measure against fires, yet the removal can have disastrous results because the chaparral acts as a check against erosion of soils during the rainy season. With the chaparral removed, soils wash or slide down hillsides during the heavy rains of winter, frequently taking homes with them.

Figure 11.3 *Typical remnant chaparral vegetation in southern Spain.* • **Why would you expect to find little "native" vegetation left in Old World Mediterranean lands?**

Brush similar to the chaparral in California is called *mallee* in Australia, *mattoral* in Chile, and *maquis* in France. In fact, the "maquis" French Underground that fought against Nazi occupation in World War II literally meant *underbrush*.

Trees that appear in the Mediterranean climate also respond to moisture conditions. Because of their drought-resistant qualities, the needle-leaf pines are among the more common species. Groves of deciduous and evergreen oaks appear in depressions where moisture collects and on the shady north sides of hills where evaporation rates are lower. Where the summer drought is more distinct, the scrub and woodlands open up to parklands of grasses and scattered oak trees. Even the great redwood forests of northern California probably could not survive without the heavy fogs that regularly invade the coastal lands in summer (Fig. 11.4). These great trees demand a unique variation of the usual Mediterranean climate regime, a climatic variation that should probably not be considered Mediterranean at all, as it has more in common with the marine west coast climate further north.

In response to the dry summers, most soils in Mediterranean regions are high in soluble nutrients, but the potential for agriculture differs widely from one region to another. The lime-rich soils around the Mediterranean Sea originally were highly productive, but destructive agricultural practices and overgrazing over thousands of years of human use have caused serious erosion problems. Today, bare white limestone is widely visible on hillsides in Spain, Italy, Greece, Crete, Syria, Lebanon, Israel, and Jordan. In the Mediterranean regions of California the soils are dense clays, gluelike when wet and hard as concrete when dry. During the Spanish period in California the clay was formed into adobe bricks and used for building material. These clay soils are hazardous on slopes, for they absorb so much water during the wet season that they can become mud flows destructive to roads and homes in areas where vegetation has been removed by fire or human interference.

In all Mediterranean regions the most productive areas are the lowlands covered by stream deposits. Here farmers have made special adaptations to climatic conditions. There is sufficient rainfall in the cool season to permit fall planting and spring harvesting of winter wheat and barley. These grasses originally grew wild in the eastern Mediterranean region. Grapevines, fig and olive trees, and the cork oak, which undoubtedly were also native to Mediterranean lands, are especially well adapted to the dry summers because

Figure 11.4 *California redwoods* (Sequoia sempervirens) *may reach heights of 100 meters (330 ft) and live for thousands of years.* • **Why is it considered unusual to find redwoods growing in a Mediterranean climate?**

of their deep roots and thick, well-insulated stems or bark (Fig. 11.5). Where water for irrigation is available, an incredible diversity of crops may be seen. These include, in addition to those already mentioned, oranges, lemons, limes, melons, dates, rice, cotton, deciduous fruits, various types of nuts, and countless vegetables. California, blessed both with fertile valleys for growing fruits, vegetables, and flowers and with snow meltwater for irrigation, is probably the most agriculturally productive of the Mediterranean regions.

Even the houses of these regions show people's adaptation to the climate. Usually white or pastel in color, they gleam in the brilliant sunshine against clear blue skies. Many have shut-

Figure 11.5 Grapes adapt well to the Mediterranean climate. This vineyard is located in California.

tered windows to cut the glare of the sun and to keep the houses cool in summer. On the other hand, much less attention is paid to keeping a place warm during the cooler winter months. This is true even in the United States, where many Midwesterners and Easterners are surprised at the lack of insulation in California homes and the small number and size of heating devices.

The following quotation describes San Francisco and some of her Mediterranean flavor. Notice the references to Italy, to an acropolis (the upper, fortified part of an ancient Greek city), to the green hills, and to the evening fog.

San Francisco put on a show for me. I saw her across the bay, from the great road that bypasses Sausalito and enters the Golden Gate Bridge. The afternoon sun painted her white and gold—rising on her hills like a noble city in a happy dream. A city on hills has it over flat-land places. New York makes its own hills with craning buildings, but this gold and white acropolis rising wave on wave against the blue of the Pacific sky was a stunning thing, a painted thing like a picture of a medieval Italian city which can never have existed. I stopped in a parking place to look at her and the necklace bridge over the entrance from the sea that led to her. Over the green higher hills to the south, the evening fog rolled like herds of sheep coming to cote in the golden city. I've never seen her more lovely.

Travels with Charley
John Steinbeck

Humid Subtropical Climate

The humid subtropical climate extends inland from continental east coasts between 15° to 20° and 40°N and S latitude (refer again to Fig. 11.1 and Table 11.1). Thus it is located within approximately the same latitudes and in a similar transitional position as the Mediterranean climate but on the eastern instead of the western continental margins. There is ample evidence of this climatic transition. Summers in the humid subtropics are similar to the humid tropical climates further equatorward. When the noon sun is nearly overhead, these regions are subject to the importation of moist tropical air masses. High temperatures, high relative humidity, and frequent convectional showers are all characteristics that they share with the tropical climates. In contrast, during the winter months, when the pressure and wind belts have shifted equatorward, the humid subtropical regions are more commonly under the influence of the cyclonic systems of the continental middle latitudes. Polar air masses can bring colder temperatures and occasional frost.

Comparison with the Mediterranean Climate. Like the inland version of the Mediterranean climate, the humid subtropical climate has mild winters and hot summers. But it has no dry season. Whereas the Mediterranean lands are under the drought-producing eastern flank of the subtropical highs, the humid subtropical regions are located on the weak western sides of the subtropical high-pressure cells. Subsidence and stability are greatly reduced or absent, even during the summer months. The warm ocean currents that are commonly found along continental east coasts in these latitudes also moderate the winter temperatures and warm the lower atmosphere, thus increasing lapse rates, which enhances instability. Furthermore, there is a modified monsoon effect (especially in Asia, but also in the southern United States), which increases summer precipitation as the moist, unstable tropical air is drawn in over the land.

As might be expected from the year-round rainfall, average annual precipitation for humid subtropical locations usually exceeds that for Mediterranean stations and may vary more widely as well. Humid subtropical regions receive anywhere from 60 to 250 centimeters (25–100 in.) a year. Precipitation generally decreases inland toward continental interiors and away from the oceanic sources of moisture. It is not surprising that these regions are noticeably drier

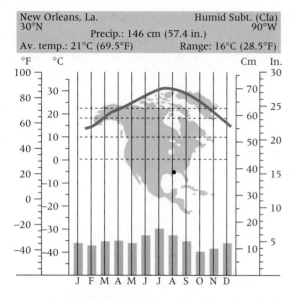

New Orleans, La.   Humid Subt. (Cfa)
30°N   90°W
            Precip.: 146 cm (57.4 in.)
Av. temp.: 21°C (69.5°F)   Range: 16°C (28.5°F)

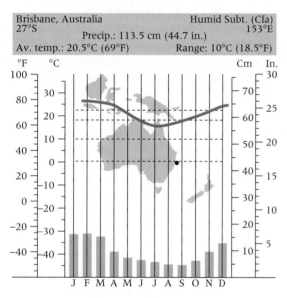

Brisbane, Australia   Humid Subt. (Cfa)
27°S   153°E
            Precip.: 113.5 cm (44.7 in.)
Av. temp.: 20.5°C (69°F)   Range: 10°C (18.5°F)

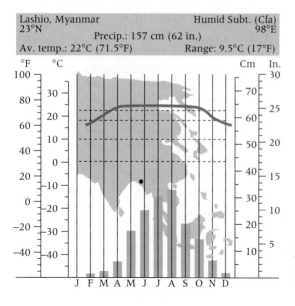

Lashio, Myanmar   Humid Subt. (Cfa)
23°N   98°E
            Precip.: 157 cm (62 in.)
Av. temp.: 22°C (71.5°F)   Range: 9.5°C (17°F)

the closer they are to steppe regions inland toward their western margins.

Both the Mediterranean and humid subtropical climates receive winter moisture from cyclonic storms which travel along the polar front. As we have noted, the great contrast occurs in the summer when the humid subtropics receive substantial precipitation from convectional showers, supplemented in certain regions by a modified monsoon effect. In addition, because of the shift in the sun and wind belts during the summer months, the humid subtropical climates are subject to tropical storms, some of which develop into hurricanes (or typhoons), especially in late summer. These three factors—the modified monsoon effect, convectional activity, and tropical storms—combine in most of these regions to produce a precipitation maximum in late summer. The climographs for New Orleans, Louisiana, and Brisbane, Australia, illustrate these effects (Fig. 11.6).

A subtype of the humid subtropical climate is found most often on the Asian continent, where the monsoon effect is most pronounced because of the magnitude of the seasonal pressure changes over this immense landmass. There, the low-sun period or winter season is noticeably drier than the high-sun period. High pressure over the continent blocks the importation of moist air, so that some months receive less than 3 centimeters (1.2 in.) of precipitation. The climograph for Lashio, Myanmar, illustrates the subtropical wet-dry climate.

Temperatures in the humid subtropics are much like those of the Mediterranean climates. Annual ranges are similar despite a greater variation among climate stations in the humid subtropical climate, primarily because the climate covers a far larger land area. Mediterranean climates record higher summer daytime temperatures, but summer months in both climates average around 25°C (77°F), increasing to as much as 32°C (90°F) as maritime influence decreases inland. Winter months in both climates average around 7° to 14°C (45°–57°F). Frost is a similar problem. The long growing season in the warmest humid subtropical regions enables farmers to grow such delicate crops as oranges, grapefruit, and lemons, but, as in the Mediterranean

Figure 11.6 *Climographs for humid subtropical climate stations.* • **What causes the atypical humid subtropical precipitation characteristic of Lashio?**

climates, farmers must be prepared with various means to protect their more sensitive crops from the danger of freezing. The growers of citrus crops in Florida are concentrated in the central lake district to take advantage of the moderating influence of nearby bodies of water.

Variation in humidity greatly affects the sensible temperatures—the temperatures we feel—in the humid subtropical and Mediterranean climates. The summer temperatures in humid subtropical regions feel far warmer than they are because of the high humidity there. In fact, summers in this climate are oppressively hot, sultry, and uncomfortable. Nor is there the relief of lower night temperatures, as in the Mediterranean regions. The high humidity of the humid subtropical climate prevents much reradiation of heat at night. Consequently, the air remains hot and sticky. Diurnal temperature ranges, in winter as well as in summer, are far smaller in the humid subtropical than in the drier Mediterranean climates. But despite the relatively mild temperatures, humid subtropical winters seem cold and damp, again because of the high humidity.

A Productive Climate. Vegetation generally thrives in humid subtropical regions, with the abundant rainfall, high temperatures, and long growing season. In the wetter portions there are forests of broad-leaf deciduous trees, pine forests on sandy soils, and mixed forests (Fig. 11.7). In the drier interiors near the steppe regions, forests give way to grasslands, which require less moisture. There is an abundant and varied fauna. A few of the common species are deer, bears, foxes, rabbits, squirrels, opossums, raccoons, skunks, and birds of many sizes and species. Bird life in lake and marsh areas is incredibly rich. Alligators inhabit the American swamps, such as the Dismal Swamp of North Carolina and the Okefenokee in Georgia.

As in the tropics, soils tend to have limited fertility due to rapid removal of soluble nutrients. However, there are exceptions in drier grassland areas, such as the pampas of Argentina and Uruguay that constitute South America's "bread basket." Whatever the soil resource, the humid subtropical regions are of enormous agricultural value because of their favorable temperature and moisture characteristics. They have been used intensively for both subsistence crops, such as rice and wheat in Asia, and commercial crops, like cotton and tobacco in the United States. When we consider that this (with its monsoon phase)

Figure 11.7 *Forest vegetation similar to this in Myakka State Park originally covered much of humid subtropical central Florida.* • **How has the physical landscape of central Florida been changed by human occupancy?**

is the characteristic climate of south China as well as of the most densely populated portions of both India and Japan, we realize that this climatic regime contains and feeds far more human beings than any other type (Fig. 11.8). The care with which agriculture has been practiced and the soil resource conserved over thousands of years of intensive use in eastern Asia is in sharp contrast to the agricultural exploitation of the past 200 years in the corresponding area of the United States. The traditional system of cotton and tobacco farming, in particular, devastated the land by exhausting the soil and triggering massive sheet and gully erosion (Fig. 11.9). Over much of the old cotton and tobacco belt, extending from the Carolinas through Georgia and Alabama to the Mississippi delta, all of the

(Continued on page 283)

Figure 11.8 *The terraced fields in this humid subtropical climatic region near Kweilin, China, are ideally suited for rice production.* **• Why is rice the preferred crop in South China?**

Figure 11.9 *Providence Canyon, Georgia, pictured here, did not exist at the time of the Civil War. It began as a system of gulleys when the original vegetative cover was removed during the most active period of farming and forestry in the post-war South. The original fertile soils are exposed near the top of the canyon walls.* **• What is being done throughout Georgia today to prevent such serious soil erosion from occurring again?**

topsoil or a significant part of it has been lost, and we see only the red clay subsoil and occasionally bare rock where crops formerly flourished. In the remaining areas, practices have had to change to conserve the soil that is left. Heavy applications of fertilizer, scientific crop rotation, and careful tilling of the land are now the rule.

Where forests still form the major natural vegetation, forest products, such as lumber, pulp, wood, and turpentine, are important commercially (Fig. 11.10). The long-leaf and slash pines of the southeastern United States have long been the world's leading source of naval stores, which are the resinous products of the pine tree (pitch, tar, resin, and turpentine). The absence of temperature and moisture limitations strongly favors forest growth. In Georgia, for example, trees may grow two to four times faster than in colder regions such as New England. This means that trees can be planted and harvested in much less time than in cooler forested regions, offering distinct commercial advantages.

The fact that living things thrive in the humid subtropical climate presents certain problems, however, for parasites and disease-carrying insects thrive along with other forms of life. African "killer bees," for example, have migrated through Mexico and now pose a serious new threat to humans and livestock in the southern United States.

Agriculture and lumbering are not the only important industries in the humid subtropical climates. In the southeastern United States, livestock raising has been increasing greatly in importance. This trend began when breeds of beef and dairy cattle were developed that could tolerate the hot summers of these regions. Tropical breeds from Africa and India were crossed with European types to produce the desired characteristics. Because of the mild climate, pastures remain green throughout the year and cattle do not need to be housed indoors. In addition, the vegetation grows so rapidly that it requires far less land to support each animal than in cooler or more arid climates.

Despite the commercial advantages of this climate, people often find it an uncomfortable one in which to live. The development and spread of air conditioning helps to mitigate this problem. Where the ocean offers relief from the summer heat, as in Florida, the humid subtropical climate is an attractive recreation and retirement region. The beauty of its more unusual features, such as its cypress swamps and forests draped with Spanish moss, has to be experienced to be fully appreciated.

Figure 11.10 *A commercial tree farm near Waycross, Georgia. Note that the pine trees are planted in orderly rows to expedite cultivation and harvest.* • **Why are tree farms common in the U.S. Southeast and Pacific Northwest but not in New England or the Upper Midwest?**

Marine West Coast Climate

Proximity to the sea and prevailing onshore winds make the marine west coast climate one of the most temperate in the world. Thus it is sometimes known as the *temperate oceanic climate.* Found in those midlatitude regions (between 40° and 65°) that are continuously influenced by the westerlies, the marine west coast climate receives ample precipitation throughout the year. However, unlike the humid subtropical climate just discussed, it has mild to cool summers. The climographs for Bordeaux, France, and Stuttgart, Germany, in Figure 11.11 are representative of the mild summer marine west coast climate, and the climograph for Reykjavik, Iceland, represents the cool summer variety.

Oceanic Influences. As they travel onshore, the westerlies carry with them the moderating marine influence on temperature, as well as much

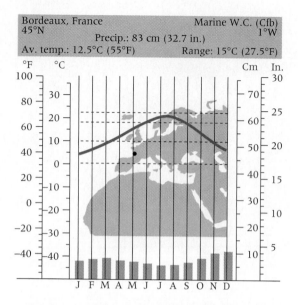

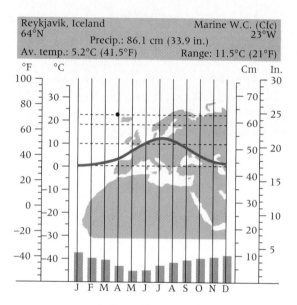

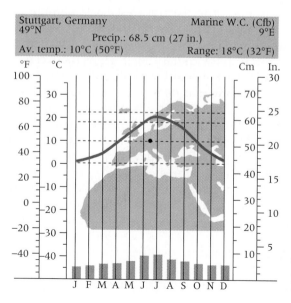

moisture. In addition, warm ocean currents, such as the North Atlantic Drift, bathe some of the coastal lands in the latitudes of the marine west coast climate, further moderating climatic conditions and accentuating humidity. This latter influence is particularly noticeable in Europe, where the marine west coast climate extends along the coast of Norway to beyond the Arctic Circle (see Fig. 11.1).

In this climatic zone the marine influence is so strong that temperatures decrease little with poleward movement. Thus, the influence of the oceans is even stronger than latitude in determining temperatures. This is obvious when we examine isotherms in the areas of marine west coast climates on a map of world temperatures (see Fig. 4.18 and Fig. 4.19). Wherever the marine west coast climate prevails, the isotherms swing poleward, parallel to the coast, clearly demonstrating the dominant marine influence.

Another result of the ocean's moderating effect is that the annual temperature ranges in the marine west coast climates are relatively small, considering the latitude. For an illustration of this, compare the monthly temperature graphs for Portland, Oregon, and Eau Claire, Wisconsin (Fig. 11.12). Though these two cities are at the same latitude, the annual range at Portland is 15.5°C, while at Eau Claire it is 31.5°C. The moderating effect of the ocean on the temperatures in Portland in both winter and summer is clearly in contrast to the effect of an inland continental position on the temperatures in Eau Claire.

Diurnal temperature ranges are also smaller than they are in other climatic regions at similar latitudes and in more arid climates. Heavy cloud cover and high humidity in both summer and winter diminish daytime heating and prevent much radiation cooling at night. Consequently, the difference between the daily maximum and minimum temperatures is small.

Of course, these climographs and climate statistics are primarily averages, and this can be misleading. Marine west coast climatic regions experience the unpredictable weather conditions associated with the polar front. Occasional invasions of a tropical air mass in summer or a polar air mass in winter can move against the general

Figure 11.11 *Climographs for marine west coast climate stations. •* **How do you explain the fact that Reykjavik has the lowest temperature range of the three stations?**

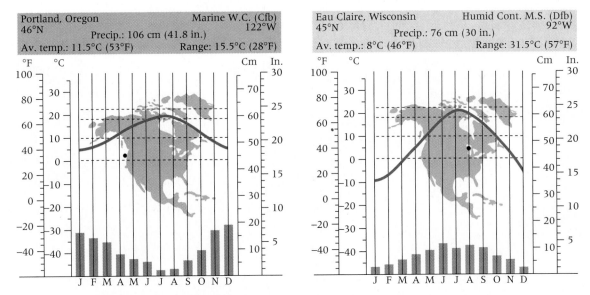

Figure 11.12 *Effect of maritime influence on climates of two stations at the same latitude. Portland, Oregon, exemplifies the maritime influences dominating marine west coast climates. Eau Claire, Wisconsin, shows the effect of location in the continental interior. The difference in temperature ranges for the two stations are significant but note also the interesting differences in precipitation distribution.* • **How do you explain these differences?**

westerly flow of air in these latitudes and produce surprising results. For example, under just such weather conditions, temperatures in Seattle, Washington, have reached a high of 38°C (100°F) and a low of −19°C (−3°F).

Despite the insulating effect of cloudy skies and high moisture content of the air, which slows heat loss at night, frost is a significant factor in the marine west coast climate. It occurs more often, may last longer, and is more intense than in other mesothermal regions. The growing season is limited to eight months or less, but even during the months when freezing temperatures may occur, only half the nights or fewer may experience them. The possibility of frost and the frequency of its occurrence increase inland far more rapidly than they do poleward, once more illustrating the importance of the marine influence.

As final evidence of oceanic influences, study the distribution of the marine west coast climate in Figure 11.1. Where mountain barriers prevent the movement of maritime air inland, this climate is restricted to a narrow coastal strip, as in the Pacific Northwest of North America and in Chile. Where the land is surrounded by water, as in New Zealand, or where the air masses move across broad plains, as in much of northwestern Europe, the climate extends well into the interior of the landmass.

Clouds and Precipitation. The marine west coast has a justly deserved reputation as one of the cloudiest, foggiest, rainiest, and stormiest climates in the world. This is particularly true during the winter season. Rain or drizzle may last off and on for days, though the amount of rain received is small for the number of rainy days recorded. Even when rain is not falling, the weather is apt to be cloudy or foggy. Advection fog may be especially common and long-lasting in the winter months, when air masses pass over warm ocean currents and pick up considerable moisture, which is then condensed as a fog when the air masses move over colder land. The cyclonic storms and frontal systems are also strongest in the winter when the subtropical highs have shifted equatorward. Conspicuous winter maximums in rainfall occur near the coasts and near boundaries with the Mediterranean climate. However, further inland a summer maximum may occur, as in Stuttgart, Germany (see Fig. 11.11).

Though all parts of this climatic type receive ample precipitation, there is much greater station-to-station variation in precipitation averages than in temperature statistics. Precipitation tends to decrease very gradually as one moves inland, away from the oceanic source of moisture. It also decreases equatorward, especially during

Figure 11.13 *The scenic Kenai Fjords of Alaska, shown here, were produced by glacial erosion during the Pleistocene ice advance.* • **In what other areas of the world are fjords common?**

summer months, as the influence of the sub-tropical highs increases and the influence of the westerlies decreases. This can bring about periods of beautiful, clear weather, something rarely associated with this climate but not uncommon in our Pacific Northwest.

The most important factor in the amount of precipitation received is local topography. When a mountain barrier like the Cascades in the Pacific Northwest or the Andes in Chile parallels the coast, abundant precipitation, both cyclonic and orographic, falls on the windward side of the mountains. Valdivia, Chile, located windward of the Andes, receives an average of 267 centimeters (105 in.) of precipitation a year. A similar location in Canada, Henderson Lake, British Columbia, averages 666 centimeters (262 in.) of rain a year, the highest figure for the entire North American continent. During the colder Pleistocene Epoch, these high precipitation amounts, falling largely as winter snow, produced large mountain glaciers. In many cases these came down to the sea, excavating deep troughs that now appear as steep-walled inlets or fjords. Fjord coasts are present in Norway, British Columbia, Chile, and New Zealand—all areas of marine west coast climates today (Fig. 11.13). In contrast, where there are lowlands and no major land-forms of high elevation, precipitation is spread

more evenly over a wide area, and the amount received at individual stations is more moderate (50 to 75 centimeters [20–30 in.] annually). This is the situation in much of the Northern European Plain, extending from western France to eastern Poland.

Two aspects of precipitation are directly related to the moderate temperatures of this climatic regime. Snow falls infrequently, and when it does, it melts or turns to slush as soon as it hits the ground. Snow is especially rare in lowland regions of this climatic zone. Paris averages only 14 snow days a year; London, 13; and Seattle, 10. In addition, thunderstorms and convectional showers are uncommon, although they occur occasionally. Even in summer, surface heating is rarely sufficient to produce the towering cumulonimbus clouds.

Resource Potential. There is little doubt that this climate offers certain advantages for agriculture. The small annual temperature ranges, mild winters, long growing seasons, and abundant precipitation all favor plant growth. Many crops, such as wheat, barley, and rye, can be grown further poleward than in more continental regions. Although the soils common to these regions are not naturally rich in soluble nutrients, highly successful agriculture is possible with the appli-

Figure 11.14 Reliable precipitation makes a diversified type of agriculture possible in the marine west coast climatic areas, with emphasis upon grain, orchard crops, vineyards, vegetables, and dairying. The village in the photograph is Iphofen, Germany. • **Although the climate is similar, why would a photo taken in a marine west coast agricultural region of the United States depict a scene that is significantly different from this one?**

cation of natural or commercial fertilizers (Fig. 11.14). Root crops, such as potatoes, beets, and turnips; deciduous fruits, such as apples and pears; berries; and grapes join the grains previously mentioned as important farm products. Grass in particular requires little sunshine, and pastures are always lush. The greenness of Ireland—the Emerald Isle—is evidence of these favorable conditions, as is the abundance of herds of fat beef and dairy cattle.

The magnificent forests that form the natural vegetation of the marine west coast regions have been a readily available resource. Some of the finest stands of commercial timber in the world are found along the Pacific coast of North America, where pines, firs, and spruces abound, commonly exceeding 30 meters (100 ft) in height (Fig. 11.15). Europe and the British Isles were once heavily forested, but most of those forests (even the famous Sherwood Forest) have been cut down for building material and have been replaced by agricultural lands and urbanization. Where European forests have been preserved through modern forestation practices, as in Germany, they are currently threatened by human-induced atmospheric pollution such as acid precipitation.

Humid Microthermal Climatic Regions

Our definition of humid microthermal includes: high enough temperatures during part of the year to have a recognizable summer; cold enough temperatures six months later to have a distinct winter; and two periods in between called spring and fall, when all life, and especially vegetation, makes preparation for the temperature extremes. Thus, in this section we will be talking about climatic regions that clearly display four readily identifiable seasons.

However, seasonality is not the only explanation for why we often use the word *variable* when describing the humid microthermal climates. As Figure 11.16 indicates, these climates are generally located between 35°N and 75°N on the North American and Eurasian landmasses. Thus they share the westerlies and the storms of the polar front with the marine west coast climate. But their position in the continental interiors and in high latitudes prevents them from experiencing the moderating influence of the oceans. In fact, the dominance of continentality in these climates is best demonstrated by the fact that they do not exist in the Southern

(Continued on page 289)

Figure 11.15 *Commercial timber is the outstanding natural resource of the marine west coast climatic region in North America.* • **Why is this not the case in other marine west coast regions?**

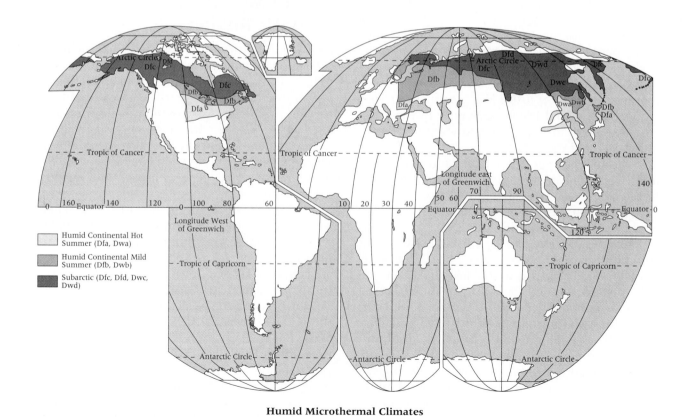

Humid Microthermal Climates

Figure 11.16 *Index Map of Humid Microthermal Climates.*

Hemisphere, where there are no large landmasses in the appropriate latitudes.

The recognition of three separate microthermal climates is based mainly on latitude and the resulting differences in the length and severity of the seasons (Table 11.2). Winters tend to be longer and colder toward the poleward margins because of latitude and toward interiors throughout the microthermal climates because of the continental influence. Summers inland are also

| TABLE 11.2 | The Microthermal Climates | | | |

Name and Description	Controlling Factors	Geographic Distribution	Distinguishing Characteristics	Related Features
Humid Continental, Hot Summer				
Warmest month above 10°C (50°F); coldest month below 0°C (32°F); hot summers; usually year-round precipitation, winter drought (Asia)	Location in the lower middle latitudes (35° to 45°); cyclonic storms along the polar front; prevailing westerlies; continentality; polar anticyclone in winter (Asia)	Eastern and mid-western U.S. from Atlantic coast to the one hundredth meridian; east central Europe; northern China, Manchuria, northern Korea, and Honshu (Japan)	Hot, often humid summers; occasional winter cold waves; rather large annual temperature ranges; weather variability; precipitation 50–115 cm (20–45 in.) decreasing inland and poleward; 140- to 200-day growing season	Broad-leaf deciduous and mixed forest; moderately fertile soils with fertilization in wetter areas; highly fertile grassland and prairie soils in drier areas; "corn belt," soybeans, hay, oats, winter wheat
Humid Continental, Mild Summer				
Warmest month above 10°C (50°F); coldest month below 0°C (32°F); mild summers; usually year-round precipitation, winter drought (Asia)	Location in the middle latitudes (45° to 55°); cyclonic storms along the polar front; prevailing westerlies; continentality; polar anticyclone in winter (Asia)	New England, the Great Lakes region and south central Canada; southeastern Scandinavia; eastern Europe, west central Asia; eastern Manchuria and Hokkaido (Japan)	Moderate summers; long winters with frequent spells of clear, cold weather; large annual temperature ranges; variable weather with less total precipitation than further south; 90- to 130-day growing season	Mixed or coniferous forest; moderately fertile soils with fertilization in wetter areas; highly fertile grassland and prairie soils in drier areas; spring wheat, corn for fodder, root crops, hay and dairying
Subarctic				
Warmest month above 10°C (50°F); coldest month below 0°C (32°F); cool summers, cold winters poleward; usually year-round precipitation, winter drought (Asia)	Location in the higher middle latitudes (50° to 70°); westerlies in summer, strong polar anticyclone in winter (Asia); occasional cyclonic storms; extreme continentality	Northern North America from Newfoundland to Alaska; northern Eurasia from Scandinavia through most of Siberia to the Bering Sea and the Sea of Okhotsk	Brief, cool summers; long, bitterly cold winters; largest annual temperature ranges; lowest temperatures outside Antarctica; low (20–50 cm/ 10–20 in.) precipitation; unreliable 50- to 80-day growing season; permafrost common	Northern coniferous forest (taiga); strongly acidic soils; poor drainage and swampy conditions in warm season; experimental vegetables and root crops

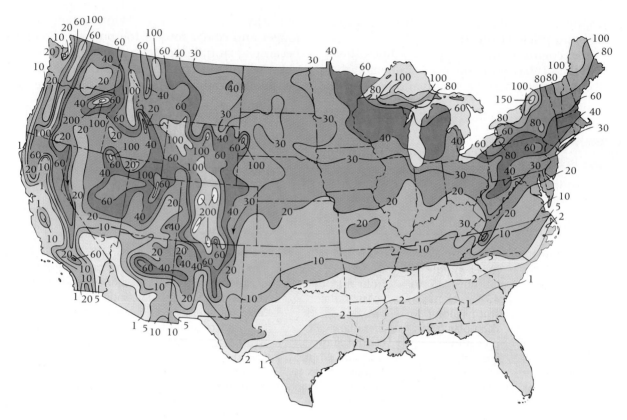

Figure 11.17 *Map of the contiguous United States showing average annual number of days of snow cover.* • **What areas of the United States average the greatest number of days of snow cover?**

inclined to be hotter, but they become progressively shorter as the winter season lengthens poleward. Thus, the three microthermal climates can be defined as humid continental, hot summer; humid continental, mild summer; and subarctic, with a cool summer and, in extreme cases, a long, bitterly cold winter.

All microthermal climates have several features in common. By definition they all experience a surplus of precipitation over potential evaporation, and, with the exception of an area in Asia that experiences winter drought because of the strong Siberian High, they have year-round precipitation. However, the greater frequency of maritime tropical air masses in summer and continental polar air masses in winter, combined with the monsoon effect and strong summer convection, produce a precipitation maximum in summer. Although the length of time snow remains on the ground increases poleward and toward the continental interior (Fig. 11.17), all three microthermal climates experience significant snow cover. This decreases the effectiveness of insolation and helps to explain their cold winter temperatures. Finally, the unpredictable and

variable nature of the weather is especially apparent in the humid microthermal climates and is present throughout all microthermal regions.

With these generalizations in mind, let's compare the microthermal climates with the mesothermal climates we have just examined. Regions with microthermal climates have more severe winters, a lasting snow cover, shorter summers, shorter growing seasons, shorter frost-free seasons, a truer four-season development, lower nighttime temperatures, greater average annual temperature ranges, lower relative humidity, and much more variable weather than do the mesothermal climatic regions.

Humid Continental, Hot Summer Climate

Unlike the other two microthermal climates, the humid continental, hot summer climate is relatively limited in its distribution on the Eurasian landmass (see Table 11.2). This is unfortunate for the people of Europe and Asia because it has by far the greatest agricultural potential and is the most productive of the microthermal climates. In the United States this climate is distributed over a wide area that begins with the eastern seaboard of New York, New Jersey, and southern New England and stretches continuously across the heartland of the eastern United States to encompass much of the American Midwest. It is one of the most densely populated, highly developed, and agriculturally productive regions in the world.

In terms of environmental conditions, the hot summer variety of microthermal climate has some obvious advantages over its poleward counterparts. Its higher summer temperatures and longer growing season permit farmers to produce a wide variety of crops. Those lands within the hot summer region that were covered by ice sheets are far enough equatorward that there has been sufficient time for most negative effects of continental glaciation to be removed, and primarily positive effects remain. Soils are inclined to be more fertile, especially under forest cover where the typical soil-forming processes are not as extreme and where deciduous trees are more common than the acid-associated pine. Of course, some advantages are matched by liabilities. The lower fuel bills of winter in the humid continental, hot summer climate are often more than offset by the cost of air conditioning during the long, hot summers not found in other microthermal climates.

Internal Variations. From place to place within the humid continental, hot summer climate,

there are significant differences in temperature characteristics. The length of the growing season is directly related to latitude. It varies from 200 days equatorward to as little as 140 days along poleward margins of the climate. In addition, the degree of continentality can have an effect upon both summer and winter temperatures and, as a result, upon temperature range. Ranges are consistently large, but they become progressively larger toward continental interiors. Especially near the coasts in this climatic region, temperatures may be modified by a slight marine influence, so that summer temperatures are not so high nor winter temperatures so low as those at inland locations at comparable latitudes. Large lakes may cause a similar effect. Even the size of the continent exerts an influence. Galesburg, Illinois, a typical station in the United States, has a significantly lower temperature range than Mukden, Manchuria, which is located at almost the same latitude but which experiences the greater seasonal contrasts of the Eurasian landmass (Fig. 11.18).

The amount and distribution of precipitation are also variable from station to station. As in all microthermal climates, the total precipitation received decreases both poleward and inland (Fig. 11.19). A move in either direction is a move away from the source regions of warm maritime air masses that provide much of the moisture for cyclonic storms and convectional showers. This decrease can be seen in the average annual

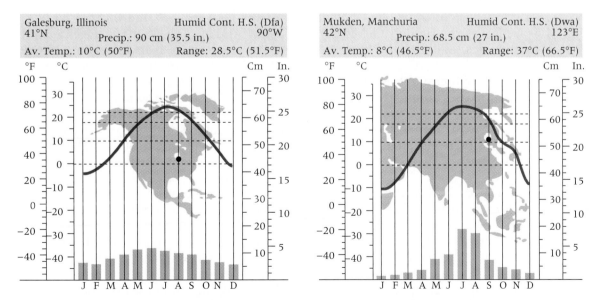

Figure 11.18 Climographs for humid continental, hot summer climate stations. • **What are the reasons for the differences in temperature and precipitation between the two stations?**

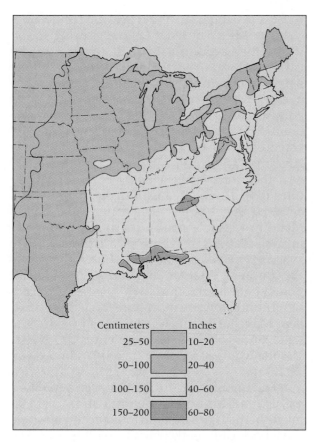

Centimeters		Inches
25–50		10–20
50–100		20–40
100–150		40–60
150–200		60–80

Figure 11.19 *Decreases of precipitation inland from coastal regions are clearly evident in this map of the eastern United States.* • **Why does precipitation decrease inland and poleward?**

precipitation figures for the following cities, all at a latitude of about 40°N: New York (longitude 74°W), 115 centimeters; Indianapolis, Indiana (86°W), 100 centimeters; Hannibal, Missouri (92°W), 90 centimeters; and Grand Island, Nebraska (98°W), 60 centimeters. Most stations have a precipitation maximum in summer, when the warm, moist air masses dominate. In certain regions of Asia, not only does a summer maximum of precipitation exist, but also the monsoon circulation inhibits winter precipitation to such a degree that they experience winter drought (see Mukden, Fig. 11.18).

As might be expected, vegetation and soils vary with the climatic elements, especially precipitation. In the wetter regions forests and forest soils predominate. At one time in certain sections of the American Midwest, tall prairie grasses grew where precipitation was sufficient to support forests, and in all the drier portions of the climate grasslands are the natural vegetation. The soils that developed under these grasslands are among the richest in the world.

Seasonal Changes. The four seasons are highly developed in the humid microthermal, hot summer climate. Each is distinct from the other three and has a character and life all its own. The winter is cold and often snowy; the spring is warmer, with frequent showers that produce flowers, budding leaves, and green grasses; the hot, humid summer brings occasional violent thunderstorms; and the fall has spells of both clear and rainy weather, with mild days and frosty nights in which the green leaves of summer turn to beautiful reds, oranges, yellows, and browns before falling to the ground.

Of course, the most significant differences are between summer and winter. In all regions the summers are long, humid, and hot. The centers of the migrating low-pressure systems usually pass poleward of these regions, and they are dominated by tropical maritime air. So-called "hot spells" can go on day and night for a week or more, with only temporary relief available from convectional thunderstorms or an occasional cold front. Asia, in particular, experiences the heavy summer precipitation associated with the monsoonal effect. Conditions are usually ideal for vigorous vegetative growth. In fact, it is said the corn grows so fast in Iowa that you can hear it! And even if this seems an exaggeration, one thing is certain. The summer heat and humidity are the ideal formula for insects: mosquitoes, flies, gnats, and bugs of all kinds abound. "Putting up the screens" is a common early summer project.

Winters are not as severe as those further poleward, but average January temperatures are usually between −5° and 0°C (23° and 32°F) or below. And once again, the averages tell only part of the story. There is invariably a prolonged invasion of cold, dry, arctic air once or twice during the winter. This often occurs just after a storm has passed and the ground is covered with snow. The sky remains clear and blue for days at a time; the temperatures will stay near −18°C (0°F) and may occasionally dip to 30° or 35°C (20° or 30°F) *below zero* at night. The ground remains frozen for long periods, and there may be snow cover for several days, even weeks, at a time. However, these characteristics do not last continuously because the greatest frequency of cyclones occurs during winter and sudden weather contrasts are common. Cold air precedes warmer air, and thaw follows freeze. Vegetation remains dormant throughout the winter season but bursts into life again with the return of consistently warmer temperatures. Throughout its early growth, vegetation is in constant danger of late spring frost.

As should be apparent from this description, the atmospheric changes within seasons are just as significant as those between seasons. The humid continental, hot summer climate is the classic example of variable midlatitude weather. This is the domain of the polar front. Cyclonic storms are born as tropical air masses move northward and confront polar air masses migrating to the south. The daily weather in these regions is dominated by days of stormy frontal activity followed by the clear conditions of a following anticyclone. Above the land is a battlefield in which storms mark the struggles of air masses for dominance. The general circulation of the atmosphere in these latitudes continuously carries the cyclones and anticyclones toward the east along the polar front. When the polar front is most directly over these regions, as it is in winter and spring, one storm and its associated fronts seem to follow directly behind another with such speed and regularity that the only safe weather prediction is that the weather will change.

Humid Continental, Mild Summer Climate

If you review the relative distributions of the humid continental, hot summer and mild summer climates in Figure 11.16, the close relationship between the two is unmistakable. Where one is found, the other is found as well, and in each situation the mild summer climate invariably lies adjacent to and poleward from the hot summer climate.

In most instances the mild summer climate is essentially a more continental or severe winter version of its equatorward counterpart. It is characterized by distinct seasonality. There is significant climatic variation, particularly with respect to precipitation, from place to place within the climate. Variable weather is the rule, and storms along the polar front provide most of the precipitation within this climatic type. Of course, there are differences between the neighboring climates. These are especially apparent when we examine certain aspects of temperature, growing season, vegetation, and human activity. A brief comparison of the humid continental, mild summer climate with the humid continental, hot summer climate should help to point out important similarities and differences.

Mild Summer–Hot Summer Comparison. In the microthermal climates, precipitation tends to decrease poleward; therefore, the humid continental, mild summer climate tends to have less precipitation than the hot summer regions closer to the equator. Precipitation continues to decrease throughout this climatic type toward the poleward margins and from the coasts toward the arid continental interiors. As in its hot summer counterpart, the monsoon effect in the mild summer climate is strong enough in Asia to produce a dry winter season (see Vladivostok, Fig. 11.20).

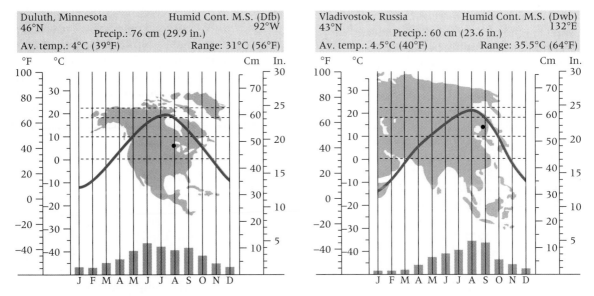

Figure 11.20 *Climographs for humid continental, mild summer climate stations.* • **What characteristics of these climographs distinguish them from the climographs of Figure 11.18?**

Figure 11.21 *People, animals, and plants living in the humid continental, mild summer regions have learned to cope with, as well as enjoy, the beauty of abundant snow, which is present continuously for many months at a time, as in this example in northern Michigan.* • **In what ways might snow cover be considered an asset for some people living in humid continental, mild summer regions?**

Winters in the mild summer climate are more severe and longer than in its neighbor to the south. Summers, on the other hand, are not as long or hot. The combination of more severe winter and shorter summers makes for a growing season of between 90 and 130 days, which is 1 to 3 months shorter than in the hot summer climate. In addition, although overall precipitation totals—50 to 100 centimeters (20 to 40 in.)—are generally lower, snowfall is greater, and snow cover is both thicker and longer lasting (Fig. 11.21).

The humid continental, mild summer regions exhibit seasonal changes just as clearly as the hot summer regions. Annual temperature

ranges are generally larger. Vigorous polar and tropical air-mass interaction makes weather change a common occurrence. However, the more poleward position of the mild summer climate brings about a greater dominance of the colder air masses and explains why temperature variability is not as abrupt or as great as it is further south. Under normal conditions, tropical air is strongly modified by the time it reaches mild summer regions: Even in the high-sun season, intrusions of warm, humid air rarely last more than a few days at a time. By contrast, winter invasions of cold arctic air periodically bring several successive days or weeks of clear skies and frigid temperatures.

As in the humid continental, hot summer climate, the wetter regions of the mild summer climate are associated with a natural forest vegetation. However, many trees common in the hot summer climate, such as oaks, hickories, and maples, find it difficult to compete with firs, pines, and spruces toward the colder, polar margins of these regions. Fortunately for agriculture, northward extensions of the grasslands and the rich soils that accompany them in the hot summer regions are found in the drier portions of the mild summer climate.

Human Activity in the Humid Continental Climates. Perhaps the greatest contrast between the hot summer and mild summer humid continental regions is exhibited in agriculture. Despite the unpredictability of the weather, the humid continental, hot summer agricultural regions are among the finest in the world. The favorable combination of long, hot summers, ample rainfall, and highly fertile soils has made the American Midwest a leading producer of corn, beef cattle, and hogs (Fig. 11.22). Soybeans, which are native to similar climatic regions in northern China, are now second to corn throughout the Midwest as feed for animals and as a raw material for the food processing, plastics, and vegetable oil industries. Wheat, barley, and other grains are especially important in European and Asian regions, and winters are sufficiently mild so that fall-sown varieties may be raised in the United States. In the mild summer climate, on the other hand, a shorter growing season imposes certain limitations on agriculture and restricts the crops that can be grown. Farmers rely more on quick-ripening varieties, grazing animals, orchard products, and root crops. Spring wheat or other spring-sown grains must be raised, and corn does not have time to mature—so, if produced, it is harvested green for fodder. Especially in Europe,

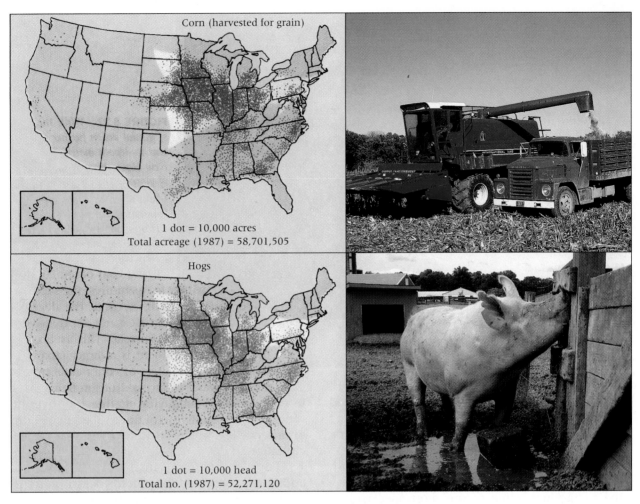

Figure 11.22 *(Top) Corn harvested in the United States; (bottom) hog production in the United States. Both maps are overlaid by boundaries of the humid continental, hot summer climate.* • **From an examination of the maps, which two states do you believe would lead the United States in production of corn and hogs each year?**

potatoes, beets, turnips, and cabbages are important. Dairy products—milk, cheese, butter, cream—are mainstays in the economies of Wisconsin, New York, and northern New England. The moderating effect of the Great Lakes or other water bodies permits the growth of deciduous fruits, such as apples, plums, and cherries.

The length of the growing season is the most obvious reason for the differences in agriculture between the two humid continental climates, but there is another climate-related reason as well. The great ice sheets of the Pleistocene Epoch have had significant, though different, effects upon mild summer and hot summer regions, especially in North America. In the hot summer regions, the ice sheets thinned and receded, releasing the enormous load of soil and solid rock debris they had stripped off the lands nearer to their centers

of accumulation. The material was laid down in a blanket hundreds of feet thick in the areas of maximum glacial advance. As the ice retreated northward, less and less debris was deposited, much being flushed away by meltwater streams. The more southerly hot summer region consequently has an undulating topography underlain by thick masses of glacial debris. The soils formed on this debris are well developed and fertile, and plant nutrients are more likely to be evenly distributed because steep slopes are lacking. The more northerly mild summer region, on the other hand, mainly shows the effects of glacial erosion (Fig. 11.23). Rockbound lakes and marshy lowlands alternate with ice-scoured rock hills. Soils are either thin and stony or waterlogged. Because of its lower agricultural potential, large sections of this area remain in forest.

Figure 11.23 Although the effect of glacial action further south was to deposit material, here in New Hampshire we see an area of glacial erosion— Newfoundland Lake, a former glacial trough. • **How might this area be considered an economic resource?**

However, because of its wilderness character and the abundance of lakes in basins produced by glacial erosion, recreational possibilities in a mild summer region far exceed those of a more subdued hot summer region. Minnesota calls itself the "Land of 10,000 Lakes," and in New England, lakes, rough mountains, and forest combine to produce some of the most spectacular scenery east of the Rocky Mountains.

Subarctic Climate

The subarctic climate is the furthest poleward and most extreme of the microthermal climates. By definition it has at least one month with an average temperature above 10°C (50°F), and its poleward limit roughly coincides with the 10°C isotherm for the warmest month of the year. As you may recall from our earlier discussion of the simplified Köppen system, forests cannot survive where at least one month does not have an average temperature over 10°C. Thus, the poleward boundary of the subarctic climate is the poleward limit of forest growth as well.

As Figure 11.16 indicates, the subarctic climate, like the other microthermal climates, is found exclusively in the Northern Hemisphere. It covers vast areas of subpolar Eurasia and North America. Conditions vary widely over such great distances. Extremely severe winter regions are located along the polar margins or deep in the interior of the Asian landmass, and climatic subtypes with winter drought are found in association with the Siberian high and its clear skies, bitter cold, and strong subsidence of air over interior Asia during winter. Other subarctic regions experience less severe winters or year-round precipitation.

Further study of Figure 11.16 suggests two additional observations. First, ocean currents tend to influence the distribution of the subarctic climate. Along the west coasts of the continents, especially in North America, the warm ocean currents modify temperatures sufficiently to permit the marine west coast climate to extend into latitudes normally occupied by the subarctic and to cause the subarctic to be found well beyond the Arctic Circle. Along east coasts where cold ocean currents help to reduce winter temperatures, the subarctic is situated further south. Second, the development of the subarctic climate is not as extensive in North America as it is in Eurasia. This is because (1) the Eurasian continent is a larger landmass, which increases the effect of continentality, and (2) the large water surface of Hudson Bay in Canada provides a modifying marine influence inland, which tends to counter the effect of continentality there.

The Effects of High Latitude and Continentality.
Subarctic regions experience short, cool summers and long, bitterly cold winters (Fig. 11.24). The rapid heating and cooling associated with continental interiors in the higher latitudes allow little time for the in-between seasons of spring and fall. At Eagle, Alaska, a station in the Klondike re-

gion of the Yukon River Valley, the temperature climbs 8° to 10°C (15°–20°F) per month as summer approaches and drops just as rapidly prior to the next winter season. At Verkhoyansk in Siberia the change between the seasonal extremes is even more rapid, averaging 15° to 20°C (30°–40°F) per month.

Because of the high latitudes of these regions, summer days are quite long and nights are short. The noon sun is as high in the sky during a subarctic summer as during a subtropical winter. The combination of a moderately high angle of the sun's rays and many hours of daylight produces some subarctic locations that receive as much insolation at the time of the summer solstice as the equator does. As a result, temperatures during the one to three months of the subarctic summer usually average 10° to 15°C (50°–60°F), though on some days they may even approach 30°C (86°F). Thus, the brief summer in the subarctic climate is often pleasantly warm, even hot, on some days.

The winter season in the subarctic is bitter, intense, and lasts for as long as eight months. Eagle has eight months with average temperatures below freezing. In the Siberian subarctic, the January temperatures regularly *average* −40° to −50°C (−40° to −60°F), and the coldest temperatures in the Northern Hemisphere—officially −68°C (−90°F) at both Verkhoyansk and Oymyakon; unofficially −78°C (−108°F) at Oymyakon—have been recorded there. In addition, the winter nights, with an average 18 to 20 hours of darkness that extend well into one's working hours, can be mentally depressing and can increase the impression of climatic severity.

As a direct result of the intense heating and cooling of the land, the subarctic has the largest annual temperature ranges of any climate. Average annual ranges in equatorward margins of the climate vary from near 40°C (72°F) to over 45°C (80°F). The exceptions are near western coasts, where warm ocean currents and the marine influence may significantly modify winter temperatures. Average annual ranges for poleward stations are even greater. The climograph for Verkhoyansk, which indicates a range of 64°C (115°F), is an extreme example.

Like subarctic temperature, subarctic precipitation is influenced by latitude and continentality. These climatic controls combine to limit annual precipitation amounts to less than 50 centimeters (20 in.) for most regions and to 25 centimeters (10 in.) or less in northern and interior locations. The low temperatures in the subarctic reduce the moisture-holding capacity of the air, thus minimizing precipitation during the occasional passage of cyclonic storms. Location toward the center of large landmasses or near lee coasts increases distance from oceanic sources of moisture. Finally, the higher latitudes occupied by the subarctic climate are dominated by the polar anticyclone, especially in the winter season. Much as the subtropical high-pressure systems restrict precipitation in the subtropical arid regions, subsidence and divergence of air in the polar

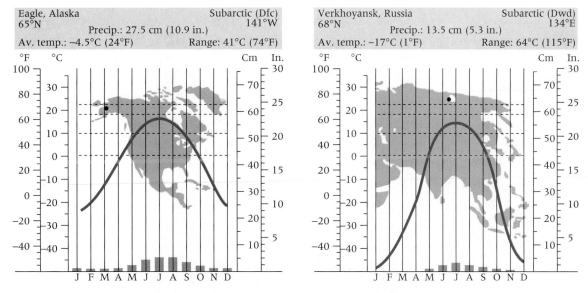

Figure 11.24 *Climographs of the subarctic climate stations.* • **Why would people settle in such severe winter climatic regions?**

anticyclone limit the opportunity for lifting and hence for precipitation in the subarctic regions. This high-pressure system also blocks the entry of moist air from warmer areas to the south.

Subarctic precipitation is cyclonic or frontal, and because the polar anticyclone is weaker and further north during the warmer summer months, more precipitation comes during that season. The meager winter precipitation falls as fine, dry snow. Though there is not as much snowfall as in less severe climates, the temperatures remain so cold for so long that the snow cover lasts for as long as seven or eight months. During this time there is almost no melting of snow, especially in the dark shadows of the forest.

A Limiting Environment.   The climatic restrictions of subarctic regions place distinct limitations on plant and animal life and on human activity. The characteristic vegetation is coniferous forest, adapted to the severe temperatures, the physiologic drought associated with frozen soil water, and the infertile soils. Seemingly endless tracts of spruce, fir, and pine thrive over enormous areas, untouched by humans (Fig. 11.25). In Russia the forest is called the *taiga,* which is the name sometimes given to the subarctic climate type itself.

The brief summers and long cold winters severely limit the growth of vegetation in subarctic regions. Even the trees are shorter and more slender than comparable species in less severe climatic regimes. There is little hope for agriculture. The growing season averages 50 to 75 days, and

frost may occur even during June, July, or August. Thus, in some years a subarctic location may have no truly frost-free season. Although scientists are working to develop plant species that can take advantage of the long hours of daylight in summer, only minimal success has been achieved, in southern parts of the climate, with certain vegetables such as cabbage and root crops such as potatoes.

A particularly vexing problem to people in subarctic regions is **permafrost,** a permanently frozen layer of subsoil and underlying rock that may extend to a depth of 300 meters (1000 ft) or more in the northernmost sections of the climate. Permafrost is present over much of the subarctic climate, but it varies greatly in thickness and is often discontinuous. Where it occurs, the land is frozen completely from the surface down in winter. The warm temperatures of spring and summer, however, cause the top few feet to thaw out. Yet because the land beneath this thawed top layer remains frozen, water cannot percolate downward, and the thawed soil becomes sodden with moisture, especially in spring, when there is an abundant supply of water from the melting snow. Permafrost poses a problem to agriculture by preventing proper soil drainage. The seasonal freeze and thaw of the surface layer above the permafrost also poses special problems for construction engineers. The cycle causes repeated expansion and contraction, heaving the surface up and then letting it sag down. The effects of this cycle break up roads, force buried pipelines out of the ground, cause walls and bridge piers to col-

Figure 11.25 Seemingly endless tracts of taiga (boreal forest) are typical of the vegetation throughout much of the Siberian Subarctic. This photo was taken between the Russian towns of Vanavara and Tura. • **Why are these virgin forests currently of little economic value?**

lapse, and even swivel buildings off their foundations.

With agriculture a questionable occupation at best, there is little economic incentive to draw humans to subarctic regions. Logging is unimportant because of small tree size. Even use of the vast forests for paper, pulp, and wood products is restricted by their interior location far from world markets. Miners occasionally exploit rich ore deposits, while isolated hunters, trappers, and fishermen, many of them native to the subarctic regions, pursue the relatively limited wildlife. Fur-bearing animals such as mink, fox, wolf, ermine, otter, and muskrat are of greatest value.

A final restriction to the human use of these regions is the relentless hordes of mosquitoes, flies, gnats, and other insects that come to life in the brief summer. Only the hardiest sportsperson will brave such a challenge to seek the moose, elk, and deer of the open subarctic forest or the abundant fish in the lakes and streams.

Polar Climatic Regions

The polar climates are the last of Köppen's humid climatic subdivisions to be differentiated on the basis of temperature. These climatic regions are situated at the greatest distance from the equator, and they owe their existence primarily to the low annual amounts of insolation they receive. No polar station experiences a month with average temperatures as high as 10°C (50°F), and hence all are without a warm summer (Table 11.3). Trees cannot survive in such a regimen, and in the regions where at least one month averages above 0°C (32°F) they are replaced by tundra vegetation. Elsewhere the surface is covered by great expanses of frozen ice. Thus there are two polar climatic types, tundra and ice-cap.

There are two important points to keep in mind in the discussion of polar climatic regions. First, these regions have a large net annual radiation loss. That is, they give up much more

TABLE 11.3 The Polar Climates

Name and Description	Controlling Factors	Geographic Distribution	Distinguishing Characteristics	Related Features
Tundra				
Warmest month between 0°C (32°F) and 10°C (50°F); precipitation exceeds potential evaporation	Location in the high latitudes; subsidence and divergence of the polar anticyclone; proximity to coasts	Arctic Ocean borderlands of North America, Greenland and Eurasia; Antarctic Peninsula; some polar islands	At least 9 months average below freezing; low evaporation; precipitation usually below 25.5 cm (10 in.); coastal fog; strong winds	Tundra vegetation; tundra soils; permafrost; swamps and bogs during melting period; life most common in nearby seas; Inuit; mineral and oil resources; defense industry
Ice-cap				
Warmest month below 0°C (32°F); precipitation exceeds potential evaporation	Location in the high latitudes and interior of landmasses; year-round influence of the polar anticyclone; ice cover; elevation	Antarctica; interior Greenland; permanently frozen portions of the Arctic Ocean and associated islands	Summerless; all months average below freezing; world's coldest temperature; extremely meager precipitation in the form of snow, evaporation even less; gale-force winds	Ice- and snow-covered surface; no vegetation; no exposed soils; only sea life or aquatic birds; scientific exploration

radiation or energy than they receive from the sun during a year, resulting in a major radiation deficiency. The transfer of heat from lower to higher latitudes to make up this deficiency is the driving force of the general atmospheric circulation. Without this compensating poleward transfer of heat from the lower latitudes, the polar regions would become too cold to permit any form of life, and the equatorial regions would heat to temperatures no organism could survive.

An equally important characteristic of polar climates is the unique pattern of day and night. At the poles, six months of relative darkness, caused when the sun is positioned below the horizon, alternate with six months of daylight during which the sun is above the horizon. Even when the sun is above the horizon, however, the sun's rays are at a sharply oblique angle, and little insolation is received for the number of hours of daylight. Moving outward from the poles, the lengths of periods of continuous winter night and continuous summer day decrease rapidly from 6 months at the poles to 24 hours at the Arctic and Antarctic Circles ($66\frac{1}{2}°$N and S). Here the 24-hour night or day occurs only at the winter and summer solstices respectively.

Tundra Climate

Compare the location of the tundra climate with that of the subarctic climate on Figure 10.4. You can see that although the tundra climate is situated closer to the poles, it is also along the periphery of landmasses, and, with the exception of the Antarctic Peninsula, it is everywhere adjacent to the Arctic Ocean. Even though temperature ranges in the tundra are large, they are not as large as in the subarctic because of the maritime influence. Winter temperatures, in particular, are not as severe in the tundra as they are inland (Fig. 11.26).

It almost seems inappropriate to call the unpleasantly chilly and damp conditions of the tundra's warmer season *summer*. Temperatures average around 4°C (40°F) to 10°C (50°F) for the warmest month, and frosts regularly occur. The air does warm sufficiently to melt the thin snow cover and the ice on small bodies of water, but this only causes marshes, swamps, and bogs to form across the land because drainage is blocked by permafrost (Fig. 11.27). And out of this soggy landscape, known as **muskeg** in Canada and Alaska, swarm clouds of black flies, mosquitoes, and gnats. The one bright note in the landscape is provided by the enormous number of migratory birds that nest in the arctic regions at this time of year and feed on the insects. However, as soon as the shrinking days of autumn approach, these birds wisely depart for warmer climates.

Winters are cold and seem to last forever, especially in tundra locations where the sun is below the horizon for days at a time. The climograph for Barrow illustrates the low temperatures of this climate. Note that average monthly tem-

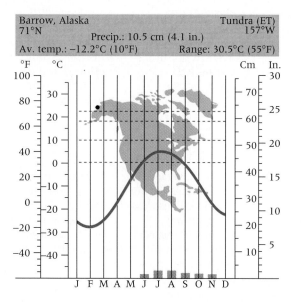

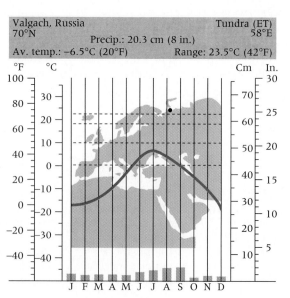

Figure 11.26 Climographs for tundra climate stations. • **Why is it not surprising that both stations are located in the Northern Hemisphere?**

Figure 11.27 *Permafrost regions, such as this area at the base of the Alaska Range, become almost impenetrable swampland during the brief Alaskan summer. Travel over land is feasible only in the winter season.* • **What is the preferred means of travel in the summer?**

peratures are *below freezing* nine months of the year. The average annual temperature is −12°C (10°F).

The tundra regions exhibit several other significant climatic characteristics. Diurnal temperature ranges are small because insolation is uniformly high during the long summer days and uniformly low during the long winter nights. Precipitation is generally low, except in eastern Canada and Greenland, because of exceedingly low absolute humidity and the influence of the polar anticyclone. Icy winds sweep across the open land surface and are an added factor in eliminating the trees that might impede their progress. Coastal fog is characteristic in marine locations, where cool polar maritime air drifts onshore and is chilled below the dew point by contact with the even colder land.

The low-growing tundra vegetation survives despite the forbidding environment. It consists of lichens, mosses, sedges, flowering herbaceous plants, small shrubs, and grasses. In particular, the plants have adjusted to the conditions associated with nearly universal permafrost, as the following passage indicates.

It seems strange that the plants . . . [willow-herb, wild rose, forget-me-not, hellebore, chives, valerian, thyme, vetches and others, which grow on the dunes of windblown sand found along the banks of some of the large rivers] should spring only from the dry sand of the dunes, but the apparent riddle is solved when we know that it is only the sand thus piled up, that becomes sufficiently warmed in the months of uninterrupted sunshine for these plants to flourish. Nowhere else throughout the tundra is this the case. Moor and bog, morass and swamp, even the lakes with water several yards in depth only form a thin summer covering over the eternal winter which reigns in the tundra, with destructive as well as with preserving power. Wherever one tries to penetrate to any depth, in the soil one comes—in most cases scarcely a yard from the surface—upon ice, or at least on frozen soil, and it is said that one must dig about a hundred yards before breaking through the ice-crust of the earth. It is this crust which prevents the higher plants from vigorous growth, and allows only such to live as are content with the dry layer of soil which thaws in summer. It is only by digging that one can know the tundra for what it is: an immeasurable and unchangeable ice-vault which has endured, and will continue to endure, for hundreds of thousands of years.

From North Pole to Equator
Alfred Edmund Brehm

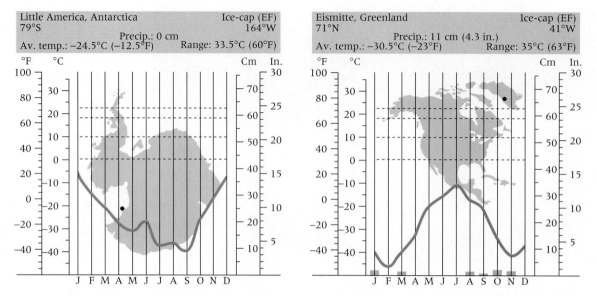

Figure 11.28 *Climographs for ice-cap climate stations.* • **If you were to accept an offer for an all-expense paid trip to visit either Greenland or Antarctica, which would you choose and why would you go?**

Ice-Cap Climate

The ice-cap climate is the most severe and restrictive climate on Earth. As Table 11.3 indicates, it covers large areas in both the Northern and Southern Hemispheres, a total of about 16 million square kilometers (6 million sq mi)—nearly twice the area of the United States. Because all average monthly temperatures are below freezing and because most surfaces are covered with glacial ice, no vegetation can survive in this climate. It is a virtually lifeless regime of perpetual frost.

Antarctica is the coldest place in the world, although Siberia sometimes has longer and more severe periods of cold in winter. Nevertheless, the world's coldest temperature, −88°C (−127°F), was recorded at Vostok, Antarctica. Consider the climographs for Little America, Antarctica, and Eismitte, Greenland, for a fuller picture of the cold ice-cap temperatures (Fig. 11.28).

The primary reason for the low temperatures of ice-cap climates is the minimal insolation received in these regions. Not only is little or no insolation received during half the year, but the sun's radiant energy that is received arrives at sharply oblique angles and consequently is spread over large areas. In addition, the perpetual snow and ice cover of this climate reflects nearly all of the incoming radiation. A further factor, in both Greenland and Antarctica, is elevation. The ice caps covering both regions rise over 3000 meters (10,000 ft) above sea level (Fig.

11.29). Naturally this elevation contributes to the cold temperatures.

The polar anticyclone severely limits precipitation in the ice-cap climate to the fine, dry snow associated with occasional cyclonic storms. Precipitation is so meager in this climate that regions within this regime are sometimes incorrectly referred to as *polar deserts*. However, because of the exceedingly low evaporation rates associated with the severely cold temperatures, precipitation still exceeds potential evaporation and the climate can be classified as humid. The annual precipitation surplus produces glaciers, which export snowfall similar to the way rivers export rainfall.

The strong and persistent polar winds are another staple of the harsh ice-cap climate. Mawson Base, Antarctica, for example, has approximately 340 days a year with gale-force winds of 15 meters per second (33 mph) or over. Katabatic winds, which are caused by the downslope drainage of heavy cold air accumulated over ice caps, are common along the edges of the polar ice. The winds of ice-cap regions can result in "whiteouts"—periods of zero visibility due to blowing fine snow and ice crystals.

Human Activity in Polar Regions

The climatic severity that limits animal life in polar regions to a few scattered species in the tundra is just as restrictive on human settlement. The celebrated Lapps of northern Europe migrate

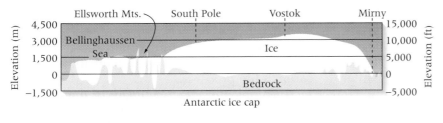

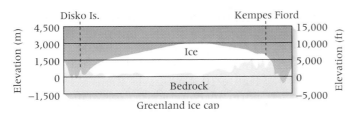

Figure 11.29 *The Antarctic ice sheet resulting from the ice-cap climate of the south polar regions is as much as 4000 meters (13,200 ft) thick locally. Where it is thinnest, it is floated by sea water to produce an ice shelf. The smaller Greenland ice sheet is about 3000 meters (10,000 ft) thick.* • **What reasons might be given for the fact that more land in Greenland than in Antarctica is free of glacial ice?**

with their reindeer to the tundra from the adjacent forest during warmer months. They join the musk ox, arctic hare, fox, wolf, and polar bear that manage to make a home there despite the prohibitive environment. Only the Inuit (Eskimos) of Alaska, northern Canada, and Greenland have in the past succeeded in developing a year-round lifestyle adapted to the tundra regime. Yet even this group relies less on the resources of the tundra than on the large variety of fish and sea mammals, such as cod, salmon, halibut, seal, walrus, and whale, that occupy the adjacent seas.

As their communication with the rest of the world has increased and as they have become acquainted with alternative lifestyles, however, the permanent population of Inuit living in the tundra have greatly diminished, and life for those remaining has changed drastically. Some have gained a new economic security through employment at defense installations or at sites where they join other skilled workers from outside the region to exploit mineral or energy resources. However, the new population centers based on the construction and maintenance of radar and missile defense stations or, as in the case of Alaska's North Slope, on the production and transportation of oil, cannot be considered permanent (Fig. 11.30). Workers depend on other regions for support and often inhabit this region only temporarily.

The ice-cap climate cannot serve as a home for humans or other animals. Even the penguins,

Figure 11.30 *One result of North Slope oil production that had a decidedly negative effect outside the polar regions. Clean-up crews work to remove oil that washed onshore after the oil tanker Exxon Valdez spilled 11 million gallons of crude oil into Prince William Sound, Alaska.* • **Considering the vulnerability of Alaska's physical environment, should development of the North Slope oil fields have been permitted?**

gulls, leopard seals, and polar bears are coastal inhabitants. It is without question the harshest, most restrictive, most nearly lifeless climatic zone on Earth. Yet, especially in Antarctica, it is of strategic importance and of great scientific interest. Scientists study Antarctic ice cores and the trapped pollen within them to help them reconstruct past climates. They also have noted that ozone concentrations have decreased over Antarctica every fall for over a decade. The result is a hole in the ozone layer above Antarctica that is as large as the continental United States. This decrease in the ozone layer is a major environmental concern. Antarctica's strategic value is so widely recognized that the world's nations have voluntarily given up claims to territorial rights on the continent in exchange for cooperative scientific exploration on behalf of all humankind.

Highland Climatic Regions

As we saw in Chapter 4, temperature decreases with increasing altitude at the rate of about 6°C per 1000 meters (3.6°F per 1000 ft). Thus, you might suspect that in highland regions there are broad zones of climate based on changes in temperature with elevation that roughly correspond to Köppen's climatic zones based on change of temperature with latitude. This is the case with one important exception: Seasons only exist in highlands if they also exist in the nearby lowland regions. For example, although zones of increasingly cooler temperature occur at progressively higher elevations in the tropical climatic regions, the seasonal changes of Köppen's middle-latitude climates are not present.

Elevation is only one of several controls of highland climates; **exposure** is another. Just as some continental coasts face the prevailing wind, so do some mountain slopes, while others are lee slopes or are sheltered behind higher topography. The nature of the wind, its temperature, and its moisture content depend on whether the mountain is (1) in a coastal location or deep in a continental interior, or (2) at a high or low latitude within or beyond the reaches of cyclonic storms and monsoon circulation. In the middle and high latitudes, mountain slopes and valley walls that face the equator receive the direct rays of the sun and are warm; poleward-facing slopes are shadowed and cool. West-facing slopes feel the hot afternoon sun, while east-facing slopes are sunlit only in the cool of the morning. This factor is known as **slope aspect** and affects where people live in the mountains and where particular crops will do best. The higher one rises in the mountains, the more important direct sunlight is as a source of warmth and energy for plant and animal life processes.

Complexity is the hallmark of highland climates. Every mountain range of significance is composed of a mosaic of climates far too intricate to differentiate on a world map, or even on a map of a single continent. Highland climates are therefore undifferentiated, signifying climate complexity. Highland climates are indicated on Figure 10.4 wherever there is marked local variation in climate as a consequence of elevation, exposure, and slope aspect. We can see that these regions are distributed widely over the Earth but are particularly concentrated in Asia, central Europe, and western North and South America.

The areas of highland climate on the world map are cool, moist islands in the midst of the zonal climates that dominate the areas around them (Fig. 11.31). Consequently, highland areas are also biotic islands, supporting a flora and fauna adapted to cooler and wetter conditions than those of the surrounding lowlands. This coolness is part of the highland charm, particularly where mountains rise cloaked with forests above arid plains, as do the Rocky Mountains and California's Sierra Nevada.

Highlands stimulate moisture condensation and precipitation by forcing moving air masses to rise over them. Where mountain slopes are rocky and forest-free, their surfaces grow warm during the day, causing upward convection, which often produces afternoon thundershowers. Mountains receive abundant precipitation and are the source area for multitudes of streams that join to form the great rivers of all of the continents.

There are few streams of significance whose headwaters do not lie in rugged highlands. Much of the stream flow on all continents is produced by the summer melting of mountain snowfields. Thus, the mountains not only wring water from the atmosphere but also store much of it in a form that gradually releases it throughout summer droughts, when water is most needed for irrigation and for municipal and domestic uses.

Peculiarities of Mountain Climates

A general characteristic of mountain weather is its changeability from hour to hour as well as from place to place. Strong convection over mountains often causes clouds to form very

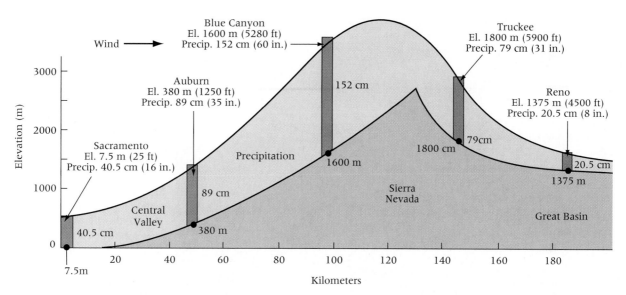

Figure 11.31 *Variation in precipitation caused by uplift of air crossing the Sierra Nevada range of California from west to east. The maximum precipitation occurs on the windward slope, since air in the summit region is too cool to retain a large supply of moisture. Note the strong rain shadow to the lee that gives Reno a desert climate.* • **Taking into consideration the locations of the recording stations, during what season of the year does the maximum precipitation on the windward slope occur?**

quickly, leading to thunderstorms and longer rains that do not affect surrounding cloud-free lowlands. Where the cloud cover is diminished, diurnal temperature ranges over mountains are far greater than those over lowlands. Since mountains penetrate upward beyond the densest part of the atmosphere, the *greenhouse effect* is less developed there than anywhere else on Earth. The thinner layer of low-density air above a mountain site does not greatly impede insolation, thus allowing surfaces to warm dramatically during the daytime. By the same token, the atmosphere in these areas does little to impede long-wave radiation loss at night. Consequently, air temperatures overnight are cooler than the elevation alone would indicate. Since the atmospheric shield is thinnest at high elevations, plants, animals, and humans receive proportionately more of the sun's shortwave radiation at high altitudes. Violet and ultraviolet radiation are particularly noticeable; severe sunburn is one of the real hazards of a day in the high country.

In the middle and high latitudes, mountains rise from mesothermal and microthermal climates into tundra and snow-covered zones. The lower slopes of mountains are commonly forested with conifers, which become more stunted as one moves upward, until the last dwarfed tree is passed at the **tree line**—the line beyond which low winter temperatures and severe wind stress eliminate all forms of vegetation except those that grow low to the ground, where they can be protected by a blanket of snow (Fig. 11.32). Where mountains are high enough, the land surface is permanently covered by snow or ice. The line above which summer melting is insufficient to remove all of the preceding winter's snowfall is called the **snow line.**

In tropical mountain regions the vertical zonation of climate is even more pronounced, and both tree line and snow line occur at higher elevations than in middle latitudes. Any seasonal change is mainly restricted to rainfall, and temperatures are stable year-round, regardless of elevation. Each climatic zone has its own particular association of natural vegetation and has given rise to a distinctive crop combination where agriculture is practiced (Fig. 11.33). In Latin America, four vertical climatic zones are recognized: *tierra caliente* (hot lands); *tierra templada* (temperate lands); *tierra fría* (cool lands), and *tierra helada* (frozen lands).

Highland Climates and Human Activity

In midlatitude highlands, soils are poor, the growing season is short, and the winter snow cover is heavy in the conifer zone, which dominates the

Figure 11.32 As in this example in the Colorado Rocky Mountains, the last tree species found at the tree line are stunted, prostrate forms, which often produce an elfin forest. Where the trees are especially gnarled and misshapen by wind stress, the vegetation is called krummholz (crooked wood).
• **What do you see in the photograph that indicates prevailing wind direction?**

lower and middle mountain slopes. Therefore, little agriculture is practiced and permanent settlements in the mountains are few. However, as the winter snow melts off the high ground just below the bare rocky peaks, grass springs into life, and humans drive herds of cattle and flocks of sheep and goats up from the warmer valleys. The high pastures are lush throughout the summer, but in early fall they are vacated by the animals and their keepers who return to the valleys. This

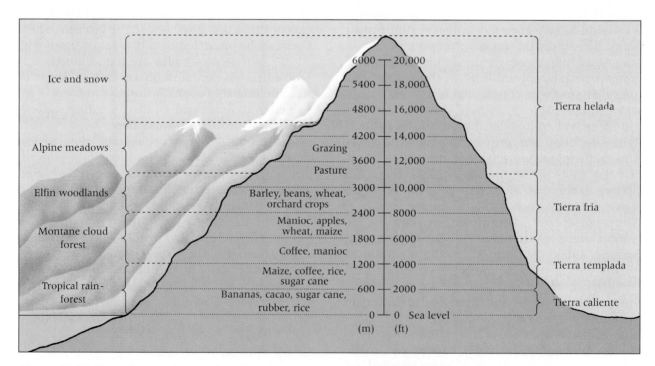

Figure 11.33 Natural vegetation, vertical climatic zones, and agricultural products in tropical mountains. Note that this example extends from tropical life zones to the zone of permanent snow and ice. There is little seasonal temperature change in tropical mountains, which allows life forms sensitive to low temperatures to survive at relatively high elevations. • **When Europeans first settled in the highlands of tropical South America, in which vertical climatic zone did they prefer to live?**

seasonal movement of herds and herders between alpine pastures and villages in the valleys, termed *transhumance,* was once common in the European highlands (the Alps, Pyrenees, Carpathian Mountains and mountains throughout Scandinavia), and still is practiced there on a reduced scale.

Otherwise, the midlatitude highlands serve mainly as sources of timber and of minerals formed by the same geologic forces that elevated the mountains, and as arenas for recreation—both in summer and in winter. Recreational use of the highlands is a relatively recent phenomenon, resulting both from new interest in mountain areas and from new access routes by road, rail, and air that did not exist a century ago.

In contrast to poleward mountain regions, tropical highlands may actually experience more favorable climatic conditions and are often a greater attraction to human settlement than adjacent lowlands. In fact, large permanent populations are supported throughout the tropics where topography and soil favor agriculture in the vertical climatic zones. Highland climates are at such a premium in many areas that steep mountain slopes have been extensively terraced to produce level land for agricultural use. Spectacular agricultural terraces can be seen in Peru, Yemen, the Philippines, and many other tropical highlands. Where the climate is appropriate and population pressure is high, people have created a topography to suit their needs, although they have had to hack it out of mountainsides.

Define and Recall

Mediterranean climate	humid continental, hot summer climate	tundra climate	exposure
sclerophyllous	muskeg	slope aspect
chaparral	humid continental, mild summer climate	ice-cap climate	tree line
humid subtropical climate			snow line
marine west coast climate	subarctic climate		
	permafrost		

Discuss and Review

1. What characteristics make the Mediterranean climate readily distinguishable from all others? What controls are responsible for producing these characteristics?

2. Summarize the special adaptations of vegetation and soils in Mediterranean regions. How have humans also adapted to these regions?

3. Compare the humid subtropical and Mediterranean climates. What are their most obvious similarities and differences?

4. What factors combine to cause a precipitation maximum in late summer in most of the humid subtropical regions?

5. What factors serve to make the humid subtropical climate one of the world's most productive? What are some of its handicaps?

6. How are temperature, precipitation, and geographic distribution of marine west coast regions linked to the controlling factors for this climate?

7. The marine west coast climate has long had its supporters and critics. Give reasons why you would or would not wish to live in such a climatic region.

8. Explain why the microthermal climates are limited to the Northern Hemisphere.

9. List several features that all humid microthermal climates have in common. How do these features differ from those displayed by the humid mesothermal climates?

10. Why is weather in the humid continental, hot summer climate so variable?

11. Describe the relationship between vegetation and climate in the humid continental, mild summer regions.

12. What are the major differences between the humid continental, hot summer and mild summer climates? Contrast the human use of regions occupied by the two climates.

13. How have past climatic changes helped to bring about differences in the configuration of the land between some hot summer and mild summer regions?

14. Refer to Figure 11.24. Using the climographs for Eagle, Alaska, and Verkhoyansk, Russia, describe the temperature patterns of the subarctic regions.

15. What factors limit precipitation in the subarctic climates?

16. How does permafrost affect human activity in both the subarctic and tundra regions?

17. Identify and compare the controlling factors of the tundra and ice-cap regions. How do these controlling factors affect the distribution of these climates?

18. What kind of plant and animal life can survive in the polar climates? What special adaptations must this life make to the harsh conditions of these regions?

19. How do elevation, exposure, and slope aspect affect the microclimates of highland regions? What are the major climatic differences between highland regions and nearby lowlands?

20. Describe and compare the vertical zonation of highland climates in the tropics and in the middle latitudes. How are the climatic zones in tropical highlands related to agriculture?

21. How does human use of highland regions in the middle latitudes differ from that in the tropics? What special human adaptations have aided utilization of these regions?

22. Do each of the following for the listed climates: Mediterranean; humid subtropical; marine west coast; humid continental, hot summer; humid continental, mild summer; subarctic; tundra; ice-cap.
    (a) Identify the climate from a set of data or a climograph indicating average monthly temperature and precipitation for a representative station within a region of that climate.
    (b) Match the climatic type with a written statement that includes one or more of the following: the statistical parameters of the climate in the modified Köppen classification; the particular climatic controls (controlling factors) that produce the climate; the geographic distribution of the climate as stated in terms of physical or political location; the unique climatic characteristics or combination of characteristics that distinguish the climate from others; types of plants, animals, and soils associated with the climate; and the human utilization typical of the climate.
    (c) Distinguish between the important subtypes (if any exist) of each climate by identifying the characteristics that separate them from one another.

Consider and Respond

1. Based on the classification scheme presented in the Appendix, classify the following climatic stations from the data provided.

	J	F	M	A	M	J	J	A	S	O	N	D	Yr.
a. Temp. (°C)	−42	−47	−40	−31	−20	15	−11	−18	−22	−36	−43	−39	−30
Precip. (cm)	0.3	0.3	0.5	0.3	0.5	0.8	2.0	1.8	0.8	0.3	0.8	0.5	8.6
b. Temp. (°C)	3	3	5	8	10	13	15	14	13	10	7	5	9
Precip. (cm)	4.8	3.6	3.3	3.3	4.8	4.6	8.9	9.1	4.8	5.1	6.1	7.4	65.8
c. Temp. (°C)	23	23	22	19	16	14	13	13	14	17	19	22	18
Precip. (cm)	0.8	1.0	2.0	4.3	13.0	18.0	17.0	14.5	8.6	5.6	2.0	1.3	88.1
d. Temp. (°C)	−27	−28	−26	−18	−8	1	4	3	−1	−8	−18	−24	−12
Precip. (cm)	0.5	0.5	0.3	0.3	0.3	1.0	2.0	2.3	1.5	1.3	0.5	0.5	10.9
e. Temp. (°C)	−4	−2	5	14	20	24	26	25	20	13	3	−2	12
Precip. (cm)	0.5	0.5	0.8	1.8	3.6	7.9	24.4	14.2	5.8	1.5	1.0	0.3	62.2
f. Temp. (°C)	9	9	9	10	12	13	14	14	14	12	11	9	11
Precip. (cm)	17.0	14.8	13.3	6.8	5.5	1.9	0.3	0.3	1.6	8.1	11.7	17	97.6
g. Temp. (°C)	−3	−2	2	9	16	21	24	23	19	13	4	−2	11
Precip. (cm)	4.8	4.1	6.9	7.6	9.4	10.4	8.6	8.1	6.9	7.1	5.6	4.8	84.8
h. Temp. (°C)	0	0	4	9	16	21	24	23	20	14	8	2	12
Precip. (cm)	8.1	7.4	10.7	8.9	9.4	8.6	10.2	12.7	10.7	8.1	8.9	8.1	111.5

2. The eight locations represented by the above data are listed below, although not in the order of presentation. Use an atlas and your knowledge of climates to match the climatic data with the following locations: Beijing, China; Point Barrow, Alaska; Chicago, Illinois; Eismitte, Greenland; Eureka, California; Edinburgh, Scotland; New York, New York; Perth, Australia.

3. Eureka, Chicago, and New York are located within a few degrees latitude of one another, yet they represent three distinctly different climatic types. Discuss these differences and identify the primary cause, or source, of these differences.

4. The precipitation recorded at Albuquerque, New Mexico (see Consider and Respond, Chapter 10), is almost twice that recorded at Point Barrow, Alaska, yet Albuquerque is considered a dry climate and Point Barrow a humid climate. Why?

5. What differentiates the *Dw* climate from the other *D* climates? Why is the *Dw* climate type only found in Asia?

6. *Csa* climatic regions and *Cfa* climatic regions are both under the influence of subtropical high-pressure cells during the summer, yet *Csa* climates are dry during the summer and *Cfa* climates are wet. Why?

CHAPTER

12

BIOGEOGRAPHY

CHAPTER PREVIEW

▶ Plants, animals, and the environments in which they live are interdependent, each affecting the other.

In what ways are plants as a group and animals as a group mutually dependent upon one another? In what ways do humans have a much greater impact on ecosystems than all other life forms?

▶ As the basic producers, autotrophs are generally considered to be the most important component of an ecosystem.

What are the other components? In what ways are the autotrophs affected by the other components?

▶ The large majority of Earth's net primary productivity is from terrestrial ecosystems, and humans use less than 1 percent for plant food.

What does this suggest concerning the oceans as a source of food? Could humans alter these figures if they changed their eating habits?

▶ The most common species in ecosystems are generalists.

Why is this so? What effect does a species' eating habits have on its ecological niche? How can the ecological niche of a generalist change from one ecosystem to another?

▶ Plant and animal distributions illustrate a mosaic on the landscape, with the most broadly distributed plant associations typically comprising the matrix of the landscape.

Why isn't a matrix based on animal distributions? What are patches and corridors within the matrix? How do successional processes relate to the concept of patches within a mosaic?

▶ Although other environmental controls may be more important on a local scale, climate has the greatest influence over ecosystems on a worldwide basis.

What is the evidence that this is the case? Which climatic factors have the greatest effect upon plants and animals? What environmental controls other than climate are important? How does climate influence the shape and size of animals and their appendages?

▶ Earth's major terrestrial ecosystems (biomes) are classified on the basis of the dominant vegetation types that occupy the ecosystems.

What are these vegetation types? Why not base the classification on animals? How does this statement relate to the second statement above?

▶ Freely floating plants called phytoplankton are the most important link in the ocean food chain.

What role do they play? Phytoplankton perform what function in relation to the atmosphere? Where in the ocean are phytoplankton concentrated?

▲ Golden aspen plant community on the slopes of the Wasatch Range, Utah.

Biogeography is the branch of physical geography, shared with the field of ecology, that specifically examines the biosphere. Of interest to biogeographers are the distributions of plants and animals, the processes that produce those distributions, and how those processes and distributions have changed over the last twenty thousand years or so. Biogeographers are also interested in the possible effects that human-induced global warming may have on plant and animal processes and distributions.

Plants, animals, and the environments in which they live are interdependent, each affecting the other. Animal life would not exist without plants as basic food, and most plants could not survive without some animals. Together, plants and animals must adapt to their environment. Humans alone have the intelligence and capacity to alter, either carelessly or deliberately, the plant-animal-environment relationship. An ecosystem is a community of organisms functioning together in an interdependent relationship with the environment they occupy. It makes good sense to examine Earth's ecosystems, in order both to understand the physical processes that created them and their geographic distributions, and to better understand the impact that human beings have upon them.

As noted in Chapter 1, there is increasing concern among scientists regarding the ability of Earth to indefinitely support future generations of humanity. What lasting effects will increased industrialization have on the water that animals in the next century will drink or on the atmosphere in which tomorrow's plants will grow? What will be the consequences for life-forms if concrete and steel replace additional square kilometers of forest? What could happen to marine life if toxic wastes accumulate in the world's oceans? Can humans learn to work with, and not against, nature in order to sustain and improve life as we know it on planet Earth today?

Organization Within Ecosystems

It can be said that ecology is an old science. The great voyages of exploration that began in the fifteenth century carried colonists and adventurers to unchartered lands with exotic environments. The more scholarly observers within each group made careful note of the flora and fauna to be found in each new part of the world. It soon became apparent that certain plants and animals were found together and that they bore a direct relationship to the climate in which they lived. As information about various world environments became more reliable and readily available, early biologists began to study plant communities and classify vegetation types. As the relationships of animals to these plant communities were recognized, naturalists in the early twentieth century began dividing Earth's lifeforms into *biotic associations*. Recently the functional relationships of plants, animals, and their physical environment have been the primary focus of attention, and the concept of the ecosystem has become widely utilized.

Our definition of an ecosystem is both broad and flexible. The term can be used in reference to the Earth system in its entirety (the ecosphere) or to any group of organisms occupying a given area and functioning together with their nonliving environment (Fig. 12.1). An ecosystem may be large or small, marine or terrestrial (on land), short-lived or long-lasting. It may even be an artificial ecosystem, such as a farmer's field. When a farmer plants crops, spreads fertilizer, practices weed control, and sprays insecticides, a new

Figure 12.1 *Relatively small ecosystems, such as the one associated with this pond in New Hampshire, were among the first to be studied because they have fairly distinct boundaries and less complicated relationships between the nonliving environment and the organisms that occupy it.* • **Why are the boundaries of ecosystems almost always arbitrary and determined by the ecologist who studies them?**

ecosystem is created, but this does not alter the fact that plants and animals are living together in an interdependent relationship with the soil, rainfall, temperatures, sunshine, and other factors that constitute the physical environment (Fig. 12.2).

As noted in Chapter 1, ecosystems are *open systems*. There is free movement of both energy and materials across their boundaries. However, they are usually so closely related to nearby ecosystems or so integrated with the larger ecosystems of which they are a part that they are not isolated in nature or readily delimited. Nevertheless, the concept of the ecosystem is a valuable model for examining the structure and function of life on Earth.

Major Components

Ecosystems may be many and varied, but the typical ecosystem has four basic components. The

Figure 12.2 Hybrid seed, fertilizers, insecticides, and, on occasion, irrigation systems may be employed by the farmer to insure the success of this artificial ecosystem in the Corn Belt. • **How does the role of humans in an artificial ecosystem differ from that in a natural one?**

first of these is the nonliving, or **abiotic,** part of the system. This is the physical environment in which the plants and animals of the system live. In an aquatic ecosystem (a pond, for example), the abiotic component would include such inorganic substances as calcium, mineral salts, oxygen, carbon dioxide, and water. Some of these would be dissolved in the water, but the majority would lie at the bottom as sediments—a natural reservoir of nutrients for both plants and animals. In a terrestrial ecosystem the abiotic component provides life-supporting elements and compounds in the soil, groundwater, and atmosphere.

The second and perhaps most important component of an ecosystem consists of the basic producers, or **autotrophs** (meaning *self-nourished*). Plants, the most important autotrophs, are essential to virtually all life on Earth because they are capable of utilizing energy from sunlight to convert water and carbon dioxide into organic molecules through the process known as photosynthesis. The sugars, fats, and proteins produced by plants through photosynthesis are the foundation for the food supply that supports other forms of life. It should be noted that some bacteria are also capable of photosynthesis and hence are classed as autotrophs along with plants. Sulfur-dependent organisms that dwell at ocean-bottom thermal vents are also classified as autotrophs.

A third component of most ecosystems consists of consumers, or **heterotrophs** (meaning *other-nourished*). These are animals that survive by eating plants or other animals. Heterotrophs are classified on the basis of their feeding habits. **Herbivores** eat only living plant material; **car-**nivores eat other animals; and **omnivores** feed on both plants and animals. Animals make an essential contribution to the Earth ecosystem of which they are a part. They utilize oxygen in their respiration and return as an end product to the atmosphere the carbon dioxide that is required for photosynthesis by plants. They can influence soil development through their digging and trampling activities, and those activities in turn may affect local plant distributions.

We might assume that plants, animals, and a supporting environment are all that are required for a functioning ecosystem, but such is not the case. Without the fourth component of ecosystems, the decomposers, plant growth would soon come to a halt. The **decomposers,** or **detritivores,** feed on dead plant and animal material and waste products. They promote decay and return mineral nutrients to the soil and sea in a form that plants can utilize.

Trophic Structure

From the discussion of the autotrophs and heterotrophs it becomes apparent that there is a definite arrangement of the major components of an ecosystem. The components form a sequence in their levels of eating: Herbivores eat plants, carnivores may eat herbivores or other carnivores, and decomposers feed on dead plants and animals and their waste products. The pattern of feeding in an ecosystem is called the **trophic structure,** and the sequence of levels in the feeding pattern is referred to as a **food chain.** The simplest food chain would include only plants and decomposers. However, the chain usually includes at least four steps: for example, grass—field mouse—owl—fungi (plants—herbivore—carnivore—decomposer). More complex food chains may include six or more levels as carnivores feed on other carnivores (zooplankton eats plants—small fish eats zooplankton—larger fish eats small fish—bear eats larger fish).

Organisms within a food chain are often identified by their **trophic level,** or the number of steps they are removed from the autotrophs or plants in a food chain (Table 12.1). Plants occupy the first trophic level, herbivores the second, carnivores feeding on herbivores the third, and so forth until the last level, the decomposers, is reached. Omnivores may belong to several trophic levels because they eat both plants and animals.

In reality, linear food chains do not operate in isolation; they overlap and interact to form a feeding mosaic within an ecosystem called a **food web** (Fig. 12.3). Both food chains and food

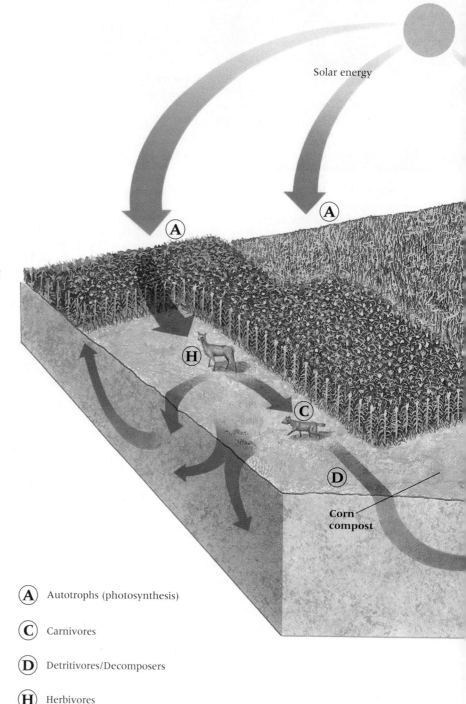

Figure 12.3 Environmental Systems: Midlatitude Ecosystems. Ecosystems are a favorite subject of study by physical geographers for several reasons. They clearly illustrate the interdependence of the variables in systems, especially the close relationships between the living components of systems (the plants and animals, particularly humans, of the biosphere) and the nonliving or abiotic components in systems (the atmosphere, hydrosphere, and lithosphere). Plants and animals are dependent upon the energy from the sun and the nutrients in the atmosphere, dissolved in the water, and in the soil for the creation of new organic material. Annual or seasonal changes in temperature and precipitation or alterations of the soil or water quality bring about changes in the energy and food supplies of plants and animals throughout the food webs of an ecosystem.

A major key to the understanding of ecosystems is revealed by an examination of energy as it flows through ecosystem food webs from one trophic level to the next. Note in the diagram the movement of solar energy as it is converted to organic material through photosynthesis by the autotrophs and then moves through successively higher levels in the ecosystem food webs—from autotrophs to herbivores to carnivores. Energy is consumed and biomass decreases at each level until the organic material is recycled by the detritivores and decomposers after death.

Solar energy

(**A**) Autotrophs (photosynthesis)

(**C**) Carnivores

(**D**) Detritivores/Decomposers

(**H**) Herbivores

Corn compost

webs merit careful study because they can be used to trace the movement of food and energy from one level to another in the ecosystem.

Energy Flow

When physical geographers study ecosystems they trace the flow of energy through the system just as they do when they study energy flow in other systems, such as streams or glaciers. And just as they do with other systems, the laws of thermodynamics apply to ecosystems. For example, as the first law of thermodynamics states, energy cannot be created or destroyed; it can only be changed from one form to another. Energy comes to the ecosystem in the form of sunlight,

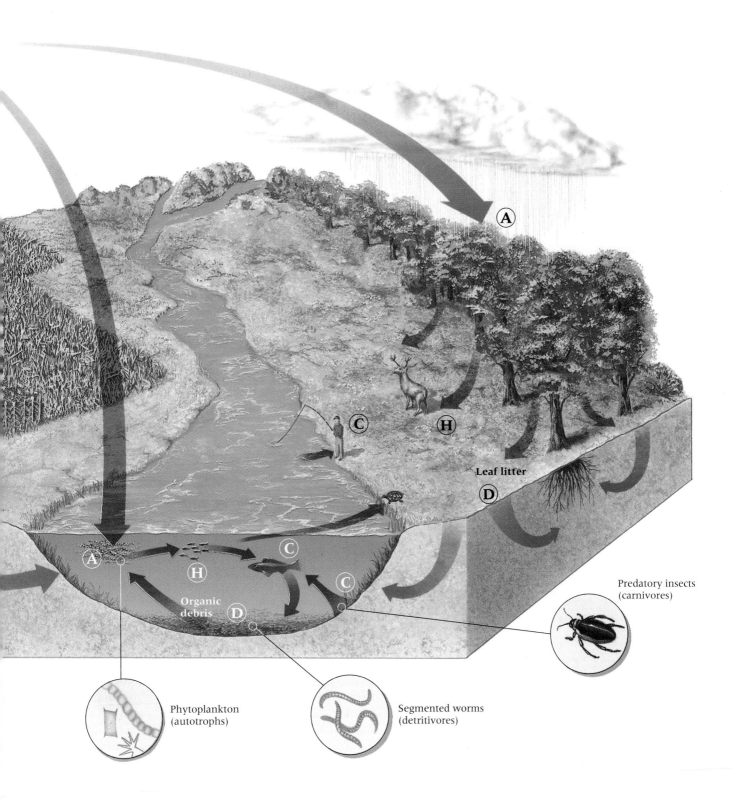

Leaf litter

Organic
debris

Predatory insects
(carnivores)

Phytoplankton
(autotrophs)

Segmented worms
(detritivores)

which is used by plants in photosynthesis. This energy is stored in the system in the form of organic material in plants and animals. It flows through the system along food chains and webs from one trophic level to the next. It is finally released from the system when oxygen is combined with the chemical compounds of the organic material through the process of oxidation.

Fire is one form of oxidation. Respiration, which involves the combination of oxygen with chemical compounds in living cells, and which can occur at any trophic level, is another.

The total amount of living material in an ecosystem is referred to as the **biomass.** Because the energy of a system is stored in the biomass, scientists measure the biomass at each trophic

TABLE 12.1	Trophic Structure of Ecosystems	
Ecosystem Component	**Trophic Level**	**Examples**
Autotroph	First	Trees, shrubs, grass
Heterotroph	Second	Locust, rabbit, field mouse, deer, cow, bear
	Third	Praying mantis, owl, hawk, coyote, wolf, bear
	Fourth, etc.	Bobcat, wolf, hawk, bear
Decomposer	Last	Fungi, bacteria

level to trace the energy flow through the system. They usually find that the biomass decreases with each successive trophic level (Fig. 12.4). There are a number of explanations for this, each involving loss of energy. The first instance occurs between trophic levels. The second law of thermodynamics states that whenever energy is transformed from one state to another there will be a loss of energy through heat. Hence, when an organism at one trophic level feeds on an organism at another, not all of the food energy is

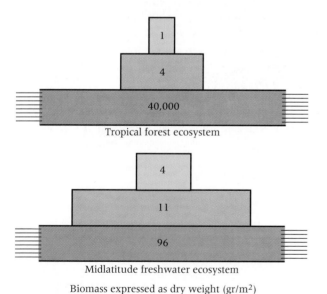

Figure 12.4 Trophic pyramids showing biomass of organisms at various trophic levels in two contrasting ecosystems. Dry weight is used to measure biomass because the proportion of water to total mass differs from one organism to another. • How can you explain the exceptionally large loss of biomass between the first and second trophic levels of the tropical forest ecosystem?

utilized. Some is lost to the system. Additional energy is lost through respiration and movement. At each successive trophic level the amount of energy required is greater. A deer may graze in a limited area, but the wolf that preys on the deer must hunt over a much larger territory. Whatever the reason for energy loss, it follows that as the flow of energy decreases with each successive trophic level the biomass also decreases. This principle also applies to agriculture. There was a great deal more biomass (and food energy) available in a field of corn than there is in the cattle that ate the corn.

Productivity

Productivity in an ecosystem is defined as the rate at which new organic material is created at a particular trophic level. **Primary productivity** refers to the formation of new organic matter through photosynthesis by autotrophs, whereas **secondary productivity** refers to the rate of formation of new organic material at the heterotroph level.

Primary Productivity. Just how efficient are plants at producing new organic matter through photosynthesis? The answer to this question depends on a number of variables. Photosynthesis requires sunlight, the amount of which is dependent upon the length of day and the angle of the sun's rays, which in turn differ widely with latitude. Photosynthesis is also affected by factors such as soil moisture, temperature, the availability of mineral nutrients, the carbon dioxide content of the atmosphere, and the age and species of the individual plants.

There have been a number of studies of productivity in ecosystems, but they have usually been concerned with measuring the net biomass at the autotroph level (Table 12.2). Wherever figures have been compiled noting the efficiency of photosynthesis, the efficiency has been surprisingly low. Most studies indicate that less than 5 percent of the available sunlight is used to produce new biomass in ecosystems. For Earth as a whole the figure is probably less than 1 percent. Despite this fact, the net primary productivity of the ecosphere is enormous. It is estimated to be in the range of 170 billion metric tonnes (a metric tonne is about 10 percent greater than a U.S. ton) of organic matter annually. Even though oceans cover approximately 70 percent of Earth's surface, slightly over two thirds of net annual productivity is from terrestrial ecosystems and less than a third is from marine ecosystems. Per-

TABLE 12.2 Net Primary Productivity of Selected Ecosystems*

Type of Ecosystem	Net Primary Productivity, g/m^2 per year	
	Normal Range	Mean
Tropical rainforest	1000–3500	2200
Tropical seasonal forest	1000–2500	1600
Midlatitude evergreen forest	600–2500	1300
Midlatitude deciduous forest	600–2500	1200
Boreal forest (taiga)	400–2000	800
Woodland and shrubland	250–1200	700
Savanna	200–2000	900
Midlatitude grassland	200–1500	600
Tundra and alpine	10–400	140
Desert and semidesert scrub	10–250	90
Extreme desert, rock, sand, and ice	0–10	3
Cultivated land	100–3500	650
Swamp and marsh	800–3500	2000
Lake and stream	100–1500	250
Open ocean	2–400	125
Upwelling zones	400–1000	500
Continental shelf	200–600	360
Algal beds and reefs	500–4000	2500
Estuaries	200–3500	1500

* From R. H. Whittaker, *Communities and Ecosystems.* 2nd ed. New York: Macmillan, 1975.

haps even more surprising is the fact that humans consume less than 1 percent of Earth's primary productivity as plant food. Biomass is also utilized by humans, however, in a variety of endeavors. Examples include the human use of lumber for construction and paper production and the use of biomass energy for feedstocks and as fodder for range animals.

Table 12.2 illustrates the wide range of net primary productivity displayed by various ecosystems. The latitudinal control of insolation and the subsequent effect on photosynthesis can easily be recognized when comparing figures for terrestrial ecosystems. There is a noticeable decrease in terrestrial productivity from tropical ecosystems to those in middle and higher latitudes. Even the tropical savannas, which are dominated by grasses, produce more biomass in a year than the boreal forests, which are found in the colder climates. Today satellites monitor Earth's biological productivity and give us a global perspective of our biosphere (Fig. 12.5).

The reasons for differences among aquatic or water-controlled ecosystems are not quite as apparent. Swamps and marshes are especially well supplied with plant nutrients and therefore have a relatively large biomass at the first trophic level. On the other hand, depth of water has the great-

est impact on ocean ecosystems. Most nutrients in the open ocean sink to the bottom beyond the depth where sunlight can penetrate and make photosynthesis possible. Hence the most productive marine ecosystems are found in the sunlit, shallow waters of estuaries, continental shelves, or coral reefs, or in areas where ocean upwelling carries nutrients nearer to the surface.

Some artificial ecosystems associated with agriculture can be fairly productive when compared with the natural ecosystems that they have replaced. This is especially true in the warmer latitudes where farmers may raise two or more crops in a year, or in arid lands where irrigation supplies the water essential to growth. However, Table 12.2 indicates that mean productivity for cultivated land does not approach that of forested land and is just about the same as that of middle-latitude grasslands. Most quantitative studies have shown that agricultural ecosystems are significantly less productive than natural systems in the same environment.

Secondary Productivity. As we have seen, secondary productivity results from the conversion of plant materials to animal substances. We have also noted that the ecological efficiency, or the rate of energy transfer from one trophic level to

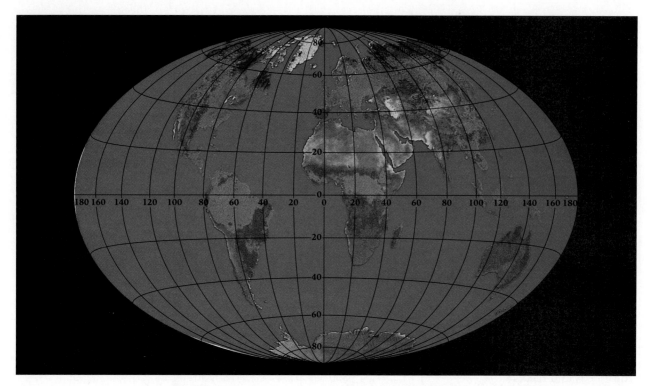

Figure 12.5 *Worldwide vegetation patterns revealed through a color index derived from environmental satellite observations. Compare this image with Figure 12.15, World Map of Natural Vegetation.*

the next, is low (Fig. 12.6). The efficiency of transfer from autotrophs to heterotrophs varies widely from one ecosystem to another. The amount of net primary productivity actually eaten by herbivores may range from as high as 15 percent in some grassland areas to as low as 1 or 2 percent in certain forested regions. In ocean ecosystems the figure may be much higher, but there is a greater loss during the digestion process. Once the food is eaten, energy loss through respiration or body movement reduces secondary productivity to a small fraction of the biomass available as net primary productivity.

Most authorities consider 10 percent to be a reasonable estimate of ecological efficiency for both herbivores and carnivores. If both herbivores and carnivores have ecological efficiencies of only 10 percent, the ratio of biomass at the first trophic level to biomass of carnivores at the third trophic level is several thousand to one. It obviously requires a huge biomass at the autotroph level to support one animal that eats only meat. As human populations grow at increasing rates and agricultural production lags behind, it is indeed fortunate that human beings are omnivores and can adopt a more vegetarian diet (Fig. 12.7).

Ecological Niche

There are a surprising number of species in each ecosystem, except for those ecosystems severely restricted by adverse environmental conditions. Yet each organism performs a specific role in the system and lives in a certain location, described as its **habitat.** The combination of role and habitat for a particular species is referred to as its **ecological niche.** A number of factors influence the ecological niche of an organism. Some species are **generalists** and can survive on a wide variety of food. The North American brown, or grizzly, bear, as an omnivore, will eat berries, honey, and fish. On the other hand, the koala of Australia is a specialist and eats only the leaves of certain eucalyptus trees. Specialists do well when their particular food supply is abundant, but they cannot adapt to changing environmental conditions. The generalists are in the majority in most ecosystems, as their broader ecological niche allows survival on alternative food supplies.

Some generalists among species occupy an ecological niche in one ecosystem that is quite different from the niche they occupy in another. As food supply varies with habitat, so varies the ecological niche. Humans are the extreme exam-

1 m² of field for 1 yr

d. Net carnivore productivity:
0.4 g (0.15% of a; 6.7% of c)

c. Net herbivore productivity:
6 g (2.2% of a; 20.7% of b)

b. Every year herbivores eat:
29 g (10.7% of a)

a. Net primary productivity
(plants): 270 g (dry weight)

Figure 12.6 Productivity at the autotroph, herbivore, and carnivore trophic levels as measured in a Tennessee field. The figures represent productivity for one square meter of field in one year. Note the extremely small proportion of primary productivity which reaches the carnivore level of the food chain. **• In this example, which group—carnivores or herbivores—is most efficient (produces the greatest percentage of energy available at the trophic level immediately below it in the food chain)?**

ple of the generalist, for in some parts of Earth they are carnivores, in some parts herbivores, and in some parts omnivores. It is also true that different species may occupy the same ecological niche in habitats that are similar but located in separate ecosystems. As an example, native horses grazed the steppes of Eurasia when bison roamed the grasslands of North America. Yet, when horses were introduced to North America they did well in the local grasslands.

Succession and Climax Communities

Up to this point we have been discussing ecosystems in general terms. In the remainder of this chapter we will note that it is the species that occupy the ecosystem that give the ecosystem its

character. At least for terrestrial ecosystems it is the autotrophs, the plant species at the first trophic level, that most easily distinguish one ecosystem from another. All other species in an ecosystem depend upon the autotrophs for food, and the association of all living organisms determines the energy flow and the trophic structure of the ecosystem. It should also be noted that the species that occupy the first trophic level are

22 people

1350 kilograms of soybeans and corn

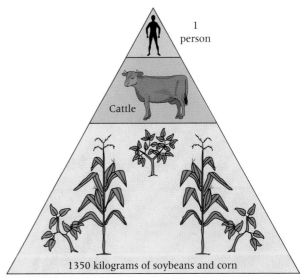

1 person

Cattle

1350 kilograms of soybeans and corn

Figure 12.7 The triangles illustrate the advantages of a vegetarian diet as we contemplate another century of rapid population growth. It is fortunate that humans are omnivores and can choose to eat grain products. The same 1350 kilograms of grain that will support, if converted to meat, only one person will support 22 people if cattle or other animals are omitted from the food chain. **• In what areas of the world today do grain products constitute nearly all of the total food supply?**

greatly influenced by climate; again we see the interconnections between the major Earth subsystems.

If the plants that comprise the biomass of the first trophic level are allowed to develop naturally without obvious interference from or modification by humans, the resulting association of plants is called **natural vegetation.** These plant associations or **plant communities** are compatible because each species within the community has different requirements in relation to major environmental factors such as light, moisture, and mineral nutrients. If two species within a community were to compete, one would eventually eliminate the other. The species forming a community at any specific place and time will be an aggregation of those that together can adapt to the prevailing environmental conditions.

Succession

Natural vegetation of a particular location develops in a sequence of stages involving different plant communities. This developmental process, known as **succession,** usually begins with a relatively simple plant community. Two types of succession, *primary* and *secondary,* are recognized. In primary succession, a bare substrate is the beginning point. No soil or seedbed exists at this point. A *pioneer* community invades the bare substrate (whether it be volcanic lava, glacially deposited sediment, or a bare beach, among others)

and begins to alter the environment. As a result, the species structure of the ecosystem does not remain constant. In time, the alterations of the environment become sufficient to allow a new plant community (a community that could not have survived under the original conditions) to appear and eventually to dominate the original community. The process continues, with each succeeding community rendering further changes to the environment. Because of the initial absence of soil, primary succession can take hundreds or even a few thousand years. Secondary succession begins when some natural process, such as a forest fire, tornado, or landslide, has destroyed or damaged a great deal of the existing vegetation. Ecologists refer to this process as **gap** creation. Even with such damage, however, soil still typically exists, and seeds may be lying dormant in that soil ready to invade the newly opened gap. Secondary succession therefore can occur much more quickly than primary succession.

A common form of secondary succession associated with agriculture in the southeastern United States is depicted in Figure 12.8. After agriculture has ceased, weeds and grasses are the first vegetative types to adapt to the somewhat adverse conditions associated with bare fields. These low-growing plants will stabilize the topsoil, add organic matter, and in general pave the way for the development of hardwood brush such as sassafras, persimmon, and sweet gum.

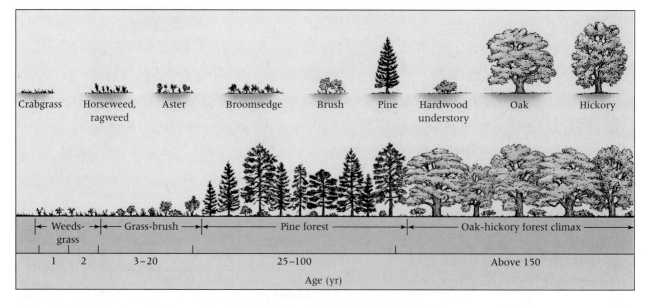

Figure 12.8 *A common plant succession in the southeastern United States. Each succeeding vegetation type alters the environment in such a way that species having more stringent environmental requirements can develop.* • **Why would plant succession be quite different in another region of the United States?**

During the brush stage the soil will become richer in nutrients and organic matter, and its ability to retain water will increase. These conditions encourage the development of pine forests, the next stage in this vegetative evolutionary process. Pine forests thrive in the newly created environment and will eventually dwarf and dominate the weeds, grasses, and brush that preceded them.

Ironically, the dominance of the pine forest leads to its demise. Pine trees require much sunlight if their seeds are to germinate. When competing with low-lying brush, grasses, and weedy annuals there is no problem in getting enough sunlight, but once the pines dominate the landscape, their seeds will not germinate in the shade and litter that their dense foliage creates. Thus, pines eventually will give way to hardwoods, such as oak and hickory, whose seeds can germinate under those conditions. These seeds may have been present and dormant in the soil, or blown into the forest, or carried into it by animals. In this example, then, the end result is an oak-hickory forest. If the changes continue uninterrupted, it is estimated that the complete succession will take 100 to 200 years. In other ecosystems, such as a tropical rainforest, this complete process may take many hundreds of years.

The Climax Community

The theory of plant succession was introduced early in the twentieth century. However, some of the original ideas have undergone considerable modification. Succession was considered to be an orderly process that included various *predictable* steps or phases and ended with a dominant vegetative cover that would remain in balance with the environment until disturbed by human activity—or until there were major changes in the environment. The final step in the process of succession has been referred to as the **climax community.** It was generally agreed that such a community was self-perpetuating and had reached a state of equilibrium or stability with the environment. In our illustration of plant succession in the southeastern United States, the oak-hickory forest would be considered the climax community.

Succession is still a useful model in the study of ecosystems. Why, then, have some of the original ideas been challenged, and what is the most recent thought on the subject? For one thing, early proponents of succession emphasized the predictable nature of the theory. One plant community would follow another in regular order as the species structure of the ecosystem evolved. But it has been demonstrated on many occasions that changes in ecosystems do not occur in such a rigid fashion. More often than not, the movement of species into an area to form a new community occurs in a random fashion and may be largely a function of chance events.

Many scientists today no longer believe that only one type of climax vegetation is possible for each major climatic region of the world. Some suggest that one of several different climax communities might develop within a given area, influenced not only by climate but also by drainage conditions, nutrients, soil, or topography. The dynamic nature of climate is now also much more fully understood than it was when the theories of succession and climax were developed, so it is now seen that by the time the species structure of a plant community has adjusted to new climatic conditions, the climate may change again. Because of the dynamic nature of each habitat, no one climax community can exist in equilibrium with the environment for an indefinite period of time. Many biogeographers and ecologists today view plant communities, and the ecosystems upon which they are based, as a *landscape* that is an expression of all the various environmental factors functioning together. They view the landscape of an area as a **mosaic** of interlocking parts, much like the tiles in a mosaic artwork. In a pine forest, for example, other plants also exist and some areas do not support pine trees. The dominant area of the mosaic—that is, the pine forest—is referred to as the **matrix.** Gaps within the matrix, resulting from areas of different soil conditions or gaps created by natural processes, are referred to as **patches** within the matrix. Relatively linear features cutting across the mosaic, including natural features such as rivers and human-created structures such as roads, fence lines, power lines, and hedgerows, are termed **corridors.** Each particular habitat is unique and constantly changing, and resultant plant and animal communities must constantly adjust to these changes. The dominant environmental influence is climate.

As alluded to above, however, climates are also changing. Over both relatively short time periods, on the scales of decades and centuries, and much longer times measured in millennia, climate changes occur. They may be subtle or they may be sufficiently drastic to create ice ages or periods of warmth between ice ages. Plant and animal communities must be able to respond to these ongoing changes or they will not survive. Many modern biogeographers are deeply involved in the reconstruction of the vegetation communities of past climatic periods, through the examination of evidence such as tree rings, pollen, insect fossils, and plant fossils. By determining how past climate changes affected and induced change in Earth's ecosystems, biogeographers hope to be able to suggest how future changes may develop as climate continues to change.

Environmental Controls

It is apparent from the preceding paragraphs that the plants and animals occupying a particular ecosystem at a given time are those species that are most successful in adjusting to the unique environmental conditions that constitute their habitat. Each species of living organism has a range within which it can adapt to environmental factors. For example, some plants can survive under a wide range of temperature conditions, whereas others have narrow temperature requirements. Biogeographers and ecologists refer to this characteristic as an organism's range of **tolerance** for a particular environmental condition. The ranges of tolerance for a species will determine where on Earth that species may be found, and species with wide ranges of tolerance will be the most widely distributed. The *ecological optimum* refers to the environmental conditions under which a species will fluorish (Fig. 12.9). As a species moves away from its ecological optimum, or as one moves from the geographic core of a plant or animal community, the environmental conditions become increasingly difficult for that species or community to survive. At the same time, those conditions may be more amenable for another species or community. The **ecotone** is the overlap, or zone of transition, between two plant or animal communities (see Fig. 12.9).

Climate has the greatest influence over natural vegetation when we observe plant communities on a worldwide basis. The major types of terrestrial ecosystems, or **biomes,** are each associated with specific ranges of temperature and critical precipitation characteristics, such as annual amounts and seasonal distribution. Climate influences leaf shape and size in trees and determines if trees will exist in a region. However, at the local scale other environmental factors can be as important as climate. A plant's range of tolerance for the acidity of the soil, the drainage of the land, or the salinity of the water may be the critical environmental factor in determining whether that plant is a part of the ecosystem. The discussion that follows serves to illustrate how the major environmental factors influence the organization and structure of ecosystems.

Climatic Factors

Of all the various climatic factors that influence the ecosystem, sunlight conditions are often the most critical. Sunlight is the vital source of energy for photosynthesis in plants and a control of life patterns for animals. The competition for light makes forest trees grow taller and limits growth on the forest floor to plants, such as ferns, that can tolerate shade conditions. Leaf sizes and shapes may reflect this variation in light reception, with large leaves in areas of limited light reception. The *quality* of light is important, especially in mountain areas, where plant growth may be severely retarded by excess ultraviolet radiation. This radiation does significant damage in the thin air at higher elevations but is effectively screened out by the denser atmosphere at lower elevations. Light *intensity* affects the rate of photosynthesis and hence the rate of primary productivity in an ecosystem. The more intense light of the low latitudes produces a higher energy input and greater biomass in the tropical forest than does the less intense light of the higher latitudes in Arctic regions. The *duration* of daylight, in association with the changing seasons, has a profound effect upon the flowering of plants, the activity patterns of insects, and the migration and mating habits of animals.

A second important climatic control of ecosystems is the availability of water. Virtually all organisms require water to survive. Plants require water for germination, growth, and reproduction, and most plant nutrients are dissolved in soil water before they are absorbed by plants. Marine plants are adapted to living completely in water (seaweed); some plants, like mangrove (Fig. 12.10) and bald cypress (Fig. 12.11), rise from coastal marshes and inland swamps; others thrive in the constantly wet rainforest. Certain tropical plants drop their leaves and become dormant during dry seasons, while others store water received during periods of rain in order to survive seasons of drought. Desert plants, such as cacti, are especially adapted to obtaining and storing water when it is available while minimizing their water loss from transpiration.

Animals, too, are severely restricted when water is in short supply. In arid regions animals must make special adaptations to environmental conditions. Many become inactive during the hottest and driest seasons, and most leave their burrows or the shade of plants and rocks only at night. Others, like the camel, can travel for great distances and live for extended periods without a water supply.

(Continued on page 324)

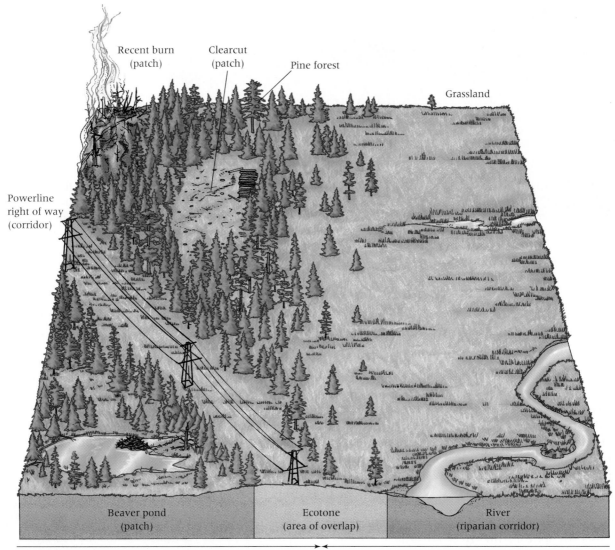

Recent burn
(patch)

Clearcut
(patch)

Pine forest

Grassland

Powerline
right of way
(corridor)

Beaver pond
(patch)

Ecotone
(area of overlap)

River
(riparian corridor)

Conditions increasingly unfavorable
for pine forest.

Conditions increasingly unfavorable
for grasslands.

Figure 12.9 The concepts of ***ecotone, ecological optimum, range of tolerance, mosaic, matrix, patch,*** *and* ***corridor*** *are illustrated in the diagram. Two ecosystems exist adjacent to each other, a pine forest on the left and a grassland on the right. Between the two is an area of overlap, where grasses and pine trees intermingle. This overlap is the* ***ecotone.*** *Toward the left of the overlap, the area would be more pine-forest-like, but with elements of grassland as well; whereas on the right of the overlap, the reverse would occur. The forest landscape achieves its* ***ecological optimum*** *some distance away from the ecotone. Within the ecotone, the pine forest (or the grassland) is reaching the outer geographic and ecological limits of its* ***range of tolerance.*** *The pine forest landscape is a* ***mosaic*** *of a number of landscape elements: the pine forest itself, which is the dominant landscape element in the area, comprises the landscape* ***matrix.*** *Smaller* ***patches*** *of non-pine forest, created by both natural and human forces, differ from the surrounding matrix. Examples include a beaver pond, a forest clear-cut, and a burned area. A* ***corridor,*** *a relatively straight-line disruption to the matrix, is created by a power-line right-of-way running through the forest. In the adjacent grassland matrix to the right of the ecotone, a* ***riparian corridor,*** *a specialized form of corridor associated with river channels, cuts through the grassland.* • **What effect would a change to drier climatic conditions throughout the area have on the relative sizes of the two ecosystems, as well as the position of the ecotone?**

Figure 12.10 Mangrove thicket along the Gulf of Mexico coast of southern Florida. • How did this vegetation influence the routes that were followed by the Spanish adventurers who first explored Florida?

Organisms are affected less by temperature variations than by sunlight and water availability. Many plants can tolerate a wide range of temperatures, although each species has optimum conditions for germination, growth, and reproduction. These functions can be impeded by excessively high and excessively low temperatures. Temperatures may also have indirect effects on vegetation. For example, high temperatures will lower the relative humidity, thus increasing transpiration. If a plant's root system cannot extract enough moisture from the soil to meet this increase in transpiration, the plant will wilt and eventually die.

Because of their mobility, animals are not as dependent upon the vagaries of climate as are plants. In spite of the great advantage afforded by mobility, however, animals are nevertheless subject to climatic stress. The geographic distribution of some groups of animals reflects this sensitivity to climate. Cold-blooded animals are, for example, more widespread in warmer climates and more restricted in colder climates. Some warm-blooded animals develop a layer of fat or fur and are able to shiver to protect themselves against the cold. In hot periods they may sweat or lick their fur in an attempt to stay cool. In extremely cold or arid regions, animals may hibernate. During hibernation the body temperature of the animal roughly changes in response to outside and ground temperatures. Cold-blooded animals such as the desert rattlesnake move in and out of shade in response to temperature change. Warm-blooded animals may migrate great distances out of environmentally harsh areas.

Among some warm-blooded animals there also exists an interesting linkage between body shape and size and variations in average environmental temperature. These adaptations have been described by biologists as *Bergmann's Rule*

and *Allen's Rule*. Bergmann's Rule states that, within a warm-blooded species, the body size of the subspecies usually increases with the decreasing mean temperature of its habitat, whereas Allen's Rule notes that, in warm-blooded species, the relative size of exposed portions of the body decreases with the decrease of mean temperature. These rules essentially boil down to the fact that members of the same species living in colder climates eventually evolve shorter or smaller appendages (ears, noses, arms, legs, and so on) than

Figure 12.11 This strand of bald cypress trees is located in extreme southern Illinois at the poleward limit of growth for this type of vegetation. • What Köppen climatic type does this site represent?

Figure 12.12 *Krummholz vegetation at the upper reaches of the subalpine zone on Pennsylvania Mountain in the Mosquito Range of the Colorado Rockies. The healthy green vegetation has been covered by snow much of the year and has been protected from the bitterly cold temperatures associated with gale-force winter winds. Note the flag trees, which give a clear indication of wind direction.* • **What type of vegetation would be found at elevations higher than the one depicted in this photograph?**

their relatives in warmer climates (Allen's Rule), and that in cold climates, body size will be larger, with more mass to provide the body heat needed for survival and for protection of the main trunk of the body where vital organs are located (Bergmann's Rule). In cold climates, small appendages are advantageous because they reduce the amount of exposed area subject to temperature loss, frostbite, and cellular disruption. In warm climates, large body sizes are not necessary for protection of internal organs, but long limbs, noses, and ears allow heat dissipation to take place in addition to that provided by panting or fur-licking.

Although most significant in areas such as deserts, polar regions, coastal zones, and highlands, wind can also serve as a climatic control. Wind may cause direct injury to vegetation or may have an indirect effect by increasing the rate of evapotranspiration. To prevent loss of water in the areas of severe wind stress, plants will twist and grow close to the ground in order to minimize the degree of their exposure (Fig. 12.12). During severe winters they are better off buried by snow than exposed to icy gales. In some coastal regions the shoreline may be devoid of trees or other tall plants, and where trees do grow they are often misshapen or swept bare of leaves and branches on their windward sides (Fig. 12.13).

Soil and Topography

In terrestrial ecosystems, the soils in which plants grow supply much of the moisture and minerals that are transformed into plant tissues. Soil vari-

ations are among the most conspicuous influences on plant distribution and often produce sharp boundaries in vegetation type. This is partly a consequence of the varying chemical requirements of different plant species and partly a reflection of other factors, such as soil texture. In a particular area clay soil may retain too much moisture for certain plants, whereas sandy soil retains too little. It is well known that pines thrive in sandy soils, grasses in clays, cranberries in acid soils, and wheat and chili peppers in alkaline

Figure 12.13 *Windswept trees along the California coast near Carmel. Strong winds from the Pacific Ocean have bowed trees inland and polished limbs on their exposed windward sides.* • **The photographer in this illustration was facing south and a little east; what is the prevailing wind direction in the area?**

soils. The subject of soils and their influence on vegetation will be explored more thoroughly in Chapter 13.

In the discussion of highland climatic regions in Chapter 11, we learned that topography influences ecosystems indirectly by providing many microclimates within a relatively small area. Plant communities vary significantly from place to place in highland areas in response to the differing nature of the climatic conditions. Some plants thrive on the sunny south-facing slopes of highland areas in the northern hemisphere; others survive on the colder, shaded, north-facing slopes. The steepness and shape of a slope also affect the amount of time water is present there before draining downslope.

Luxuriant forests tower above the well-watered windward sides of mountain ranges such as the Sierra Nevada and Cascades, while semi-arid grasslands and sparse forests cover the leeward sides. Spatial variations in precipitation drainage and resulting vegetation differences would not exist were it not for the presence of the topographic barrier inducing orographic uplift. Each major increase in elevation also produces a different mixture of plant species that can tolerate the lower ranges of temperature found at the higher elevation.

Natural Catastrophes

The distributions of plants and animals are also affected by a diversity of natural processes frequently termed *catastrophes*. It should be noted, however, that the term catastrophe is applied from a strictly human perspective. What may be catastrophic to a human, such as a hurricane, tornado, forest fire, landslide, or avalanche (Figure 12.14), is simply a natural process operating to produce openings (gaps) in the prevailing vegetative mosaic of a region. The resulting successional processes, whether primary or secondary in nature, produce a variety of patch habitats within the broader regional matrix of vegetation. The study of natural catastrophes and the resulting patch dynamics among the plant and animal residents of an area is a topic of strong interest and ongoing research in modern landscape biogeography.

Biotic Factors

Although they might tend to be overlooked as environmental controls, other plants and animals may be the critical factors in determining whether a given organism is a part of an ecosystem. Some interactions between organisms may

be beneficial to both species involved, whereas others may have an adverse effect on one or both. Because most ecosystems are suitable to a wide variety of plants and animals, there is always competition between species and among members of a given species to determine which organisms will survive. The greatest competition occurs between species that occupy the same ecological niche, especially during the earliest stages of life cycles when organisms are most vulnerable. Among plants, the greatest competition is for light. Those trees that become dominant in the forest are those that grow the tallest and partially shade the plants growing beneath them. Other competition occurs underground where the roots compete for soil water and plant nutrients.

Interactions between plants and animals and competition both within and among animal species also have significant effects upon the nature of an ecosystem. Animals are often helpful to plants during pollination or the dispersal of seeds, and plants are the basic food supply for many animals. The simple act of grazing may help to determine the species that make up a plant community. During dry periods, herbivores

Figure 12.14 The vegetation mosaic of this area is coniferous forest, but frequent snow avalanches passing down the gully keep rigid-stemmed conifers from invading the patch of open, low shrubs and grasses. The effects of attempted colonization are seen at right, where a swath of conifers have been destroyed by a recent snow avalanche. A bridge abutment is visible in the foreground. • **What other environmental factors might account for the limited amount of coniferous vegetation on the left, south-facing slope? What could happen to the bridge in the event of a massive snow avalanche?**

may be forced to graze an area more closely than usual, with the result that the taller plants are quickly grazed out. Plants that grow close to the ground, that are nonpalatable, or that have the strongest root development are the ones that survive. Hence, grazing is a part of the natural selection process, but serious overgrazing rarely results under natural conditions because wild animal populations increase or decrease with the available food supply. To be more precise, the number of animals of a given species will fluctuate between the maximum number that can be supported when its food supply is greatest and the minimum number required for reproduction of the species. For most animals, predators are also an important control of numbers. Fortunately, when the predator's favorite species is scarce, it will seek an alternative species for its food supply, and it would be rare for an animal to be excluded from an ecosystem through predation.

Human Impact on Ecosystems

Throughout human history we have modified the natural development of ecosystems. Except in regions too remote to be altered significantly by civilization, humans have eliminated much of Earth's natural vegetation. Farming, fire, grazing of domesticated animals, afforestation and deforestation, road building, urban development, dam building and irrigation, raising and lowering of water tables, mining, and the filling in or drainage of wetlands are just a few ways in which humans have modified the plant communities around them. Overgrazing by domesticated animals can seriously harm marginal environments in semiarid climates. Trampling and compaction of the soil by grazing herbivores may reduce the soil's ability to absorb moisture, leading to increased surface runoff of precipitation. In turn, the decreased absorption and increased runoff may respectively lead to *land degradation* and gully erosion.

It should be noted that ecosystems are not the only victims as humans alter natural environments; the changes can often produce long-term negative effects on humans themselves. The desertification of large tracts of semiarid portions of east Africa resulted in widespread famine in the 1980s in countries such as Ethiopia and Somalia. Elsewhere, the widespread destruction of wetlands not only eliminates valuable plant and animal communities, it often seriously threatens the quality and reliability of the water supplies for the people who drained the land (see The Environment: Restoring Our Priceless Wetlands).

We have, in fact, so changed the vegetation in some parts of the world that we can characterize classes of cultivated vegetation cared for by humans—for example, flowers, shrubs, and grasses to decorate our living areas, and grains, vegetables, and fruits that we raise for our own food and to feed the animals we eat. Our focus in the remainder of this chapter, however, will be on the major ecosystems of Earth as we assume they would appear without human modification.

Classification of Terrestrial Ecosystems

The geographic classification of natural vegetation is as difficult as the classification of any other complex phenomena influenced by a variety of factors. However, plant communities are among the most highly visible of natural phenomena, so they can be categorized on the basis of form and structure or gross physical characteristics. Of course, the composition of the natural vegetation changes from place to place in a gradational manner, just as temperature and rainfall do, and although distinctly different types are apparent, there may be broad transition zones (ecotones) between them. Nevertheless, over the world there are distinctive recurring plant communities, indicating a consistent botanical response to systematic controls that are essentially climatic. It is the dominant vegetation of these plant communities that we recognize when we classify the Earth's major terrestrial ecosystems (biomes).

All of Earth's biomes can be categorized into one of four easily recognized types: forest, grassland, desert, and tundra. Because vegetation adapted to cold climates may occur at high elevations at any latitude as well as in high-latitude regions in general, the last of the major types is often referred to as arctic and alpine tundra by biogeographers. However, the forests of the equatorial lowlands are an entirely different world from those of Siberia or of New England, and the original grasslands of Kansas bore little resemblance to those of the Sudan or Kenya. Hence, the four major types of ecosystems can be subdivided into distinctive biomes, each of which is an association of plants and animals of many different species. Pure stands of particular trees, shrubs, or even grasses are extremely rare and are limited to small areas having peculiar soil or drainage conditions.

Earth's major biomes are mapped in Figure 12.15 on the basis of the dominant associations

(Continued on page 330)

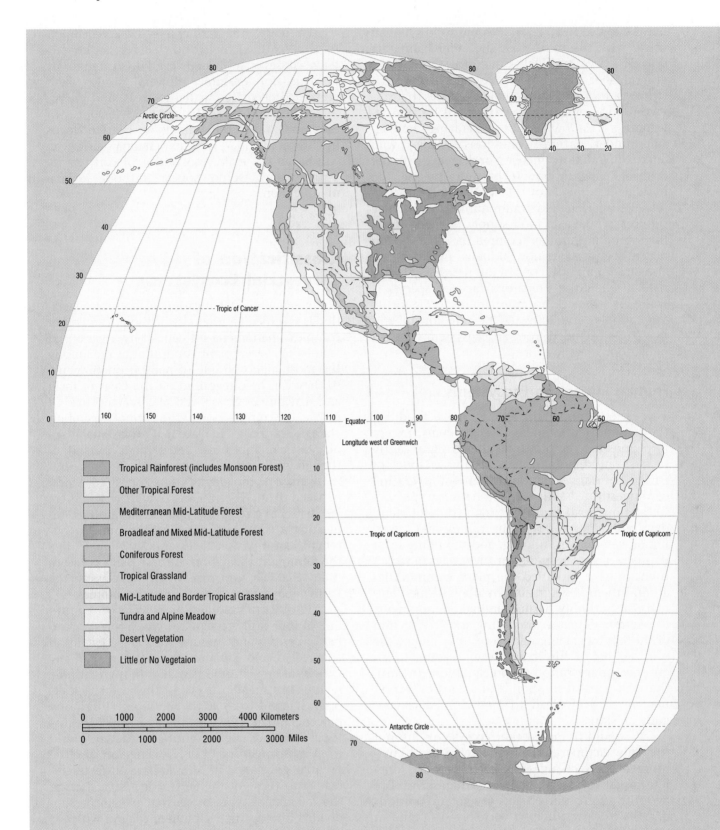

Figure 12.15 World Map of Natural Vegetation.

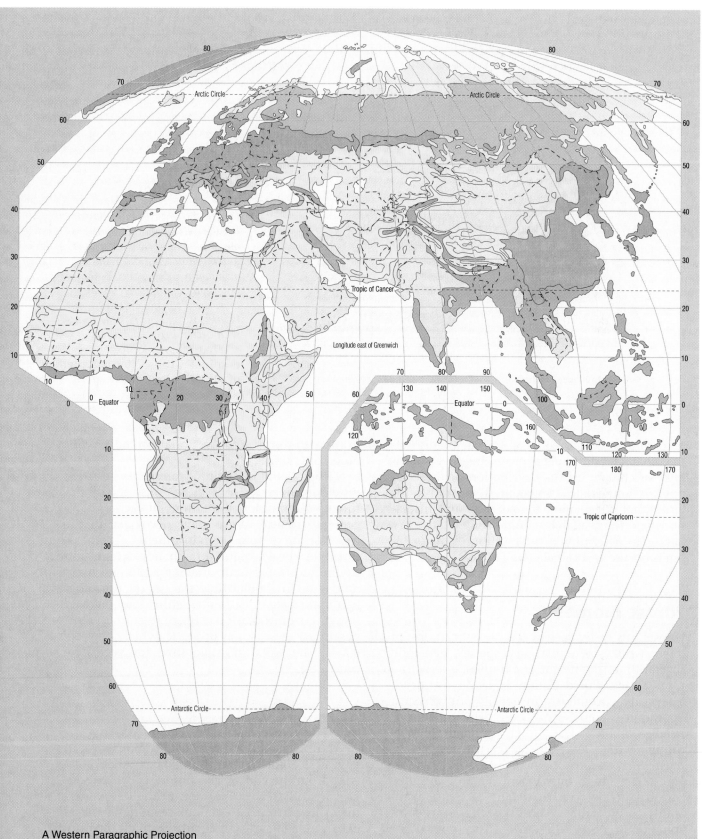

A Western Paragraphic Projection
developed at Western Illinois University

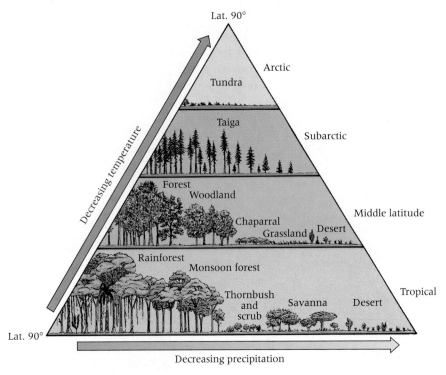

Influence of latitude and moisture on distribution of biomes

Figure 12.16 Schematic diagram showing distribution of Earth's major biomes as they are related to temperature (latitude) and the availability of moisture. Within the tropics and middle latitudes there are distinctly different biomes as total biomass decreases with decreasing precipitation.
• What major biome dominates the wetter margins of all latitudes but the Arctic?

of natural vegetation that give each its distinctive character and appearance. The direct influence of climate on the distribution of these biomes is immediately apparent. Temperature (or latitudinal effect upon temperature and insolation) and the availability of moisture are the key factors in determining the location of major biomes on the world scale (Fig. 12.16).

Forest Biomes

Forests are easily recognized as associations of large, woody, perennial tree species, generally several times the height of a human, and with a more or less closed canopy of leaves overhead. They vary enormously in density and physical appearance, some being evergreen and either needle- or broad-leaf, others being deciduous, dropping their leaves to reduce moisture losses during dry seasons or when soil water is frozen. Forests are found only where the annual moisture balance is positive—where moisture availability considerably exceeds potential evapotranspiration in the growing season. Thus, they occur in the tropics, where either the ITC or the monsoonal circulation brings plentiful rainfall, and in the midlatitudes, where precipitation is associated with cyclonic depressions along the polar

front, with summer convectional rainfall, or with orographic uplift.

Tropical and midlatitude forests have evolved different characteristics in response to the nature of the physical limitations in each area. In general, tropical forests have developed in the less restrictive forest environments. Temperatures are always high, though not extreme, in the humid tropics, encouraging rapid and luxuriant growth. Midlatitude forests, on the other hand, must adapt to combat either seasonal cold (ranging from occasional frosts to subzero temperatures) or seasonal drought (which may occur at the worst possible time for vegetative processes).

Tropical Forests

The forests of the tropics are far from uniform in appearance and composition. They grade poleward from the equatorial rainforests, which support Earth's greatest biomass, to the last scattering of low trees that overlook seemingly endless expanses of tall grass or desert shrubs on the tropical margins. We have subdivided the tropical forests into three distinct biomes: the tropical rainforest, the monsoon rainforest, and other tropical forest types, primarily thornbush and scrub. Of course, there are gradations (ecotones) between the different types as well as distinctive variations that are found in individual localities only.

Tropical Rainforest. In the equatorial lowlands dominated by Köppen's tropical rainforest climate, the only physical limitation for vegetation growth is competition between adjacent species. The competition is for light. Temperatures are high enough to promote constant growth, and water is always sufficient. Thus, we find forests consisting of an amazing number of broad-leaf evergreen tree species of rather similar appearance, since special adaptations are not required (Fig. 12.17). A cross section of the forest often reveals concentrations of leaf canopies at several different levels. The trees composing the distinctive individual tiers have similar light requirements—lower than those of the higher tiers but higher than those of the lower tiers. Little or no sunlight reaches the forest floor, which may support ferns but is often rather sparsely vegetated. The forest is literally bound together by vines, **lianas,** that climb the trunks of the forest trees and intertwine in the canopy in their own search for light. Aerial plants may cover the limbs of the forest giants, deriving nutrients from the water and the plant debris that falls from higher levels. Light and variable wind conditions in the rainforest (associated with the latitudinal belt of the doldrums) preclude wind from being an effective agent of seed and pollen dispersal, so large colorful fruits and flowers, designed to attract animals who will unwittingly carry out seed and pollen dispersal, prevail.

The forest trees commonly depend upon grotesquely flared or buttressed bases for support, since their root systems are shallow (see Fig. 13.11). This is a consequence of the richness of the surface soil and the poverty of its lower levels. The rainforest vegetation and soil are intimately associated. The forest litter is quickly decomposed, its nutrients released and almost immediately reabsorbed by the forest root systems, which consequently remain near the surface. Tropical soils that maintain the amazing biomass of the rainforest are fertile only as long as the forest remains undisturbed. Clearing the forest interrupts the crucial cycling of nutrients between the vegetation and the soil; the copious amounts of water percolating through the soil leach away its soluble constituents, leaving behind only inert iron and aluminum oxides that cannot support forest growth. The present rate of clearing threatens to wipe out the worldwide

Figure 12.17 *A rainforest near El Salto Falls in eastern Mexico. The dense nature of the forest canopy effectively conceals the vast number of different evergreen tree species and relatively open forest floor.* • **How might this rainforest differ from the rainforests of the Pacific Northwest of the United States?**

tropical rainforests within the foreseeable future. The largest remaining areas of unmodified rainforests are in the upper Amazon Basin, where they cover hundreds of thousands of square kilometers (see again The Environment: Destruction of the Amazon Rainforest, pages 254–255).

Because of the darkness and extensive root systems present on the forest floor, animals of the tropical rainforest are primarily arboreal. A wide variety of species of tree-dwelling monkeys and lemurs, snakes, tree frogs, birds, and insects characterize the rainforest. Even the large herbivorous and carnivorous mammals, such as sloths and jaguars respectively, are primarily arboreal.

Within the areas of rainforest and extending outward beyond its limits along streams are patches or strips of jungle. Jungle consists of an almost impenetrable tangle of vegetation contrasting strongly with the relatively open nature (at ground level) of the true rainforest. It is often

(Continued on page 334)

RESTORING OUR PRICELESS WETLANDS

The low-lying marshes, swamps, shallow lakes, and pools of nearly stagnant water that occupy floodplains and border rivers or coastlines throughout the United States have long been considered by many people to be useless wastes of the land resource. As obstacles to land development for agricultural, industrial, and residential purposes, these wetlands have been efficiently and methodically eliminated by dredging and the channelization of rivers, drainage through the construction of ditches and canals, and the filling of lowlands.

Recently we have come to realize the true value of our wetland resources. Wetlands offer a stabilizing effect on the shorelines of seas, lakes, and rivers. Their loss drastically increases the erosion rates around these water bodies. We have also discovered that wetlands have a great capacity to filter and purify water. The saturated soil beneath the wetland vegetation naturally breaks down pollutants into harmless substances. As wetlands have disappeared, the quality of municipal water supplies has rapidly deteriorated.

Of major significance is the loss of some of the most complex and important ecosystems in the world. Scientists, environmentalists, government officials, and concerned citizens have become determined not only to preserve the nation's remaining wetlands but also to restore former wetland areas as nearly as possible to their original natural state. The stories of two river floodplains in quite different regions of the United States can serve as encouraging examples of what cooperative effort and tireless dedication to an important cause can accomplish.

The Kissimmee River

The Kissimmee River and its floodplain are integral parts of a much larger wetland ecosystem that originally covered a major portion of southern Florida and also includes Lake Okeechobee and what remains of the Everglades. The entire ecosystem has undergone traumatic alteration since land reclamation began in the nineteenth century, and today the region suffers from loss of water supplies, unnatural water flow that from time to time either bypasses or floods living organisms in the wetlands and soils, impure water quality, and salt water intrusion. One major part of the problem was the channelization of the Kissimmee River, which took place as part of a flood control project. As a result, river flow that formerly had been carried by a meandering river channel and broad floodplain was confined to a narrow excavated canal. The floodplain marshes were drained and replaced to a large extent by pastures for high-quality beef cattle. The ability of the river system to store and release excess water slowly to Lake Okeechobee and its ability to purify the runoff it received had been largely destroyed.

It must have seemed an impossible task to convince the state and the federal governments, which a few years before had spent millions of dollars to channelize the Kissimmee, that it was now essential to spend millions more in an attempt to return the river to its natural state. But this is exactly what a small band of determined environmentalists set out to do. **The fact that they are succeeding is important evidence that when the truth is known and is presented with courage and persistence, government will respond.**

On the basis of studies conducted by a council created to coordinate restoration of the Kissimmee, dechannelization was recommended in 1983. A demonstration project that returned water flow to a small portion of the floodplain followed the recommendation and proved that the floodplain and its ecosystems could be restored if large sections of the

Restored wetlands along the original course of the Kissimmee River.

Sign identifying the location of the Kissimmee restoration project.

river were dechannelized. The estimated cost of the 15-year Kissimmee project will exceed $372 million, with funding divided equally between Florida and the federal government. For those who understand the true value of the project, the cost is small by comparison with the benefits to humans and the natural environment.

The Cache River

The rolling hills of southern Illinois are a long way from the flat plains of south Florida, but the rivers that drain the two regions have some important things in common. Although the Cache River of Illinois is a small stream compared with the Kissimmee, it, too, supports irreplaceable wetlands as well as valuable forest ecosystems throughout its drainage basin.

By the end of the nineteenth century, much of the virgin forest had been removed by commercial logging, and in the early twentieth century, humans turned their attention to the wetlands. Ditches were dug, swamplands were drained, and new outlets to the Ohio River provided steeper gradients that robbed the lower half of the Cache basin of much of its flow. Fields and crops had replaced swamps and marsh in many of the wetland areas by the 1970s.

The future of the river and its resources appeared bleak indeed, but unknown to most, other forces were already at work in the 1960s to save the Cache. Through the efforts of a few dedicated individuals who believed in the preservation of natural areas, the Illinois Nature Preserves Commission Act became law in 1963. Wetland areas along the Cache were high on the list of those that should be acquired as Illinois nature preserves. Purchase of the first tract along the upper Cache occurred in 1969, and the move toward conservation and away from exploitation had begun.

Progress toward preservation of one of Illinois' most valuable wetland areas will continue for years to come, but it reached a high water mark when the U.S. Fish and Wildlife Service authorized a National Wildlife Refuge along the Cache River in 1990. Within the refuge over 35,000 acres of land are to be purchased or protected while still in private ownership. Authorization of the refuge will allow preservation of nearly 13,000 acres of wetlands in their natural state and restoration of an additional 10,000 wetland acres. **The Cache River project should encourage people from throughout the United States to do for their rivers and wetlands what the citizens of southern Illinois are doing for the Cache.** As the president of the citizens group to save the Cache said at the dedication of the wildlife refuge, "It's like a dream come true—I'll tell you, it almost brings tears to your eyes after working that long, and seeing all that happen."

Rapid downcutting of the upper Cache River and its tributaries in southern Illinois is a direct result of the ditching and lowering of the water table that has destroyed valuable wetlands throughout the river's drainage basin.

Eroded stream bank in southern Illinois showing exposed tree roots.

CRITICAL THINKING ▼

(1) Is all the time and money being expended in Florida and Illinois to preserve wetlands worth the effort? Would you support similar projects if you had the opportunity? **(2)** What areas of the environment in addition to wetlands should be preserved? **(3)** What areas of the United States do you believe will be preserved in a reasonably natural state by the year 2050?

composed of secondary growth that quickly invades the rainforest where a clearing has allowed light to penetrate to the forest floor. Jungle commonly extends into the drier areas beyond the forest margins along the courses of streams. There it forms a gallery of vegetative growth closing over the watercourse and hence has been called **galeria forest.**

Monsoon Rainforest. In areas of monsoonal circulation there is an alteration between the dry monsoon season, when the dominant flow of air is from the land to the sea, and the wet monsoon season, when the atmospheric circulation reverses, bringing moist air onshore along tropical coasts. The wet monsoon season rainfall may be very high, even hundreds of centimeters where air is forced upward by topographic barriers. In any case, it is sufficient to produce a forest that, once established, remains despite the dry monsoon season. Monsoon forests themselves may have discernible tiers of vegetation related to the varying light demands of different species, and they are included with the tropical rainforest in Figure 12.15. However, the number of species is less than in the true rainforest, and the overall height and density of vegetation are also somewhat less. Some of the species are evergreen, but many are deciduous.

Other Tropical Forests, Thornbush, and Scrub. Where seasonal drought has precluded the development of true rainforest, or where soil characteristics prevent the growth of such vegetation, variant types of tropical forests have developed. These tend to be found on the subtropical margins of the rainforests and on old plateau surfaces where soils are especially poor in nutrients. The vegetation included in this category varies enormously but is generally low-growing in comparison to rainforest, without any semblance of a tiered structure, and is denser at ground level. Commonly it is thorny, indicating defensive adaptation against browsing animals, and it shows resistance to drought in that it is generally deciduous, dropping its leaves to conserve moisture during the dry winter season. Ordinarily grass is present beneath the trees and shrubs. As we move away from the equatorial zone, we find the trees more widely spaced and the grassy areas becoming dominant. Along tropical coastlines, a specially adapted plant community, known as mangrove, thrives (see again Fig. 12.10). Here trees are able to grow in salt water.

Midlatitude Forests

The forest biomes of the midlatitudes differ from those of the tropics because the dominant trees have evolved mechanisms to withstand periods of water deprivation due to low temperatures and annual variations in precipitation. Evergreen and deciduous plants are present, equipped to cope with seasonal extremes not encountered in tropical latitudes.

Mediterranean Sclerophyllous Woodland. Surrounding the Mediterranean Sea and on the

Figure 12.18 The distinctive sclerophyllous evergreen vegetation type encountered wherever hot, dry summers alternate with rainy winters. Oaks commonly occupy relatively damp sites, such as the gullies and windward slopes seen here in southern California. • **What are the general characteristics of sclerophyllous vegetation?**

southwest coasts of the continents between approximately 30° and 40°N and S latitude, we have seen that a distinctive climate exists—Köppen's mesothermal hot and dry summer type (Mediterranean). Here annual temperature variations are moderate and freezing temperatures are rare. However, there is little or no rainfall during the warmest months, and plants must be drought-resistant. This requirement has resulted in the evolution of a distinctive vegetation that is relatively low-growing, with small, hard-surfaced leaves and roots that probe deeply for water. The leaves must be capable of photosynthesis with minimum transpiration of moisture. The general look of the vegetation is a thick scrub plant community, called *chaparral* in the western United States and *maquis* in the Mediterranean region (see Fig. 11.3). Wherever moisture is concentrated in depressions or on the cooler north-facing hill slopes, deciduous and evergreen oaks occur in groves (Fig. 12.18). Drought-resistant needle-leaf trees, especially pines, are also part of the overall vegetation association. Thus, the vegetation is a mosaic related to site characteristics and microclimate. Nevertheless, the similarity of the natural vegetative cover in such widely separated areas as Spain, Turkey, and California is astonishing. (Note the location of Mediterranean midlatitude forest, Fig. 12.15.)

Broad-Leaf Deciduous Forest. The humid regions of the middle latitudes experience a seasonal rhythm dominated by warm air in the summer and invasions of cold polar air in the winter. To avoid frost damage during the colder winters and to survive periods of total moisture deprivation when the ground is frozen, trees whose leaves have large transpiring surfaces drop these leaves and become dormant, coming to life and producing new leaves only when the danger period is past (Fig. 12.19). A large variety of trees have evolved this mechanism: certain oaks, hickory, chestnut, beech, and maples are common examples. The seasonal rhythms produce some

Figure 12.19 The appearance of hardwood forests in midlatitude regions with cold winters changes dramatically with the seasons. The green leaves of summer change to reds, golds, and browns in fall and drop to the forest floor in winter. Leaf dropping in areas of cold winters, such as this example in western Illinois, is a means of minimizing transpiration and moisture loss when the soil water is frozen. • **What length of growing season (frost-free period) is associated with the climate of western Illinois?**

beautiful scenes, particularly during the periods of transition between dormancy and activity, with the sprouting of new leaves in the spring and the brilliant coloration of the fall as chemical substances draw back into the plant for winter storage.

The trees of the deciduous forest may be almost as tall as those of the tropical rainforests and, like them, produce a closed canopy of leaves overhead or, in the cold season, an interlaced network of bare branches. However, lacking a multi-storied structure and having lower density as a whole, the midlatitude deciduous forests allow much more light to reach ground level. Forests of this type are the natural vegetation in much of western Europe, eastern Asia, and eastern North America. To the north and south they merge with mixed forests composed of broad-leaf deciduous trees and conifers. (Broad-leaf deciduous forests and mixed forests are included together on Figure 12.15 as mixed midlatitude forest.) Both the broad-leaf deciduous and mixed forests have been largely logged off or cleared for agricultural land, and the original vegetation of these regions is rarely seen.

Broad-Leaf Evergreen Forest.

Beyond the tropics, broad-leaf trees remain evergreen (active throughout the year) in significant areas only in certain Southern Hemisphere locations. Here the mild maritime influence is strong enough to prevent either dangerous seasonal droughts or severely low winter temperatures. Southeastern Australia and portions of New Zealand, South Africa, and southern Chile are the principal areas of this type. In the Northern Hemisphere, broad-leaf evergreen forest may once have been significant in eastern Asia, but it has long since been cleared for cultivation. Limited areas occur in the United States in Florida and along the Gulf Coast as a belt of evergreen oak and magnolia.

Mixed Forest.

Poleward and equatorward, the broad-leaf deciduous forests in North America, Europe, and Asia gradually merge into mixed forests, including needle-leaf coniferous trees, normally pines. In general, where conditions permit the growth of broad-leaf deciduous trees, coniferous trees cannot compete successfully with them. Thus, in mixed forests the conifers, which are actually more adaptable to soil and moisture deficiencies, are found in the less hospitable sites: in sandy areas, on acid soils, or where the soil itself is thin. The northern mixed forests reflect the transition to colder climates with increasing latitude; eventually conifers become dominant in this direction. The southern

mixed forests are more problematic in origin. In the United States they are transitional to pine forests situated on sandy soils of the coastal plain. In Eurasia they coincide with highlands dominated by conifers during a stage in the plant succession that began with the change of climatic environments at the end of the ice ages, some 10,000 years ago.

Coniferous Forest.

The coniferous forests occupy the frontiers of tree growth. They survive where most of the broad-leaf species cannot endure the climatic severity and impoverished soils. The hard, narrow needles of coniferous species transpire much less moisture than do broad leaves, so that needle-leaf species can tolerate conditions of physiologic drought (unavailability of moisture because of excessive soil permeability, a dry season, or frozen soil water) without defoliation. Pines, in particular, also demand little from the soil in the form of soluble plant nutrients, especially basic elements such as calcium, magnesium, sodium, and potassium. Thus, they grow in sandy places and where the soil is acid in character. As a whole, conifers are particularly well adapted to regions having long, severe winters combined with summers warm enough for vigorous plant growth. Because all but a few exceptions retain their leaves (needles) throughout the year, they are ready to begin photosynthesis as soon as temperatures permit without having to produce a new set of leaves to do the work.

Thus we find a great band of coniferous forests (the **boreal forests,** or **taiga**) dominated by spruce and fir species, with pines on sandy soils, sweeping the full breadth of North America and Eurasia northward of the fiftieth parallel of latitude, approximately occupying the region of Köppen's subarctic climate (see Fig. 10.4). Conifers differ from other trees in that their seeds are not enclosed in a case or fruit but are carried naked on cones. All are needle-leaf and drought-resisting, but a few are not evergreen. Thus a large portion of eastern Siberia is dominated by larch, which produces a deciduous, coniferous forest. In this area, January mean temperatures may be $-35°$ to $-51°C$ ($-30°$ to $-60°F$). Here we encounter the most severe winter climate in which trees can maintain themselves, and even needle-leaf foliage must be shed for the vegetation to survive. Hardy broad-leaf deciduous birch trees share this extreme climate with the indomitable larches.

Extensive coniferous forests are not confined to high-latitude areas of short summers. Higher elevations in midlatitude mountains of the Northern Hemisphere have forests of pine, hem-

Figure 12.20 As this photograph in the Colorado Rocky Mountains indicates, evergreen coniferous forest is the characteristic vegetation of higher-elevation regions in the middle latitudes. The needle-leaf trees are an adaptation to the physiologic drought of the winter season, which is longer and more severe at higher elevations. • Why are needle-leaf trees better adapted to physiologic drought than broad-leaf trees?

lock, and fir (Fig. 12.20), with subalpine larch and specially adapted pine species characterizing the harshest and highest forest sites. The forests along the sandy coastal plain of the eastern United States are there in part because of the sandy soils present, but they may also reflect a stage in plant succession that in time will lead to domination by broad-leaf types. Similarly, a temporary stage in the postglacial vegetation succession of the Great Lakes area included magnificent forests of white pine and hemlock that were completely logged off during the late nineteenth century.

A more maritime coniferous forest occupies the west coast of North America extending from southern Alaska to central California. It is made up of sequoias, Douglas fir, cedar, hemlock, and, farther north, Sitka spruce. Many of the California sequoias are thousands of years old and more than 100 meters (330 ft) high. The southern regions experience summer drought, and farther north, sandy, acidic, or coarse-textured soils dominate.

Grassland Biomes

Grasses, like conifers, appear in a variety of settings and are part of many diverse plant communities. They are, in fact, an initial form in most plant successions. However, there are enormous, continuous expanses of grasslands on our Earth. In general, it is thought that grasses are dominant only where trees and shrubs cannot maintain themselves either because of excessive or deficient moisture in the soil. On the global scale grassland biomes are located in continental interiors where most, if not all, of the precipitation falls in the summer. Two great geographic realms of grasslands are generally recognized: the tropical and the midlatitude grasslands. However, it is difficult to define grasslands of either type using any specific climatic parameter, and geographers suspect that human interference with the natural vegetation has caused expansion of grasslands into forests in both the tropical and middle latitudes.

Tropical Savanna Grasslands

The tropical grassland biome differs from grassland biomes of the midlatitudes in that it ordinarily includes a scattering of trees; this is implied in the term **savanna** (Fig. 12.21). In fact, the demarcation between tropical scrub forest and savanna is seldom a clear one. The savanna grasses tend to be tall and coarse, with bare ground visible between the individual tufts. The related tree species generally are low-growing and wide-crowned forms, having both drought- and

Figure 12.21 *The savanna biome of eastern Africa. This classic landscape of grasses and scattered trees supports dwindling herds of grazing animals as land use patterns change.* • **What human activities have replaced much of the eastern Africa savanna?**

fire-resisting qualities, indicating that fires frequently sweep the savannas during the drought season. Indeed, the role of fire in maintaining savannas (suppressing tree invasions) and the role of humans in creating or expanding the savannas at the expense of the forest has been a topic of continuing discussion among geographers. Savannas occur under a variety of temperature and rainfall conditions but generally fall within the limits of Köppen's tropical savanna type. Commonly they occur on red-colored soils leached of all but iron and aluminum oxides and which become bricklike or slaggy when dried. They likewise coincide with areas in which there is a dramatic fluctuation in the level of the water table (the zone below which all soil and rock pore space is saturated by water). The up-and-down movement of the water table may in itself inhibit forest development, and it is no doubt a factor in the peculiar chemical nature of savanna soils. Large migratory herds of grazing animals and associated predators, responding to the periodically abundant grasses followed by seasonal drought, characterized the savanna prior to widespread human disruption.

Midlatitude Grasslands

The zone of transition between the midlatitude deserts and forests is occupied by the midlatitude grasslands. On their dry margins, they pass gradually into deserts in Eurasia and are cut off westward by mountains in North America. However, on their humid side, they terminate rather abruptly against the forest margin, again raising questions as to whether their limits are natural or have been created by human activities, particularly the intentional use of fire to drive game animals. The midlatitude grasslands of North America, like the African savannas, formerly supported enormous herds of grazing animals—in this case antelope and bison, which were the principal means of support of the American Plains Indians.

Like the savannas, the midlatitude grasslands were diverse in appearance. They, too, consisted of varying associations of plant species that were never uniform in composition. In North America the grasses were as much as 3 meters (10 ft) tall in the more humid sections, as in Iowa, Indiana, and Illinois, but only 15 centimeters (6 in.) high on the dry margins from New Mexico to western Canada. Thus the midlatitude grassland biomes are usually divided into tall-grass and short-grass prairie, often with a zone of mixture recognized between them. Unlike growth in the tropical savannas, the germination and growth of midlatitude grasses are attuned to the melting of winter snows, followed by summer rainfall. Although the grasses may be annuals that complete their life cycle in one growing season or perennials that grow from year to year, they are dormant in the winter season. Also, unlike in the savannas, the soils beneath these grasslands are extremely rich in organic matter and soluble nutrients. As a consequence, most of the midlatitude grasslands have been completely transformed by agricultural activity. Their wild grasses have been replaced by domesticated varieties—wheat, corn, and barley—so they have become the "breadbaskets" of the world.

Tall-Grass Prairie. The tall-grass prairies were an impressive sight; in some better-watered areas they made up endless seas of grass moving in the breeze, reaching higher than a horse's back. Flowering plants were conspicuous, adding to the effect. Unfortunately, this **tall-grass prairie** scene, which inspired much vivid description by those first encountering it, is no longer visible anywhere. The tall-grass prairies, which once reached continuously from Alberta to Texas, have been destroyed. Their tough sod, formed by the dense grass root network, defeated the first wooden plows, which had served well enough in breaking up the forest soils. But the steel plow, invented in the 1830s, subdued the sod and was aided by the introduction of subsurface tile for draining the nearly flat uplands and by the simultaneous appearance of well-digging machinery and barbed wire. These four innovations transformed the tall-grass prairie from grazing land to cropland.

In North America the tall-grass prairie pushed as far eastward as Lake Michigan. Why trees did not invade the prairie in this relatively humid area remains an unanswered question. Farther west, shallow-rooted grass cover is fully understandable because the lower soil levels, to which tree roots must penetrate for adequate support and sustenance, are bone dry. In such areas trees can survive only along streams or where depressions collect water.

In Eurasia, tall-grass prairies were found on a large scale in a discontinuous belt from Hungary, through Ukraine, Russia, and central Asia, to northern China. In South America the grasslands are known as the pampas of Uruguay and Argentina and in South Africa as the Veldt. Today, all of these areas have been changed by agriculture. The factor that seems to account best for the tall-grass prairie—precipitation that is both moderate and variable in amount from year to year—is the principal hazard in the use of these regions as farmland. However, this hazard becomes much greater in the areas of short-grass prairie.

Short-Grass Prairie. West of the hundredth meridian in the United States and extending across Eurasia from the Black Sea to northern China, roughly coinciding with the areas of Köppen's middle-latitude steppe climate, are vast, nearly level grasslands composed of a mixture of tall and short grass, with short grass becoming dominant in the direction of lower annual precipitation totals (Fig. 12.22). On the Great Plains between the Rocky Mountains and the tall-grass prairies, the **short-grass prairie** zone more or less coincides with the zone in which moisture rarely penetrates more than 60 centimeters (2 ft) into the soil, so that the subsoil is permanently dry. Moving toward the drier areas, the grassland vegetation association dwindles in diversity and, more conspicuously, in height to less than 30 centimeters (1 ft). This is a consequence of reduction in numbers of tall-growing species and higher abundance of shallow-rooted and lower-growing types. The total amount of ground cover also declines toward the drier margins as the deeply rooted, sod-forming grasses of the prairie grassland give way to bunchgrass species (so called because, instead of forming a continuous grass cover, they occur in isolated clumps or bunches). In their natural conditions the short-grass prairies of North America and Eurasia supported higher densities of grazing animals—bison and antelope in the former, wild horses in the latter—than did the tall-grass prairies. Indeed, it is suspected that the specific plant association of the short-grass regions may have been a consequence of overgrazing under natural conditions. The short-grass prairies cannot be cultivated without the use of irrigation or dry farming methods, so they remain primarily the domain of wide-ranging grazing animals; however, today's animals are domesticated cattle, not the thundering self-sufficient herds of wild species that formerly made these plains one of Earth's marvels.

Figure 12.22 Short-grass prairie vegetation of the Nebraska Sand Hills. Although the tall-grass prairies have been completely transformed by humans, vast areas of short-grass prairie remain because of their low and unpredictable precipitation, which makes them hazardous for agriculture. • **What obviously would have been different if a photograph had been taken here 200 years ago?**

Desert

Eventually, lack of precipitation can become too severe even for the hardy grasses. Where evapotranspiration demands greatly exceed available moisture throughout the year, as in Köppen's desert climates, either special forms of plant life have evolved or the surface is bare. Plants that actively combat low precipitation are present, equipped to probe deeply or widely for moisture, to reduce moisture losses to the minimum, or to store moisture when it is available. Other plants evade drought by merely lying dormant, perhaps for years, until enough moisture is available to assure successful growth and reproduction. The desert biome is recognized by the presence of plants that are either drought-resisting or drought-evading (Fig. 12.23). In extremely dry deserts, only a few plants can survive and ground cover is much more uncommon (Fig. 12.24).

Plants that have evolved mechanisms to combat drought are known as *xerophytes*. They

Figure 12.24 *The absence of ground cover in this photograph of the Namib Desert in Africa is a clear indication of the extremely low annual rainfall experienced in this region. The animal pictured in the photograph, a gemsbok (oryx antelope), indicates exotic water and vegetation nearby.*

are perennial shrubs whose root systems below ground are much more extensive than their visible parts or that have evolved tiny leaves with a waxy covering to combat transpiration. They may have leaves that are needlelike or trunks and limbs that photosynthesize like leaves or that have expandable tissues or accordionlike stems to store water when it is plentiful (the succulent cacti). They may be plants that can tolerate excessively saline water or shrubs that shed their leaves until sufficient moisture is available for new leaf growth. The nonxerophytic vegetation consists mainly of short-lived annuals that germinate and hurry through their complete life cycle of growth—leaf production, flowering, and seed dispersal—in a matter of weeks when triggered by moisture availability. Like other species, these ephemeral plants also require days of a certain length, so they appear only in particular months; therefore, the month-to-month and year-to-year variation in form and appearance of desert vegetation is enormous. Animals of the deserts are primarily nocturnal to avoid the searing heat of the daytime, and many have evolved long ears, noses, legs, and tails which allow for greater blood circulation and cooling. The similar life-forms and habits of the different plant and animal species found in the deserts of widely separated continents are a remarkable display of repeated evolution to insure survival in similar climatic settings.

Figure 12.23 *Ocotillo in bloom in the Shavers Valley of the Mojave Desert, California.* • **What drought-resistant adaptations by this ocotillo plant can be observed easily?**

Arctic and Alpine Tundra

Proceeding upward in elevation and poleward in latitude, we finally come to regions in which the growing season is too brief to permit tree growth.

Nearing the poles, we enter a vast realm dominated by subfreezing temperatures and thin snow cover much of the year, so that the ground is frozen to depths of hundreds of meters. Only the top 36 to 60 centimeters (15–25 in.) thaw during the short summer interval. Still, vegetation survives here and in fact forms a nearly complete cover over the surface. Such vegetation must be equipped to tolerate frozen subsoils (permafrost), icy winds, a low sun angle, summer frosts, and soil that is waterlogged during the short growing season. The result is **tundra**—a mixture of grasses, flowering herbs, sedges, mosses, lichens, and occasional low-growing shrubs. Most of the plants are perennials that produce buds close to or beneath the soil surface, protected from the wind. Many of the plants show xerophytic adaptation—such as small, hard leaves—in response to extreme physiologic drought resulting from wind stress. This is particularly true in areas of alpine tundra, where extended periods of waterlogging of soils are uncommon. The effect of wind is evident from the fact that the less-exposed valleys within the tundra region are often occupied by coniferous woodlands.

In a band of varying width reaching across northern Alaska, Canada, Scandinavia, and northern Russia, several types of tundra are recognized: *bush tundra,* consisting of dwarf willow, birch, and alder, which grow along the edge of the coniferous forest; *grass tundra,* which is hummocky and water-soaked during the summer (Fig. 12.25); and *desert tundra,* in which expanses of

bare rocks may be covered by colorful lichens. In a few ice-free valleys of Antarctica, only desert tundra occurs.

Alpine conditions are not exactly like those in the arctic latitudes, for the deeper snow cover of the high mountains prevents the development of permafrost and the summer sun results in considerably more evaporation. However, many high areas are swept clean of snow by wind and are thus extremely exposed, supporting only desert tundra. Microclimate becomes an important control of vegetation because of the varying exposures to sun and wind. Nevertheless, the short growing season and severe wind stress produce an overall plant community similar to that in the arctic regions (Fig. 12.26).

Animals of the tundra obviously must cope with long periods of extreme cold as well as darkness. Many animals hibernate through the long Arctic winters (or the cold and windy Alpine winters); others, such as the caribou of Alaska and Canada, migrate into the boreal forest to escape the extreme cold exacerbated by Arctic winds. Year-round residents, such as the polar bear and musk ox, must have extremely large fat reserves around their large chests, in addition to extremely efficient fur. Many insects, such as the ubiquitous mosquito, cope with the extreme climate by emerging from eggs in the spring, maturing and laying eggs, and dying within one short Arctic summer. Enough eggs survive, buried under insulating snow, to continue the cycle in the following year.

Figure 12.25 Tundra fireweed in bloom during the brief arctic summer in Alaska. The photograph was taken near Atigun Pass in the Brooks Mountain Range.
• **Why might alpine tundra be of more use to humans than arctic tundra?**

Figure 12.26 Close examination of the tundra biome of the mountainous western United States reveals a mixture of grasses, sedges, and rushes. • Why does all vegetation, including the occasional shrub or stunted tree, grow so close to the ground?

Marine Ecosystems

The living organisms of the ocean can be divided into three groups according to where or how they live in the ocean. The first group is called **plankton.** Plankton is made up mostly of the ocean's smallest—usually microscopic—plants **(phytoplankton)** and animals **(zooplankton).** These tiny plants and animals float freely with the movements of ocean water, are true "drifters," and form the basis of the oceanic food chain. The second group is composed of the animals that swim in the water. This group is called **nekton** and includes fish, squid, marine reptiles, and marine mammals such as whales and seals (Fig. 12.27). The third group is comprised of the plants and animals that live on the ocean floor. This group is called the **benthos.** It includes corals, sponges, and many algae; such burrowing or crawling animals as the barnacle, crab, lobster, and oyster; and attached plants such as turtle grass and kelp.

Life in the ocean depends on the sun's energy and on the nutrients available in the water. The phytoplankton are the most important link in the ocean food chain. They take the dissolved

Figure 12.27 Elephant seals basking in the nutrient-rich and cool upwelled waters off the San Benitos Islands, Mexico. • To which group of ocean organisms do these elephant seals belong?

nutrients in the water and, through the process of photosynthesis, produce oxygen and foods needed by zooplankton and by the smallest nekton. Phytoplankton are the only food source for these animals, which in turn form the food source for larger carnivorous fish and marine mammals. These in turn are prey for still larger animals.

The marine *food chain* just described is part of a full *food cycle* in the ocean, for through the excretions of animals and the decomposition of both plants and animals in the ocean, chemical nutrients are returned to the water and are again made available for transformation by phytoplankton into usable foods. The phytoplankton also play an essential role in the production of oxygen for Earth's atmosphere.

The uneven distribution of nutrients in the oceans and the fact that sunlight can penetrate only to a depth of about 120 meters (400 ft), depending on the clarity of the water, means that the distribution of marine organisms is also variable. Most organisms are concentrated in the upper layers of the ocean, where the most solar energy is available. In deep waters, where the ocean floor lies below the level to which sunlight penetrates, the benthos organisms depend on whatever nutrients and plant and animal detritus filter down to them. For this reason, benthos

animals are scarce in deep and dark ocean waters. They are most common in shallow waters near coasts, such as coral reefs and tide pools, where there is sunlight and a rich supply of nutrients and where phytoplankton and zooplankton are abundant as well.

The waters of the continental shelf have the highest concentration of marine life. The supply of chemical nutrients is greater in waters near the continents, where nutrients are washed into the sea from rivers. Marine organisms are also concentrated where deeper waters rise (upwelling) to the surface layers, where sunlight is available. Such vertical exchanges are sometimes the result of variations in salinity or density. A similar situation occurs where convection causes bottom layers of water to rise and mix with top layers, as is the case in cold polar waters and in midlatitude waters during the colder winter months. Marine life is also abundant in areas where there is a mixing of cold and warm ocean currents, as there is off the northeastern coast of the United States (Fig. 12.28).

During the 1977 dives of the manned submersible vessel *Alvin* off the East Pacific Rise, scientists for the first time observed abundant sea life on the floor of the ocean at depths of over 2500 meters (8100 ft). It was previously presumed that these cold (2°C/46°F), dark waters were a

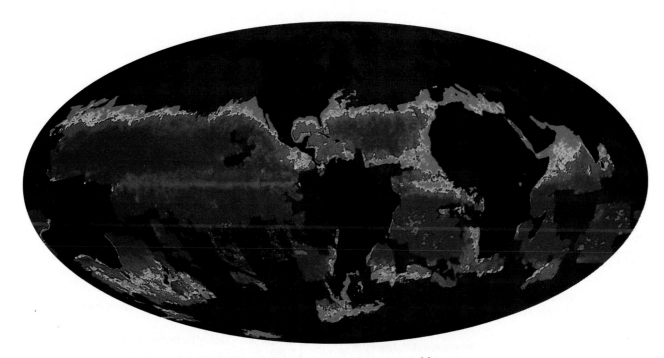

Figure 12.28 Distribution of chlorophyll-producing marine plankton mapped by space observations of ocean color. Purple to yellow to red: increasing chlorophyll concentration.

virtual biological desert. The undersea volcanic mountain range produces vents of warm, mineral-rich waters, which nourish bacteria and large colonies of crabs, clams, mussels, and giant 3-meter (10-ft) red tube worms. This entirely new deep-ocean ecosystem, which exists without the benefit of sunlight, has caused scientists to re-think old theories about the ocean and its chemistry. This unusual "chemosynthetic" vent community has more recently been observed at several other oceanic ridge sites in the Pacific and Atlantic Oceans.

Define and Recall

abiotic	biomass	mosaic	savanna
autotroph	primary productivity	matrix	tall-grass prairie
heterotroph	secondary productivity	patch	short-grass prairie
herbivore	habitat	corridor	arctic tundra
carnivore	ecological niche		alpine tundra
omnivore	generalist	tolerance	
detritivore		ecotone	phytoplankton
decomposer	natural vegetation	biome	zooplankton
trophic structure	plant community		nekton
food chain	succession	liana	benthos
trophic level	gap	galeria forest	
food web	climax community	taiga	

Discuss and Review

1. Give reasons why the study of ecosystems is important in the world today. Give several examples of natural ecosystems within easy driving distance of your own residence.

2. What are the four basic components of an ecosystem?

3. How can an organism belong to more than one trophic level? Give an example.

4. How does the biomass at one trophic level usually compare in weight with that at the next level?

5. What is the difference between primary and secondary productivity? Which is most important to the ecosystem?

6. What factors are most critical in affecting the net primary productivity of a terrestrial ecosystem?

7. Why are generalists in the majority in ecosystems?

8. In what ways has the original theory of succession been modified?

9. How do the terms mosaic, matrix, patch, corridor, and ecotone relate to each other and to a vegetation landscape?

10. What are the important environmental controls of ecosystems? Which are the most important on a worldwide basis?

11. What are the four major types of Earth biomes?

12. What important climatic characteristics are related to each of the major biomes?

13. Describe a true tropical rainforest. How does such a forest differ from jungle?

14. What are the distinctive features of chaparral vegetation? What climatic conditions are associated with chaparral?

15. What are the major differences between tropical and midlatitude forests?

16. What conditions of climate or soil might be anticipated for each of the following in the middle latitudes: broad-leaf deciduous forest; broad-leaf evergreen forest; needle-leaf coniferous forest?

17. What factors contribute to the development and maintenance of savannas?

18. How have xerophytes adapted to desert climatic conditions?

19. What are the chief characteristics of tundra vegetation? What adaptations to climate does tundra vegetation make?

20. Which link in ocean food chains is most important?

21. The highest concentrations of marine life (Fig. 12.28) are found in which parts of the oceans? Why?

Consider and Respond

1. Arrange the following organisms into six logical trophic levels in order from the first to the last.
   (a) owl; fungi; vulture; grass; bobcat; rabbit
   (b) herbivorous zooplankton; small fish; phytoplankton; shark; bottom bacteria; large fish

2. Refer to Table 12.2. Based on the means, which is more important for terrestrial ecosystem productivity between the Equator and the polar regions: annual temperatures or the availability of water? Use examples from the table to explain your choice.

3. What broad groups might you use if you were to classify Earth's vegetation on the basis of latitudinal zones? Why, do you suppose, did your textbook authors choose not to do this, electing instead to identify broad groups based on major terrestrial ecosystems (biomes)?

4. Refer to Figure 12.15. There is considerable variation in the natural vegetation of the United States between 30° and 40° north latitude. Describe the broad changes in vegetation that occur within that latitudinal band as you move from the east coast to the west coast of the United States.

5. Examine the climatic variation (see Fig. 10.4) within the same latitudinal band described in statement 4. Does there appear to be a relationship between natural vegetation and climate? If so, describe this relationship.

CHAPTER

13

SOILS AND SOIL DEVELOPMENT

CHAPTER PREVIEW

▶ Soil serves as an outstanding example of the integration of Earth's subsystems.

Which subsystems are involved? Why is soil such a good example of interaction between subsystems?

▶ Soil water is essential to vegetation, partly because it is the means by which plants receive the dissolved nutrients required for growth.

Why is gravitational water such an effective solution agent? How is capillary water important during periods of drought?

▶ Fertility of a soil depends on many factors, and a soil that is fertile for one crop may not be for another.

What factors determine a soil's fertility? How are acidity and alkalinity related to soil fertility? How is the nature of the vegetation involved?

▶ On a global scale, climate exerts the major influence over soil formation.

How do temperature and precipitation affect the development of soils? How does climate affect soils indirectly through vegetation? Through the nature of the moisture regime?

▶ Soils are most productive when no one soil-forming regime is dominant.

Why is this so? What often results when natural vegetation is removed or sufficient time permits a single soil-forming process to proceed unchecked?

▶ Although the National Resources Conservation Service (NRCS) soil classification system was devised so that soil scientists and agriculturalists could identify differences among soils in limited areas, it is still of major importance to physical geographers.

How is the system important to geographers? Why is there a general correlation among the distributions of NRCS soil orders, climatic classes, and vegetation types on a worldwide scale?

▶ Soils are among the world's most important, least understood, and most widely abused natural resources.

What determines a "natural resource"? Why is there so much ignorance concerning soils? How are soils abused? What can be done about it?

▲ Pineapple plants growing in the andisols of Maui, Hawaii.

I n an urban society, few people outside the fields of science and agriculture give much thought to their nation's heritage of varied, fertile soils. This is a serious oversight in an age of environmental awareness and concern. Composed primarily of weathered minerals and varying amounts of water, oxygen, and organic materials, soil covers most land surfaces with a fragile mantle that, along with water and air, ranks as an indispensable resource.

Soil is a dynamic natural body capable of supporting a vegetative cover. It contains chemical solutions, gases, organic refuse, and an active fauna. The complex physical, chemical, and biological processes that take place among the various components of soil are an integral part of its dynamic character. As an active body, soil responds to changes in climate (especially to temperature and moisture), to land surface configuration, to vegetative cover and composition, and to animal activity. The result is that soil evolves and matures as it changes in response to any alterations in its environment. Thus soil serves as an outstanding example of the integration of Earth's subsystems.

The formation of soils depends on a large number of factors. But the dominant influence of climate on soils is unmistakable when viewed on a worldwide scale. The climate-soil relationship and the association of soils with climate-controlled vegetation were both obvious as climatic regions were considered in Chapters 10 and 11.

Principal Soil Components

What actually makes up soil? What does the scoop of a bulldozer contain when it shovels up a load of soil? What does it take to support Earth's varied vegetation?

Inorganic Materials

Soil is made up of both insoluble mineral material—that is, minerals that will not dissolve in water to form a solution—and soluble minerals or chemicals in solution. The most common minerals found in soils are combinations of the most common elements of Earth's crust: silicon, aluminum, oxygen, and iron. In fact, most of the known chemical elements are found in soils in some form. Some of these occur in chemical combinations; others are found in the air and water that are also part of soil. We have already learned that a large number of these elements are necessary to sustain the flora and fauna of Earth's ecosystems. Important among these, in addition to the four elements listed previously, are carbon, hydrogen, nitrogen, sodium, potassium, zinc, copper, and iodine.

The elements and chemical compounds that form a part of soil material come from many sources. Some are derived from the weathering of underlying rocks or from accumulations of loose sediments. Others enter in solution in water or as a part of the chemical structure of water itself. Still others are part of the air found in soils or are derived from organic activity, some of which helps to disintegrate rocks, release gases, or create new chemical compounds.

Because plants need many chemical elements for growth, knowledge of a soil's mineral and chemical content is necessary to determine its productive potential. Frequently we can rectify a deficiency in a specific element through fertilization and thereby increase the productivity of a given soil.

Soil Water

Plants need air, water, and minerals to function, live, and grow. They depend upon the soil in which they are rooted for much of their supply of these necessities (Fig. 13.1). Soil water is not pure water; rather, it is a solution bearing traces of many soluble nutrients. Not only does soil water supply the moisture necessary for the chemical reactions that sustain life, but it also provides the nutrients in such a form that they can be extracted and used.

The original source of soil moisture is, of course, precipitation. When precipitation falls on the land surface, it is either absorbed or it runs off downslope to a stream that eventually channels it into a larger body of water. The water absorbed by the soil washes over and through various soil materials, dissolving some of these materials and carrying them through the soil.

Water is found in soil in different circumstances. The water that percolates downward through the soil, pulled by the force of gravitation, is called **gravitational water.** Gravitational water moves downward through the spaces between the individual soil particles and the clumps of soil toward the *water table*—the level below which all the spaces between the soil particles and clumps are filled with water. As a consequence, the water cannot percolate any further.

As we might expect, the amount of gravitational water in a soil is related to several conditions, including the amount of rainfall that has occurred, the length of time since the last rainfall, the ease with which the water moves through the soil, and the amount of space available for water storage.

Gravitational water functions in the soil in several ways (Fig. 13.2). First, as gravitational water moves down through a soil, the water takes with it the finer particles from the upper layer of soil. This removal of soil components from the topsoil is called **eluviation.** Eventually, as the gravitational water percolates downward, it begins to deposit some of the clay and silt particles it removed from the topsoil. Such deposit in the subsoil of soil components is called **illuviation.** Gravitational water therefore serves as an agent

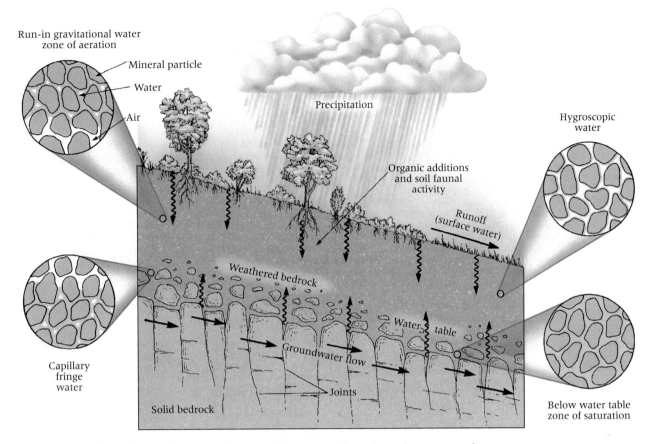

Figure 13.1 *The relation of soil to environmental factors.* • How does the source of gravitational water differ from the source of capillary water?

of transportation and mixing as it moves soil particles from one level to another. The result of eluviation is that the texture of the topsoil becomes coarser as the finer particles are removed. Consequently, the topsoil's ability to retain water is reduced, while illuviation enhances the subsoil's water retention capability. This process may be carried to such extremes that the subsoil becomes extremely dense and compact, forming a clay **hardpan.**

Gravitational water also affects the chemical composition of a soil and, as a result, its color, texture, structure, and ability to provide plant nutrients. As gravitational water moves downward, it dissolves the soluble inorganic soil components and carries them into the deeper levels of the soil, perhaps to the zone where all open spaces are saturated. This depletion of the nutrients in the upper soil is called **leaching.** When extensive, as it is under conditions of heavy precipitation, leaching can rob a topsoil of all but the most inert substances.

The processes of leaching and eluviation are a major cause of the characteristic stratification found in soils. The upper portions are composed of coarse material and are somewhat impoverished in soluble nutrients. Both fine material and some of the substances dissolved from the upper soil come to rest in the lower portion, which becomes dense and sometimes is strongly colored by accumulated iron compounds.

Some soil water is held to the surface of the individual soil particles and soil clumps by surface tension (the same property that causes small droplets of water to form rounded beads instead of spreading out in a thin film). This soil water, called **capillary water,** serves as the storage supply of water for plants. Capillary water can move in all directions through soil, its migration determined by its tendency to move from areas with more water to areas with less. Thus, during the periods between rainfalls when there is no gravitational water flowing through the soil, capillary water can move upward or horizontally to supply plant roots with the necessary moisture and dissolved nutrients.

When capillary water moves upward in a soil toward the surface, it carries with it minerals from the subsoil. If this water is evaporated in the upper soil layers, the minerals are left behind as

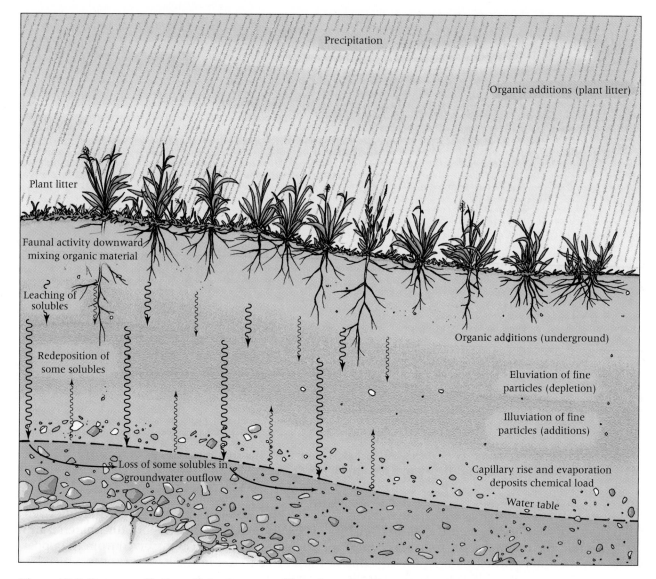

Figure 13.2 *Processes affecting soil development.* • **How does deposition by capillary water differ from deposition (illuviation) by gravitational water?**

alkaline or saline deposits in the topsoil. Such deposits can be detrimental to the plants and animals existing in the soil. Lime deposits formed in this way may produce a cement-like layer, called *caliche,* which, like clay hardpan, prevents the downward percolation of rainwater.

Soil water is also found in a very thin film, invisible to the naked eye, which is bound to the surface of all soil particles. This water, called **hygroscopic water,** is held by strong electrical forces to the surfaces of the soil particles. Because it does not move through the soil, it cannot supply plants with the water they need.

Soil Air

A large part of soil, in some cases as much as 50 percent, is made up of the voids between individual soil particles and between clumps or aggregates of soil particles. When not filled with water, these spaces are filled with air. Soil air is much like the air of the atmosphere above Earth's surface, though it is likely to have less oxygen, more carbon dioxide, and a fairly high relative humidity because of the presence of capillary and hygroscopic water.

For most of the microorganisms and plants that live in the soil, soil air supplies the oxygen and carbon dioxide necessary for life processes. Thus, the problem with water-saturated soils is not so much the excess water itself but the fact that, with all the pores filled with water, there is no supply of air. It is because of the lack of air, then, that many plants and animals find it difficult to survive in water-saturated soils.

Organic Matter

In addition to various chemical compounds (both soluble and insoluble), air, other gases, and water, soil also contains organic matter. The decayed remains of plant and animal material, partially transformed by bacterial action, are called **humus.** Humus is important to soils in several ways, but it is most important for its role as a catalyst in the chemical reactions by which plants extract nutrients from the soil and for its role in restoring minerals to the soil. Humus also improves soil structure, making it more workable and increasing its capacity to retain water. Humus serves, too, as a source of food for the enormous variety of microscopic organisms that live in soil.

Living organisms in soil range from countless microscopic bacteria to good-sized earthworms, rodents, and other burrowers. Many of these animals are useful in the development and enrichment of soils. They are important in the creation of humus from inert plant litter and in mixing organic material deeper into the soil. In addition, we cannot ignore the chemical and mechanical roles of the plants and their root systems, which are an integral part of the soil system.

Each of the soil constituents is important in determining the characteristics of a particular soil variety. Soils vary from place to place both locally and over Earth, as do the proportions of their constituents. For example, soils in midlatitude grasslands normally have a very high proportion of organic debris or humus; those in deserts are baked dry, have very little water, and are rich in soluble constituents such as lime and salt; tropical soils have a noticeably high iron and aluminum oxide content. Knowledge of a soil's water, mineral, and organic components and their proportions can help determine its potential productivity and what the best use for that particular soil might be.

Characteristics of Soil

Soils have many physical properties that are useful in describing and differentiating among them. The most important include color, texture, structure, acidity or alkalinity, and capacity to hold and transmit water and air.

Color

Color, if not the most important physical attribute of soil, is at least the most visible. Many laypersons who know next to nothing about the constituents of soil or its formation processes are aware of the variations in color that exist from place to place. The well-known red clay of Georgia is not far from Alabama's belt of black soil. Soil colors vary from black to brown to reds, yellows, grays, and near-whites, and each of these colors offers a clue to the physical and chemical characteristics of the particular soil.

For example, humus or decomposed organic matter is black or brown, and soils that are high in humus content are generally black or dark brown. As the humus content of soil decreases, because of either low organic activity or leaching, the color gradually fades to light brown or gray. A large proportion of humus in a soil usually indicates that the soil is highly fertile, for humus acts as a catalyst in the complex chemical reactions that allow plants to obtain nutrients from the soil. For this reason dark brown or black soils are spoken of as *rich* and are considered superior soils. It should be noted, however, that this is not always the case, for there are some black or dark brown soils with little or no humus content that get their dark color from other factors.

The red and yellow colors of soils are usually due to the presence of iron compounds. In moist climates a light gray or white soil indicates that iron has been removed and oxides of silicon and aluminum are present, while in dry climates the same color usually indicates a high proportion of salts.

Soil colors are useful in providing clues to the physical and chemical characteristics of soils, as well as in making the job of soil differentiation easier. But of course, color alone does not answer all of the important questions about a soil's qualities or potential for use.

Texture

Soil texture varies according to the size of the particles that make up the soil. In **clay** soils, the particles have diameters that are less than 0.002 millimeters (soil scientists universally use the metric system). The particles of **silt** soils are defined as being between 0.002 and 0.05 millimeters. **Sandy** soils have particles with diameters between 0.05 and 2.0 millimeters. Individual soil particles with diameters larger than 2.0 millimeters are regarded as inert gravel or rock fragments and technically are not soil particles. Since no soil is made up of particles of uniform size, the proportion of particles in various size ranges determines the texture of the soil.

For example, a soil composed of 50 percent silt-sized particles, 45 percent clay, and 5 percent

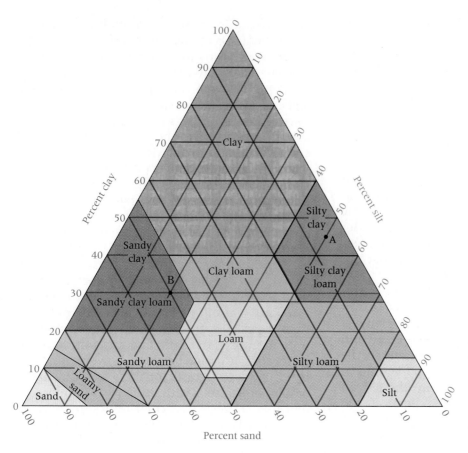

Figure 13.3 *The texture of any soil can be represented by a dot on this diagram. Texture is determined by sieving the soil to determine the percentage of particles falling into the size ranges for clay, silt, and sand.*

sand would be considered a silty clay. A triangular diagram (Fig. 13.3) has been developed to show different classes of soil texture and the percentage ranges of each **soil grade** (as sand, silt, and clay are called) within each class. Point A within the silty clay class represents the example just given. A second soil sample (B) that is 20 percent silt, 30 percent clay, and 50 percent sand would be referred to as a sandy clay loam. The **loam** soils, which occupy the central areas of the triangular diagram, are those in which no one of the three grades of soil particles dominates over the other two. It is interesting to note that the loam soils are those best suited to the support of plant life.

Soil texture is important because it helps determine the capacity of a soil to retain moisture and air, both of which are necessary for plant growth. Soils with a greater proportion of larger particles are well aerated and allow water to pass through (or infiltrate) the soil more quickly—sometimes so quickly, in fact, that plants are unable to make use of the water. Clay soils present the opposite problem: They transmit water very slowly, become waterlogged, and are deficient in air. An important part of cultivation is the aera-

tion of soil. This is accomplished by plowing, disking, harrowing, or shoveling the soil in order to open up its structure and allow it to breathe.

Structure

In most soils the individual mineral particles aggregate into larger distinctive masses or clumps, known as **peds,** which give the soil a particular structure. The structure of a soil is an important factor in its workability. Structure also influences a soil's **permeability** (the rate at which fluids such as water pass through the soil) and its **porosity** (the amount of space in a soil that may contain fluids). Permeability, which is usually greatest in sandy soils, and porosity, which is usually greatest in clayey soils, control soil drainage as well as the amount of moisture available to plants and organisms living in the soil. As a further complication, soils of similar textures may have different structures. Consequently, one may be more productive than another.

Soil structure may be influenced by such outside factors as the moisture regime and the nature of the nutrient cycle by which plants and the soil constantly interchange chemicals, keeping certain ones in the system while others are leached away. We have all seen the structural change in certain soils from when they are wet to when they have been baked dry in the sun. Soil structure can also be influenced by such human actions as plowing, cultivation, irrigation, and even fertilization. The addition of certain fertilizers, as well as the existence of lime or decayed organic debris in a soil, affects structure through chemical means that encourage first the clumping of individual soil particles and then the maintenance of those groups or clumps. Excess sodium and magnesium work the other way, causing clay soils to be structureless "glue" when wet, or concretelike when dry. On the other hand, the development of a definite structure is hindered by the absence of some of the smaller soil particles. It is for this reason that sandy beaches and deserts, made up as they are of the larger soil particles, have no apparent soil structure. This also explains why some soil has more structure below the layers closest to the surface, for the smaller particles of the topsoil have been removed to lower layers by water traveling down through the soil.

Scientists have classified soil structure according to the various forms assumed. These range from columns, prisms, and angular blocks, to nutlike spheroids, laminated plates, crumbs, and granules (Fig. 13.4). For our purposes it is most important to know merely that both massive and fine structures are less useful than aggregates of intermediate size and stability, which permit ideal drainage and aeration.

Acidity and Alkalinity

What makes a soil fertile or infertile is the many complex chemical processes and exchanges that take place in soils and plant systems. The general nature of the soil chemistry is usually expressed as the degree to which the soil departs from chemical neutrality toward either acidity or alkalinity (baseness).

Soil acidity or alkalinity is important because it helps determine the availability of nutrients to plants and ultimately controls plant growth. Plants receive virtually all of their necessary nutrients in solution. That is, a plant is unable to absorb nutrients unless they are dissolved in liquid. However, when the soil moisture lacks some degree of acidity, the soil water has little ability to dissolve minerals and release their nutrients. As a result, even though the nutrients are in the soil, plants may not have access to them. To correct this alkalinity, which is more common in arid soils than in any others, and to make the soil more productive, a farmer can flush the soil by irrigating under conditions of good drainage.

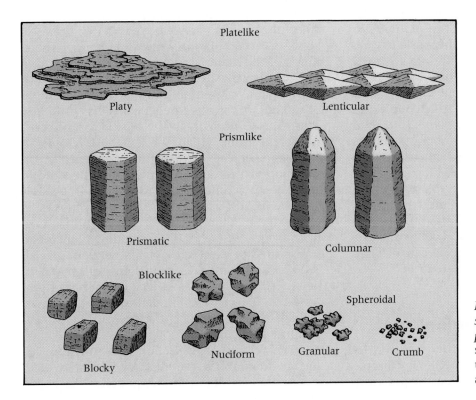

Figure 13.4 Classification of soil structure on the basis of soil peds. • **How does soil structure affect a soil's usefulness or suitability for agriculture?**

As might be expected, a strongly acidic soil is also detrimental to plant growth. In a soil that is too acidic, the soil moisture dissolves nutrients, which become leached away before they can be obtained by plant roots. This makes the soil still more acid and accentuates the problem. It is, moreover, a situation occurring in soils and climates that are often otherwise suitable to plant growth. Luckily it can be corrected by the addition of lime to the soil.

The acidity or alkalinity of a soil is measured on a scale of 0 to 14, called the **pH scale.** This actually is a measure of the concentration of highly reactive hydrogen ions present in the soil moisture. The scale is logarithmic, meaning that each change in a whole number on the scale represents a tenfold change in the hydrogen present. It is also inverted, meaning that the lower the pH value, the greater the amount of hydrogen present. Low pH values indicate acid soil moisture; high values indicate alkaline conditions. With increased rainfall, leaching increases, gradually replacing soil elements such as sodium (N_2), potassium (K), magnesium (Mg), and calcium (Ca) with hydrogen. This effect is enhanced by the fact that rain falling through the atmosphere is slightly acidic ($H_2O + CO_2 = H_2CO_3$ [carbonic acid]). The result of these processes is that desert soils tend to be alkaline whereas humid soils tend to be acidic.

Soil scientists have shown that most complex plants will grow only in soils whose pH is between pH 4 and pH 10. The optimum pH for plant productivity varies with the plant itself. Vegetation has evolved around the world in a variety of climates and is in equilibrium with the soil environment in which it is native. Thus barley can tolerate alkaline soils, but camellias and rhododendrons prefer more acid conditions. Those who bring a diversity of plants into their gardens must be prepared to produce soils of varying pH values for their specimens. This is achieved through **fertilization.**

In addition to affecting plant growth through the availability of nutrients, the acidity or alkalinity of a soil also affects the microorganisms functioning in the soil. Like plants, the microorganisms are highly sensitive to a soil's pH, and each has its optimum situation.

Soil Profile Development

Soils begin to develop when either rocks or deposits of loose material are colonized by simple plant and animal life. Once the organic processes of life and death begin to take place among mineral particles or disintegrated rock, differences begin to develop from the surface down through the soil parent material. This vertical differentiation comes about originally from such simple factors as the gradual accumulation of organic matter at the surface and the removal of fine particles and dissolved matter from the top layers by water percolating downward, followed by the deposition of these materials at a lower level. As climate, vegetation, animal life, and steepness of slope affect soil formation over time, this vertical differentiation becomes more and more apparent. Often, especially in middle latitudes, fully developed soils exhibit a vertical zonation into distinct layers or **horizons** that are distinguished by their different physical and chemical properties.

In most soils in the middle latitudes there are several generally recognizable layers. At the surface in humid regions with sufficient vegetation and moderate rates of organic decomposition is the O horizon. This is a layer of loose organic debris and raw humus. Immediately below this layer is the A horizon. Typically this horizon is a dark horizon due to a concentration of organic matter and decomposed humus. It is often referred to as "topsoil." Immediately below this horizon in some soils is a lighter colored horizon termed the E horizon, for the eluvial processes that dominate. The lighter color is the product of enhanced leaching, and can even become white in humid midlatitude soils when all of the colorful soil bases are removed. Below this is a zone of accumulation, the B horizon, where much of the material removed from the A and E horizons is deposited. Except in soils that have a high organic content and in which there is a lot of mixing, the B horizon generally has little humus. The C horizon is the weathered **parent material** from which the soil is developed—fragments of the bedrock directly beneath, or transported and deposited material. The C horizon does not reflect the movements of matter and the organic activity in the higher zones. The lowest layer, sometimes called the R horizon, is unchanged bedrock or unmodified material transported to the site by water, wind, or glacial action. Particular horizons in some soils may not be as well developed as others, and certain horizons may be missing altogether. In other cases definite vertical differentiation can be noted within a particular horizon.

The vertical cross section of a soil from its surface down to the parent material from which it is formed is called the **soil profile** (Fig. 13.5).

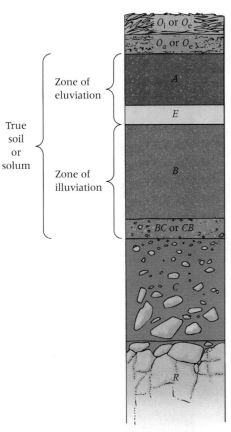

O_i or O_e — Loose leaves and organic debris

O_a or O_e — Partly decomposed organic debris

Zone of eluviation

A — Topsoil; dark in color; rich in organic matter

E — Zone of intense leaching or eluviation

True soil or solum

Zone of illuviation

B — Zone of accumulation
CaCO$_3$ or CaSO$_4$ in mollisols
deeper colored zone of
maximum accumulation
of colloids

BC or CB — Transition to C

C — Partly weathered parent material

R — Regolith or rock layer

Figure 13.5 *Soils are categorized by the degree of development and the physical characteristics of their horizons.* • **In which soil profiles of Figure 13.6 is it easiest to recognize the different layers or horizons?**

The differences among the infinite variety of soils that exist are apparent in an examination of their profiles. For this reason soil scientists have grouped and classified soils in large part on the basis of differences exhibited in the soil profiles and their horizon development (Fig. 13.6).

Factors Affecting Soil Formation

Many agents and influences are involved in the formation of soil. Because of the variations that exist among soil components, as well as the effects wrought by the changing character of the various agents and controls of soil formation, no two soils are identical in all their characteristics.

Perhaps it would be useful at this point to draw a parallel between soil and the atmosphere, as both are dynamic systems acted upon by many interrelated agents. The atmosphere is made up of various gases, water vapor, dust, and so on; likewise, soil is composed of organic and inorganic materials, water, and air. We note changes in the state of the atmosphere from time to time and place to place by noting changes in the elements of weather. Variations in soil appear as lateral changes in soil characteristics or in the vertical development of the soil profile. Weather patterns are controlled by the climatic factors in much the same way that soil characteristics are affected by the agents and controls of soil formation that we are now going to examine. The factors controlling the formation and distribution of different types of soils are parent material, organic processes, climate, land surface configuration, and time. Of these, the parent material is distinctive as the raw material. The other factors influence the type of soil that is formed from this raw material.

Parent Material

All soil is derived from the weathered fragments of rock material. If this weathered material accumulates *in place* through rock decay—that is, through the physical and chemical breakdown of the bedrock directly beneath the soil—we refer to the weathered fragments as **residual parent material.** If the rock fragments from which a soil is formed have been carried to the site by streams, waves, winds, or glaciers to form a new deposit, this mass of sediment, which will eventually develop a surface soil, is called

Oxisol, central Puerto Rico

Vertisol, Lajas Valley, Puerto Rico

Ultisol, North Carolina Piedmont

Alfisol, southern Michigan

Spodosol, northern New York

Inceptisol, northern Alaska

Figure 13.6 Soil profiles representing a variety of climatic and other soil-forming factors.

transported parent material. It is the development and action of organic matter through the life cycles of living things that are primarily responsible for differentiating soil from its fragmentary rock source or parent material, which will always be present beneath it.

Parent material is one of the agents or factors of soil formation that help to determine a soil's characteristics. It differs from the other factors, however, because it is the original substance with which the whole process of soil formation

begins. The parent material is the raw material on which the processing factors of climate, organic activity, surface configuration, and time are imposed to manufacture a soil.

Parent material varies in the degree to which it influences the characteristics of the soil derived from it. Some parent materials, such as sandstone, which contains mainly the extremely stable mineral quartz, are far less subject to weathering and change than others. Soils that develop from these parent materials demonstrate a high

Mollisol, central Iowa

Mollisol, southeastern South Dakota

Mollisol, eastern New Mexico

Mollisol, eastern Colorado

Aridisol, southern New Mexico

Aridisol, central Nevada

level of similarity with their parent source. On the other hand, some chemically reactive bedrock materials are easily weathered, and the soils that develop from them, like those on transported material, are apt to show a greater correlation with soils of similar climate than with those of similar parentage but different climate. Thus, it is common to find that one or more of the other soil-forming agents or controls may have a far greater influence on the soil than does the parent material. In fact, on a worldwide ba-

sis, climate and the associated plant communities cause stronger and larger variations in soil characteristics than do parent materials. The differences among soils based on variations in parent material are visible on a local level, however, and are of more than casual interest to the soil scientist and agriculturist.

The influence of a soil's parentage is affected by time. As a soil evolves and matures, the influence of parent material on its characteristics declines. Thus a younger or less mature soil,

Figure 13.7 *Despite strong leaching under a wet tropical climate, Hawaiian soils remain high in nutrients because their parent material is of recent volcanic origin.* • **What other parent materials provide the basis for continuously fertile soils in wet tropical climates?**

provided the other agents such as climate are similar, will show more similarity to its parent material than will a more mature one.

Both residual and transported parent material affects the soil that develops from it in very specific ways. First, the chemicals and nutrients available to the plants and animals living in a soil are derived from the soil's parent material (Fig. 13.7). Thus, a parent material deficient in calcium will produce a soil deficient in calcium. Its natural fauna and plant cover will be of a type that requires little calcium. The artificial addition of lime (calcium carbonate) can correct such a deficiency. Likewise, a parent material rich in aluminum, such as granite or basalt, will produce a soil also rich in aluminum. In fact, the main source of metallic aluminum is the bauxite ore found in tropical soils, where it has been concentrated by the intense leaching that readily removes the other bases.

The size of the particles that result from weathering of the parent material is a prime fac-

tor in the determination of soil texture and structure. A rock material such as sandstone, which contains little clay and is weathered into relatively coarse fragments, will make up a soil of coarse texture. The parent material therefore exerts an important influence on the availability of air and water to a soil's living population.

Organic Activity

Plants and animals affect soil formation in many ways. The life processes of the dominant plants are as important to the soil as its microorganisms—the microscopically small plants and animals that abound in most soils.

At the most general level, the completeness of the vegetative cover affects erosion rates. A forest of any type, because it provides a protective canopy over the soil and produces a mulch of litter on the surface, keeps rain from beating on the soil surface and increases the proportion of rainfall entering the soil rather than running off its surface. Vegetation can also affect the evaporation rate. A scanty vegetative cover will allow greater evaporation of soil moisture than will thick protective vegetation. This evaporation in turn affects the movement of capillary water toward the surface. Furthermore, the nature of the plant community determines the nutrient cycles that are a part of soil formation. Certain elements are absorbed by plants and then returned to the soil after the plants die and are decomposed. These exchanges vary among types of plants, as some use more of certain chemicals than others. Soluble nutrients that are not used by plant cover are soon lost in the leaching process, which impoverishes the soil (Fig. 13.8). The roots of larger plants affect the soil structure as well by making it more porous and by absorbing water and various plant nutrients from the soil.

All parts of vegetation (their leaves, bark, branches, flowers, and root networks) contribute to the organic content of a soil when the plants die. As is logical, the organic content of soil varies with the nature of the associated plant life. A grass-covered prairie is able to supply a far greater abundance of organic matter than the incomplete surface cover found in a desert region. There is some question, however, as to whether forests or grasslands (with their thick root network) furnish the soil with greater organic content. There is no question that many of the grassland regions of the world, like the American prairies, provide some of the richest and most fertile soils for cultivation, in part because of the high amount of organic debris that is naturally present.

The process of decay, aided by bacterial action, transforms organic matter into the jellylike mass called humus. As we noted earlier, humus is important to soils in many ways. For one, it serves as the primary food supply for the microorganisms in the soil. Humus also affects soil structure by enhancing the water retention capabilities and workability of the soil. As it is further acted upon by microorganisms, humus returns to the soil organic and inorganic materials necessary for further plant life. Consequently, in most soils there is a direct correlation between humus content and fertility.

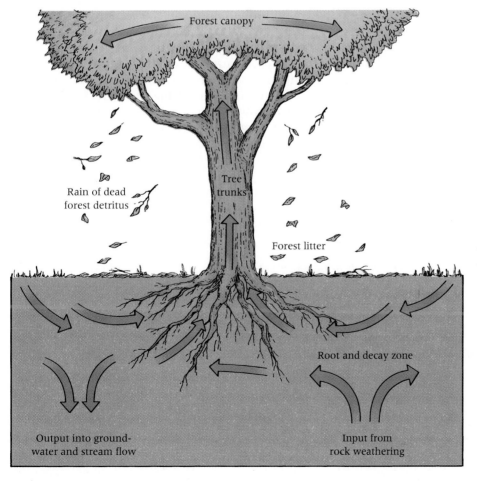

Figure 13.8 The nutrient cycle in a forest. In a wet climate, if trees do not take up soluble nutrients in the soil, they are flushed away in groundwater and lost permanently. The more demanding the forest vegetation, the richer the resulting soil. Pines are notoriously undemanding of soluble bases; consequently, in pine forests bases are lost by leaching and are not replaced by the vegetation, resulting in soil deterioration. • In what way does the destruction of tropical rainforests offer a notorious illustration of the nutrient cycle?

Bacteria, which are microscopic life-forms, are perhaps the most important of the microorganisms that live in the soil and contribute to its formation. Bacteria feed on the organic matter and humus of the soil and by this process break down the debris of living things into their organic and inorganic components, allowing the formation of new organic compounds all of which are then available for the promotion of further plant life.

It is difficult to estimate the number of bacteria, fungi, and other microscopic plants and animals that live in the soil, though some have suggested as many as one billion per gram of soil. Whatever the number, it is enormous. It is no surprise, then, to learn that the activities and remains of these microorganisms, minute though they are individually, add considerably to the organic content of a soil.

In addition to the microscopic animals already referred to, earthworms, nematodes, ants, termites, wood lice, centipedes, burrowing rodents, snails, and slugs stir up the soil, mixing mineral components from the lower levels with organic components from the upper portion. Earthworms are especially important in soil formation because they take soil in, pass it through their digestive tracts, and excrete it in casts. The process not only helps to mix up the soil, but also changes the texture, structure, and chemical quality of the soil. In the late 1800s, Charles Darwin estimated that earthworm casts produced each year would equal as much as 10 to 15 tons per acre. As for the number of earthworms themselves, sampling suggested that the weight of the earthworms beneath a pasture in New Zealand equaled the weight of the sheep grazing above them.

Climate

The discussions of climatic regions in Chapters 10 and 11 have clearly demonstrated that, on a world scale, climate is a major factor in soil formation. There are, of course, many variations among soils that are apparent on a smaller-than-global basis. The differences that are apparent at a more local level show the influence of such other factors as parent material, land surface configuration, vegetation type, and time.

Temperature clearly affects the activity of soil microorganisms. This activity in turn affects the rate of decay and decomposition of organic matter. In the hot equatorial regions the great activity of soil microorganisms precludes any thick accumulation of organic debris or humus. Figure 13.9 shows that the amount of partially decomposed organic matter and humus in a soil increases as we move into the middle latitudes from the equatorial zone. In the Köppen mesothermal C and microthermal D climates, the activity of soil microorganisms slows enough to allow the accumulation of rich layers of decaying organic matter and humus. However, moving further poleward into colder regions, the combination of retarded microorganism activity and limited plant growth results in the accumulation of only thin layers of undecomposed or partially decomposed organic matter.

Temperature also affects the rates of the chemical reactions that take place in soil, many of which make available nutrients necessary for plant growth. Chemical activity tends to increase and decrease directly with temperature. As a result, the parent material of soils in the hot equatorial regions is transformed to a far greater de-

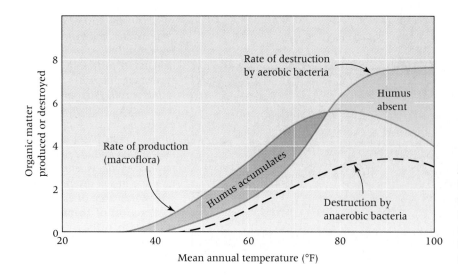

Figure 13.9 *Relationship of temperature to production and destruction of organic matter in the soil.* • **What range of mean annual temperatures is most favorable for the accumulation of humus?**

gree by chemical means than is the parent material in colder zones.

Temperature affects soil indirectly through its influence on the nature of vegetation that develops in a particular region. We know that particular vegetation associations are adapted to certain temperature regimes. The soil in which this vegetation grows often reflects the character of the plant cover as a result of nutrient cycles that keep both the vegetation and the soil in chemical equilibrium.

Moisture conditions affect the development and character of soils more clearly than any other factor. Without precipitation—and, consequently, soil water and the chemicals dissolved therein—plant life is impossible. The absence of plant life greatly diminishes the organic content and thereby the fertility of a soil.

We have already discussed the effects of both gravitational and capillary water on soil structure, texture, color, and development. As precipitation is the original source of all soil water (disregarding the minor contribution of dew), the amount of precipitation received by a soil affects the rate and degree of leaching, eluviation, and illuviation that occur, and thereby the rate of soil formation and horizon development.

When considering the effect of precipitation on the quantity and movement of soil water, we should note that the evaporation rate is a factor as well. Thus, salt and gypsum deposits from the

upward migration of capillary water are more extensive in hot, dry regions—like the southwest United States, where the evaporation rate is high—than in colder, dry regions (Fig. 13.10).

Just as temperature affects soil development indirectly through its effects on vegetation, so too does the moisture regime of a region. Where the warm season is dry, the soils are alkaline. When the summers are warm and extremely wet, with drier winters following, intense chemical weathering and fluctuating water tables produce a peculiar type of soil crust, known as *laterite*. This crust is common in the tropics, where it is quarried for building material (see again Fig. 10.13).

Land Surface Configuration

The slope of the land, as well as its aspect (the direction in which it faces), affects soil development both directly and indirectly. Steep slopes are generally better drained than more gentle ones. They are also subject to more rapid runoff of surface water. As a consequence there is less infiltration of water on steeper slopes, which retards the soil-forming processes. This retardation inhibits the development of mature layered soils, sometimes to the extent that no soil at all will develop over the parent material. In addition, rapid runoff on steep slopes often erodes their surfaces as fast or faster than soil can develop on them. On more gentle slopes, because there is less runoff and more infiltration, more water is available for soil-forming purposes and for the support of vegetation, and erosion is not as extensive as on steeper slopes. In fact, the rate of erosion on gently rolling hills is often just enough to offset the ongoing production of soil from the underlying parent material. It is often gently rolling land that allows mature soils of the most ideal characteristics to develop.

Valley floors and flatlands often are very poorly drained. When this is the case and the water level in the soil is near the surface, gravitational water is unable to percolate downward, and capillary water may in fact move salt and alkaline substances to the surface in harmful concentrations. This condition is a constant danger in the irrigation of flatlands because irrigation tends to raise the water table. Artificial drainage ditches must be provided to lower high water tables in such instances.

The direction a slope faces affects its microclimate. Thus, a north-facing slope in the Northern Hemisphere has a microclimate that is colder and wetter than a south-facing exposure, which receives the sun's rays more directly and, as a

Figure 13.10 Harmful concentrations of chemical substances are frequently encountered in arid regions. Shown here is a surface salt crust resulting from the upward capillary rise of water from a water table that is saline and close to the surface. Evaporation at the surface deposits the salt. • **What negative soil effects can result when humans practice irrigated agriculture in arid regions?**

consequence, is warmer and drier. Local variations in soil depth, texture, and profile development result directly from such differences between microclimates.

Land surface configuration also affects the development of soils indirectly through its effect on vegetation. Where a steep slope prevents the development of a mature soil to support abundant vegetation, the poor quality of the plant cover diminishes the amount of organic debris available for the soil.

Time

We have spoken of the evolution of a soil toward a state of dynamic equilibrium with its environment. A soil is mature when it has reached such a state of equilibrium. Young soils are still in the process of alteration to achieve equilibrium. The mark of a mature soil is a well-developed and stable horizon structure. A young or immature soil, on the other hand, generally has poorly developed horizons or none at all. Very old soils may have horizons that are so well developed as to present problems. Such soils frequently contain dense pans or crusts in their *B* horizons. These may consist of eluviated clay or chemically concentrated calcium carbonate (lime), silica, or oxides of iron and aluminum. Soils on ancient flat surfaces in the tropics, where soil formation is rapid, frequently present such problems. To use old soils agriculturally, farmers may need to break up their crusts with dynamite or, less dramatically, by deep plowing.

Another effect of time is that the more mature soil is decreasingly influenced by its parent material and increasingly reflects its climatic environment and vegetative cover. In fact, as we have previously noted, the influence of climate is usually the most apparent of all controls on soil development at a global scale, provided the soil has had sufficient time to reach maturity.

The importance of time in soil formation is especially clear in the case of soils developed upon transported parent materials. Depositional surfaces are in most cases quite young and have not been exposed to the effects of the atmosphere as long as have those parts of Earth's surface that have been worn down by erosion over periods of tens of millions of years. Recent deposits of transported materials have not yet been leached of their soluble nutrients, nor has their soil developed undesirable characteristics. Deposition is occurring today in a variety of settings: on the floodplains of large rivers, where the material accumulating is known as *alluvium;* downwind from dry areas, where dust settles out of the atmosphere to form deep blankets of *loess;* and in volcanic regions that are occasionally showered by ash and veneered by lava. Ten thousand years ago, glaciers withdrew from vast areas, leaving behind a veneer of *till* or *outwash,* some of it plastered over the land surface by the advancing ice and other portions let down or washed out in great sheets as the ice melted. In terms of agricultural productivity, the world's best soils are found on young alluvium, loess, volcanic ash and lava, and certain types of glacial debris. Somewhat older deposits of similar materials are less productive. They have already been leached of vital nutrients and have developed unfavorable structures, similar to those of the soils on older erosional surfaces that constitute most of Earth's land surface.

There is no fixed amount of time that it takes for a soil to become mature. This is because of the number and variability of the factors that affect soil formation. Generally, though, it takes hundreds to thousands of years for a soil to reach maturity.

Soil-Forming Regimes

By now it should be clear that there are infinite possible combinations of the factors that function together to produce soils of all descriptions. Nevertheless, an examination of the world's soils reveals that they can be separated into a limited number of general types. The characteristics that differentiate these major types can be attributed to their different soil-forming regimes, each resulting from a combination of different processes. The differences between these soil-forming regimes are primarily the result of climatic differences and indirectly the result of differences in plant cover.

At the broadest scale of generalization, there are three primary soil-forming regimes that relate to climatic differences. These are laterization, podzolization, and calcification (Fig. 13.11). On a local scale other processes become important, but they are of lesser significance in the worldwide distribution of soils, and for that reason we will briefly mention only two of them. Soils may be formed by any one of these processes or by a combination of two or more, primarily depending on the climate but also on surface configuration, vegetation, and parent material.

Laterization

Laterization is a soil-forming process that occurs in humid tropical and subtropical climates

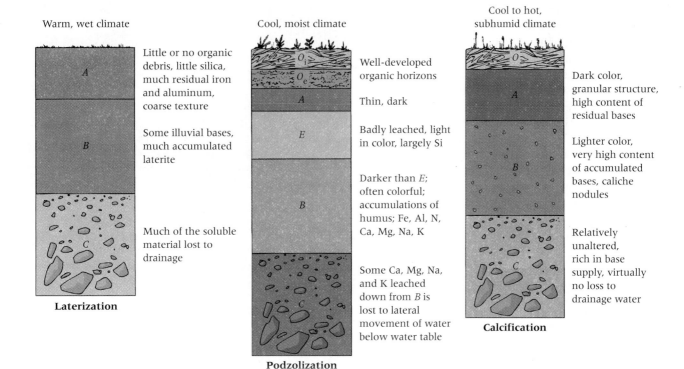

Warm, wet climate — Little or no organic debris, little silica, much residual iron and aluminum, coarse texture — Some illuvial bases, much accumulated laterite — Much of the soluble material lost to drainage — **Laterization**

Cool, moist climate — Well-developed organic horizons — Thin, dark — Badly leached, light in color, largely Si — Darker than *E*; often colorful; accumulations of humus; Fe, Al, N, Ca, Mg, Na, K — Some Ca, Mg, Na, and K leached down from *B* is lost to lateral movement of water below water table — **Podzolization**

Cool to hot, subhumid climate — Dark color, granular structure, high content of residual bases — Lighter color, very high content of accumulated bases, caliche nodules — Relatively unaltered, rich in base supply, virtually no loss to drainage water — **Calcification**

Figure 13.11 Profile development in the three major soil-forming regimes. • **How are these three generalized soil profiles related to Figure 13.9?**

as a result of the high temperatures and abundant precipitation. These effects produce rapid breakdown of rocks and decomposition of nearly all minerals. Despite the dense vegetation that is typical of these climates, little humus is incorporated into the soil because of the rapid decomposition of plant litter and enormous numbers of microorganisms in the soil. Because of the abundance of moisture, eluviation and leaching of all but iron and aluminum oxides play dominant roles in the formation of humid tropical soils. There is no *O* horizon, and fine soil particles as well as most minerals and bases are removed from the *A* horizon except for iron and aluminum compounds, which are insoluble in the soil solution primarily because of the absence of organic acids from humification. As a result, the topsoil is reddish in color, coarse in texture, and tends to be porous. In contrast to the *A* horizon, the *B* horizon has a heavy concentration of illuviated materials. Where the tropical forest vegetation remains, the soluble nutrients released in the weathering process are quickly absorbed by the vegetation, which eventually returns them to the soil where they are reabsorbed by plants. This rapid nutrient cycle prevents the total leaching away of the bases, so the soil is only moderately acidic. Removal of the vegetation permits total leaching of bases, resulting in the formation

of crusts of iron and aluminum compounds (laterites), as well as accelerated erosion of the *A* horizon.

Laterization can take place year-round because of the lack of distinct seasonal variation in temperature or precipitation. This continuous activity, and the strong weathering of the parent material, causes some humid tropical soils to develop to depths of as much as 8 meters (25 ft).

Podzolization

Podzolization occurs in its purest form in the high middle latitudes, where the climate is moist with short, cool summers and long, severe winters. The typical coniferous forest of this climate is an integral part of the process of podzolization.

Where temperatures are low much of the year, the activity of microorganisms is reduced enough that humus is allowed to accumulate; however, because of the small number of animals living in the soil, there is little mixing of this humus below the soil surface. Leaching and eluviation by a strongly acidic soil solution remove the soluble bases and aluminum and iron compounds from the *A* horizon. The remaining silica gives a distinctive ash-gray color to the *A* horizon (*podzol* is derived from a Russian word meaning *ashy*). Because most coniferous trees require a minimum amount of bases, they return little basic material to the soil. The needles they drop are chemically acidic. This contributes to the acidic quality of the soil. Indeed, it is difficult to say whether the soil is acidic because of the vegetative cover, or whether the vegetative cover is adapted to the acidic soil.

Podzolization can take place beyond the typical cold, moist climate when the parent material is highly acidic, as on the sands common along the east coast of the United States. The pine forests that can grow in such acidic conditions return acids to the soil, promoting the process of podzolization.

Calcification

The third distinctive soil-forming process is called **calcification.** In contrast to both laterization and podzolization, which require humid climates, calcification demands climates where evapotranspiration exceeds precipitation. In areas of low precipitation the air is often loaded with alkali dusts such as calcium carbonate ($CaCO_3$). When calm conditions prevail or when it rains, the dust settles across the landscape and accumulates in the soil. The rainfall is just sufficient to translocate these materials to the B horizon of soils. Over hundreds to thousands of years, the $CaCO_3$-enriched dust concentrates in the B horizons of soils, forming hard layers of *caliche* or the much thicker *calcretes* (Fig. 13.12). These accumulations can be enhanced by the upward (capillary) movement of dissolved alkaline salts in groundwater when the water table is near the surface.

Calcification becomes important in the climatic regimes where moisture penetration is shallow. The subsoil is too dry to support tree growth, and shallow-rooted grass is the primary form of vegetation. Calcification is enhanced by the fact that grass uses calcium, drawing it up from the lower soil layers and then returning it to the soil when the grass dies. The grasses and their dense root networks provide large amounts of organic matter, which typically is mixed deep into the soil by the numerous animals found there. Thus midlatitude grassland soils are rich both in bases and in humus, and are the world's most productive agricultural soils.

It is interesting that a soil transect across the United States from Illinois to Colorado shows a gradual westward *decrease* in soil humus content and thickness of the A horizon, while it shows an *increase* in the prominence of calcium carbonate at the base of the B horizon. The most

Figure 13.12 *An especially thick calcrete layer caps Mormon Mesa in southern Nevada. Note that calcification has formed a layer that measures twice the height of the truck standing above it.* • **What precipitation characteristics are associated with the calcification soil-forming process?**

productive soils are found about midway in this transect, where both humus and base status are relatively high. The deserts of the Far West have no humus, and the rise of capillary water leaves not only calcium carbonate but even more soluble sodium chloride (salt) at the surface.

Regimes of Local Importance

Two additional soil-forming processes are important enough to merit consideration. Both are characteristic of areas with locally poor drainage, although they occur under strikingly different climatic conditions. The first, **salinization,** occurs in stream valleys, interior basins, or other low-lying areas in desert regions that have high groundwater tables. The high groundwater can be the result of adjacent mountain ranges or stream flow originating outside a desert region (Fig. 13.13), but increasingly it is the result of extensive irrigation. Rapid evaporation of this water leaves behind the high concentration of soluble salts in the soil that characterizes the soil-forming regime and gradually destroys the agricultural productivity of the area. An extreme example of salinization is the fertile crescent of Mesopotamia (Iraq), where thousands of years of irrigated agriculture in the desert has led to soils too salty to cultivate today (see Fig. 13.10 for such an example). The second, **gleization,** occurs in poorly drained areas under cold and wet environmental conditions. This process is usually associated with peat bogs, where the soil is an accumulation of humus layers overlying a blue-gray layer of thick, gummy, water-saturated clay. (Un-

Figure 13.13 Salt deposits associated with high groundwater levels along the Amorgosa River in the area of Death Valley. • **What causes salinization and why is it often associated with agriculture?**

Figure 13.14 Peat has been harvested for fuel for hundreds of years in bogs such as this one in northern Scotland. Note the spade markings where peat has been removed and the spade-sized peat blocks (in the background) cut by hand and awaiting transport to a nearby farm where they will be stored, dried, and used as fireplace fuel. • **What are the similarities and differences between peat and various grades of coal?**

reduced iron in the early stages of decomposition imparts the blue-gray color to the soil.) In formerly glaciated, poorly drained regions such as northern Russia, Ireland, Scotland, and Scandinavia, the peat has long been harvested and used as an important source of energy (Fig. 13.14).

Soil Classification

Soils, like climates and other phenomena that vary spatially, can be classified. The agency in the United States that is responsible for soil classification (termed **soil taxonomy**) is the Soil Survey Division of the Natural Resources Conservation Service (NRCS), a branch of the Department of Agriculture. As with any classification system, the methods and categories are continually being updated.

Soil classifications are published in **soil surveys.** Soil surveys show the distribution of soils in a region, usually at the county level. These documents are available for most parts of the United States, and are very useful reference sources for factors such as soil fertility, irrigation, and drainage.

The NRCS Soil Classification System

The current classification system is based on the nature of the soil horizons. The largest classification of soils is the **soil order,** of which eleven are recognized. Subdivisions of the soil orders are: suborders, great groups, subgroups, families, and series. More than 10,000 soil series have been recognized in the United States. Some soil orders reflect regional climatic conditions because they have developed during the Holocene under rela-

tively stable climates. During this interval, the dominant soil-forming processes have produced similar types of soil horizons for regions with similar climates and vegetation cover. However, other soil orders reflect the recency or type of parent material present, and thus the distribution of these soils does not conform to climatic regions.

An entirely new vocabulary of names, derived from root words of classical languages like Latin, Arabic, and Greek, was developed to label the many different soils in the system. The names, like the system, are both precise and consistent since they were chosen to describe the particular characteristics that distinguish one soil from another, thus assuring that a soil will be classified in the proper category.

When examining a soil for classification under the NRCS system, particular attention is paid to certain horizons or layers that characterize the soil. Some of these horizons are situated below the surface **(subsurface horizons),** and some, called **epipedons,** are surface layers that usually exhibit the darker shading associated with the presence of organic material (humus). Examples of some of the more common horizons, which illustrate how names were chosen to represent actual soil properties, may be found in Table 13.1.

NRCS Soil Orders

Seven of the eleven soil orders largely reflect Holocene climates. They are discussed here in a

365

TABLE 13.1 Selected Common Horizons in the NRCS Soil Classification System*

Oxic horizon (from *oxygen*)
A subsurface layer that contains hydrated oxides of iron and aluminum. Found in tropical and subtropical climates at low elevations.

Argillic horizon (from Latin: *argilla*, clay)
Usually formed beneath the *A* horizon by illuviation, this layer contains a high percentage of accumulated silicate clays.

Ochric epipedon (from Greek: *ochros*, pale)
A surface horizon that is light in color and very low in organic matter or very thin.

Albic horizon (from Latin: *albus*, white)
Commonly the *A2* horizon overlying a spodic horizon, this layer is usually sandy and very light in color due to the removal of clay and iron oxides.

Spodic horizon (from Greek: *spodos*, wood ash)
Usually underlying the *A2* horizon, the spodic horizon is dark in color because of the illuviation of humus, aluminum oxides, and often, iron oxides.

Mollic epipedon (from Latin: *mollis*, soft)
Very high in content of basic substances (calcium, magnesium, potassium), the mollic epipedon is a relatively thick and dark-colored surface layer.

Calcic horizon (from *calcium*)
A subsurface horizon rich in accumulated calcium carbonate or magnesium carbonate.

Salic horizon (from *salt*)
A layer of soil material at least 6 inches thick and containing at least 2 percent salt. Common in desert basins.

Gypsic horizon (from *gypsum*)
A subsurface soil horizon rich in accumulated calcium sulfate (gypsum).

* This table includes only some of the more common horizons.

sequence that reflects increasing moisture levels. Frequent comparison of Figure 13.15 with Figure 10.4 will help to illustrate the relationship between soils and climate within the United States.

Aridisols are soils of desert regions, and they develop primarily under conditions where precipitation is less than half of potential evaporation. Consequently, most Aridisols reflect the calcification process. Where groundwater tables are high, evidence of salinization may be common. Although ordinary horizon development is weak because of lack of water movement in the soil, there is often a subsurface accumulation of calcium carbonate (calcic horizon), salt (salic horizon), or calcium sulfate (gypsic horizon). Soil humus is minimal because vegetation is lacking in deserts; thus desert soils are often light colored. Aridisols are usually alkaline, but because few nutrients have been leached, they can produce bountiful harvests if irrigated in a way to reduce the pH and excess salts. Geographically,

Aridisols are the most common soils on Earth because deserts cover such a large portion of the land surface.

Mollisols are most closely associated with grassland regions and are among the best soils for sustained agriculture. Because they are located in semiarid climates, Mollisols are not heavily leached, and thus they have a generous supply of bases, especially calcium. The characteristic horizon of a Mollisol is a mollic epipedon, a thick, dark-colored surface layer rich in organic matter from the decay of the abundant root material. Grasslands and associated Mollisols served as the grazing lands for countless herds of antelope, bison, and horses. Before the invention of the steel plow, the thick root material made this soil nearly uncultivable in the United States, and thus led to the widespread public image of the Great Plains as a "Great American Desert." Today, Mollisols support most of the grain production from domesticated grasses (wheat, rye, oats, corn, barley, and sorghum). In regions of adequate precipitation, such as the tall-grass prairies of the American Midwest, the combination of soils and climate is unexcelled for agriculture. In areas of lesser precipitation, periodic drought is a constant threat, and the temptation of fertile soils had been the downfall of many farmers prior to the advent of center-pivot irrigation.

Vertisols are typically found in regions of strong seasonal precipitation, such as tropical wet and dry climates. In the United States, they are most common where clay-rich soils are produced by the weathering of shale. The combination of clayey soils in a wet and dry climate leads to the drying of the soil and consequent shrinkage and formation of deep cracks during the dry season, followed by the expansion of the soil during the wet season. The constant shrink-swell action disrupts horizon formation to the point that soil scientists often describe Vertisols as "self-plowing" soils. Anything solid constructed atop or in heaving soils, such as highways, sidewalks, and basements, may be disrupted. Vertisols are dark in color, are high in bases, and contain considerable organic material derived from the grasslands or savanna vegetation with which they are normally associated. Although they harden when dry and become gummy and difficult to cultivate when swollen with moisture, Vertisols can be agriculturally productive. The use of modern mechanized farm machinery has permitted the black Vertisol belt of Texas to become one of the world's leading cotton-producing regions.

Alfisols occur in a wide variety of climatic settings in the United States. They are characterized by a subsurface clay horizon (argillic *B* horizon), a medium to high base supply, and a light-colored ochric epipedon. The five suborders of *Alf*isols reflect climate types and exemplify the hierarchical nature of the classification system: Aqu*alf*s are seasonally wet and can be found in mesothermal areas such as Louisiana, Mississippi, and Florida; Bor*alf*s are found in moist, microthermal climates such as Montana, Wyoming, or Minnesota; Ud*alf*s are common in both microthermal and mesothermal climates that are moist enough to support agriculture without irrigation, such as Wisconsin, Ohio, and Tennessee; Ust*alf*s are found in mesothermal climates that are intermittently dry, such as Texas and New Mexico; and Xer*alf*s are found in California's Mediterranean climate, which is characterized by wet winters and long, dry summers. Because of their abundant bases, Alfisols can be very productive agriculturally if the local deficiencies

are corrected: irrigation for the dry suborders, properly drained fields for the wet suborders.

Spodosols are most closely associated with the podzolization soil-forming process. They are readily identified by their strong horizon development. There is often a white or light gray *E* horizon (albic horizon) covered with a thin, black layer of partially decomposed humus and underlain by a colorful *B* horizon enriched in relocated iron and aluminum compounds (spodic horizon). Spodosols are generally low in bases and form in porous substrates such as glacial drift or beach sands. In New England and Michigan, Spodosols are also acidic. In these regions, as well as in similar regions in northern Russia, Scandinavia, and Poland, only a few types of agricultural plants, such as cucumbers and potatoes, can tolerate the microthermal climates and sandy, acidic soils. Consequently, the cuisine of these regions directly reflects the Spodosols that dominate the areas.

Ultisols, like Spodosols, are also low in bases because they develop in moist or wet regions. Ultisols are characterized by a subsurface clay horizon (argillic horizon) and are often yellow or red colored because of the accumulation of residual iron and aluminum oxides in the *A* horizon. In the eastern United States, the Ultisols are most closely associated with the old Confederacy. When first cleared of forests, these soils can be agriculturally productive for several decades. But a combination of high rainfall and the associated runoff and erosion from the fields decreases the natural fertility of the soils. Ultisols remain productive only with the continuous application of fertilizers. Today, forests cover many former cotton and tobacco fields of the old Confederacy because of the reduction in soil fertility and extensive soil erosion (see again Fig. 11.9).

Oxisols have developed over long periods of time in regions with high temperatures and heavy annual rainfall. They are almost entirely leached of soluble bases and are characterized by a thick horizon of iron and aluminum oxides (oxic horizon). The soil consists mainly of minerals that resist weathering (for example, quartz, kaolin, hydrated oxides). Oxisols are most closely associated with the humid tropics, but they also extend into savanna and tropical thorn forest regions. In the United States, Oxisols are present only in Hawaii. Oxisols are dominated by laterization and retain their natural fertility only as long as the soils and forest cover maintain their delicate equilibrium. The bases in the tropical rainforests are stored mainly in the vegetation.

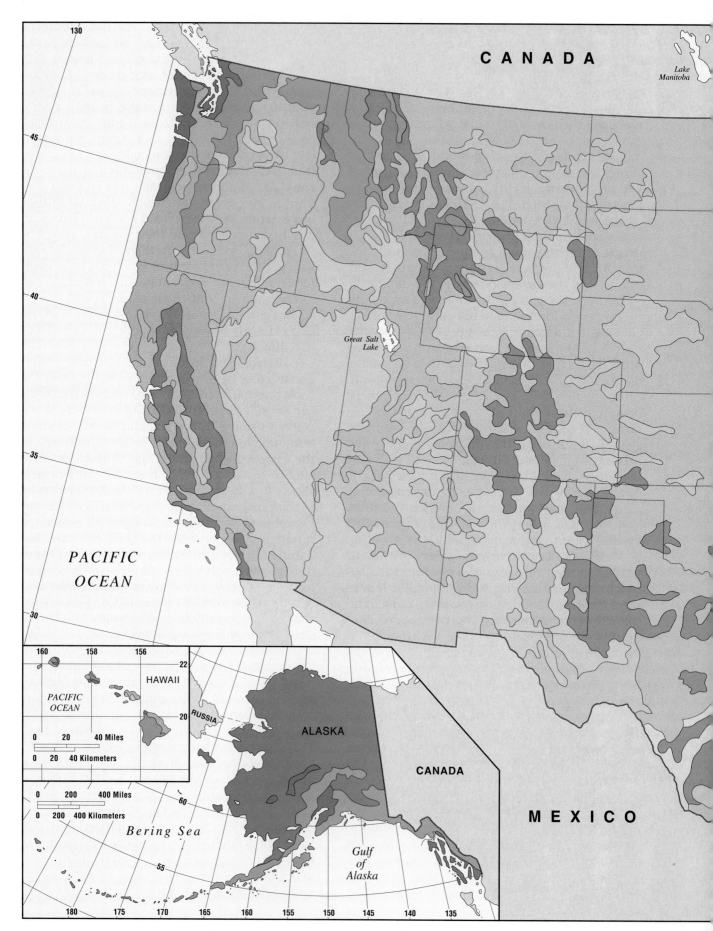

Figure 13.15 *Distribution of NRCS Soil Orders in the United States.*

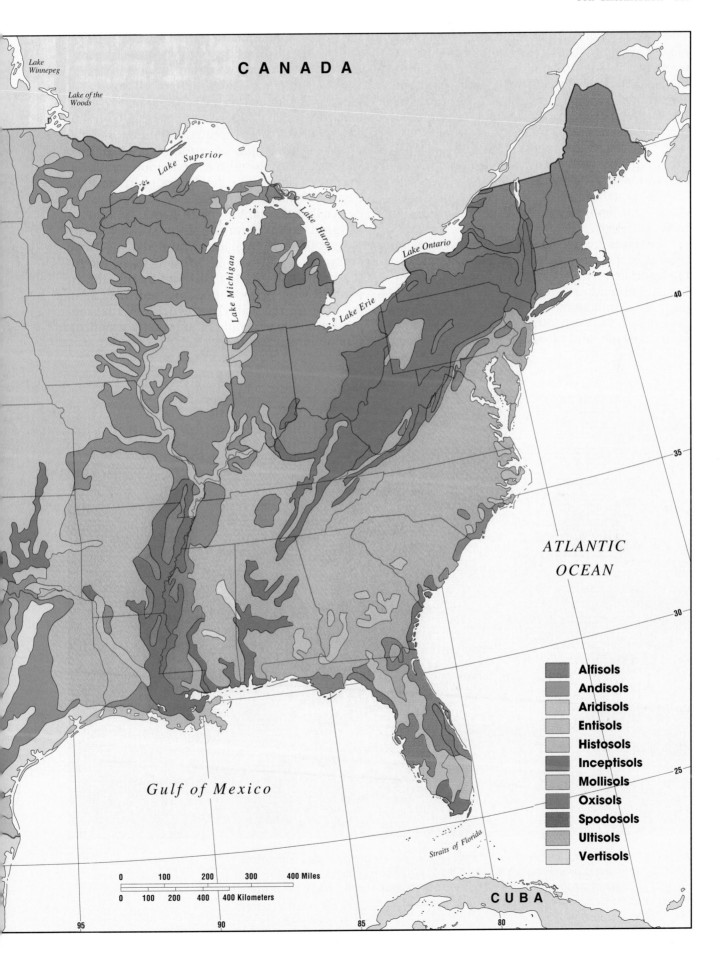

Alfisols
Andisols
Aridisols
Entisols
Histosols
Inceptisols
Mollisols
Oxisols
Spodosols
Ultisols
Vertisols

ATLANTIC OCEAN

CANADA

Lake Winnepeg

Lake of the Woods

Lake Superior

Lake Michigan

Lake Huron

Lake Ontario

Lake Erie

Gulf of Mexico

Straits of Florida

CUBA

0   100   200   300   400 Miles

0   100   200   400   400 Kilometers

When a tree dies, the bases must be recycled rapidly by epiphytes and insects before the heavy rainfall leaches them from the system. The burning of vegetation associated with "slash-and-burn agriculture" in rainforests releases the nutrients necessary for crop growth but quickly results in their loss from the ecosystem. Many tropical Oxisols that once supported lush forests are now heavily dissected and only support a combination of weeds, shrubs, and grasses.

The remaining four soil orders are limited in extent, and are not closely associated with climatic patterns:

Entisols are soils that lack horizons, usually because of their recent development. They are often associated with the constant erosion of sloping land in mountain regions or the frequent deposition of alluvium on alluvial fans or river floodplains. They can also occur in areas of heavy sand accumulation (Nebraska sand hills) where the porous and mobile substrate greatly decreases the rate of horizon development.

Inceptisols are young soils with weak horizon development. The processes of A horizon depletion (eluviation) and B horizon deposition (illuviation) are just beginning, usually because of a very cold climate, repeated floodplain inundation, or a high rate of erosion. In the United States, Inceptisols are most common in Alaska, the lower Mississippi River floodplain, and the western Appalachians. Globally, Inceptisols are especially important along the lower portions of the great river systems of South Asia, such as the Ganges-Brahmaputra, the Irrawaddy, the Chao Phraya, and the Mekong. In these areas, the Inceptisols are constantly enriched by the silts associated with periodic flooding, and they form the basis for the paddy-rice agriculture that supports millions of people in an otherwise agriculturally limited region.

Figure 13.16 *Gully erosion is one of the more spectacular examples of poor agricultural practices. It produces permanent alteration of the landscape and guarantees that the original productivity of the land cannot be regained.* • **What should have been done to prevent this example of land mismanagement?**

Histosols develop in poorly drained areas, such as swamps, meadows, or bogs, as a product of gleization. They are largely composed of slowly decomposing plant material. The waterlogged conditions of the soil deprive bacteria of the oxygen necessary to prevent accumulation of organic matter. Although Histosols may be found in low areas with poor drainage at all latitudes, they are most common in tundra areas or in recently glaciated, subpolar locations such as Canada, Ireland, and Scotland. Histosols in the subpolar latitudes are commonly acidic and only suitable for special bog crops such as cranberries. Histosols are important as the primary source of peat, which is a fuel source in regions where Histosols are common.

Andisols are the newest soil order in the classification, having been subdivided mostly from the Inceptisols. Andisols are soils that develop atop volcanic parent materials, usually volcanic ash. They often have a low bulk density as well as substantial proportions of glassy minerals and the weathering products of extrusive igneous rocks. Because many of the soils are constantly replenished by eruptions, they are often fertile compared to those in surrounding regions. Intensive agriculture atop Andisols supports the dense populations of the Philippines and West Indies. However, in tropical climates the intensive cultivation, without proper terracing of the fields, can lead to severe erosion and dissection of the landscape. In the United States, Andisols are most common on the slopes of, and downwind from, the Cascade volcanoes and, to a lesser extent, in Hawaii and Alaska.

Regardless of their composition, origin, or state of development, Earth's soils remain one of our most important and vulnerable resources. Even the word *fertility,* so often associated with soils, has a meaning that takes into consideration the usefulness of these soils to humans. Soils are fertile in respect to their effectiveness in producing specific vegetation. Some soils may be fertile for corn and others for potatoes. There are soils that retain their fertility only as long as they remain in delicate equilibrium with their vegetative cover. But in every instance the significance of the soil's fertility is only of consequence to those human beings who would use the soil resource.

It is clearly the responsibility of all of us who enjoy the agricultural end-products of farm, ranch, and orchard, or who simply appreciate the beauty of forest and field, to recognize and help protect our valuable soils. Although space does not permit a thorough examination of the problems of soil erosion, soil depletion, and land mismanagement, we should be conscious of their existence in the world today (Fig. 13.16). At the same time we should be aware that for each of these problems there are reasonable solutions (Fig. 13.17). Maintaining soil fertility and usefulness is a serious challenge to humanity and one of the essentials in our continuing struggle to protect the resources of our natural environment.

Figure 13.17 *The contour farming techniques used on this farm are excellent examples of conservation methods designed to preserve the soil resource.* • **What other soil conservation practices are often used to preserve the soil resource?**

Define and Recall

gravitational water	permeability	podzolization	Vertisol
eluviation	porosity	calcification	Alfisol
illuviation	pH scale	salinization	Spodosol
hardpan	fertilization	gleization	Ultisol
leaching			Oxisol
capillary water	soil horizon	soil taxonomy	Entisol
hygroscopic water	soil profile	soil survey	Inceptisol
humus	residual parent material	subsurface horizon	Histosol
	transported parent	epipedon	Andisol
loam	material	Aridisol	
soil ped	laterization	Mollisol	

Discuss and Review

1. Why is soil an outstanding example of the integration of Earth's subsystems?

2. Describe the different circumstances in which water is found in soil.

3. What is the effect of eluviation if carried to an extreme? What is the effect of illuviation if carried to an extreme?

4. Under what conditions does leaching take place? What is the effect of leaching on the soil and consequently on the vegetation it supports?

5. How can capillary water contribute to the formation of caliche? What is the effect of caliche on drainage?

6. How might soil air differ from air in the atmosphere? What is the effect on life when air is excluded from water-saturated soils?

7. How is humus formed? What relation does humus have to soil fertility?

8. What conclusions can you draw from the color of the soil in your area? How might color relate to fertility?

9. How is texture used to classify soils? Describe the ways scientists have classified soil structure.

10. What pH range indicates soil suitable for most complex plants?

11. What are the general characteristics of each horizon in a soil profile? How are soil profiles important to scientists?

12. What factors are involved in the formation of soils? Which is most important on a global scale?

13. How does transported parent material differ from residual parent material? List those factors that help to determine how much effect the parent material will have on the soil.

14. What are the most important effects of parent material on soil?

15. List a number of ways in which humus is important to soils.

16. How does the presence of earthworms alter soil?

17. Describe the various ways in which temperature and precipitation are related to soil formation.

18. The Bonneville Salt Flats in Utah are well known as a natural soil formation that provides a perfect surface for auto racing. How do you suppose these salt flats were formed?

19. Describe the three major soil-forming regimes.

20. Which soil orders of the NRCS Soil Classification System have the most agricultural potential? Why?

Consider and Respond

1. Refer to Figure 13.3. Using the texture triangle, determine the textures of the following soil samples.

	Sand	Silt	Clay
(a)	35%	45%	20%
(b)	75%	15%	10%
(c)	10%	60%	30%
(d)	5%	45%	50%

What are the percentages of sand, silt, and clay of the following soil textures? (Note: Answers may vary, but they should total 100 percent.)

   (e) Sandy clay   (f) Silty loam

2. Refer to Figure 13.5 and associated pages in the text.
   (a) What horizons make up the zone of eluviation?
   (b) What are two processes that occur in the zone of eluviation?
   (c) The various *B* horizons are in what zone?
   (d) Weathered parent material is the major constituent of what horizon?
   (e) Partly decomposed organic debris makes up which horizon?

3. Refer to Table 13.1.
   (a) What materials accumulate in an argillic horizon?
   (b) Which would generally be better suited for agriculture—a soil with an ochric epipedon or a mollic epipedon? Why?
   (c) What name would be given to a 7-inch-thick horizon that contained at least 2 percent salt?

4. Refer to Figure 13.15 and the world map of Figure 10.4 and use an atlas to respond to the following:

   What soil order or orders and associated climate or climates are most often found together in the Corn Belt located principally in Illinois, Indiana, Iowa, and Ohio? In the northern Great Lakes Region (Lake Huron, Lake Michigan, and Lake Superior)? In the grasslands astride the fiftieth parallel of latitude in North America? In most of Nevada, southeastern California, and southern Arizona? In the Old South between the Appalachians and the Coastal Plain?

5. Where would you rank soils among a nation's environmental resources? Give your opinion of the overall value of soils in the United States and the extent to which these soils are preserved and protected.

CHAPTER
14

EARTH'S INTERIOR, EARTH'S CRUST, AND PLATE TECTONICS

CHAPTER PREVIEW

▶ The opposing sets of forces that affect Earth's landforms are locked in an "eternal struggle," with no clear beginning and no definite end in sight.

What does this mean? What is its significance?

▶ Gradation and tectonism act simultaneously and not in isolation from one another.

How is this important to an understanding of the opposing processes? How does it affect an explanation of the origin of a particular landform?

▶ We have less real knowledge of Earth's interior than we have of outer space.

Why? What does this tell us about specific priorities? Will this change?

▶ Only eight elements account for almost 99 percent of Earth's crust by weight, and the most common minerals are combinations of these same elements.

What does this suggest about mineral classification? What does it suggest concerning the most common rocks in Earth's crust?

▶ Knowledge of the origin, nature, and structure of bedrock is important to physical geographers because all three properties may influence the character and appearance of the surface environment.

What surface characteristics does bedrock influence? How can topography be influenced by bedrock? How are humans affected by bedrock properties?

▶ The theory of plate tectonics has had an all-encompassing relationship to the Earth sciences, similar to that which evolution has to the life sciences.

What is the importance of such a theory? Why is science so concerned with theory?

▶ For a scientific theory to be fully accepted, it is not sufficient merely to describe what happens; the scientist must also explain *how* it happens.

How is this statement related to "continental drift"? What processes explain plate tectonics?

▲ Mt. Evans in the Colorado Rockies.

The study of the origin and development of landforms, called *geomorphology,* is an important subdivision of physical geography. Landforms are the surface expression of the lithosphere and owe their development to processes originating from both within and outside Earth's surface. We have already distinguished among the large subsystems that make up the total Earth system: the atmosphere, the hydrosphere, the biosphere, and the lithosphere. Up to this point we have primarily examined the atmosphere and, to a lesser extent, the biosphere. In the following nine chapters, we will concentrate our attention on the lithosphere and, particularly in Chapter 21, on the interactions between the lithosphere and the oceans. It will become increasingly obvious as we proceed that all the subsystems are interrelated—we cannot focus on one without noting the influences upon it of all the others. For example, as we identify typical landforms and the forces that have produced them, as well as how they are currently being changed, we will quickly recognize the roles of the atmosphere, hydrosphere, and biosphere in helping to create them.

Landforms and Geomorphic Processes

Though humans have been and continue to be an effective force in altering the shape of the land, the work of humans is minute and recent in contrast to the natural forces that have been active over the whole surface of Earth throughout the immense span of geologic time. Thus, in order to understand and explain such different landforms as Death Valley, Mount St. Helens, the Grand Canyon, Cape Cod, the Mississippi Delta, and the glacial plains of Illinois, we must understand something of the nature and arrangement of the materials composing them, or their structure. It is also necessary to understand the processes by which these materials have been shaped into distinctive landforms (known as *geomorphic processes*). Finally, because the geomorphic processes are evolutionary, we must know what stage of development a particular landform has achieved. Knowing the stage helps us understand what a landform was like in the past and what it will probably be like in the future.

At one time scientists believed that Earth's landscapes were created in great cataclysms—that the Grand Canyon, for example, split open one violent day and has remained that way ever since. This theory is called **catastrophism.** For almost two centuries, however, geographers, geologists, and other Earth scientists have accepted the theory of **uniformitarianism**—the idea that forces are operating today in the same fashion as they have for millions of years. They may, however, vary in their geographic location and geologic intensity.

The formation of landforms is a worldwide continuous struggle between the forces that elevate, disrupt, and create inequalities in Earth's surface and the processes that wear down, fill in, and tend to level the surface. The processes that roughen Earth's surface gain their energy from Earth's interior and are called **tectonic processes** (from the Greek: *tekton*, carpenter, builder). The leveling processes are known as **gradational processes.** Gradation includes the fragmentation and chemical breakdown, or **weathering,** of Earth materials, which makes them removable; **mass wasting**—the movement downslope of the weathered materials due to the pull of gravity; the actual removal, or **erosion,** of the materials and their transportation by agents such as wind, water, and ice; and their **deposition** at lower elevations. The opposing forces, one roughening and the other smoothing Earth's surface, are constantly in conflict.

Though we will consider first the tectonic processes and then the processes of gradation, the shape of the land is actually the result of continuous interaction between the internal tectonic forces and the external gradational processes. The Grand Canyon, for example, is the result of the lifting of a plateau by tectonic processes, combined with the gradational action of water on the land, which has carved out the canyon itself. As a ranger at Grand Canyon National Park once suggested, "Not only is the knife (the river) cutting down through the slab of butter (the rocks), but the slab of butter is also being pushed up against the knife."

When we see the Grand Canyon today, we must remember that this is only an early stage in its evolution. In time the water will remove more and more of the uplifted land, the shape of the canyon will change, and perhaps additional uplifting will occur as well. In some regions, there is a seemingly endless cycle of uplift and wearing down, of elevation and leveling. Elsewhere the tectonic forces have long remained dormant, the leveling process has dominated, and gradation is almost complete.

The Structure of Earth

Scientists still know relatively little about the interior of Earth. Increased knowledge of its structure, its composition, and the processes going on within will help answer questions about crustal motion, earthquakes, volcanic eruptions, the formation of mineral deposits, and the origins of the continents and of Earth itself.

Earth has a radius of about 6400 kilometers (4000 mi), but scientists have been able to penetrate, and examine directly, only its thin outer skin. Through direct means such as mining and drilling, we have gained a very limited knowledge of Earth's interior. The lure of gold has taken prospectors to a depth of 3.2 kilometers (2 mi) in South African mines, and drilling for oil and gas has penetrated to several times that distance. These explorations have been helpful in providing knowledge about Earth's uppermost layers, but they are really only scratches in the surface of Earth.

Most of what we have learned about the interior structure and composition of Earth has been deduced through indirect means. The most important tool that scientists have used to gain such indirect knowledge is the behavior of vibratory earthquake waves and other shock waves (usually generated by manmade explosions) as they pass through Earth. These vibratory or seismic waves can be recorded by an instrument called a **seismograph.**

There are two major types of seismic waves, which travel at various speeds in materials of different densities and states. These are usually labeled P (primary) waves, which travel fastest and arrive first at the seismograph recording a quake,

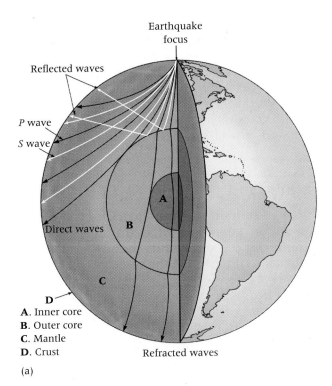

A. Inner core
B. Outer core
C. Mantle
D. Crust

(a)

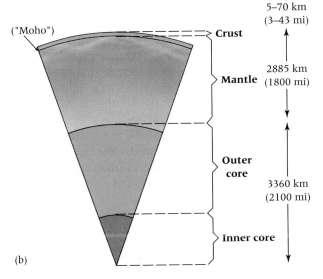

(b)

***Figure 14.1** The internal structure of Earth as revealed by seismic waves. **(a)** The presence of internal changes is revealed by the refraction of P (primary) seismic waves and the inability of S (secondary) waves to pass through Earth's liquid outer core. **(b)** Cross section through Earth's internal structural zones.* • How does the thickness of the crust compare with the thickness of the core and mantle?

and S (secondary) waves, which travel more slowly. Repeated patterns of the P and S waves on the seismograph suggest that these waves are refracted, or bent, as they meet marked changes of density in the materials within Earth's interior. The waves speed up in denser material and slow down in less dense material. Such information, supplemented by studies of Earth's magnetism and gravitational pull, suggests a series of layers, or zones, in Earth's internal structure. These zones, from the innermost to the surface, are known as the core, the mantle, and the crust (Fig. 14.1).

Earth's Core

The core forms one-third of Earth's mass and has a radius of about 3360 kilometers (2100 mi). It is believed to be composed primarily of iron. The core material is under enormous pressure—several million times the atmospheric pressure at sea level. The **outer core** is 2400 kilometers (1500 mi) thick. Because it blocks the passage of the seismic S waves that will not travel through liquids, Earth scientists assume that the outer core is molten in spite of the very high pressure that it must be under. Estimates of internal temperatures are 4800°C (8643°F) at the core-mantle boundary, increasing to 6900°C (12,423°F) at the very center of Earth.

However, the **inner core** of Earth appears to be solid iron. Scientists explain the solid state of the inner core in this way: The melting point of a material depends not only on temperature but on pressure as well. The pressure on this innermost part of Earth is so great that the inner core remains solid; that is, its melting point has been raised to a temperature above even the high temperatures found there. The outer core, on the other hand, though its temperatures are lower, is under less pressure and can exist in a molten state.

Density is estimated at 10 grams per cubic centimeter at the core-mantle boundary, increasing to 13 g/cm^3 at the center of Earth. These high core densities balance the lower density of Earth's mantle (3.3 to 5.5 g/cm^3) and crust (2.7 to 3.0 g/cm^3), and give Earth an overall density of 5.5 g/cm^3. This high density of Earth's core is one reason that scientists believe that iron is its primary component, and the density and composition of meteorites also support this theory.

Earth's Mantle

The **mantle** is about 2885 kilometers (1800 mi) thick and constitutes about two-thirds of Earth's mass. Earthquake waves that pass through the mantle indicate that this zone of Earth's interior is for the most part a dense solid, in contrast to the molten outer core that lies beneath it. Although the mantle is solid due to its extremely high temperatures and pressure, this solid mantle material is plastic and may actually "flow" a few centimeters per year. Scientists agree that the most common mineral of the mantle is probably olivine, an iron magnesium silicate. Small crystals of this mineral are common in lavas that have erupted in oceanic areas, such as Hawaii.

Despite its overall solid character, the mantle contains layers, or zones, of differing strength and rigidity. The uppermost rigid layer of the mantle, together with the crust, forms the **lithosphere.** The term *lithosphere* has traditionally been used to describe the entire solid Earth (see page 3). In recent decades, however, Earth scientists have used the term in this more precise way to describe the material of the crust and upper rigid mantle (Fig. 14.2). Located immediately beneath the lithosphere, at a depth of about 100 to 700 kilometers (62–435 mi), is a thick layer of plastic mantle called the **asthenosphere** (Greek: *asthenos,* weak). The plastic quality of the asthenosphere permits its material to move both vertically and horizontally, dragging the lighter and more rigid lithospheric plates with it. Many Earth scientists now believe that the energy for tectonic forces comes from movement within the asthenosphere produced by thermal convection currents originating deep within the mantle and heated by the decay of radioactive elements.

The upper level of the mantle, at its interface with the crust, appears to be marked by a significant change of density, or a discontinuity, indicated by an abrupt increase in the velocity of earthquake waves at this internal boundary. Scientists have labeled this zone the **Mohorovicic discontinuity,** or **Moho** for short, after the Serbian geophysicist who first detected it in 1909. The Moho does not lie at a constant depth around Earth. In fact, it tends to be the mirror image of surface topography, being deepest under mountain ranges and rising to within 8 kilometers (5 mi) of the ocean floor (see Fig. 14.2). In the early 1960s, the United States sponsored Project Mohole, which was an attempt to drill through Earth's crust to the Moho and beyond to obtain a sample of the material of the mantle. Only experimental drilling in the ocean floor, where the crust is thinnest, had been done when, in 1966, federal funds were cut off and the project had to be abandoned. During the 1980s, the former Soviet Union drilled over 13 kilometers (8 mi) deep in the Kola Peninsula near Finland in unsuccessful attempts to penetrate Earth's crust and recover samples of the mantle.

Earth's Crust

The only portion of Earth that Earth scientists have direct knowledge of is the **crust,** which forms about 1 percent of its mass. Not only are we in daily contact with the surface of this layer, but we have also been able to penetrate and sample it to depths of several kilometers. Earth's crust is the outer portion of the lithosphere and is of primary importance in understanding landforms. While Earth's deep interior (the core and mantle) are of concern to physical geography because they are responsible for, and can help to explain, changes in the lithosphere, it is the crust on which we live and which forms the ocean floors and continents. It is the rocks and debris of the crust from which soils are formed and which we penetrate in search of energy and mineral resources.

Earth's crust is less dense than either the core or the mantle. It is a thin outer layer that has been compared to an eggshell. It varies in thickness from 5 kilometers (3 mi) in the ocean basins to as much as 70 kilometers (43 mi) under some continental mountain systems. The average

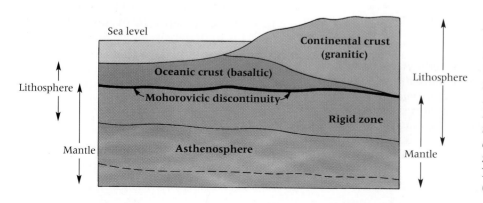

Figure 14.2 There are two distinct types of crust: oceanic and continental. Together with the rigid upper zone of the mantle, these form the lithosphere. Below the lithosphere is the plastic mantle zone called the asthenosphere. • **What two zones are separated by the Mohorovicic discontinuity (Moho)?**

thickness of the continents is about 32 to 40 kilometers (20–25 mi).

Earth's crust is relatively cold, rigid, and brittle compared to the mantle. It responds to stress by fracturing, wrinkling, and raising or lowering into domes and basins. Two layers can be distinguished in Earth's crust. The lower layer is **oceanic crust** composed of material similar to a dark-colored, fine-grained rock termed **basalt.** This layer is more basic chemically than the upper layer that is exposed on the continents and has a higher density (3.0 g/cm^3). It constitutes the deep ocean floors around Earth, and its most common minerals are compounds of silica (Si) and magnesium (Mg). Basalt itself also appears in great lava outflows on all of the continents.

The upper layer, which is the **continental crust,** is the material of the continents. It is less dense (2.7 g/cm^3), more acidic, and lighter in color than the oceanic crust. Although every known rock type of every geologic age may be found on the continental masses, the density of the continental crust is similar, on average, to the density of **granite,** a common coarse-grained rock underlying all continents. Thus, the continental crust is regarded as granitic, in contrast to the basaltic oceanic crust (see Fig. 14.2).

The Composition of Earth's Crust

The tectonic and gradational processes that work to create the variety of landforms over Earth's surface produce vastly different results from place to place. Earth's crust is composed of a variety of rocks and minerals that respond in different ways and at different rates to the Earth-shaping processes. Therefore, the physical geographer should know the different types of rocks and their primary characteristics, especially their responses to the tectonic and gradational processes.

Rocks and Minerals

A rock is an aggregate (a whole made up of parts) of mineral particles. Each mineral in a rock remains separate and retains its own distinctive properties, which together determine the properties of the rock itself. Though there are usually several different minerals in a rock, as in the case of both granite and basalt, a few rocks, such as limestone or quartzite, may be composed totally of particles of a single mineral. Rocks are the materials of the lithosphere. They are lifted, pushed down, and deformed by the tectonic processes; they are weathered and eroded by the gradational forces, to be deposited as sediment elsewhere.

Solid rock that underlies surface material is called **bedrock.** Above the bedrock is usually a layer of decomposed rock called **regolith.** Above this regolith may be soil. A trench cut into Earth's surface does not always reveal these three layers. On a mountain slope, for example, running water may remove weathered material as fast as it forms, so that the bedrock is left exposed. Such exposed bedrock is called an **outcrop.**

Mineral Classification

The most common elements of Earth's crust (and so of the minerals and, accordingly, the rocks that make up the crust) are oxygen and silicon, followed by aluminum and iron, and the bases: calcium, sodium, potassium, and magnesium. As you can see in Table 14.1, a mere eight chemical elements—out of the more than 100 known—account for almost 99 percent of the weight of Earth's crust. The most common minerals are combinations of these eight elements.

Minerals are naturally occurring inorganic substances. They are well-defined combinations of atomic elements, and each can be characterized by its unique chemical formula. Each mineral has other distinctive qualities as well: a particular color, luster, hardness, tendency to fracture, and specific gravity. Minerals are usually crystalline in nature, although this may be evident only when they are viewed through a microscope. Thus, in addition to other uniform characteristics, mineral crystals have consistent geometric forms that express their atomic structure (Fig. 14.3). The atomic elements composing each mineral are held together by electrical bonds; to be stable, each mineral must have a

TABLE 14.1	Most Common Elements in the Earth's Crust
Element	Percentage of the Earth's Crust by Weight
Oxygen (O)	46.60
Silicon (Si)	27.72
Aluminum (Al)	8.13
Iron (Fe)	5.00
Calcium (Ca)	3.63
Sodium (Na)	2.83
Potassium (K)	2.70
Magnesium (Mg)	2.09
Total	98.70

Source: J. Green, "Geochemical Table of the Elements for 1953," *Bulletin of the Geological Society of America* 64 (1953).

(a)

(b)

Figure 14.3 *The geometric arrangement of atoms determines the crystal form of a mineral.* **(a)** *Pyrite crystals, also known as "fool's gold."* **(b)** *Quartz crystals.*
• What is a mineral?

balance between positive and negative charges. Accordingly, minerals are latticelike structures held together either by covalent atomic bonding or by the electrostatic attraction between oppositely charged ions of the elements included (for example: Na^+ and Cl^- combine to produce common salt). Bonding affects the breakdown of minerals, and thereby of rocks. Minerals whose internal bonds are weakest are most easily altered. Ions may leave or be traded within their structure, producing physical changes. These characteristics of bonding are the chemical basis of weathering.

Because of the ease with which certain elements combine with a variety of others, there are many discrete families of minerals. The most active of these elements are silicon, oxygen, and carbon. Consequently, the most common mineral groupings are the silicates, oxides, and carbonates. The **silicates** are by far the largest and

most important group, constituting 92 percent of Earth's crust. They are created by the cooling of magma (a molten mass containing all the elements from which minerals form), which causes the crystallization of certain minerals at successively lower temperatures. All of these are compounds of oxygen and silica and of one or more metals and/or bases. Olivine is one of the first silicate minerals to crystallize (at high temperatures), and quartz is one of the last (at relatively low temperatures). This relationship parallels their relative stability in a rock. In the weathering of granite, which may include a variety of silicates, olivine is one of the first minerals to decompose and quartz is one of the last.

The **oxides** do not form masses of rock. Oxides formed by crystallization occur in veins, sometimes large veins (one example is magnetite, an iron ore). The more common oxides are actually the product of weathering: Oxygen that combines with other substances, such as iron or aluminum, is introduced by water entering the structure of an iron- or aluminum-bearing mineral. Oxides, such as goethite (iron rust) and some hematite, are a product of mineral alteration.

The **carbonates** are of both organic and inorganic origin. Carbon's ability to form complex compounds makes this element an important building block in nature. In combination with oxygen and calcium, carbon produces calcite, the mineral composing limestone, one of the more common rock types. Calcite added to magnesium produces the mineral dolomite, which itself forms rock masses. The carbon of limestone and dolomite is frequently derived from organic material in the form of microscopic marine organisms.

The only mineral groups in which oxygen is not an important constituent are halides and sulfides. In the halides, which may form rock masses, chlorine plays the role usually played by oxygen, combining with the base sodium to form halite (common salt). In the sulfides, which occur only as veins, sulfur acts like oxygen, combining with iron, for example, to form pyrite ("fool's gold"). Many of Earth's deep metallic mineral deposits form sulfide ores, whereas those formed near the surface have been changed to oxides by proximity to atmospheric oxygen.

The only remaining group of minerals is comprised of those consisting of single elements uncombined with any others. Some of these are rare and valuable and include gold, silver, copper, sulfur, and carbon (in the form of graphite and diamonds). Most metals, however, occur as oxides, carbonates, or sulfides—ores that must be refined at considerable cost.

Figure 14.4 *Basalt, a fine-grained extrusive igneous rock, forms these fresh lava flows on the island of Hawaii.*

Figure 14.5 *Granite, a coarse-grained intrusive igneous rock, forms the mountainous terrain of the Sierra Nevada in California.* • **What is the difference between granite and basalt?**

The Classification of Rocks

Although the number of minerals making up most of the rocks of the lithosphere is limited, they are combined in so many different ways that the variety of rock types is enormous. Nevertheless, all rocks can be categorized as one of three major types, based on their origin. These rock types are igneous, sedimentary, and metamorphic.

Igneous Rocks. **Igneous rocks** are formed when molten rock-forming material cools and solidifies. Below Earth's surface this melt is called **magma.** The igneous material with which we are most familiar is **lava,** the surface form of magma (Fig. 14.4). This molten material is spewed forth by volcanoes at the surface at temperatures of as much as 1090°C (2000°F).

Molten material that emerges at Earth's surface and solidifies is called **extrusive** igneous rock, or volcanic rock (after Vulcan, the Roman god of fire). If rising magma does not break through to the surface but solidifies *within* Earth below the surface, the resulting internally cooled and crystallized rock is known as **intrusive** igneous rock. When intrusive igneous masses have cooled deep beneath the surface, they are sometimes referred to as **plutonic** rocks (after Pluto, the Roman god of the underworld). Generally, intrusive igneous rocks are more resistant to the gradational processes than are extrusive igneous rocks.

Different igneous rocks have little in common except their method of formation: crystallization and solidification from molten material. Igneous rocks vary in chemical composition, tendency to fracture, texture, crystalline structure, and the presence or absence of layering. Never-

theless, they may be grouped or classified in terms of their crystal size or texture as well as their chemical composition.

Extrusive rocks, and some intrusive rocks that have pushed close to the surface, undergo rapid cooling and solidification. This forms fine-grained igneous rock, such as basalt or rhyolite (see Fig. 14.4). Only small crystals are produced under conditions of rapid cooling, because there is little time for crystal growth prior to solidification. In some instances cooling and degassing are so rapid that the resulting rock has a glassy texture, as in obsidian (volcanic glass).

Plutonic rocks cool far more slowly because the surrounding masses of rock retard the loss of heat from the molten magma. Slow cooling allows more time for larger crystal formation prior to solidification. Rocks formed in this manner are coarse-grained, with crystals often as long as three centimeters or more. Granite and gabbro illustrate this coarse texture (Fig. 14.5).

The chemical composition of igneous rocks varies from acidic (rich in light minerals, especially silica) to basic (low in silica and rich in heavy minerals, such as compounds of iron and magnesium—the *ferromagnesium minerals*). Granite, an acidic, coarse-grained, plutonic rock, has the same chemical and mineral composition as rhyolite, a fine-grained extrusive rock. Likewise, gabbro is the coarse-grained, plutonic version of the fine-grained basalt, both having been formed from the same chemically basic magma.

Many igneous rocks are jointed. **Joints** are cracks within a rock structure, which may develop as an igneous rock shrinks in volume during its formation. Basalt is famous for its hexagonal columnar jointing. Devil's Postpile in California and Devil's Tower in Wyoming are formed of huge, hexagonally jointed basalt

(a)

(b)

Figure 14.6 *Hexagonal columnar jointing forms basalt columns. (a) Devil's Postpile National Monument, California. (b) Devil's Tower National Monument, Wyoming.* • **What does the devil have to do with these unusual landforms?**

columns (Fig. 14.6). Some jointing may also be caused by regional tectonic stress.

Sedimentary Rocks.   As their name implies, **sedimentary rocks** are derived from accumulated sediments—unconsolidated mineral material that has been deposited as a result of gradational processes. After having been deposited in layers, the sediment is compacted by the pressure of the material above it, expelling water and reducing pore space. Cementation also occurs when silicon dioxide, calcium carbonate, or iron oxide accumulates in the remaining pores between the particles of sediment. Together the processes of compaction and cementation transform (lithify) the sediment into a solid, coherent layer of rock. The sedimentary materials (cobbles, pebbles, sand, silt, or clay) are debris particles eroded from any previously existing rock and transported and

deposited on a riverbed, on the surface as a beach or sand dune, on a lake bottom, or on the ocean floor. Sedimentary rocks formed from such debris are called **clastic** rocks (Latin: *clastus,* broken). By far the most common clastic rocks are those formed on the ocean floors from slowly sinking and settling continental and marine sediments.

Common clastic rocks are conglomerate, sandstone, siltstone, and shale. Conglomerate is a solid mass of cemented boulders, cobbles, pebbles, or gravel, often with sand filling in the spaces between large particles. It is a hard rock relatively resistant to weathering. Most sandstone is formed by the cementation of fine grains of quartz. It is usually granular, porous, hard, and also resistant to weathering. However, its qualities are largely determined by the cementing material. When sandstone is cemented by substances other than silica (such as calcite or iron oxide), it is more easily weathered. Siltstone is similar to sandstone, though much finer textured, being formed of the much smaller particles of silt. Shale is produced from the compaction of very fine-grained sediments, especially clay. The resulting sedimentary rock is finely bedded, smooth-textured, and nonporous. It is also brittle and easily cracked, broken, or flaked. These clastic sedimentaries may be further classified as marine or terrestrial sedimentaries, depending on their origin. Thus, marine sandstones may have been formed originally in coastal zones, while some terrestrial sandstones have originated in desert conditions on land.

Sedimentary rocks may also be formed from the remains of organisms, both plants and animals. Such rocks are **organic sedimentaries.** Coal, for example, was formed through the accumulation and compaction of decayed vegetation in acid, swampy environments where water-saturated ground prevents oxidation. The initial transformation of such material produces peat, which, when subjected to burial and further compaction, is lithified to produce coal. Unlike other types of rock deposits, most of the world's greatest coal deposits originated during a geologic time some 300 million years ago known as the Pennsylvanian Period (called the Carboniferous Period in Europe—see Table 14.2).

Other organic sedimentary rocks have been formed from organisms growing in lakes and seas. As the skeletal remains of shellfish, corals, and microscopic drifting organisms called plankton sank to the bottom of such water bodies, they became cemented and compacted together to form *shell and coral limestones.* Frequently, however, these calcium carbonate ($CaCO_3$) materials

Figure 14.7 *The White Cliffs of Dover along the English Channel are made of limestone from the skeletal remains of miscroscopic marine organisms.* • **What special type of limestone is this very white variety?**

were dissolved in the water and later were precipitated or separated out of the water. In this way secondary limestones were formed, including the chalk seen in the White Cliffs of Dover along the English Channel (Fig. 14.7).

Limestone, therefore, may vary from a jagged and cemented complex of visible shells or skeletal material to a smooth-textured rock of building-stone quality. Where magnesium is an important constituent along with calcium carbonate, the rock is called dolomite.

Other mineral salts that have been precipitated out from evaporating seawater or lake basins have formed a variety of sedimentary deposits often useful to humans. These include gypsum (or alabaster in its purest form), halite (common salt), and borates, which are important in hundreds of products from fertilizer, fiberglass, and pharmaceuticals to detergents and photographic chemicals.

Because sediments accumulate over time in layers, sedimentary rocks display distinctive layering, or **stratification.** For example, a period of time during which coarse-textured sands have been deposited may be followed by a period during which fine clays are laid down. At other times organic deposits may predominate. Each different type of sediment may become a different rock type, resulting in differentiated **strata,** or beds. The **bedding planes,** or boundaries between the differing sedimentary strata, indicate changes in the nature of the deposits but no real break in the sequence of deposition (Fig. 14.8). Where a

marked mismatch occurs between beds, the surface of contact between the rocks is called an **unconformity.** This indicates an interruption in the sequence of deposition during which erosion may have removed earlier deposits before deposition was renewed.

Within some sedimentary rocks, especially sandstones and shales, even finer "microbedding" may occur. Thus we may find **lamination planes,** parallel to the major bedding planes, in laminated sandstones or shales (Latin: *lamina,* thin leaf). Lamination planes are evidence of minute variations in deposition with each tide on a shore, for example, or during flood periods of a river's flow. Another form of microbedding is called **cross-bedding,** characterized by a pattern of zigzags at an angle with the main bedding, often reflecting shifts of direction by waves along a coast or winds over desert dunes (Fig. 14.9).

Many sedimentary rocks are *jointed* or cracked during the formation process or when subjected to stress as part of Earth's crust. Limestone especially may be massively jointed, and the impressive slabs of rock at Arches National Park, Utah, owe their form to vertical joints in great beds of sandstone (Fig. 14.10).

Structures such as bedding planes, lamination planes, and joints are important in the formation of different physical landscapes, since these structures are weak points in the rock, where weathering and erosion can attack. Joints can allow water to penetrate deeply into some rock masses, causing them to be removed at a faster rate than the surrounding rock.

Figure 14.8 *Different types of sedimentary rocks may be laid down at various time periods. Each different type is considered a strata separated by bedding plains, such as these at Grand Canyon National Park, Arizona.* • **Where would the youngest strata be located in this photo?**

Figure 14.9 Cross-bedding in sandstone at Zion National Park, Utah. • This zigzag pattern usually indicates what type of condition during deposition?

Metamorphic Rocks. The word **metamorphic** means "changed form," and that is just what metamorphic rocks are. Enormous heat and pressure deep in Earth's crust, often associated with tectonic activity, can totally reconstitute rock, changing it to a new product. Usually the resulting rock is harder and more compact, has a crystalline structure, and is more resistant to weathering than before.

Metamorphism occurs most commonly where crustal materials are forced down to lower levels by tectonic processes or where molten magma is rising through the crust, giving off heat and also solutions and gases that can modify the rock nearby. The major effect of metamorphism is to either wholly or partially fuse or melt the affected rock, so that it can be deformed, or *flow* slightly, but without becoming molten magma. This process causes mineral crystals to recrystallize with an orientation that reflects the direction of flow. Such metamorphism produces rocks whose minerals are segregated in wavy bands, an effect known as **foliation.** Where the banding is very fine, the individual minerals have a flattened, "platy" structure, and the rocks tend to flake along these bands. Such rocks are called *schists*.

Where the bands are broad, the rock is extremely sound and is known as *gneiss* (pronounced "nice"). Coarse-grained rocks such as granite generally recrystallize as gneiss, whereas fine-grained rocks may produce schists. Some shale produces a more massive metamorphic rock known as *slate,* which exhibits a tendency to break apart or cleave along flat surfaces.

Rocks originally composed of one dominant mineral are not foliated by metamorphism. Limestone is merely reconstituted into much denser *marble*; the impurities in the limestone produce

a beautiful variety of colors (Fig. 14.11). Silica-rich clastic rocks such as sandstone are fused in solid sheets of quartz, known as *quartzite*. Quartzite is brittle but almost inert chemically. Thus, it is virtually immune to chemical weathering and commonly forms cliffs and rugged mountain peaks after less resistant rocks have been chemically altered and removed by erosion.

The Rock Cycle. Like landforms, rocks do not remain in their original form indefinitely but instead are always in the process of transformation. There is, in fact, a cycle of rock formation that has no obvious beginning and no predictable end (Fig. 14.12). When magma is cooled, igneous rocks are formed. Igneous rocks can return to a molten condition (magma) through the addition of heat, or they can be changed into metamorphic rock through the application of heat, pressure, and/or chemical action, or their weathered particles may form the basis of sedimentary rocks. Sedimentary rocks can be formed from the weathered particles of all three basic rock types. Metamorphic rocks can be created out of either igneous or sedimentary rock. In addition, metamorphic rocks can be heated sufficiently to become magma.

Plate Tectonics: A Unifying Theory

Scientists in all disciplines are constantly searching for broad explanations that shed light on the

Figure 14.10 Vertical jointing of sandstone strata at Arches National Park, Utah. • What would cause joints such as the one in the upper strata in this photo?

Figure 14.11 The marble cliffs in Little Cottonwood Canyon in Alta, Utah. • **What is the origin of marble?**

detailed facts, recurring patterns, and interrelated processes that they observe and analyze. The theories of evolution and natural selection proposed by Charles Darwin and his scientific colleagues in the mid–nineteenth century comprised just such an explanation. Those theories ultimately revised the thinking of scientists in many fields and still are the model for biologists and other life scientists today.

But what are the underlying explanations for the Earth sciences? We have already mentioned the general acceptance of uniformitarianism—the idea that Earth today is only the current product of orderly processes that have been operating in a similar fashion for millions of years. Are there similar hypotheses that explain how and why these processes work? Is there one broad theory that can help to explain such diverse subjects as the growth of continents, the movement of solid rock beneath Los Angeles, the

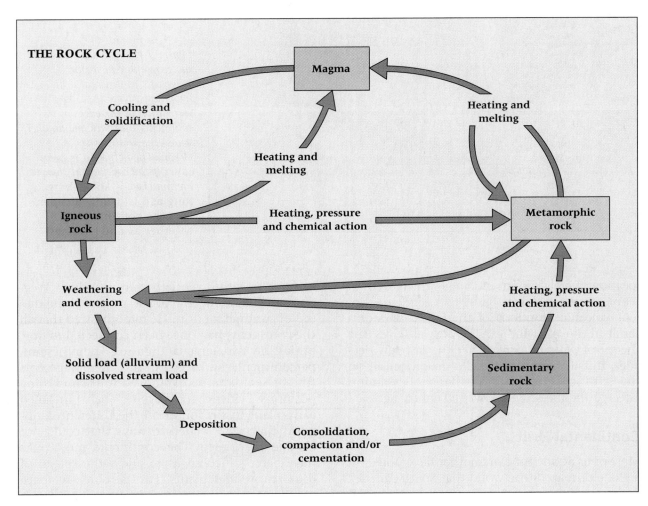

Figure 14.12 The Rock Cycle: "No trace of a beginning, no prospect of an end," *according to the eminent British geologist James Hutton (1726–1797).* • **What conditions are necessary to change igneous rock to metamorphic rock, and metamorphic rock to igneous rock?**

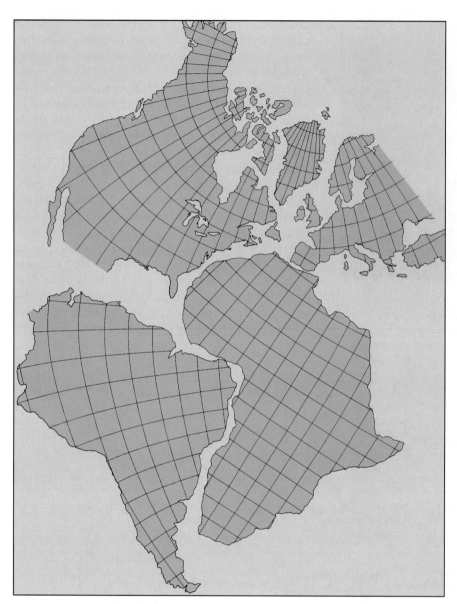

Figure 14.13 *This figure shows the geographic basis for Wegener's continental drift hypothesis. Note the close correlation of the edges of the continents that face one another across the width of the Atlantic Ocean. The actual fit is even closer if the continental slopes are matched.* • **About how long ago is it now assumed that the continents fit together as one landmass?**

location of great mountain ranges, the pattern of temperatures in ocean basin rock, and the violent volcanic eruptions of the island of Montserrat in the West Indies? The answer is yes: the theory of *plate tectonics*. Over the last few decades, this theory has created a major impact on the Earth sciences just as the theory of evolution affected the life sciences over a century ago.

Continental Drift

Most of us at one time or another have noted on a globe or map of the world that South America and Africa look as if they fit together. In fact, if a globe were made into a spherical jigsaw puzzle, several of the widely separated large landmasses could be made to fit without large gaps or overlaps (Fig. 14.13). Is there a scientific explanation?

In the early twentieth century, Alfred Wegener, a German meteorologist, proposed the theory of continental drift. He hypothesized that all the continents had once been connected in one, or possibly two, large landmasses. Then these supercontinents broke apart, and their fragments (the present continents) moved to their existing positions. Evidence for Wegener's belief included early plant fossils found on the different continents that were related in ways that could not be due to chance. The continents must once have been joined so as to allow the spread of those early land plants. Furthermore, evidence of climatic changes (for instance, glacial evidence in the Sahara Desert and tropical fossils in Antarctica) could be explained best by the movement of the large landmasses from one climatic zone to another.

The reaction of the scientific community to Wegener's proposal was one ranging from skepticism to outright ridicule. A major objection to his hypothesis, which Wegener admitted, was that he could provide no explanation for the breakup, nor could he provide an acceptable explanation for the energy that would have been needed to move huge landmasses through the rigid crust and across vast oceans.

Wegener did propose that perhaps the crust was not so rigid as had been believed. If instead it were plastic under certain conditions, and if the lighter material of the continents "floated" on the heavier rocks in which they are imbedded, then the movements of the continents away from each other might be possible. Nevertheless, Wegener was still unable to find a convincing and scientifically acceptable movement mechanism for continental drift.

Supporting Evidence for Continental Drift

More than half a century passed before Wegener's seemingly fanciful proposal concerning the movement of continents received serious consideration by a majority of Earth scientists. During that time, World War II had brought rapid advances in science and technology that often accompany an all-out war effort. Research in oceanography and geophysics in particular had been aided by the development of sonar and radioactive dating, by the improvement of magnetometers, and by the coming of the computer age. Earth scientists were flooded with new evidence that portions of the lithosphere (including the continents) had indeed been on the move.

As one example, scientists were originally unable to explain the unusual orientation of magnetic fields in many crustal rocks that had cooled from the molten state millions of years before. Minute iron minerals within molten rock orient themselves like tiny compass needles pointing to the magnetic pole. This orientation is locked into the rock as it solidifies and is known as **paleomagnetism.** Rocks of different ages, or on different continents, show magnetic orientations at a variety of angles to Earth's magnetic field as it is today. At first it was assumed that Earth's magnetic poles had wandered, but this could not explain varying paleomagnetism in rocks of the same age.

A far better explanation emerged when a model of the continents was projected backward in time so that the magnetic orientation of the rocks was made to coincide with Earth's magnetic field during past periods of Earth history. The an-cient magnetism indicated an almost perfect fit of the continental jigsaw puzzle some 200 million years ago.

Supporting evidence for crustal movement came from a variety of additional sources. New fossil discoveries indicated that certain reptiles and land plants, earlier known to have been found in Australia, India, South Africa, and South America, were to be found in Antarctica as well. The plants or animals found in each instance were so similar and specialized that they could not have developed without land bridges between their now separate locations. Continental fit was discovered to be even better a few hundred meters below sea level along the continental slopes. Mountain ranges on the different continents were carefully matched by radioactive-dating techniques and were shown to be continuous when the continents were joined. Even climatology continued its contribution. Evidence of ancient glaciation in Brazil and South Africa or of tropical forest climate (coal deposits) in Alaska and Antarctica could only be explained by crustal movement.

Convection Currents and Sea Floor Spreading

The real key to the theory of plate tectonics was to be found on the ocean floor. First, mapping of the ocean basins revealed a system of midocean ridges (also called oceanic ridges or rises) that had configurations remarkably similar to the outlines of the continents. Second, it was discovered that parallel bands of rocks with similar magnetic properties extended one after the other in identical fashion on both sides of the ridges in the Atlantic and Pacific oceans. Third, scientists made the surprising discovery that although some continental rocks were nearly 4 billion years old, rocks on the ocean floor are all geologically young—they have been in existence less than 200 million years. Fourth, the oldest ocean floor rocks are under the deepest ocean waters and nearest to the continents, while the youngest ocean floor rocks are under shallower ocean depths and nearest the midocean ridges. Finally, temperatures of rocks on the ocean floor vary significantly; they are highest near the ridges and become progressively lower farther away.

Only one logical explanation emerged to fit all of the new evidence. It became apparent that hot new oceanic crust is being formed at the mid-ocean ridges while older, cooler, and more dense oceanic crust is being destroyed near the margins of the ocean basins. Most now believe that the emergence of this new oceanic rock is associated with the movement of great sections or plates of lithosphere away from the midocean ridges. Earth scientists label this phenomenon **sea floor spreading.** The plates move laterally at an average rate of 2–5 centimeters (1–2 in.) per year above the plastic asthenosphere of the mantle. Ocean floors are not permanent, and the young age of ocean basin rock is the result of crustal renewal at the ridges and the movement of this crustal material as part of lithospheric plates toward crustal destruction at the margins of the ocean basins.

Plate tectonics, the modern version of continental drift, also includes a plausible explanation of the mechanism involved in sea floor spreading and the movement of the lithosphere. The mechanism is **convection,** the transfer of hot mantle material upward toward Earth's surface. New molten material rises toward the surface and is expelled at the midocean ridges as part of huge subcrustal convection cells. The cellular motion is continued as cooling crustal material moves away from the ridges and is completed as the older and denser oceanic crust is consumed in the great trenches that are often near the boundaries where continental and ocean basin plates meet.

Tectonic Plate Movement

The theory of plate tectonics suggests that the lithosphere (the crust and the rigid upper mantle) consists of as many as 20 rigid plates (Fig. 14.14). All plates move as distinct units—in some places traveling away from each other (diverging), in other places sliding past each other (moving laterally), and elsewhere coming together (converging). Seven are considered major plates and are of continental or oceanic proportions. Five are of minor size, although they have apparently maintained their own identity and direction of movement for some time. Several other plates are quite small and are all in the active zones marking the boundaries betwen major plates. Although the largest plate, the Pacific plate, is primarily oceanic, all other major plates are composed of both continental and oceanic crustal material.

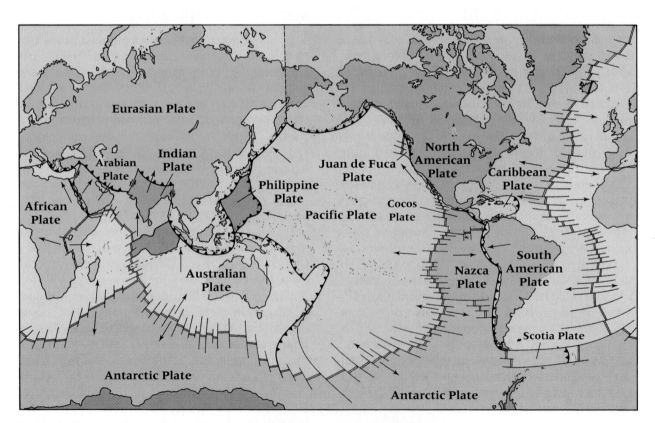

Figure 14.14 World map of the major tectonic plates and their general directions of movement. Most tectonic activity occurs along plate boundaries, where plates separate, collide, or slide past one another. • Do the edges of continents and plate boundaries correlate closely? Do all the plates have a continent located on them?

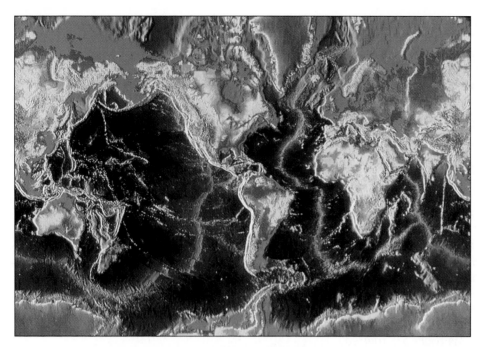

Figure 14.15 *A computer-generated relief map of the surface of Earth. This global view of the major landform features of both the continental and oceanic crusts helps confirm plate tectonics.* **• Can you see the outline of any of the major plate boundaries (compare this relief map to Figure 14.14)? How would this type of modern data have helped Alfred Wegener when he first proposed his theory of continental drift?**

Once set in motion, the shifting of the larger tectonic plates relative to one another provides the explanation for many landform features on Earth's surface. This is of particular interest to the physical geographer because it is now possible through tectonic plate theory to understand the *paleogeography* of the planet and the worldwide distributional patterns and close relationships among such diverse phenomena as earthquakes, volcanic activity, zones of crustal movement, and major landforms (Fig. 14.15). Let us briefly examine the three ways in which tectonic plates relate to one another as a result of movement along their boundaries.

Plate Divergence. Tectonic **plate divergence,** or pulling apart along the oceanic ridges, is a direct result of sea floor spreading (Fig. 14.16). Tension-producing forces cause the oceanic crust to thin out and weaken. Shallow earthquakes are often associated with the crustal stretching, and magma (molten rock beneath Earth's surface) wells up from the asthenosphere, forming new crustal ridges and ocean floor as the plates move away from each other. The formation of new crust in these divergence or spreading areas gives the label *constructive plate margins* to these zones. Thus, the Atlantic Ocean floor was formed as the

South American and African continents were moved apart. Occasional "oceanic" volcanos like those of Iceland, the Azores, and Tristan da Cunha mark such boundaries.

Though most plate divergence is along oceanic ridges, sometimes this process is associated with continents. The best example is the well-known *rift valley* system of East Africa, stretching from the Red Sea to Lake Malawi. The entire system, including the Sinai Peninsula and the Dead Sea, is characterized by a series of crustal blocks which have moved downward with respect to the crust on either side, with lakes frequently occupying the depressions. Measurable widening of the Red Sea suggests that it may indeed be a "proto-ocean" between Africa and the Arabian Peninsula, similar to the young Atlantic between Africa and South America less than 200 million years ago (Fig. 14.17).

Plate Convergence. Tectonic **plate convergence** causes a maximum of crustal activity. Despite the exceedingly slow rate of plate movement, the incredible energy involved causes the brittle crust to crumple as one plate overrides the other. The denser plate is forced deep below the surface in a process called **subduction.** This usually occurs where oceanic crust meets continental crust, and the denser oceanic crust descends below the lighter continental crust (Fig. 14.16). Such is the situation along the Pacific coast of South America, where the oceanic Nazca plate subducts beneath the South American plate, or in Japan, where the Pacific plate dips under the Eurasian plate. Crustal rocks are being lost in these subduction zones, known as *destructive plate margins.*

Deep oceanic trenches form where the crust is dragged downward toward the mantle, as in the Peru-Chile trench and the Japanese trench. Frequently hundreds of meters of sediments, eroded from the adjacent landmasses, are carried into these trenches, later to form sedimentary rock. When the rock becomes squeezed and contorted between the colliding plates, it is heavily folded and metamorphosed. Many great mountain ranges, such as the Andes, have been formed by such processes at convergent plate margins.

A subducting plate is heated as it plunges downward into the mantle. Its rocks are melted, and the resulting hot magmas begin to migrate upward along fissures and zones of weakness in the overriding plate. Where the magma reaches the surface, it forms a series of volcanic peaks in these same mountain areas, like the Cascades in Washington and Oregon or the Andes in South America. Where two oceanic plates meet, the older, denser one will subduct below the younger, less-dense oceanic plate, and volcanoes may form major *island arcs* separating the continents from the ocean trenches. The Aleutians, the Kuriles, and the Marianas are all examples of island arcs near oceanic trenches that border the Pacific plate.

As the subducting plate grinds downward, enormous friction is produced, which explains the occurrence of major earthquakes in these regions. Called Benioff zones, these are named after the seismologist Hugo Benioff, who first plotted the existence of earthquakes occurring at a steep angle along the descending edge of the subducting plate (see Fig. 14.16 on pages 392–393).

Where two continental plates collide, massive folding and crustal block movement occur, rather than volcanic activity. This causes crustal thickening, and again major mountain ranges may form. The Himalayas, the Tibetan Plateau, and other high Eurasian ranges were formed in this way where the continental Indian plate collided with Eurasia some 40 million years ago, and the Alps were formed when the African plate pushed into the Eurasian plate (see Fig. 14.16).

Thus, the zones of plate convergence mark the locations of the more spectacular landforms on our planet: huge mountain ranges, volcanoes, and deep ocean trenches. Their geographic locational patterns can be understood within the framework of tectonic plate theory.

Lateral Movement. A third type of plate contact occurs when plates neither pull apart nor converge but instead slide laterally past each other as they move in opposite directions. Such a plate boundary exists along the famous San Andreas Fault in California (Fig. 14.18). In fact, Mexico's Baja California peninsula and southern California, an area west of the fault, are actually a part of the Pacific plate. The land to the east of the fault, on the other hand, is part of the North American plate. In the area along the fault, the Pacific plate is moving laterally northwestward in relation to the North American plate at a rate of about 8 centimeters (3 in.) a year. If this movement continues, in a few million years Los Angeles will eventually move to the position of San Francisco on its way to final disappearance in the Aleutian Trench.

Another type of lateral plate movement occurs on the ocean floor in areas of plate divergence. As the plates pull apart, they usually do so along a series of fracture zones that tend to form at right angles to the major zone of plate contact. These cross-hatched plate boundaries along which lateral movement takes place are called *transform faults.* Such transform faults or fracture zones are common along midocean ridges, but examples can also be seen elsewhere, as on the sea floor off the coast of the Pacific Northwest between the Pacific and Juan de Fuca plates (see Fig. 14.18). Transform faults are probably caused by variable speeds of plate motion, causing stress and eventually movement of small portions of the plate. The most rapid plate spreading is on the East Pacific rise, where plate motion is over 17 centimeters (5 in.) per year.

Growth of Continents

The origin of continents themselves remains a mystery. It is clear that the individual continents tend to have a core area of very old igneous and metamorphic rock. These cores are usually areas of relatively low relief that are far from active plate boundaries and lack major tectonic disturbances over an immense period of time. These ancient crystalline rock areas are called **continental shields.** The Canadian, Scandinavian, and Siberian shields are outstanding examples. Around the peripheries of the exposed shields, flat-lying, younger sedimentary rocks indicate the continued presense of a stable and rigid mass beneath, as in the American Midwest, western

Figure 14.17 A Space Shuttle view of the Sinai Peninsula looking west along the Mediterranean coast of Egypt. North is to the right in this image. To the east of the Sinai is the rift system that forms the Gulf of Aqaba. To the west is the Red Sea rift. The irrigated valley of the Nile River is clearly visible as it flows northward across the desert into the Mediterranean Sea. • What two major landmasses are separated by the Red Sea?

Siberia, and much of Africa. Most continents appear to grow outward by mountain building around the sediment-covered margins of the ancient shield areas. This process is clearly related to the plate tectonic concept, for the descent of lithosphere in an ocean trench around a continent generates new molten material that forms deep-lying rocks and volcanoes on the continental edge. Simultaneously, the continental edges tend to be buckled by the subduction process. Oceanic rocks are peeled off and piled against the continental margin forming ranges such as the Coast Ranges of California and Oregon and the Olympic Mountains of Washington. Except where two continents have collided to become one, mountain building is generally restricted to those plate boundaries where subsurface injections, surface volcanism, and the squeezing up of marine formations all add new material to the continental mass.

Earth scientists have also suggested that continents grow by **accretion**—that is, by adding numerous chunks of crust to the main continent by collision. North America may have grown in this manner over the last 200 million years by adding chunks of crust, known as **microplate terranes,** as it moved westward over the Pacific and former oceanic plates. Though their exact

origin is still speculative, many parts of western North American from Alaska to California may have originated from south of the equator. Terranes, which have their own distinct geology from that of the continent they are now joined to, may have originally been offshore oceanic island arcs, undersea platforms, or continental fragments, such as New Zealand or Madagascar.

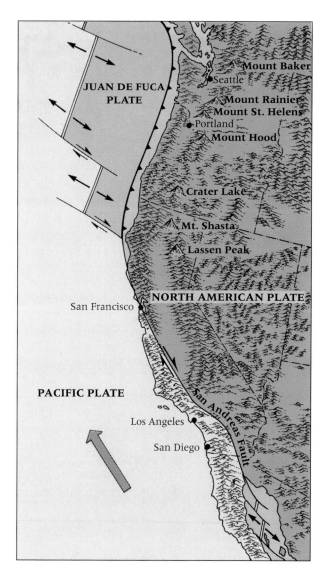

Figure 14.18 Diagram of a lateral plate boundary. A lateral plate boundary occurs where the Pacific plate moves along the edge of the North American plate in western North America. The San Andreas Fault system is formed along this boundary. The area west of the fault is moving northward in relation to the rest of North America. This movement occasionally produces major earthquakes as the plates slip past each other. • Note that north of San Francisco the boundary type changes. What type of boundary is located there, and what type of surface features indicate this change?

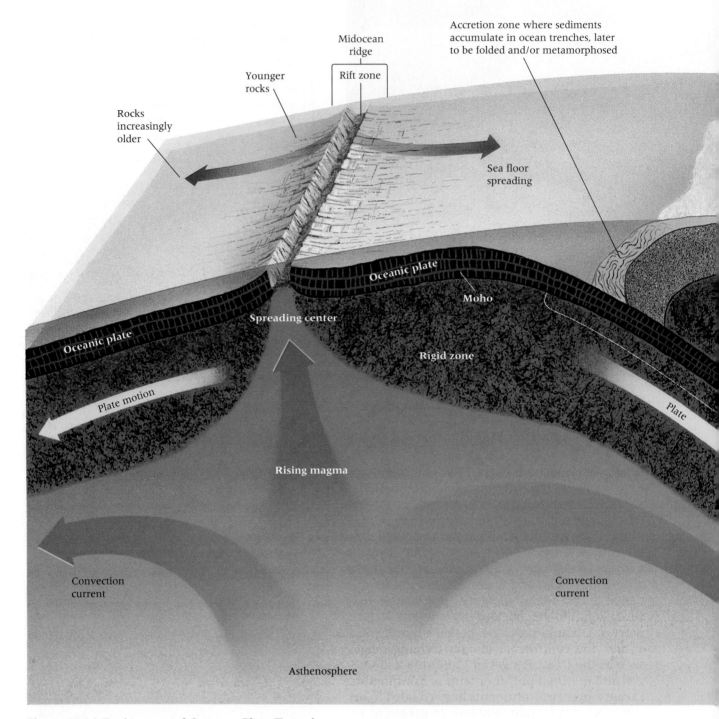

Midocean ridge

Accretion zone where sediments accumulate in ocean trenches, later to be folded and/or metamorphosed

Younger rocks

Rift zone

Rocks increasingly older

Sea floor spreading

Oceanic plate

Moho

Oceanic plate

Spreading center

Rigid zone

Plate motion

Plate

Rising magma

Convection current

Convection current

Asthenosphere

Figure 14.16 Environmental Systems: Plate Tectonic Movement. *Unlike the other major systems discussed in this book, the plate tectonics system does not derive its energy from the sun. Instead, the movements of Earth's lithosphere are caused and controlled by energy derived from Earth's interior. Exactly how deep the convection cells of heat energy move within the mantle and the outer core of Earth is still debated. The most common theory is that the convection cells flow within the mantle's asthenosphere, and the rigid upper mantle and crust above are carried along with these currents at a speed that approximates human fingernail growth.*

Earth's crust is broken into giant slabs, known as plates. As these plates move, they interact with adjoining plates, forming different boundary types, each having distinct landform features. This diagram shows three major plate boundary types: spreading centers, subduction zones, and collision zones. The spreading center boundary, on the left of the diagram, is a constructive boundary as it brings new crustal material to the surface along an active rift zone. Over time, newer material pushes older rock progressively away from the active volcanic rift zone in both directions. Earth's longest spreading center boundary is the midocean ridge

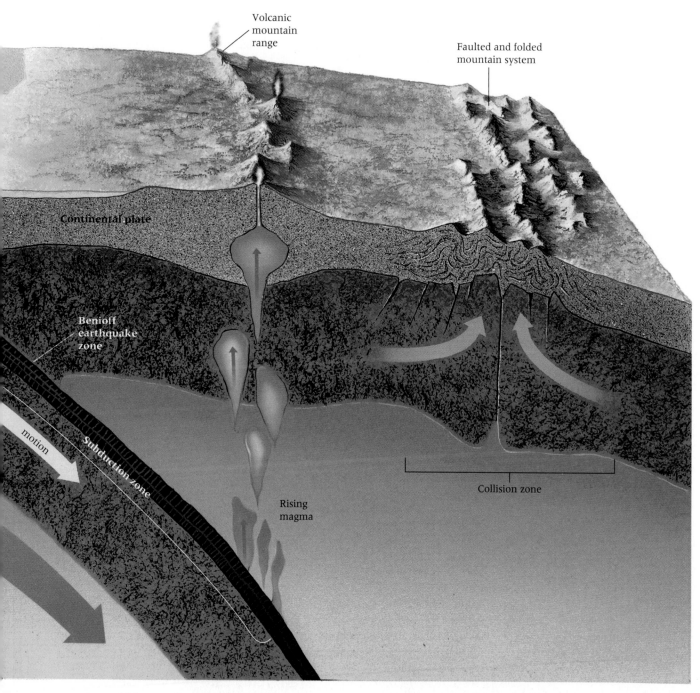

Volcanic
mountain
range

Faulted and folded
mountain system

Continental plate

Benioff
earthquake
zone

motion

Subduction zone

Rising
magma

Collision zone

system, which runs through all the major oceans. The best known rift system on a continent is the Great Rift Valley of East Africa.

Subduction zones, as shown in the middle of the diagram, occur where two plates are converging. This is a destructive boundary where crustal material is returned to the interior of Earth. The denser oceanic plate is forced by gravity and plate movement to subduct beneath the less dense continental plate. Surface features common to subduction zones are deep ocean trenches and volcanic mountain ranges or island chains. The best examples of subduction are found around the Pacific Ring of Fire, in

volcanic eruption and earthquake-prone sites such as those in Japan, Chile, New Zealand, and the northwest coast of the United States.

Collision zones, as shown on the right, are found where two continental plates collide. Massive mountain building occurs as the crust is compressed and thickened. Unlike in the other two boundary types, volcanoes are lacking here. The world's highest mountains, the Himalayas, formed when the Indian plate collided with Eurasia. The Alps were formed in a similar manner in a collision between the African and Eurasian plates.

TABLE 14.2			*Timetable of Earth's Geologic History*			
ERA	**PERIOD**	**EPOCH**	**DISTINCTIVE FEATURES**	**MILLIONS OF YEARS AGO**	**PERCENT OF EARTH'S HISTORY**	**MOUNTAIN-BUILDING EPOCHS (OROGENIES)**
Cenozoic	Quaternary	Holocene	Modern humans	.01	.04%	Alpine and Cascadian
		Pleistocene	Early humans; glaciation	1.6		
	Tertiary	Pliocene	Large carnivores	5	1.3%	
		Miocene	Abundant grazing mammals	24		
		Oligocene	Large running mammals	37		
		Eocene	Modern types of mammals	58		
		Paleocene	First placental mammals	65		Laramide (Rocky Mountains)
Mesozoic	Cretaceous		First flowering plants; climax of dinosaurs, followed by extinction	144	3.6%	
	Jurassic		First birds, first true mammals; many dinosaurs	208		
	Triassic		First dinosaurs; abundant cycads and conifers	245		Appalachian
Paleozoic	Permian		Extinction of many kinds of marine animals including trilobites; continental glaciation in Southern Hemisphere	286	8.0%	Hercynian (Europe)
	Carboniferous — Pennsylvanian		Great coal swamps, conifers; first reptiles	320		
	Carboniferous — Mississippian		Sharks and amphibians; large-scale trees and seed ferns	360		Acadian
	Devonian		First amphibians; fishes very abundant	408		
	Silurian		First terrestrial plants	440		Caledonian Taconic (Taconian)
	Ordovician		First fishes; marine invertebrates	505		
	Cambrian		First abundant record of marine life; trilobites and brachiopods dominant	570		
PROTEROZOIC	Precambrian		Limited evidence of abundant algae	2.5 Billion	87.0%	Archaean
ARCHEOZOIC			Oldest rocks	4 Billion		

Paleogeography

The study of past geographic environments is known as **paleogeography.** The goal of paleogeography is to try to reconstruct the past environment of a geographic region based on geologic and climatic evidence. For the student of physical geography, it seems that the present is complex enough without trying to know what the geography of the past was like. However, peering into the past may actually help us forecast and prepare for change in the future.

The immensity of geologic time over which major Earth events (such as plate tectonics, ice ages, or the formation and erosion of mountain ranges) took place is difficult to picture in the human time scale of days, months, and years. The geologic timetable is a calendar of Earth history (see Table 14.2). It is divided into *eras,* the largest units of time, such as the Mesozoic Era (which means "middle life"), and shorter *periods,* such as the Jurassic Period. *Epochs,* such as the Pleistocene Epoch (recent ice age), are the smallest time units and are used within the Cenozoic Era ("recent life") where geologic evidence is more abundant.

If we took a 24-hour day to represent the whole of the approximately 4.6-billion-year history of Earth, the first 21 hours would be consumed by Precambrian time, an era of which we know very little. The most recent, or Quaternary Period, would take less than 35 seconds, and human beginnings, over about the last 4 million years, less than a single second.

If we look at evidence for the paleogeography of the Mesozoic Era (which ended 65 million years ago), for instance, we would find a much different physical geography than today. This was a time when the supercontinent Pangaea was gradually coming apart and the ocean floors widened, separating the supercontinent into the continents familiar to us today. A paleogeographic map of North America during the Jurassic Period (208 million years ago) shows that the Appalachian Mountains were in existence, as were the ancestral Rockies (Fig. 14.19). Most of the Great Plains and the present Pacific Coast states were beneath shallow seas, while the Midwest was a lowland region.

The Mesozoic climates were much different as well, as North America drifted northwest. By the Jurassic Period, much of the present United States was centered at about 30°N latitude, still in warmer climates than today. Fern and conifer forests were common, and the animal life was totally unlike what it is today. The Mesozoic was the "age of the dinosaurs," the large reptiles that ruled the land and the sea. Other life also thrived, including marine plants and invertebrates, insects, and the earliest birds and mammals.

The Mesozoic Era ended with an episode of great extinctions, including the end of the dinosaur age. Geologists, paleontologists, and paleogeographers are not in agreement as to what caused these great extinctions. Some of the strongest evidence points to a large meteorite striking Earth 65 million years ago, disrupting global climate. Other evidence points to plate tectonic changes in the distribution of oceans and continents, which in turn caused rapid climatic change and increased volcanic activity, as possible causes of mass extinctions.

Each era and period of Earth's history has had a unique paleogeography with its own distribution of land and sea, climate regions, plants, and animal life. For instance, the life of early humans was under different environmental conditions than today. How did this have an influence on the present distribution of human population? (See Fig. 1.13, World Map of Population Density.) As time passes and more evidence is collected, paleogeographers may be able to fill in the empty spaces on those maps of the past, which are so unfamiliar to us.

Landforms as Part of the Earth System

We have already seen that the distribution and shapes of landmasses influence climate significantly. Likewise, major landforms such as mountain ranges affect climate by forcing moisture-carrying winds to rise on windward slopes, producing heavy rainfall and creating dry conditions on the leeward side. This in turn affects vegetation, soils, and animal life. Another example of the interrelationship between landforms and climate is the effect of slope and terrain on soil development. It is difficult for soils to mature on steep mountainsides. Very flat terrain also affects soil development by causing poor drainage.

There are innumerable ways in which landforms and people are interrelated. These include both the influences landform configurations have on human life and the many ways people in turn affect the landforms on which they live.

Landforms are one of the major aspects of the Earth system with which human beings must interact on a daily basis. The character and quality of the land and its resources (slope, elevation, relief, soil, minerals, and water) are significant influences on the character and quality of human life. The distribution of population over Earth's

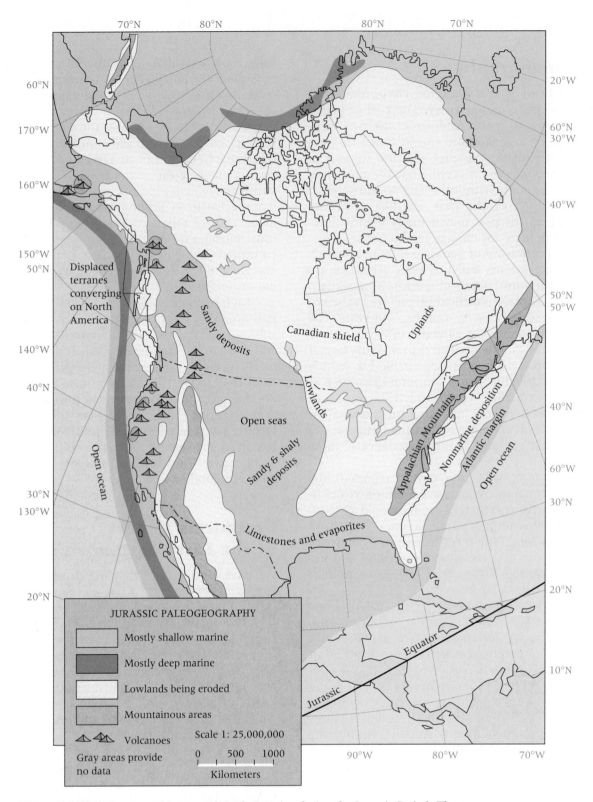

Figure 14.19 *Paleogeographic map of North America during the Jurassic Period. The continent had a very different geography 180 million years ago. Its location, size, landforms, and coastlines were completely different from today.*

surface shows the importance of land-surface configuration to people (see again Fig. 1.13). All over the world people have tended to congregate along coastlines that allow easy access to the sea, especially where there are good harbors, and on plains where climate and soils have allowed agri-cultural development. They have avoided moun-tainous areas, remote hill lands, and isolated tablelands because of the restrictions these land-forms impose on agriculture, commerce, and communication.

Humans in turn, particularly in today's technologic world, also have made changes in the landforms. Agriculture has significantly altered the world's landscapes, from paddy terraces in the Orient to the drained marshes (polders) of the Netherlands. Construction and urbanization completely change landforms, and in many cases the natural terrain is totally obscured. Mining has created huge pits and slag piles, and strip mining for coal has devastated entire regions (as in some parts of Appalachia). We also use landforms for waste disposal or alter landforms by filling canyons, as in Los Angeles, or filling productive bays and coastal regions. All of these activities also affect the other Earth systems as well, by changing soils, vegetation, animal life, air quality, and water supplies.

Define and Recall

geomorphology	asthenosphere	igneous rocks	metamorphic rock
catastrophism	Mohorovicic	magma	foliation
uniformitarianism	discontinuity (Moho)	lava	
tectonic process	crust	extrusive	continental drift
gradational process	oceanic crust	intrusive	paleomagnetism
weathering	basalt	plutonic	sea floor spreading
mass wasting	continental crust	joint	plate tectonics
erosion	granite	sedimentary rock	convection
deposition		clastic	plate divergence
	bedrock	organic sedimentary rock	plate convergence
seismograph	regolith	stratification	subduction
outer core	outcrop	strata (beds)	continental shield
inner core	mineral	bedding plane	accretion
mantle	silicates	unconformity	microplate terrane
lithosphere (crust and	oxides	lamination plane	paleogeography
upper mantle)	carbonates	cross-bedding	

Discuss and Review

1. What are the differences between tectonic and gradational processes?

2. Identify the major zones of Earth from the center to the surface. How do these zones differ from one another? What is the special significance of the asthenosphere?

3. Define the differences between continental crust, oceanic crust, lithosphere, and asthenosphere.

4. List the most common elements in Earth's crust. What is a mineral? What is a rock?

5. Describe the three major classifications of rock and the means by which they are formed. Give an example of each.

6. Describe the rock cycle. What is magma and how is it related to the rock cycle?

7. What evidence did Wegener rely on in the formulation of his theory of continental drift? What evidence did he lack? What evidence has since been found to support his theory?

8. What type of plate boundary is found near the Andes, along the San Andreas Fault, in Iceland, and near the Himalayas?

9. Define paleogeography. Why are geographers interested in this topic?

Consider and Respond

1. Explain why the eastern United States has relatively little tectonic activity in comparison to the western United States.

2. List the four major plate boundary types. Name a country which is located along each one of the plate boundary types.

3. List several types of human activities that have a major impact on landforms. Give some examples of human activities which are geomorphic agents in your local area.

4. How have landforms affected world human population distribution? Cite some specific examples of densely populated regions and sparsely populated regions and how they relate to the landforms there.

CHAPTER

15

LANDFORMS AND TECTONIC PROCESSES

CHAPTER PREVIEW

▶ Although the capacity of humans to influence life-forms on Earth is almost limitless, the changes wrought by humans on Earth's crust are miniscule compared to those wrought by tectonic forces.

Can humans expect to control these natural forces? How should humans deal with these forces and the landforms they create?

▶ Natural catastrophes such as volcanic eruptions and earthquakes can lead to great human suffering and economic loss, but each has a direct cause and rational explanation well understood by Earth scientists.

How does this relate to superstition and religion? How is this linked to disaster prediction? Why do humans often avoid clear warning signs?

▶ Although individual examples of mountain building and other crustal uplift may seem to be a direct product of one particular tectonic process, such as folding, faulting, or volcanism, these processes rarely work in isolation of one another and their geographic distributions follow strikingly similar worldwide patterns.

What does this statement suggest concerning a common underlying explanation for these phenomena? How is this important to human beings?

▶ The locations of tectonic plate boundaries have a direct correlation with the distribution of some of Earth's great "hazard zones."

In what ways are volcanism and faulting involved in this relationship? Why do humans persist in living in or near areas subject to volcanic and earthquake activity?

▲ Mt. Rainier, Washington.

As we noted in Chapter 14, the tectonic processes are those that create major irregularities in Earth's physical landscapes. Therefore, although the tectonic forces operate from within Earth, they are still dominant controls of topography, the appearance of Earth's land surface. Tectonic landforms can be produced by deep-seated Earth processes or from materials which are ejected from the interior onto Earth's surface. Tectonic landforms may also be produced by both the movement of molten materials and the deformation of solid rock through tectonic activity.

Some of the world's largest and most spectacular landform features are the product of tectonic forces. Mountains in particular are primarily a result of tectonic activity. It would be hard to imagine Earth without the Himalayas, Alps, Andes, Rockies, and Appalachians. Not only do the tectonic forces build large-scale mountain systems, but they also produce many smaller-scale landforms which we will study in this chapter.

The tectonic processes, especially those involved in volcanic eruptions and earthquakes, are responsible for many serious natural hazards. Physical geography is concerned not only with the origins and description of tectonic landforms, but also with their distribution and those potential hazards that are related to the tectonic processes.

Figure 15.1 The two major types of lava use Hawaiian terminology. Hotter, more fluid lava forms pahoehoe, *or ropy lava (lower portion of the photo). Viscous pasty lava forms block lava called* aa *(upper portion of the photo).*

The Magmatic Processes

Volcanism refers to the rise of magma and its cooling above Earth's surface. It also includes the extrusive rocks and landforms created from this surface activity. **Plutonism** is a general term that refers to igneous processes that occur below Earth's surface. It also includes the intrusive rocks and rock masses formed from the magma.

Volcanoes

There is no spectacle in nature as awesome as an explosive volcanic eruption. The most violent eruptions are so cataclysmic that thousands of human lives may be lost and the landscape completely destroyed.

Volcanic eruptions vary greatly in character. Consequently, the volcanic landforms that result are extremely diverse. The variation in eruptive style and in resulting landforms is mainly the result of chemical differences in the magma feeding the eruption. Some magmas have cooled before eruption, so that heavy minerals have crystallized while light minerals, those high in silica, remain dissolved. In such cases, a great deal of gas may already have separated from the magma, building up pressure in the magma chamber below the surface. The explosive release of these gases produces a violent eruption. If, on the other hand, the magma is very hot, crystal formation will not yet have begun and the gases will still be dissolved in the magma itself. If such material leaks or is forced to the surface, the eruption is not explosive, although enormous amounts of highly fluid lava may be produced. Fluid layers become **pahoehoe,** or ropy lavas, while the more viscous, pasty lavas are called **aa,** or blocky lavas; both terms are of Hawaiian origin (Fig. 15.1).

In addition to lava flows and gaseous materials, most volcanic eruptions hurl into the air molten material that hardens at the surface and solids of various sizes. These **pyroclastic materials** (Greek: *pyros,* fire; *clastus,* broken) vary from huge volcanic "bombs" to cinders, ash, and fine volcanic dust. In the most explosive erup-

(a)

(b)

*Figure 15.2 Composite cones are formed from layers of lava and pyroclastic material. **(a)** Cotopaxi in Ecuador is a classic composite cone formed in the Andes. **(b)** Mt. Shasta is a composite cone in the Cascade Range of northern California. Shastina is a secondary crater to the right of the main summit cone.* • *Along what type of tectonic plate boundary are both these volcanoes located?*

(a)

(b)

(c)

Figure 15.3 *Mount St. Helens, in the Cascade Range of Washington. **(a)** A view of Mount St. Helens and Spirit Lake prior to the eruption. **(b)** On May 18, 1980, at 8:32 A.M., Mount St. Helens erupted violently. **(c)** A view of Mount St. Helens after the eruption. The massive landslide and blast removed over 4.2 cubic kilometers (1 cu mi) of material from the mountain's north slope, leaving a crater more than 400 meters (1300 ft) deep. The blast cloud and monstrous mudflows destroyed the surrounding forests and lakes and took 60 human lives.*
• **What type of volcano is Mount St. Helens?**

tions, the volcanic dust may be hurled high into the atmosphere. The 1991 eruptions of Mount Pinatubo in the Philippines ejected a volcanic aerosol cloud that circled the globe. The suspended material caused spectacular sunsets and may have lowered global temperatures slightly.

There are four basic types of volcanoes: composite cones, shield volcanoes, plug domes, and cinder cones. Many volcanoes consist of a mix of lava flows and pyroclastic materials and therefore are known as **composite cones.** Most of the well-known volcanoes of the world are of this type: Fujiyama in Japan, Cotopaxi in Ecuador, Vesuvius in Italy, Mount Rainier in Washington, and Mount Shasta in California (Fig. 15.2). Composed of great volumes of both lava—particularly andesite lava—and ash, composite cones, also known as *stratovolcanoes* (consisting of stratified pyroclastic material and lava), are potentially dangerous. The andesite lava produces explosive eruptions, such as those that occur in the Andes of South America.

On May 18, 1980, residents of the Pacific Northwest—and, soon after, people throughout the United States—were shocked by the realization that every age in history is a volcanic age. Mount St. Helens, a stratovolcano in southwestern Washington that had been smoldering and venting steam and ash for several weeks, exploded with incredible force. A menacing bulge had been growing on the side of the mountain, and Earth scientists warned of a possible major eruption, but no one could forecast the magnitude of the blast. Within minutes nearly 400 meters (1300 ft) of the mountain's north summit had disappeared into the sky and down the mountainside (Fig. 15.3).

Much of the explosion blew pyroclastic debris laterally outward from where the bulge had been. An incredible storm cloud of intensely hot steam, noxious gas, and volcanic ash, traveling at a speed of more than 320 kilometers (200 mi) per hour, obliterated forests, lakes, streams, and camping sites for a distance of nearly 32 kilometers (20 mi). Monstrous mud-flows of ash and melted ice choked streams and valleys and added to the devastation by engulfing everything in their paths. Over 500 square kilometers of magnificent forest and recreation land were destroyed. Hundreds of homes were buried or badly damaged; choking ash several centimeters thick covered nearby cities; untold wildlife was killed; and more than 60 human beings lost their lives in the eruption. It was a minor event in Earth's history, but a sharp reminder of the awesome power of natural forces.

Some of Earth's worst natural disasters have occurred in the shadows of composite cones. Mount Vesuvius, in Italy, killed over 20,000 people in the cities of Pompeii and Herculaneum in 79 A.D. Mount Etna, on the Italian island of Sicily,

(a)

(b)

Figure 15.4 Shield volcanoes of Hawaii Volcanoes National Park. *(a)* The gentle slopes surrounding the active Kilauea Crater. *(b)* The active Hawaiian craters emit an extremely hot, fluid, basaltic lava. As the lava cools, it changes from its molten orange color to a black pahoehoe basalt. • **Why are Hawaiian volcanoes less explosive than those of the Cascades or Andes?**

destroyed 14 cities in 1669, killing over 20,000 people. Today, Mount Etna is active much of the time and remains Europe's most dangerous volcano. The greatest volcanic eruption in recent history was the explosion of Krakatoa in the Dutch East Indies (now Indonesia) in 1883. The violent eruption killed over 36,000 persons, many by the subsequent tsunamis (seismic sea waves) as they swept the coasts of Java and Sumatra. In 1985 the Andean composite cone Nevado del Ruiz, in the center of Colombia's coffee-growing region, erupted and melted its snowcap, sending torrents of mud and debris down its slopes and burying cities and villages, resulting in a death toll in excess of 23,000. The 1991 eruption of Mt. Pinatubo in the Philippines killed over 300 people and caused suspected climatic impacts for years following the eruption (see The Environment: Volcanic Mudflow Hazards in the Cascades, Chapter 16). Beginning in 1995, the Soufriere Hills volcano began a series of violent eruptions, eventually destroying over one-half of the Caribbean island of Montserrat.

Another type of volcano is the **shield volcano,** composed of lavas and relatively little ash or cinders (Fig. 15.4). The gentle cones of Hawaii illustrate best this type of volcano. They emit basaltic lava with temperatures of over 1090°C (2000°F). There is some minor escape of gases and steam, which hurls the lava into the air a few hundred meters, and some buildup of cinders (lava clots that congeal in the air), but the major feature is the outpouring of very fluid basaltic lava in flows only a meter or so deep. The accumulation of flow on top of flow builds broad structures with very gentle slopes.

By contrast, **plug dome** volcanoes extrude extremely stiff acidic lava that fills the initial pyroclastic cone without flowing beyond it (Fig. 15.5). They are characteristically steep-sided volcanoes with broad summits composed of solid lava. Spikes of hard lava often project at the summit. These volcanoes' vents repeatedly jam with congealed lava and then are cleared in cataclysmic explosive eruptions. In 1903 Mount Pelée, a plug dome on the French West Indies island of Martinique, destroyed in a single blast all but one resident of a town of 30,000. Mt. Lassen, in California, is a large plug dome that has been active in this century. Other plug domes have erupted in Japan, Guatemala, and the Aleutian Islands.

The smallest type of volcano, known as a **cinder cone,** produces little lava and consists largely of pyroclastics. Cinder cones are illustrated by Craters of the Moon and Menan Buttes,

Figure 15.5 Plug dome volcanoes extrude stiff acidic flows, which form very steep slopes. Mt. Lassen, located in northern California, is a plug dome and the southernmost volcano in the Cascade Range. It was last active between 1914 and 1921. • **Why are plug dome volcanoes considered dangerous?**

Idaho (see Map Interpretation: Volcanic Landforms), and Sunset Crater, Arizona (Fig. 15.6). In 1943 the remarkable cinder cone called Paricutín grew from a crack in a Mexican cornfield to a height of 92 meters (300 ft) in five days, and to over 360 meters (1200 ft) in a year.

Occasionally an explosive composite volcano may expel so much material in a violent eruption that its summit collapses into its empty magma chamber, producing a vast crater known as a **caldera.** This caldera may subsequently fill with water. The best known caldera in North America is Crater Lake in Oregon, a circular body of water 10 kilometers (6 mi) across and almost

610 meters (2000 ft) deep, surrounded by near-vertical cliffs as much as 610 meters (2000 ft) high. The present rim crests at about 2440 meters (8000 ft). A new cone has built up from the floor of Crater Lake to peak above the surface as Wizard Island (Fig. 15.7). Valles Caldera in New Mexico is another excellent example, and Yellowstone National Park is the site of three ancient calderas. Krakatoa in Indonesia and Santorini (Thera) in the Greek Islands are also examples of calderas. Other calderas can be found in the Philippines, the Azores, Japan, Nicaragua, Tanzania, and Italy, most of them occupied by deep lakes.

Distribution of Volcanism

Most volcanoes are concentrated in several well-defined zones along plate boundaries, while a few volcanic regions are distributed more sporadically (Fig. 15.8). The midocean ridge system where crustal plates are diverging is entirely volcanic, as are the oceanic islands on the midocean ridge, such as the Azores and Iceland. Volcanoes also occur where continental plates are spreading; examples along the East African rift valleys include Mt. Kilimanjaro and Mt. Kenya.

A ring of volcanoes almost completely circles the Pacific Ocean and is known as the "Pacific Ring of Fire." These explosive volcanoes erupt andesite lava. When the oceanic crust descends below the continental crust and into the oceanic trench systems, it melts and is reconstituted into magma. The magma moves up to the surface under the continental borders to produce plutons (see following section) and volcanism above them. In fact, wherever there is descent of crustal material into a deep-ocean trench system, volcanoes occur in the vicinity. This process produces the Pacific ring of volcanoes, which includes the Andes, the Cascades, and the Aleutians; the Kuril Islands and the Kamchatka Peninsula in Russia; and Japan, the Philippines, New Guinea, Tonga, and New Zealand.

Another volcanic belt marks the line of collision between the northward-moving Southern Hemisphere crustal plates and the Eurasian plate. On this line are located the volcanoes of the Mediterranean region, Turkey, Iran, and Indonesia.

The movement of Earth's major crustal plates controls the distribution of most volcanism, as it does the tectonic processes of folding and faulting, which we will discuss next. However, some volcanic areas do not occur near plate boundaries but appear randomly within plates. The Hawaiian Islands, the Galapagos Islands, and

Figure 15.6 Sunset Crater, Arizona, is typical of the smallest type of volcano, the cinder cone. • **What are pyroclastics?**

(Continued on page 407)

VOLCANIC LANDFORMS

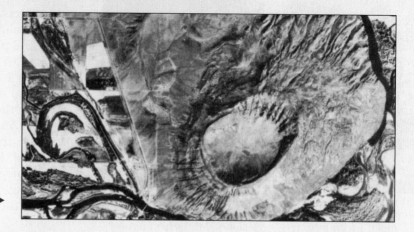

Vertical aerial photo of ▶
Menan Buttes, Idaho

The Map

The Menan Buttes map area is located on the upper Snake River Plain in eastern Idaho. The Snake River Plain is a region of recent lava flows that extends across southern Idaho. It is part of the vast volcanic Columbia Plateau, which covers over 520,000 sq km (200,000 sq mi) of the northwestern United States. The lava originated from fissure eruptions that spread vast amounts of fluid basaltic lava across the landscape, accumulating to a thickness of several thousand feet. Most of the Snake River Plain has an elevation between 900 and 1500 m (3000 and 5000 ft). Rising above the basalt plain are numerous volcanic peaks, including Menan Buttes and Craters of the Moon. The Snake River flows westward across the region.

   The Snake River Plain has a semiarid or steppe climate. Due to the moderately high elevation of the plain,

temperatures are cooler than in nearby lowlands, with an average annual temperature of about 10°C (50°F). Precipitation is between 25 and 50 cm (10 and 20 in.). It is unpredictable from year to year, and there are thunderstorms during the summer months. The low rainfall total is mainly a result of a rainshadow effect. Moisture-producing storms from the Pacific Ocean are blocked from reaching the region by the Cascade Range and the Idaho Batholith section of the northern Rockies. The upper Snake River Plain and the Menan Buttes area are left with dry, adiabatically warmed air descending from the lee slopes of the mountains. The vegetation cover in the area is sparse, mainly characterized by sagebrush and bunch grasses.

Interpreting the Map

1. As what type of volcano are Menan Buttes classified? Upon what landform characteristics is your decision based?

2. Define a butte. Are these landforms really buttes?

3. What is the local relief of the northern Menan Butte? What is the depth of the crater of each butte?

4. Do you think that these volcanoes are active at present? What evidence from the map and aerial photograph indicates activity or a period of inactivity?

5. Sketch an east-west profile across the center of northern Menan Butte from the railroad tracks to the channel of Henry's Fork. Is this profile typical of a volcanic summit and crater? What is the horizontal distance of the profile?

6. Slope ratio can be calculated by dividing the relief into the horizontal distance. For example, a 3000-foot-high mountain slope with a horizontal distance

of 9000 feet would have a slope ratio of 1:3. What is the slope ratio for the western slope of northern Menan Butte from the crater ridge down to the railroad tracks at the foot of the butte?

7. This is a shaded relief topographic map and differs from most of the other topographic maps in this book. What is the major advantage of this shaded relief mapping technique? Are there any disadvantages, compared with the regular contour topographic maps?

8. Compare the southern Menan Butte on the map with the vertical aerial photograph on this page. Why would it be useful to have both a map and an aerial photograph when studying landforms? What is the chief advantage of each?

Menan Buttes, Idaho ▶
Scale 1:24,000
Contour interval = 10 ft
U.S. Geological Survey

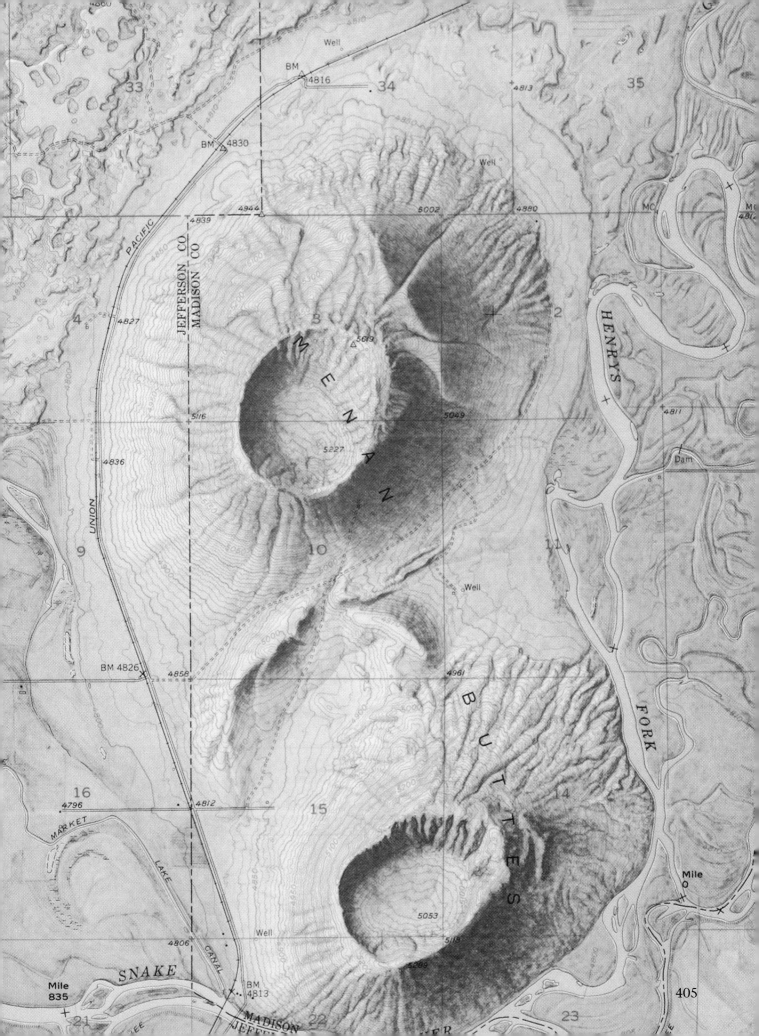

405

(a)

(b)

Figure 15.7 *Crater Lake National Park, Oregon.* *(a)* *Crater Lake is the best-known caldera in North America. It formed when a violent eruption about 6000 years ago of Mt. Mazama caused the summit to collapse forming a deep crater.* *(b)* *Later the crater filled with water, forming the 610-meter-deep (2000 ft) Crater Lake. Wizard Island is a secondary volcanic peak formed in the caldera.* **• Do you think other Cascade volcanoes could repeat this event?**

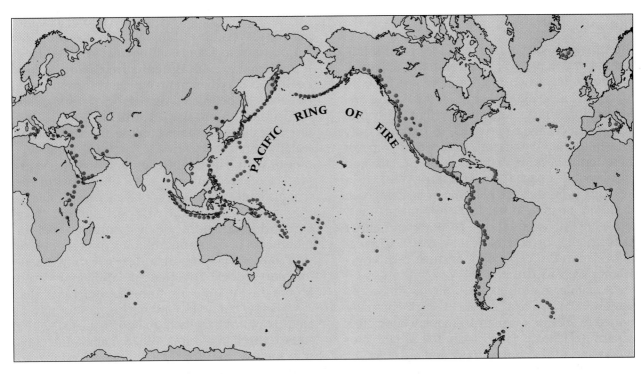

Figure 15.8 *Map of major volcanic regions of the world. Note the strong correlation with the plate boundaries in Figure 14.14. Intraplate "hot spots," such as Hawaii, are the only volcanic areas not near plate boundaries.* • **Which plate has the greatest number of volcanoes around it?**

Yellowstone National Park are examples of these intraplate "hot spots." Some may be associated with a plume of magma rising from the mantle.

Not all volcanism involves erupting volcanoes. Some continental areas are covered with enormous accumulations of lava, called **lava plateaus** or **flood basalts,** that consist of hundreds of overlapping flows. In past geologic periods, basaltic magmas welled up quietly through many separate **fissures** and flowed across the landscape, often engulfing it in depths of thousands of meters. The Columbia Plateau in Washington, Oregon, and Idaho, covering 520,000 square kilometers (200,000 sq mi), is a major example of such a lava plateau (Fig. 15.9), as is most of India's Deccan Plateau.

Figure 15.9 *Columbia Plateau flood basalts in southwestern Idaho. River erosion has exposed the uppermost thick layers of basalt underlying the Columbia Plateau.*

Plutons

The variety in shapes, sizes, and forms of solidified magma that result from plutonic activity is enormous, but when first formed most have no effect on the shape of Earth's surface. **Plutons** refer to those forms of intrusive igneous bodies that are especially deep-seated, so that it is only after millions of years of erosion that some of these forms become exposed at the surface and thus become a part of the landscape. What usually happens is that the plutonic form, composed of granite or some other intrusive igneous rock, is eventually exposed by erosion and either stands higher or is reduced lower than the materials around it, depending on their relative resistance (Fig. 15.10).

A smaller pluton exposed at the surface by erosion is known as a **stock.** A stock is usually limited in area to less than 100 square kilometers (40 sq mi). Plutons larger than 100 square kilometers are termed **batholiths** and are complex masses of solidified magma, usually granite. A batholith is composed of many individual plutons that push aside some of the rocks of the crust while melting and digesting others. Batholiths vary in size; some are as much as several hundred kilometers across and thousands of meters thick. Batholiths form the core of many major mountain ranges, primarily because uplift has caused older covering rocks to be eroded away. These older rocks are preserved where the degree of uplift has been less. The Sierra Nevada, Idaho, Rocky Mountain, Coast, and Baja California batholiths cover areas of hundreds of thousands of square kilometers of western North America.

When magma forces its way into cracks and between layers, disturbing but not digesting the surrounding rock and coming near to the surface, it is referred to as an **igneous intrusion.** A **laccolith** occurs when molten magma forces its way horizontally between layers of rock, forming a solidified mushroom-shaped structure. A laccolith resembles a mushroom because the structure is usually connected with a source of magma by a pipe or stem. The resulting mound on Earth's surface is often compared to a blister, with the magma beneath the surface layers comparable to the water beneath the skin of a blister. Like batholiths, laccoliths often form the core of mountains or hills after millions of years of erosion have worn away the surface covering of sedimentary rocks. The Henry, LaSal, and Abajo Mountains in Utah are formed from laccoliths.

Smaller but no less interesting landforms created by plutonic activity may be exposed at the surface through erosion. When magma forces its way toward the surface, it is sometimes able to intrude into a vertical fracture in the crust across the general trend of the surrounding rocks. If it solidifies, a wall-like sheet of magma will be formed, known as a **dike** (Fig. 15.11). When exposed by erosion, dikes appear as flat walls of resistant igneous rock rising above the countryside. At Shiprock, in New Mexico, dikes many kilometers in length rise sharply to over 90 meters (300 ft) above the plateau. Sometimes magma intrudes between and parallel with the layers of rock and solidifies in a horizontal sheet, called a **sill** (Fig. 15.12). The Palisades, along New York's Hudson River, are an example. Shiprock itself is a **volcanic neck,** which is the exposed (formerly

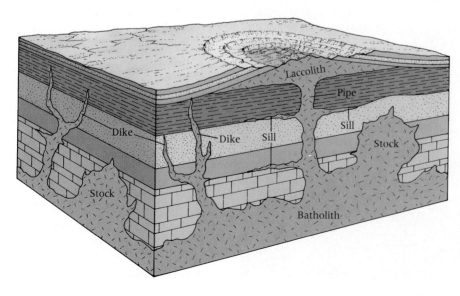

Figure 15.10 The major plutonic features. These intrusive igneous features crystallize below the surface. Later these resistant forms may be exposed to erosion.
• **Of what type of rock are many of these plutons composed?**

Figure 15.11 *An exposed dike in central Colorado. This resistant rock was intruded into softer sandstone. Later the sandstone was eroded away, leaving the resistant dike exposed.*

Figure 15.13 *Shiprock, New Mexico, is a volcanic neck. Volcanic necks form when the resistant intrusive pipe of the volcano is exposed by erosion.* • **Why do you think this feature is called Shiprock?**

subsurface) pipe that fed a volcano situated above it about 30 million years ago (Fig. 15.13).

Solid Tectonic Processes

Solid tectonic processes include the bending, folding, warping, and fracturing of the Earth's crust that are largely a product of the conflicting motions of Earth's crustal plates. Most of our in-

Figure 15.12 *Sills form where magma intrudes between parallel layers of surrounding rocks. This photo in Yellowstone National Park, Wyoming, shows two sills intruded within layers of volcanic ash.* • **What is the major difference between a dike and a sill (see Fig. 15.11)?**

formation about such deformation comes from direct observation of the rocks of the crust, particularly their structure and arrangement. Sedimentary rocks are especially useful to study, since we know that when they are first formed the layers in these rocks are nearly horizontal and the youngest layers are always on top of older layers. If the layers of sedimentary rocks appear tilted or bent, or displaced, then we can assume some kind of deformation has taken place.

Geologists describe the tilting of the beds as the **dip.** Dip is measured as an angle from the horizontal. The **strike** of such beds is their compass direction at right angles to the dip. Thus we might say that certain layers of outcropping rocks have a dip of 50° west and a north/south strike (Fig. 15.14).

Earth's crust seems to have been subject to tectonic pressures and tensions throughout its history, though during some periods the stresses seem to have been greater than at others. The crust has responded to these stresses by wrinkling, warping, and sometimes fracturing, and by being pushed up or by sinking down. Most of these changes have occurred slowly over millions of years; some have been rapid and cataclysmic. The varied responses, often appearing in combination with one another, have yielded a variety of complex structural configurations. In addition, the gradational processes act on these structures as soon as they begin to appear, working to wear them down and fill them in, to level the landscape.

Tectonic movements may involve uplift or depression of large sections of the crust, or may

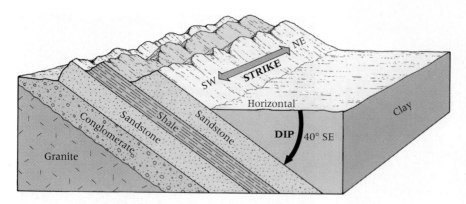

Figure 15.14 Dip and strike in sedimentary rocks. Dip is the angle and direction that rock layers tilt, measured from the horizontal. Strike is the compass direction of the line marking the intersection of the rock layers with Earth's surface. The strike is always at right angles to the dip. • **What is the dip and strike of the upper strata of sandstone in this diagram? (Assume that north is toward the top of the diagram.)**

result from either compression, which tends to shorten and thicken the crust, or tension, which tends to stretch and thin the surface. Generally speaking, compression usually results in wrinkling of the crust, which may eventually shear or slide over itself horizontally if the forces are great enough. Tension causes the crust to crack or fracture, and it may collapse because of the loss of support.

Warping of Earth's Crust

The broad and gentle deformation of Earth's crust over large areas is called **warping.** The state of Florida and Mexico's Yucatán Peninsula were raised to their present levels from below the ocean by crustal warping. Evidence for this warping in Florida is found in the number of marine fossils in the sedimentary rocks there. The extensive Colorado Plateau has been uplifted thousands of feet in similar fashion. The North Sea has been created by the slow downward warping of the crust and the consequent inundation of low plains by the sea. The crust beneath Hudson Bay is rising slowly, so that eventually the bay will disappear. A final example occurs around Lake Superior, where the north shore is rising while the south shore is slowly sinking. The last two examples are a consequence of the disappearance of continental ice sheets, whose sheer weight depressed these areas hundreds of meters. With disappearance of the ice, the areas are rebounding to their former levels.

This type of movement is related to **isostasy,** the concept that Earth's crust is floating in hydrostatic equilibrium on the denser material of the mantle. The effect of isostasy is that the addition of weight (in the form of sediment, glacial ice, or a large body of water) causes the crust to sink slightly, whereas the removal of weight (by erosion, deglaciation, or disappearance of large bodies of water) allows the crust to rise (isostatic rebound) or to float a bit higher. While much of the broad crustal warping occurring today is clearly related to isostasy, the kind of warping that is raising the Colorado Plateau may be due to deep-seated upwelling.

Crustal warping is an important process of landform change, especially since it affects broad parts of Earth's surface. Although it is a form of solid tectonic movement, it is unrelated to tension or the compression of Earth's crust; rather, it proceeds from vertical movements probably related to lateral transfer of material in the asthenosphere.

Folding

The wrinkling of Earth's crust, known as **folding,** usually occurs in response to slow lateral compression. If you spread a cloth on a table and push it from one side, the folds of various kinds that result are much like the folds in Earth's crust that result from certain kinds of pressure exerted on it. Placing both hands some distance apart on the cloth and moving one hand toward you and the other away produces an even larger number of folds. Thus, if underlying rocks move laterally past one another, compressional folding occurs.

Folds produced naturally in rock may be very small, covering an area of a few centimeters, or they may be enormous, with vertical distances between the crests and troughs measured in kilometers. Folds can be tight or broad, symmetrical or asymmetrical (Fig. 15.15). Much of the Appalachian Mountain system is an example of simple folding on a large scale (see Fig. 16.18). In contrast, the Alps were formed by highly complex folding, in which folds are overturned, sheared off, and piled on top of one another. In any case, almost all mountain systems exhibit some degree of folding, though the folding is usu-

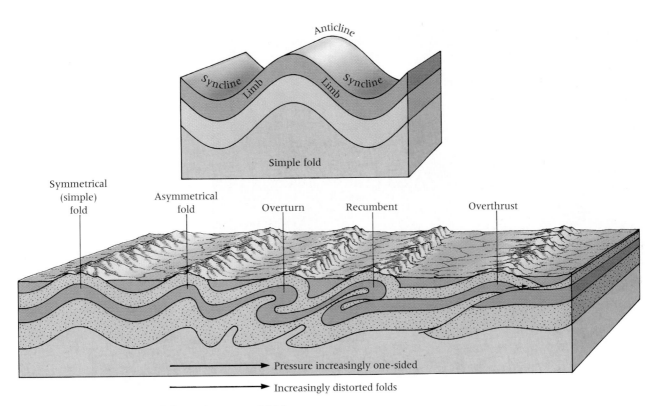

Figure 15.15 *A diagram of the major types of folds.*

ally found in connection with some faulting (fracturing of the crust).

In a series of simple folds, the upfolds are called **anticlines** and the troughs are called **synclines** (Fig. 15.16). The flanks of the folds between the anticlinal crests and synclinal troughs are the fold limbs. Folds may be symmetrical where the compressional forces are somewhat even from both sides. As these forces become stronger from one side, the folds become more and more asymmetrical, until they are overturned. Some may be so lopsided that the fold lies horizontally; these are known as **recumbent folds.** Ultimately the bending rocks will not yield further, and the upper part of the fold shears or breaks and slides over the lower. Such a development is called an **overthrust** (thrust *fault*). Major overthrusts occur along the northern Rocky Mountain front and the southern Appalachians. Recumbent folds and overthrusts are important in the formation of complex mountains such as the Andes, Canadian and Montana Rockies, Alps, and Himalayas (Fig. 15.17).

Faulting

Some rocks are too rigid to bend into folds and rupture instead. Both tension and compression cause rock masses to fracture and move differentially with respect to one another. When slippage or displacement of the crust occurs along a fracture plane, the fracture is called a **fault** (Fig. 15.18). The instantaneous movement along a

Figure 15.16 *Folded sedimentary strata at Ecola Beach, Oregon.* • Identify the anticlines and synclines in this photo.

(Continued on page 413)

Figure 15.17 The Canadian Rockies are complex mountains where folds have been broken and sheared, forming spectacular overthrusts.

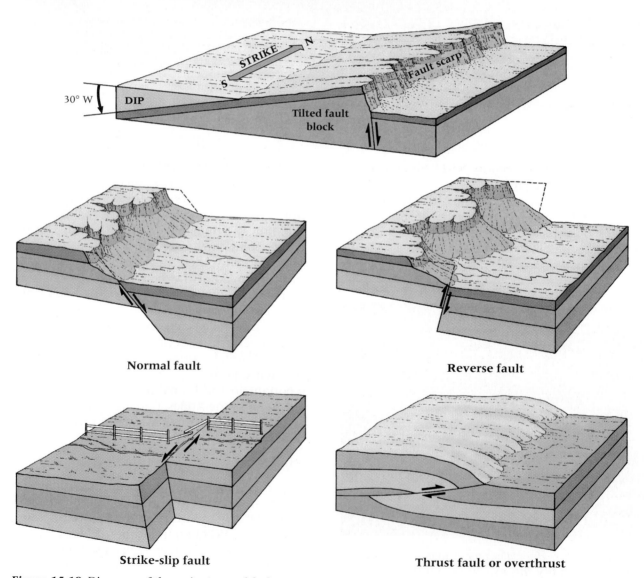

Normal fault

Reverse fault

Strike-slip fault

Thrust fault or overthrust

Figure 15.18 Diagrams of the major types of faults.

fault during an earthquake varies from fractions of a centimeter to several meters. The maximum horizontal displacement along the San Andreas Fault in California during the 1906 San Francisco earthquake was over 6 meters (21 ft). A vertical displacement of over 10 meters (33 ft) occurred during the Alaskan earthquake in 1964. The cumulative displacement along a major fault over millions of years may be tens of kilometers vertically and hundreds of kilometers horizontally, though the vast majority of faults show offsets of much less magnitude.

Vertical displacement along a fracture occurs when one part of the crust moves up or drops down in relation to another. Movement is parallel to the *dip* of the fracture plane extending into Earth. Thus, this type of movement is known as **dip slip.** Faults that have vertical displacement resulting from tension are called **normal faults;** those resulting from compression are called **reverse faults.** When compression pushes one crustal slab up and over another, the displacement is called a **thrust fault** (overthrust).

The steep escarpment of the higher block where vertical displacement occurs is called a **fault scarp** or **fault escarpment.** Such fault scarps account for some of the most spectacular mountain walls, especially in the western United States. The east face of the 645-kilometer-long (405 mi) Sierra Nevada of California is a classic example of a fault scarp rising steeply 3350 meters (11,000 ft) above the desert. In contrast, the west side of the Sierra Nevada (the "back slope") descends very gently over a distance of 100 kilometers (60 mi) through rolling foothills. The Sierra Nevada is a great tilted fault block presenting its steep fault scarp to the east (Fig. 15.19). The equally dramatic Teton Range of Wyoming rises in the same way out of the plains to the east (Fig. 15.20). The Colorado Plateau steps down to the Great Basin by a series of similar fault scarps, this time facing westward in southern Utah and northern Arizona.

Let us consider for a moment the effects on other physical systems of such uplifting as occurred in the Sierra Nevada. Stream erosion is accelerated by the increase in slope. Precipitation on the windward slope is increased because of orographic lifting. The steep lee side of the tilted fault block is drier than before, because it now lies in the rain shadow of the range. Increased precipitation and the lower temperatures at higher elevations combine to change the climate of the raised land significantly, which then changes vegetation, soils, and animal life as well. Soils are also affected by an increase in surface runoff and erosion. The process of uplift extended over several million years, so these changes have been very gradual. The Sierra Nevada are continuing to rise rapidly,

Figure 15.19 The east front of the Sierra Nevada in California is a steep fault scarp.

Figure 15.20 *The steep fault scarp of the Tetons in Wyoming.* • **What is unique about a fault scarp?**

Figure 15.21 *The San Andreas Fault across the Carrizo Plain in central California. The area to the west of the fault is moving northwest, while the area to the east is moving relatively to the south. Note the extensive erosion that is taking place in the weakened rock along the fault zone.* • **What type of fault is the San Andreas?**

geologically speaking—perhaps a centimeter each year. In addition, it should be remembered that the gradational agents (water, ice, wind, and gravity, alone or in combination) begin working to level the land as soon as lifting begins. The Sierra Nevada, for instance, are much different today than they once were, for they have been carved and etched by thousands of years of glaciation, stream erosion, and gravitational mass movement of weathered material.

The displacement along some faults is horizontal rather than vertical. All such faults descend steeply into Earth's crust, but the slippage is parallel to the surface trace, or *strike,* of the fault. This horizontal movement creates a **strike-slip fault.** It is visible in the horizontal displacement of roads, railroad tracks, fences, and stream beds. The San Andreas Fault, which runs through much of California, is a strike-slip fault (Fig. 15.21).

Often a series of parallel faults is encountered. Where the block between two faults is elevated above the land to either side because it has been pushed up or because the land to either side has dropped down, the raised portion is called a **horst.** The central European Highlands north of the Alps contain many horsts, as does the Basin and Range region of the southwestern United States. The Basin and Range region extends eastward from California to Utah and southward from Oregon to New Mexico, including all of Nevada. The great Ruwenzori Range of East Africa is a horst, as is the Sinai Peninsula between the fault troughs in the gulfs of Suez and Aqaba (see again Fig. 14.17).

The opposite of a horst is a **graben,** a depression between two facing fault scarps that may occur when the block between two faults drops down or when the land to either side of the faults is uplifted. Classic examples of grabens are Death Valley, California (Fig. 15.22), and the middle Rhine Valley in Germany. Several interconnected grabens form major lowlands called **rift valleys.** Examples of rift valleys are the Rio Grande rift of New Mexico, the Great Rift Valley of East Africa, in which are located Lakes Tanganyika and Malawi, and the large Jordan rift valley, where

Figure 15.22 *Death Valley, California, is a classic example of a topographic graben. The valley floor is 86 meters (282 ft) below sea level—the lowest elevation in North America.*
• **What is the difference between a graben and a horst?**

the Dead Sea lies at a level some 390 meters (1280 ft) below the Mediterranean Sea, which is only 64 kilometers (40 mi) away. Rift valleys are also found along the centers of oceanic ridges.

Immense regions may be affected by crustal shattering along faults, producing a mosaic of horsts, grabens, and tilted fault blocks. The outstanding example is the U.S. Southwest's Basin and Range region.

Faulting, like folding, is an important part of landform development, especially mountain building. In fact, the two—faulting and folding—often occur in the same areas, sometimes in the vicinity of volcanism, the other important mountain builder. These processes are ongoing ones, as attested to by associated continuing earthquake activity.

Earthquakes

The most prominent evidence of present-day solid tectonic movement, **earthquakes** are vibrations of Earth that occur when the accumulating strain of slow crustal deformation is suddenly released by displacement along a fault. An earthquake does not cause displacement. Rather, the shock waves of an earthquake signify the release of energy caused by the movement of crustal blocks past one another. It is the earth-

quake's surface waves, which pass along the crustal surface, that cause the damage and loss of life that we associate with major earthquakes. The point at which an earthquake originates is the **focus,** which may occur anywhere from the surface to a depth of 700 kilometers (435 mi). The earthquake **epicenter** is the point on Earth's surface that lies directly above the focus; this is where the strongest shock is normally felt.

Most earthquakes are so slight that we cannot feel them, and they produce no visible damage. Usually they occur deep within Earth; no displacement is visible at the surface. Others are strong enough to rattle a few dishes, while a few are so strong as to topple buildings, break power and gas lines and water pipes, and trigger rockslides and landslides. Aftershocks may follow the main quake as the crustal movement settles. Geophysicists are currently investigating the possibility that foreshocks may alert us to major earthquakes, though evidence is at present inconclusive.

The scale of **earthquake magnitude,** developed by Charles F. Richter in 1935, is based on the energy released and is a measure of the intensity of ground motion as recorded on seismographs. Every increase of one number in magnitude (for example, from 6 to 7) means the ground motion is 10 times greater. (The actual energy released may be 30 times greater.) The extremely destructive 1906 San Francisco earthquake was rated at 8.3, but the strongest earthquake in North America (8.6) occurred in Alaska in 1964. The 1989 Loma Prieta earthquake near San Francisco measured only 7.1, but it was still responsible for the deaths of over 60 people. Most were crushed in their automobiles when an elevated freeway collapsed on the streets below. Factors other than magnitude also play a role in the loss of human life during an earthquake. These include depth of the epicenter, time lapse of the quake, and population density and building construction in the affected area.

The Los Angeles area earthquake in 1994 killed about 60 people, collapsed freeways, and caused commuter traffic jams for months afterward (Fig. 15.23a). The earthquake epicenter magnitude of 6.7 at Northridge, a San Fernando Valley community, was not considered the "big one" that Californians have been warned is com-

(a)

(b)

Figure 15.23 Earthquake damage along the Pacific rim. (a) The San Fernando Valley earthquake occurred on January 17, 1994. The epicenter was below Northridge, California, and the quake killed about 60 people in the Los Angeles area. (b) Exactly one year later, on January 17, 1995, an earthquake struck Kobe, Japan. The deadly tremor killed over 5000 people and destroyed over 50,000 buildings.

ing (see The Environment: California: Living in a Hazard Zone). Residents of the Pacific Northwest have also been warned about an impending "big one" as the North American plate slips over the subducting Juan de Fuca plate.

The location of an earthquake epicenter in relation to population density greatly influences its impact on humans. A strong tremor that struck the sparsely settled Mojave Desert communities of California in 1992 (magnitude 7.5) caused little loss of life. Yet a magnitude 6.9 quake in 1988 in Armenia (in the former Soviet Union) destroyed two cities, with a death toll of 25,000. Other relatively recent earthquake tragedies have occurred in Peru in 1970 (magnitude 7.8), where 65,000 lives were lost when enormous avalanches obliterated entire mountain villages; in Mexico in 1985 (magnitude 8.1), where there was a death toll of over 9000, mostly in the high-rise sections of Mexico City; and in Iran in 1990 (magnitude 7.7), where at least 50,000 died in remote areas of the country. More recently, in January 1995, a 7.2 magnitude quake struck the densely populated port city of Kobe, Japan. It killed more than

5000 people and destroyed more than 50,000 buildings (Fig. 15.23b). The death toll from a May 1995 earthquake (magnitude 7.5) on Russia's Pacific Rim island of Sakhalin was almost 2000. This quake destroyed the oil town of Neftegorsk, killing over one-half the town's residents. In 1998, two major earthquakes rocked northeastern Afghanistan, killing over 8000 people.

Although most earthquakes are related to major known faults and are in mountainous regions, the one most widely felt in North America was not. It occurred as one of a series during the period from 1811 to 1812, was centered near New Madrid, Missouri, and was felt from Canada to the Gulf of Mexico and from the Rocky Mountains to the Atlantic Ocean. Fortunately, the area was not densely settled at that stage in history. Though they are not common, recent small earthquakes have occurred in New England, New York, and the Mississippi Valley. Probably no region on Earth is what could be called "earthquake safe." The map in Figure 15.24 shows world regions where earthquakes are most common. Note the especially strong correlation between

(Continued on page 420)

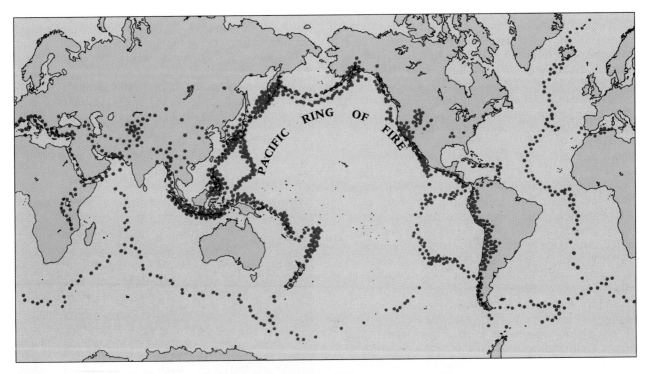

Figure 15.24 *Map of the locations of earthquake epicenters. Compare this map with Figure 14.14, which shows the major tectonic plates and demonstrates the strong relationship of earthquakes to plate boundaries. Also compare this map with Figure 15.8, the map of the world volcanic regions.* • **Is there a strong correlation between volcanic and earthquake locations? What parts of the world are susceptible to both tectonic hazards?**

CALIFORNIA: LIVING IN A HAZARD ZONE

In early November 1993, bone-dry Santa Ana winds were blowing out of the desert and through the canyons of southern California. The sustained winds reached 40 miles per hour, and in some areas 80-miles-per-hour winds whipped the dry California chaparral-covered hills. Several years of drought and fire prevention had allowed the buildup of great amounts of flammable tinder under the chaparral and scrub. Soon, numerous fires started, some from natural causes, some from arson. The Santa Ana winds whipped small fires into massive, fast-moving firestorms with temperatures in excess of 1700°F. From north of Los Angeles to the Mexican border, 14 major fires were burning out of control. In Malibu and Laguna Beach, the flames reached the beautiful Pacific coast.

On television, Americans watched as exhausted and helpless firemen tried to gain control of the fires and save homes. When it was all over, 800 homes were lost, with damage estimates over $500 million. Two communities suffered the most: in Laguna Beach, 318 homes were burned, and in Altadena, 118 homes were destroyed. In recent years, other devastating fires have struck Santa Barbara and Oakland, destroying hundreds of homes and taking human lives.

Fire is part of the natural cycle of the coastal California ecosystem. It rejuvenates the native vegetation. **However, the millions of people attracted to the sunny coastal region do not always realize that they have moved into a danger zone.**

From January through March 1998, Californians were again facing another natural hazard—this time floods. Pacific storms, powered by warmer than normal El Niño waters, slammed into the state, dumping record amounts of rainfall. Some areas received between 25 and 38 cm (10 and 15 in.) of rainfall in 24 hours, more than much of the region receives in a year. Rivers jumped their banks, flooding downtown areas of cities, washing away highways, collapsing a freeway bridge, and destroying numerous homes. Mudslides buried other homes and many streets. Miraculously, only a few people were killed by the dangerous flood waters and mudslides. Damage was estimated in the billions. Much of California's agricultural crop in the Central Valley was lost or damaged by the floods and winds.

As if fires and floods were not sufficient challenges, Californians must also face the hazard of a shifting ground. A deadly 6.7-magnitude earthquake struck beneath the San Fernando Valley, north of Los Angeles, on January 17, 1994. People were literally shaken out of their beds by the early morning quake. Fortunately, the quake struck at 4:31 A.M. and not during the morning commuter rush hours. Even though approximately 60 people were killed, hundreds of lives could have been lost had the quake struck at a later time. Apartment buildings collapsed, freeways crumbled, and natural-gas lines ruptured, setting off fires. Aftershocks scared residents for many days, hundreds of people were homeless during the cleanup, and commuting was a nightmare for months. The earthquake caused $20 billion in damage, the most expensive natural disaster in the history of the United States.

Earthquakes are also part of the California environment, and new faults are still being discovered. In fact, the epicenter of the quake that struck near Los Angeles was from a fault unknown to scientists. The Los Angeles Basin has been rocked by numerous past quakes. Over 60 people were killed by another quake, which struck the San Fernando Valley in 1971. Nearby Whittier, east of Los Angeles, was struck by a deadly quake in 1987.

Then there is San Francisco—a jewel of a city with its hills and cable cars and bridges. Like the Los Angeles Basin, the San Francisco Bay area is home to several million people. Yet the beautiful skyscrapers of San Francisco are built within sight of the infamous San Andreas Fault and its numerous branches.

During the 1989 baseball World Series, millions of fans across the country were glued to their TV sets for the third game between San Francisco and neighboring Oakland, when the ground shook again. Vivid television coverage brought views of heroic rescue operations, cars hanging from the Bay Bridge, spectacular fires in the Marina District, a destroyed downtown Santa Cruz, and grisly scenes of excavating bodies of commuters crushed in their automobile tombs beneath a two-level freeway in Oakland. The 7.0-magni-

Wildfires burn Laguna Beach home.

418

Building destruction from the Northridge earthquake on January 17, 1994.

still to come! According to studies, both regions have the potential to be struck by earthquakes with magnitudes above 7.5. Estimates are that the Los Angeles area is struck about every 140 years, while the Bay area is shaken every 100 years.

Certainly California does not have a monopoly on the hazard zones of the world. Nor does California have a monopoly on people who choose to live in areas of high risk. However, the "golden state," with its large urban population, its earthquake-prone terrain, and a variable wet and dry climate, does offer a classic example of the growing tendency to build and live in hazard zones.

It seems apparent that there is little scientists can do to convince people not to build in hazard zones, and even less that they can do to convince them to move out once they have settled there. There is also little that can be done to avert or prevent natural disasters such as brushfires, floods, and earthquakes. **Therefore, to minimize loss of life and property we must focus our attention on a better understanding of the environment itself and on better preparation for the inevitable natural disasters.** Local governments can prevent development of the most dangerous areas by hazard zoning. Building codes can be enforced to prevent the rapid spread of fires or structural collapse during moderate earthquakes. In fact, California has enacted some of the strictest earthquake construction legislation in the world. Although it is expensive, individuals can buy earthquake and disaster insurance. Communities can organize to have disaster plans for future emergency situations. The benefits of all these preparations will not prevent future natural disasters from striking, but they will help save lives and property.

tude quake had an epicenter on the Loma Prieta fault in the Santa Cruz Mountains south of San Francisco. It was the strongest temblor since the 1906 quake that destroyed San Francisco.

Scientists have warned residents of both the Los Angeles and San Francisco areas that the "big one" is

CRITICAL THINKING ▼

(1) How does California's Mediterranean climate contribute to the catastrophic flood and fire events in that state? **(2)** Along what type of plate boundary is coastal California located? **(3)** If you were a regional planner in California, what types of terrain and landforms would you recommend be zoned as open space?

419

earthquakes and plate boundaries (see Fig. 14.14 for comparison).

Solid Tectonic Landforms

The original structural forms that result from solid tectonic activity vary from microscopic joints to major mountain ranges. However, none of these remain in original form because of the leveling action of the agents of gradation. It is important to note the difference between land-forms and the structural configurations mentioned here. For instance, an anticline may form a surface ridge but frequently does not because of erosional activity. For example, Nashville, Tennessee, occupies a topographic valley, yet it is sited in the remains of a structural dome. Only where the anticline is composed of extremely resistant rock does it persist as a ridge. The important distinction is that *mountain* or *ridge* refers to surface configuration, whereas the term *anticline* or *horst* signifies a geologic structure.

Define and Recall

volcanism	fissure	warping	fault scarp (escarpment)
plutonism	pluton	isostasy	strike-slip fault
pahoehoe (ropy lava)	stock	folding	horst
aa (blocky lava)	batholith	anticline	graben
pyroclastic materials	igneous intrusion	syncline	rift valley
composite cone	laccolith	recumbent fold	earthquake
shield volcano	dike	overthrust	focus
plug dome	sill	fault	epicenter
cinder cone	volcanic neck	dip-slip fault	earthquake magnitude
caldera		normal fault	
lava plateau (flood basalts)	dip	reverse fault	
	strike	thrust fault	

Discuss and Review

1. What are the four basic types of volcanoes? Give an example of each.

2. Which is the most dangerous type of volcano? Explain why.

3. What factors affect the geographic distribution of volcanoes?

4. How do plutonism and volcanism differ from each other?

5. What is a batholith? What is the difference between a dike and a sill?

6. What are the major differences between folding and faulting? What causes this type of crustal activity?

7. Draw a diagram of folding, showing anticlines, synclines, and thrusts.

8. What causes an earthquake? What is the relationship between the focus and the epicenter of an earthquake?

9. What is the major difference between volcanism and solid tectonic activity? What structural and landform features are the results of each of these two types of tectonic activity?

Consider and Respond

1. Name the tectonic activity that would be most responsible for the following:
   Sierra Nevada
   Cascades
   Basin and Range region
   Ridge and Valley section of the Appalachians

2. Name several areas in the United States that are highly susceptible to natural hazards from tectonic activity. Give examples of populated locations that may someday face potential disasters from earthquakes and/or volcanic activity.

3. List five countries that lie along the "Pacific Rim of Fire." Name a few countries that also face tectonic hazards but that are not in the Pacific region.

4. Can you recall any recent earthquakes or volcanic eruptions in the news? If so, where did these occur?

5. Assume you are a regional planner for an urban area in the western United States. What hazards must you plan for if the region has active fault zones and volcanoes? What recommendations would you make as far as land use and settlement patterns are concerned to lessen the danger from these tectonic hazards?

CHAPTER

16

GRADATION, WEATHERING, AND MASS MOVEMENT

CHAPTER PREVIEW

► The cycle of tectonism and gradation is endless because the magmatic and solid tectonic processes invariably create new landforms to be reduced before gradation has been completed.

What evidence in Earth's crust and on Earth's surface supports this statement? What does the statement suggest concerning the appearance and distribution of past and future landforms?

► The theoretical end product of gradation is a land surface reduced to base level (sea level).

Why has this not happened? What are the sources of the energy that produces gradation?

► Weathering and mass movement are the initial steps in the overall process of gradation.

Why are they essential? Why are they initial? How are they related to erosion and deposition?

► Processes such as sedimentation, volcanism, faulting, and folding sow the seeds of destruction within the rock structures and landforms they create.

How is this so? How are these processes related to joints and fractures?

► Different rock types weather at differing rates; these rates are especially influenced by climatic conditions and the type of weathering involved.

In what ways does climate influence weathering? How can differential weathering influence landforms?

► Despite the spectacular nature of many types of mass movement, it is the imperceptible downhill motion of soil and regolith termed creep that is most effective.

How can this be so? What factors facilitate creep?

► Despite the personal risks and potential economic loss involved, people still persist in living in many of the world's greatest "hazard zones."

Why do they continue to live in such areas? What can be done to reduce the risks?

▲ The weathered "Red Rocks" at Sedona, Arizona.

*I*n Chapters 14 and 15 we summarized the theories, materials, processes, and structures associated with the creation of topographical irregularities on Earth's surface. Now we will begin to examine the gradational processes that are simultaneously at work removing these irregularities. Gradation fills in depressions in the land and wears down areas of higher elevation. In some places gradation also helps build up the land surface by creating depositional landforms such as alluvial fans, deltas (Chapter 18), sand dunes (Chapter 20), and glacial moraines (Chapter 19). The gradational forces have their origin and action at Earth's surface. In this sense they can be contrasted to the tectonic forces that originate below Earth's surface.

Although the ultimate result of gradation is a leveling and smoothing of the surface, the various stages between the original uplift by the tectonic forces and the final result of gradation may include landforms even more irregular than those produced by tectonism alone. For example, the surface of a plateau shortly after uplift may be relatively smooth, although its elevation has been raised thousands of feet. Gradation, working through its primary agent — running water — gradually begins to cut up the land; valleys develop, separated by mountains or hills. Eventually, as the moving water wears away more and more of the raised portions of the land, the irregularities in the surface originally created by water erosion are reduced. Thus, many variably shaped landforms are actually the result of gradational reductions in differences in elevation.

Gradation and Tectonism

We have been speaking here of a cycle of landform development in which the land is raised by tectonic processes and then worn down or leveled by gradation. To complete the cycle, the land surface must be reduced to a level and form beyond which no further modification by the gradational processes is possible. This is, however, a hypothetical model of what actually happens. In reality, gradation is an ongoing process, and its agents are constantly at work in all places where there are differences in elevation. Furthermore, the forces of solid tectonism and volcanism do

not wait for gradation to be completed before they act again; they may occur at any time, creating new landforms to be acted on by the gradational processes. Thus we see many of Earth's subsystems at work: tectonic action producing new landforms, such as fault scarps, folded mountains, and volcanoes, on which gradational forces work to create reduced slopes which themselves will again be altered by future tectonic movements, renewing the cycle once again.

The Gradational Processes

Gradation includes the picking up and removal of loose material, its transportation to another location, and its deposition there. **Erosion**—the wearing down of the land—and the transportation of material together act to lower the elevation of the land. **Deposition,** on the other hand, is the filling in of depressions or the raising of land elevation. The combined result of erosion and deposition is the gradational reduction of irregularities in elevation (Fig. 16.1).

The ultimate effect of gradation is to reduce the land surface to **base level**—a surface so flat that erosional forces can no longer affect it. The ultimate base level is the level of the sea, below

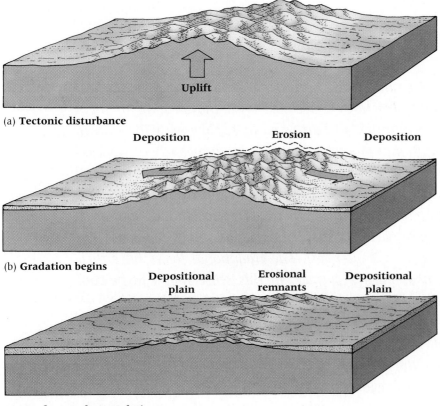

(a) **Tectonic disturbance**

Deposition   Erosion   Deposition

(b) **Gradation begins**

Depositional plain   Erosional remnants   Depositional plain

(c) **Nearly complete gradation**

Figure 16.1 The process of gradation. (a) tectonic uplift; (b) gradation begins; (c) nearly complete gradation. • What is the ultimate effect of the gradation process?

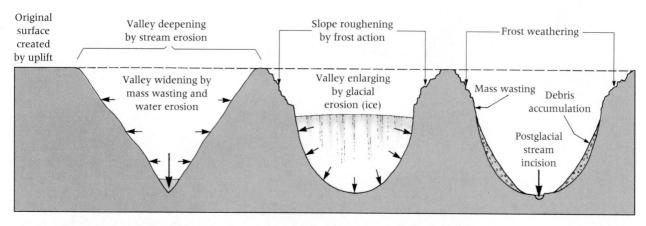

Figure 16.2 *The scenery of most high mountain regions has been formed by at least three separate phases of gradational development: stream cutting and valley development during tectonic uplift; glacial enlargement of former stream valleys, with intense frost weathering above the level of the ice; and postglacial weathering and mass wasting with recent stream cutting.* • **How does the profile of a valley change at each phase?**

which the land surface cannot be lowered by the normal gradational forces. If there were no further tectonic activity, the surfaces of the continents would eventually be reduced to sea level by gradation.

The removal of material from one place and its deposition in another place is accomplished by gradational agents, the most important of which are water, wind, and ice. Chapters 17 and 18 discuss how water below the land surface and in streams acts as a gradational agent. Chapter 20 covers wind as a gradational agent. Chapter 19 includes an examination of ice as an agent of gradation, and Chapter 22 includes a discussion of gradation by waves in coastal zones. There are also other gradational agents, such as humans, animals, and plants.

Gravity and solar radiation provide the energy for gradational agents. Gradation may also be accomplished by the force of gravity acting alone. Gravity may be the primary cause of the movement of rock and weathered debris downslope to lower elevations. Such removal and deposition is called **mass movement** or **mass wasting.** In the final part of this chapter we will consider some of the varieties of mass movement.

The individual gradational agents usually do not work alone on a particular landform. More often they combine, sometimes sequentially, to level the land. For example, the spectacular scenery of the northern Rockies is to a large extent the result of ice action, but stream erosion and mass wasting have also been involved (Figs. 16.2 and 16.3). However, in order to simplify the

explanation of the gradational forces and processes and their resultant landforms, each agent and its major landforms will be discussed separately.

Weathering

Weathering includes various processes by which rocks at or near Earth's surface are disintegrated or decomposed in preparation for their removal and transportation. Weathering eventually breaks down even Earth's most massive rocks into components small enough to be moved downslope by one or more of the gradational agents.

Some Earth scientists include weathering in the concept of gradation. However, weathering of Earth's rocks can occur without gradation. Weathering does not by itself level the land; it only fragments rocks so they can be moved, thus allowing gradation to take place. The definition of weathering is the breakdown of the surface rock material *in place,* involving little or no movement of the rock material by the agent of weathering itself. However, it is important to note that practically no gradation will result without initial preparation by weathering.

Earth scientists usually divide weathering into two basic types. **Physical** or **mechanical weathering** disintegrates rocks without altering their chemical composition. **Chemical weathering** decays rock by a variety of chemical reactions. Both types of weathering normally take

Figure 16.3 *An aerial view across the summits of the northern Rockies. The mountains in the foreground are the White Cloud Peaks in the Sawtooth Range of Idaho.* • **Can you identify evidence of the three phases shown in Figure 16.2?**

place at or near the surface of Earth, though evidence of weathering has been found as far as 185 meters (600 ft) below the surface. Variations in the depths to which weathering will occur, as well as variations in the type and rate of weathering, depend primarily on (1) the structure and composition of the rocks, (2) climate, (3) the configuration of the land surface, and (4) the vegetative cover. Of particular importance to geographers is the relationship of climate, especially temperature and moisture conditions, to rates and types of weathering (Fig. 16.4). In field observations, physical and chemical weathering processes cannot always be separated.

Physical Weathering

The fragmentation or mechanical disintegration of rocks by physical weathering is important to gradation in two ways. First, the smaller rock pieces are more easily removed and transported by one of the gradational agents. In this way mechanical disintegration aids gradation. Second, the fragmentation of a large rock into smaller ones encourages chemical weathering by increasing the rock's surface area and thus exposing more of the rock to possible chemical decomposition.

The joints and fractures that develop in igneous, sedimentary, and metamorphic rocks are areas of weakness that encourage physical weathering by making fragmentation easier. Chemical weathering also proceeds faster along joint planes or zones of stress. Jointing is especially common in certain sedimentary rock types such as limestone, but it can be found in any rock that has been subjected to the stresses of volcanism and solid tectonic activity (Fig. 16.5). Thus, the

THEORETICAL WEATHERING REGIONS

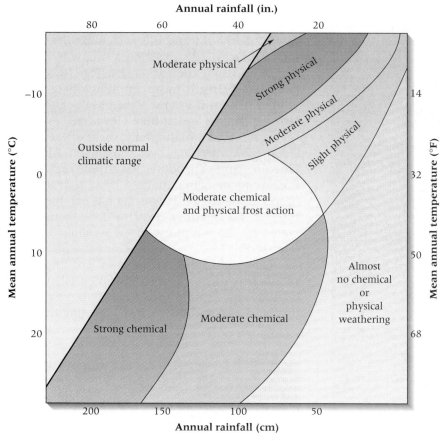

Annual rainfall (in.)

Moderate physical
Strong physical
Moderate physical
Slight physical

Outside normal climatic range

Moderate chemical and physical frost action

Almost no chemical or physical weathering

Strong chemical

Moderate chemical

Mean annual temperature (°C)

Mean annual temperature (°F)

Annual rainfall (cm)

Figure 16.4 *This diagram of weathering regions shows the relationship between the climate and the weathering processes. Physical weathering is most active where temperature and rainfall are both low, while chemical weathering is most active in regions of high temperature and rainfall. Most regions of the world have a combination of both physical and chemical weathering.* • **In what weathering region would an area that had an annual mean temperature of 5°C and annual rainfall of 100 centimeters be located?**

Figure 16.5 *Jointed and weathered granite near the summit of Mt. Whitney in the Sierra Nevada of California. The closer the spacing of joints in any rock type, the more rapid the weathering, both physical and chemical.* • **Why do high mountain regions such as this one have such severe weathering?**

processes of mountain building and rock formation work with physical and chemical weathering to produce gradation.

Plants, animals, and humans all contribute to the mechanical disintegration of rocks. Plant roots wedge into cracks, and as their roots grow, they exert pressure on the rocks, eventually fragmenting them (Fig. 16.6). Burrowing animals, such as gophers and ground squirrels, weaken rocks and cause fragmentation and disintegration. The actions of humans also encourage physical weathering. Bulldozers, trail bikes, bombs, mining, quarrying, and even hikers all do their bit to shatter rocks into smaller fragments.

The freezing of water in cracks is one of the most important means of physical weathering. When water freezes it becomes less dense and expands in volume by 9 percent, exerting tremendous outward force. Drivers who fail to fill their radiators with antifreeze before winter or before driving to the mountains on a ski trip find this

Figure 16.6 *Physical weathering in Olympic National Park, Washington. As tree roots grow, the rock is wedged apart.* • **Besides trees, what other organisms cause organic weathering?**

Figure 16.7 *Frost wedging of a rock at Devil's Lake, Wisconsin. As water freezes in rock joints, it expands, wedging further into the rock.* • **What is the reason that frost wedging can split such a large rock?**

an expensive lesson. Plumbing pipes in homes in cold-winter areas may burst if they are not well insulated.

Similarly, if water fills the cracks in exposed rock surfaces, the expanding ice will form a wedge and split the rock apart (Fig. 16.7). Such **frost wedging** is most common where freezes are frequent and intense—in middle- and high-latitude regions and at high elevations. Even the "ageless" 60-foot-high granite faces of Presidents George Washington, Thomas Jefferson, Abraham Lincoln, and Theodore Roosevelt on Mount Rushmore are being attacked by this weathering process. The freeze-thaw cycle has enlarged fractures in the granite since the faces were completed by the sculptor Gutzon Borglum in 1941 (Fig. 16.8).

Although frost wedging is an important weathering process throughout high latitudes, it is most active in high mountains above the tree line. The exposed bedrock of high mountains is

Figure 16.8 *Even the granite face of President Lincoln at Mount Rushmore is aging due to the freeze-thaw cycle and the work of frost wedging.* • **Will he need a facelift someday?**

Figure 16.9 *Talus cones in the Canadian Rockies formed by weathering and rockfalls at the base of steep slopes.* • **Why does this rock debris form a cone shape?**

splintered by frost wedging, and when shattered rock breaks loose, the force of gravity causes it to tumble downslope. The weathered rock material that accumulates at the base of mountain slopes (called **talus**) often takes a fan-shaped form termed a **talus cone** (Fig. 16.9).

Mechanical disintegration or wedging also results from the growth of salt crystals. This commonly occurs in sandstone or fractured igneous rocks in arid and semiarid areas. After rain falls, capillary moisture moves outward to the surface of porous sandstone. Evaporation of this water leaves behind deposits of salt crystals. As these crystals grow, they wedge out sand grains. Repeated over time, this action can create multitudes of niches and shallow caves in large sandstone cliffs. **Salt weathering** or **salt wedging** is also an important factor in the breaking down of rocks in the splash zone along rocky marine coasts (Chapter 22).

Weathered caves or overhangs were important living space for prehistoric Native American communities in the American Southwest—for example, at Mesa Verde, Colorado, or Canyon de Chelly, Arizona (Fig. 16.10). When the Native American cliff-dwelling villages were facing toward the south, they were shaded when the sun was high in summer and warmed by the low-angle sun's rays in winter.

Granular disintegration breaks rock down into its constituent grains or particles, especially when rocks are composed of a variety of minerals. Granite, composed of quartz, feldspar, and mica, is one example. These minerals expand with heat and contract with cold at different rates, setting up stresses and strains within the rock that eventually may cause it to disintegrate.

It was once believed that extreme diurnal changes of temperature in desert regions were a *primary* cause of rock disintegration there. Laboratory studies now refute this. Fire, however, can be an effective weathering agent. Since rock is a poor conductor of heat, the fire expands only the rock's outer shell. In addition, the expansion fracturing is compounded by the pressure of steam formed from the water in the rocks. Thus, bedrock and boulders exposed to forest and brush fires often undergo peeling and granular disintegration.

Exfoliation refers to the *spalling*, or breaking off, of concentric or curved rock shells parallel to the surface. It is especially common in granite and was once thought to be caused by temperature changes. However, exfoliation may actually be caused by both chemical and physical weathering processes. Shallow exfoliation producing sheets or spalls a millimeter or so thick is closely associated with hydration, a chemical weathering process. Apparently there is also some validity to the theory that the unloading of pressure on rocks can result in exfoliation. When upper layers of rock are removed through

Figure 16.10 *Prehistoric Native American ruins of dwellings built into the cliffs at Montezuma Castle National Monument, Arizona.* • **Can you think of reasons why these Native Americans chose these cliffs for their dwellings?**

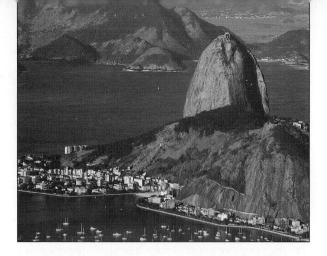

Figure 16.11 *Sugar Loaf, Rio de Janeiro, Brazil, is a granite exfoliation dome.* • **Why is granite so susceptible to exfoliation?**

erosion, pressure on the rocks below is lessened, so the rocks can minutely expand. As they do so, fractures develop, creating concentric shells of rock that are often compared to the layers of an onion. This characteristic fracturing is especially apparent in the massive intrusive rocks like granite, which crystallized under very high pressure. Also, bedding planes do not exist in granite to absorb the expansion as the overlying rocks are removed. As rock layers are peeled away from large, solid rock formations, immense dome-shaped masses, called **exfoliation domes,** are left. Yosemite Valley in California contains some of the world's best examples of exfoliation domes, especially Half-Dome. Sugar Loaf, overlooking Rio de Janeiro's bays, is a similar exfoliated granite dome (Fig. 16.11).

Chemical Weathering

Chemical weathering, or **decomposition,** prepares rocks for removal and transportation by the gradational agents in three ways. First, chemical alteration forms new minerals that are softer and/or finer and therefore less resistant to erosion. Second, chemical weathering may create substances—through the addition of water or through chemical change—that are greater in volume than the original rock material. This expansion can fracture the rock, increasing the rate of both physical and chemical weathering by increasing the amount of surface area and by weakening the rock mass. Third, chemical weathering may dissolve minerals in a chemically active water solution (weak acid), making them easy to remove and transport. As more and more minerals are removed, the number and size of pore spaces and other openings in the rock are increased, allowing the rate of weathering to increase.

Almost without exception, chemical weathering can take place only in the presence of wa-

ter. Therefore, the rate of chemical weathering is always increased by the addition of water, and chemical weathering is most rapid in humid climates. Even arid climates, however, have enough moisture to allow some chemical weathering to take place, evidence of which is found in the rounded boulders showing the exfoliation and granular disintegration common in those arid regions where granite is present (Fig. 16.12).

Chemical reactions are also more rapid at high temperatures. Consequently, tropical regions are subject to more chemical weathering than are places with lower temperatures. Subarctic and polar climates are subject to chemical weathering, but the effects of physical weathering are much more visible.

Thus, in hot, humid regions, such as the tropical rainforest, savanna, and monsoon climates, chemical weathering is more significant in affecting the shape of landforms than is physical weathering. The landforms and rocks of warm climates show the importance of chemical weathering in their rounded shapes. In contrast, the landforms and rocks of drier and cooler regions, where physical weathering is more important, are sharp, angular, and jagged (refer back to Fig. 16.4).

The principal processes of chemical weathering are oxidation, hydration and hydrolysis, carbonation, and solution. The most important agents performing these processes are water, oxygen, and carbon dioxide, all of which are common elements found in soil, rock, precipitation, groundwater, and air.

Figure 16.12 *Block disintegration of granite rocks, Mojave Desert. The blocks of rock became rounded or "spheroidal" where angles were attacked by both physical and chemical weathering.* • **Can you see where the original joints were located?**

Oxidation. The chemical union of oxygen with another substance to form a new product is called **oxidation** (Fig. 16.13). The chemical compounds formed by oxidation are usually softer and finer than the original substances and often have a greater volume. One of the most common of the *oxides,* as these compounds are called, is iron rust, derived from the combination of iron and oxygen. Rust is composed of two iron oxides: the minerals hematite (red) and limonite (yellow). The rusty stains visible on many rocks are a combination of hematite and limonite, both of which are softer and more easily removed from rocks than the original iron-bearing minerals from which they were formed.

Hydration and Hydrolysis. Though both involve the addition of water to a chemical substance, hydration and hydrolysis are distinctly different processes. In **hydration**, water molecules are attached to the molecules of another substance without any chemical change occurring. Hydration produces expansion, which can in turn result in either granular disintegration or exfoliation. Hydration also weakens minerals, making them more susceptible to other chemical weathering processes—particularly oxidation and hydrolysis—as well as to physical weathering.

Unlike hydration, **hydrolysis** does involve a chemical change through the union of water with another substance. Many common rock minerals are susceptible to hydrolysis, particularly the silicate minerals that form igneous rocks. In many cases hydrolysis, like hydration, results in expansion that can lead to exfoliation. Hydrolysis is not limited to rock exposed at the surface. In warm, humid climates it can decay rock 30 meters or more below the surface.

Solution. Though it does not actually involve chemical change, solution is an important process of chemical weathering. Once minerals are dissolved in water, they are easily removed and transported by groundwater or soil water, or surface water flow. This is known as *chemical erosion.*

Some rock minerals, such as sodium chloride (salt) and calcium sulfate (gypsum), are soluble

Figure 16.13 *The Painted Desert in Petrified Forest National Park, Arizona.* • **The red shales result from chemical weathering of what element?**

in neutral or acidic water. Mineral salts that are immediately soluble in water are called **evaporites,** since they precipitate when the water becomes saturated with them, as during the evaporation process. Other minerals, the products of previous chemical weathering, are also soluble. Even silica becomes soluble in hot, wet climates. Removal of silica is a major feature of the laterization process (see Chapter 13).

Many minerals that are insoluble or only slightly soluble in pure water are easily dissolved in slightly acidic water solutions. Rain and soil water, by absorbing carbon dioxide from the atmosphere and from decaying vegetation in the soil, often form a weak solution of carbonic acid, which is capable of dissolving a variety of minerals, notably calcite or calcium carbonate, the ingredient of common limestone (Fig. 16.14). When acted on by carbonic acid, calcium

Figure 16.14 *Weathered limestone near the shores of Lake Boringo, Kenya. Exposed limestone quickly discolors and becomes pitted by solution in wet tropical climates.* • **Why is limestone so easily weakened under these conditions?**

Figure 16.15 The Parthenon in Athens, Greece. • Why have the last several decades been so damaging to this ancient historic structure?

ble (calcite), there is a growing concern about the damage to these treasures. The Parthenon in Greece, the Taj Mahal in India, and the Great Sphinx in Egypt are examples where chemical solution and salt buildup are damaging and rotting away monument surfaces (Fig. 16.15).

Carbonic acid is not the only acid active in chemical weathering by solution. Other acids, derived primarily from the decay of organic matter, are also present in groundwater and are capable of dissolving other rock minerals. Although most organic weathering takes place below the soil surface, it can also affect exposed rock. Such decay is associated with the chemical activity of lichens and mosses, which secrete acid substances that assist rock disintegration and thus produce nutrients for these simple life forms as they colonize bare rock.

Differential Weathering

Different rock types are affected by different weathering processes and breakdown, and are removed at different rates. Granite is resistant to physical weathering but is altered by oxidation, hydration, and hydrolysis. Accordingly, it is usually covered by a deep weathered regolith. Limestone is removed quite rapidly by carbonation and solution. Shale is chemically inert but is mechanically weak and is susceptible to hydration, which converts it back to clay, its original material. Sandstone is only as strong as its

carbonate forms calcium bicarbonate, a salt that is soluble in water. Thus, the solution of limestone involves both carbonation (creation of calcium bicarbonate) and solution. In humid regions, the leaching (or removal) of salts such as calcium bicarbonate can severely weaken rock by greatly increasing the size of pore openings.

In recent decades the phenomena of acid rain and acid fogs, caused by air pollutants dissolved into atmospheric moisture, have speeded up chemical weathering. Since many of the great monuments and sculptures of the world are composed of limestone (calcium carbonate) or mar-

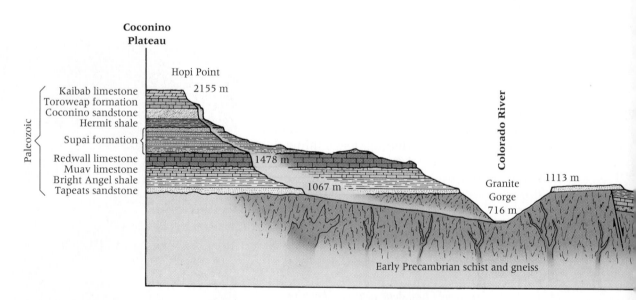

Figure 16.17 The Grand Canyon of the Colorado River is a classic example of differential weathering and erosion in an arid climate. This cross section shows the relationship between rock type and surface form in the canyon. • **What surface features do the more resistant rocks, such as sandstone, form?**

Figure 16.16 *Differential weathering of the Grand Canyon, Arizona. Two-billion-year-old schist forms the steep inner gorge. It is overlain by the Tapeats sandstone, which forms the Tonto Platform immediately above the inner gorge.*

cement, which varies from soluble calcite to inert silica.

Wherever a number of different rock types are associated in a landscape, some will be relatively *strong,* or resistant to weathering, and others will be *weak,* or easily altered and removed. A rock that is strong in contrast to other rocks in one area may be weak relative to different rocks in another area. Likewise, rocks that are resistant in a climate dominated by chemical weathering may be weak where physical weathering processes dominate. In general, the more massive the rock (the fewer the joints and bedding planes), the more resistant it is to all types of weathering. The process by which certain rock types weather more rapidly than others, resulting in an uneven landscape, is known as **differential weathering.**

Differences in rock resistance to weathering are highly visible in the landscape. Resistant rocks stand out as cliffs, ridges, or mountains, while weak rocks are eroded away to form valleys, subdued hills, and gentle slopes. One of the most outstanding examples of *differential weathering and erosion* is the scenery at Arizona's Grand Canyon (Figs. 16.16 and 16.17). In Arizona's dry climate, limestone is a resistant rock, as are sandstones and conglomerates, while shale is relatively weak. Thus, the canyon is a landscape of cliffs and ledges composed of limestone, sandstone, and conglomerate, separated by gentler slopes of shale. The cliffs are undermined as the shale beneath is removed. At the base of the canyon, resistant and ancient metamorphic rocks have produced a steep-walled inner gorge.

An excellent regional example of differential weathering and erosion may be seen in the Appalachian Ridge and Valley region of the eastern

(Continued on page 435)

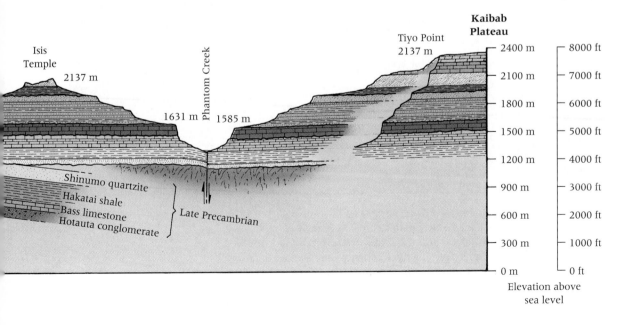

Isis Temple
2137 m

Phantom Creek

1631 m   1585 m

Tiyo Point
2137 m

Kaibab Plateau

Shinumo quartzite
Hakatai shale
Bass limestone
Hotauta conglomerate
Late Precambrian

— 2400 m   — 8000 ft
— 2100 m   — 7000 ft
— 1800 m   — 6000 ft
— 1500 m   — 5000 ft
— 1200 m   — 4000 ft
— 900 m   — 3000 ft
— 600 m   — 2000 ft
— 300 m   — 1000 ft
— 0 m   — 0 ft

Elevation above
sea level

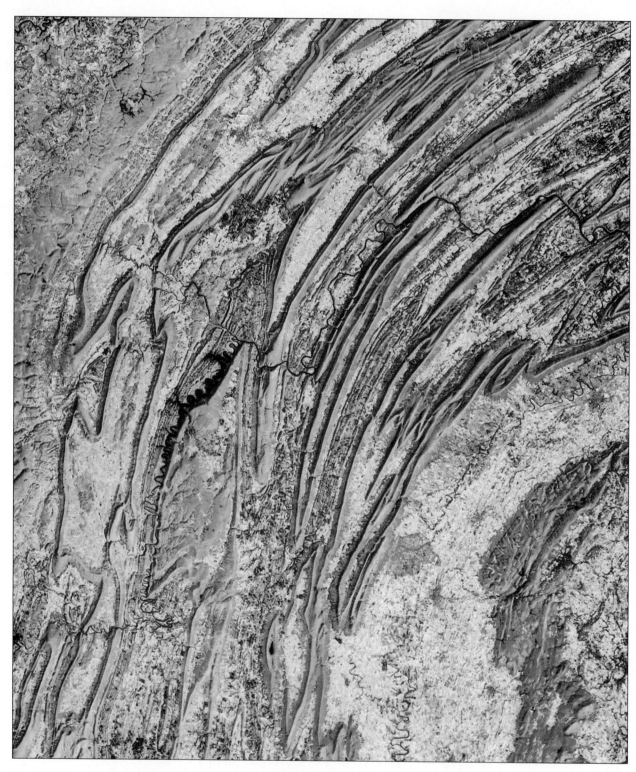

Figure 16.18 *A color-enhanced satellite image of Pennsylvania's Ridge and Valley section of the Appalachians. The image shows the effects of weathering and erosion on rock layers of different resistance. The most resistant rocks form the ridges, while weaker rocks form the valleys. The image also shows the pattern of anticlines and synclines formed from a collision between the North American plate and the African plate. The Susquehanna River can be seen cutting across the ridges at the northern part of the image. The river flowed across the region prior to the development of the present surface. • **Can you see how the topography of the Ridge and Valley section influences human settlement patterns?***

United States (Fig. 16.18). The geologic structure here consists of sandstone, conglomerate, shale, and limestone folded into anticlines and synclines. These have been eroded so that the edges of the steeply dipping rock layers are exposed at the surface. In this humid area, forested ridges composed of resistant sandstone and conglomerate stand up to 2000 feet above agricultural lowlands excavated out of shale and soluble limestone.

We have used the expression *differential weathering and erosion* as a single term because, although it is weathering that breaks up rocks at differential rates, it is erosion, or removal, that produces the landforms we see. Thus, weathering and erosion are closely related in discussions of the origin of landforms.

The Importance of Weathering

The fragmented material prepared by both chemical and physical weathering not only is available for erosion—either by direct gravitational transfer or as part of the load carried by rivers, glaciers, or wind—but it also forms the raw material for soil development. The regolith formed by weathering is the major source of the inorganic portions of soils, without which most vegetation could not grow. It is also the major source of nutrients for ocean waters as rivers carry the regolith to the coastlines.

Chemical weathering plays an important role in breaking down rocks into their mineral components and in creating new compounds. In this way valuable ores of iron, aluminum, tin, manganese, and uranium may be concentrated over long periods of time. This effect usually occurs under humid, tropical conditions as part of the laterization process (see Chapter 13). Such secondary ores are to be distinguished from primary metallic ores, which are concentrated by magmatic processes.

Mass Movement

Mass movement is a collective term for all the downslope movements of surface materials that take place in direct response to the pull of gravity. All over Earth's surface, and at all times, gravity pulls objects toward Earth's center. On sloping surfaces the result of this force is a general downslope movement of loose material. Any material that does not have the stability to resist the force of gravity responds by rolling, falling, sliding, or flowing downslope, stopping at the bottom of the slope or at a point where there is

Figure 16.19 *The scar and debris pile from the rockslide at Frank, Alberta, in 1903. In less than two minutes, millions of tons of rock slid down the steep slopes of Turtle Mountain and buried the town of Frank.* • **What might trigger such a catastrophic event?**

enough support to resist gravitational pull—the **angle of repose** for the material in question.

Mass movement occurs in a wide variety of ways. A single rock rolling and tumbling downhill is a form of gravitational transfer, as is an entire hillside flowing hundreds or thousands of feet downslope, burying homes, cars, and trees. Some mass movement is catastrophic in scale and violence, as in Figure 16.19. However, the more common type of mass movement is so slow as to be imperceptible, though tilted telephone poles, gravestones, and fenceposts indirectly reveal its presence.

The combination of all forms of mass movement rivals the gradational work of water as a modifier of physical landscapes, because the pull of gravity is always present. Wherever there is regolith or soil on a slope of almost any inclination, gravity will be able to effect some movement downslope. Because gravitational force is stronger on steeper slopes, it pulls down the regolith faster there. Consequently, the regolith on steep slopes is apt to be thinner than it is on more gentle slopes, and bedrock may be exposed.

Gravity is obviously the main cause of mass movement. However, other factors are also important, especially in providing the initial instability leading toward movement. Water is the most important additional factor encouraging mass movement. Water fills pore spaces, greatly increasing the weight of the material into which it has soaked. Water also lubricates material, helping particles to slip over each other more easily. Thus, especially in saturated material, water acts to minimize friction between particles of weathered material and reduce the resistance to gravitational pull.

The shaking produced by earthquakes, explosions, and even by the movements of heavy trucks or trains can be enough to shake material loose from a supported position, triggering movement. The undercutting of slopes by streams, waves, or bulldozers is an especially conspicuous trigger of mass movement. Groundwater, meltwater, and alternate periods of freezing and thawing or wetting and drying can contribute to mass movement, especially to slow mass movement. Even plant root systems and burrowing animals promote mass movement.

Forms of Slow Mass Movement

Creep is the name given the slowest downhill movement of soil and regolith. Creep is so slow as to be imperceptible to the observer. Yet creep is the most widespread, persistent, and effective of all forms of mass movement, for it is going on at all times on nearly all slopes where there are weathered materials and soil available for movement.

For the most part, creep does not produce distinctive landforms. Instead, it gradually wears away slopes and deposits material at the base to be carried away by one of the gradational agents, usually running water. Sometimes the material that creeps downslope is not removed but accumulates in the basin, filling it in and leveling the bottom of the slope.

Several factors facilitate creep; in most locations these factors act in combination to move the regolith and soil slowly downslope. When soil water freezes, it expands in volume, pushing soil particles upward. When the ice thaws, the soil contracts. Soil particles sink back down, but the force of gravity causes them to be displaced downslope of their original position, as shown in Figure 16.20. The repeated cycles of freezing and thawing, which cause repeated lifting and downslope sinking of soil particles, result in downslope creep of the mass of the soil. The rate of movement is very slow, usually less than a few centimeters per year.

Alternate periods of wetting and drying, which result in the alternate expansion and contraction of soil or regolith, also produce a net effect of downslope movement. This effect occurs because expansion tends to be greater on the downslope side of the material, in response to the ever-present pull of gravity. Likewise, there is less contraction on the downslope side.

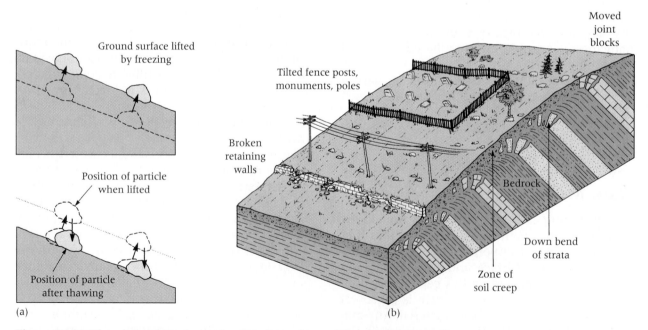

Figure 16.20 *The relationship of creep to soil volume change. Diagram (a) shows the freeze-thaw mechanism that moves particles downslope. Diagram (b) presents some common visible effects of creep on natural and human features.* • **What other human structures might be damaged by creep?**

Small burrowing animals such as gophers and ground squirrels are very effective soil movers. The tunnels they dig help push soil downhill, and the excavated material they bring to the surface also tends to fall downslope. Plant roots, too, push soils outward. Even the trampling of soil and regolith by humans or herds of grazing animals tends to push surface material downhill. Every step up or down a steep slope or even across it pushes some surface material downslope to a slightly lower position. Because most of the elements that encourage soil creep act near the surface, the rate of creep is greater there than below the surface. Therefore, telephone poles, fence posts, and even trees, all of which are anchored at a level below the surface, exhibit a downslope tilt.

Solifluction is the relatively slow downslope movement of soil and/or regolith that is saturated with water. Solifluction, which means "soil flow," is most common at high latitudes or elevations, where there is a layer of permafrost beneath the surface. During summer, when temperatures are higher, only the top few meters of the soil thaw. However, the layer of permafrost beneath the surface prevents the downward percolation of thawed water. As a consequence, the unfrozen part of the soil becomes watersoaked, creating a soggy mass that sags slowly downslope in response to gravity until the next freeze arrives. The annual movement may amount to only several centimeters. Solifluction is encouraged by the sparseness of vegetation found in tundra regions. Plant roots on gentle slopes act to retard mass movements, but where permafrost exists, roots are very shallow. The portion of the soil that freezes and thaws annually and that moves when saturated is called the *active layer*.

Evidence of solifluction can be seen in tundra landscapes, where irregular lobes of soil produce hummocky, or mounded, slopes (Fig. 16.21). Sometimes these slopes exhibit fractures and wrinkles from compression and tension during downslope movement. The general effect of both creep and solifluction is to produce rounded hillcrests and a landscape free of sharp angular features—the subdued landscapes usually associated with humid climates.

Forms of Rapid Mass Movement

There are several varieties of rapid mass movement. Some of these are landslides, rockslides, rockfalls, slumps, earthflows, and mudflows. They are distinguished from the slower types in that their movement is visible and their effect

Figure 16.21 *Solifluction flow near Suslositna, Alaska.*
• **Solifluction is commonly found in what climate type?**

on surface configuration is more dramatic. Of course, the speed of movement varies in each particular situation, depending on such variables as the quantity and composition of the material being moved, the steepness of slope, the amount of water involved, the nature of the vegetative cover, and the triggering factor.

Rapid mass movements usually leave a visible scar where material has been removed upslope and leave a deposit below consisting of debris that has slid, fallen, flowed, or rolled downslope. Despite the often spectacular nature of rapid mass movement both in force and in effect on the land, slow mass movement, especially creep, has a far greater cumulative effect on surface configuration.

Landslide. Though the term is often used to refer to any form of rapid mass movement, Earth scientists give the name **landslide** only to the rapid downslope movement of a mass of material that moves as a unit and carries with it all loose material, and possibly great masses of bedrock as well. The movement is sudden and the moving mass often attains very high velocities.

(Continued on page 440)

VOLCANIC MUDFLOW HAZARDS IN THE CASCADES

Between northern California and southern British Columbia is a chain of spectacular volcanic mountains known as the Cascades. It is a range of great beauty, offering vistas of glacier-covered volcanic peaks, deep gorges, waterfalls, lakes, and some of the nation's deepest winter snowfalls. The range also forms a climatic barrier. It separates the rainy, evergreen-forested western slopes from the semiarid eastern slopes.

Until Mount St. Helens' eruption in 1980, most residents and visitors to the Pacific Northwest were not aware of the danger that lurks in these beautiful mountains. It should be of particular concern, as two large urban areas are close to the volcanic peaks. Within sight of massive Mt. Rainier (elevation 4323 m/14,410 ft) is Washington's Puget Sound lowland. This urban region, centered on Seattle-Tacoma, is home to over 2.5 million people and two of the nation's busiest ports. To the south, in the shadow of Mt. Hood (elevation 3372 m/11,329 ft), is Portland, Oregon, an urban area of over a million people. Other Cascade peaks, especially Mt. Shasta and Mt. Baker, could also threaten local populations.

A mammoth eruption, such as Mount St. Helens produced in 1980, is always a possibility. However, the Cascade volcanoes could produce another danger, a massive volcanic mudflow or *lahar*. Lahar is the Indonesian word for such a traumatic event. These rapid flows are a slurry mix of rock debris, mud, glacial ice, and water. In 1985, a volcanic mudflow killed 23,000 people in the Colombian town of Armero. The deadly rapid lahar destroyed and then buried the town located in a valley 50 kilometers (30 mi) from the Andean volcanic summit of Nevado del Ruiz. A major eruption is not necessary to set off such a deadly event. It can be triggered by a minor eruption, which heats the ice and snow at the summit, or an earthquake, which destabilizes the steep, volcanic rock-, ash-, and ice-covered slopes.

Due to its hazard potential, Mt. Rainier was targeted by scientists as one of several high-risk world volcanoes for intensive research under the International Decade for Natural Disaster Reduction (IDNDR). Mt. Rainier was chosen because it is potentially more dangerous than any other Cascade volcano for several reasons. First, Mt. Rainier is the highest and steepest of the peaks. Second, due to corrosive volcanic gases seeping from its active summit area, the rocks have been weakened and the unstable clay-rich debris could slip. Third, Rainier is second only to Mount St. Helens in earthquake activity, and one of these could jar loose massive amounts of material. Fourth, there is more water stored in Rainier's summit ice cap, its dozen major glaciers, and deep winter snow cover than in the ice and

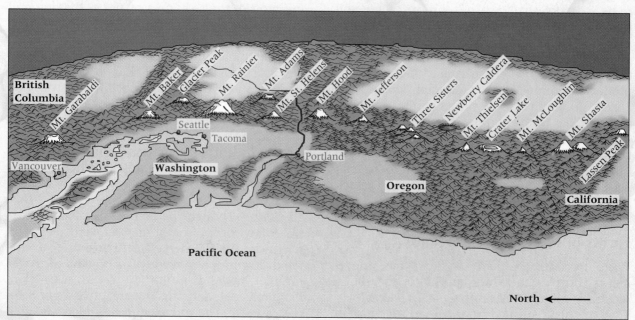

Map of the Cascade Range.

snow of all the other Cascade volcanoes combined. Last, adding to the danger is the rapid growth of the urban area at the base of its western slopes.

Mt. Rainier also has a chronology of past lahar activity. About 5700 years ago a lahar known as the Osceola mudflow removed the summit of Rainier. Scientists have noted that the summit was probably 620 meters (2000 ft) higher prior to the top sliding off the steep volcano. This lahar changed river courses and buried river deltas and valleys with as much as 31 meters (100 ft) of material. The flow reached as far as 100 kilometers (62 mi) from the summit and covered an area estimated at 505 square kilometers (195 sq mi). After the flow, it took the volcano 2000 years of lava flows to build up the new rounded summit cone. About 550 years ago, another lahar, known as the Electron mudflow, buried the Puyallup Valley under 3 meters (10 ft) of mud.

During the 10-year IDNDR program, scientists will use remote sensing, field mapping, and volcanic gas sampling to learn more about Mt. Rainier. Earth scientists hope to improve their ability to monitor the mountain, predict eruptions, and give ample warning of the potential for a catastrophic lahar roaring downslope toward the populated Puget Sound lowlands. The volcano's hazardous history should play a role in future land use planning and in disaster preparation. Not only are lives at stake, but such a lahar could cause hardship and economic havoc to the region's economy and international trade. As one scientist commented, "I think a lot of people don't realize the hazard potential that Mt. Rainier has. They just look up and see a beautiful mountain."

Mount St. Helens, lahar flow of 1982.

CRITICAL THINKING ▼

(1) Along which type of plate boundary are the Cascades located? **(2)** Mt. Rainier and the other Cascades are classified as what type of volcano? A lahar is classified as what type of mass movement? **(3)** If you were a planner in the volcanic Pacific Northwest, what type of sites would you recommend be classified as hazard zones?

Aerial view of Mt. Rainier, Washington. Note the steep slopes, unstable rockslide areas, and vast amounts of glacial ice surrounding and capping the 4323 m/ 14,410-foot volcanic summit.

Large landslides are rare but are often newsworthy because of their destructive qualities.

Landslides tend to occur more frequently on steep slopes in mountain regions during or after periods of heavy rain because of the additional weight of water and its lubricating qualities. They are a particular hazard in the clay-rich soil areas of the northern Appalachians and New England. Many landslides have been triggered by earthquakes. During the 1994 Northridge earthquake, thousands of landslides were triggered in the local Santa Susanna Mountains. They are also more frequent on slopes that are undercut by streams. More than 25 fatalities occur each year in the United States due to landslides.

Whereas some landslides involve only the surface regolith, larger movements may detach huge masses of rock. These **rockslides** are truly enormous, with volumes measured in cubic kilometers. Anything in their path is obliterated. More important, they may dam up valleys, which soon become filled with lakes. When the lakes become deep enough, they may wash out the dams, producing catastrophic sudden floods in the valleys downstream. Thus, immediately after such a major rockslide, workers must clear out the resulting dam and control the outlet of water trapped above it. This was done successfully after the Hebgen slide in southwestern Montana

in 1959 (Fig. 16.22). This slide, one of the largest in North American history, was triggered by an earthquake and killed 28 people camped along the Madison River.

Major rockslides result from rock mass instability related to geologic structure and stream or glacial undercutting of slopes. Like all landslides, major rockslides usually occur during exceptionally wet periods when the rock mass, or a sliding plane at its base, is well lubricated. Earthquakes have been associated with many, but not all, large historic rockslides. Today there are many locations in mountain regions where enormous slabs of rock supported by weak materials are poised on the brink of detachment, waiting only for an unusually wet year or a jarring earthquake to set them in motion.

Rockfall. When an individual rock or several rocks fall downslope, clattering over other loose rock and debris, the movement is called a **rockfall.**

The rocks or rock fragments involved in rockfalls can vary greatly. A rockfall may consist of tiny granular particles skittering downslope or a huge boulder bounding downhill and fragmenting along the way. In steep, rocky, mountainous areas where rockfalls are common, cone-shaped accumulations of rock fragments build up at the base of the cliffs, forming the talus slopes or cones mentioned earlier (see Fig. 16.9).

Rockfalls are particularly common in spring, when a traveler can encounter rocks scattered on mountain roads. Meltwaters and alternate periods of freezing and thawing at that time disturb precariously balanced rock masses, loosening them from their previously secure positions.

The steep granite cliffs of Yosemite Valley, California, have experienced some very large rockfalls. The most recent, in July 1996, killed one hiker and injured numerous others. The rockfall, estimated to have been moving downslope at over 200 kilometers per hour (160 miles per hour), produced a compression air blast in front of the moving rocks. The dusty air blast destroyed trees hundreds of meters from the base of the cliff as the rockfall crashed to the valley floor (Fig. 16.23).

Slump and Earthflow. Sometimes a mass of soil on a slope slips or collapses in a backward rotation, down at the top and up at the base, so that what was once a portion of the hill slope ends up tilting backward (Fig. 16.24). The curved backward rotation of such a **slump** distinguishes it from a landslide. Slumps are likely to occur where

Figure 16.22 Looking down the valley from the site of the 1959 earthquake-caused slide on the Madison River, Montana. This rapid fall of rock, soil, and trees completely blocked the river valley, creating the lake in the right foreground. The massive slide killed 28 people in a valley campground.

Figure 16.23 *The scar from the Happy Isles rockslide that occurred in Yosemite Valley on July 10, 1996. The rockslide killed one hiker and injured numerous others. The slide was so rapid (estimated at over 250 kilometers per hour or 160 miles per hour) and large that it created an air blast that set off a giant dust cloud which rose from the valley floor.*

moisture is concentrated at the base of a water-soaked mass of clay-rich soil. Slumps are conspicuous phenomena during exceptionally wet winters in California where they frequently damage expensive hillside homes.

Earthflows are linear movements of moist (almost liquefied), clay-rich regolith. Slumps often change into earthflows in a downslope direction. Some California earthflows are a mile or more long, and they generally have a tongue-like shape. Active earthflows move at rates of a few meters per hour, per day, or per month. Thus, they are somewhat more rapid than solifluction but slower than landslides.

Debris Flow. The channeled movement of water and rock debris mixed in various proportions is called a **debris flow** or **mudflow.** It is more fluid than an earthflow and moves much more rapidly, its high water content helping to speed it along. In general, debris flows follow valleys rather than flow as sheets downslope.

A debris flow is the most fluid of all mass movements. Sometimes it is difficult to distinguish between a runny debris flow and a muddy stream. Debris flows result from torrential rainfall on steep, poorly vegetated slopes in regions with distinct wet and dry seasons, such as coastal California, Oregon, and Washington (Fig. 16.25). The rains flush weathered debris into canyons, where it is mixed with flood waters. The result is a torrent of mud, gravel, boulders, plant debris, and water that can knock out bridges and destroy buildings and homes. (See The Environment: California: Living in a Hazard Zone, in Chapter 15.) Such torrents frequently occur in wet winters following dry summers in which fires have destroyed hillside vegetation and left the soil vulnerable to removal by water runoff.

Probably the most dangerous mudflow hazard occurs in regions with active volcanoes. Here,

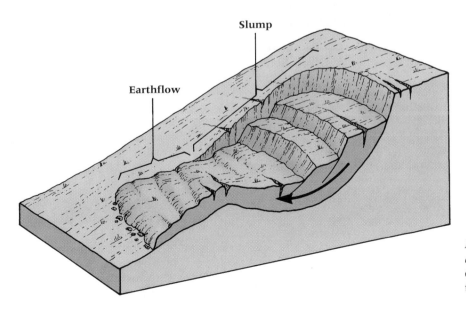

Figure 16.24 *Diagram of slump and earthflow.* • **What is the difference between these two types of mass movement?**

Figure 16.25 A debris flow in Laguna Beach, California. During torrential rains from El Niño storms in 1998, many debris flows of mud, soil, vegetation, and human structures occurred in the mountains of southern California.

steep slopes may be covered with hundreds of feet of loose volcanic ash. During volcanic eruptions, emitted steam, cooling and falling as rain, may saturate the ash, sending down dangerous and fast-moving mudflows, known as **lahars.** Of particular concern are high volcanic peaks capped with glaciers and snow fields. Should an eruption melt the ice, rapid and catastrophic lahars rush down the mountains with little warning and bury whole valleys and towns. In 1985, over 23,000 people were buried by mud in a valley below the Andean volcanic peak Nevada del Ruiz in Colombia. In the United States, there is a concern over some of the high Cascade volcanoes in the Pacific Northwest. Mounts Rainier, Baker, Hood, and Shasta all have a potential for such an event (see The Environment: Volcanic Mudflow Hazards in the Cascades).

Weathering, Mass Movement, and the Landscape

In this chapter we have discussed only two phenomena associated with gradation: weathering and mass movement. Although both weathering and the slower forms of mass movement rarely attract human attention, they vitally affect the landscape. They can create distinctive landforms without help from other more familiar processes, and every slope reflects the local nature of weathering and mass wasting processes. The effect that rock structure has on the landscape is influenced by the weathering characteristics of the minerals involved, and both the nature of the weathering processes and the way that weathered regolith is affected by gravity are major determinants of how the landscape looks. Steep slopes reflect slow weathering; gentle slopes result from rapid weathering; rounded forms suggest chemical weathering and slow but general mass movement (creep or solifluction); and angular slopes are associated with physical weathering and mass movements in the form of rockfalls, rockslides, and debris flows. In the following chapters we will consider the forms produced by the more dynamic and generally visible agents of gradation: running water, groundwater, wind, glacial ice, and coastal waves.

Define and Recall

erosion	chemical weathering	decomposition	solifluction
deposition	frost wedging	oxidation	landslide
base level	talus	hydration	rockslide
mass movement (mass wasting)	talus cone	hydrolysis	rockfall
	salt weathering (wedging)	evaporites	slump
		differential weathering	earthflow
weathering			debris flow (mudflow)
physical (mechanical) weathering	exfoliation	angle of repose	lahar
	exfoliation dome	creep	

Discuss and Review

1. How does weathering differ from the transporting agents of gradation?

2. How are the joints and fractures in a rock related to the rate at which weathering takes place?

   How does physical weathering encourage chemical weathering in rock?

3. What are several ways in which expansion and contraction can affect the weathering of rock?

4. Why is chemical weathering more rapid in humid climates than in more arid climates?

5. In what ways does chemical decomposition prepare rock for gradation? Give examples of oxidation, hydration, and solution.

6. Compare the ways in which hydrolysis and carbonation work below Earth's surface.

7. What types of rocks best resist all types of weathering? How visible are they in the landscape, compared to less resistant rocks?

8. What function does weathering perform in shaping landforms?

9. What two factors encourage mass movement the most? How do they work together?

10. What factors facilitate creep?

11. Describe the general effects of solifluction on the landscape. With what climates is solifluction usually associated?

12. What conditions encourage landslides and rockslides?

13. State the primary differences between a lahar and a debris flow. What causes the development of each?

14. Suggest ways in which weathering and mass movement affect human lives.

Consider and Respond

1. Based on Figure 16.4, in what weathering region would your local area be classified?

2. Based on Figure 16.4, in what weathering regions would the following geographic regions be located?
   Brazil's Amazon Basin
   The North Slope of Alaska
   The summit of Pike's Peak, Colorado
   The Mojave Desert of southern California
   The Appalachian Mountains of Pennsylvania

3. What would you recommend as a solution to prevent the loss of valuable historical monuments due to weathering processes?

4. If you were an urban planner located in a city with numerous steep slopes, what would be the major hazards you would have to plan for? What recommendations would you make to lessen these dangers to the community?

CHAPTER
17

UNDERGROUND WATER AND KARST LANDFORMS

CHAPTER PREVIEW

▶ Because of its usefulness and unique properties, water is considered by many scientists to be Earth's most important resource and the most distinctive feature of this planet's environment.

What are the peculiar properties of water that make it so important? What role does it play in making Earth a unique planet in the solar system?

▶ Groundwater and surface water are closely interrelated parts of a much larger system known as the hydrosphere.

How are they related? What is the movement of water through the hydrosphere called? What is the source of the power that drives the system?

▶ The depth (location) of the water table is the most important factor in the availability of groundwater.

How does the water table reflect precipitation? Surface landforms? How do humans affect the water table?

▶ The removal of bedrock by groundwater produces readily recognizable features both above and below Earth's surface that together characterize karst topography.

What processes are involved in the development of karst topography? What unusual surface and subsurface features result from these processes?

▶ Karst landscapes are found primarily in regions where limestone bedrock lies at or near the surface.

Why is this so? What areas of the United States and of the rest of the world have such landscapes?

▲ Karst topography in Guilin, a limestone region in southern China.

Although fresh water on or below Earth's surface, apart from that locked up as snow and ice, is most visible in streams and lakes, the greatest amount of this resource is actually found beneath the surface as **groundwater** (also referred to as **underground water**). In this chapter we will look at the significance of groundwater for human beings, especially as it affects their domestic, agricultural, and industrial water supplies. And, in keeping with our continuing examination of those processes that help to shape the land, we will look at the gradational effects of groundwater. Although of minor gradational importance when compared with streams, groundwater gradation has a significant impact on landforms in certain areas.

Figure 17.1 The Groundwater System. The groundwater system is really a subsystem of Earth's hydrologic cycle. The major input source for groundwater is precipitation when rainwater or snowmelt percolates into the ground. Some water may also become groundwater via surface runoff when streams and lakes contribute to the groundwater system. The major output or natural loss of groundwater occurs when it is released at the ground surface. This usually occurs through springs which may flow directly out of bedrock or surface material.

The groundwater system is usually divided into three parts: the zone of aeration, the zone of saturation, and, between them, the fluctuating water table. Water percolates downward through the zone of aeration, where air fills most of the openings between soil and rock particles. Eventually a depth is reached below which all openings are filled or saturated with water; this is the zone of saturation. The depth where the two zones meet is the water table. The fluctuating water table varies due to changes in precipitation. During dry seasons or years, the water table drops; during wet seasons or years, the water table rises.

Humans depend heavily upon groundwater systems. In the United States, for instance, over 50 percent of Americans receive their drinking water from groundwater. In the more arid western United States, groundwater is the major source of irrigation water. Overpumping of wells has caused the water table to drop drastically in much of this region. Although groundwater is harder to pollute than surface water, because it moves and recharges so slowly, it is also more difficult to clean up. Some major sources of groundwater pollution are leaking septic systems, animal feed lots, leaking underground storage tanks, and seepage related to disposal of toxic materials.

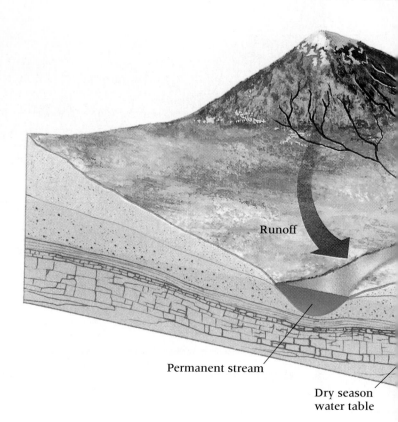

Runoff

Permanent stream

Dry season
water table

Occurrence and Supply of Groundwater

By far the largest proportion of groundwater is derived from atmospheric sources and represents Earth's primary supply of fresh water. It is brought to the surface by springs and wells, and it contributes significantly to standing bodies of water, such as lakes, and to running water in streams.

A small portion of groundwater is so deep below Earth's surface that it has probably never been part of the hydrologic cycle. Other groundwater was once involved in the hydrologic cycle, but because of changes in Earth's surface, it has been locked out of that cycle for a long period of time. It is trapped in layers of sediment laid down by ancient rivers or seas. Future changes in the lithosphere could release these trapped waters

and return them to the hydrologic cycle. Through volcanic activity these waters could be released in the form of geothermal energy (steam and hot water). The most obvious evidence of this activity is in the form of hot springs and geysers.

Groundwater Zones and the Water Table

Groundwater includes water found within the soil as well as in the loose rock and bedrock below. Under conditions of modest precipitation and good drainage, as water sinks into the ground it first passes through a level where air as well as water fills the spaces and pores within the soil and rock. This level is called the **zone of aeration.** Further movement downward brings the water to a second level, called the **zone of sat-**

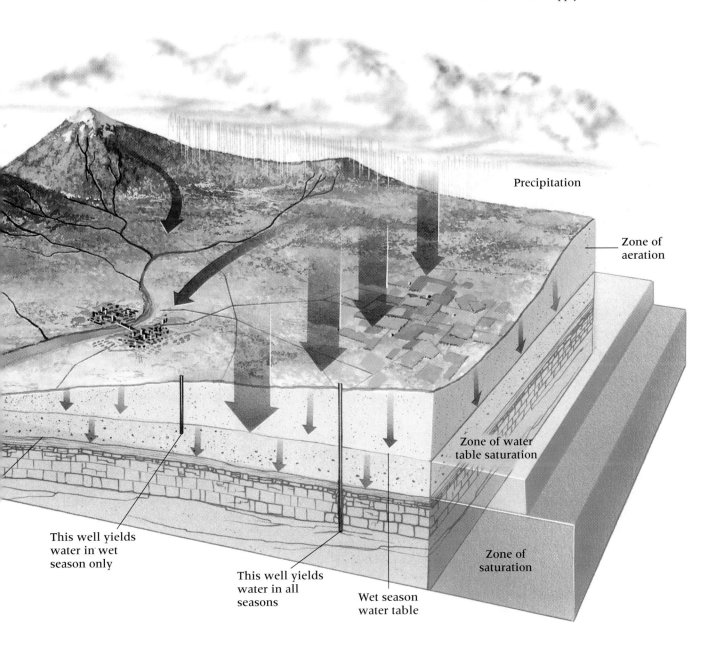

Precipitation

Zone of
aeration

Zone of water
table saturation

Zone of
saturation

This well yields
water in wet
season only

This well yields
water in all
seasons

Wet season
water table

uration, where all the openings are filled with water. Between the two levels, and marking the upper limit of the zone of saturation, is an undulating surface called the **water table.** The water table does not remain fixed in position but fluctuates with the quantity of recent precipitation. After heavy precipitation, the water table will be higher. Since the water table usually reflects the precipitation amount, it generally lies closer to the surface in humid regions than in arid regions.

In humid regions there are actually three groundwater zones (Fig. 17.1). The lowest zone is always saturated. The upper zone is almost

never saturated. Between these two is a zone that is saturated under conditions of ample precipitation and not saturated under conditions of low precipitation. It is through this middle zone that the water table fluctuates. Obviously, a well or spring originating within the permanently saturated zone will always have water, but one originating in the intermediate zone of fluctuation will run dry when the water table falls below it.

In many desert regions there is no saturated zone at all. Soon after rainstorms, water evaporates at or just below the surface. If groundwater is present, it may be very old, having accumulated during a past period of greater precipitation.

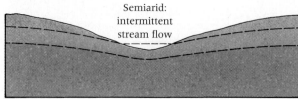

(a) **Effluent condition**

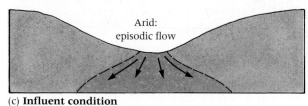

(b) **Influent condition**

Arid:
episodic flow

No water table except
by seepage from stream

(c) **Influent condition**

Figure 17.2 The influence of the water table on stream flow. *(a)* In humid regions, groundwater flows into major stream channels throughout the year, providing streams with continuous flow. This is known as an effluent condition. *(b)* In semiarid regions, with a dry season, the water table may drop below the stream bed, so that the stream dries up until the next wet season. *(c)* In desert regions, the absence of a water table means that stream runoff occurs only during rains, and flow diminishes downstream due to continuous seepage loss. This is known as an influent condition.

When such water is extracted in wells, it is not replaced from the atmosphere and the water table falls rapidly.

The water table in humid regions is not level but instead tends to follow the general contours of the land. It is higher under hills and other high surfaces and lower beneath valleys and depressions. It is usually closer to the surface under low places than under high places. Because water tends to seek its own level and because it is affected by gravitational force, the groundwater under higher land surfaces tends to flow downslope to a lower level as would a stream on the surface.

In humid regions of low relief the water table is often so high that it intersects the ground surface, producing lakes, ponds, or marshes such as those common in New England and along the Gulf Coast from Louisiana to Florida. Where the landscape is one of hills and narrow valleys, the lowest points on the water table are controlled by the position of valley floors. As a stream deepens its valley, it may cut below the level of the water table. Groundwater then moves downslope in the subsurface and enters the stream. This *effluent* condition keeps the stream flowing between rains.

Streams in semiarid and arid regions flow only seasonally or immediately after rains. In semiarid regions, the water table lies below the stream bed during dry periods and rises to inter-

sect the stream bed during wet periods. The stream is fed by groundwater only during wet seasons but loses water by seepage during dry *influent* periods. In desert regions, this influent condition is continuous, since there is no groundwater available to feed streams and there is surface water flow only during and immediately after rains. Much of this downstream flow is eventually lost by seepage into the dry ground under the stream channel (Fig. 17.2).

Factors Affecting the Distribution of Groundwater

The amount of groundwater in an area depends on a variety of factors. Most fundamental is the amount of precipitation that falls in a given area and in areas that drain into it. Second is the rate of evaporation. A third factor is the amount and type of vegetation cover on the land. Although dense vegetation transpires great amounts of moisture back to the atmosphere, it prevents rapid runoff of rainfall, encourages percolation of water into the ground, and prevents rapid evaporation by providing shade. Thus, the effect of forests in a humid region is to increase the supply of groundwater.

A fourth factor affecting the distribution of groundwater is the porosity of the soil and rocks. *Porosity* refers to the amount or proportion of

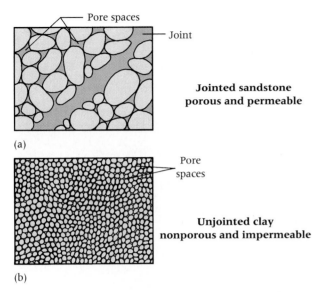

Pore spaces

Joint

**Jointed sandstone
porous and permeable**

(a)

Pore
spaces

**Unjointed clay
nonporous and impermeable**

(b)

Figure 17.3 *The relationship between porosity and permeability in two different sedimentary rocks.* ***(a)*** *Jointed sandstone, porous and permeable, and* ***(b)*** *unjointed clay, nonporous and impermeable. Which of these two rock types is the ideal aquifer?*

space between the particles that make up the soil or rock. Thus, some gravels, consisting of coarse particles that do not completely interlock, are exceedingly porous and can hold large amounts of water. Dense, interlocking crystalline rocks like granite, with virtually no pore space, can hold little water.

On the other hand, some granite may indeed hold water within joints, which allow the passage of this water rather freely. Thus jointed granite can be described as permeable. The *permeability* of a material—its ability to allow the passage of water through it—is related to the number and size of the spaces within the rock, which may be large pore spaces or joints, bedding planes, or other fractures. Porosity and permeability are not synonymous terms. A coarse sandstone, large-pored and jointed, will typically be porous and permeable; on the other hand, a finely porous and unjointed clay may contain significant amounts of water but at the same time be impermeable, since the water is locked around the tiny particles by soil moisture tension (Fig. 17.3). Porosity therefore affects the storage of groundwater and permeability affects the groundwater movement. Both of these factors affect the availability of water for wells and springs.

A rock layer of sandstone or limestone, which is porous and permeable, or a gravel and sand bed can act as a container and transmitter of water and is called an **aquifer** (Latin: *aqua,*

water; *ferre,* to carry). Any rock layer that is relatively impermeable and very finely porous, such as slate or shale, will restrict the passage of water and limit its storage, and therefore is called an **aquiclude** (Latin: *claudere,* to close off).

Sometimes an aquifer will be found between two aquicludes. In this case water flows in the aquifer much as it would in a water pipe or hose. Water can pass through the aquifer and cannot escape outward through the aquicludes. Furthermore, soil water percolating downward may be prevented from reaching the zone of saturation by an aquiclude. An accumulation of such water above an aquiclude is called a **perched water table** (Fig. 17.4). Careless drilling can puncture the aquiclude supporting a perched water table so that the water drains out. The well must then be deepened to reach the true water table.

Springs are surface outflows of groundwater. They are caused by the landform configuration, the level of the water table, and the relative position of various types of aquicludes. For example, a spring may occur along a valley wall where a stream or river has cut through the land to a level lower than a perched water table. There an impermeable layer of rock prevents further downward penetration of groundwater and thus forces the water to move horizontally. Such a spring is permanent if the water table always remains at a level above the valley floor; otherwise the spring is intermittent, flowing only when the water table is at a high level.

Availability of Groundwater

Groundwater is an extremely vital resource to many areas of the United States and most of the world. In fact, half of the American population derives its drinking water from groundwater. In some states, such as Florida, over 90 percent of the drinking water is from groundwater. Today, over two-thirds of the groundwater used in the United States is for irrigation. One of the largest aquifers is the Ogallala aquifer; it underlies the Great Plains from west Texas to South Dakota and supplies over 30 percent of all the irrigation groundwater in the United States.

Groundwater also plays a major role in supporting many wetlands and forming shallow lakes in land depressions. These bodies of water, fed by groundwater, are extremely important for thousands of resident and migratory birds. The flow of groundwater is vital for the survival of the Everglades in southern Florida. This "river of

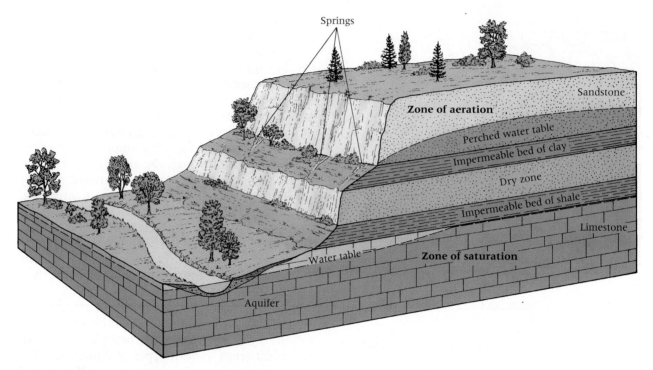

Figure 17.4 *A perched water table occurs when impermeable rock strata are present. In this example, the perched water table is underlain by a dry zone, between impermeable beds of shale and clay. Below the dry zone is the regional water table and the zone of saturation through which the water flows toward the nearby river.* • **Is the perched water table a reliable source of groundwater?**

grass" and its great variety of birds and other animals are totally dependent upon the continued southward movement of groundwater flow through the region.

Wells

Wells are artificial openings in the land surface dug or drilled below the water table. Water is extracted from wells by lifting devices ranging from simple rope-drawn water buckets to pumps powered by gasoline or electricity. In shallow wells, the supply of water depends on fluctuations in the water table. In contrast, deeper wells that penetrate into lower aquifers provide more reliable sources of water and are less affected by seasonal periods of drought.

In an area where there are many wells, the removal of groundwater may exceed the intake of water that replenishes the supply. In most areas irrigated from wells, the water table has fallen far below the reach of the original wells (Fig. 7.5). Progressively deeper wells must be dug (or the old ones extended) in order to reach the lower supply of water. In northern India in the Ganges Valley, the development of modern wells, deeper

than the traditional hand-dug wells, has resulted in the lowering of the water table. In the southern High Plains of Texas, Oklahoma, and Nebraska, the drawdown of the Ogallala aquifer, mainly for the irrigation of crops, is of serious concern.

In extreme cases of high groundwater demand, *subsidence,* or sinking of the land, occurs as the water pressure is reduced due to pumping. The Central Valley of California, Mexico City, and Venice, Italy, all have subsidence problems. In southern California, groundwater is artificially replaced by diverting rivers over permeable deposits. This process is known as artificial recharge.

Because groundwater filters down through many layers of soil and rock before it reaches an aquifer, it is free of sediment but often carries a large chemical load dissolved from the materials through which it passes. Groundwater is thus said to be *hard* in comparison with *soft* rainwater. Moreover, just as increases in population, urbanization, and industrialization have resulted in the pollution of some of our surface waters, they have also resulted in the pollution of some of our groundwater supplies. The most recent danger to groundwater supply has been from toxic waste

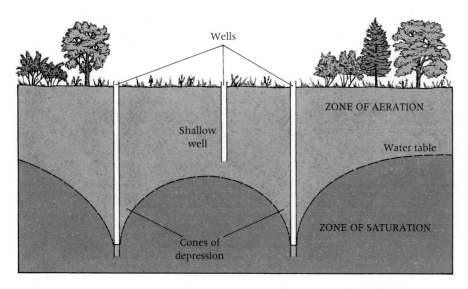

Figure 17.5 Cones of depression in the water table caused by pumping wells. In areas with many wells, adjacent cones of depression intersect, lowering the regional water table, causing shallow wells to go dry. • **In extreme cases, what other condition may occur?**

Wells

Shallow well

ZONE OF AERATION

Water table

ZONE OF SATURATION

Cones of depression

seepage. In coastal regions, where groundwater pumping has lowered the water table, salt water from the sea seeps in and replaces the fresh water since the pressure is no longer there to hold back the salt water. This salt water replacement has occurred in many localities, notably in southern Florida, Long Island (New York), and Israel.

Artesian Wells

In an *artesian* system, water under pressure rises to a level above the water table. In **artesian wells,** water rises to the surface and flows out under its own pressure, without pumping. Certain conditions are prerequisite to an artesian structure (Fig. 17.6). First, a porous aquifer such as sandstone must be exposed at the surface in an area of high precipitation or infiltration. This aquifer must absorb water at the surface, must incline downward hundreds of meters below Earth's surface, and must be confined between impermeable layers that prevent escape of the water. These conditions cause the aquifer to act as a pipe that conducts water through the subsurface. Provided with no exit, the water in the "pipe" is under pressure from the water above it (closer to the area of intake at the surface). As a consequence of this pressure, water will move toward any available outlet. If that outlet happens to be a well drilled through the impervious layer and into the aquifer, the water will rise in the well, sometimes gushing out at the surface. The height to which the water rises depends on the amount of pressure exerted on the water. Pressure in turn depends on the quantity of water in the aquifer (more water, more pressure), on the

angle of incline (steeper slope, more pressure), and on the number of other outlets, usually wells, available to the water (more outlets, less pressure). Sandstone exposed at the surface in Colorado and South Dakota transmits artesian water eastward to wells as far as 320 kilometers (200 mi) away. Other well-known artesian systems are found in Olympia, Washington, the western Sahara, and eastern Australia's Great Artesian Basin, which is the largest artesian formation in the world. The word *artesian* was derived from the Artois region of France, where such systems were once common.

Land Sculpture by Underground Water

The runoff of surface water is the major process that shapes landforms, but groundwater is also an agent of gradation. It is, of course, vital in the subsurface chemical weathering process, and, like surface water, it removes, transports, and deposits material prepared by weathering.

The principal mechanical effect of groundwater is to encourage mass movement by lubricating weathered material and soil, producing slumps, earthflows, and landslides. But it is by chemical action that groundwater contributes most to gradational processes. Through the removal of rock materials by solution and the deposition of those materials elsewhere, groundwater is an effective land-shaping agent, especially in areas where limestone is present: Because limestone is easily dissolved by acidic

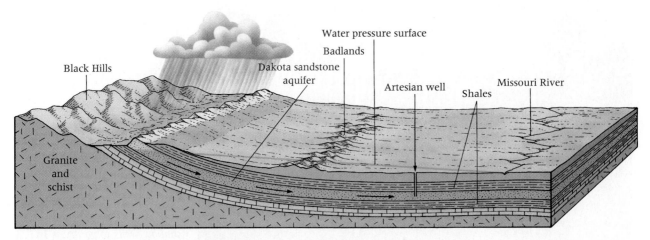

Figure 17.6 *Conditions producing an artesian system. The Dakota sandstone, an aquifer which averages 30 meters in thickness, transmits water from the Black Hills to over 320 kilometers eastward beneath South Dakota.* • **What is unique about artesian wells?**

groundwater, distinctive landscapes evolve wherever groundwater can act on limestone.

Karst or Limestone Landscapes

The distinctive features of a limestone landscape occur where limestone strata lie at or near the surface. The eastern Mediterranean region in particular exhibits features of limestone solution on a large scale. These are most clearly developed on the Karst Plateau along Croatia's scenic Dalmatian Coast. Any landforms developed by solution in limestone are called **karst** landforms after this classic locality. Other karst regions are located in central France, the southeastern United States, southern China, Laos, Mexico's Yucatan Peninsula, and the larger Caribbean islands.

Since limestone is a common sedimentary rock type, features created by the action of surface water or groundwater are found in many parts of the world. However, a true karst landscape, in which solution of limestone has been the dominant agent of land formation, is rather uncommon because of the special circumstances required.

A humid climate with plenty of precipitation is most conducive to karst development. Karst features are not developed in arid climates. However, some arid regions have karst features which originated during a period when the climate was much wetter than it is today.

A second important factor in the development of karst landforms is active movement of groundwater; this allows water saturated with dissolved calcium carbonate to be quickly replaced by unsaturated water. Vigorous movement of

groundwater occurs when an outlet at a lower level is available, such as a deeply cut stream valley or a tectonic depression.

Limestone landscapes are typically marked by an absence of surface streams, since the permeability of the rock encourages the disappearance of surface water underground. Frequently the surface junction between impermeable clays and shales and permeable limestones is marked by a series of *swallow holes* down which streams and rivers are "lost," or disappear. These seepage points may be widened and deepened by solution into great vertical *potholes,* very often tens or even hundreds of meters deep.

Surface outcrops of limestone are pitted and pockmarked by chemical solution, especially along the joints, forming large, flat, furrowed exposures of limestone pavements. As this surface solution tends to be concentrated at joint intersections, circular depressions, called **sinkholes,** gradually develop. Often hundreds of these may be conspicuously developed, as in the Yucatan of Mexico, in Jamaica, and in the southeastern United States—especially Missouri, Florida, Alabama, Kentucky, Tennessee, and southern Indiana (Fig. 17.7). Sinkholes that have become plugged with clay or that have eaten through to a layer of insoluble shale produce lakes and ponds such as those found in central Florida (see Map Interpretation: Karst Topography).

In time, such sinkholes, enlarged by continuing solution, may merge to form larger depressions called **dolines** (Fig. 17.8). Meanwhile, the slowly circulating water below the surface, traveling along joint and bedding planes, is also dissolving the limestone, so **caves** and **caverns**

(Continued on page 457)

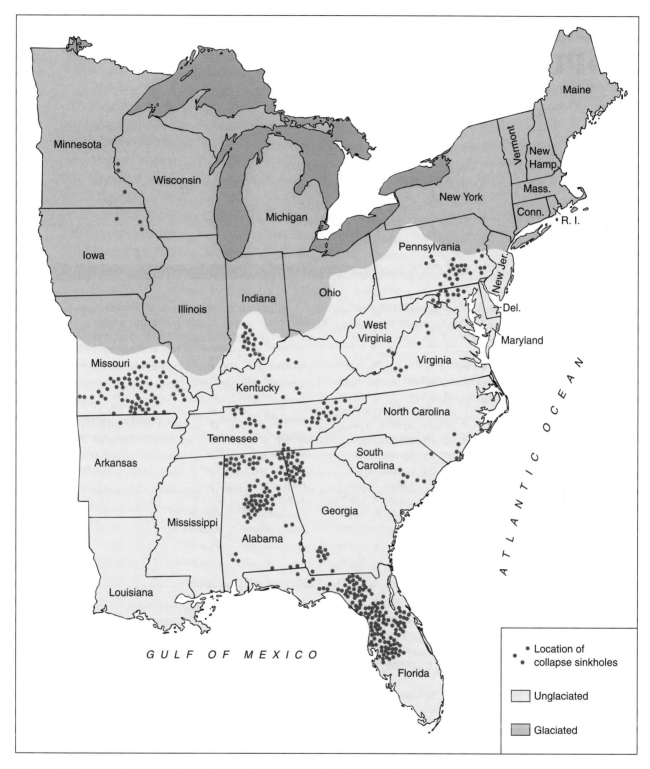

Figure 17.7 *Map of recent sinkhole sites in the eastern United States.* • **Which states have the greatest concentration of sinkholes?**

KARST TOPOGRAPHY

Putnam Hall, Florida
Scale 1:24,000
Contour interval = 10 ft
U.S. Geological Survey ▶

The Map

The Interlachen map area is located in northern central Florida. The Florida peninsula is the emerged portion of a wide anticlinal ridge, known as the Peninsular Arch. The arch encompasses virtually the whole state of Florida and includes the wide continental shelf off the Gulf Coast of Florida. It also includes the Bahamas Shelf off the Atlantic Coast of Florida. The submerged part of the arch is more than three times wider than Florida itself. The region is underlain by thousands of feet of marine limestone and shale. Much of the great thickness of marine sediments originated in the Mesozoic Era, when Florida was a marine basin. As the arch rose, Florida became a shallow shelf and eventually became elevated above sea level. Today, Florida has the lowest average elevation of any state. Most of the state has a humid subtropical climate, and millions of tourists flock to the state seeking its warm winter weather.

Although outsiders think of Florida mainly as a state with magnificent beaches, it is also a state with hundreds of lakes dotting its center. The largest, Lake Okeechobee, has an average depth of less than 46 m/15 feet. Most of the lakes, such as those in the Interlachen map area, are much smaller. The lake region is formed on the Ocala Uplift, a gentle arch of limestone that reaches to 46 m/150 feet above sea level.

Florida is an ideal area to study karst topography, especially the central lake region. Not only do the surface features express the geomorphic effects of groundwater, but the subsurface also does. Recent discoveries indicate great and complex subsurface cavern systems hundreds of feet beneath the surface. Much of the state's runoff is channeled through huge aquifers, and groundwater springs are quite common.

Interpreting the Map

1. As what type of major landform surface would this area be classified? What is its average elevation?

2. What is the contour interval of this map? Why do you think the cartographers chose this interval?

3. On what type of bedrock is the map area situated? Do you think the climate has any influence on the landforms in the area? Explain.

4. From the landform features on the map, how do you know this is a karst region?

5. What are the round, steep depressions called? Why are some of the depressions occupied by lakes?

6. Locate Clubhouse Lake on the full-page map (scale 1:62,500) and the smaller map (1:24,000). What is the elevation of Clubhouse Lake? What is its maximum width?

7. What is Levys Prairie (on the west margin of the map) and the area north of Lake Grandin?

8. What is the approximate elevation of the water table? (*Note:* You can determine this from the elevation of the lakes.)

9. Underground, the water flows through an aquifer. Define an "aquifer" and list the characteristics an aquifer must have. What is the general direction of groundwater flow in the aquifer underlying the Interlachen area?

10. Since much of central Florida is rapidly urbanizing, what problems and hazards do you foresee for this karst area?

Interlachen, Florida
Scale 1:62,500
Contour interval = 10 ft
U.S. Geological Survey ▶

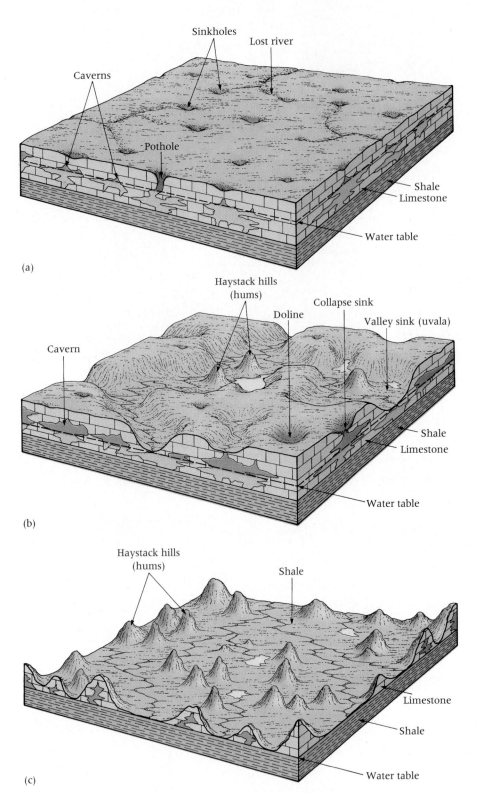

(a)

(b)

(c)

Figure 17.8 *An idealized sequence in the development of Karst topography.* **(a)** *Surface solution at joint intersections in limestone causes the development of sinkholes. Caverns form by groundwater solution along joint patterns. Cavern ceilings may collapse, causing larger and deeper sinkholes. Surface streams may disappear into sinkholes to join the groundwater flow.* **(b)** *Eventually the layer of limestone may be divided by the merging of sinkholes, forming large valley sinks called dolines, and longer karst valleys, called uvalas. Surface drainage now flows on insoluble rocks below the limestone.* **(c)** *In the latest stage, limestone remnants, called haystack hills or hums, are isolated above the new land surface of insoluble rocks.* • **Overall, what has been the general effect on the area of solution by groundwater?**

DIANE MULVILLE-FRIEL

University Education: *Bachelor of Science, Geography/Land-use Planning, Northern Michigan University. Master of Science, Geography, Western Illinois University.*

Early Interests: *Forestry, conservation.*

Diane Friel is an environmental scientist for Ayres Associates in Tampa, Florida. As a student, her interest in conservation led to further studies in geography and a focus on land-use controls and environmental regulations which has allowed her to become a more effective advocate for the environment. At Ayres, Diane is something of an "environmental detective," conducting on-site contamination investigations and reviewing historical documents to determine if environmental hazards exist on properties. These investigations are called Environmental Site Assessments (ESAs).

There is legislation called the Comprehensive Environmental Response Compensation and Liability Act, known as Superfund, that says if you purchase property contaminated by chemicals or hazardous wastes, you can be held liable for cleaning it up, whether or not you caused the contamination. This law created the need to investigate land prior to purchasing it. So now when someone wants to buy property, they hire a consultant like me to conduct an ESA. Historical research and land-use reconstruction is a major component of the ESA process.

As a consulting firm, most of the ESAs we prepare are confidential documents that are not seen or reviewed by regulatory officials. However, we have an ethical responsibility to notify regulatory agencies if we come across something that is an immediate threat to the public. Ayres has conducted more than 30 ESAs on over 140,000 acres of rural agricultural and ranch lands that the South Florida Water Management District is purchasing for conservation purposes.

Some common environmental hazards in rural Florida include the use of underground storage tanks that held gasoline and petroleum products; agricultural land use that leads to contamination by pesticides and other chemicals; asbestos in aging building materials; and vats where ranchers used to dip cattle in arsenic to kill ticks and other disease-carrying insects.

My training in geography and historical research is of great use in my current work. We use topographic maps, plats, land records, aerial photographs, and current environmental regulatory listings to obtain information on sites. These are the tools of geographic research. Geographers are quite comfortable in the field, and these skills allow me to navigate over and around a 20,000-acre parcel in undeveloped Florida.

It is both dangerous and exciting work. You really have to be very alert while in the field and never put yourself into a life-threatening situation. If you are a geographer and you get up to speed on the environmental regulations, you can step right into this kind of work because it really relies heavily on map and aerial photography analysis. You must also be able to do the background research. Writing ability is also very important. But most importantly, you have to have an investigative bent to have a job like mine. The good news for female students out there is that I don't really feel like I have had to overcome being a female to get a job in my field."

develop, until the massive limestone complex becomes eaten into like Swiss cheese. Collapsing subsurface cave roofs cause **collapse sinks** (Fig. 17.9), and the coalescing sinks and potholes and dolines create larger and larger depressions, often linearly arranged along the former underground water courses, to form **valley sinks** and **uvalas.** (The terms *doline* and *uvala* are derived from Slavic languages used in what was formerly Yugoslavia.)

Sudden collapse from the formation of sinkholes is a significant hazard that can cause severe property damage and human injury. These rapidly forming sinkholes are usually caused by excessive groundwater withdrawal for human use during drought periods. This lowers the water

Figure 17.9 This large sinkhole formed in Winter Park, Florida, when limestone caverns collapsed during a severe drought in May 1981.
• **What human activities might contribute to such hazards?**

table, causing a loss of buoyant support for the ground above, followed by collapse. Rapid sinkhole collapse has damaged roads and railroads and has even swallowed buildings.

In a mature karst landscape, a complex underground drainage system all but replaces the normal surface patterns. The landscape may show large valleys with no streams at their bases; such valleys have been excavated by surface streams that were later diverted to underground paths by the development of swallow holes on the valley floors. This process is characteristic of the Mammoth Cave area in Kentucky. Some of the "lost rivers" may reemerge as springs where they have cut down to impervious beds below the limestone.

In the late stage of karst development, especially in the wet tropics, only limestone remnants are left standing above the insoluble rock below. These remnants are usually in the form of small steep-sided and cave-riddled karst towers called **haystack hills** (or **hums**) (see again Fig. 17.8). Examples of this spectacular landscape are found in southern China (Fig. 17.10) and adjacent southeast Asia and on the islands of Puerto Rico, Cuba, and Jamaica. These have been described as "egg-box" landscapes, since an aerial view of the pits and hums resembles the designs of cardboard egg boxes.

Underground caverns are the most spectacular forms created by the solution of limestone.

Groundwater, sometimes flowing as streams underground, carves out networks of caverns that later become filled with air and decorated with depositional forms. Examples of these in the United States are Carlsbad Caverns in New Mexico, Mammoth and Colossal Caves in Kentucky, and Luray and Shenandoah Caverns in Virginia. In fact, 34 states have caverns open to the public. Some are quite extensive, with rooms over 30 meters (100 ft) high and with kilometers of connecting passageways.

One factor necessary for cavern development in karst landscapes is the jointing of limestone strata. Groundwater solution works along jointing planes; the joint pattern is evident in the rectangular pattern of the caverns that are created. Streams do not run through all caverns, although the dry caverns often show evidence of previous stream activity, such as sedimentation of clay and silt on the cavern floor.

The depositional features of limestone caverns produce some of the most beautiful forms found in nature. Where water drips from the ceiling, carbon dioxide gas is vented from the water into the cave air. This makes the water less acidic and thus less able to hold the lime in solution. The dripping water leaves behind a deposit of calcium carbonate called *travertine* or *dripstone*. As these travertine deposits grow downward, they form icicle-like spikes called **stalactites** that hang from the ceiling. Calcium-saturated water

Figure 17.10 Karst towers (hums) of Guilin, a limestone region in southern China.

dripping onto the floor of a cavern builds up similar but more massive structures called **stalagmites.** Stalactites and stalagmites often meet and continue growing to form columns or pillars (Fig. 17.11). A great variety of other depositional forms (**speleothems**) are also found, many of them very delicate in structure.

Not all caverns are alike. Some are well decorated with speleothems; some are not. Many have several levels and are almost spongelike in structure, while others are linear in pattern. The great variation in cavern forms indicates variations in mode of origin. Most large caverns appear to have developed either at the water table,

Figure 17.11 Cavern formations of dripstone in Carlsbad Caverns National Park, New Mexico. • **How do you explain the presence of such huge caverns in such a dry climate?**

where the rate of solution is most rapid, or below the water table in the zone of saturation. Subsequently, a decline in the level of the water table caused by the incision of surface streams, climatic change, or tectonic uplift has replaced the cavern water with air, allowing speleothem formation to begin. Underground rivers deepen some air-filled caverns, and collapse of their ceilings enlarges them upward. Many smaller caverns appear to have been formed by percolating water in the zone of aeration entirely above the water table.

Cavern development is a very complex process, involving such variables as rock structure, groundwater chemistry, hydrology, and regional tectonic and erosional history. The science of **speleology** (cavern studies) is a particularly challenging one, especially if we appreciate the fact that all that is known about caverns has been discovered by men and women "spelunkers" crawling through mud and water in dark passages beset by unknown obstacles hundreds of meters underground. Recent cave exploration and mapping of the deep, water-filled caves below the surface in Florida have involved scuba diving, an extremely risky operation.

Reference here should be made to another unique set of groundwater-related phenomena, namely **hot springs** and **geysers**. Here the water is heated at its source and hence is referred to

Figure 17.12 One of the world's most famous geysers is "Old Faithful" in Yellowstone National Park, Wyoming.

Figure 17.13 Hot spring calcareous deposits in Yellowstone National Park, Wyoming.
• **Explain how these unique deposits are formed.**

as geothermal water. It probably has been heated by contact with hot solidified rocks deep below the surface and is accompanied by steam. Where the water bubbles out fairly continuously, this is a hot spring; where it is an intermittent and somewhat eruptive process, like Old Faithful in Yellowstone National Park, it is a geyser (Fig. 17.12). Geyser is an Icelandic term for these stream eruptions so common on that volcanic island. Geysers appear to erupt when the steam pressure below reaches a critical level and forces the column of superheated water and steam out of the fissure in an explosive manner.

Both hot springs and geysers contain significant amounts of minerals in solution, which become deposited in various forms, often as terraces or cones around the vent. These calcareous *tufa* and siliceous *geyserite* deposits are colorful and impressive forms (Fig. 17.13).

Geothermal activity is usually associated with areas of tectonic instability, especially along plate boundaries. This geothermal energy has been harnessed to produce electricity in areas such as California, Mexico, New Zealand, Italy, and Iceland.

Define and Recall

groundwater
  (underground water)
zone of aeration
zone of saturation
water table
aquifer
aquiclude

perched water table
spring
well
artesian well

karst
sinkhole

doline
caves (cavern)
collapse sink
valley sink (uvala)
haystack hill (hum)
stalactite
stalagmite

speleothem
speleology
hot spring
geyser

Discuss and Review

1. What is the water table? How is it related to climate?

2. What is porosity? What is permeability? How are these two characteristics related to groundwater?

3. Define and describe the conditions necessary for an aquifer.

4. Explain how an artesian well works.

5. Define karst. What conditions are necessary for a region to have karst landforms?

6. Describe a sinkhole. What is the formation process involved?

7. Explain how haystack hills would be formed. What type of climate would be most conducive to their formation?

8. Describe how caverns form and name some common cavern features.

9. What is a geyser? What is the energy source for a geyser? Where are geysers most often located?

Consider and Respond

1. If you were a water resources geographer and had to plan for the development of groundwater resources for a community, what would be your major considerations?

2. Describe the major landform features in a region of karst topography. Over time, how will the region be changed?

3. What are some human impacts and environmental problems related to the use of groundwater?

CHAPTER

18

FLUVIAL LANDFORMS AND PROCESSES

CHAPTER PREVIEW

▶ Surface water, through sheet wash and stream action, is the most important single agent of erosion and does more work than all of the other gradational agents combined.

What does this tell us about Earth's landforms? How is this significant to humans?

▶ Moderate increases in either the volume or velocity of stream flow can cause major increases in the load-carrying capacity of a stream.

What does this tell you about stream erosion and deposition? About a stream at flood stage? What are the major characteristics of stream erosion, transportation, and deposition processes?

▶ There are significant differences among the upper, middle, and lower courses of streams in terms of gradational processes and the landform features that result from these processes.

What features most usually characterize the upper, middle, and lower courses of streams? What gradational processes are most important in the three different stream courses?

▶ Careful analysis of a stream system provides an excellent opportunity to illustrate the concepts of energy budget, water budget, and equilibrium as applied to natural systems.

What is meant by a budget? Why study these budgets? In what other systems is equilibrium an important concept?

▶ Although it may seem incongruous, running water is the chief agent of gradation in dry regions, and more arid landforms are gradational products of water action than wind action.

How can this be? What features in arid regions are the product of water action?

▶ Because water is only an occasional agent of gradation in desert regions, the widespread evidence of landforms created by running water in such regions is often attributed to earlier wetter climatic conditions.

What is the evidence of climatic change? How might the most recent rainy period be associated with glaciation?

▶ Exotic streams such as the Nile, Indus, and Colorado exert great influence on the human use of desert regions.

Why is this so? In what ways have humans and their civilization been influenced by exotic streams?

▶ Although a wide variety of landforms results from both fluvial erosion and deposition, landforms of fluvial deposition at the base of mountain regions in desert areas are often the most attractive for human settlement.

What combination of factors causes this to be true? What regions of the United States are excellent examples? What are the most common landforms of arid fluvial erosion and deposition?

▲ Fluvial erosion in the White Mountains of New Hampshire.

Running water exerts a greater influence in shaping the landforms of our planet than any other gradational process. Through both erosion and deposition, running water modifies existing landforms, and directly creates others, as it flows over land surfaces or in confined channels. With few exceptions, nearly every region of the world has been shaped by the power of water flowing downslope under the influence of gravity. This is true for both humid and arid regions, although a major exception would be the perennially frozen polar regions. The long-term impacts of running water, whether dominated by erosion or deposition, can be quite dramatic. Two prime examples in the United States illustrate the effectiveness of water in creating landforms: the Grand Canyon (Fig. 18.1a), carved by long-term *erosion* on the Colorado River, and the Mississippi River delta (Fig. 18.1b), where *deposition* is building new land into the Gulf of Mexico.

The Stream System

Streams are bodies of water that flow in well-defined channels (except during times of flood). In the Earth sciences, **stream** is a general term for water flowing in a natural channel, although in general usage we describe large streams as *rivers,* and there are many other local terms for streams (creek, brook, run, draw, bayou). Processes associated with the work of streams are known as **fluvial processes** (from Latin: *fluvius,* river).

The flow of water on the land surface is called **runoff,** which occurs when more precipitation falls than is intercepted by vegetation, evaporated into the air, or infiltrated into the ground. Runoff intensity depends on several factors. Among these are the amount and rate of precipitation, the evaporation rate, permeability of surface materials, density and type of vegetation, and slope angle. As in any natural system, these factors are interrelated. For example, after a given rainfall amount, less vegetative cover will mean more runoff, and any factor that impedes infiltration will increase runoff (and vice versa). Human activities can alter many of these variables, and in some places the natural runoff has been greatly modified by urbanization, dam building, mining, logging, and agriculture.

For short distances, runoff may occur in sheets (as **sheet wash** or unconcentrated flow), or it may form tiny channels called rills. Eventually, overland flow develops a pattern of well-defined channels, which join to form larger channels and eventually a stream system.

Each major stream and its **tributaries**—the smaller streams that feed it—make up a **stream system.** A stream system has three major subsystems: a *catchment system* that collects and directs runoff into the *transport system;* a network of channels that move water and sediment; and a *depositional or dispersal system*—the location to which water and sediment are eventually transported. The land surface drained by a stream system (or one or more of its tributaries) is called a **drainage basin, watershed,** or **catchment basin** (Fig. 18.2). Drainage basins are the areas that feed runoff into a stream; they exist for every

(a)

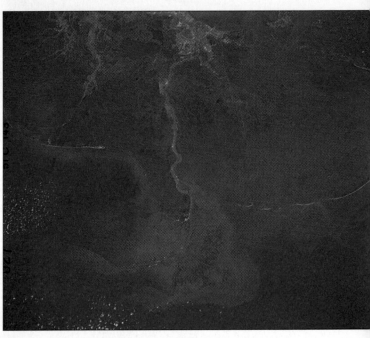

(b)

Figure 18.1 (a) *The Grand Canyon has been carved through an uplift in the Colorado Plateau by long-term fluvial erosion by the Colorado River (assisted by weathering and mass movement). The Canyon is 1.6 km (1 mile) deep, 466 km (277 miles) long, and averages nearly 16 kilometers (10 miles) in width. This orbital photograph shows the snow-covered, forested Kaibab Plateau (north of the canyon, right) and the Coconino Plateau (south of the canyon, left). The view is toward the west and the river is flowing toward the right corner. (b) Through the process of fluvial deposition at its mouth, the Mississippi River has been building its delta outward into the Gulf of Mexico for thousands of years. The flow of the river slows as it enters the Gulf, resulting in the deposition of muddy sediments that the stream has transported from its drainage basin. This is a Landsat false-color composite, a digital image, produced by the Multispectral Scanner (MSS). Muddy water is light blue, and clearer, deeper water is dark blue.*

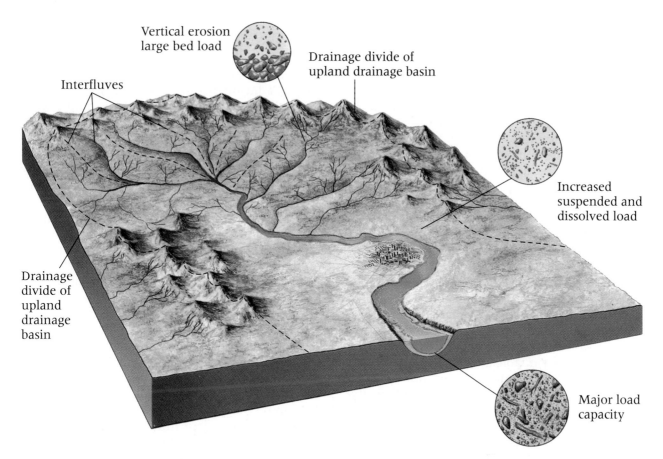

Vertical erosion
large bed load

Drainage divide of
upland drainage basin

Interfluves

Increased
suspended and
dissolved load

Drainage
divide of
upland
drainage
basin

Major load
capacity

Figure 18.2 Environmental Systems: The Hydrologic System—Streams. *The stream
or surface runoff system is a subsystem of the hydrologic cycle. Its major water input is
from precipitation. However, groundwater may also contribute to the steam system in
regions with a high water table. The major water output for most stream systems returns
water where the stream flows into the ocean. Output or loss of water also occurs by
evaporation back to the atmosphere and by infiltration into the groundwater system.
Stream systems are generally divided into regions known as drainage basins. Drainage
basins are separated from each other by higher elevations known as divides. Interfluves are
higher areas that separate streams flowing within the same drainage basin.*

*A stream system is a complex system of moving water that also involves energy
transfers and the transport of a variety of surface materials. Energy enters the system with
precipitation. As the runoff flows downslope, gravity aids in increasing the amount of
energy available to the stream system for cutting and eroding channels. Most of the energy
is lost to heat from frictional flow, while some goes to the sea with the stream system
waters.*

*Materials, known as load, enter the stream system by erosion and mass movement
from the banks, particularly in the upper headwaters of the drainage basin. Much of the
surface material eroded here consists of large particles, including boulders. Coarse material
is carried along the channel bottom and is called bed load. As the number and size of
tributaries grow downstream, the amount of load carried by the stream system generally
increases dramatically. This is especially true for finer materials suspended in water
(suspended load) and dissolved minerals (dissolved load). Materials leave the stream
system when carried and deposited in the sea at the mouth of the river. Some materials are
also deposited along the normal stream channel, as a river flows over its banks during
flood periods. Human activities can change the amount of load available for stream
systems by building dams, altering land with construction projects, overgrazing pasture
land, and clearing forests. These activities may also affect water quality downstream,
where communities may depend upon the stream system for their water supply.*

channel. Large river systems drain extensive watersheds that consist of many sub-basins, one for each tributary. Drainage basins are open systems involving both inputs and outputs of water, materials, and energy. Knowing the boundaries of a drainage basin and its sub-basins is critical in properly managing the water resources of a watershed. For example, pollution discovered in a river generally comes from a location in its drainage basin, either at the point where the pollutant was detected or from somewhere upstream. This knowledge helps us detect and correct sources of pollution.

The higher land between two tributary valleys of a drainage basin is referred to as an **interfluve** (from Latin: *inter,* between; *fluvius,* river). On an interfluve that separates two drainage basins there is an imaginary line called a **divide.** Some divides are ridges, but in many cases the higher land is not necessarily ridge-shaped. For example, drainage divides can exist on high plains, on tablelands, or along the crests of rounded hills. Surface runoff flows into a stream system (or sub-basin) on one side of the divide, while on the opposite side, runoff flows toward another stream system. The *Continental Divide* separates North America into a western region where runoff flows to the Pacific Ocean, and an eastern region where runoff flows to the Atlantic Ocean. Generally, the Continental Divide follows high ridges in the Rocky Mountains, but in some locations it also lies on gently sloping high plains.

Water Flow in Streams

Stream flow varies considerably around the world, responding to short-term storms, seasonality of precipitation or snow melt, the size and nature of drainage basins, and regional climatic patterns. Not all streams flow during the entire year. In arid and semiarid regions, streams often flow only after a heavy rain or during the rainy season, if there is one. These are called **intermittent** streams. In contrast, **perennial** streams flow all year, though not always with the same volume or at the same velocity. In times of high flow, many streams overflow their banks, flooding the flat lowlands on either side called **floodplains.**

Total stream flow, or **discharge,** is the volume of water flowing past a cross section of the channel in a given unit of time (given in either cubic meters, or cubic feet, per second). There are several reasons why *stream discharge* data are so important to collect and analyze. First, discharge

can be used to indicate and compare the size of stream systems. Also, because a stream's discharge changes in response to runoff—during storms, floods, or droughts—a stream system continually makes adjustments to these changes in flow.

Discharge is calculated by the simple equation $Q = AV$. *Discharge (Q) is equal to the stream cross-sectional area (A—channel width times channel depth) times the average stream velocity (V).* Conceptually *understanding* this *equation* is easy yet also very important. Because it is an equation, both sides must always be equal, so a change on one side must be accompanied by a change on the other. If discharge increases, the velocity and/or the size of the channel must also increase, and this is what happens during a flood. The discharge equation ($Q = AV$) illustrates the natural balance of water flow in a stream system. Discharge also is important in understanding changes in a stream's energy, ability to erode its channel, and the amount and/or size of sediment that a stream can transport.

The United States Geological Survey maintains over 6000 gauging stations to measure stream discharge in the United States (Fig. 18.3). In the past, the data from these stream gauges had to be collected at the site, but today many gauging stations operate on solar energy and beam their data to a satellite that relays them to a receiving station. With this system, stream flow changes can be monitored at the exact time that they occur, which is very beneficial in issuing flood warnings.

Comparing the largest streams worldwide, the Amazon has by far the greatest discharge, while the Mississippi-Missouri system is ranked fourth (Table 18.1).

Erosion by Streams

Water flowing in a stream is affected by the force of gravity in two main ways. First, gravitational force causes water to flow downslope, and second, gravity causes water to exert pressure on the stream bed. The strength of these forces, translated into flow velocity and the pressure exerted on the stream bed, depends on (1) the steepness of the stream gradient and (2) the depth of water in the channel. The volume of water in the stream is related to the depth, provided the channel width remains constant. Every drop of water is a unit of potential energy for gravity to use. Thus, the greater the stream flow, the greater the amount of energy available to shape the land.

About 95 percent of a stream's energy is consumed in overcoming external friction against channel walls and internal friction between water molecules, eddies, and currents. The remaining energy, probably less than 5 percent, is used for vertical or lateral erosion of the stream bed, channel, and banks, and for moving its **load**—the sediment that it is transporting downstream.

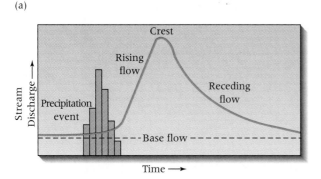

(a)

(b)

(c)

Streams continually undergo changes in an effort to sustain a balance among these energy uses. *Dynamic equilibrium* is maintained by adjustments among channel slope, shape, and roughness; amount of load eroded, carried, or deposited; and particularly, the velocity and discharge of stream flow. These last two factors can change very rapidly. This is one reason why velocity and discharge (which is also partly a function of velocity) are such significant aspects of a stream's behavior, regulating its ability to erode and to transport sediment (Fig. 18.4). Friction along the bottom and sides of a channel slows down the flow, so the maximum velocity in a stream is usually found in the center of the channel slightly below the surface. Anyone who boats, rafts, canoes, or floats on inner tubes on rivers or streams knows that not only does the flow vary seasonally, but the velocity also varies with its position in the channel.

The ability of a stream to pick up and carry materials is largely determined by the degree of *stream turbulence*. Turbulence is chaotic flow that mixes and churns the water and is controlled by channel roughness and flow velocity. A rough

*Figure 18.3 (a) The U.S. Geological Survey (USGS) monitors the flow of streams and rivers using gauging stations like the one illustrated here. When the level of the river rises or falls, so does the water in the lower part of the station, connected to the channel by intake pipes. A gauging float moves up and down with the water level, and this motion is measured and recorded. Until recent years, a person had to come to the station to retrieve the data, or to see how much water flowed in the channel during a flood. Today, stream gauging stations electronically beam flow data to a satellite that transfers the information to a receiving station. Now it is possible to obtain data from many stations at once and to instantly monitor how a stream responds to a storm at the same time that a flood event is occurring. • **What advantages are there to receiving flow data automatically at a receiving station distant from the stream channel?** (b) A stream hydrograph is a graph of changes in flow and water depth, recorded over a period of time by a gauging station. This hydrograph example shows the rise in a river in response to flood runoff, the flood peak, and the recession of flow waters following the end of a storm. Note that there is a lag between the time that precipitation starts and the rise in the river. • **Why would such a time lag occur between the rainfall and rise in the river?** (c) USGS stream gauging station. These are commonly located where highway bridges cross rivers or streams. Note the antenna for sending stream flow data through a satellite link to the USGS receiving station.*

TABLE 18.1	Ten Largest Rivers of the World*					
	Length		Area of Drainage Basin		Discharge	
	km	mi	sq km (×1000)	sq mi	1000 m³/s	1000 cfs
Amazon	6276	3900	6133	2368	112–140	4000–5000
Congo (Zaire)	4666	2900	4014	1550	39.2	1400
Chang Jiang (Yangtze)	5793	3600	1942	750	21.5	770
Mississippi– Missouri	6260	3890	3222	1244	17.4	620
Yenisei	4506	2800	2590	1000	17.2	615
Lena	4280	2660	2424	936	15.3	547
Paraná	2414	1500	2305	890	14.7	526
Ob	5150	3200	2484	959	12.3	441
Amur	4666	2900	1844	712	9.5	338
Nile	6695	4160	2978	1150	2.8	100

*Adapted from Morisawa: *Streams: Their Dynamics and Morphology.* New York, McGraw-Hill Book Company, 1968.

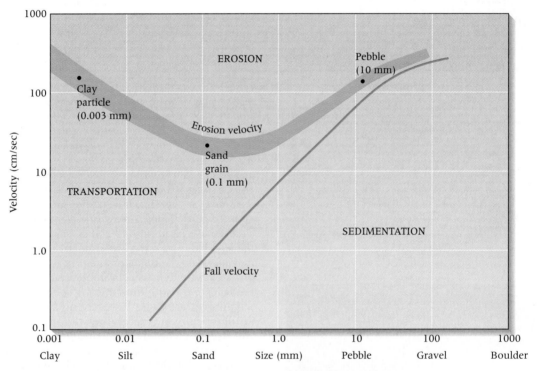

Figure 18.4 *This graph shows the relationship between stream flow velocity and the ability to erode or transport material of varying sizes (inability to erode or transport particles of a particular size or larger will result in deposition). Note that small pebbles (particles with a diameter of 10 millimeters, for example) need a high stream flow velocity to be moved because of their size and weight. The fine silts and clays (smaller than 0.06 millimeter) also need high velocities for movement because they stick together cohesively. Sand-sized particles (between 0.06 and 2.0 millimeters) are relatively easily eroded and transported, compared to clays or gravel (a mix of any particles larger than 2.0 millimeters).*

channel bottom increases the intensity of turbulent flow. A small increase in velocity can result in a significant increase in turbulence, which in turn greatly increases the rate of erosion as well as the load-carrying capacity of the stream (as shown in Fig. 18.4).

Volume of stream flow and resistance of the stream bed also affect erosion rates. Where rocks are highly resistant, little erosion can take place. Because erosion rates are affected by so many variables (velocity, turbulence, volume of stream flow, resistance of rocks), erosion in a stream varies greatly from place to place. All streams, however, have a **base level** below which they cannot erode. At the mouth of a stream—its outlet into an ocean or lake—either sea level or the level of the lake forms the stream's base level. The ultimate base level for virtually all stream gradation is sea level. Channel and valley deepening (downward erosion) can occur as long as there is enough slope to allow flow toward the outlet.

A stream that begins to flow on land that is significantly above its base level will gradually deepen, widen, and lengthen its channel through erosion. Fluvial erosion is accomplished in several ways. First, **hydraulic action** occurs as turbulent river currents wedge under rock slabs on the bed and pound away at the river banks or below waterfalls. **Plunge pools** at the base of waterfalls reveal the power of stream erosion when it is directed toward a localized point (Fig. 18.5). Hydraulic action also displaces loose particles from the stream bed or channel walls and carries the material along in the current. The sediments that a stream may transport range in size from clay to silt, sand, pebbles, cobbles, and boulders. Another term used for sediment is gravel, which is a mix of particle sizes larger than sand.

As rock particles bounce, scrape, and drag along the bottom and sides of a stream channel, they break off additional rock fragments. This form of erosion is called **stream abrasion** (or **corrasion**). Under certain conditions, stream abrasion makes round depressions called **potholes** in the rock of the stream bed (Fig. 18.6). Potholes are formed only in special circumstances, such as below waterfalls or swirling rapids, or at points of structural weakness, such as joint intersections in the stream's bed. They range in diameter and depth from a few centimeters to many meters. If you peer into a pothole, you can often see one or more round stones at the bottom. These are the *grinders*. Swirling whirlpool movements of the stream water cause such stones to grind the bedrock and enlarge the

Figure 18.5 Plunge pool at the base of Vernal Falls in Yosemite National Park, California. • **Why is a deep plunge pool found at the base of most waterfalls?**

Figure 18.6 These potholes were cut into a stream bed of solid rock during times of high flow. Potholes result from stream abrasion and the swirling currents in a fast-flowing river. • **Explain how a pothole is carved into bedrock on the stream bottom.**

pothole, while finer sediments are carried away. As potholes widen and join, the stream bed is gradually deepened.

Some fluvial erosion occurs by stream water chemically dissolving the bedrock. This chemical process called **corrosion** has a limited effect on many rocks but may be significant in limestone areas.

Over distance, the particles transported by a river gradually are reduced in size and their shape changes from angular to rounded. This process is called **attrition**—the wear and tear on sediments as they tumble and bounce against each other and against the stream channel. Attrition explains why rounded pebbles are found in stream beds and why the sediment load in the lower reaches of most large rivers is composed primarily of silt- or clay-sized particles and dissolved materials.

The slope of a stream bed is its **gradient**—the vertical drop over a horizontal distance (meters per kilometer, or feet per mile). In general, steeper gradients produce greater velocity and increased turbulence in the stream, which result in increased erosion. Stream gradients are usually steepest at the headwaters and in new tributaries, and diminish in a downstream direction. However, the greater volume of water farther downstream accelerates flow velocity even more than the high slope upstream, so the ability to pick up and carry materials is equivalent or even increased downstream.

As erosion continues, an idealized gradient is reached in which the frictional loss of energy is counterbalanced by energy produced by the stream's flow. A stream in this condition is said to be in a graded condition. A **graded stream** has just the velocity necessary to transport the load eroded from the drainage basin. The graded stream is a theoretical balanced state averaged over a period of years. Near the mouth, stream erosion erodes to a graded condition much sooner than at the headwaters, where the discharge is lower. As the stream gradient is progressively reduced, grade is reached farther and farther upstream, eventually producing a smooth profile (Fig. 18.7).

Stream erosion widens and lengthens stream channels and the valleys that they occupy. Lengthening occurs primarily at the source through **headward erosion,** accomplished partly by surface runoff flowing into a stream and partly by springs undermining the slope. The lengthening of a river's course in an upstream direction is particularly important where erosional gullies are rapidly dissecting agricultural land.

Such gullying may be counteracted by soil conservation practices to reduce erosional soil loss. Channel lengthening also occurs as the stream becomes more sinuous and thus decreases its gradient. Streams flowing on low-gradient floodplains do not generally flow in straight lines but *meander* from side to side in the valley. **Mean-**

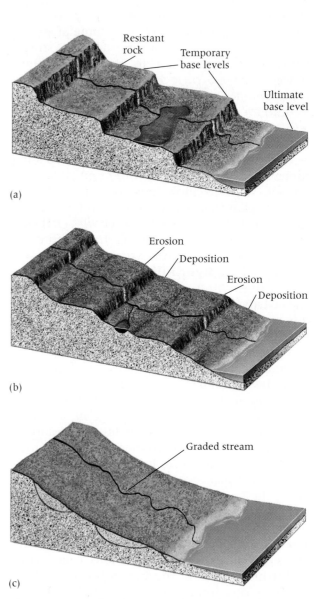

Figure 18.7 Idealized concepts of a graded stream and base level. (*a*) *An ungraded stream with many temporary base levels.* (*b*) *The stream profile becomes smoother as the stream is graded toward ultimate base level through erosion and deposition. Temporary base levels control the levels of the river upstream from resistant bedrock obstructions.* (*c*) *Eventually streams that flow to the ocean tend to develop a smooth profile graded to sea level (ultimate base level) that flattens downstream and is steeper in the headwaters.* • **What would account for the rapids and waterfalls in the initial gradient?**

Figure 18.8 *Valley widening is accomplished by meandering, as shown here by the Tazlina River, Alaska. This is due to the increased stream velocity on the outside turn of each meander, eroding and undercutting the valley side.* • **What occurs along the inside turn of a meander?**

ders are broad, sweeping bends in a channel that develop on most streams where they flow in low gradients on floodplains.

Valley widening is accomplished by lateral erosion on the outside of bends and commonly accompanies the outward expansion of meanders (Fig. 18.8). Lateral undercutting encourages slumping of the banks into the channel. Channel widening is greatly encouraged by mass movement of material down the side slopes of a stream valley. Tributaries flowing into a larger stream also help to widen the valley through which the trunk stream flows.

Stream Transportation

Some fluvially transported sediment is directly eroded by the stream, but a far greater proportion of the sediment is brought to the stream by surface runoff and mass movement. As previously noted, these materials transported by fluvial processes are called the stream load. Streams transport their load in several ways (Fig. 18.9). Some minerals are dissolved in the water—and carried in **solution.** The finest solid particles are carried in **suspension,** buoyed by vertical turbulence. Such particles can remain suspended as long as the force of upward turbulence is stronger than the downward settling tendency of the particles. Sediment particles, too large to be carried in suspension, slide and roll along by **traction.** When large particles hop and bounce along the channel bottom, the process is called **saltation** (from French: *sauter,* to jump).

The load of a stream is measured by the weight of the material it is transporting. There are three main types of stream load: In most streams the largest portion of sediment load is the **suspended load**—the materials being carried in suspension; **bed load** consists of the particles that roll or saltate along the stream bed; **dissolved load** is that portion that is held in solution.

The relative proportion of each type of load varies with flow rates and the nature of the drainage basin (vegetative cover, slope, rock and soil types). Some rivers transport huge suspended loads; the highest is that of the Huang He in

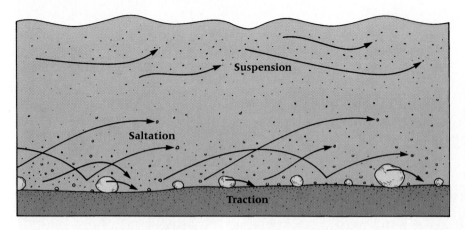

Figure 18.9 *Transport of solid load in a stream. Clay and silt particles are carried in suspension, sand travels by suspension and saltation, and larger material from pebbles to boulders move by traction.*
• **What is the difference between traction and saltation?**

northern China, known as the "Yellow River" because of the color of its suspended load. The Huang He carries a silty suspended load of over one million tons per year. Compared to the Mississippi River, the Huang He carries five times the suspended sediment load with only one-fifth the discharge.

Two terms are used to convey the relationship between load and a stream's ability to transport this material: **Capacity** refers to how *much* load a stream can carry; a stream's **competence** is determined by the diameter of the largest particle it can carry as bed load. Both capacity and competence can significantly increase in response to a relatively small increase in velocity. A stream that doubles its velocity during a flood may increase its sediment load six to eight times. The huge boulders seen in many mountain streams arrived there during some past high-velocity flow that greatly increased stream competence; they will be moved again when a flow of similar magnitude occurs. Rapid increase in discharge can temporarily upset the equilibrium in a stream, causing it to pick up more load through erosion in order to reassume a balance. In fact, rivers do most of their heavy earth-moving work during short periods of flood.

Stream Deposition

Since a stream's capacity to carry material depends primarily on velocity and discharge, a reduction in either of these elements will cause a stream to reduce its load through deposition. Deposition most commonly occurs in locations where the velocity is slowed, such as on the insides of bends in the channel, on floodplains (or valley floors), at river mouths (deltas), and where gradient abruptly flattens (such as where streams leave mountain areas). **Alluvium** is the general

name given to fluvial deposits, regardless of the type or size of material. Alluvium is usually recognized by the characteristic sorting and/or rounding of sediments that streams perform. Velocity changes cause a stream to sort particles by size, transporting the sizes that it can and depositing the larger ones. As velocity fluctuates, particle sizes that can be picked up, transported, and deposited vary accordingly. The velocity of a stream varies with the changes in discharge that occur between wet and dry seasons or in response to storm events. The alluvium deposited by a stream with fluctuating discharge will exhibit velocity changes by alternating layers of coarse and fine material.

When a stream has a heavy bed load in relation to its discharge, it deposits much of its load as sand and gravel bars in the stream bed. These obstructions break the stream into strands that interweave, separate, and rejoin, giving a braided appearance to the channel. Such a pattern is called a **braided stream** (Fig. 18.10). This pattern may develop wherever the coarse sediment input into a stream is extremely high owing to banks of loose sand and gravel or the presence of a nearly unlimited, coarse bed load such as that found downstream from glaciers and also in many desert areas. Braided streams are common on the Great Plains (for example, the Platte River), in the desert Southwest, in Alaska, and in Canada's Yukon.

Land Sculpture by Streams

One way to examine some of the variety of landform features resulting from fluvial gradational processes is to examine a typical river course from its headwaters to its mouth. In the following discussion of upper, middle, and lower river courses

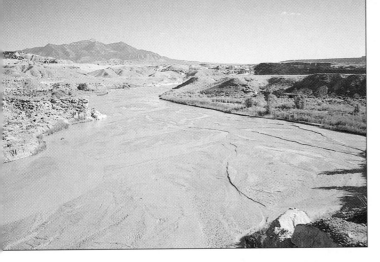

Figure 18.10 *The braided stream channel of the Fremont River, Utah. Braiding results from an abundant bed load of coarse sediment that obstructs flow and separates the main stream into numerous strands.*
• **Under what circumstances might a braided stream develop?**

you should note that erosion tends to be more significant in the upper course, whereas deposition is more important in the lower course.

Features of the Upper Course

In the headwaters, in the upper course of a river, the gradient is steepest. Here vertical erosion, particularly corrasion, works most strongly as the stream seeks base level through down-cutting. Erosion in the upper course generally creates a steep-sided valley, a gorge, or a ravine as the stream channel in the bottom of the valley cuts deeply into the land. No floodplain is present, and the steep valley sides encourage mass movement of material directly into the flowing stream. Valleys of this type, dominated by the down-cutting activity of the stream, are often called **V-shaped valleys** because of their shape (Fig. 18.11).

Because the stream in the upper course is generally in contact with bedrock, the effects of *differential erosion* are significant. Differential erosion occurs where rivers cut through rock layers of varying resistance. Typically, rivers flowing over resistant rock have a steeper gradient than where they encounter weaker rock. A steep gradient gives the stream more energy. Rapids and waterfalls indicate the resistant materials and steep gradients most common in the upper course of streams and their tributaries (see again Fig. 18.11). Where rocks are particularly resistant to weathering and erosion, valleys will be narrow, steep-sided gorges or canyons, and where rocks are less resistant, valleys will be more spacious.

Many streams spill from lake to lake in their upper courses, either over open land (like the Niagara River at Niagara Falls, between Lake Erie and Lake Ontario), or through gorges. In either case, the lakes are eventually eliminated as stream erosion lowers their outlets, or deposition fills the lake by building deltas at the inlet point.

Features of the Middle Course

When a stream has reduced its gradient, smoothed out its channel bed, and begun to approach its base level, vertical erosion becomes less significant and lateral erosion of the channel sides assumes a more important role. Combined with mass movement, this effect displaces the channel laterally, eroding valley walls to produce a wider valley floor with a smoother stream gradient. Lateral erosion on the outside of river bends eventually produces a narrow floodplain along the banks. Vertical erosion is minimized because the stream has little or no excess energy available

Figure 18.11 *Near the headwaters of a stream, valleys typically have a characteristic "V" shape. The stream flows in a steep-walled valley, with rapids and waterfalls, as shown here in Yellowstone Canyon, Wyoming.* • **Describe the gradient of the Yellowstone River (compare with Fig. 18.7).**

to pick up and transport more material. Consequently, streams that have reached such a balance or equilibrium between load, velocity, and discharge erode laterally, picking up material on one side and dropping it on the other.

It is in the middle course of most rivers that a graded condition exists. Here the river valley contains a floodplain but still maintains definite valley walls. The river becomes sinuous, forming wide bends or meanders on the valley floor. Such a stream erodes the **cut bank** on the outside of a meander loop and deposits material on the **point bar** on the inside of the meander (Fig. 18.12). Centrifugal force accelerates stream velocity and forces water against the outside of these bends, allowing sediment to be picked up, and decelerates velocity on the inside, allowing sediment to be deposited. Over time, this activity increases the size of the stream meanders, further widening the floodplain.

The *meander core,* the land between the curves of a stream's meanders, is relatively flat

and covered with alluvium, particularly during periods of falling velocity as floods subside. As lateral erosion continues and the meander size is increased, the area of flat land covered with rich alluvium increases. Though flooding of such land is always a potential hazard, the richness of the floodplain's alluvial soils offers an irresistible lure for farmers.

Features of the Lower Course

In the lower course of a river, deposition becomes more significant in shaping the landscape. The lower valley of a major river is wider than the width of the meander belt and shows evidence of shifts in the river's course (Fig. 18.13). In most cases this great valley width results from the lower course being a vast floodplain of alluvium rather than an erosional plain (see Map Interpretation: Fluvial Landforms). In other words, the alluvium filling the valley is much deeper than the stream channel. These giant lower valleys are

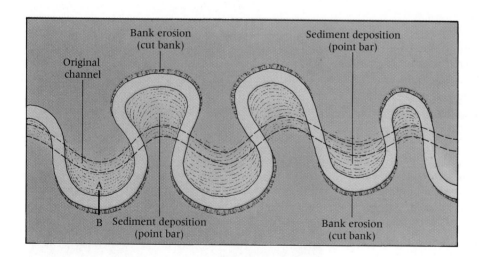

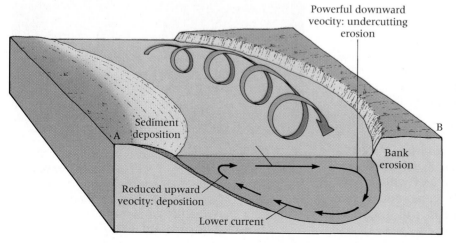

Figure 18.12 Characteristics of a meandering river channel. Note that water flowing in a channel has a tendency to flow downstream in a "corkscrew" fashion, which moves water against one side of the channel and then to the opposite side. The up and down motion of the water contributes to the processes of erosion and deposition. • Why do the meanders become more sinuous as time passes?

actually filled marine **estuaries** (river mouths drowned by the sea where salt and fresh water mix)—as in the case of the Mississippi River—or subsiding tectonic depressions, like the Amazon and Congo Basins.

During floods, the floodplain of a river's lower course is inundated with sediment-laden

Figure 18.14 *During floods, low areas adjacent to the river are inundated with sediment-laden water that flows over the banks to deposit alluvium, mainly silts and clays, on the floodplain. This is the Missouri River floodplain at Jefferson City, Missouri, during the 1993 Midwest flood.* • **What would the river floodwaters leave behind in flooded homes after the water recedes?**

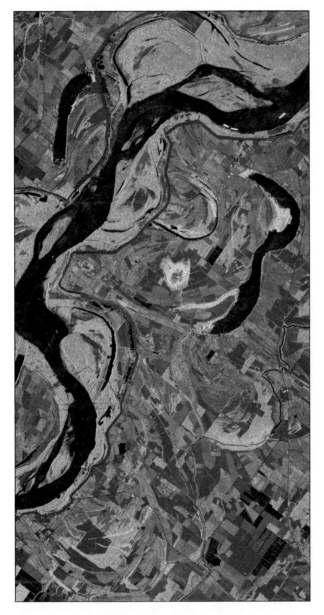

Figure 18.13 *A colorized radar image shows part of Mississippi River floodplain along the Arkansas-Louisiana-Mississippi state lines. Images like this help us to assess flood potential and to learn much about the geomorphological history of the river and its floodplain. The colors are used to enhance various landscape features like water bodies (dark), field patterns, and forested areas (green). Note how the river has changed its channel position many times leaving oxbow lakes and meander scars on the floodplain.*

water that deposits alluvium over it. These plains accordingly are called **alluvial plains** (Fig. 18.14).

A common landform of the lower course, the **oxbow lake,** provides evidence of the river's shifting position over time. Especially during flood periods, oxbow lakes form when a portion of the channel is cut off as the stream seeks a shorter, steeper, and straighter path. The cut-off section of the channel forms the oxbow lake (Fig. 18.15).

During flooding, a sharp reduction in a stream's velocity caused by friction at the channel's edge results in heavy sedimentation along the riverbanks. Over time, the banks are built up into **natural levees** through successive deposition that occurs most frequently on the immediate banks of the river. Natural levees along the Mississippi River rise up to 5 meters (16 ft) above the rest of the floodplain.

When a river slows down during times of reduced discharge, deposition will occur in the channel. Thus, in its lower course, a river can sometimes raise the level of its channel bed. In some instances, as in China's Huang He and the Yuba River in Northern California, the stream channel may be raised above the surrounding floodplains, although the natural levees tend to hold the water in the channel. Flooding is an even greater danger in this situation because the

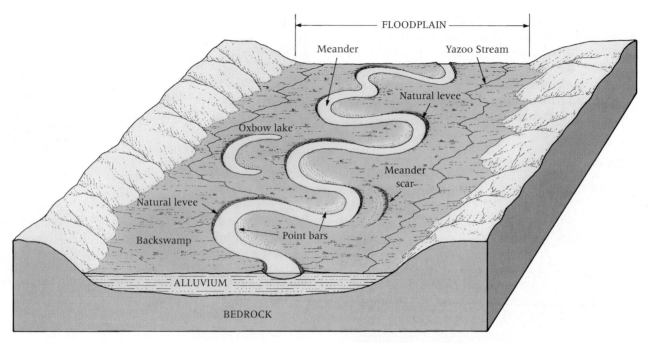

Figure 18.15 Features of a large floodplain common in the lower courses of major rivers. Low marshy or swampy parts of the floodplain, generally at the water table, are called back swamps. • **What is the origin of an oxbow lake?**

so-called "bottom lands" of the floodplain actually lie below the level of the river. Sometimes humans attempt to control a river by building up the levees artificially in order to keep the river in its channel. Unfortunately, artificial levees tend to make the river higher and deeper, so that when floodwaters go over the levees, they may be even more extensive and destructive. This happened during the 1993 Mississippi River floods in the Midwest.

The presence of levees—both natural and artificial—prevents tributaries in the lower course from joining the main stream. Smaller streams are forced to flow parallel to the main river until a convenient junction can be found. These parallel tributaries are called **yazoo streams,** named after the Yazoo River, which parallels the Mississippi River for over 160 kilometers (100 mi) until it finally joins the larger river near Vicksburg, Mississippi.

When a stream flows into a body of water such as a lake or an ocean, both flow velocity and load capacity are reduced. Consequently, the stream deposits its load—larger particles first and then finer ones. The resulting landform is known as a **delta,** after the Greek letter Δ, which some deltas, like that of the Nile River, resemble when seen on a map, or from space (Fig. 18.16a). Most deltas, such as the Niger and Nile deltas, have a triangular arc shape.

Multiple channels flowing away from the main stream are called **distributaries,** a common channel form on deltas. Natural levees build up along the sides of these distributary channels, even developing outward from the coast into the larger body of water. Continued deposition and delta formation can build new land far out into the water. Rich alluvial deposits and the abundance of moisture allow vegetation to quickly become established on this fertile new land and further secures its position. Delta lands like those of the Mekong, Indus, and Ganges rivers form important agricultural areas that feed the dense populations of many parts of Asia.

In general, deltas can attain great size only where the sediment supply is high, where the continental shelf does not drop off sharply, and where waves, currents, and tides are not strong.

The *bird's-foot delta* of the Mississippi is a special type that results from subsidence after delta formation, leaving only the natural levees of the distributary channels above sea level (looking like a bird's foot). The Mississippi delta (Fig. 18.16b) has changed rapidly over the years as new distributaries deposit alluvium that subsequently sinks, leaving only the levee crests extending seaward along the edge of the channel.

Another type of delta fills in at the heads of some marine *estuaries.* As previously noted, estuaries are river valleys that have been drowned by

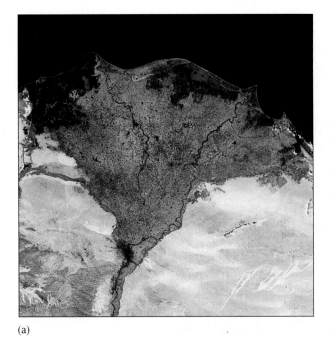

(a)

(b)

(c)

Figure 18.16 *Satellite views of river deltas.* ***(a)*** *The Nile delta, the classic arcuate river delta, as it enters the Mediterranean Sea.* ***(b)*** *The delta of the Colorado River where it enters the Gulf of California. The deposition of sediments occurs where streams enter estuaries.* ***(c)*** *A false-color satellite view of upper Chesapeake Bay. Chesapeake Bay was formed by submergence of the lower valley of the Susquehanna River as sea level rose at the end of the last glaciation. The Potomac River enters the main estuary at the bottom left of the image, while numerous minor tidal estuaries are found along the margins of the bay.* • **What change in a river's equilibrium causes the development of a delta?**

the sea either because the sea level has risen or because the land has subsided (Fig. 18.16c). Deltas form slowly in estuaries because of the counteraction of tides and sometimes because of low sediment input caused by a gentle gradient.

Rejuvenation

Changes in the base level of a stream can occur as a result of either tectonic processes or climatic change. Landmasses may rise in response to fault-

ing, or perhaps sea level may drop in response to the growth of glaciers. In either case, the result will be similar: streams flowing in that region (or, in the case of sea level change, worldwide) will be above their base level and will now have the energy to regrade (downcut) their channels to the new base level. Valleys with waterfalls or rapids may develop as these channels are deepened by erosion. The landscape and its stream are then said to be **rejuvenated.**

If new uplift occurs gradually, after the formation of large stream meanders, these meanders

Figure 18.17 The "Goosenecks" of the San Juan River in Utah provide classic examples of entrenched meanders. • **Explain how rejuvenation could form such spectacular features.**

become **entrenched** as the rejuvenated stream deepens its valley (Fig. 18.17). Now, instead of eroding the land laterally, with meanders migrating across a plain, the rejuvenated stream's primary activity is vertical incision.

It is important to note that all rivers reaching the sea were rejuvenated during the Pleistocene as a consequence of the lowering of sea level that resulted from continental glaciation. The accumulation of water on the land in the form of glacier ice caused sea level to drop as much as 120 meters (400 ft), thus lowering the base level for all coastal streams, which consequently cut deep valleys for their channels. Subsequent melting of the ice elevated sea level—or base level—causing the streams to deposit their loads and filling valleys with sediment. These times of deposition produced broad, flat, alluvial floodplains above buried channels cut far below sea level.

While a drop in base level causes downcutting and a rise causes deposition, an upset of stream equilibrium resulting from sizable increases or decreases in discharge or load can have similar effects on the landscape. Variations in sea level, tectonic movements, and changes in stream equilibrium can cause rejuvenation in stream valleys, so that the valley is slightly deepened and the old valley floor preserved in "stair-stepped" banks along the walls of the valley. These older valley floors are **stream terraces.** Some terraces are a consequence of successive periods of rejuvenation and deposition (Fig. 18.18). Stream terraces provide a great deal of evidence about the geomorphic history of the river and its surrounding region.

Stream Patterns

When viewed from the air or on maps, the channel networks of a stream system form distinct **drainage patterns** (also called **stream patterns**). Two primary factors that influence the stream pattern are geologic structure and the nature of the land surface. A **dendritic** (treelike) stream pattern (Fig. 18.19) is an irregular branching pattern with tributaries joining larger streams at acute angles (less than 90°). A dendritic stream pattern is the most common type, in part because water flow in this pattern is highly efficient. Dendritic patterns form where the underlying geologic structure does not strongly control the position of stream channels. Hence they tend to develop in areas where the rocks have a roughly equal resistance to weathering and erosion. In contrast, a **trellis** stream pattern consists of long, parallel streams linked by short, right-angled segments. A trellis pattern is usually evidence of folding, where parallel outcrops of erodible rocks form valleys between more resistant ridges, such as in the ridge and valley section of the Appalachians. A **radial** pattern develops where streams flow away from a common high point on cone- or dome-shaped geologic structures,

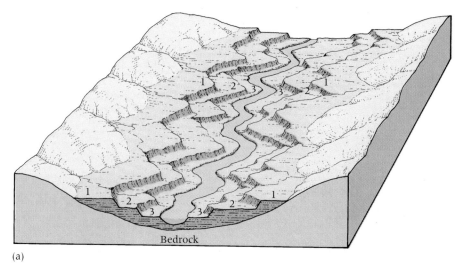

(a)

(b)

Figure 18.18. *(a) Diagram of stream terraces formed by three phases of change in the stream that may have developed through rejuvenation (changes in base level through rise or fall of sea level, or uplift/subsidence of the land surface). Major changes in the equilibrium of streams (from erosion-dominated to deposition-dominated or vice versa) can also form terraces. The terrace development process has produced three successively lower valley floors along the river in this diagram. The oldest terrace is the highest level (1); the youngest is the lowest level (3); (b) River terraces in the Tien Shan Mountains of China.* • **How many terraces can be clearly identified in this photo?**

such as volcanoes and domal uplifts. The opposite pattern is **centripetal,** with the streams converging on a central area, as in a basin of *interior drainage.* **Rectangular** patterns occur where streams follow sets of intersecting fractures to produce a blocky network of straight channels with right-angle bends.

Most stream patterns that drain a large region generally require a considerable amount of geologic time to become well established (varying from tens of thousands to millions of years). In areas that were covered by extensive glacier ice until about 10,000 years ago, streams flow on low gradients, wandering between marshes and small lakes in a pattern (actually a chaotic pattern) called **deranged drainage.** These streams have yet to develop a "normal" drainage pattern because, given the gradient, there has not been enough time (since deglaciation) for a stream network to become well established.

Many streams follow the "grain" of the topography or structure, eroding valleys in weaker rocks and flowing between divides formed on resistant rocks. Many examples also exist, however, of streams that flow across (transverse to) the structure, cutting a gap or canyon through mountains or ridges. These **transverse streams** can be puzzling, giving rise to questions such as how can a stream cut a gorge through a mountain range, and how did the stream get from one side of a mountain range to the other? Such systems are probably either *antecedent* or *superimposed.* Antecedent streams existed before the formation of the mountains that they flow

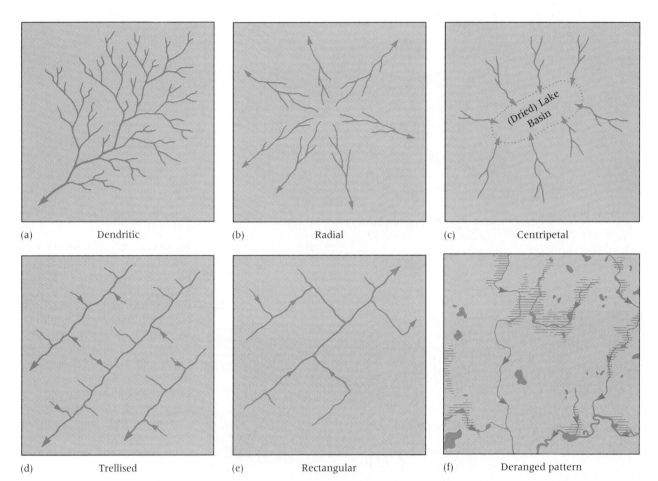

Figure 18.19 Drainage pattern often reflects geologic structure. **(a)** A dendritic pattern is found on homogeneous rocks. **(b)** A radial pattern indicates a domed upland or volcano. **(c)** A centripetal pattern indicates a structural lowland or basin. **(d)** A trellised pattern indicates parallel valleys of weak rock between ridges of resistant rock. **(e)** A rectangular pattern indicates linear joint patterns in the bedrock structure. **(f)** A deranged pattern typically results following the retreat of continental ice sheets, and is characterized by a chaotic arrangement of channels connecting small lakes and marshes.

through, maintaining their courses by cutting an ever-deepening canyon as gradual mountain building took place across their paths. The Columbia River Gorge through the Cascade Range in Washington and Oregon and many of the great canyons in the Rocky Mountain region (Royal Gorge and the Black Canyon of the Gunnison, both in Colorado, and Flaming Gorge in Utah/Wyoming) probably originated in this way. Other rivers, such as those of the central Appalachians, have cut gaps through mountains in a very different manner. These streams originated on earlier rock strata, since stripped away by erosion, so that the streams have been superimposed on the rocks beneath. This sequence would explain why, in many instances, the rivers flow *across* the folds, creating water gaps through

mountain ridges. Examples include the Cumberland Gap and the gap formed by the Susquehanna River in Pennsylvania, both of which were important as travel routes for the first American settlers crossing the Appalachians.

Quantitative Fluvial Geomorphology

Many studies in geomorphology are based on statistical analyses of data measured through instrumentation in the field or from remotely sensed imagery. Measurements and quantitative data allow geomorphologists to understand and scientifically test relationships concerning how landforms develop and the processes responsible for their formation. Although scientific and quantitative methods are important in studying virtually all aspects of the Earth system and are used by climatologists, biogeographers, geomorphologists, and hydrologists, some examples are given here that apply to rivers. Because of the importance of streams to human existence, extensive efforts have been undertaken to gather the data necessary to help us understand how streams work, to predict their behavior, to compare streams objectively, and to describe their nature. Since streams are dynamic systems in a constant state of change, many types of quantitative data must be developed, tested, and applied for descriptive and analytical purposes. An understanding of fluid mechanics and energy flow may also be integrated with that of the geomorphic processes and landform descriptions; part of this approach was discussed earlier in the section on equilibrium in streams (for example, $Q = AV$).

An example of the value of the use of quantitative data can be illustrated with these questions: How can we objectively compare streams in terms of their size or importance as a stream system? Do stream systems develop in a logical manner or is the process completely random? If a channel network is a part of a stream system, and a system is a set of interacting parts, how do parts of the stream system interact with each other? A first step to finding the answers to these questions is to objectively compare streams within a river basin, or to compare similar river basins. The problem, though, is how to determine what "similar" means in an objective and quantitative manner. A solution was suggested in the 1940s by a hydrologist, Robert Horton, who proposed a method (since modified by others) for determining **stream order.** First-order streams are the first headwater channels, formed without tributaries, that can be seen in the field or mapped on large-scale topographic maps. When two or more first-order streams join, the channel becomes larger, and the stream is classified as a second-order stream. When two or more second-order streams intersect, the stream is classified as a third-order stream, and so on. Generally, higher order means larger and longer channels, but fewer channel segments exist as the stream order rises. Drainage basins can also be classified in this manner, according to the highest-order stream in the basin, which is called the *trunk stream* (Fig. 18.20). To answer our questions at the start of the paragraph, the size or importance of a stream or basin can be objectively compared using the stream order system—the Mississippi River would be classed as a *tenth-order stream,* and the Ama-

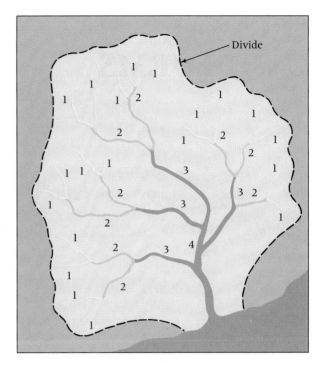

Figure 18.20 *The hierarchy of stream ordering is illustrated by this fourth-order watershed. Note that when two streams of a lower order flow together, the downstream segment will become the next higher order (note also that lower number tributaries flowing into a higher one does not change the order of the downstream segment). Each segment of the stream has only one order following this system.* • **What is the highest order stream called?**

zon would probably be a *thirteenth-order stream.* Further, stream systems do seem to develop in a logical manner and their parts (tributaries, subbasins) are generally well adjusted to each other. Higher order streams, for example, tend to have larger basins, longer channels, larger channels, and greater discharge than lower order streams (and vice versa).

Drainage density studies provide us with information on the degree of erosional dissection that exists in parts of a basin area. The formula used is $D = L/A$: *Drainage density (D) is equal to the total length of all stream channels in the basin (L) divided by the area of the drainage basin (A).* The easily eroded Dakota Badlands have an extremely high drainage density of over 125 (km of channel per one km of land), while very resistant granite hills may have a drainage density of only 5 (5 km of stream channel for every square kilometer of basin area). High drainage density is associated with steep slopes, sparse vegetation, weak rocks, high erosion rates, and high runoff.

(Continued on page 484)

FLUVIAL LANDFORMS

LANDSAT false-color image of Louisiana. ▶

The Map

The Campti map area is located in northwestern Louisiana in the Gulf Coastal Plain region. The Gulf Coastal Plain extends from northern Florida to the Texas-Mexico border. The region extends inland for over 200 miles in some areas and actually extends offshore under the Gulf of Mexico as a shallow continental shelf. Most of the coastal plain is at a very low elevation. The elevation gradually increases from just above sea level along the coast to an elevation of several hundred feet far inland. The region is underlain by gently southward-dipping sedimentary rock layers. Much of the surface material is dominated by alluvial plain deposits from the many rivers crossing the region, especially those of the Mississippi drainage system. This is a land of river meanders, natural levees, and bayous.

The Red River's headwaters originate in the semiarid steppe plains of the Texas Panhandle. The main stream forms much of the Texas-Oklahoma boundary as it flows eastward toward an increasingly more humid climate and high-volume tributaries. About midcourse, the Red River enters a humid subtropical climate region, which supports rich farmland and dense forests. In southern Louisiana, about 161 kilometers (100 mi) downstream from the map area, the Red River flows into the Mississippi River. The Red River is the most southern major tributary of the extensive Mississippi drainage basin.

Louisiana is a region of mild winters but hot and humid summers. Annual rainfall totals for Campti, Louisiana, average about 127 centimeters (50 in). The nearby warm waters of the Gulf of Mexico supply vast amounts of atmospheric energy, producing a high frequency of thunderstorms, tornadoes, and, on occasion, hurricanes that strike the Gulf Coast.

Interpreting the Map

1. How would you describe the general topography of the Campti map area? What is the local relief? What is the elevation of the land bordering the Red River?

2. What is the name of the dominant flat landform feature that extends across the map area? Is this mainly an erosional or depositional landform?

3. In what general direction does the Red River flow? Is the direction of flow easy or hard to determine just from the map area? Why?

4. Does the Red River have a gentle or steep gradient? Why is it difficult to determine the river gradient from the map area? Is the Red River in the map area a low- or high-velocity river?

5. What is the origin of Smith Island? Adjacent to Smith Island is Old River; what is this feature called?

6. Explain the stippled brown areas in the meanders south of the city of Campti. Are these areas on the inside or the outside of the meander turn? Explain why.

7. How would you describe the features labeled a "bayou"?

8. Is this map area more typical of the upper, middle, or lower course of a river?

9. Although this is not an active tectonic region, what do you think would happen to the region if it were uplifted? What would be the major geomorphic process change?

Campti, Louisiana
Scale 1:62,500
Contour interval = 20 ft
U.S. Geological Survey ▶

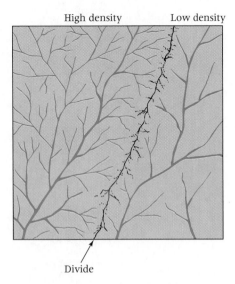

High density   Low density

Divide

Figure 18.21 *Drainage density the length of channels per unit area varies according to several factors. For example: other factors being equal, soft, highly erodible, and impermeable rocks tend to have higher drainage density than areas dominated by strong, resistant, and or permeable rocks. Slope and vegetation cover also can affect drainage density.* • **What kind of drainage density would you expect in an area of steep slopes and sparse vegetation cover?**

To understand this concept, think about what would happen on a hillside if the vegetation was burned off in a fire. Erosion would rapidly cut gullies, creating more channels where fewer existed before. In other words, the drainage density would increase, and we could use the quantitative measure of $D = L/A$ to determine precisely the immediate change and monitor the change over time. Figure 18.21 illustrates drainage density.

Future quantitative studies not only will help us better understand the origins and formational processes of landforms and landscapes, but will also allow us to apply quantitative and scientific methods to help us predict water supplies and flood hazards, estimate soil erosion, and trace sources of pollution.

The Importance of Surface Waters

Streams

Historically, people have used streams and rivers for a variety of purposes. The settlement and growth of the United States would have been very different without the Mississippi River and its far-reaching tributary system that drains most of the United States between the Appalachians and the

Rockies. The Mississippi, like many other rivers, has been used for exploration, migration, and settlement, and the number of major cities along it (Minneapolis/St. Paul, St. Louis, Memphis, and New Orleans, to name a few) provides evidence of people's tendency to settle along rivers. We have also used the Mississippi for the inexpensive transportation of bulk cargo. Today, our navigable rivers still compete successfully with railroads and trucks as carriers of grain, lumber, and mineral fuels.

Rivers and streams have generated power for lumber and paper mills, and, more recently, for hydroelectricity. Rivers provide irrigation water, and the rich alluvial soils of the floodplains and along their banks are often productive agricultural lands. We also use streams as sources of food and water, and for recreation (boating, fishing, waterskiing, swimming), and as a depository for wastes.

There are many benefits to living near streams and rivers. However, settlement along a river has its risks, particularly in the form of floods. Variability of stream flow constitutes the greatest problem to life along rivers and is also an impediment to their use. Stream channels can generally carry (and contain) the maximum flows that are estimated to occur once every year or two. The maximum flows that are probable over longer periods of 5, 10, 100, or 1000 years spill out of the channel and inundate the surrounding land, sometimes with disastrous results. Similarly, exceptionally low flows may produce crises in water supply and bring river transportation to a halt.

Because stream flow can be so variable, and in some cases unreliable, few communities today use only water supplies that are not regulated in some way by humans. Many river systems now consist of a series of **reservoirs**—artificial lakes impounded behind dams. The dams contain potentially devastating floodwaters and store the discharge of wet periods to make the water available during dry seasons or drought years (Fig. 18.22). An outstanding example of such human river-basin management is the Tennessee River Valley, which was "tamed" during the 1930s. A look at a hydrographic map will demonstrate that most of our great rivers, such as the Missouri, Columbia, and Colorado, have been transformed by dam construction into a string of reservoirs. Unfortunately, the life of these reservoirs, like that of any lake, will only be a few centuries at most, because they will gradually fill with sediment carried by the inflowing streams. To protect the natural flow and environments of the few remain-

Figure 18.22 The multipurpose Lookout Point Dam on the Middle Fork of the Willamette River, Oregon.
• **Name several functions that multipurpose dams serve.**

Figure 18.23 The majority of lakes in the world are of glacial origin, as shown here by Stanley Lake in the Sawtooth Range of Idaho. • **Why are lakes considered temporary features?**

ing undeveloped streams and rivers in the United States, Congress enacted the Wild and Scenic Rivers Act in 1968.

Lakes

Lakes are standing bodies of inland water, and the water in most lakes is runoff that is held in temporary storage. Most of the world's lakes, such as Lake Superior, Lake Tahoe, and Lake Victoria, contain fresh water. However, some lakes, such as the Caspian Sea, Dead Sea, and Great Salt Lake in Utah, are salty, because they exist in a closed basin, where evaporation recycles water to the atmosphere, leaving dissolved minerals to accumulate in the lake.

Natural lakes form wherever the water supply is adequate and geologic or topographic processes have created depressions on the land surface. The majority of the world's lakes, such as North America's five Great Lakes and Minnesota's "10,000 lakes," are products of glaciation. Rivers, groundwater, tectonic activity, volcanism, and human activities also produce lakes. Lake Baikal in Russia is the world's deepest lake. This long, narrow Siberian lake is over 1525 meters (5000 ft) deep and was caused by a fault depression. Crater Lake in Oregon, North America's deepest lake, occupies a caldera formed by the collapse of a volcano.

Lakes and ponds (small shallow lakes) are temporary features on Earth's surface. Few have been in existence for more than 10,000 years, and most are very recent in terms of geologic time (Fig. 18.23). As soon as they are formed, their destruction begins through several processes, including sedimentation, biological activity, and perhaps down-cutting of the outlet by a stream.

Streams that flow into lakes deliver sediment—sand, silt, and gravel—to form deltas, which eventually fill the entire lake. Plants and animals, using nutrients brought by streams, occupy the shore and build communities on the bottom, helping the shoreline advance farther into the lake area. As the lake becomes smaller and shallower, plants take root on its bed and it becomes a marsh. Eventually the marsh is filled and becomes a meadow. This inevitable process is accompanied by progressive changes in the lake's flora and fauna. The first occupants are life-forms adapted to cold, deep water. As the process continues, there is a shift toward organisms adapted to progressively shallower and warmer water. Eventually, toads hop where pike and trout formerly swam and humans fished.

Lakes are important to humans for more than their scenic appeal and their value for fishing or recreational activities. Like oceans, lakes affect the nearby climates, particularly by reducing daily and seasonal temperature ranges and by increasing humidity. Lakes can also cause a downwind increase in precipitation—snow or rain generated by the "lake effect," in which storms either pick up more moisture from the lake or undergo uplift as they move over a relatively warm body of water. Major fruit-producing areas have formed near lakes in Florida, New York, Michigan, and Wisconsin because of the moderating temperature effects of lakes.

The benefits of lakes are such that humans have produced tens of thousands of artificial lakes through the construction of dams. Reservoirs are some of humankind's most ambitious construction projects. However, because lake water tends to stratify by temperature, it does not mix as well as rivers or the oceans. Poor circulation makes lakes easily susceptible to destruction by the chemical, thermal, and biological pollution often resulting from human activities. The Great Lakes—Lake Erie in particular—provide instructive examples of the damage that can be done to a large, complex natural system by human misuse over a short period of time.

Water in the Desert

The scenery of desert regions is unlike that of most other environments. Desert landforms are generally angular, with exposures of bare bedrock (Fig. 18.24). In the desert, landforms may dominate the landscape as opposed to the sparse desert vegetation. Actually, however, it is the absence of continuous vegetative and soil cover, rather than the direct influence of arid climates, that gives desert landforms their unique character. Without sufficient vegetation to break the force of raindrops, and without the binding effects of root networks, a blanket of moisture-retentive soil cannot accumulate on slopes. Grains of rock loosened by weathering may be swept away by the next rainfall. Slopes in more humid areas tend to have a mantle of weathered rock and soil that gives them a subdued, rounded form. If the vegetative cover were removed, erosion would accelerate, the permeable soil cover would be lost, and a desertlike landscape would be created, regardless of the climate.

Although desert landscapes strongly reflect a deficiency of water, the effects of running water are widely evident, on slopes as well as in valley bottoms. Much of the rain that falls, sparse as it is, encounters soil-free surfaces and runs off immediately, eroding the land. A long-established fallacy is that most desert landforms are produced by wind erosion. Wind erosion does occur in deserts, but it is much less important than water erosion. The effects of wind erosion are mainly confined to moving poorly consolidated loose materials and to removing fine, dust-sized particles from desert regions. Nevertheless, since sparse vegetation permits wind gradation to op-

Figure 18.24 *This stream flows through a deep gorge on the east side of the Atlas Mountains in Morocco. This stream is on the arid, leeward rain-shadow side of the mountains, facing the Sahara to the east. The stream shown here loses water by infiltration and evaporation, and finally ends by disappearing into the Sahara. Note the steeply dipping, folded rocks of the Atlas, and the thin line of vegetation along the stream channel.* • **Do you think that the gorge was eroded by the stream with this amount of flow, and if not, what factors might have produced more discharge in the stream to erode the deep canyon?**

erate on a far grander scale in deserts than in any other environment, we will examine wind-produced, or *eolian,* landforms in detail in Chapter 20.

Climate and Vegetation

The importance of water as a gradational agent in arid regions is not only related to the climate of those areas today, but also to the climates of the past. Evaporation and precipitation are important factors in determining the amount of runoff available for stream flow and gradational work. In addition, these same climatic factors affect the density of vegetative cover in arid regions. Where little or no vegetation exists, water is extremely effective in shaping the land.

A characteristic of desert climates is the small amount of precipitation received. Sometimes years may pass without any rain in certain desert areas, though such situations are exceptional. Most desert locations receive some precipitation each year, although the frequency and amount are highly unpredictable. The rains that do fall are often brief and limited in their coverage, although they can also be intense. As mentioned earlier, the most important impact of desert rainfall on landform development is that, when it does occur, much of it falls on impermeable surfaces, producing rapid runoff, generating flash floods, and performing as a powerful agent of erosion.

Although running water is a highly effective agent of landform development in deserts today, it only operates occasionally. In most desert regions, running water is active only during and shortly after rainstorms.

Paleogeographic evidence, however, indicates that many deserts have not always had the arid climates they have today. Geomorphologists studying arid regions have found certain desert landforms that are incompatible with the present climate, and attribute these to the work of water under an earlier, wetter, climate. A majority of desert areas were wetter in the past, most recently during the Pleistocene Epoch. While glaciers were advancing in the high latitudes and in mountain regions, precipitation increased in the middle and subtropical latitudes where deserts are found today. Evidence of *pluvial* (rainy) periods in many of today's deserts include lake deposits, wave-cut shorelines (see Chapter 9), and large canyons occupied by streams that today are too small to have eroded such large valleys.

Desert Streams

Most desert streams are *intermittent;* that is, they flow only during and shortly after a particularly rainy period. The rest of the time, the beds of these streams lie exposed and dry. There is no groundwater input to sustain a **base flow** between periods of surface runoff. The Mojave River in California is an excellent example of an intermittent stream that loses water volume by seeping into the ground rather than gaining volume from groundwater inflow as do most streams in humid regions.

Since desert streams seldom have enough discharge to flow to the sea before evaporating or infiltrating into their dry beds, they often terminate in interior depressions, where they form temporary lakes. These, too, eventually evaporate and disappear, and then reappear when rains provide adequate inflow. In arid regions there are many closed basins that have not been filled with water since the pluvial times of the Pleistocene Epoch.

Where surface runoff drains into closed desert basins, sea level does not govern erosional base level as it does in streams that flow into the ocean. Low areas characterized by this pattern of intermittent streams that are controlled by a local base level are known as **interior drainage basins.** When deposition raises the surface of the

basin, the local base level also rises. If tectonic activity lowers the basin, the local base level is also lowered and stream erosion is rejuvenated. Some basins of interior drainage have floors below sea level, as in Death Valley, California, the Dead Sea Basin, the Turfan Basin in western China, and Australia's Lake Eyre (see Map Interpretation: Desert Landforms).

Most streams found in deserts originate in nearby humid regions or cooler, wetter, mountain areas. Many streams in arid regions, however, have insufficient discharge to sustain flow across a large desert. Without inflow provided by tributaries or groundwater to replenish water losses caused by evaporation and underground seepage, the streams dwindle and finally disappear. The Humboldt River in Nevada is an outstanding example, because, after flowing 465 kilometers (290 mi), the river disappears into the Humboldt Basin, a closed depression. Only a few large streams that originate in humid uplands have sufficient volume to survive the long journey across hundreds of kilometers of desert to the sea (Fig. 18.25). **Exotic streams,** rivers that flow from humid regions through the desert to reach the ocean, erode toward a base level governed by sea level. Other desert streams—during their infrequent flows—erode toward the base level set by the lowest elevation of their basin. Classic examples of exotic streams include the Nile (Egypt and Sudan), Tigris-Euphrates (Iraq), Indus (Pakistan), Murray (Australia), and Colorado (United States and Mexico) Rivers.

Water As a Gradational Agent in Arid Lands

When rain falls in the desert, sheets of water run down unprotected slopes, picking up sediment. Channels become filled with flooding streams of muddy water. The material removed by runoff and surface streams is transported, just as in humid lands, until velocity and/or volume of stream flow decreases sufficiently for deposition to occur. Eventually these streams disappear because their seepage and evaporation losses exceed their flow. Huge amounts of debris are deposited along the way as the stream loses volume and velocity. The processes of erosion, transportation, and deposition by running water are the same in both arid and humid lands. However, the resulting landforms differ, partially because of the intermittent nature of desert runoff, the presence of interior drainage that does not flow to the sea, and the lack of vegetation to protect surface materials against rapid erosion.

Figure 18.25 A false-color satellite image of the Nile River meandering across the Sahara in Egypt. The dark-red irrigated croplands contrast with the barren, sandy, and rocky terrain of the surrounding desert. The Nile is an exotic stream. Its headwaters are in the wetter climates of the Ethiopian Highlands and lakes in the east African rift zone, which support its northward flow across the Sahara to the Mediterranean Sea. • Can you name three other exotic streams?

Landforms of Arid Fluvial Erosion

Among the most common desert landforms created by surface runoff and erosion are the channels of intermittent streams. Created by the rushing surface waters, these vertical-walled stream channels, usually cut into unconsolidated alluvium, are called **arroyos** or **washes** in the southwest United States, **barrancas** in Mexico, and **wadis** in North Africa and southwest Asia (Fig. 18.26). Arroyos are prone to flash floods that make them potentially dangerous areas for desert camping. Though it may sound bizarre, many

people have drowned in the desert during flash floods.

Where steep slopes are underlain by weak clays and shales, rapid runoff produces a dense network of barren ridges dissected by V-shaped gullies and dry ravines. Early fur trappers in the Dakotas called such areas "bad lands to cross" (Fig. 18.27). The phrase stuck, and those regions are still called the Badlands. Similar area of rugged, highly dissected, barren topography, with extremely high drainage density, are now called **badlands.** Other examples of badlands topography can be seen in Death Valley National Park, California, and Big Bend National Park, Texas. Badlands generally do not form naturally in humid climates because the vegetation there slows runoff and erosion, decreasing drainage density. Removing the vegetation from clay or shale areas by overgrazing, mining, or logging, however, will cause badlands topography to develop even in humid areas.

Plateaus are regions of generally low relief and high elevation, often dominated by a structure of horizontal layers. Many striking plateaus exist in the deserts and semiarid regions of the world. In the United States, the Colorado Plateau is an excellent example, centered on the Four Corners area of Utah, Arizona, New Mexico, and Colorado. In tectonically uplifted desert plateau regions such as this, streams and their tributaries respond to uplift by cutting steep-sided canyons. Where the canyon walls consists of horizontal layers of alternating resistant rocks and erodible rocks, differential weathering and erosion may cause the canyon walls to retreat quite rapidly. The walls of canyons in these areas tend to be stair-stepped, with near-vertical cliffs marking the resistant layers (ordinarily sandstone or limestone), and weaker rocks (often shales) forming the slopes. The distinctive appearance of the walls in the Grand Canyon have this appearance, with cliffs forming where the rocks are strong and slopes developing where the rocks are weaker. The rim of the Grand Canyon forms a flat-topped cliff produced by a **caprock,** a resistant horizontal layer that forms the top of the stair-stepped valley walls. Caprocks not only form the rims of canyons, but they also form the summits of other kinds of flat-topped landforms that are found in other climate regions; however, they are most common and best developed in deserts.

The processes of erosion and weathering will eventually reduce the extent of a caprock until only flat-topped, steep-sided **mesas** remain (mesa means "table" in Spanish). Mesas are relatively common landscape features in the Colorado

(Continued on page 492)

Figure 18.26 *Dry stream beds such as this one in southern Utah are caused by runoff channeled after rains in arid regions. These stream-cut channels are called arroyos in the southwestern United States and wadis in the Middle Eastern and North African deserts (see Fig. 18.24).* • **Why do desert stream channels have a high flash-flood risk?**

Figure 18.27 *The Badlands of South Dakota. Impermeable clays that lack a soil cover produce rapid runoff, leading to intensive gully erosion and a high drainage density.* • **Was this rugged terrain named appropriately?**

DESERT LANDFORMS

LANDSAT image, Death Valley, California. ▶

The Map

The Furnace Creek map area is located in Death Valley National Park, California. Death Valley is located in the Basin and Range region, which consists of north-south trending tilted fault block mountains (ranges), separated by downfaulted valleys (basins). As the young rugged ranges erode, the basins are filling with debris carried down by effective but rare flash floods during desert rains.

The mountain block forming the western slopes (in the map area) of Death Valley is the Panamint Range. The highest summit, Telescope Peak, reaches an elevation of 3368 meters (11,049 ft). To the east of Death Valley is the Amargosa Range. Two peaks in the Amargosa Range, Coffin Peak and Funeral Peak, are an indication of the infamous history Death Valley had with early prospectors and miners who came to this place for gold and borax. Death Valley's lowest elevation of −86 meters below sea level (−282 ft) is located midvalley at Badwater, which is appropriately named!

The present climate of the Basin and Range region is arid, except for the high mountain summits, which may receive snow and support pine forests. Death Valley has the most extreme aridity in the region. The average annual rainfall for Death Valley is 3.5 centimeters (1.7 in.), and the potential evaporation may exceed 380 centimeters (150 in.)! Death Valley summer temperatures commonly exceed 40°C (104°F), and the record maximum is 57°C (134°F). In winter, much of the Basin and Range region has freezing temperatures at night.

During the Pleistocene Epoch, lakes several hundred feet deep occupied numerous basins in the region. From observing the map and satellite photo, it is hard to believe that several thousand years ago, Death Valley was occupied by Lake Manly. This Pleistocene lake had a depth of 183 m. Today, only traces of Lake Manly's shoreline can be seen, as Death Valley has lost its water and is now a classic site to study young desert landforms.

Interpreting the Map

1. Based on the general location of Death Valley, why is the area so arid?

2. Describe the general topography of the map area.

3. What is the lowest elevation on the map? What is significant about the elevation of Death Valley?

4. What specific type of arid landform is the blue striped feature located in the depression? Why are the edges of this feature shown with blue dashed lines?

5. Of what types of surface materials is the feature in Question 4 composed? Explain how this feature was formed.

6. What specific type of arid landform is indicated on the map by the curved parallel contours at the base of the mountains? What do the blue —··— ··— lines crossing the contours represent?

7. Of what types of surface materials is the feature in Question 6 composed? Explain how this feature was formed.

8. Note that the large, curved landforms at the base of the mountains coalesce into similar features. What is the specific landform name for such broad features?

9. What evidence from the map indicates that this is an interior drainage basin?

10. Sketch a general east-west profile from the benchmark at Devil's Speedway to the 2389-foot benchmark on the mountain straight to the west. Label the following landforms: mountain front, pediment, alluvial fan, and basin.

Furnace Creek, California
Scale 1:62,500
Contour interval = 80 ft
U.S. Geological Survey ▶

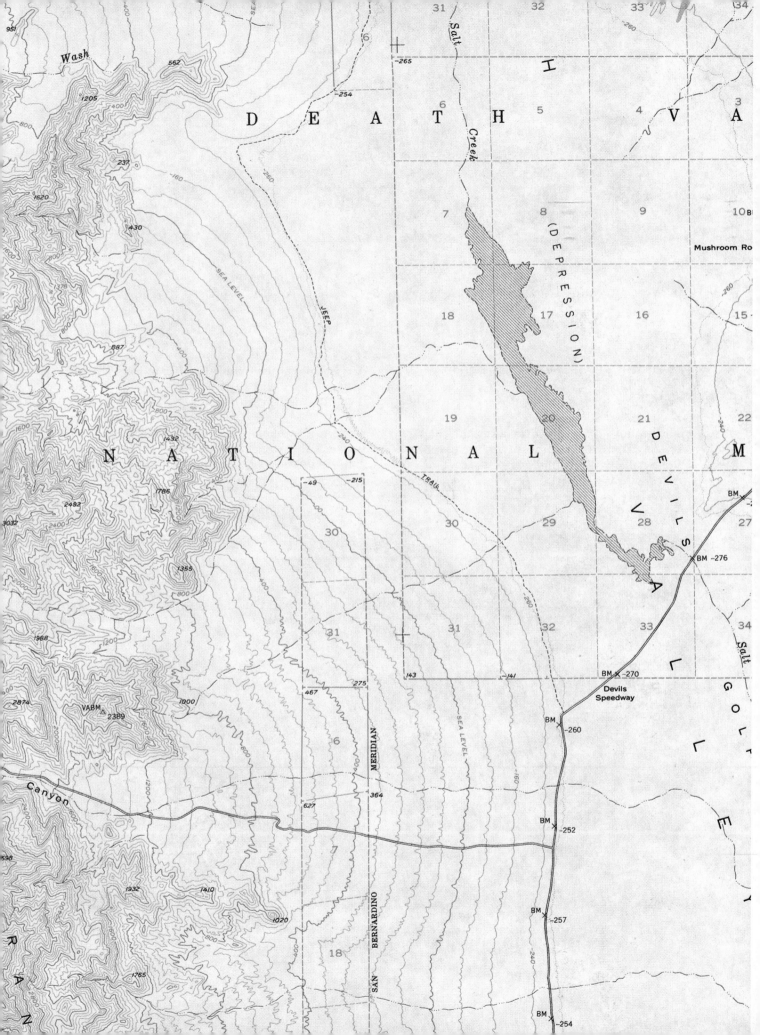

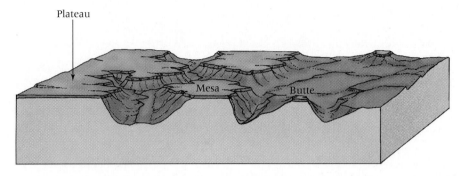

Figure 18.28 *Diagram of landforms developed through weathering and erosion in an area of horizontal layers of rocks with a resistant cap rock (such as the Colorado Plateau of Arizona, Colorado, New Mexico, and Utah). (Compare this diagram to the photo in Figure 18.29.)*

Plateau region. After additional erosion of the caprock from all sides (Fig. 18.28), a mesa will be reduced to a smaller flat-topped remnant, called a **butte.** Mesas and buttes in a landscape are generally evidence that uplift occurred in the past and that fluvial erosion of the uplifted land has been extensive since that time. Monument Valley in the Navajo Tribal Reservation on the Utah-Arizona state line is an exquisite example of such a landscape formed with a caprock that is particularly thick, contributing to the distinctive scenery (Fig. 18.29). Many famous Hollywood western movies have been filmed in Monument Valley and in nearby areas of the Colorado Plateau because of the striking, photogenic desert landscape.

Figure 18.29 *The landscape of Monument Valley, Arizona, with prominent buttes and mesas. The cap rock here is particularly thick, and represents a single layer of rocks that once covered the entire region. The buttes and mesas are erosional remnants of that layer.*
• How were these landforms produced?

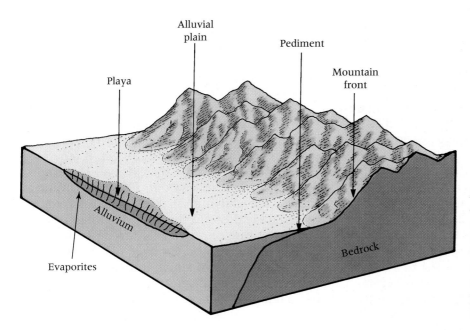

Playa

Alluvial
plain

Pediment

Mountain
front

Alluvium

Evaporites

Bedrock

Figure 18.30 Pediments are erosion surfaces cut into bedrock at the bases of mountain slopes in arid regions. At many locations, pediment surfaces are covered with a veneer of alluvial deposits. • **How can a pediment be distinguished from an alluvial fan?**

The erosion of slopes fringing a desert basin or plain results from sheetwash and gully development. As these slopes gradually undergo erosional retreat (aided by weathering), a gently sloping bedrock surface, called a **pediment,** is formed (Fig. 18.30). Characteristically in desert areas, there tends to be a sharp break in slope between the steeper hills or mountains, which rise at angles of 20 to 30 degrees or steeper, and the gentle pediment, whose slope is usually only 2 to 7 degrees.

Geomorphologists do not agree on how pediments are formed, perhaps because different processes may be responsible for their formation in various regions of the world. However, there is general agreement that most pediments are erosional surfaces created by the action of running water. In some areas, however, weathering may also have played a strong role in the development of pediments.

Landforms of Arid Fluvial Deposition

Deposition is as important as erosion in creating landform features of the desert, and in many areas the deposition carried on by water does as much to level the land as does erosion. Many desert areas have wide expanses of alluvium deposited either in closed basins that contain the sediments or at the base of mountains by streams as they lose water in the arid environment. As the flow of a stream diminishes, so does its *capacity*—its ability to transport load. Most landforms formed by fluvial deposition in arid lands are not exclusive to desert regions, but are common and visible in any dry environment barren of soil and sparse in vegetative cover.

Alluvial Fans. Where streams, particularly intermittent ones, pass abruptly from narrow canyons to open plains, their channels may flare out to become wide and shallow. The greater external friction of the shallow channel quickly reduces stream velocity and carrying power, and the discharge decreases as water seeps from the channel into the coarse alluvium below. As a result, most of the sediment load carried by the stream is deposited at the base of the highland.

Upstream, the channel is constrained by valley walls in the highlands, but as it flows out of the canyon, the channel is free to shift its position. As deposition occurs on the plain area, the channel, acting much like a braided stream, shifts, subdivides, and continues to deposit material. The location at which the stream exits the canyon mouth serves as a pivotal point. Deposition takes place radially around that point in the shape of an Oriental fan (or sometimes a cone), called an **alluvial fan.**

An important characteristic of an alluvial fan is the sorting that generally occurs on its surface. Coarse sediments like boulders and pebbles are deposited near the fan's apex, where the slope is steepest. The alluvium becomes increasingly finer downslope from the apex. Alluvial fans are especially common in arid regions because of intermittent flow and the heavy sediment loads carried during brief periods of runoff. In some regions many alluvial fans are composed largely of debris-flow deposits (a mix of rock and soil), which are not sorted like deposits produced by running water.

Alluvial fans achieve their greatest development where intermittent streams laden with debris flow out of a mountainous region onto arid plains. In the western United States, alluvial fans are especially common in landscapes consisting of fault-block mountains and basins, as in the Great Basin of California, Nevada, and Utah. Here streams periodically rush from canyons in the uplifted fault-block mountains and deposit their load of debris in the adjacent basins (Fig. 18.31).

There are several reasons for the development of alluvial fans in such settings. First, highland areas in desert regions are subject to intense erosion, primarily because of their thin vegetation cover, steep slopes, and the orographically intensified downpours that can occur over mountains. In addition, the streams in arid regions probably carry a larger load (in comparison to the discharge) than comparable streams in more humid regions. As the streams flow from confined channels in mountain canyons into desert basins, they deposit most of their coarse sediment near the mouth of the canyon. As streams flow out into the desert basin, their depth decreases and their volume is significantly reduced through seepage into the dry ground. Not far from the canyon, the stream itself may disappear. (Note: In humid areas, streams generally have enough discharge to continue flowing in restricted valleys after leaving highland areas. The water loss is not nearly as significant because groundwater inflow and entering tributaries sustain the stream flow. As a result, most highland streams that flow into the lower lands of humid regions do not create large alluvial fans.)

Where alluvium has been deposited on a pediment, the surface appearance of this *erosional* form may closely resemble a water-*deposited* alluvial fan. In some situations, in fact, it may not be possible to determine the existence of an underlying pediment without either excavating through the alluvium at the surface or finding the surface exposed in the walls of washes or gullies. In the case of an alluvial fan, the alluvium will be tens or even hundreds of meters thick. In contrast, the layer of alluvium overlying a pediment is only a relatively thin veneer, no more than a few meters deep and sometimes much less.

Along the bases of many mountains in arid regions, adjacent alluvial fans are so large that they join together to form an undulating, ramp-like surface called a **bajada** (Fig. 18.32). A bajada consists of adjacent alluvial fans forming an "apron" around the mountain base. Where the associated streams are large, fans tend to be less steep, less permeable, and larger in area. Where extensive fans coalesce they form a **piedmont alluvial plain,** like the area surrounding Phoenix, Arizona.

Many piedmont alluvial plains have rich soils and a potential to be transformed into highly productive agricultural lands. The major

Figure 18.31 Alluvial fans form at the base of a mountain range where streams generally lose their carrying power. The apex of a fan forms at the mouth of the canyon or wash where the stream encounters more gently sloping land. Fans are particularly common landforms in the arid Basin and Range region of the western United States. For scale, note that a road runs across the lower portion of this alluvial fan in Death Valley, California.
• **Where would the coarser deposits be found on an alluvial fan?**

Figure 18.32 *A bajada is formed when a series of alluvial fans join, forming a continuous alluvial slope along the front of an eroding mountain range. This example is from the Mojave Desert of California.* • **Why would a series of alluvial fans have a tendency to eventually join to form a bajada?**

obstacle in accomplishing this goal is inadequate water to grow crops in an arid environment. In many areas of the world, arid alluvial plains are irrigated with water diverted from mountain regions or obtained from reservoirs on exotic streams. The fertile alluvial farmlands near Phoenix are a good example, producing citrus fruits, dates, cotton, alfalfa, and vegetables.

Playas. Desert basins of interior drainage surrounded by mountains are sometimes called **bolsons.** Most bolsons were formed by the same faulting that uplifted the fringing highlands. The lowest part of a bolson commonly is occupied by a dry lake bed, known as a **playa.** Floods occasionally transform the playa into a shallow pond,

called a **playa lake,** in the space of a few hours. The lake may be gone the next day, or it may persist for weeks (Fig. 18.33). Playa lakes lose most of their water to evaporation in the desert air, although seepage plays a role. Repeated cycles of inflow and evaporation leave behind thick mineral deposits crystallized from evaporating brines: chlorides, sulfates, and carbonates (for example, rock salt, gypsum, and lime, respectively). When the playa lake dries out, its bed may be encrusted with sparkling white salt deposits, although some playas form a salt crust from groundwater loss through surface evaporation. Saline playas are known as **salt flats** or **salinas.** Most playas have a clay surface, baked hard by the desert sun when it is dry but extremely sticky and slippery when

Figure 18.33 *Badwater, in Death Valley, California. The playa surface, seen just behind the edge of the water, is the lowest elevation in North America at 86 meters (282 ft) below sea level. Badwater, the small lake, is mainly fed by groundwater from the surrounding mountains moving below the surface through alluvium deposits. The water quickly evaporates in this extremely arid environment.* • **Why was this small lake called Badwater?**

Figure 18.34 *A false-color satellite image of Death Valley, California, which is a desert bolson in the Basin and Range region. Death Valley lies between the Amargosa Range to the northeast and the Panamint Range to the southwest. Alluvial fans dominate the edges of the valley.* • **What do you think the white areas are in the center of the valley?**

it is wet. Playas are among the flattest landforms on Earth.

Playas are useful in several ways. For one, we mine the rich deposits of evaporite minerals, including such important industrial chemicals as potash, salt, borates, and sodium nitrate that have been deposited in playa beds. Also, the extensive, flat surfaces of some playas make them suitable as racetracks and airstrips. Utah's famous Bonneville Salt Flats are the bed of an extinct Pleistocene lake formed during pluvial times. The western portion of the Salt Flats, where world land speed records are set, still floods to a depth of one or two feet, usually in the winter season. The hard, flat surface of Edwards Dry Lake, in California's Mojave Desert, serves as a favored landing site for the space shuttle. Occasionally this playa also floods, and the shuttle must be diverted to an alternative landing location.

Landscape Development in Deserts

Landscape development in arid climates is similar in many ways to that of humid climates. The major differences in arid regions are the great expanses of exposed bedrock, the lack of continuous water flow, and the more active role of the wind. As an arid landscape develops over time, the lowering of uplands and filling of lowlands is accomplished mainly by fluvial erosion and deposition. The presumed final product is a pediplain—an extensive low-relief erosional area formed by the coalescence of pediments.

An excellent example of desert landscape development is found in the Basin and Range region of the western United States. The region extends from west Texas to eastern Oregon and includes all of Nevada and large portions of New Mexico, Arizona, Utah, and eastern California. Here over 200 mountain ranges, with basins between them, dominate the topography. Much of this area is part of the Great Basin subregion, characterized by interior drainage and centered on Nevada (Fig. 18.34). However, many individual basins also have interior drainage, such as Death Valley and the Coachella Valley in southern California.

In an early stage of development, the recently uplifted fault-block ranges (horsts) rise thousands of feet above the desert basins (grabens). These high mountains, such as the Guadalupe Mountains, Sandia Mountains, Warner Mountains, and the Panamint Range, encourage orographic rainfall. Thus, fluvial processes begin eroding the flat blocks to form arroyos and pediments (Fig. 18.35). Fluvial deposition develops alluvial fans at the base of the range. As time

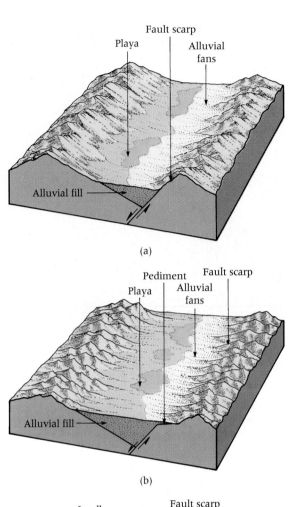

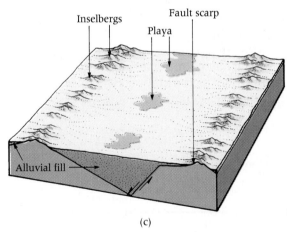

Figure 18.35 *The sequence of desert landscape evolution in a desert region characterized by fault-block mountains and intervening basins, such as the Basin and Range region. (a) Alluvial fans develop at the base of uplifted fault-block mountains as streams downcut and erode the young ranges. (b) After mountain uplift ceases, the range fronts erode back, producing pediments, and the basins continue to fill with eroded material, forming bajada and playa surfaces. (c) Eventually, over millions of years, erosion may reduce the mountains to isolated remnants called inselbergs, which rise above flat desert plains.*
• **What do you think would happen to the landscape if vertical faulting (uplift) resumed in the region?**

Figure 18.36 *Ayers Rock is a striking sandstone inselberg, an erosional remnant, rising above the arid, flat interior of the Australian Outback. The Olgas, in the background, formed as another erosional remnant divided into many bedrock hills by the widening of joints.*
• **Explain how inselbergs form.**

jave Desert in California and parts of the Sonoran Desert in Arizona, however, have localities where uplift along faults has been inactive long enough for erosion and deposition to dominate the landscape. Generally, over time relief is reduced to landscapes dominated by desert plains and extensive bajadas interrupted by a few isolated, resistant erosion remnants called **inselbergs** (island mountains) as reminders of earlier, more tectonically active and mountainous landscapes. The ancient and geologically stable deserts of Australia and the Sahara also have areas of inselbergs surrounded by extensive desert plains (Fig. 18.36).

Although fault-block mountains and fault-block basins dominate the geologic structure of many regions of the American West and other arid locations around the world, it is important to know that desert landscapes are as varied as those of other environments. Desert regions exist where the landscape is dominated by nearly every imaginable geologic setting, including volcanoes, ash deposits and lava flows, folded rocks forming ridges and valleys, horizontal layers of rock, and exposures of massive intrusive rocks. The geologic structures and geomorphic processes found in desert areas are, for the most part, the same as those found in humid regions. It is the variation in the effects and power of these processes that makes the desert landscape distinctive. *Viva la difference,* for deserts are places with a beauty all their own. One only needs to count the number of national parks, national monuments, and other scenic attractions in the arid southwestern United States to find ample support for this contention.

passes, erosion and deposition subdue relief, resulting in lower mountains and increased valley filling as alluvial fans coalesce into bajadas.

Most areas of the Basin and Range region and its spectacular counterparts in western China and Iran are too tectonically active to reach a later stage of development. The landscapes of the Mo-

Define and Recall

stream	base level	suspended load	delta
fluvial processes	hydraulic action	bed load	distributary
runoff	plunge pool	dissolved load	rejuvenated stream
sheet wash	stream abrasion	stream capacity	entrenched meander
tributary	(corrasion)	stream competence	stream terrace
stream system	pothole (stream)	alluvium	dendritic stream pattern
drainage basin	corrosion	braided stream	trellis stream pattern
(watershed, catchment	attrition		radial stream pattern
basin)	gradient (stream)	V-shaped valley	centripetal stream
interfluve	graded stream	cut bank	pattern
divide	headward erosion	point bar	rectangular stream
intermittent stream	meander	estuary	pattern
perennial stream	solution	alluvial plain	deranged stream pattern
floodplain	suspension	oxbow lake	transverse streams
stream discharge	traction	natural levee	stream order
stream load	saltation	yazoo stream	drainage density

base flow
interior drainage basin
exotic stream
wash (arroyo, barranca, wadi)

badlands
plateau
cap rock
mesa
butte

pediment
alluvial fan
bajada
piedmont alluvial plain
bolson

playa
playa lake
salt flat (salina)
inselberg

Discuss and Review

1. On a worldwide basis, what is the most important gradational process operating to shape the landscapes of the continents?

2. What dominant stream process formed the Grand Canyon? What was the dominant stream process that formed the delta of the Mississippi River?

3. What factors affect the discharge of a stream?

4. Why is it important to monitor stream gauging stations and the stream flow data that they provide?

5. Which variables affect the erosion rate of a stream bed? What determines the base level to which a stream bed will erode?

6. What constitutes the largest portion of a stream's load? State the ways a stream can transport its load.

7. How does a stream sort alluvium? Explain the relationship of this sorting to velocity changes in a stream.

8. Where in the course of a stream is the gradient steepest? How do rapids and waterfalls in stream beds relate to rock resistance and steepness of gradient?

9. Describe the development of natural levees and the relationship of levees to yazoo streams.

10. What would cause rejuvenation of a stream? What unique features would distinguish a rejuvenated stream?

11. What are stream terraces? How are they formed? How did changes in sea level during the Pleistocene cause stream terraces to form on land?

12. Name the two primary variables that influence a stream system's pattern. How do the various patterns reflect these variables?

13. Discuss the positive and negative aspects of building dams.

14. What kinds of processes help form lakes? Why are lakes such short-lived features of Earth's surface?

15. What are some general characteristics that illustrate how landforms in arid regions differ from landforms in humid areas?

16. How does an exotic stream differ from an intermittent stream? Name three examples of exotic streams.

17. What factors contribute to the development of badlands topography?

18. What is the difference between a mesa and a butte?

19. Why do streams deposit their load to form alluvial fans? In what ways does the formation of a delta resemble that of an alluvial fan?

20. Why do temporary lakes form in basins of interior drainage?

Consider and Respond

1. One aspect of the concept of equilibrium in streams is that a stream continually adjusts several factors to match the discharge that it is carrying (velocity, discharge, load capacity, load competence, channel width and depth). How would each of these factors change in response to a flood event during each of the three flood phases—rising, cresting, and receding? What effects would these changes have on the landscape?

2. In the study of a drainage basin, what types of geographic observations and quantitative data might prove useful in planning for flood control and water supply?

3. Why are there so many national parks and monuments in the arid parts of the American Southwest? What landform characteristics and other factors of desert landscapes draw people to these locations?

CHAPTER

19

GLACIAL SYSTEMS

CHAPTER PREVIEW

▶ A glacier serves as an excellent example of an open system.

Why is a glacier considered an open system? What materials enter and leave the system? How does energy move through the system?

▶ There are two major forms of glaciation which, although similar in some respects, have their own impact on the landscape and originate in different areas in different ways.

What are the two forms and how do they originate?

▶ Much of the stark grandeur and scenic beauty of mountain regions is due to alpine glaciation.

What characteristics of most high mountains are a product of alpine glaciation? How might human use of mountains be influenced by alpine glaciation?

▶ The ice budget of a glacial system is determined by two factors: the amount of snow (ice) that accumulates each year at the source of the glacier and the amount of ablation (melting, evaporation, sublimation) near the snout of the glacier.

What happens to the glacier when accumulation exceeds ablation? When ablation exceeds accumulation? What happens when dynamic equilibrium has been achieved?

▶ The most visible effects of continental glaciation in North America are from the Wisconsinan, the last stage of Pleistocene glaciation.

Why are the effects of Wisconsinan glaciation so prominent in the physical landscape? How did Pleistocene glaciation change the physical appearance of North America?

▶ Although both are produced by continental glaciation, there are major differences between erosional and depositional glacial landforms.

What are the major differences? How did they develop through glaciation? How has human use been affected by these landscapes?

▲ Susitna Glacier and Mt. Hayes, Alaska.

One can hardly imagine greater natural beauty than in the Swiss Alps, the Canadian Rockies, or the coastal fjord country of Norway and Alaska. Rugged mountain peaks rise high above deep, lake-filled valleys or narrow, deep sea lanes, creating the ultimate in scenic appeal for many people. Masses of moving ice, known as **glaciers,** have transformed the appearance of high mountains, as well as large portions of continents, into unique landscapes. These great icy currents are one of the most effective and spectacular geomorphic agents on Earth's surface.

Glacier Formation and the Hydrologic Cycle

Glaciers are moving masses of ice that have accumulated on land in areas where more snow falls during a year than melts. The snow falls as hexagonal crystals of intricate beauty and variety. Once on the land surface, however, it is soon transformed into a more compacted mass of smaller, rounded grains. As the air space around them is lessened by compaction and melting, the grains become more dense. With further melting, refreezing, and increased weight from newer snowfall above, the snow reaches a granular recrystallized stage between flakes and ice known as **firn.** With additional time, pressure, and refrozen meltwater from above, the firn will be recrystallized into glacial ice. The small firn granules will become larger, interlocked blue ice crystals. When the ice is thick enough, usually over 30 meters (100 ft), the weight of the snow and firn above will cause the crystals to become plastic and flow outward or downward from the area of snow accumulation (see pages 508–510, "How Does a Glacier Flow?").

Glaciers are open systems, with snow entering the system and mainly meltwater leaving the system in a constant cycle. The glacial system is controlled by two basic climatic conditions: precipitation in the form of snow, and freezing temperatures. First, there must be sufficient snowfall to exceed the annual loss through melting, evaporation, sublimation, and **calving,** which occurs when the glacier loses solid chunks as icebergs to the sea or to large lakes. Mountains along middle-latitude coastlines and even at the equator can support glaciers due to heavy orographic snowfall, despite intense sunshine and warm surrounding climates. Yet some very cold polar regions in subarctic Alaska and Siberia, and a few valleys in Antarctica, have no glaciers due to a dry climate.

A second climatic condition is temperature. Summer temperatures must not be high for too long or all the snowfall from the previous winter will melt. Surplus snowfall is essential, since it allows for the pressure of accumulated snow over the years to transform older buried snow into firn and glacial ice and to create depths great enough for the ice to flow. Glaciers are sometimes classified by temperature as more active temperate glaciers or colder, slower-flowing polar glaciers.

Glaciers are part of Earth's hydrologic cycle and are second only to the oceans in the total amount of water contained. About 2 percent of Earth's water is currently frozen as ice. Two percent may be a deceiving figure, however, since over 80 percent of the world's *fresh* water is locked up as ice in glaciers, with the majority of it in Antarctica. The total amount of ice is even more awesome if we estimate the water released upon the hypothetical melting of the world's glaciers. Sea level would rise about 60 meters (200 ft). This would change the geography of the planet considerably. In contrast, should another ice age occur, sea level would drop drastically. During the last ice age, sea level dropped about 120 meters (400 ft).

When snow falls on high mountains or in polar regions, it may become part of the glacial system. Unlike rain, which returns rapidly to the sea or atmosphere, the snow that becomes part of a glacier is involved in a much more slowly moving system. Here water may be stored in ice form for hundreds or even hundreds of thousands of years before being released again into the liquid water system as meltwater. In the meantime, however, this ice is not stagnant but is moving slowly across the land with tremendous energy, carving into even the hardest rock formations. The glacier reshapes the landscape as it engulfs, pushes, drags, and finally deposits rock debris in places far from its original location, so that even long after the land is released from its icy entombment, a tremendous variety of glacial landforms remains as a reminder of the energy of the glacial system.

Throughout most of Earth's history, glaciers did not exist. However, when a period of time occurs during which significant areas of Earth are covered by glaciers, we call it an *ice age*. At the present time about 10 percent of Earth's land surface is covered by glaciers. Present-day glaciers are found in Antarctica, in Greenland, and at high elevations on all the continents except Australia. In the recent past, from about 2.4 million to about 10,000 years before the present, nearly a third of Earth's land area was periodically covered by ice thousands of meters thick. In the much more distant past, other ice ages have occurred.

Types of Glaciers

There are two major types of glaciers: **alpine glaciers** and **continental ice sheets.** Alpine glaciers have their source in mountain areas, usually accumulating in depressions initiated by stream erosion. Those that are confined by the rock walls of the valley they occupy may also be called **valley glaciers** (Fig. 19.1).

Another variety of alpine glacier is the **piedmont glacier** (Fig. 19.2). This variety forms where valley glaciers coalesce and spread out like pancake batter over flatter land at the base of a mountainous region. Whereas piedmont glaciers result from glacial flow beyond the limits of confining valleys, some alpine glaciers do not reach the valleys below the zone of high peaks. Instead,

Figure 19.1 A valley glacier in Alaska's Kenai Peninsula. A clearly defined firn line separates the white upper accumulation zone from the blue-toned lower ablation zone. • **Explain what firn is.**

Figure 19.2 *The upper portion of the Malaspina Glacier, Alaska. This large piedmont glacier lies at the flat base of the St. Elias Mountains.* • **How do you think this type of glacier forms?**

they occupy distinctive high amphitheaters, called **cirques,** and are known as **cirque glaciers** (Fig. 19.3).

Alpine glaciers create the characteristic rugged scenery of the high mountains of the world. They can currently be found in the Rockies, Sierra Nevada, Cascades, and the Olympic and Coast Ranges, and in numerous Alaskan ranges of North America. They are also found in the Andes, in New Zealand, in the Alps, and in the Himalayas, Pamirs, and other large Asian mountain ranges. Small glaciers are even found on tropical mountains in New Guinea and East Africa, on Mounts Kenya and Kilimanjaro, and in the Ruwenzori Range. The largest alpine glaciers in existence today are found in Alaska and the Himalayas, where some reach lengths of over 100 kilometers (62 mi).

The second and larger type of glacier is the continental ice sheet, a far more significant formation than the valley glacier. At one time, continental ice sheets covered as much as 30 percent of the land. They still blanket Greenland and Antarctica, and smaller ice caps are present on some Arctic islands. In contrast to alpine gla-

ciers, continental ice sheets are unconfined, flowing over even the higher portions of the land. Thus, they are not elongated but flow outward in all directions from their source area.

Features of an Alpine Glacier

Alpine glaciers can be divided into two parts: a zone of accumulation and a zone of ablation (Fig. 19.4). The upper portion of the glacier, where snowfall exceeds **ablation** (losses through melting, evaporation, and sublimation), is termed the **zone of accumulation.** The lower portion of the glacier, where ablation exceeds snowfall, is termed the **zone of ablation.** This lower zone is able to develop only because, under the stress of gravity, glacial ice flows constantly from the zone of accumulation down to the zone of ablation. Downslope movement is seen in the development of a great crack or crevasse around the head of the glacier. This crack is known as the *bergschrund* (see Fig. 19.3). It shows that the ice mass is pulling away from the confining rock walls of the cirque and is moving outward and downward. The end of a glacier—its *terminus* or *snout*—marks the lowest advance of the zone of ablation. If the snout reaches the sea, it will calve icebergs.

The **firn line,** which separates the two zones of a glacier, represents an equilibrium point between snowfall and ablation. The firn line tends to coincide with the snow line. Several factors influence the level of the snow line (see center of photograph, Fig. 19.1). Latitude and elevation, which are temperature controls, are important factors. Equally important is the amount of snowfall received during the winter. In general, the colder the temperature and the greater

Figure 19.3 *A cirque glacier in Alaska. A bergschrund, or great crevasse, can be seen along the upper margin of the glacier.* • **What does such a large crevasse indicate about this glacier?**

the snowfall, the lower the elevation of the snow line. Other factors cause variation in the level of the snow line, such as the amount of insolation. A shady mountain slope, or one where there is a high percentage of cloud cover, will have a lower snow line than one that receives more insolation. Wind is another factor, since it produces snow drifts on the lee side of the mountain ranges. Thus, in the midlatitudes of the Northern Hemisphere, the snow line is lower on the north (shaded) and east (leeward) slopes of mountains. Consequently, the most significant glacier development is on these slopes.

Some alpine glaciers are not caused primarily by snowfall on the glacier itself, but by the accumulation of snow blowing and drifting into protected areas. The Colorado Rockies and the Ural Mountains in Russia provide good examples of alpine glaciers formed by snow drift.

Equilibrium and the Glacial Budget

When the snout of a glacier does not move, the glacier is said to be in a state of equilibrium— that is, a balance in the system has been achieved between accumulation and ablation of ice and snow. As long as this rare condition is maintained, the glacier's snout will remain in the same location, although the glacial ice continues to flow forward, like the stairs of an escalator.

Let us assume that for several years, exceedingly heavy amounts of snow are received and the glacier's equilibrium is upset. Under the pressure of this surplus snow accumulation, more ice will be produced and the snout will advance until it reaches a new point of equilibrium where receipt of ice and snow equals wastage. A deficit of snow in the glacial budget will cause the snout to recede, by melting, until an equilibrium between receipt and wastage is again achieved. An increase or decrease in temperature will also cause advancing and receding of the ice mass.

However, even in a retreating glacier, the ice is constantly flowing forward. Forward movement stops only when the ice becomes too thin to flow at all. Glaciers fluctuate constantly (Fig. 19.5). Since about 1890 to the present, most Northern Hemisphere glaciers have been receding. This overall receding may be an indication of global warming, yet each individual glacier has its own balance. For example, in 1986 Alaska's Hubbard Glacier advanced so rapidly that it cut off Russell Fjord from the sea and trapped many seals and porpoises. Yet just a few hundred miles away, the giant Columbia Glacier is rapidly receding and calving increased numbers of icebergs into Prince William Sound. Scientists and the

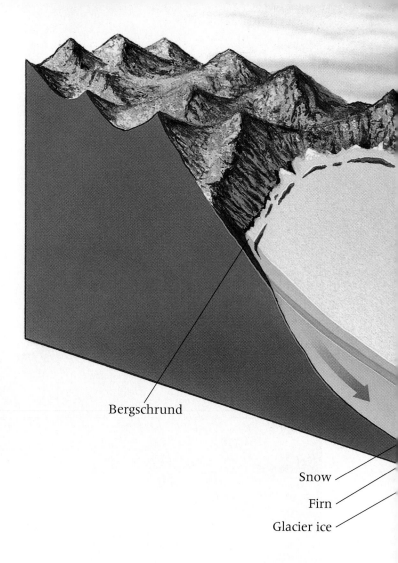

Bergschrund

Snow

Firn

Glacier ice

Coast Guard are concerned, since further receding will increase the hazard from icebergs in the nearby oil tanker lanes from the TransAlaska Pipeline.

How Does a Glacier Flow?

The mechanics by which an alpine glacier flows are complex, and explanations are still theoretical. Most of the movement, however, is caused by gravity. It is believed that a glacier flows because of a combination of factors, one of which is basal slip over its rock floor. Steep slopes, which allow gravity to work, and meltwater, which reduces friction by lubricating, must both be involved. Basal slip is particularly important in middle-latitude glaciers and during summer, when much of the glacier is near its melting point and meltwater is available. During winter and in cold-climate glaciers, with little meltwater available, the glacier may freeze onto the bedrock and basal slip may be prevented.

(Continued on page 508)

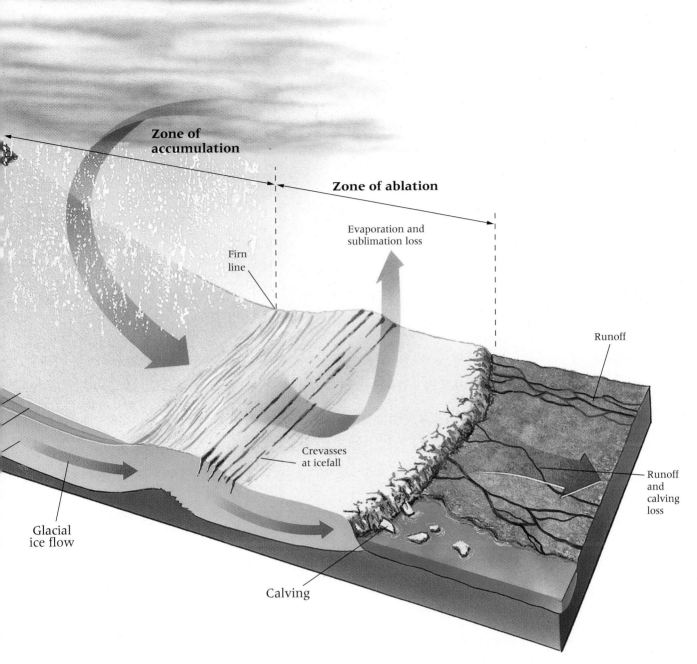

Figure 19.4 Environmental Systems: The Hydrologic System—Glaciers. *Glacial systems are mainly controlled by two major climatic factors: the input of snowfall (accumulation) and the loss of snow and ice (ablation) due to evaporation or melting at temperatures above the freezing point. Most of the glacial growth is in the zone of accumulation, where temperatures permit freezing much of the year and annual snowfall exceeds melting. The zone of accumulation in alpine regions is mainly in the middle elevations, where temperatures are not too cold for snowfall, and in protected basins, where strong mountain winds cannot remove the snow. The highest elevations will remain relatively ice-free due to the severe cold and wind. Over the years in areas of accumulation, snow becomes buried under newer snowfall, changes into granular firn, and much later transforms into glacial ice. As the thickness increases, the buried glacial ice will move downslope.*

Most loss from a glacier is in the zone of ablation, or the area below the firn line where summer melting exceeds snowfall. Here the glacier loses its mass from direct evaporation and sublimation at the glacier's surface, from runoff below the snout due to melting throughout the ablation zone, and from calving of icebergs at the snout into the sea or lakes.

A glacial system is in a state of dynamic equilibrium if a balance is achieved between the input in the zone of accumulation and the output of the zone of ablation. If accumulation exceeds ablation for several years, the glacier's snout will advance; if ablation exceeds accumulation, the glacier will lose mass and the glacier will recede. At present most of the world's alpine glaciers are receding, possibly due to a global warming trend. Of concern to many scientists are the potential effects of global warming on glacial systems, especially the continental ice sheets of Antarctica and Greenland. With increasing world temperatures, glacial systems could be changed, with more rapid ice loss due to melting and calving. These changes in turn could affect world sea level and could cause local increases in flooding hazards.

Figure 19.5 *The receding Columbia Glacier, Alaska. As this large glacier retreats, it calves icebergs into Prince William Sound.* • **Why is the rapid receding of this glacier of particular concern to scientists?**

Internal plastic deformation is another means by which ice flows. At depth the tremendous weight of the ice causes the individual ice crystals to arrange themselves in parallel layers and then slide over each other, much like a deck of cards. This type of flow seems to occur only below about 30 meters (100 ft) of ice, in an area known as the *zone of plastic flow.*

The upper surface of the glacier is brittle and does not have plastic deformation. Motion of the ice in this upper *zone of brittle flow* occurs by fracturing and faulting. Here the ice breaks and cracks as a solid material. These cracks, called **crevasses,** are common where the ice mass is stretched, particularly along the margins and snouts of alpine glaciers (Fig. 19.6). Where glaciers drop over steep slopes, they form an **ice fall.** Here, intersecting crevasses break the ice into a morass of unstable ice blocks. This is an extremely dangerous area for mountain climbers and scientists who venture onto the ice. Where the snout of the glacier meets seawater can also be dangerous, as large blocks of ice calve off and topple into the water, creating large waves.

The speed of glacial flow varies from an imperceptible fraction of a centimeter per day to as much as 30 meters (100 ft) per day. In addition, the speed of an individual glacier may change from time to time because of changes in the dynamic equilibrium and from place to place because of changes in gradient or even in the amount of friction encountered with adjacent rock. The speed of glacial flow is greatest where there is a steep slope, where the ice is thickest, and where temperatures are warmest. For example, the Nisqually Glacier, on the steep slopes of Washington's Mt. Rainier, moves 38 centimeters (16 in.) per day in summer. As a general rule, temperate alpine glaciers flow much faster than the cold polar Antarctic Ice Sheet.

Even within a small section of a glacier, glacial flow varies. In the middle of the upper surface, where friction is least, speed of flow is greatest. On the sides and bottom surfaces of the glacier, friction slows the rate of flow.

Sometimes a glacier's velocity will increase by many times its normal rate, causing the glacier to travel hundreds of meters per year. The reasons for such enormous *glacial surges,* as these velocity increases are called, are not completely clear, although lubrication of the glacial bed by pockets of meltwater explains some of them.

Glaciers as Agents of Gradation

As a glacier moves over the land, whether as an ice sheet or as an alpine glacier, it scrapes along the land surface, picking up and carrying along with it boulders and rock fragments. This erosive process of lifting and incorporating rock and soil into the glacial ice is called glacial **plucking** or **quarrying.** Plucking is encouraged by weathering processes, particularly the freezing of water in joints and fractures in the bedrock, which breaks rock fragments loose.

Glaciers also drag rock fragments along their undersides. Many of the rock fragments traveling with a glacier become tools of erosion themselves as they scrape and gouge stationary surface rocks. The **striations** (gouges, grooves, and scratches) produced by such **glacial abrasion,** as this erosion process is called, are often cited as evidence of previous glaciation in areas devoid of glaciers today (Fig. 19.7). Striations indicate the direction of flow long after a glacier has disappeared. Rock surfaces subjected to intense glacial abrasion are typically smoother and more rounded than those

produced mainly by plucking or quarrying. The glacially produced landform of such rock surfaces is the **roche moutonnée**—a bedrock hill that is smoothly rounded on the upstream side most subject to abrasion, with some plucking evident on the downstream side (Fig. 19.8).

Plucking and abrasion provide glaciers with a load of rock fragments of all sizes, from the finest-ground rock flour to giant boulders. Much of the load is concentrated near the surfaces of the glacier that are in contact with land surfaces—the source of most of a glacier's load. More material, particularly in the case of alpine glaciers, is transported on the upper glacial surface, where it has been eroded from the valley sides. Even more material joins the glacier from mass movement of mechanically weathered materials due to ice wedging on the high, unglaciated slopes above. Where valley glaciers join, these

(Continued on page 511)

(a)

(b)

Figure 19.6 *Crevasses form in the upper surface of glaciers. **(a)** A large crevasse on the Aleyska Glacier, Alaska. **(b)** Crevasses are particularly numerous where the ice flow speeds up, especially at icefalls, such as this one on the Chilkat Glacier, Alaska.*

Figure 19.7 *Glacial abrasion produces smooth rock surfaces with scratches and grooves parallel to the direction of ice movement.* • **Can you see evidence of the direction of ice flow from this photograph in Mount Rainier National Park, Washington?**

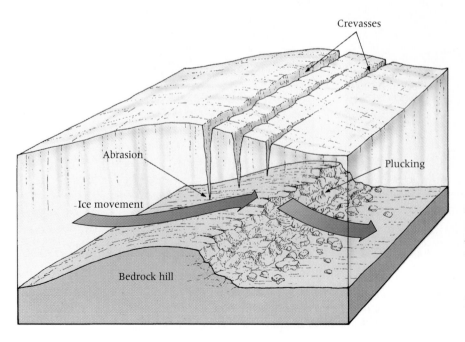

Figure 19.8 *The characteristic asymmetric form produced by glacial erosion of a bedrock hill. Arrows indicate the direction of ice flow. This glacial landform is known as a roche moutonnée, or sheep-backed rock.* • **Why would crevasses form above this feature?**

Figure 19.9 *Where valley glaciers join, the eroded materials from glacial erosion, weathering, and mass movement form characteristic stripes as shown here in the Alaska Range.*
• **Explain how these stripes can be found in the middle of a valley glacier.**

materials cause the characteristic stripes on the glacier surface (Fig. 19.9). The volume and velocity of a glacier do not determine the size of the particles it can erode and transport, as is the case with running water. As a consequence, the sorting of sediment by size that occurs as running water changes speed does not occur with flowing ice.

Ice is especially aggressive in removing non-resistant rock, often producing erosional depressions. The bottom ice currents (but not the ice surface) may move obstructions even if the ice has to move uphill for a while to drag material out of the depression. The fact that ice currents, unlike flows of water, can move upward means that after the ice has melted, a glaciated surface may have many depressions. After the retreat of a glacier these depressions are often filled with water to form lakes.

Erosional Landforms Due to Alpine Glaciation

Ice thickness and erosion are greatest at the head of alpine glaciers. Glacial erosion works headward even as the glacier flows downslope. The headward erosion of a glacier produces a valley head shaped like a steep-sided bowl or amphitheater. Glacial undercutting of the rock walls above the ice level, combined with mechanical weathering, causes mass movement and increases the steepness of the bowl's walls, forming a cirque. If a

lake forms in the cirque depression after the ice melts, it is called a **tarn** (Fig. 19.10a).

Often two or more cirque glaciers will form near the top of the same mountain. As the cirques of two such glaciers are enlarged, the wall of rock between them will be shaped into a jagged, saw-tooth spine of rock, called an **arête** (Fig. 19.10b). Where three or more cirques meet at a mountain summit, they form a characteristic pyramid-like peak called a **horn** (Fig. 19.10c). The Matterhorn, in the Swiss Alps, is the world's classic example. A **col** is a pass formed by the headward erosion of two cirques that have intersected to produce a low saddle in a high mountain ridge or arête.

Unlike streams, which initially erode V-shaped valleys, glaciers erode characteristically steep-sided, U-shaped valleys called **glacial troughs.** In addition, a glacier's tendency to move straight ahead rather than to meander causes it to straighten out the original valley.

By eroding weak rock on the valley floor, the glacier often creates a sequence of rock steps and excavated basins. During glaciation, ice falls will be present at the steps. The result is a "glacial stairway." When the ice recedes, rock-bound lakes may fill the basins, often looking like beads connected by a glacial stream flowing down the glacial trough. Such lake chains are called **paternoster lakes.**

A large glacier often has tributaries that merge with it. These tributary glaciers, like the main ice stream, carve U-shaped channels. However, because they have less volume than the

(a)

(b)

(c)

(d)

Figure 19.10 *Erosional landforms due to alpine glaciation. (a) A cirque and tarn lake formed by headward erosion of a glacier in the Canadian Rockies. (b) Jagged sawtooth spines of rock, such as these in the French Alps, are called arêtes. (c) Mt. Assiniboine in the Canadian Rockies is a classic example of a horn. Horns are formed when several glaciers cut headward into a mountain peak. (d) Glaciers carve steep-sided U-shaped valleys called glacial troughs. Yosemite Valley, California, is a classic example of a glacial trough.*

main glacier, the rate of erosion in these tributaries is less rapid. As a result, their troughs are smaller and not as deep as those of the main glacier. Nevertheless, during the peak glacial phases, the top surface of the ice that flows from the smaller glacier is at the same level as the ice in the larger glacier. Not until the two glaciers begin to wane does the difference in height between their trough floors become apparent. The higher trough of the old tributary glacier is called a

hanging valley. A stream that flows down such a channel will drop down to the lower channel by a high waterfall or a series of steep rapids. Yosemite Falls and Bridalveil Falls in Yosemite National Park are excellent examples of hanging valley waterfalls. Yosemite Valley itself is a classic example of a glacial trough (Fig. 19.10d).

Once a landscape has been eroded by an alpine glacier, it shows a sharp contrast between the glacial trough scoured smooth by ice flow and

512

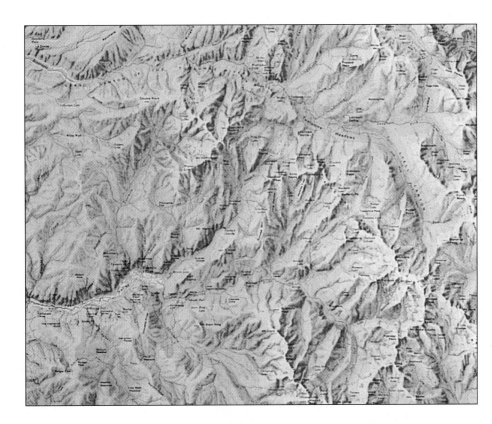

Figure 19.11 Map of Yosemite National Park and the central Sierra Nevada in California, showing rugged alpine topography.

the jagged peaks above the former level of the ice. The rugged quality of these upper surfaces is caused primarily by mechanical weathering above the ice surface and by undercutting at the head of the ice mass. California's Sierra Nevada (Fig. 19.11) and Glacier National Park, Montana, are excellent sites of such rugged alpine glacial terrain (see Map Interpretation: Alpine Glaciation). The sequence in development of alpine glacial landforms is illustrated in Figure 19.12.

In areas where mountainous regions lie near the coast, alpine glaciers may reach the sea and calve icebergs (see page 508). This situation is found at present along the coasts of British Columbia, southern Alaska, Chile, Greenland, and Antarctica, and was formerly characteristic of Scotland, Norway, Iceland, and New Zealand. When such a glacier then recedes landward, the sea invades the former glacial trough, creating a deep, inland finger of the sea called a **fjord**. This invasion by the sea is possible primarily because—unlike streams, which can erode only to base level—glaciers can erode far below sea level. Because of its density, ice must be nine-tenths submerged before it will float. Thus, the sea may enter deep glacial channels once the ice has melted. In addition, fjords that were formed during periods of large-scale glaciation were later submerged as sea level rose with the melting of

the glaciers (see Figs. 11.13 and 22.11). Most of the deep, narrow channels of Washington's Puget Sound were carved into bedrock by glacial erosion and later invaded by the sea.

Depositional Landforms Due to Alpine Glaciation

Alpine glaciers carry debris on their surfaces, frozen in their interiors, and dragged along at the bottom. As mentioned previously, the material they carry includes boulders, rocks, and fragments plucked by the glaciers themselves from their channel sides and floor as well as the smaller fragments and particles produced by abrasion. Rockfall from steep trough walls may also supply a glacier with debris. Glaciers deposit huge chunks of bedrock, fine rock flour, layers of pollen, dead trees and other plants, soil, and volcanic dust.

All glacial deposits are included within the general term **drift**, whether they are unsorted and unstratified ice deposits or the orderly deposits of meltwater streams issuing from the glacier. To differentiate these two types of deposits, the term **till** is applied to unsorted drift laid down by ice. **Glaciofluvial deposits** refer to more orderly meltwater deposits.

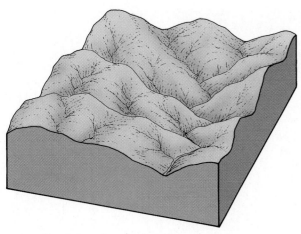

(a) **Preglacial fluvial topography**

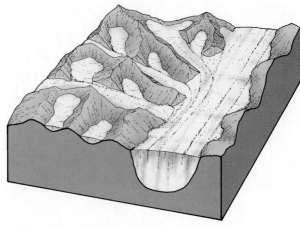

(b) **Maximum glaciation**

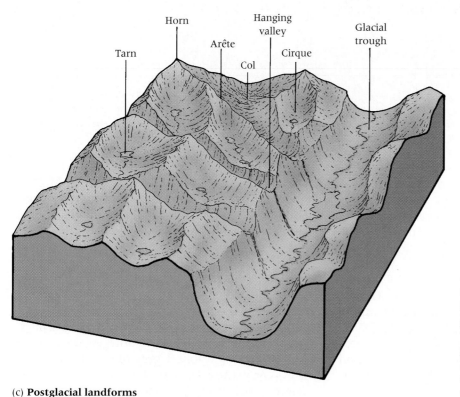

Tarn Horn Arête Col Hanging valley Cirque Glacial trough

(c) **Postglacial landforms**

Figure 19.12 *The development of glacial landforms in an alpine region.* *(a) Mountain topography prior to glaciation.* *(b) Maximum valley glaciation.* *(c) Major postglacial landforms.* • **How have the valley profiles changed from preglacial to postglacial times?**

Glaciers deposit a portion of their load when their capacity is reduced. Glacial deposits called **moraines** occur along the margins of glaciers. The ridgelike deposits laid down along the side margins are called **lateral moraines** (Fig. 19.13a). When two glaciers join together, their interior lateral moraines form a **medial moraine** in the center of the new glacier. At the snout of the glacier, all the debris carried forward by the "conveyor belt" of ice and pushed ahead of the glacier is deposited in a jumbled heap of rocks and fine material, called an **end moraine.**

End moraines marking the farthest advance of the snout are **terminal moraines.** End moraines deposited as a consequence of a halt in snout receding, followed by a stabilization of the ice front prior to further receding, are called **recessional moraines.** Glaciers also deposit a great deal of till along the floor of their channel as they retreat, particularly near the snout, where melting is greatest, where the ice becomes thinner, and where load capacity is consequently reduced. This glacial till deposit is called **ground moraine.**

(a)

(b)

Figure 19.13 *Depositional landform features due to alpine glaciation. (a) Moraines on the west slope of the Wind River Range, Wyoming.* • **Can you identify the lateral and end moraines in this photo?** *(b) Meltwaters from glaciers in the Chugach Mountains of Alaska formed these braided streams from deposits of glacial outwash.* • **What is the source of all this material?**

Braided streams of meltwater, laden with sediment, commonly issue from the glacier terminus. The sediment, called **glacial outwash,** is deposited beyond the terminal moraine, with larger rocks and debris deposited first and then the progressively finer particles. Often resembling an alluvial fan confined by valley walls, this depositional form is called a **valley train.** Valleys in glaciated regions may be filled to depths of a few hundred meters by outwash, producing extremely flat valley floors (Fig. 19.13b).

Continental Ice Sheets

Continental ice sheets differ from alpine glaciers, especially in size and shape. However, as agents of landform gradation, ice sheets are similar in most ways to alpine glaciers, and much of what we have discussed about alpine glaciers applies as well to continental ice sheets. The differences in the gradational activities of the two types of glaciers are primarily differences in scale, attributable to the enormous disparity in size between the two.

Existing Ice Sheets

Glacial ice currently covers about 10 percent of Earth's land area, and alpine glaciers can be found in mountain regions on most continents. However, in area and mass, alpine glaciers are insignificant compared to the ice caps of Greenland and Antarctica, which account for 96 percent of the area covered by glaciers today. Ice caps resembling those of Greenland and Antarctica, but on a much smaller scale, are also present in Iceland, in the arctic islands of Canada and Russia, and in Alaska and the Canadian Rockies.

The Greenland ice cap covers the world's largest island with a layer of ice that is over 3000 meters (10,000 ft) thick in the center. The only land exposed in Greenland is a narrow, mountainous strip along the coast (Fig. 19.14). Where the ice does reach the sea, it usually does so through fjords. These ice flows to the sea resemble alpine glaciers and are called **outlet glaciers.** When the ice reaches the sea, huge chunks are broken off by melting and the action of waves and tides. The resulting **icebergs** are a hazard to vessels in the North Atlantic shipping lanes south

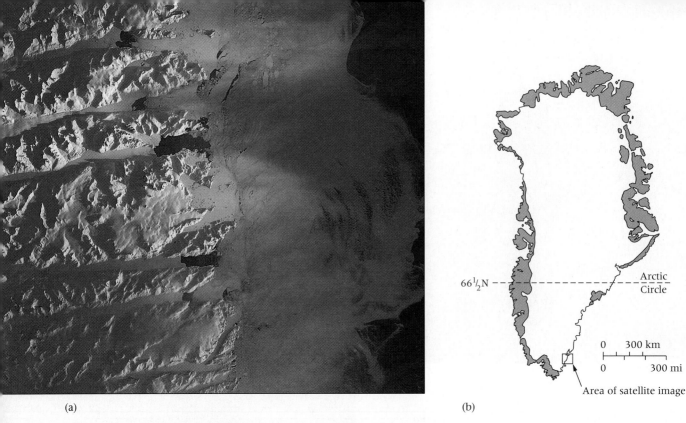

(a)

(b)

66½N - - - - - - - - - - - Arctic Circle

0   300 km

0   300 mi

Area of satellite image

Figure 19.14 *The Greenland Ice Sheet. **(a)** Except for the mountainous edges, the Greenland Ice Sheet almost completely covers the world's largest island. Ice thickness is over 3000 m (10,000 ft) and depresses the bedrock below sea level. In this satellite view, several outlet glaciers flow seaward from the ice sheet to the east coast of the island. **(b)** Map showing the extent of the Greenland Ice Sheet.*

of Greenland. Tragic maritime disasters, such as the sinking of the *Titanic,* have been caused by collisions with these huge irregular chunks of ice, which are nine-tenths submerged and thus mostly invisible to ships. Today, with radar, satellites, and the ships and aircraft of the International Ice Patrol, these sea disasters are minimized.

The Antarctic ice cap covers some 13 million square kilometers (5 million sq mi), an area almost seven-and-one-half times the size of the Greenland ice cap, which covers 1.7 million square kilometers (650,000 sq mi). Little is known about the land beneath the thick layer of ice in Antarctica (Fig. 19.15). Like Greenland, little land is exposed in Antarctica, and the weight of the 4500 m (14,000 ft) thick ice in some interior areas has depressed the land well below sea level. Where the ice reaches the sea, it floats in enormous flat-topped plates called **ice shelves.** These are the source of icebergs in Antarctic waters. They do not have the irregular shape of Greenland's icebergs, and because they do not float into heavily used shipping lanes, these huge tabular-shaped Antarctic icebergs are not as much of a hazard to navigation (Fig. 19.16). They do, however, add to the problem of access to Antarctica

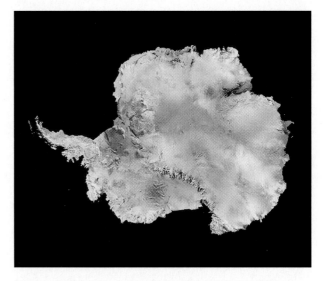

Figure 19.15 *The Antarctic Ice Sheet. The world's largest ice sheet covers an area larger than the United States and Mexico and has a thickness over 4500 m (14,000 ft). This infrared satellite mosaic image covers the whole south polar continent. Notice that most of Antarctica is ice-covered (white and blue on the image). The only rocky areas (darker areas on the image) are the Antarctic Peninsula and the Transantarctic Mountains. Large ice shelves flow to the coastline, the largest being the Ross Ice Shelf. The image is oriented with the Greenwich Meridian at the top.*

516

Figure 19.16 The Antarctic Ice Sheet calves enormous flat-topped tabular icebergs. • What portion of the iceberg is hidden below the sea surface?

for scientists. The huge wall of the ice shelf itself, as well as the broken-up and melting sea ice and the extreme climate, combine to make Antarctica inaccessible to all but the hardiest individuals and equipment. This icy continent serves, however, as a natural laboratory for scientists from many countries to study Antarctic glaciology, climatology, and ecology.

Pleistocene Ice Sheets

The Antarctic and Greenland ice sheets are remnants of the Pleistocene. As discussed in Chapter 9, the Pleistocene Epoch, or ice age, is the name given that time period of recent geologic history during which there were numerous major advances and retreats—and a vastly greater number of minor movements—of continental ice sheets over large portions of the world's land (Fig. 19.17). Scientists believe the Pleistocene Epoch began about 2.4 million years ago. Since that time, ice sheets have spread outward over the land from centers in Canada, Scandinavia, and eastern Siberia, and also from high mountain ranges, to cover nearly a third of Earth's land surface. At the same time, sea ice expanded

(Continued on page 520)

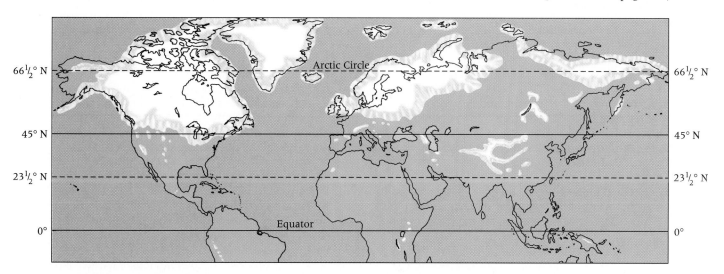

Figure 19.17 The maximum extent of glacial ice coverage in the Northern Hemisphere during the Pleistocene. Much of North America and Eurasia was covered by ice up to several thousand meters thick. • Why were some very cold areas, such as portions of interior Alaska and Siberia, ice-free during this time?

ALPINE GLACIATION

An oblique aerial photo of Glacier National Park. ▶

The Map

The Chief Mountain, Montana, map area is located in Glacier National Park. Glacier National Park adjoins Canada's Waterton Lakes National Park located across the border. Alpine glaciation has produced the spectacular scenery of this region of the Northern Rocky Mountains. Most of the glaciated landscape was produced during the Pleistocene. Today, about 50 glaciers still exist in Glacier National Park, but due to rapid melting in recent decades, it is estimated that there may be none within 70 years. Prior to glaciation, this region was severely faulted and folded during the formation of the Northern Rocky Mountains.

The oblique aerial photo clearly shows the rugged nature of the terrain of this map area. The steep slopes, glaciers, lakes, and sharp arête ridges are obvious landform characteristics of alpine glaciated regions.

Temperature is a primary control of highland climates such as the ones in this map area. As you would expect, the rapid decrease in temperature with increasing elevation results in a variety of climates within alpine regions. Elevation is only one control of highland climates; exposure is another factor. West-facing slopes receive the warm afternoon sun, while east-facing slopes are sunlit only in the cool of the morning.

Interpreting the Map

1. What is the approximate local relief of this map?

2. From looking at this mountain region on the topographic map and in the aerial photograph, do you think most of the landscape was produced by glacial erosion or deposition?

3. Locate Grinnell, Swiftcurrent, and Sperry glaciers. What type of glaciers are they?

4. What evidence indicates that the glaciers were once larger and extended further down the valleys?

5. Note that most of the existing glaciers are located to the northeast of the mountain summits. Explain this orientation. At what elevation are the glaciers found?

6. What types of glacial landform are the following features?
   a. The features occupied by Helen, Iceberg, and Ipasha lakes.
   b. The features occupied by Lake Sherburne and Josephine Lake.
   c. Pyramid Peak, Mt. Gould, and Mt. Wilbur.
   d. The series of lakes that occupy the valley of Swiftcurrent Creek.

7. Along the high ridges runs a dashed line labeled "Continental Divide." What is its significance?

8. If you were to hike southeast from Auto Camp (on McDonald Creek) up Avalanche Creek to the base of Sperry Glacier, how far would you travel and how much elevation change would you encounter?

Chief Mountain, Montana
Scale 1:125,000
Contour interval = 100 ft
U.S. Geological Survey ▶

equatorward. In the Northern Hemisphere, sea ice was present along the coasts as far south as Delaware in North America and Spain in Europe. Between each glacial advance a warmer *interglacial* occurred, during which the enormous continental ice sheets and sea ice retreated and almost completely disappeared. An examination of glacial deposits has determined that within each major glacial advance there were many minor retreats and advances which reflect smaller changes in global temperature and precipitation.

The gradational effects of the last major glacial advance, known as the Wisconsinan stage, which ended about 10,000 years ago, are the most visible in landscapes today. The glacial landforms created during the Wisconsinan stage are relatively new and have not been destroyed to any great extent by the other gradational agents. Consequently, we are able to derive a fairly clear picture of the extent and actions of the ice sheets at that time.

The land covered during the major glacial advances in North America and Eurasia is depicted in Figure 19.17. Continental ice sheets in North America extended as far south as the Missouri and Ohio Rivers and covered nearly all of Canada and much of the northern Great Plains, the Midwest, and the northeastern part of the United States. In New England, the ice was thick enough to overrun the highest mountains, including Mount Washington, with an elevation of 2063 m (6288 ft). The ice was more than 2000 m (6500 ft) thick in the Great Lakes area. In Europe, glaciers spread over what is now most of Great Britain, Ireland, Scandinavia, Germany, Poland, and western Russia. In much of Siberia and interior Alaska, it was too cold and dry (as it is today) to allow such massive snow and ice accumulation as occurred further south in North America and northwestern Europe.

During each advance of the ice sheets, alpine glaciers were much more extensive and massive in highland areas than they are today. In fact, it was the examination of landforms in the Swiss Alps that led Louis Agassiz in 1840 to publish the theory of past glaciations. Though at first considered radical, the theory today is acknowledged as fact and is supported by countless detailed studies of glacial deposits and erosional forms in various parts of the world.

Where did the water locked up in all the ice and snow come from? Its original source was the oceans. During the periods of glacial advance there was a general lowering of sea level, exposing large portions of the continental shelf and forming land bridges across the present-day North, Bering, and Java Seas. The most recent melting and glacial retreat raised the oceans a similar amount—about 120 meters (400 ft). Evidence for this rise in sea level can be seen along many submerging coastlines around the world.

Movement of Continental Ice Sheets

It is a popular misconception that all continental ice sheets originate at the poles and spread toward the equator. Actually, the great centers of Pleistocene glacial accumulation (aside from highland areas, Antarctica, and Greenland) were in the upper midlatitudes, in the vicinity of Hudson Bay in Canada, on the Scandinavian Shield, and in eastern Siberia. These huge ice sheets also covered some fairly rugged terrain, such as New York's Adirondacks, New England's White Mountains, and the highlands of Scandinavia and Scotland.

Continental ice sheets began to flow outward in all directions from a central zone of accumulation when the ice was thick enough to allow for plastic flow. As with valley glaciers, the initial flow direction of advancing ice sheets was determined largely by the path of least resistance, found in preexisting valleys and belts of softer rock. Thus the radial expansion is analogous to the spreading of pancake batter in a frying pan. If enough batter is poured into the center of the pan, it will eventually spread outward to cover the entire bottom of the pan. Like an alpine glacier, an ice sheet flows outward from a zone of accumulation to a zone of ablation. Like alpine glaciers, ice sheets advance and recede with small changes in temperature and snowfall, and will melt and disappear when ablation exceeds accumulation.

Ice Sheets and Erosional Landforms

Ice sheets erode the land in ways similar to alpine glaciers but on a much larger scale. As a result, landforms created by ice-sheet erosion are far more extensive than those formed by alpine glaciation, stretching over millions of square kilometers of North America, Scandinavia, and Russia. As ice sheets flowed out over the land, they gouged Earth's surface, enlarging valleys that already existed, scouring out rock basins, and smoothing off existing hills. The eroding ice sheets removed most of the soil and then attacked the bedrock itself.

Today these ice-scoured plains are areas of low, rounded hills, lake-filled depressions, and wide exposures of bedrock. Because the ice sheets plowed through and totally disrupted the former stream patterns, and because glaciation has been so recent that new drainage systems have not had time to form, ice-scoured plains are characterized by thousands of lakes, marshes, and areas of muskeg (poorly drained areas in cold climates grown over with vegetation). The major characteristic of glacially eroded lands, such as those found in Canada and Finland, is the great expanse of exposed gouged bedrock and standing water.

Ice Sheets and Depositional Landforms

Again, scale makes ice-sheet deposition different from that of alpine glaciers. Though terminal and recessional moraines, ground moraines, and glaciofluvial deposition are found, these deposits form significantly larger features in the land-scapes caused by ice sheets (Fig. 19.18 and Map Interpretation: Continental Glaciation).

Terminal and Recessional Moraines. Terminal and recessional moraines form belts of low hills and ridges deposited on the land in glaciated areas. These features are rarely more than 60 m (200 ft) high (Fig. 19.19). The last major Pleistocene glacial advance through New England left a terminal moraine running the length of New York's Long Island and formed the offshore islands of Martha's Vineyard and Nantucket. Cape Cod and Lake Michigan's rounded southern end were formed by recessional moraines. End moraines are usually arcs convex toward the direction of ice flow. Their pattern and placement indicate that the ice sheets did not maintain an even front but spread out in tongue-shaped lobes channeled by the previous terrain (Fig. 19.20). The positions

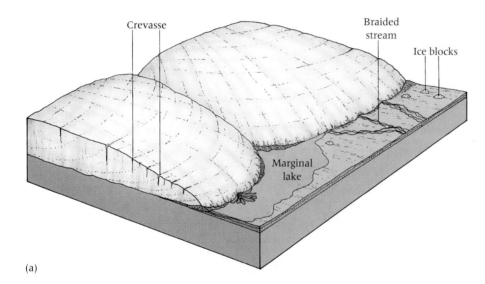

(a)

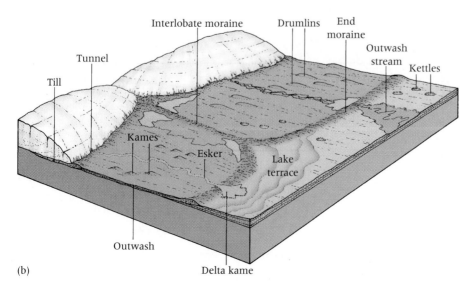

(b)

Figure 19.18 Landscape alteration by ice-sheet deposition. (a) Features associated with ice stagnation at the edge of the ice sheet. (b) Landforms due to further modification by glacial meltwater as the ice sheet retreats. • **What is the direct cause of most of these landscape features?**

Figure 19.19 *The hilly topography of an end moraine due to a continental ice sheet. This aerial photo is of an end moraine on the Waterville Plateau in eastern Washington.* • **How do you account for the more bumpy terrain at the left of the photo compared to the smoother surface of the plain at the right? Refer to Figure 19.18(b) to help explain your answer.**

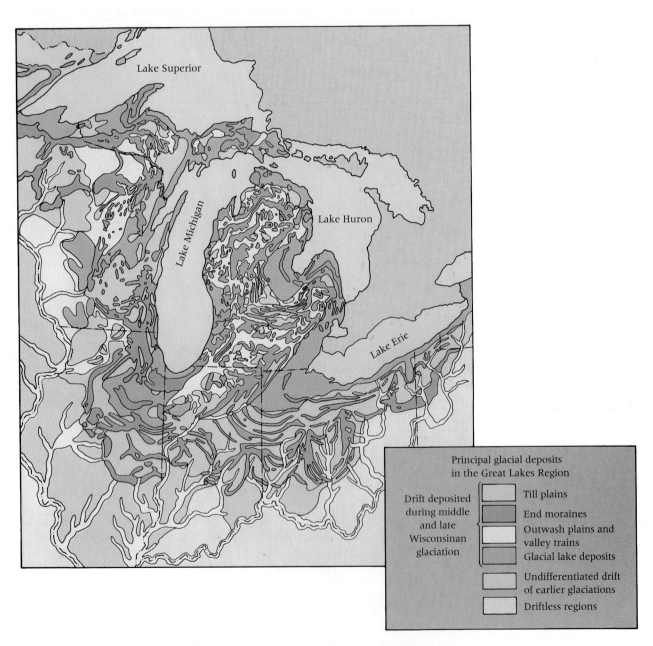

Principal glacial deposits
in the Great Lakes Region

Drift deposited
during middle
and late
Wisconsinan
glaciation

Till plains

End moraines

Outwash plains and
valley trains

Glacial lake deposits

Undifferentiated drift
of earlier glaciations

Driftless regions

Figure 19.20 *Glacial deposits in the Great Lakes region.* • **Why do the many end moraines have such a curved pattern?**

of the terminal and recessional moraines give evidence of more than simply the direction of ice flow. In addition, the character of the deposited material can be examined to detect the pattern and sequence of advances and retreats of each successive ice sheet.

Till Plains. In the zone of ice-sheet deposition, massive amounts of unsorted glacial till accumulated, often to depths of 30 m (100 ft) or more. Because of the uneven nature of the deposition, the configuration today of the land covered by till varies from place to place. In some areas the till is too thin to hide the original contours of the land; in other regions the thick deposits of the till form broad, rolling plains of low relief. Small hills and slight depressions, some filled with water, characterize most **till plains** and reflect the uneven glacial deposition. Some of the best agricultural land of the United States is found on the gently rolling till plains of Illinois and Iowa. The young, dark-colored, grassland soils (mostly mollisols) that developed on the till are extremely fertile.

Outwash Plains. Beyond the belts of hills that are the terminal and recessional moraines lie the **outwash plains.** These are extensive, relatively smooth plains covered with the sorted deposits carried forward by the meltwater from the ice sheets. The outwash plains, which may cover hundreds of square kilometers, are analogous to the valley trains of alpine glaciers.

Some outwash plains, till plains, and moraines are marked by occasional water-filled pits, called **kettles,** which were formed when blocks of ice became detached from the glacial terminus and were buried in the glacial deposits. Eventually, where the blocks of ice melted, holes remained. Today countless kettles form lakes in the glaciated landscapes.

Drumlims. A **drumlin** is a streamlined hill, usually 0.5 km (0.3 mi) in length and less than 50 m (160 ft) high, molded in glacial drift on the till plains (Fig. 19.21a). Drumlins are oddly shaped, resembling half an egg or the convex side of a teaspoon, and are usually found in swarms, with as many as a hundred or more clustered together. Glaciologists do not yet understand exactly how drumlins were formed. The most conspicuous feature is their elongated shape, which follows the direction of ice flow. Their broad, steep noses face the direction from which the ice advanced, while their gently sloping, narrow tails point in the direction of ice flow. Thus their geometry is the reverse of that of roches moutonnées. They are well developed in Ireland and in the states of New York and Wisconsin. Boston's Bunker Hill, a drumlin, is one of America's best known historical sites.

Eskers. An **esker** is a narrow, winding ridge composed of glaciofluvial gravels (Fig. 19.21b). Sometimes eskers are as long as 200 kilometers (130 mi), but usually they do not exceed several

(Continued on page 526)

(a)

(b)

*Figure 19.21 Drumlins and eskers. **(a)** Porcupine Islands, a series of submerged drumlins at Acadia National Park, Maine. Drumlins are streamlined hills elongated in the direction of ice flow. **(b)** An esker near Albert Lea, Minnesota. Eskers are depositional ridges formed by meltwater deposits in a tunnel under the ice.*
• **What economic importance do eskers have?**

CONTINENTAL GLACIATION

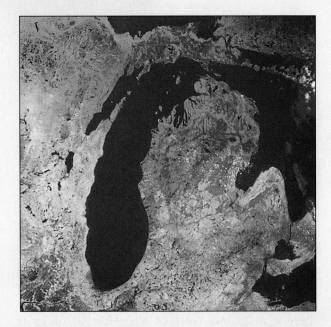

LANDSAT mosaic image of Michigan and the surrounding Great Lakes. ▶

The Map

The Jackson, Michigan, map area is located in the Great Lakes section of the Central Lowlands. It is a region that was covered by the continental ice sheets during the Pleistocene Epoch. Several thousand years ago, the ice melted back, exposing a totally new topography completely different from that of preglacial times. Today evidence of the work of the ice is found throughout the region. The most obvious glacial topographic features are the thousands of lakes (including the Great Lakes), the knobby topography, and moraine ridges. The shapes of the Great Lakes themselves are dominated by the moraines piled up by the advancing and retreating tongue-shaped ice lobes that extended outward from the main continental ice sheet (see satellite image). Other, smaller scale fea-

tures were also created by the ice sheet and its meltwaters. Many of these local landform features are well illustrated on the Jackson, Michigan, topographic map.

The map region has a humid continental, mild summer climate. The summers are pleasant. Excessively warm and humid air seldom reaches this region for more than a few days at a time. Instead, cool but pleasant evening temperatures tend to be the rule. However, winters are long and often harsh. Snow can be abundant and can be on the ground continuously for many weeks or even months at a time. The annual temperature range is significant, and precipitation occurs year round but is provided primarily by storms moving along the polar front.

Interpreting the Map

1. Describe the general topography of the map area.

2. What is the local relief?

3. Does the topography of this region indicate glacial erosion or glacial deposition?

4. Does the area appear to be well drained? What are the three main hydrographic features that indicate the drainage conditions?

5. What type of glacial landform is Blue Ridge? What are its dimensions: length and average height?

6. How is a feature such as Blue Ridge formed? What economic value might it have?

7. What is the majority of the surface material in the map area? What is the term for this type of surface cover?

8. What probably caused the many small depressions and rounded lakes? What are these depressions called?

9. The Jackson, Michigan, map area is located in the south-central part of the state. Describe how the topography of the area appears on the satellite image.

10. What is the advantage of having satellite image and topographic map coverage of an area of study?

Jackson, Michigan
Scale 1:62,500
Contour interval = 10 ft
U.S. Geological Survey ▶

kilometers. It is believed that most eskers were formed by streams of meltwater flowing in ice tunnels at the base of ice sheets. Eskers are a prime source of gravel and sand for the construction industry. Being natural embankments, they are frequently used in marshy, glaciated landscapes as highway and railroad beds. Eskers are especially well developed in Finland, Sweden, and Russia.

Kames. Conical hills composed of sorted glacio-fluvial deposits are called **kames.** They are presumed to have formed in contact with glacial ice when sediments accumulated in ice pits, in crevasses, and among jumbles of detached ice blocks. Like eskers, kames are excellent sources for mining sand and gravel and are especially common in New England. **Kame terraces** are landforms resulting from accumulations of glaciofluvial sand and gravel along the margins of ice lobes melting in valleys in areas of hilly relief. Examples of kame terraces can be seen in New England and New York.

Erratics. Scattered in and on the surface of the glacial drift may be boulders that differ from the local bedrock. Often the source regions of such rocks, called **erratics,** can be identified, indicating the direction of ice flow during the Pleistocene. In Illinois, erratics may be from a source region as far away as Canada. Before the theory of an ice age was proposed, many hypotheses were developed to explain the existence of erratics. Among these explanations was one based on the belief that the Biblical flood transported rocks from one place to another. The term *drift* originated in connection with the flood hypothesis. However, a flood would not account for the striations present on the erratics and could not move large boulders hundreds of kilometers. Also, such boulders would be found worldwide, not just in the middle and high latitudes.

Glacial Lakes and Periglacial Landscapes

The ability of ice sheets to form lake basins has already been mentioned. Millions of glacier-created lakes are still in existence due to the continental ice sheets that once covered much of North America and Eurasia. The ice sheets created some lakes by scooping out deep elongated basins along former stream valleys. If the glacier deposited a moraine dam at one end of such a basin, a meltwater lake would remain after the retreat of the ice sheet. New York's beautiful Finger Lakes are the classic example of moraine-dammed lakes in ice-deepened elongated basins. Alpine glaciers can produce similar results, as in the case of Washington's Lake Chelan and Lakes Maggiore, Como, and Garda of the Italian Alps. Although many of the lakes formed by the ice sheets no longer exist, the **glaciolacustrine** (from *glacial,* ice; *lacustrine,* lake) deposits that they laid down prove their former existence and size.

Some lakes formed where the disruption of drainage by glacial deposition prevented depressions from being drained of meltwater. This situation usually developed where water was trapped between a large end moraine and the ice front, or where the land sloped toward, instead of away from, the ice front. In both situations, temporary **ice-marginal lakes** were formed from meltwater. They were drained and ceased to exist when the retreat of the ice front uncovered an outlet route.

During their existence, the floors of these ice-marginal lakes accumulated layers of fine sediment. The result of this sedimentation can be seen in the extremely flat surfaces that characterize most glaciolacustrine plains. The outstanding example of such a plain is the valley of the Red River in North Dakota, Minnesota, and Manitoba. This plain is the flattest landscape in North America and is of great agricultural significance, since it is well adapted to growing wheat. The Red River flows northward where it eventually flows into Lake Winnipeg. This plain is the result of deposition in a vast Pleistocene lake held between the front of the receding continental ice sheet on the north and the morainic dams and higher topography to the south. This lake has been called Lake Agassiz, after the Swiss scientist who espoused the theory of an ice age. Lake Winnipeg is the last remnant of Lake Agassiz, lying in the deepest part of the ice-scoured and sediment-filled lowland.

Another ice-marginal lake produced much more spectacular landscape features but not in the area of the lake itself. In northern Idaho a glacial lobe, moving southward from Canada, blocked the valley of a major tributary of the Columbia River and caused the formation of an enormous ice-dammed lake known as Glacial Lake Missoula. This lake covered almost 7800 sq km (3000 sq mi) and was 610 m (2000 ft) deep at the ice dam. On occasions when the ice dam failed, Lake Missoula emptied in stupendous

ANNA DAVILA

High School Education: *San Dimas High School, 1989, San Dimas, California*

University Education: *Bachelor of Arts, Geography/Environmental Analysis, 1997, California State University, Fullerton*

Early Interests: *Hiking, photography*

Prior Work Experience: *Volunteer tour guide for National Geographic exhibition at Fullerton Museum Center*

Anna Davila is a cartographer in the Member Services department of AAA in Costa Mesa, California. Along with other cartographers, she prepares and updates all of the street, guide, and county maps covering southern California for auto club members. This involves a great deal of research in addition to the physical aspect of making maps by hand and with the help of computers. Anna talks about her geography experience, the map-making process, and the other avenues her geography education has opened up for her.

"I work with a team of editors and cartographers who research, create, edit, and revise maps. The work is very detailed and time-consuming—in fact, I've been working on two maps for four-and-a-half months now. Most maps are updated every two years, though some are revised yearly.

There are three main types of maps that we produce. Street maps contain listings of streets as well as some point features like schools, parks, malls, etc., in adjoining cities. Guide maps are much more feature-oriented and would show things such as campgrounds, national parks, and recreation areas. County maps provide information on roads and mileage from point A to point B, but may also include point features.

New construction and other factors make it necessary to keep maps up-to-date so it's important that we gather as much information as possible when we're creating a new map or revising an existing one. The information comes from various sources. Our research coordinator may write to city planners requesting tracts that list any changes to buildings, streets, or plots of land.

We also use other companies' maps for comparison. We use the Postal Services' Zip Plus, for example, to check for discrepancies. If we find any, we have to actually go to the site and determine the correct information. This type of research can literally take months to complete.

Once we've gathered the necessary information, it's time to pull the research together and put it on paper. Some of our maps are now done on computers using Adobe Illustrator, which makes it easier to manipulate the figures and do color separations. But a large part of our cartography work is still done manually.

The various elements of a map are transferred to Mylar-coated plates. Generally there is a separate plate for each individual element—one for drawings of streets and roads, one for type, one for city or county borders. We use scribers, which are like needles of varying widths, to draw or inscribe the map elements on the Mylar surface. The Mylar is also coated with a red rubbery substance called rubylithe that peels off and allows for color separations when we send the plates to the printer.

My studies in geography have given me the kind of knowledge that I've needed to do my job, but they've also prepared me for other things. I think that tools like GIS really are the future. Geography, especially if you have an interest in the environment, is not a narrow field. At some point, I'm planning to continue my education, hopefully to communicate my knowledge of geographic and environmental issues through photography or even documentaries."

floods that engulfed much of eastern Washington. The racing floodwaters scoured the basaltic terrain, producing Washington's channeled scablands consisting of intertwining steep-sided troughs **(coulees),** dry waterfalls, scoured-out basins, and other features quite unlike those associated with normal stream erosion.

The Great Lakes of the eastern United States and Canada make up the world's largest lake system. Lakes Superior, Michigan, Huron, Erie, and Ontario lie in former river valleys that were vastly enlarged and deepened by glacial erosion. All of the lake basins except that of Lake Erie have been gouged out to depths below sea level and have irregular bedrock floors lying beneath thick blankets of glacial till.

The history of the Great Lakes is very complex, resulting from the back-and-forth movement of the ice front which produced many changes of lake levels and overflow in varying directions at different times (Fig. 19.22). The earliest lakes appear to have emerged near the southern tip of the Lake Michigan Basin (Pleistocene Lake Chicago) and the western end of the Erie Basin (Pleistocene Lake Maumee). These lakes drained westward to the Mississippi through the Illinois and Wabash Rivers.

Ice retreat exposed the southern fringe of the Ontario Basin (Pleistocene Lake Iroquois), which was occupied by a lake with an eastern outlet through New York's Mohawk and Hudson Valleys. By this time the basin of Lake Huron was emerging, with an outlet westward across Michigan through the Grand River to Lake Michigan. This outlet also channeled the overflow from the Erie Basin as the Wabash route ceased to function.

Further ice receding exposed the western portion of the Superior Basin (Pleistocene Lake Duluth), which overflowed westward to the Mississippi through the St. Croix River in Minnesota. Lakes Michigan, Huron, Erie, and Ontario were now linked to overflow eastward to the Mohawk and Hudson Valleys, with Lakes Michigan and Huron spilling along the ice front into Lake Iroquois rather than following their present route through Lakes St. Clair and Erie.

About 9000 years ago, the St. Lawrence outlet was exposed after the ice receded northward, and the Great Lakes formed a single system emp-

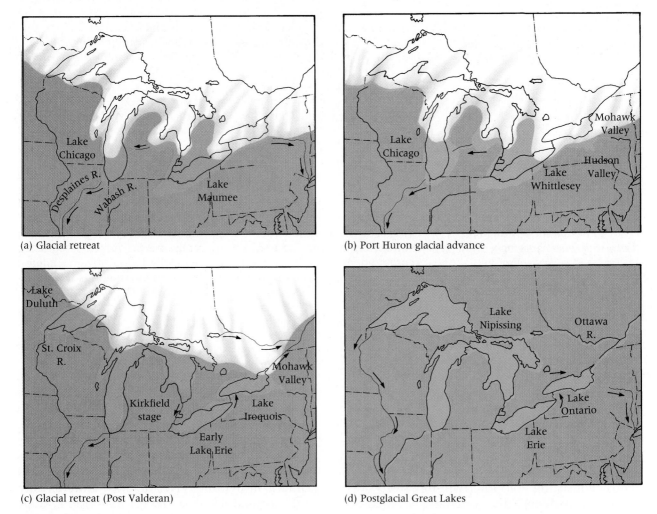

(a) Glacial retreat

(b) Port Huron glacial advance

(c) Glacial retreat (Post Valderan)

(d) Postglacial Great Lakes

Figure 19.22 *Stages in the formation of the Great Lakes of North America as the ice sheet receded at the close of the Pleistocene Epoch.*

Figure 19.23 *A satellite image mosaic of the modern Great Lakes system.* • **Name and locate the five Great Lakes.**

tying to the east—the upper lakes entering the St. Lawrence by way of the Ottawa River. This low outlet permitted the lakes to diminish well below their present levels and areas. However, complete deglaciation caused Earth's crust to slowly rise and raised the outlet. This process so enlarged the lakes that the old Illinois River outlet of Lake Michigan began to function again, and a new connection formed between Lakes Huron and Erie through Lake St. Clair. Eventually the St. Lawrence outlet was lowered, the Illinois river link was abandoned, and crustal rise terminated the Ottawa River link between the upper and lower lakes. With these developments the modern lake system was finally established (Fig. 19.23).

Beyond the ice sheets themselves, other changes occurred. The cool, Pleistocene midlatitude arctic and subarctic conditions created **periglacial landscapes** (*periglacial* means "near the ice"). The permafrost conditions caused landforms, such as ice wedges, patterned ground, and smoothed hill slopes due to solifluction, that are peculiar to tundra climates and areas where frost action dominates (see Chapter 16).

The weight of the ice depressed the land surface several hundred meters. As the ice receded, the land started to rise slowly in a process known as **isostatic rebound.** This is what helped form today's Great Lakes, and areas such as Hudson Bay and the Baltic Sea may someday emerge above sea level. Measurable isostatic rebound raises elevations of areas such as Sweden, Canada, and some of eastern Siberia by up to 2 centimeters (1 in.) per year. Should Greenland and Antarctica lose their ice sheets someday, their depressed central land areas would also rise to reach isostatic balance.

Define and Recall

glacier	crevasse	drift	ice shelf
firn	ice fall	till	till plain
calving	glacial plucking	glaciofluvial deposit	outwash plain
alpine glacier	(quarrying)	moraine	kettle
continental ice sheet	striation	lateral moraine	drumlin
valley glacier	glacial abrasion	medial moraine	esker
piedmont glacier	roche moutonnée	end moraine	kame
cirque	tarn	terminal moraine	kame terrace
cirque glacier	arête	recessional moraine	erratic
	horn	ground moraine	glaciolacustrine
ablation	col	glacial outwash	ice-marginal lake
zone of accumulation	glacial trough	valley train	coulee
zone of ablation	paternoster lakes		periglacial landscapes
firn line	hanging valley		isostatic rebound
internal plastic	fjord	outlet glacier	
deformation		iceberg	

Discuss and Review

1. Describe the two major types of glaciers and explain how each is formed.

2. Diagram and label the characteristic parts of an alpine glacier.

3. How does a glacier maintain its budget in a state of equilibrium?

4. Explain how a glacier moves.

5. How does a glacier accumulate its load? Cite examples of evidence of glacial erosion and movement.

6. What is the difference between *till* and *outwash*?

7. Explain how one can distinguish glacial valleys from stream valleys in mountain areas.

8. How are hanging valleys formed?

9. Describe the major types of deposition by alpine glaciers.

10. Where are the two major existing ice sheets located? How do they compare in area and thickness?

11. How has ice-sheet erosion altered the landscape? What kinds of landscapes are produced after continental glaciers deposit their loads and recede?

12. What are the differences between eskers, drumlins, and kames?

13. Explain the relationship of glaciation to the origin and history of the Great Lakes.

14. How does the present extent of continental ice-sheet coverage compare with the maximum extent of the Pleistocene ice-sheet coverage?

Consider and Respond

1. Using your textbook and lecture notes, give the correct glacial landform term for each of the following definitions.
   (a) Amphitheater-shaped head of a glacial valley.
   (b) Peaked mountain summit formed by erosion of valley glaciers on several sides.
   (c) Glacial deposits located along the sides of glaciated valleys.
   (d) Steep-sided "sawtoothed ridge" between two glacial valleys.
   (e) Material deposited by meltwater beyond the leading edge of the glacier.

2. Name five coastal areas of the world where fjords can be found. What type of climate do most of these coastal regions have?

3. Assume that the maximum advance of the Pleistocene ice sheet is occurring today. Describe what the following present-day locations would be like in terms of climate and terrain.
   (a) Upper Michigan
   (b) Coastal New Jersey
   (c) Eastern Washington
   (d) Central interior Alaska

4. Describe the ice budget of a glacial system. What two major factors control this budget? How is this related to the movement of glaciers?

5. Glaciation is truly an interdisciplinary topic. Besides physical geographers, what other scientists are involved in the study of glaciers? Why are they so interested in ice?

CHAPTER

20

EOLIAN LANDFORMS

CHAPTER PREVIEW

▶ Over most land surfaces, wind exerts little influence over the development of landforms, but under certain conditions it can be an important agent of gradation.

Why is wind so limited as a major agent in landform development? Under what conditions is the wind an important agent of gradation?

▶ Wind gradation is most effective in arid regions, but it can be an important factor in humid regions as well.

Why is wind gradation so closely associated with arid regions? What areas in humid regions are most likely to be subject to wind gradation?

▶ In many ways the processes of surface water and wind gradation are similar, but in some important ways they are not.

In what ways are they similar? What are the differences and why do they exist?

▶ Although most people associate deserts with sand dunes, in most desert areas the surface is covered by sparse vegetation, loose rock, or gravel, or where bare bedrock is exposed.

Under what conditions are the various types of sand dunes formed? What type of landscape is found in the majority of the world's deserts?

▶ Deposition of loess by the wind has had a significant positive impact on humans in a number of major world regions.

Why is this so? What regions are involved and what were the sources of the loess?

▲ Sand dunes in Africa's Namib Desert.

As a force in leveling the land, wind is less effective than running water, waves, groundwater, moving ice, and mass movement. However, under certain conditions, wind can be a significant modifier of topography that contributes visibly to the shaping of Earth's surface. Landforms—whether in the desert or elsewhere—that are created by wind are called **eolian landforms** (after Aeolus, the god of the winds in classical Greek mythology). Note that outside of North America, eolian is usually spelled aeolian.

Wind As a Gradational Agent

The two primary conditions necessary for wind to become an effective gradational agent are the absence of complete vegetation cover and the presence of dry, fine material at the surface. These two conditions are most widespread in arid regions and beaches, though they are also found on or adjacent to dry lake beds, areas of recent alluvial or glacial deposition, newly plowed fields, and overgrazed lands.

A continuous vegetation cover reduces wind velocity near the surface, absorbs the force of the wind and prevents it from being directed against the land surface, and holds materials in place with its root network. Without such a protective cover, fine-grained and sufficiently dry surface materials are subject to removal by any strong gust of wind that occurs. However, if surface particles are damp, they will adhere together in wind-resistant aggregates.

The gradational activities of the wind are similar in most ways to those of running water. Running water and wind detach and remove materials by similar means. Both gradational agents transport material by traction, saltation, and suspension. However, unlike running water, wind cannot transport material in solution. Furthermore, wind has hardly any lateral or vertical limitations on movement. The result is that the dissemination of material by the wind can be far more widespread than that by streams.

A similarity between the gradational properties of wind and running water is that the size of the particles they can pick up and carry (their *competence*) is controlled by their velocity. However, because air is a much less dense fluid medium than water, its competence is greatly reduced to only sand-size or finer material. The result is that wind erosion selects the finer particles for transportation and leaves behind the larger, coarser particles that the wind is incapable of lifting. Likewise, wind deposits are stratified according to changes in velocity.

Wind Erosion

Strong winds frequently blow through arid regions, whipping up loose surface materials and transporting them within turbulent air currents. Wind removes or erodes surface materials by two processes (Fig. 20.1). The first of these is **deflation,** which is similar to the hydraulic force of running water. As wind velocity increases, the first particles to be affected are the finest ones present at the surface—microscopic bits of clay and silt—essentially, fine dust. The finest particles transported by the wind are carried in suspension, buoyed by vertical currents. Such particles will remain in suspension as long as the strength of the upward currents of air exceeds the tendency of the particles to fall to the ground. If the wind velocity surpasses 16 kilometers (10 mi)

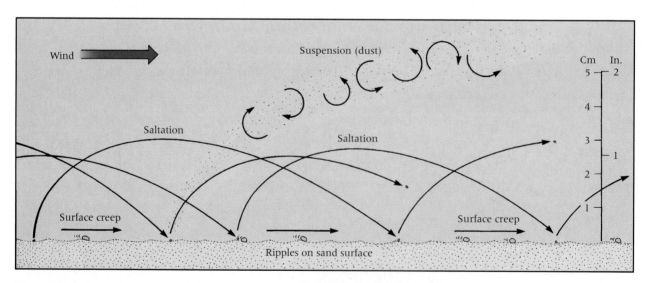

Figure 20.1 *The movement of sand by wind saltation and surface creep due to impact of saltating grains. When the wind is strong, the forward ripple movement of sand is rapid enough to be seen.* • **Why are only sand-sized or smaller particles moved by the wind?**

per hour, surface sand grains will be put into motion. The particles that are too large to be carried in suspension are bounced along the ground by the transportation process of *saltation.* As these particles are bounced along, they dislodge other particles that are then added to the wind's load or are driven forward on the ground as surface sand creep. The particles carried in suspension by the wind make up its suspended load. The particles that bump and jump along the ground make up the wind's bed load.

The second way in which the wind erodes is by **abrasion.** This process is analogous to the abrasion produced by the transported load of streams, breaking waves, and glacial ice, but it operates on a much more limited scale. In order for the wind to remove surface materials by abrasion, it must carry cutting tools that are harder than the particles of the abraded surface. As these materials hit a surface of weaker material, they break off bits of the weaker rock. Quartz sand is the most effective of the wind's abrasive agents. Yet sand grains are relatively large and heavy and rarely are lifted higher than 1 meter (3 ft) above the surface. Consequently, the effect of natural sandblast is limited to a zone close to ground level.

Where fine dust particles predominate on the land surface, they will be picked up and carried in suspension by the strong winds. The result is a thick, dark, swiftly moving cloud of dust swirling over the land. **Dust storms** can be so severe that visibility drops to nearly zero and almost all sunlight is blocked. They can also be highly destructive, removing layers of surface materials and depositing them elsewhere, sometimes in a thick, choking layer, all within a matter of a few hours. **Sandstorms** occur in areas where sand is the dominant surface material. Because sand is heavier than dust, most sandstorms are confined to a low level near the surface. Evidence of the restricted height of desert sandstorms can be seen on automobiles that have traveled through the desert: After a sandstorm, the pitting, gouging, and abrading effects of natural sandblast are often more damaging to the lower portions of the vehicle.

Erosion by deflation can produce hollows or depressions in a barren surface of unconsolidated materials. These depressions, which vary in size from a few centimeters to a few kilometers in diameter, are called **deflation hollows** or **blowouts.** In the Kalahari Desert, deflation hollows collect rainwater and attract animals and their hunters, the Bushmen. Often blowouts form where there was already a slight depression in the surface, such as in silty playa deposits.

Deflation has been thought to produce a close-fitting mosaic of rock fragments called **desert pavement** (or *reg* in North Africa and *gibber* in Australia), common in many desert regions. Because erosion by deflation is selective, the smaller clay and silt particles from an area of materials of mixed sizes are selectively eroded and transported to another location; sometimes even the coarser and heavier grains of sand are removed as well. Such selective removal leaves behind the largest particles, pebbles, and rock fragments, which together form the cobbled surface of desert pavement (Fig. 20.2). This erosional feature is widespread in parts of the Sahara, interior Australia, the Gobi in central Asia, and the American Southwest. Some recent research indicates that desert pavement may also be a product of unchanneled running water. Regardless of its origin, desert pavement is important for the protection it affords the material below the top layer of coarse pebbles and rocks. Pavement formation stabilizes desert surfaces by preventing continuous wide-scale erosion. Unfortunately, off-road recreational vehicles may disturb this stability, thus damaging desert ecological systems.

Where abrasion or sandblast action is at work on rocks of varying resistance, differential erosion results in the etching away of softer sections

Figure 20.2 *Desert pavement in the Mojave Desert of California. Removal of fine materials by wind and unchanneled running water leaves a surface of large particles, pebbles, and rock fragments, forming a desert pavement.* • **Is desert pavement a surface indestructible to human activities? Why?**

of rock while the more resistant rock remains. The rock face may become honeycombed or latticed in intricate designs.

Eolian abrasion also produces **ventifacts** (wind-fashioned rocks). A ventifact is commonly a rock fragment that has been trimmed flat on one side by sandblast, after which erosion of its support has caused it to turn over, so that a second side is faceted by eolian abrasion. Often three flat faces, which meet along sharp edges, are produced. This highly distinctive rock form is an indication of sandblast. Although not extremely common, ventifacts are plentiful where wind conditions are ideal for their formation.

Another feature often attributed to wind abrasion is the pedestal or balancing rock, commonly and incorrectly thought to form where sandblast attacks the base of an individual rock so that the larger top part appears balanced on a thinner pedestal below. Actually, such forms result from salt crystallization and weathering processes in the damper environment at the base of an outcrop and are unrelated to sandblast (Fig. 20.3).

Where the land surface has been eroded to bedrock, wind abrasion will polish the rock surface but will not significantly erode it. The speed of eolian erosion in arid regions depends primarily on the character of the materials exposed.

Figure 20.3 *Mushroom Rock in Death Valley, California, is a pedestal rock caused by desert weathering processes.* • **What would account for such an unusual shape?**

Only where these are mechanically weak can abrasion and deflation be effective.

Wind Deposition

All material transported by wind is deposited somewhere in some characteristic manner. The coarser material is often deposited in drifts in the shape of hills or ridges, called **dunes.** The finer silty material, called *loess,* settles far from its source area, blanketing and sometimes modifying the existing topography.

Sand Dunes

To many people the word *desert* evokes the image of endless sand dunes, blinding sandstorms, a blazing sun, mirages, and an occasional palm oasis. Although there *are* such deserts, particularly in Arabia and North Africa, many others have rocky or gravelly surfaces, some scrubby vegetation, and no sand dunes. Nevertheless, sand dunes are certainly the most spectacular feature of wind deposition, whether they occur as "sand seas," or *ergs,* in the Sahara, or as hills behind a Cape Cod or Oregon beach (Fig. 20.4).

Dune topography is almost infinite in its variation. For instance, dunes in the great ergs of the Sahara and Arabia look like continuously rolling sea waves. Others are shaped like individual crescent rolls sitting on a plate of sand. Sometimes eolian sand forms "sand sheets" with no dune formation at all. Laboratory experiments indicate that the different formations are the result of the amount of sand available, the strength and direction of the dominant winds, and the amount of vegetation cover. As wind carrying sand encounters obstacles or changes in the topography that decrease its velocity, sand drops out and piles up in drifts. These formations further decrease wind velocity so that the dunes grow larger, until an equilibrium is reached between dune size and the ability of the wind to feed sand to the dune.

Sand dunes may be classified as *active* or *stabilized* (Fig. 20.5). Active dunes change their shape *and* advance downwind. Dunes may change their shape with changing wind direction and/or wind strength. Dunes move forward as the wind erodes their windward slope. This causes sand ripples to migrate up to the crest and deposit their load on the steep leeward slope, or **slip face,** which is at the angle of repose (rest) for dry sand (about 35 degrees, which is the steepest slope dry sand can retain without slipping or

(Continued on page 538)

Figure 20.4 *Dunes along the Oregon coast.*
• **Why are coastlines such good locations for dune formation?**

(a)

(b)

Figure 20.5 *Active and stabilized dunes. **(a)** Active dunes usually have sharp crests. The gentle back slopes are erosional and face windward, while the leeward advancing slip faces are steep and depositional. The slip face is at the angle of repose for dry sand, as shown by these active dunes in Death Valley, California. **(b)** If plants can establish themselves in a dune area, they bring up moisture beneath the dunes, which stabilizes the dune movement. Stabilized dunes tend to have more rounded crests, as do these in the Vizcaino Desert of Baja California, Mexico.*

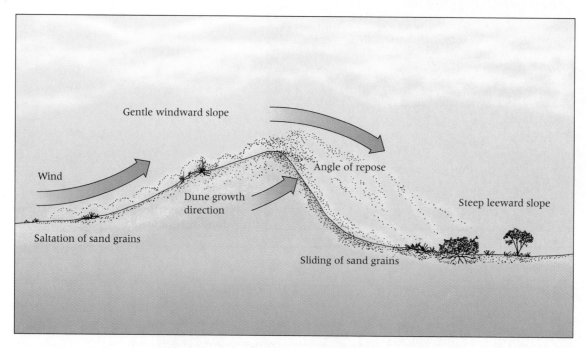

Figure 20.6 *Active dunes move downwind. This is done by the wind moving sand up the back slope (windward) of the dune toward the sharp crest. The sand then slides down the steeper slip face (leeward) of the dune, causing the dune to advance. Sand supply and wind speed and duration are major factors in the speed of the dune advance.*

falling). When wind direction and velocity are relatively constant, a dune can move forward while maintaining its form (Fig. 20.6). The speed at which active dunes move downwind varies. Large dunes travel extremely slowly; smaller ones may move up to 40 meters (130 ft) a year. During sandstorms, a dune may migrate over one meter in a single day.

A dune whose shape and position are maintained over time is said to be stabilized. Dunes are normally stabilized by vegetation, by the position of a wind-breaking obstacle, or by the back-and-forth movement of the crest under the influence of opposing winds. Where vegetation lies in the path of an active dune, the dune may move over plants and drown them in sand (Fig. 20.7). However, depending on the size and extent of the vegetation, dune movement may be impeded. Vegetation is able to stabilize a sand dune if plants can gain a foothold and send roots down to moisture beneath the dune. This task is difficult for most plants, because the sand itself offers little in the way of nutrients or moisture. In places where a sufficient cover of vegetation has been able to develop, a sand dune invasion may be halted. One such place is in the Sand Hills of Nebraska, where giant dunes, formed during a drier interglacial period, have been stabilized by a cover of grasses that now serve as grazing lands

Figure 20.7 *Plants invading a sand dune area on Padre Island, Texas. In coastal dune regions, there is almost always constant change due to dunes moving inland and plants invading the dune area.* • **Explain how plants can stabilize dunes.**

Figure 20.8 The rolling hills and grazing lands of the Sand Hills in central Nebraska were once a major sand dune region.

(Fig. 20.8). Similar stabilized dunes are found along the southern edge of the Sahara, which clearly extended farther toward the equator in the recent geologic past. Both instances involve changes in climate that affected sand supply and wind patterns (see Map Interpretation: Eolian Landforms).

Types of Sand Dunes. Many dunes are similar enough so that some basic types can be described (Fig. 20.9). **Barchans** (see Fig. 20.9a) are crescent-shaped isolated dunes. Their windward slope is the convex curve of the crescent, a gentle slope up which sand is moved. The steeper concave lee-ward slope is at the angle of repose. The two

(Continued on page 542)

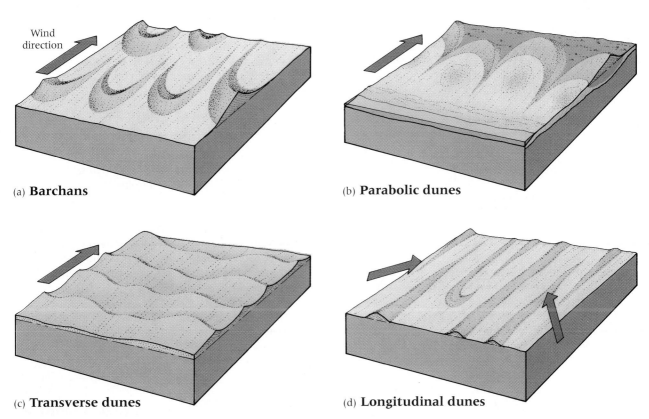

(a) **Barchans**

(b) **Parabolic dunes**

(c) **Transverse dunes**

(d) **Longitudinal dunes**

Figure 20.9 The four principal sand dune types: *(a)* barchan, *(b)* parabolic, *(c)* transverse, and *(d)* longitudinal. The wind direction is similar in all figures, as indicated by the arrow. • **What factors play a role in which type of dune will be found in a region?**

EOLIAN LANDFORMS

A location map of the Nebraska Sand Hills which cover almost one-third of the state. ▶

The Map

The Nebraska Sand Hills are located on the Great Plains, but do not resemble the typical flat Great Plains topography. The Sand Hills region was formed by eolian processes and is the largest expanse of sand in North America. The region covers over 52,000 square kilometers (20,000 square miles) of central and western Nebraska (see map on this page).

About 5000 years ago the Sand Hills region was part of a huge North American desert with giant active sand dunes dominating the terrain. The giant dunes reached over 120 meters (400 ft) high, and inundated post-glacial peat bogs and rivers. As the climate became wetter, vegetation was able to establish on the dunes and stabilized them into sand hills. Underlying the Sand Hills is the Ogallala formation, a sedimentary rock layer that forms the largest groundwater aquifer in the United States. The many lakes found between the dunes are supported by the high water table of the Ogallala aquifer.

The Sand Hills area has a middle-latitude steppe climate (BSk) and receives an average of about 50 cm (20 in.) of precipitation annually. Temperatures have a great annual range from freezing winters to very hot summers. The region is also subject to severe storms. During summer the region is often pelted by hail from large thunderstorms, while during the winter the area is subjected to blinding blizzards.

The Sand Hills region of the Great Plains is one of the windiest areas in the United States. Evidence of continuing eolian processes can be seen in huge blowouts (deflation hollows) carved into some of the sandy hills.

Vegetation is dominated by bunch grasses which can gain a foothold in the dry unstable sandy and hilly slopes. Some species of bunch grasses have extensive root systems that may extend more than one meter (3 feet) into the sandy ground. Surrounding the lakes and marshes, which are located in the interdunal valleys, is a marsh plant community which supports thousands of migratory and local birds. Populations of deer and rabbits feed on the local vegetation, and even bison graze in a wildlife refuge. The main land use of the region is cattle grazing. Some Earth scientists are now forecasting that if global warming continues, the Sand Hills area will lose its protective grass cover and once again be dominated by giant migrating desert sand dunes.

Interpreting the Map

1. What is the approximate relief between the dune crests and the interdunal valleys?

2. What is the general linear direction of the dunes and valleys?

3. Which side of the dunes has the steepest slopes?

4. If the slip face is the steepest slope of the dunes, what was the prevailing wind direction when the dunes were active?

5. Based on your answers to the previous three questions, determine what type of sand dune formed the Sand Hills.

6. What is the general direction of groundwater flow in the aquifer beneath the Sand Hills? (*Note:* Use the elevation of the lakes to determine the water table elevation.)

7. Sketch a north-south profile across the middle of the map from School No. 94 to the eastern end of School Section Lake. Label the following landform features: dune crests, dune slip faces, interdunal valleys, and lakes.

8. What cultural features on the map indicate the major land use for the region?

Steverson Lake, Nebraska ▶
Scale 1:62,500
Contour Interval = 20 ft
U.S. Geological Survey

Figure 20.10 *Barchan sand dunes at Bahia Magdalena, Baja California, Mexico.* • **What is the dominant wind direction in this photo (see Fig. 20.9a)?**

horns of the crescent point downwind (Fig. 20.10). Barchans are formed in areas of low sand supply where moderate winds blow from a prevailing direction. Although they form as isolated dunes, barchans often appear in swarms.

Parabolic dunes (Fig. 20.9b) are somewhat similar to barchan dunes but have a reverse orientation. Here the points of the curving dune trail behind, while the crest advances most rapidly. Such dunes often resemble hairpins that are open in the upwind direction. Often their "tails" are stabilized by vegetation. Parabolic dunes most commonly occur along beaches.

The *blowout dune* is most commonly formed on beaches where sand supply is abundant, where winds are moderate and blow from a prevailing direction, and where vegetation has partially stabilized the sand. The shape of the blowout dune is that of an elongated sand hill with a deflation hollow on the windward side. Where vegetation has insufficiently fixed a blowout dune, its leeward side gradually encroaches upon the land as a parabolic dune.

Transverse dunes (Fig. 20.9c) form where light to moderate winds blow from a constant direction and there is an abundant supply of sand covering the entire landscape. Transverse dunes take the shape of a series of crests and troughs whose peaks run perpendicular to the direction of prevailing winds (hence the name *transverse*). These dunes look like sea waves. The windward

slope of transverse dunes is gentle like that of the barchans, while the steeper leeward slope is at the angle of repose.

Longitudinal dunes (Fig. 20.9d) are long, narrow, parallel dunes that are aligned with the prevailing wind direction. A small sand supply and strong winds are important factors contributing to the formation of longitudinal dunes. There is no consistent distinction between the back slopes and slip faces of these dunes, and their summits may be either rounded or sharp. Longitudinal dunes cross vast areas of interior Australia, where they are known as *sand ridges*. A type of longitudinal dune called a **seif** (from the Arabic for "sword") is found in the deserts of Arabia and North Africa. Seifs are huge, sharp-crested dunes, sometimes hundreds of kilometers long, whose troughs are almost clear of sand (Fig. 20.11). They may reach 180 meters (600 ft) in height.

Dune Protection.   To many who visit the desert or the beaches, dunes are one of nature's most beautiful landforms. However, dune areas are also very attractive to recreationists, and they are particularly exciting for drivers of ORVs (off-road vehicles). Although dunes appear to be indestructible or rapidly changing environments that do not damage easily, this is far from the truth. Dunes are actually fragile environments with easily impacted ecologies. Since dune regions are really the result of environmental balance between moving dunes and the plants trying to stabilize them, nature's equilibrium is easily upset. Many of the most spectacular dune areas in the United States have special protection as national monuments or national seashores, such as White Sands, New Mexico; Great Sand Dunes, Colorado; Indiana Dunes, Indiana; and Cape Cod, Massachusetts. Many dune areas, however, do not have special protection and environmental degradation is a constant threat.

There are many practical reasons for dune preservation, especially in coastal zones. Dunes play an important role in coastal protection and are sometimes the last defense of our coastal communities from storm waves (Fig. 20.12). They are particularly important along the low-lying Gulf and Atlantic coasts of the United States, where occasional hurricanes and "Northeasters" batter the coastlines and erode the beaches in front of the dunes. In nations such as the Netherlands, coastal dunes are even more important because the land behind them is below sea level and a breach through the dunes could mean disaster.

(Continued on page 544)

Figure 20.11 *A satellite view of longitudinal dunes in the Sahara. The area in the photograph is approximately 160 km (100 mi) across.*
• **Estimate the length of the dunes in this satellite image.**

Figure 20.12 *A sign indicating protected dune area along the New Jersey Coast.* • **Why do you think some dune areas need to be protected from human activities such as "dune buggies" and recreational vehicles?**

Coastal dune regions also play an important role as wildlife habitats, especially for many bird species.

Loess Deposits

The wind can carry dust-sized particles of clay and silt, resulting from deflation, for hundreds or thousands of kilometers before depositing them. Eventually these particles settle down to form a tan or gray blanket of **loess** that covers the existing topography over widespread areas. These deposits vary in thickness from a few centimeters to over 100 m (330 ft). In northern China, on the margins of the Gobi Desert, the loess is 30 to 90 m (100–300 ft) thick (Fig. 20.13).

Loess may originate from deserts, dry-river floodplains, or other unvegetated surfaces. The extensive loess deposits of the American Midwest and Europe were derived from the glacial deposits of retreating continental ice sheets. As winds blew across the barren glacial plains, they picked up a large load of fine sediment that formed the loess deposits of downwind regions.

Certain interesting characteristics of loess affect the shape of the land where it forms the surface material. (See again the unusual physical landscape in Fig. 20.13.) For example, though

Figure 20.13 A steep gully eroded into the thick loess deposits of northern China.
• Where is the origin of these loess deposits?

fine and dusty to the touch, loess maintains vertical walls when cut through naturally by a stream or artificially by a road. Sometimes slumping will occur down these steep faces. This slumping gives a step-like profile to many loess bluffs. Furthermore, loess is easily eroded because of its fine texture and unconsolidated character. As a result, loess-covered plains that are unprotected by vegetation often become gullied. Where loess covers hills, both gully erosion and slumping are conspicuous. A particularly severe erosion problem is currently causing the collapse of the high loess bluffs along the Mississippi River at Vicksburg, Mississippi (Fig. 20.14).

Due to its high calcium carbonate content and young unleached characteristics, loess is the parent material for many of Earth's most fertile agricultural soils. Extensive loess deposits are found in northern China, the Pampas of Argentina, the North European Plain, Ukraine, and Kazakhstan. In the United States, the Midwestern Plains, the Mississippi Valley, and the Palouse region of eastern Washington are underlain by rich loess soils (Fig. 20.15). All of these areas are extremely productive grain-farming regions (Fig. 20.16). Although winds exert only minor influence over the development and appearance of Earth's landforms, they have produced land-surface characteristics that greatly impact the lives of humans who occupy a wide variety of physical landscapes.

Figure 20.14 *The steep and unstable loess bluffs of Vicksburg, Mississippi.*

Figure 20.15 *The rich wheat fields of the Palouse region of eastern Washington.* • **Why are loess soils associated with wheat-growing regions?**

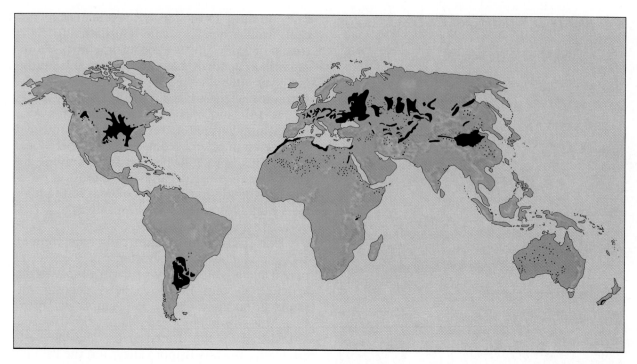

Figure 20.16 Map of the major loess regions of the world. Most loess deposits are
peripheral to deserts and recently glaciated regions.

Define and Recall

eolian	deflation hollow	dune	longitudinal dune (sand
deflation	(blowout)	slip face	ridge)
abrasion	desert pavement (reg,	barchan	seif
dust storm	gibber)	parabolic dune	loess
sandstorm	ventifact	transverse dune	

Discuss and Review

1. What are eolian landforms? Give some examples.

2. What two conditions are necessary for wind to be an effective gradational agent?

3. Explain the two processes by which wind removes or erodes surface materials.

4. Why are most sandstorms confined to near the surface? How do they differ from dust storms?

5. Describe the formation of a ventifact.

6. What is deflation? How is it related to the formation of desert pavement?

7. What is the difference between a barchan and a transverse dune?

8. Why are coastlines conducive to dune formation?

9. Why is loess an important natural resource? What problems might you encounter if you tried to cultivate loess?

10. Name several regions in the world where loess deposits are located. What was the probable origin of the loess for each of these regions? What is the major type of economic activity for these loess regions today?

Consider and Respond

1. What type of sand dunes might be expected to develop under the following conditions?
   a. Moderate winds from a constant direction and abundant sand over a large desert area.
   b. Low sand supply and moderate winds from a constant direction.
   c. Abundant sand, moderate winds, and sparse vegetation along beaches.
   d. Strong winds and a small sand supply.

2. Discuss several reasons why some sand dune regions need environmental protection from human activities.

3. Give some reasons why regions of loess are more common in the Northern Hemisphere than in the Southern Hemisphere.

CHAPTER
21

THE GLOBAL OCEAN

CHAPTER PREVIEW

▶ Although the oceans cover approximately 70 percent of Earth's surface, oceanography is among the youngest branches of science and we have just begun to investigate its associated content, concepts, and unanswered questions.

Why is oceanography such a young science? What does the global ocean mean to you as a student? What are the major divisions of the global ocean?

▶ There are significant surface and depth differences in the salinity, temperature, density, and pressure of seawater.

What are the most important differences? Why does salinity vary in different geographic locations? What effects does water pressure have on humans?

▶ Contrary to early scientific opinion, there are topographic irregularities on the ocean floor comparable to those that exist on the exposed continents.

What did early scientists believe the ocean floor looked like? Why? What are the major features associated with the continental margins and the deep ocean floor?

▶ Islands are landmasses that are smaller than continents and surrounded by water.

What are the three major types of islands? How are they formed? What are coral reefs?

▶ The surface movements of the global ocean have a major impact on humans and their activities.

What are the most important forms of ocean movement? How do they effect humans?

▲ A wall of waves crashes upon the north shore of Oahu in Hawaii.

nowledge of the oceans and seas is of major importance to an understanding of physical geography. Not only do the oceans cover most of Earth's surface geography, they also play a major role in the physical systems of the planet. For instance, the oceans are the world's major solar collector, and they absorb, store, and distribute much of the world's energy. The oceans play a significant role in Earth's hydrologic cycle and in the global circulation system, and they are a major control of weather and climate. The oceans are a significant producer of the world's oxygen and absorber of carbon dioxide. The ocean's floors contain two-thirds of the world's geology, and its submarine geological processes are still being discovered. Finally, the oceans are a source for food, minerals, and energy resources that (it is hoped) will serve humankind for centuries to come.

Introduction to the Oceans

The oceans contain over 97 percent of Earth's water and cover about 71 percent of Earth's surface. In fact, the name for our planet, "Earth," is a misnomer. It should probably be called "Oceanus" or "Hydro," since it is the only water planet in our solar system. Yet our knowledge of the oceans is not extensive because they present a hostile environment to humans as land-dwelling creatures. Consequently, we have been slow to learn about them. Only since the voyage of the *H.M.S. Challenger* from 1872 to 1876 have scientists seriously begun to explore the ocean's complexity. Only during the twentieth century did oceanography become an important science. Most of our knowledge of the ocean has developed since World War II with the refinement of sonar devices (which use sound waves to determine the topography of the ocean floor), deep-diving vehicles, coring devices, and satellites.

The oceans actually form a single, large, continuous body of water that surrounds all the landmasses of Earth. The "global ocean" is divided into three or four separate oceans that are distinguished largely on the basis of their geographic locations, although they do differ somewhat in certain other characteristics as well.

The three principal oceans are the Pacific, the Atlantic, and the Indian Oceans. The approximate area of the Pacific Ocean is 166 million sq km (64 million sq mi). The Atlantic is about half that size, or 83 million sq km (32 million sq mi), and the Indian Ocean is about 73 million sq km (28 million sq mi). In comparison, the entire continent of North America is only about 23 million sq km (9 million sq mi) in area. In fact, the total land area of Earth—149.5 million sq km (57.5 million sq mi)—is smaller than the area of the Pacific Ocean, which is the largest geographic feature on Earth's surface.

The oceans vary greatly in depth as well as in size. The Pacific Ocean has an average depth of 4200 m (14,000 ft), while the Atlantic and Indian Oceans have average depths of 3900 m (13,000 ft). However, these figures are somewhat misleading, as the ocean floors are not flat plains but have mountains, trenches, and basins that vary considerably in depth. For instance, the Pacific reaches a maximum depth of over 11,000 meters (36,000 ft).

The Arctic Ocean is far smaller, at 13 million square kilometers (5 million sq mi), and shallower, averaging 930 meters (3250 ft), than the other three oceans. For this reason, it is sometimes referred to as a *sea.* Seas are saltwater bodies that are smaller than oceans and somewhat enclosed by land. Unlike lakes, which can be of either fresh or salt water, seas always interchange water with the oceans. Some major seas include the Mediterranean, Baltic, Bering, Caribbean, Coral, North, Black, Yellow, and Red seas. Some salty lakes, such as the Aral, Caspian, and Dead "Seas," are incorrectly named because they are actually landlocked (see page 485).

Characteristics of Ocean Waters

The world ocean is 96.5 percent water by weight, but seawater is a dilute solution of many dissolved solids, dominated by the salts. On average, 3.5 percent (35 parts per thousand [ppt]) is dissolved solid matter. Of this matter, the most common substance by far is sodium chloride, or common table salt (Table 21.1). Other major constituents of seawater are magnesium, sulfur, calcium, and potassium. These major constituents are uniformly proportional to each other throughout the world's oceans and major seas. The oceans also contain the dissolved gases of the atmosphere, especially oxygen and great amounts of carbon dioxide.

There is little doubt that ocean waters contain virtually every element found on land. Most, however, are trace elements found only in concentrations of a few parts per billion (ppb). For example, there are about 40 pounds of gold and 200 pounds of lead per cubic mile of ocean water. We now extract salt (sodium chloride) and magnesium from the sea, and the future extraction of other substances may also prove worthwhile.

The measurement of all dissolved solids in seawater is referred to as **salinity,** and it varies throughout Earth's oceans and seas. Although, as

TABLE 21.1	**Composition of Dissolved Solids in Seawater**	
Element	Parts per Thousand of Seawater	% of Dissolved Solids
Chlorine (Cl)	18.98	55.0
Sodium (Na)	10.56	30.6
Magnesium (Mg)	1.27	7.7
Sulfur (S)	0.88	3.7
Calcium (Ca)	0.40	1.2
Potassium (K)	0.38	1.1

already mentioned, the average salinity is 3.5 percent (35 ppt), this can vary in the open ocean from 3.2 to 3.8 percent (32–38 ppt). The salinity is even greater in enclosed or partially enclosed seas.

Several factors affect salinity—primarily the amount of precipitation and the rate of evaporation. In humid regions, where precipitation is high, fresh water tends to dilute the seawater and reduce salinity. Moreover, the flow of major rivers into the sea tends to lower the concentration of dissolved salts. On the other hand, in arid and semiarid areas, where precipitation is low and few rivers exist, evaporation is high, salts are more concentrated, and salinity is therefore higher.

Temperature in the oceans varies and, together with salinity, affects density. Density is the mass per unit volume of a substance and is measured in kilograms per cubic meter or pounds per cubic foot. The maximum density of fresh water occurs at about 4°C (39°F) and is 1000 kilograms/cubic meter (62.4 lb/cu ft). Salinity and density are directly proportional—the greater the salinity, the denser the water. The salinity of ocean water raises water density to about 1025 kilograms/cubic meter (64 lb/cu ft). Above 4°C/39°F in fresh water (lower in salty water), an increase in temperature results in a decrease in density. The density of ocean waters affects circulation because differences in density cause gravitational displacement of the water. Since warm water is less dense than cold, it tends to float above colder, denser water. Cold or highly saline water near the surface sinks and is replaced by warmer or less saline water. This effect produces patterns of surface-water movement that vary with the seasons. More important is the fact that these density differences cause deep-ocean currents (thermohaline currents) that circulate even the ocean's deepest waters.

Surface Variations

The ocean surface shows significant variations in salinity. Salinity is highest in subtropical high-pressure regions near 30°N and S because of the low precipitation, scant stream flow, and high evaporation rate in that area. Salinity decreases toward the equator because of the abundant rainfall, heavy stream flow, and lower evaporation rate due to increased cloud cover. For example, salinity is only 2 percent (20 ppt) near the equatorial coast of Brazil because of the diluting influence of the Amazon River as well as the abundant rainfall and reduced evaporation rate. The lowest salinity is found in polar regions because

of the extremely low evaporation rate and the fresh meltwater inflow during the warm season, whose source is the fresh water stored on land in the form of snow and glacial ice. Salinity is higher in middle latitudes than in polar regions due to warmer ocean temperatures and greater evaporation. The highest salinities are found in semienclosed seas in dry, hot regions, where the evaporation rate is high and the stream flow low. The Red Sea, for example, has a salinity of more than 4 percent (40 ppt). Desert basin lakes have the highest salinity; the Dead Sea is so saline—23.8 percent (238 ppt)—that only primitive life-forms can survive in it (thus its name). The Great Salt Lake in Utah has a maximum salinity of 22 percent (220 ppt).

Variations with Depth

The vertical temperature distribution in the oceans is due to insolation heating the surface water. The heat is mixed in the seawater by conduction and wave action. The result is a surface layer of generally warmer temperatures below which there is a transition zone to the uniformly cold (average 3.5°C/37°F) waters of the deep ocean. The change from the heated surface layer to the cold deeper water occurs in a well-defined transition zone called the **thermocline.** The thermocline is most apparent in the tropics and during the warm season in middle latitudes, when the surface water is most strongly heated. There may be no thermocline in polar waters as seawater temperatures are often as cold at the surface as they are at significant depth.

Pressure increases with depth. The effect is comparable to the increase in atmospheric pressure as we descend down a mountain slope to a lower elevation. However, water is over 800 times denser than air and water pressure increases by the weight of 1 atmosphere (14.7 lb/in.^{2}) for every 10 meters (33 ft) of water depth. At a depth of 300 meters (1000 ft), water pressure is 445 pounds per square inch. At 600 meters (2000 ft), it is 892 pounds per square inch, and at 1800 meters (6000 ft), it is 2685 pounds per square inch. At the bottom of the Mariana Trench, it exceeds 16,000 pounds per square inch. Water pressure strains air-filled human organs such as eardrums and lungs, even during shallow descents into the ocean. For this reason, the deepest scuba dives have been only to about 100 meters (330 ft). However, deep-diving vehicles, such as bathyscaphes and submersibles, can descend thousands of meters while maintaining a more normal pressure for humans inside (Fig. 21.1).

Figure 21.1 *The deep-diving submersible* Alvin *has taken scientists to the ocean bottom to study the oceanic ridges and make numerous scientific discoveries. Inside* Alvin, *crew members are under normal atmospheric pressure, while outside, the submersible is under tremendous pressure from the ocean depths.* • **Does all undersea exploration require manned submersibles? Why or why not?**

The Ocean Floor

Only in the twentieth century were scientists first able to explore the ocean floor and discover the details of its topography. The invention and continuing refinement of sonic depth-finding devices (sonar) that can make continuous recordings of ocean depths using reflected sound waves have helped enormously, as have specialized research ships designed for deep-sea drilling, such as the *Glomar Challenger,* and deep-diving vehicles like *Alvin* (see Fig. 21.1). Recent advances in marine technology have produced materials and engineering designs able to withstand the enormous pressures found far below the ocean surface.

It was not until the 1960s that the first physiographic diagram of the floor of the oceans was produced by Bruce Heezen and Marie Tharp (Fig. 21.2, see pages 554–555). Today knowledge of the topography of the ocean floor, known as **bathymetry,** has advanced, with thousands of echo soundings by ships and the use of modern computer-imaging techniques to produce even more accurate bathymetric maps. Amazingly, the sea bottom can also be mapped from satellites, based on the measurement of sea level. Variations in gravity cause the sea to bulge over high areas such as submarine mountains and to dip over deep trenches (Fig. 21.3).

At present we know more about the surface of Mars than we do about our own seabed. However, recent innovations will help us to explore the deep-sea terrain. Undersea research laboratories will allow scuba-diving scientists to stay in the shallower reaches of the ocean for days at a

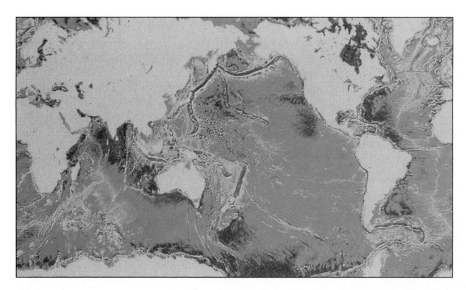

Figure 21.3 *The sea floor can now be mapped by using spaceborne sensors on satellites. The sea floor topography in this image was obtained from data collected by radar altimeters aboard the Geosat satellite. The altimeters can measure the seawater surface accurately to within a few centimeters. Variations in gravity cause the seawater to bulge slightly over seamounts and oceanic ridges, as the water is pulled toward them, creating a small mound of water at the sea surface. Sea level dips slightly over trenches. Thus the ocean bottom can be "seen" from space.* • **Identify the Mid-Atlantic Ridge and the trenches around the Pacific rim from this image. What colors represent these features?**

time. New side-scanning sonar can now record swaths of sea floor several kilometers wide with great accuracy and can produce three-dimensional bathymetric contour maps. Deep-tow camera systems can photograph the deepest sections of the ocean floor. Though manned submersibles, such as *Alvin,* will continue to take scientists to the ocean floor, robotics will probably be the way of future ocean exploration. Unmanned submersibles, such as *Argo-Jason,* which helped locate the site of the *Titanic* and can accompany manned submersibles, are the safest and most

likely deep-ocean exploration vehicles for the future.

Before the most recent explorations of the sea, it was believed that most of the ocean floor consisted of flat plains. We now know that this is far from the truth. Beneath the ocean waters are irregularities on the seabed surface greater in size and extent than those found on the continents (Fig. 21.4). Mountains, basins, plains, volcanic cones, escarpments, canyons, and trenches all are present beneath the ocean (see Figs. 21.2 and 21.3).

(Continued on page 556)

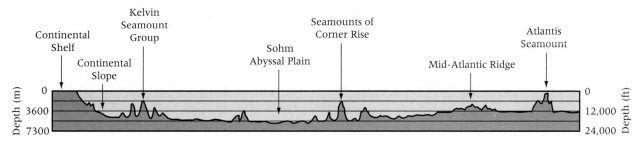

Figure 21.4 *The major features of the ocean floor are shown in this profile of the North Atlantic Ocean from the east coast of the United States to the Mid-Atlantic Ridge.* • **What is the deepest part of this section of the North Atlantic Ocean?**

Figure 21.2 *The "Floor of the Ocean" map by Heezen and Tharp in the 1960s was the first detailed relief map of the world ocean floor. It still serves well for a general view of the major ocean floor features.* • **Locate the deepest point in the global ocean.**

The ocean waters actually spill over the rim of their basins and onto the edges of the continents. Thus, much of the shallow sea floor is geologically a part of the continents. As a result, there are two major topographic subdivisions of the ocean basins: (1) the continental margins, which consist of the continental shelf and continental slope and are part of the continental crust, and (2) the deep-ocean floor, which is formed by oceanic crust (see Fig. 21.2).

Features of the Continental Margins

Continental Shelf. The **continental shelf** consists of the rims of the continents flooded by ocean water. This part of the sea floor slopes gently from sea level to depths of about 200 meters (656 ft) below sea level. The continental shelf varies a great deal in width. In some places it is almost nonexistent; in others it is as wide as 1300 kilometers (800 mi). Generally, where broad continental plains slope to meet the ocean (passive margins), the continental shelf is quite wide. Where mountains are found along the edge of the continent (active margins), the shelf is narrow. These two extremes are found on the two coasts of the United States. Beyond the "passive" plains of the Atlantic and Gulf coasts, the continental shelf extends outward as much as 480 kilometers (300 mi). On the Pacific Coast, with its "active" mountainous margins, the continental shelf is narrow. It is, of course, not surprising that the character of the continental shelf is related to the character of the nearby land, especially when we remember that the shelf is an extension of that land and geologically is not a part of the ocean basins.

The continental shelf area is the "treasure chest of the sea." The amount of sunlight and the supply of nutrients washed into these waters from the continents allow marine plants and animals to grow and thrive. Thus it is in the shallow waters above the continental shelf that 90 percent of the fish we eat are caught. Virtually all the lobster, crab, shellfish, and shrimp live in these rich waters. Vast untapped quantities of gas and oil are stored in the shelf sediments, as are other minerals, including diamonds, tin, and gold.

The topography of the continental shelf is not as smooth as we might expect from the effects of constant wave motion, settling of sediments (sand, silt, and clay) brought from the adjacent land, and sea-level changes. Actually, though it is a *relatively* smooth, sloping plain, the shelf has ridges, depressions, hills, valleys, and canyons. Some of the higher features break the surface of the water and appear as islands, such as New York's Long Island and Massachusetts' Martha's Vineyard. During periods of lower sea level during the Pleistocene Epoch, when much

Figure 21.5 *Monterey Canyon off the central California coast is an example of a submarine canyon that is deeper than the Grand Canyon of the Colorado River. The image was created from high-resolution satellite imagery and high-tech digital lithography. This region is now designated the Monterey Bay National Marine Sanctuary.*
• **What is the probable origin of submarine canyons such as this one?**

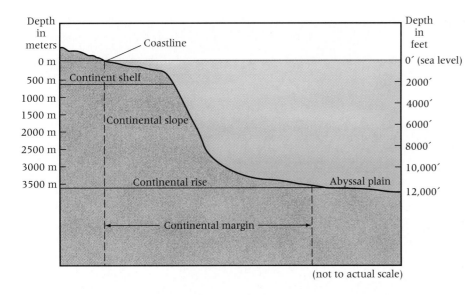

Figure 21.6 *A general profile of the major features of the continental margin.* • **Where is the true boundary between the continents and the deep-ocean basins?**

water ordinarily stored in the seas was held on the land as glacial ice, the continental shelves formed land-bridges connecting Alaska to Siberia and Great Britain to continental Europe. The teeth and bones of grazing Pleistocene land mammals are occasionally dredged up by fishermen from shelves previously exposed as land areas.

Submarine canyons are probably the most striking features of the continental margin. These canyons are usually steep-sided erosional valleys cut into the sediment and rock of the shelf and slope. They resemble the canyons cut by rivers on land (Fig. 21.5) and some are larger than the Grand Canyon. V-shaped in cross section, these submarine canyons are often found opposite the mouths of major rivers, such as the Congo or Hudson. This distribution has led to the hypothesis that rivers have helped form the canyons. Yet there are canyons that are not located opposite either present-day or ancient river mouths. Many theories have been advanced as to how the submarine canyons were formed. The most widely accepted explanation is that they were produced by **turbidity currents**—periodic massive submarine flows of sediment and water (a slurry) that move from the continental shelves down to the deep-ocean floor. Apparently such flows have a high capacity for eroding the ocean floor. The upper portions of the canyons may have developed by normal erosional processes during the Pleistocene Epoch, when low sea levels exposed much of the shelf areas. The lower portions of the canyons may be cut thousands of meters below sea level and have fans of sediment, such as beach sand, at their mouths on the deep-ocean floor.

Continental Slope. Marking the outer edge of the continental shelf is a relatively steep drop, usually 3000 to 3600 meters (10,000–12,000 ft), to the ocean floor (Fig. 21.6). This drop, called the **continental slope,** forms the boundary between the adjacent continent and the ocean basin. The landward boundary of the continental slope, where the land drops off abruptly, usually occurs where the waters are somewhere between 120 and 180 meters (400–600 ft) deep.

The continental slope is actually not a steep incline, since it descends at an angle of 15 degrees at the most, though it definitely slopes more sharply than does the shelf. What is most characteristic of the continental slope is its great descent, usually some 3600 meters (12,000 ft) but sometimes as much as 9000 meters (30,000 ft), to the deep-ocean floor or to the trenches.

Far less sediment is deposited on the continental slope than on the shelf because of the slope's greater incline and increased distance from the continents. Sediment is transported across the slope by turbidity currents in submarine canyons.

Sometimes a gently sloping surface, known as the **continental rise,** forms at the base of the continental slope. It has an average descent of less than one degree. Although the continental rises are well developed along the passive margins of the Indian and Atlantic Oceans, they are almost nonexistent along the active Pacific margins due to the presence of trenches at the edges of the continental margins. The continental rise is a depositional feature consisting of muddy sediments from turbidity currents and slumped materials from the shelf and slope.

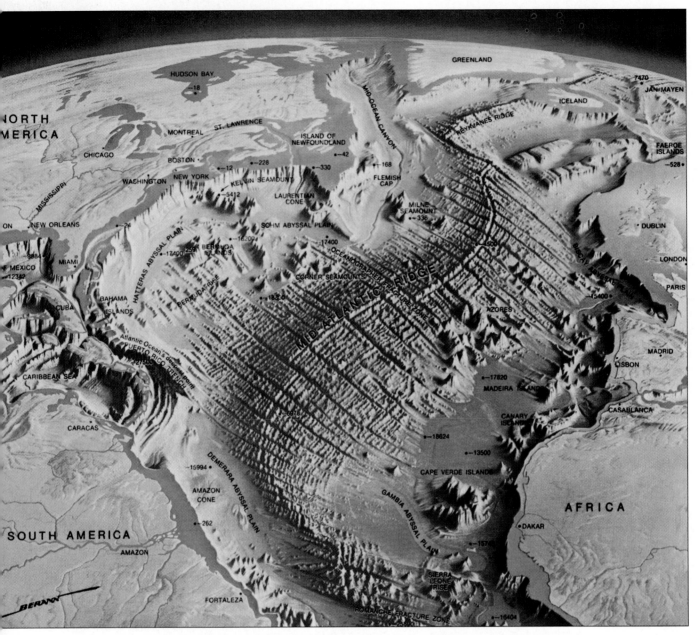

Figure 21.7 Relief map of the sea floor of the North Atlantic Ocean, showing such features as the Mid-Atlantic Ridge, the Puerto Rico Trench, the abyssal plains, numerous seamounts, and the continental shelf and slope of the surrounding continents. • What is the most obvious submarine topographic feature in the North Atlantic Ocean?

Features of the Deep-Ocean Floor

The deep-ocean floor lies at an average depth of 3600 to 3900 meters (12,000–13,000 ft) below sea level. Until recently it was believed to be a relatively smooth plain. Now, with sophisticated hydrographic profiling devices, we have discovered that the topography of the deep-ocean floor is almost as irregular as that of the land. The ridges and depressions of the ocean floor rival and sur-

pass those found on continents, both in size and complexity of pattern. On the other hand, large areas blanketed by marine sediments are virtually featureless plains.

Smoothing and leveling agents are important in shaping land above sea level (see Chapters 16–20). Running water, glaciers, wind, weathering, and gravity all work to smooth and level the land. They do this by wearing down landforms that are higher than their surroundings and by filling in depressions that are lower than their surroundings. Under the sea, these leveling agents are either missing entirely or much less active than they are on land. Consequently, many of the submarine landforms created by volcanic action and by breakage and bending of Earth's crust remain basically in their original form. The two most impressive submarine features of the deep-ocean floor are the oceanic ridges and the trenches. Other major features of deep-ocean submarine topography are abyssal plains, seamounts, and guyots.

Oceanic Ridges. The **oceanic ridges** (or **mid-ocean ridges**) are interconnected chains of mountains found in all three major oceans (see Figs. 21.2 and 21.3). The best known of these is the Mid-Atlantic Ridge (Fig. 21.7), which extends from Iceland almost to Antarctica before it swings eastward from Africa toward the Indian Ocean. The Mid-Indian Ridge forms an inverted Y shape that extends into the Red Sea (see Fig. 21.2). Its southerly arms link the Mid-Atlantic Ridge to the Pacific Ridge and on to the East Pacific Rise. This continuous chain of mountains 64,000 kilometers (40,000 mi) long averages about 1600 kilometers (1000 mi) wide and rises an average of 1500 to 3000 meters (5000–10,000 ft) above the ocean floor. In some places its highest peaks rise above the surface of the water as islands. The Azores, Ascension Island, and Iceland are high peaks of the Mid-Atlantic Ridge. For example, the Azores rise 8100 meters (27,000 ft) above the ocean floor.

From the results of recent oceanographic research, especially ocean-floor drilling and direct observation of the oceanic ridges from the manned U.S. submersible *Alvin,* Earth scientists now know that new material is being added to Earth's oceanic crust by undersea volcanic activity along the oceanic ridges.

The topography of the oceanic ridges is very rugged, and the whole range consists of volcanic rocks, chiefly basalts forming "pillow lavas." Pillow lavas are globular formations formed by lava cooling rapidly under seawater. Numerous fractures and faults add to the complexity of the oceanic ridges. Running parallel through the middle of the oceanic ridges is a central rift valley which may reach several kilometers in width. It is volcanically active and the center of earthquake activity. Most dramatic are the undersea volcanoes called "black smokers" that spew out hydrothermal fluids (hot water mixed with fine grains of iron and zinc sulfides). The central rift valleys and oceanic ridge axis are offset by **frac-**

Figure 21.8 *The drill ship* JOIDES Resolution *explores the ocean bottom sediments under the Ocean Drilling Program (ODP). A computer-controlled positioning system supported by two main shafts and 12 side thrusters maintains the ship's position over a specific sea bottom site while it is drilling in water depths up to 8000 meters.* • **Why are drill cores of ocean sediments so important to scientists?**

ture zones that cut perpendicularly across the oceanic ridges. These fracture zones are actually transform faults, where sections of oceanic plates slide past each other. They usually form long fault scarps that may extend several thousand kilometers beyond the oceanic ridges. Where they cross the rift valleys, the greatest concentration of seismic activity occurs.

The Deep Sea Drilling Project (DSDP) operated the drill ship *Glomar Challenger* between 1968 and 1983. *Glomar Challenger* drilled over 1000 cores in all the major oceans. The most remarkable discovery was the confirmation of seafloor spreading. The addition of new crustal rock material along the midocean ridges is pushing the older parts of the crust apart. Those drilling in the ocean floor now seek more detailed data on the hidden three-quarters of Earth's geology. The Ocean Drilling Program (ODP) continues to explore the ocean bottom with the newer drill ship *JOIDES Resolution* (Fig. 21.8).

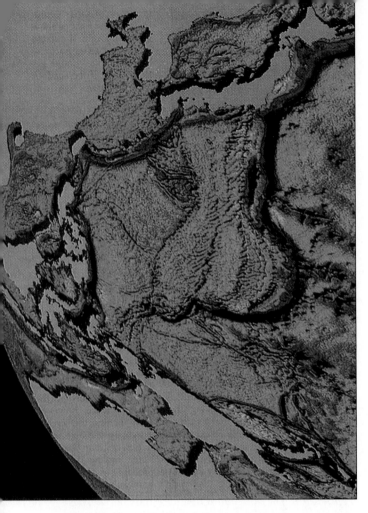

Figure 21.9 An image of the western Pacific Ocean from Geosat gravity data shows the major trenches off Asia. Note the great depths of the Japan, Mariana, and Philippine trenches. • **What caused the formation of these trenches?**

Trenches. **Trenches** represent the deepest parts of the ocean. Usually long, narrow, arc-shaped, and steep-sided, these depressions are aptly named. Most trenches are found not in the middle of the ocean basins, but near their active margins. Trenches are usually found adjacent to zones that have a concentration of volcanic and earthquake activity and are most common on the seaward (convex) side of curving chains of volcanic islands (called **island arcs**), such as the Aleutians.

Most trenches, including all of the deepest ones, are found around the Pacific "Ring of Fire." Challenger Deep, in the North Pacific's Mariana Trench, is the deepest known part of the ocean, reaching 10,915 meters (35,810 ft) below sea level. In 1960 Jacques Piccard and Donald Walsh descended in the bathyscaphe *Trieste* to the bottom of that trench. Placed in that trench, Mount Everest, the highest mountain on Earth, would still have a mile of seawater above its summit. Other major trenches in the Pacific are the Kuril,

Japan, Philippine, Tonga, and Peru-Chile Trenches (Fig. 21.9). The deepest part of the Atlantic is the Puerto Rico Trench, at 8648 meters (28,374 ft), while the Java Trench, at 7125 meters (23,377 ft), is the Indian Ocean's deepest point (see Figs. 21.2 and 21.3).

As with the oceanic ridges, the trenches play a major role in Earth's geologic evolution. As noted in Chapter 14, Earth scientists believe that oceanic plates are descending below continental plates and are being recycled to Earth's interior by way of the trenches. This process, known as *subduction,* is a key concept in the theory of plate tectonics. Ocean-floor drilling has shown that no ocean-floor rock is older than 200 million years. The youngest oceanic crustal rock is being created at the oceanic ridges, while the oldest oceanic crustal rock is being destroyed in the

Figure 21.10 The Seven Sisters section of the chalk-white cliffs of Dover along the English Channel. • **What is the origin of the cliffs of Dover?**

560

trenches by subduction. Thus, the oceanic crust is completely recycled over a period of a few hundred million years.

Abyssal Plains. **Abyssal plains,** or basins of low relief, lie at depths between 3000 and 6000 meters (10,000–20,000 ft) and cover about 40 percent of the ocean floor. Most are covered by thick masses of marine sediments that bury the ocean-floor relief. Much of this sedimentary blanket consists of fine brown and red clays, contributed by turbidity flows from the continental slopes, wind deposition, and volcanic eruption. A significant portion also consists of the remains of microscopic marine organisms. Such organic deposits are known as **ooze.** Britain's famous white chalk cliffs of Dover are uplifted ancient ocean-floor ooze deposits (Fig. 21.10). In regions of slow deposition, manganese nodules form on the abyssal plains. These potato-sized concentrations of manganese, iron, copper, and nickel may someday be an important source of minerals.

Seamounts and Guyots. One of the most common features of ocean topography is the **seamount,** a submarine volcanic mountain. Unlike the continuous chains of mountains that form the oceanic ridges, seamounts are relatively isolated mountains or groups of mountains, usually with an elevation above the ocean floor of 900 meters (3000 ft) or more. Often steep-sided with small summits, seamounts are volcanic peaks that grow from the deep-ocean floor. Though most seamounts are not nearly as impressive in size as the midocean ridges, some are high enough to break the ocean surface and appear as oceanic islands, such as the Canary Islands, Tahiti, and Hawaii (Fig. 21.11).

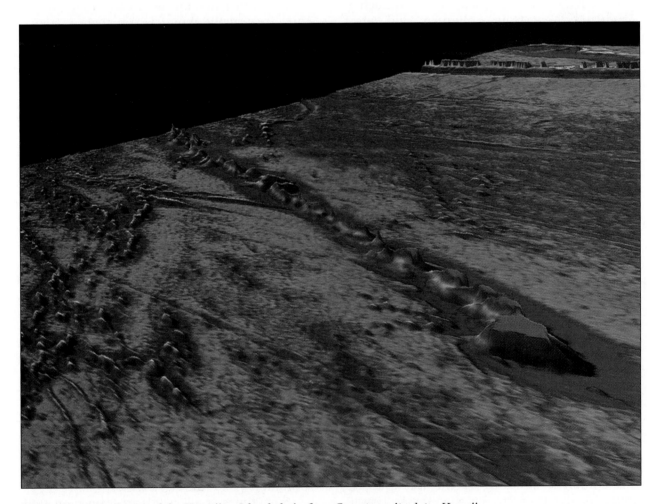

Figure 21.11 *An image of the Hawaiian Island chain from Geosat gravity data. Hawaii, the largest and youngest island, is located over an active "hot spot" in the Pacific plate. The other islands and seamounts of the Hawaiian chain become inactive and sink into the oceanic crust as the Pacific plate drifts to the northwest.* • **What is the difference between an oceanic island and a seamount?**

Guyots, discovered during World War II, are seamounts with flat instead of peaked tops, found at depths of a few thousand meters below sea level. The origin of their flat tops at such depths is not certain. Where they have been studied, research indicates that they are volcanoes whose summits have been planed off by wave erosion. Later they subsided to their present depths, possibly during lateral movement of the sea floor away from the oceanic ridges and "hot spots," where new volcanic material rises to the oceanic crust surface, anchored in the mantle.

Islands and Coral Reefs

There are three basic types of islands—continental, oceanic, and coral. **Continental islands** are usually found on the continental shelves. They are geologically part of the continent but are separated from it because of sea-level changes or tectonic activity. The world's largest islands, such as Greenland, New Guinea, Borneo, and Great Britain, are continental. Smaller continental islands include Washington's San Juan Islands, New York's Long Island, and California's Channel Islands. The hundreds of barrier islands along the Gulf and Alantic coasts of the United States are also continental. A few large continental islands, such as New Zealand and Madagascar, are "continental fragments" that separated from continents millions of years ago.

Oceanic islands are volcanoes that rise from the deep-ocean floor. They are not geologically related to the continents. Most of the oceanic islands, such as the Aleutians, Tonga, and the Marianas, occur in *island arcs* along the edges of the trenches. Others, like the Azores, are peaks of the oceanic ridges rising above sea level. Many oceanic islands occur in chains or lines, such as the Hawaiian Islands. These **island chains** are caused by the oceanic crust sliding over a stationary "hot spot" in the mantle. The exact cause of these hot spots is not known. Yet we can predict, for instance, that the Hawaiian Islands will move with the Pacific plate toward the northwest, and that the islands will slowly sink deeper into the thin oceanic crust. Hence, the islands to the northwest will submerge to become seamounts. A new volcanic island, named Loihi, will form to the southeast (Fig. 21.12). Evidence of this motion is indicated by the fact that the youngest islands of the Hawaiian chain, Hawaii and Maui, are to the southeast, while the older islands, such as Kauai and Midway, are located to the northwest.

Coral reefs are shallow, wave-resistant structures formed by an accumulation of skeletal remains of tiny sea animals called polyps that secrete a limy skeleton of calcium carbonate. Many other organisms, including algae, sponges, and mollusks, add material to the reef structure. Reef corals need special conditions to grow—clear and well-aerated water, water temperatures above 20°C (68°F), plenty of sunlight, and normal salinity. These conditions can be found in the shallow waters of tropical regions such as Hawaii, the West Indies, Indonesia, and the Queensland coast of Australia. Today, increasing pollution, dredging, souvenir coral collecting, and other man-made stresses threaten the survival of many coral reefs.

A **fringing reef** is a coral reef attached to the coast (Fig. 21.13a). Fringing reefs tend to be wider where there is more wave action to bring a continuous supply of well-aerated water and additional nutrients for increased coral growth. They are usually absent where there is a river mouth because the corals cannot grow where the waters are laden with sediment or where river water lowers the salinity of the marine environment.

Sometimes the coral forms a **barrier reef,** which lies offshore, separated from the land by a shallow lagoon (see Fig. 21.13b). Most barrier reefs form in association with slowly sinking oceanic islands, growing at a pace that keeps them above sea level. Other barrier reefs, such as Australia's Great Barrier Reef, the Florida Keys, and the Bahamas, were formed on continental shelves and grew upward as sea level rose after the Pleistocene Epoch. The Great Barrier Reef of Australia is over 1930 kilometers (120 mi) long.

An **atoll** is a ring of coral reefs encircling a lagoon that has no inner volcanic island (see Fig. 21.13c). Figure 21.14 indicates the manner in which atolls develop. In response to the subsidence of volcanic islands, the fringing reef grows upward as fast as subsidence occurs, becoming a barrier reef and finally an atoll. This explanation of atolls was developed by Charles Darwin in the 1830s. It was proved correct when the U.S. Navy drilled into several Pacific atolls in the 1950s. Drilling evidence indicates that there has been as much as 1200 meters (4000 ft) of subsidence and an equal amount of reef development in the past 60 million years (see Fig. 21.13c).

Sometimes small islands are formed on atolls from storm wave action, which erodes the reef and heaps the coral debris above sea level. Such atoll islands are most common in the Pacific

Ocean. Atolls pose severe challenges as environments for people. First, they have a low elevation above sea level and provide no defense against huge storm waves that can inundate the entire atoll, drowning all its inhabitants. Possible future sea-level rise from global warming may also threaten the people of these low islands. Second, there is little fresh water available on the porous coral limestone surface. Third, little vegetation can survive in the lime-rich rock and soil of

(Continued on page 565)

(a)

(b)

Figure 21.12 *The island of Hawaii was formed over a "hot spot." (**a**) Two large shield volcanoes, Mauna Loa and Mauna Kea, dominate the island. Mauna Loa and Kilauea on its eastern slope are still very active. To the south of the island is Loihi seamount, an active submarine volcano, which may reach above sea level in 50,000 years.* • **Do all islands form in this manner? Why or why not?** *(**b**) A submarine view of the steep slopes near the summit of Loihi, at a depth of 980 meters (3200 ft). The yellow areas are iron-rich minerals spewed from the volcanic vents. The white areas are bacterial mats, which grow around the active volcanic vents. The bacteria chemosynthetically feed on the chemicals spewed from the hydrothermal vents.*

(a)

(b)

(c)

Figure 21.13 *The major types of coral reefs are evident in the Society Island chain of French Polynesia. (a) Moorea is a rugged young oceanic island with a fringing reef. (b) The older island of Bora Bora is subsiding and now has a fringing reef around it. (c) Tetiaroa Atoll has no surface evidence of its former volcanic core. The individual island portions of an atoll are known as "motu" in the South Pacific.*

which the atoll islands are composed. The coconut palm is one exception to this rule, and therefore the coconut is depended upon to fulfill many of the needs of atoll-island inhabitants in Polynesia and Micronesia.

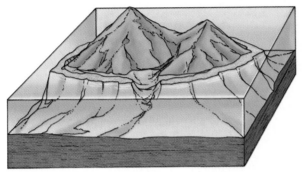
Fringing reef

(a)

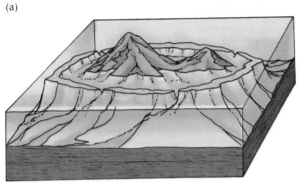
Barrier reef

(b)

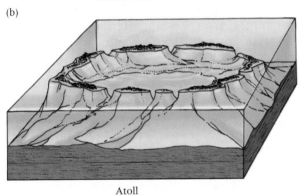
Atoll

(c)

Figure 21.14 The theory of coral reef development around oceanic islands is based on an explanation by Charles Darwin. Three successive reef forms develop from island subsidence and coral reef upbuilding. (a) First, a fringing reef forms along the shore. (b) Later, as the island erodes and subsides, a barrier reef forms. (c) Further subsidence causes the coral to build upward while the volcanic core of the island is completely submerged below a central lagoon, forming an atoll.
• **Explain why the island subsides while the coral grows upward.**

Tides and Waves

The oceans exhibit three major forms of movement. First are the major ocean currents, discussed in Chapter 6, which are the movements of surface water pushed by the winds. Second are the tides, which are the periodic rise and fall of sea level caused by the gravitational forces of the moon, sun, and Earth. Third are the waves, which are the oscillatory motions on the ocean surface that are mainly generated by the force of the winds. Coastlines (which will be discussed in Chapter 22) are primarily a product of the work of waves, especially storm waves. The tides allow the waves to break closer or further from shore and produce local coastal currents that move sediments and circulate water in bays and estuaries.

Tides

The periodic rise and fall of ocean waters in response to gravitational forces acting on them are called the **tides.** For the waters to rise in one place as a high tide, they must be pulled away from another part of Earth, resulting in a low tide there. The gravitational pull of the moon and sun, and the force produced by rotation of the Earth-moon system, are the major causes of the tides (Fig. 21.15). Tides are complex phenomena; in order to explain them we must first consider the relationship between Earth and the moon.

Since the moon revolves around Earth every 29.5 days (one "moonth," or month), one might assume that the center of its revolution would be at the center of Earth. However, both the moon and Earth actually revolve around a common center of gravity that lies within Earth and that is always on the side of Earth that faces the moon. Because of the rotation of the Earth-moon system around this axis, centrifugal force causes Earth and the moon to tend to fly away from each other. However, this tendency is blocked because the centrifugal force is exactly balanced by the gravitational attraction between Earth and the moon. Centrifugal force causes Earth's fluid hydrosphere to bulge out on the side opposite the moon. This effect is one of the external forces that raises the water surface enough to produce a high tide. The ocean waters on the side of Earth facing the moon respond to the moon's gravitational force by bulging toward the moon. This bulge of water is directly opposite the bulge produced by centrifugal force, and Earth rotates through two high tides daily. Because they are 180° of longitude apart, they are separated by about 12 hours (half a day, because 180° is half

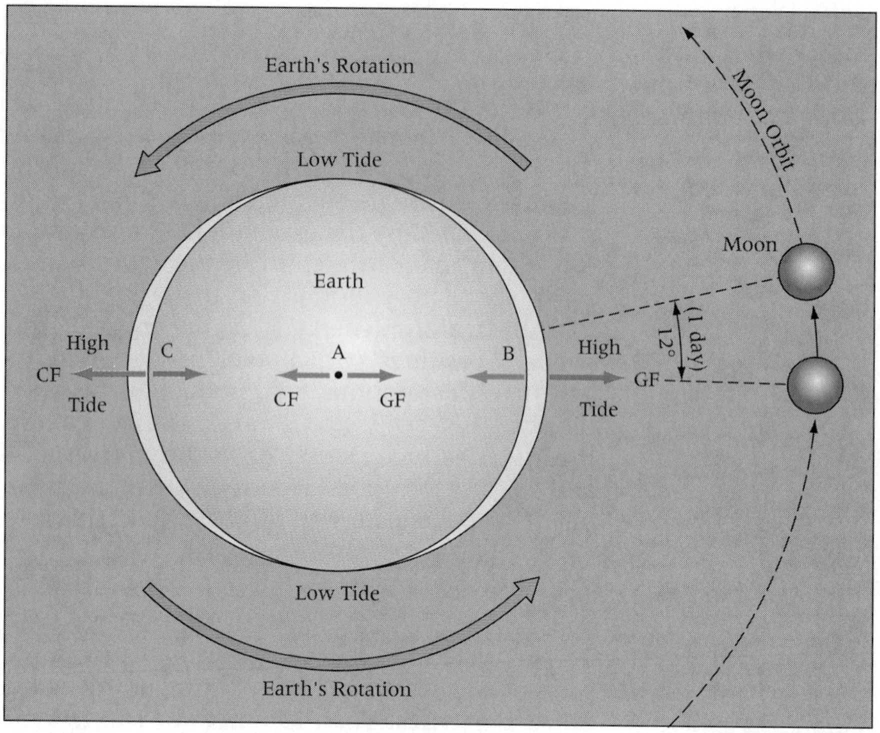

A. Gravitational force (GF) and centrifugal force (CF) are equal. Thus separation between earth and moon remains constant.

B. Gravitational force exceeds centrifugal force, causing ocean water to be pulled toward moon.

C. Centrifugal force exceeds gravitational force, causing ocean water to be forced outward away from moon.

Figure 21.15 *The tides are a response to gravitational attraction of the moon (periodically reinforced or opposed by the sun), which pulls a bulge of water toward it while the centrifugal force of rotation of the Earth-moon system forces an opposing mass of water to be flung outward on the opposite side of Earth. Earth rotates through these two bulges each day. Actually, a "tidal" day is 24 hours and 50 minutes long.* • **Explain why a tidal day is slightly longer than an Earth day. How many high tides and low tides are there during each tidal day?**

Earth's circumference). Between these two tidal bulges, the waters recede as they are pulled toward the areas of high tide. Accordingly, there are two low tides midway (90° of longitude) between the high tides.

Earth's 24-hour rotation, together with the moon's daily movement of 12° eastward along its monthly 360° path around Earth, means that theoretically coastlines will experience two high tides and two low tides every 24 hours, 50 minutes (the length of a tidal day). The time between the two high tides is called the **tidal interval,** and it averages 12 hours, 25 minutes. However, this ideal tidal pattern does not occur everywhere, though the most common tidal pattern does approach the ideal model of two high tides and two lows in a tidal day. This *semidiurnal* tidal regime is characteristic along the Atlantic coast of the United States. In bodies of water that have restricted access to the open ocean, such as the Gulf of Mexico, tidal patterns of only one high and one low tide occur during a tidal day. This type of tide is called *diurnal* and is not very com-

mon. A third type of tidal pattern consists of two high tides of unequal height and two low tides, one lower than the other. The waters of the Pacific coast of the United States exhibit this *mixed pattern.*

Tidal regimes are extremely complex and variable, even over short distances. We have described the tides as though they were bulges of water through which Earth rotates. In actuality, this "wave of equilibrium" theory does not account for the many variations in tidal regimes. A somewhat different theory explains the tides as an oscillatory movement of the hydrosphere set up as a consequence of the forces we have discussed. This "stationary-wave theory" describes the tides as a back-and-forth sloshing of water, like the waves produced if a tray of water is tipped first one way and then the other. A circular motion, due to Earth's rotation, is superimposed on this up-and-down or rocking movement.

The difference in sea level at high tide and low tide is called the **tidal range.** Tidal range varies from place to place in response to factors such as the shape of the coastline, the depth of the water, access to the open ocean, and the topography of the ocean floor. The average tidal range along open-ocean coastlines like the Pacific coast of the United States is 2 to 5 meters (6–15 ft). In restricted or partially enclosed seas, such as the Baltic or Mediterranean Sea, the tidal range is usually 0.7 meter (2 ft) or less. Funnel-shaped bays off major oceans, such as the Bay of Fundy on Canada's east coast, are apt to produce extremely high tidal ranges. The Bay of Fundy is famous for its enormous tidal range, which averages 15 meters (50 ft) and may reach as much as 21 meters (70 ft) (Fig. 21.16). Other narrow, elongated coastal inlets that exhibit great tidal ranges are Cook Inlet, Alaska; Puget Sound, Washington; and the Gulf of California in Mexico.

Spring and Neap Tides. The sun acts as a secondary tidal influence on the ocean waters, but because it is so much farther away, its tidal effect is less than half (46 percent) that of the moon. When the sun, moon, and Earth are lined up, as they are when there is a new or full moon, the additional influence of the sun on the ocean waters causes abnormally high and low tides, increasing the tidal range. This situation occurs every two weeks and is called **spring tide** (*spring* here does not refer to the season). A week after a spring tide, when the moon has revolved a quarter of the way around Earth, its gravitational pull on Earth is exerted at an angle of 90 degrees to that of the sun. At this time the forces of the sun and moon tend to counteract one another. The moon's attraction, at the time of the first-quarter and last-quarter moons, is diminished by the counteracting force of the sun's gravitational pull. Consequently, the high tides are not as high and the low tides are not as low. This moderated situation, which also occurs every two weeks, is called **neap tide** (Fig. 21.17).

Using astronomical information, long periods of observation of tidal patterns, and up to 40 local factors, tide tables can be prepared years in advance. Tide-predicting machines, ingenious devices of gears and pinions, have been used since the late 1800s. Today, computers provide tide tables of local coastal areas; these are essential to safe ship navigation and useful to sailors, fishermen, scuba divers, surfers, beach joggers, and sea shell collectors.

Tidal Currents. The tidal movement that we have been describing is a vertical rise and fall of ocean waters. There are also horizontal currents, especially in bays or sounds, called **tidal currents.**

The *flood tide* is the incoming current that accompanies rising tide; the outgoing *ebb tide* accompanies falling tide. The time period between flood and ebb, with no current, is called *slack water.* The velocity of tidal currents may be high enough to affect shipping, erode the shoreline, and move fine sediments. The tidal current through the Golden Gate, the entrance to San Francisco Bay, is 4 knots (4.6 mph) during both flood and ebb tides. In some of the channels between islands of the Inside Passage of the Alaska–

Figure 21.16 *The extreme tidal range of the Bay of Fundy in Nova Scotia, Canada. Note the difference between the low tide in the top photo and the high tide, $6\frac{1}{2}$ hours later, in the bottom photo.* • **Why does the Bay of Fundy have such a great tidal range?**

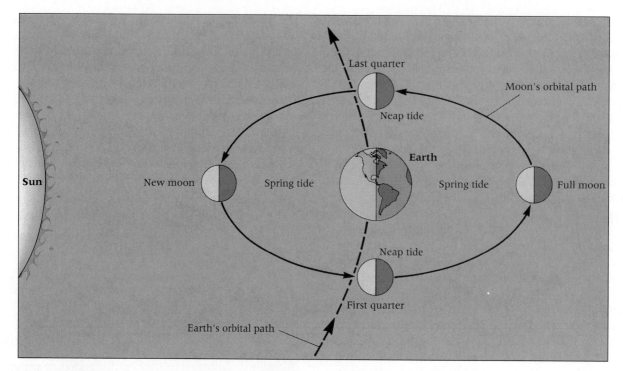

Figure 21.17 *The maximum tidal ranges of spring tides occur when the moon and sun are aligned on the same side of Earth, or on opposite sides of Earth, at new moon or full moon respectively. The minimum tidal ranges of neap tides occur when the gravitational forces of the moon and sun are acting at right angles to each other, at first- and last-quarter moons.* • **How many spring tides and neap tides occur each month?**

British Columbia Coast, there are tidal currents of up to 10 knots (11.5 mph). These currents can be hazardous to ships trying to navigate between the islands.

When incoming flood tides are opposed by the strong currents of large rivers, the tide may develop a steep, wavelike front, known as a *tidal bore*. A few rivers famous for their tidal bores are the Amazon, Yangtze, and Seine. In exceptional instances, as at the mouth of the Amazon, the tidal bore may be over 5 meters (16 ft) in height.

Waves

Waves are undulations at the surface of bodies of water. Contrary to appearance, waves do not transport water horizontally from one place to another, except where they break as surf along a coastline. Rather, the form or shape of waves and their energy are transmitted *through* water. The movement of waves is similar to the movement of stalks of wheat as wind blows across a wheat field, causing wavelike ripples to roll across its surface. The wheat returns to its original position after the passage of each wave. Water, too, returns to its original position (or close to it) after transmitting a wave. Another image that may serve as an analogy for the idea of a wave moving through water is the movement of a wave transmitted along the lengths of a snapped rope.

Most natural water waves are initiated by wind; the few that are not are produced by seismic or volcanic activity (tsunamis) or are an effect of tidal activity (tidal bores). When wind blows across water, friction arises. Some of the wind's energy is thus transmitted to the water. Waves are the result of this energy transfer.

In a sufficiently deep, open body of water, waves appear as undulations on the surface. In Figure 21.18 you can see what happens to individual parts of the surface water during the passage of a wave. First, the upward movement of water produces a **wave crest.** The subsequent sinking of the water surface produces a **wave trough.** Water particles rise and fall, producing an endless series of waves passing along. The actual movement of water particles is circular, so that there is a small amount of forward movement during each rise. This circular or oscillatory

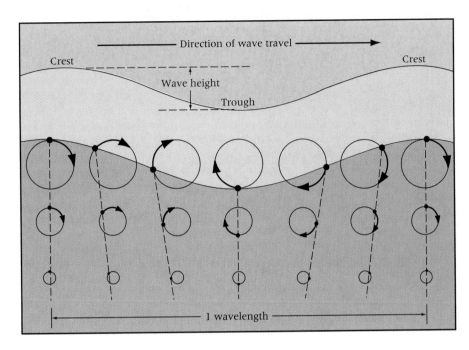

Figure 21.18 Orbital paths of water particles cause oscillatory wave motion in deep water. Wave height is the diameter of the surface orbit, which is the vertical distance from trough to crest. **• Wavelength is the measurement of what distance?**

pattern of movement is the reason why deep-water ocean waves are called *waves of oscillation* (see Fig. 21.18). The height of a wave is measured vertically from trough to crest. The horizontal distance between two wave crests is the **wave length.** A **wave period,** measured in seconds, is the time it takes for successive crests to pass a fixed point. A "sea state" report would read something like this: Three- to four-foot waves from the northwest, at 10-second intervals. This type of report is useful to sailors, fishermen, and surfers.

The factors that determine wave size are (1) velocity of the wind, (2) duration of the wind, and (3) fetch. **Fetch** refers to the distance of open water across which the wind can blow without interruption. An increase in any of these three factors produces an increase in the size (height and wavelength) of waves.

Gentle breezes form *ripples* on the sea surface, providing roughness necessary for the wind to "grip" the water. If the wind increases, the ripples are soon transformed into larger waves. In a storm, steep, choppy, chaotic waves called "sea" occur. They become *whitecaps* if the wind is strong enough to break the tops off the waves (Fig. 21.19).

When the waves travel out of the generation area or the wind dies out, the chaotic sea is transformed into more gentle, longer-period waves called *swells.* Swells can travel thousands of kilometers and completely cross an ocean. For instance, large southwest swells arriving on the Cal-ifornia coast in summer have been generated by winter Antarctic storms occurring south of New Zealand. The regular rhythm of the swells is the type of wave motion usually associated with sea-sickness.

Catastrophic Waves

Tsunamis are a series of waves caused by undersea seismic activity, such as earthquakes and volcanic eruptions, or undersea landslides. In deep-ocean water they travel at speeds of up to 725 kilometers (450 mi) per hour and may pass beneath a ship unnoticed. However, they are extremely dangerous, rising over 30 meters (100 ft) as they meet the shore, engulfing and devastating entire coastal settlements and often causing great loss of life and property (see The Environment: Tsunami Warning!).

In 1946, an Alaskan earthquake caused a tsunami in Hilo, Hawaii, that attained a height of over 10 meters (33 ft) and killed over 150 people. When Krakatoa erupted in 1883, it generated a tsunami that reached over 40 meters (130 ft) and killed over 37,000 people in the nearby Indonesian Islands. Tsunamis are sometimes called "tidal waves," even though they have no relationship to tidal activity.

Similarly, **storm surges,** which are sometimes called "storm tides," are not related to tides, although they may coincide with high tide. Storm surge is the name given to the rise in sea level beneath a severe storm (low air pressure)

(Continued on page 572)

TSUNAMI WARNING!

For coastal inhabitants, probably one of the most terrifying, though rare, natural hazards is the tsunami. *Tsunami* means "harbor wave" in Japanese. The news media, however, continue to incorrectly call tsunamis "tidal waves," even though these hazardous, seismic-produced waves are not related to the tidal forces.

Tsunamis are a series of long-period ocean waves mainly generated by submarine earthquakes, especially strong-magnitude quakes along the subduction zones of the Pacific Ocean. Destructive tsunamis have also occurred in the seismically active Mediterranean and Caribbean Seas. Volcanic eruptions and massive submarine landslides can also set them off. In 1883, the volcanic island of Krakatoa along the Java Trench erupted, causing waves over 40 meters (130 ft) high. The death toll on the neighboring islands of Java and Sumatra was over 37,000 people.

As these seismic sea waves travel across the deep-ocean basins, they may be 15 minutes to an hour apart but only a few feet high. At great ocean depths tsunamis lose very little energy, and their speeds may exceed 450 miles per hour. A ship's crew in midocean cannot even detect them passing beneath their hull.

As a tsunami reaches shallower coastal water, however, friction slows it down, compressing the waves, which increases their height. When it hits land the waves may be 15 meters (50 ft) high or more and be slowed to speeds of 30 to 60 miles per hour. **A tsunami has the potential to destroy coastal towns and override peninsulas and low islands.**

Tsunamis can also travel considerable distances up rivers, fjords, and estuaries, and may increase in height as they move a massive volume of water into a more constricted area upstream and inland. If a tsunami is generated in or enters a narrow coastal waterbody, such as Lituya Bay, Alaska, or Puget Sound, Washington, it may actually rock back and forth. This deadly motion is termed a *seiche*.

When a tsunami arrives, there may be anywhere from a single wave to over a dozen. The tsunami wave crest does not always arrive first; instead, the first arrival may be the wave trough. This causes the sea to rapidly recede far offshore, exposing the bottom of a bay, coral reefs, or fish helplessly stranded in the mud. In 1946, the people of Hilo, Hawaii, learned this dangerous lesson when Hilo Bay suddenly drained of wa-

(a)

(b)

Tsunami damage aftermath: (a) Coastal cliff formed by erosion from the 1992 tsunami that struck Flores Island, Indonesia. (b) tsunami destruction on Okushiri Island, Japan, in 1993.

ter. Within a few minutes, however, the giant wave crests, generated by an Alaskan earthquake off the Aleutian Islands, crashed into the city, destroying the waterfront. This tsunami killed 159 Hawaiians, 96 in Hilo itself! In 1960, a powerful earthquake off Chile caused another tsunami that struck Hilo, this time taking 61 lives.

One of the most dangerous aspects of a tsunami is the great amount of debris carried with the waves, including trees, trucks, trains, boulders, docks, boats, and almost anything imaginable. This mass of water and materials will destroy everything in its path. Alaska's 1964 Good Friday earthquake, one of this century's most powerful, produced a tsunami that struck the southern Alaskan port city of Seward at an estimated speed of 60 miles per hour. The impact of the waves ruptured the harbor petroleum tanks, causing a large oil spill. Another tsunami crest arriving later struck the city as a flaming wall of water 12 meters (40 ft) high, igniting the city of Seward. Not only did this earthquake generate tsunamis in Alaska, but it also sent deadly waves southward to coastal Washington, Oregon, and California, taking more lives.

In recent years, tsunamis have continued to take lives. During 1992, Nicaragua and Flores Island in Indonesia were struck. The latter tsunami killed over 700 people. In 1993, Okushiri Island, off northern Japan, was struck by a deadly tsunami. The death toll was 165. During 1994, a tsunami hit Russia's Kuril Islands, and another struck Mindoro Island in the Philippines, destroying several coastal villages. In 1998, a tsunami took over 2000 lives along the north coast of Papua New Guinea.

Today, people of the seismically active Pacific region have some warning protection from these destructive waves. In 1948, the Pacific Tsunami Warning System was established by the U.S. National Oceanic and Atmospheric Administration (NOAA). The headquarters, known as the Pacific Tsunami Warning Center (PTWC), is at the Honolulu Observatory, and the system comprises a transoceanic network of seismic stations and tidal gauges from Pacific rim coastal cities and Pacific islands. Japan also maintains a national tsunami warning system.

The PTWC will issue a *tsunami watch* after detecting any earthquake in the Pacific region with a magnitude severe enough to produce a tsunami. The epicenter is quickly located, and computers provide estimated times of arrival (ETAs) for potential tsunamis in the Pacific region. A "watch" status is similar to tornado and hurricane watches issued by NOAA, indicating that a possibility of a danger exists. It is recommended that coastal populations be alert and stay tuned to local radio and television stations for possible emergency bulletins.

If a tsunami is actually sighted, the Honolulu Observatory will issue a *tsunami warning* for all threatened coastal regions and islands. Although the height and

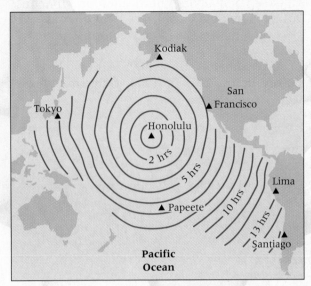

The PTWC locates the epicenter of the earthquake and provides estimated times of arrival for potential tsunamis in the Pacific region.

number of tsunami waves cannot be forecasted, the ETA is quite accurate, as the open ocean speed of the waves is predictable. Local officials are notified to undertake emergency procedures, including sirens and announcements to clear beaches, harbors, and other potentially dangerous areas. Evacuation of threatened coastal towns and cities may also be undertaken if time allows.

Although the system is not perfect and damage cannot be prevented, the Pacific Tsunami Warning System has saved lives by taking the element of surprise away from these terrifying and deadly seismic sea waves. However, because tsunamis do not occur often, the general public lacks experience with and knowledge of their destructive power. Officials are concerned that thousands of people who have recently moved to coastal or island locations may have no idea of the danger of such natural hazards. The states of California, Oregon, Washington, Alaska, and Hawaii have formed a National Tsunami Hazard Mitigation Program to improve detection, educate the public, and improve civil disaster planning.

CRITICAL THINKING ▼

(1) Although the Pacific Ocean has a tsunami warning system, why has none been established for the Atlantic Ocean? **(2)** Why should you never go down to the coast to watch for a tsunami? **(3)** Why is it possible for an earthquake in Chile to cause damage and take lives in Hawaii, the Philippines, and Japan?

and to the waves driven onshore by the strong winds accompanying hurricanes or typhoons. During Hurricane Camille in 1969, the sea level was over 8 meters (25 ft) above normal along the Gulf Coast.

Like tsunamis, these high seas may be enormously destructive. In 1900, Galveston, Texas, was destroyed and several thousand lives were lost as a result of a hurricane-produced storm surge. In 1970, a storm surge produced by a typhoon in the Indian Ocean drowned over 200,000 people in the Ganges Delta area of Bangladesh, and in 1985, thousands more lives were lost in a similar storm there.

Figure 21.19 *A U.S. Coast Guard vessel in storm waters in the Pacific off the Oregon coast. Within a storm, chaotic waves called "sea" occur. They become whitecaps when the wind is strong enough to blow the tops off the waves and become huge breakers when they enter the shallow water near shore.* • **What factors would determine the height of such storm waves?**

Define and Recall

salinity
thermocline

bathymetry
continental shelf
submarine canyon
turbidity current
continental slope
continental rise
oceanic ridge (midocean
  ridge)

fracture zone
trench (ocean)
island arc
abyssal plain
ooze
seamount
guyot
continental island
oceanic island
island chain

coral reef
fringing reef
barrier reef
atoll

tide
tidal interval
tidal range
spring tide
neap tide

tidal current
wave
wave crest
wave trough
wave length
wave period
fetch
tsunami
storm surge

Discuss and Review

1. How do the oceans play a major role in the physical systems of Earth?

2. List the three major oceans. Which is the largest? What is the difference between an ocean and a sea?

3. List the major constituents of seawater. Which ones are economically recoverable for human use?

4. Why and how do salinity and temperature affect the density of water? How is density related to oceanic circulation?

5. At which latitudes does seawater have the highest salinity? What are some factors which cause this high salinity?

6. Describe the major topographic features of the deep-ocean floor. Why have these features remained relatively unchanged since their formation?

7. How are the features of the continental shelf related to the adjacent landmasses? What are some of the differences between the continental shelf and the continental slope?

8. What is the difference between a submarine canyon and a trench? How is each formed? Where are they generally located?

9. What is the relationship between the oceanic ridges and the trenches? How do they relate to the theory of plate tectonics?

10. What is a coral reef and how does it form? What sea conditions are necessary for coral reef formation?

11. Describe where most of the world's coral reefs are located. Name several islands or island groups which have coral reefs. Name two continental coastlines which have coral reefs.

12. Describe the major factors that produce tides. What are some of the variations in tidal patterns?

13. What force causes most ocean waves? What characteristics of that force affect wave height?

Consider and Respond

1. Many scientists consider the oceans to be Earth's last true frontier and a great resource base for the future. So why is the ocean the least explored part of our planet?

2. Refer to Figures 21.2 and 21.3. Describe how the major topographic features of the deep-ocean floor differ from the major continental features. Why is there such a difference, and what geomorphic forces are dominant in each case?

3. Differentiate between oceanic and continental islands. List several examples of each.

4. Assume you were asked to plan for the construction of a major tourist resort on a beautiful tropical atoll. What would be the attractions of such a resort location? What limitations and concerns would you have to evaluate before construction of the resort?

CHAPTER

22

COASTAL LANDFORMS

CHAPTER PREVIEW

▶ The dynamic coastal regions of Earth are greatly impacted by human activities and need protection from these impacts.

Why are the coasts such dynamic environments? Why are the coastal zones the most polluted ocean waters?

▶ Waves are the dominant agent of erosion and deposition in the narrow coastal zones where the land and water meet.

What are the major erosional and depositional features of gradation by waves? What would be the ultimate result of wave action on shorelines if gradation continued without interruption?

▶ Beaches are the most visible evidence of coastal deposition, and they reflect a balance between input and removal of material by waves and currents.

Of what types of materials are beaches made? How do coastal currents form major coastal landforms?

▶ Coasts are dynamic and complex systems, and they are hard to classify due to their change-able nature.

Based on global plate tectonics, what are the two major types of coastlines? What are the differences between coastlines of emergence and coastlines of submergence?

▲ Ecola Beach, Oregon.

The most changeable environment in the world is the coastal zone. This is the zone where the hydrosphere, lithosphere, and atmosphere meet. It is a zone of continuous interaction between Earth's physical spheres. The coastlines of the world offer a tremendous variety of landforms, weather patterns, and marine and coastal life-forms. The coastal waters of the oceans are not only dynamic from a geomorphic viewpoint, they are also biologically dynamic with a great variety of niches for organisms to occupy.

To many people, the coasts are the most scenic and interesting landscapes in the world. The coastal regions of the world attract more tourism than any other natural environment. Places such as the Mediterranean's Riviera, Hawaii, Florida, southern California, and Australia compete for the hundreds of millions of tourists who head for the coast for their vacations each year. The coastal zones, however, often include our most polluted waters, since they are heavily impacted by urban growth, ports, offshore oil development, tourism, and agricultural runoff. The coasts need protection from these human impacts, and by understanding the natural processes which operate there, we will be helping to solve future problems. In this chapter we will look at the processes that shape our coastal zones and the magnificent coastal landforms produced by these forces.

Coastal Landforms and Processes

Waves and their resulting local currents are the gradational agents associated with large bodies of water: the oceans, seas, and major lakes. Their effects are felt in a narrow, dynamic zone where land, air, and water meet. The line of this meeting is called the **shoreline.** Landform changes are continuous both landward and seaward of the shoreline. The position of the shoreline fluctuates with tides, storms, and long-term rises and falls in sea (or lake) level, as well as with tectonic movements. The **coastal zone** includes the dynamic region on land as well as areas currently submerged under water, through which the shoreline boundary fluctuates.

Waves, like streams, erode and deposit materials. By removing, transporting, and depositing material, waves constantly work on the narrow strip of land with which they are in contact. Although the gradational effects of waves are multiplied many times during storms, when the largest waves develop, all waves that reach the

Figure 22.1 Large breakers pounding the shore at Point Lobos, California. • **What is the origin of such large breakers?**

coastal zone do some work in shaping the land (Fig. 22.1). The changing tidal levels allow the waves to extend their effects over a wider zone, while fluctuating world sea levels do the same in a longer time frame.

As waves pound away at the shore, their general effect is to straighten and smooth the shoreline. Peninsulas or headlands that extend further into the water than other parts of the land are gradually cut back by wave action. Bays and inlets, on the other hand, are gradually filled in.

The Breaking of Waves

As long as the water is deep, waves of oscillation can roll along without disturbing the bottom and the ocean (or lake or sea) floor will remain unaf-

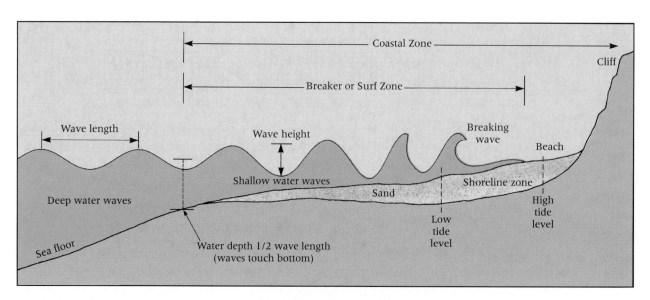

Figure 22.2 Waves begin to "feel" bottom when the water depth becomes half the distance between wave crests. Then the wave velocity and wavelength decrease while the wave height and steepness increase until breaking occurs. • **Why don't the waves break in deeper water?**

fected. However, when the water depth is about half the wave length, the waves begin to "feel" the bottom and are retarded by friction with sand and rock (Fig. 22.2). These shallow water waves (waves of translation) slow down and their wavelength decreases while both wave height and steepness increase. Eventually wave height reaches a point of instability, and the waves curl and collapse, producing surf, or breakers. If the waters far from shore are shallow, the waves will break at some distance offshore. If, however, water remains deep adjacent to the coast, the waves are forced to break against the land. The tidal range will also influence how far from shore the waves will break.

If a wave crashes directly against solid material, the impact is often sufficient to break the material apart, whether it is jointed rock or a man-made structure. The direct impact of storm waves erodes back coastal cliffs and headlands. More commonly, the waves break before striking the land directly; then the water surges forward as foaming **swash** or *uprush,* which picks up sand and carries it onto a beach. The force of gravity eventually overcomes the waning momentum of the water, which finally reverses direction and drains back down the beach slope as **backwash,** taking some sand with it. The material returned to the water in the backwash is flung back onto the shore during the breaking of the next wave.

When the waves are large, chunks of rock and sand serve as abrasive tools similar to those carried as stream bed load. Thus abrasion is an important factor in wave erosion, just as it is in fluvial erosion. The power of waves, combined with the buoyancy of water, enables large storm waves to carry large rocks, even boulders weighing tons. Such pieces of rock, when thrown against cliffs, act like cannonballs. Wave erosion of the coastal zone is also carried on by the process of solution, as is the case with surface water and groundwater, as well as by hydraulic action from the sheer physical force of the pounding water and the explosive effect of the air compressed between breaking waves and cliff faces. Expansion of salt crystals from evaporating ocean spray also helps detach mineral grains from cliffs, helping to push back the land.

Wave Refraction

Wave refraction refers to the bending of waves as they approach a shore. An important consequence of wave refraction is that wave energy becomes concentrated along some parts of the shoreline and is greatly reduced in others. To see how this happens, imagine an irregular shoreline of bays and headlands (Fig. 22.3). Offshore waves are essentially parallel to each other, approaching the shore either directly or obliquely.

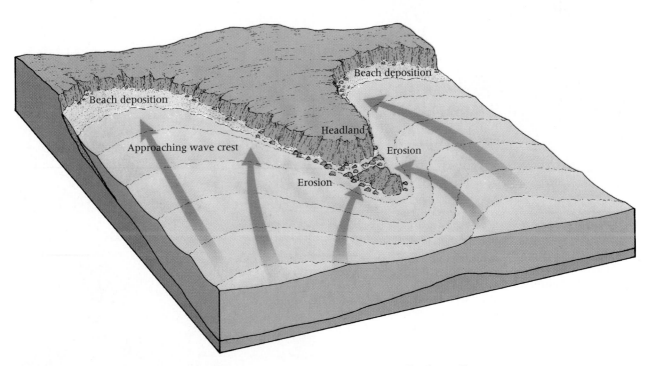

Figure 22.3 Wave refraction causes wave energy to be concentrated on headlands, eroding them back, while in bays, wave deposition causes beaches to grow seaward. • How will this coastline change over a long period of time?

However, as they approach land, the waves reach shallow water in some places sooner than in others. As a rule, the shallow waters off headlands will be reached by a wave before the shallow waters of a bay. As a result, the part of a wave that approaches a headland will be slowed down and forced to break before the part that is approaching the bay. Because one part of the wave is slowed down before another part, the wave is bent or refracted.

Consequently, wave energy is directed more toward the headlands and less toward bays and

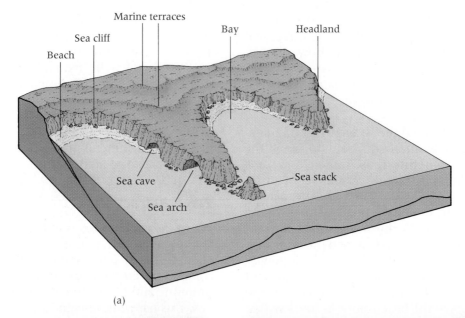

(a)

(b)

Figure 22.4 **(a)** *Diagram of the major coastal erosional landforms due to wave activity.* **(b)** *The rugged sea cliffs along the wave-eroded and uplifted Washington coastline.* **(c)** *Sea caves in the steep limestone sea cliffs on the Mediterranean island of Malta.* **(d)** *Sea arches, such as this one at Cabo San Lucas at the tip of Mexico's Baja California peninsula, form as sea cliffs are eroded completely through a headland.* **(e)** *A sea stack such as this one off Santa Barbara Island in Channel Islands National Park, California, forms when sea cliffs retreat, leaving a resistant pillar of rock standing above the waves.*

(c)

coastal indentations. Because wave energy is concentrated at headlands, wave erosion is more intensive there. The debris produced by wave erosion at coastal headlands and cliffs joins with fluvial material brought to the coast by streams and is shunted toward areas where there is less wave energy, accumulating in the bays between headlands. This effect is the reason waves tend to reduce shoreline irregularity: The concentra-

tion of erosion on headlands wears them back, while beach deposition in bays fills them seaward.

Coastal Erosional Landforms

As waves erode the coastal zone, they create distinctive landforms (Fig. 22.4a; also see Map Interpretation: Active-Margin Coastlines). As is the

(Continued on page 582)

(d)

(e)

ACTIVE-MARGIN COASTLINES

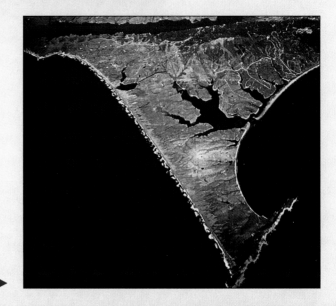

High-altitude oblique aerial photo of Point Reyes. ▶

The Map

Point Reyes National Seashore is located north of San Francisco on the rugged California coast. Point Reyes itself consists of resistant bedrock that is being cut back by the forces of the sea. Marine life is abundant, including many marine birds and marine mammals (note Sea Lion Cove off the southern tip of Point Reyes). The hilly area, Punta de los Reyes, separates Drakes Bay and Point Reyes from Tomales Bay (just showing in the far northeast corner of the map).

The oblique aerial photo of Point Reyes was taken from a high-altitude NASA aircraft. The color infrared film makes vegetation appear red. The view is looking northeast. Trending across the upper portion of the photograph is the San Andreas Fault, which forms the linear Tomales Bay. The San Andreas Fault also separates the Pacific plate, on which Point Reyes is located, from the North American plate, which underlies the area to the east of the fault.

Point Reyes is moving northwest (to the left), while inland California is moving southeast (to the right). Active-margin coastlines, such as California's, generally indicate young coastal features and tectonic controls such as fault zones.

The Point Reyes map area has a Mediterranean climate, strongly influenced by the cool California current flowing offshore. The sea here is not for swimming due to the uncomfortably cool seawater, large surf, and dangerous rip currents. The proximity of the sea and the prevailing onshore westerly winds create a truly temperate climate. Although very hot and very cold temperatures are rarely experienced, the area is one of the windiest and foggiest coastlines in the United States. It is an area of rugged natural beauty with wind-sculpted pines, grasslands, rocky sea cliffs, and long sandy beaches.

Interpreting the Map

1. Which area of the coast is most exposed to wave erosion? What features indicate this type of high-energy activity?

2. Which area of the map is under the influence of strong longshore currents? What is the general direction of flow and what coastal feature would indicate this flow?

3. Looking into the future, what may happen to Drakes Estero? What would Limantour Spit become?

4. Which area of the map appears to have wind activity? What would indicate this? What do you think the prevailing wind direction is?

5. Locate examples of the following coastal erosional landforms:
   a. Headland.
   b. Sea stack.
   c. Sea cliffs.

6. Note that the bays and esteros (Spanish for estuary) have mud bottoms. Since the creeks and streams are very small in the region, what probably is the source and cause for the movement of the mud?

7. Note the offshore bathymetric contours (blue isolines). Which has a steeper gradient, the Pacific coast or Drakes Bay? Why do you think there is such a great difference?

Point Reyes, California
Scale 1:62,500
Contour interval = 80 ft
U.S. Geological Survey ▶

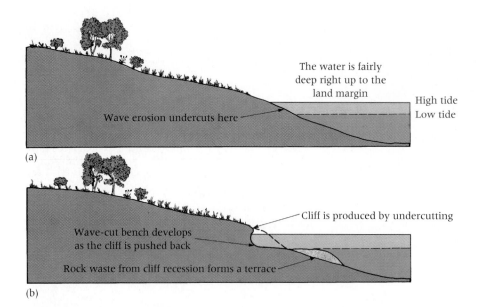

(a)

The water is fairly deep right up to the land margin

High tide
Low tide

Wave erosion undercuts here

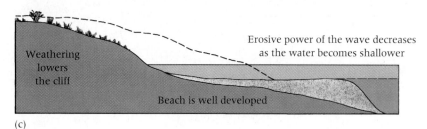

(b)

Cliff is produced by undercutting

Wave-cut bench develops as the cliff is pushed back

Rock waste from cliff recession forms a terrace

(c)

Weathering lowers the cliff

Erosive power of the wave decreases as the water becomes shallower

Beach is well developed

Figure 22.5 *Diagram of the origin of sea cliffs, wave-cut benches, and wave-built terraces.* • **What is the difference between a wave-cut bench and a wave-built terrace?**

case with *differential weathering and erosion* by water inland, rock resistance to erosion by waves is a major factor in the formation of landforms along coasts. **Sea cliffs** are created where waves pound against steeply sloping land or highly resistant headlands. The erosive action of the waves gradually undercuts the land, creating steep coastal cliffs (Fig. 22.4b). Where the rocks are well jointed but cohesive, wave erosion may create **sea caves** along lines of weakness (Fig. 22.4c). **Sea arches** result where two caves meet from either side of a headland (Fig. 22.4d). When the top of an arch collapses, or a sea cliff retreats and a resistant pillar is left standing, it is called a **sea stack** (Fig. 22.4e).

Waves frequently cut notches in a cliff by erosive action. Notches are particularly common where limestone cliffs border the sea. The seawater dissolves the limestone just as groundwater does on land, creating notches. However, notches are also cut in other rocks by abrasion and are an important factor in cliff retreat. The rate of coastal erosion is controlled by both wave action and rock type, and is accelerated during severe storms.

The presence of a wave-cut cliff implies removal of a large mass of material. The cliff is the vertical portion of a notch cut into the land. The horizontal portion of the notch is below water level and consists of a broad abrasion platform or **wave-cut bench** (Fig. 22.5). The material eroded from the sea cliff and wave-cut bench

Figure 22.6 *A marine terrace on the Palos Verdes Peninsula on the southern California coast.* • **How does a marine terrace form?**

comes to rest in the beaches of the coastal indentations. Some of the eroded material is then carried to sea by backwash and coastal currents. Because the turbulence important in transport of debris is less seaward, coarse debris accumulates and is battered into smaller particles close to the land. Finer particles move out into the sea. Thus marine sedimentation, like fluvial deposition, is organized by size and weight, with the finest particles being laid down farthest from shore. As these deposits accumulate, a **wave-built terrace** is created just seaward of the wave-cut bench. Should tectonic activity uplift these wave-cut benches and wave-built terraces above sea level, they are then called **marine terraces** (Fig. 22.6).

Beaches and Coastal Deposition

Beaches are the most visible evidence of coastal deposition. They reflect the balance between input of material by swash and removal by backwash, which *combs* the beach. Though sand-sized material is the most common, not all beaches are made of sand. Some are formed from gravel, cobble, and even silt (Fig. 22.7). Granite, basalt, shale, conglomerate, and coral all result in beaches that differ in type, color, and texture. Where particle sizes are large and wave energy is high, beaches will be much steeper than where only fine material is present and wave energy is low. Since wave height due to storms is greater

(a)

(c)

(b)

(d)

Figure 22.7 Beaches are the most visible evidence of wave deposition and may be made of any material deposited by waves. *(a)* Sand beach at Laguna Beach, California. *(b)* Cobble and gravel beach at Kopachuck State Park, on Washington's Puget Sound. *(c)* Fine-grained, light-colored sand beach, common on tropical islands with coral reefs, such as Huahine in French Polynesia. *(d)* Black Sand Beach on the island of Hawaii, formed from wave deposits of volcanic material.

in winter than in summer in the midlatitudes, winter beaches are generally narrower, steeper, and composed of coarser material than are summer beaches. Winter storm waves are relatively destructive, whereas the smaller summer waves are constructive. On the Pacific coast of the United States, beach deposits may be removed entirely by destructive winter storm waves. Summer beaches are generally temporary accumulations of sand deposited over winter beach materials. On the Atlantic and Gulf coasts of the United States, where there is a summer hurricane season, this pattern may be reversed.

Coastal Currents

The movement of water and material by obliquely approaching waves along a shoreline is called a **longshore current.** Where waves come in at an angle and break, the swash pushes material ahead of it toward shore. Though the waves and consequent swash strike the shore at an angle, the backwash, responding to gravity, moves directly downslope perpendicular to the shoreline (Fig. 22.8). The result is that material pushed diagonally up the beach by swash is not returned to its original position by backwash. Repeated over time, the result is the mass transport of materials along the shore. Longshore currents move tons of material along U.S. coastlines throughout the year.

Rip currents are strong, seaward-moving currents found near the shore and are caused by the channeled return of water and sediments from large waves that have broken against the land. Rip currents (also known as *undertow* and *rip tides*) can be dangerous, for the rapidly moving water can pull swimmers out to sea. Rip currents are usually visible as streaks of foamy, turbid water flowing perpendicular to the shore.

Beach systems are in equilibrium when income and outgo of sand are in balance. An increase in the size of a beach can be accomplished by building an obstruction to the longshore current. This will prevent sand removal while sand input remains the same. The obstruction may be accomplished by constructing a **groin** or **jetty,** a concrete or rock wall perpendicular to the beach. Of course, this obstruction starves the next beach area, which now has no input but still has the usual rate of removal (see Fig. 22.8). Beach deposition is frequently engineered to keep harbors free of sediment or to encourage recreational beach growth. However, every change in the natural process has effects that extend well beyond the area of concern, with results that are seldom beneficial in the long run.

Beaches are transitory by nature. When humans upset the natural sand supply by damming rivers or building groins to slow beach migration, the beach's temporary nature may be accelerated. In Florida and New Jersey, hundreds of millions of dollars are being spent to replenish the sand beaches. The beaches not only serve the obvious recreational needs but are also necessary to protect coastal settlements from storm waves.

Figure 22.8 *Longshore currents carry beach material parallel along the coast. The building of a jetty traps sand on one side, while it starves the beach on the other side, as evidenced by this jetty at East Hampton, New York.* • **What direction does the longshore current move in this aerial photo?**

Depositional Features

Wave erosion of nonresistant materials produces a great deal of sediment and a variety of coastal depositional landforms (Fig. 22.9a). Where rivers flow into the sea, they often bring additional sed-

iment. Material not carried out into deeper waters or deposited on the wave-built terrace is transported along the shore from the area of intense wave action to a more protected area, where deposition can take place. When a shoreline is relatively straight except for a bay, the material

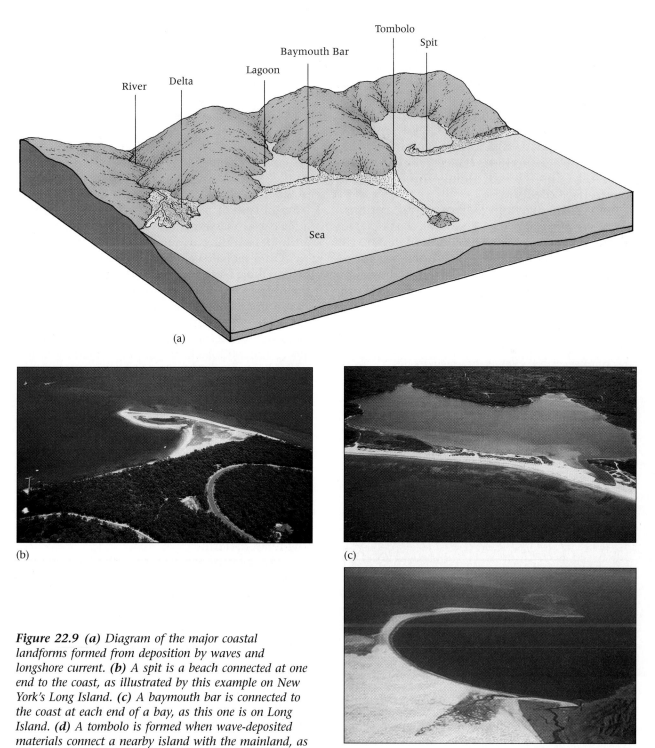

(a)

(b)

(c)

(d)

Figure 22.9 (a) *Diagram of the major coastal landforms formed from deposition by waves and longshore current. **(b)** A spit is a beach connected at one end to the coast, as illustrated by this example on New York's Long Island. **(c)** A baymouth bar is connected to the coast at each end of a bay, as this one is on Long Island. **(d)** A tombolo is formed when wave-deposited materials connect a nearby island with the mainland, as shown by this example in Baja California, Mexico.*

carried by longshore currents will be deposited as a **spit** that continues along the shoreline into the bay mouth (Fig. 22.9b). If the spit is joined by one extending from the opposite side of the bay, or if the spit grows completely across the bay, the result is called a **baymouth bar** (Fig. 22.9c).

The formation of a baymouth bar changes the bay into a protected lagoon. The salinity of such a lagoon will vary from that of the open sea, depending on such factors as river flow and climate. The salinity in turn will affect the marine life of the lagoon. A river that flows into the waters of a bay protected from wave erosion by a spit or bar will be able to build up a delta. Eventually, the entire lagoon may be transformed into salt marsh or coastal wetland by river deposits on one side trapped behind wave deposits on the other. If wave deposition connects the mainland to islands close to the shore, **tombolos** are formed (Fig. 22.9d).

On gently shelving coasts or along coastlines where waves are forced to break at considerable distance from the shore, the churning motion of the water builds submerged **offshore bars** parallel to the coast. As these bars grow and emerge above sea level they form barrier beaches and the larger **barrier islands** (Fig. 22.10; also see Map

Interpretation: Passive-Margin Coastlines). Usually there is a sand-dune zone and a shallow lagoon between the barrier island and the mainland. Barrier islands are the most common type of beach deposit on low-relief coastlines. The Atlantic and Gulf coasts from New York to Texas are dominated by barrier islands. Some excellent examples of large barrier islands are Fire Island (New York), Cape Hatteras (North Carolina), Cape Canaveral and Miami Beach (Florida), and Padre Island (Texas).

Types of Coasts

Because coasts are such dynamic and complex systems, influenced by tectonics, sea-level changes, storms, and marine and terrestrial geomorphic processes, there is no universally accepted classification system. However, there are a few different coastal classifications that do contribute to an understanding of such complex systems.

On a global scale, coastal classification may be based on plate tectonics. Two major types are noted: passive-margin coasts and active-margin coasts. **Passive-margin coasts** are best represented by the Atlantic Ocean. Here, most major tectonic activity is in the center of the ocean along the Mid-Atlantic Ridge. The coast is tectonically passive, with little volcanic and mountain-building activity. Generally, passive-margin coasts have low relief with broad coastal plains and wide continental shelves. Most have been modified by marine deposition and some subsidence. The East coast of the United States is a good example of a passive-margin coast (see Map Interpretation: Passive-Margin Coastlines). There are exceptions among some of the younger passive-margin coasts, such as those of the Red Sea and the Gulf of California, that do not show this low-relief pattern.

Active-margin coasts are best represented by the Pacific Ocean. Here, most tectonic activity is around the margins of the ocean, along plate subduction or transform boundaries. High-relief coasts with narrow coastal plains and narrow continental shelves are the rule; mountain building and volcanic activity are common. The coasts tend to be erosional and spectacular in nature, with limited time for the development of marine or terrestrial depositional features. The West coast of the United States is a good example of an active-margin coast (see Map Interpretation: Active-Margin Coastlines).

Figure 22.10 *Barrier islands form from wave deposition along gentle shelving coasts, such as these in a satellite view of Cape Hatteras, North Carolina.* • **Name several other barrier islands along the U.S. Atlantic and Gulf coasts.**

On a regional scale, coastlines may be classified as coastlines of emergence or coastlines of submergence. **Coastlines of emergence** occur where the water level has been lowered or the land has risen in the coastal zone. In either case, land emerges that was once covered by seawater, and features created by wave action, such as marine terraces, sea cliffs, stacks, and beaches, are found above the level of the present shoreline. The position above the present shoreline of wave-created landforms serves as evidence of emergence. Coastlines of emergence were probably common during Pleistocene glaciation, prior to 10,000 years ago, because the formation of glaciers would have lowered the level of the oceans over 120 meters (400 ft). Features of emergence are at present best developed along active-margin coasts such as those of California, Oregon, and Washington, where marine terraces are found as much as 370 meters (1200 ft) above sea level. Less spectacular emergent coastlines also occur where isostatic rebound has raised the land after the retreat of the continental ice sheets, such as around the Baltic Sea and Hudson Bay.

As the Pleistocene glaciers melted, sea level gradually rose, creating **coastlines of submergence** in which features of the coastal zone were submerged beneath the seas all around the world. Coastlines of submergence also occur where the level of the land has been lowered by tectonic forces, as is the case in San Francisco Bay. Sediment compaction as in coastal Louisiana may also cause coastal submergence. The features of a new coastline of submergence are related to the character of the coastal lands prior to submergence. Plains, for instance, will produce a far more regular shoreline than will a mountainous region. Two special types of submerged shorelines are ria shorelines and fjord shorelines. **Rias** are created where river valleys are "drowned" by a relative rise in sea level or a sinking of the coastal area. The valleys become narrow bays, and the interfluves form peninsulas. The coasts of Greece and western Turkey are outstanding examples of such coasts. *Fjords,* which are drowned glacial valleys, form scenically spectacular shorelines (Fig. 22.11). A fjord shoreline is highly irregular with deep, steep-sided arms of the sea penetrating far inland in the troughs originally deepened by glaciers. Tributary streams cascade down the canyonlike fjord walls, which may be several thousand feet high. Fjord coastlines are found in Norway (where the term originates), Chile, New Zealand, Greenland, and Alaska. Canada, however, has more fjords than any other coastal nation.

When areas of low relief and soft sedimentary rocks are submerged, barrier islands form with enlarging bays and lagoons behind them, such as those found almost continuously along the Atlantic and Gulf coasts of the United States.

Some coastlines, such as those formed by coral reefs and river deltas, cannot be classified as either submerging or emerging. Actually, most shorelines show evidence of more than one type of development, largely because the level of the land and the level of the seas have changed many times during the geologic history of Earth. For this reason, most coastlines are characterized by features of both submerged and emerged shorelines.

Since coastlines are really shaped by both terrestrial and marine geomorphic processes,

(Continued on page 590)

Figure 22.11 *An aerial photo of an Alaskan fjord with an active "tidewater" glacier.* • **How did this fjord form?**

PASSIVE-MARGIN COASTLINES

False-color LANDSAT image of
Long Island–New Jersey area. ▶

The Map

The Eastport, New York, map area is on the south shore of Long Island about 70 miles east of New York City. It is located in the far northeast corner of the satellite image. Long Island is part of the Atlantic Coastal Plain, which extends from Cape Cod, Massachusetts, to Florida. Much of the Atlantic Coastal Plain in the northeastern United States is embayed by water bodies such as Delaware Bay, Chesapeake Bay, and Long Island Sound. The map area has low relief, and its humid continental climate is moderated by the nearby ocean waters.

Although this is a coastal region, its recent glacial history plays an important role in the landforms that exist today. Long Island was formed by two east-west trending glacial terminal moraine ridges deposited during the Pleistocene ice age. The southern moraine is known as the Ronkonkoma moraine. Between the coast and the Ronkonkoma moraine is a sandy glacial outwash plain that forms the higher elevations at the very northern part of the map area. As the Pleistocene ice melted, sea level began to rise, submerging the lowland that now forms Long Island Sound (it is the body of water in the upper center of the satellite image). This separated the island from the mainland coastal plain.

The United States Atlantic and Gulf coasts have 295 barrier islands with a combined length of over 4400 kilometers (2600 mi). New York, especially the coastal zone of Long Island's south shore, has 15 barrier islands with a total length of over 240 kilometers (150 mi). Among these are Fire Island National Seashore, Jones Beach, and the Hamptons. The coastal barrier islands are important for protection of the mainland from storm waves. They are also important because they form rich marshlands, which are a critical habitat for fish, shellfish, and birds. Barrier islands play an important role in the economy of coastal regions, such as Long Island's south shore, as they are prime recreation and vacation destinations.

Interpreting the Map

1. On the map, what is the long, narrow feature called that forms the straight shoreline labeled Westhampton Beach? From observing the satellite image, do you see similar features along the south shore of Long Island and the New Jersey coast?

2. What is the highest elevation on the linear coastal feature? What do you think forms the highest portions of this feature?

3. What forms the outer seaward portion of the landform in Question 1?

4. Behind the linear feature is a water body. Is it deep or shallow? Is it a high-energy or low-energy environment? What is a water body such as this called?

5. Is the Eastport coastline erosional or depositional? What features support your answer? Is this a coastline of submergence or emergence?

6. Note Beaverton Creek in the middle of the map area. Does it have a steep or gentle gradient?

7. Based on its gradient, does Beaverton Creek flow seaward continuously? If not, what would influence its flow?

8. Looking into the future, how will this map area be modified by geomorphic agents? Which area will probably be modified the most? Explain.

9. What type of natural disaster is this coastal area of Long Island most susceptible to? Why?

Eastport, New York
Scale 1:24,000
Contour interval = 10 ft
U.S. Geological Survey ▶

another regional classification system recognizes only two types of coasts: primary and secondary coastlines. **Primary coastlines** are formed mainly by land erosion and deposition processes. Such formation is due to tectonic activity or sea-level changes that cause the shoreline to change its level too rapidly for the marine processes to shape the coast. The major types and good examples of primary coastlines are drowned river valleys (Delaware and Chesapeake Bays), glacial erosion coasts (southeastern Alaska, Maine, and Puget Sound in Washington), glacial deposition coasts (Cape Cod, Massachusetts, and the north shore of Long Island in New York), river deltas (Mississippi delta), volcanic coasts (Hawaii), and faulted coasts (California).

Secondary coastlines are those formed mainly by marine geomorphic agents, especially waves, and by marine organisms. Marine erosional coasts are dominated by such features as sea cliffs, arches, stacks, and sea caves (Oregon). Marine depositional coasts have such features as barrier islands, spits, and bars (North Carolina). One example of coasts built by marine organisms is the coral reef (Florida Keys). Mangrove trees and salt-marsh grasses also trap sediments to build new land areas in shallow coastal water.

Regardless of the classification type, coastlines are one of the most spectacular and dynamic regions on Earth. They are the true meeting place of all Earth's spheres—hydrosphere, lithosphere, atmosphere, and biosphere.

Define and Recall

shoreline
coastal zone
swash
backwash
wave refraction

sea cliff
sea cave

sea arch
sea stack
wave-cut bench
wave-built terrace
marine terrace

beach
longshore current

rip current
groin (jetty)
spit
baymouth bar
tombolo
offshore bar
barrier island

passive-margin coast
active-margin coast
coastline of emergence
coastline of submergence
ria
primary coastline
secondary coastline

Discuss and Review

1. What is the difference between a shoreline and the coastal zone?

2. How do ocean waves change when they enter shallow coastal waters? What is the main factor causing this change in the waves?

3. What is wave refraction? How is wave refraction related to the shape of the coastline?

4. Describe how sea cliffs form. Name three other coastal erosional features which may form in sea cliff areas.

5. Describe a marine terrace. What is it evidence of?

6. What are the differences between longshore currents and rip currents?

7. Explain the relationship between the sorting of coastal marine sediment and the development of beaches. Why do beaches change seasonally?

8. Why do baymouth bars develop? What effect do they have on bays?

9. What is the difference between an offshore bar and a barrier island? Name several examples of barrier islands. Where are most barrier islands located in the United States?

10. What are the major differences between active-margin coastlines and passive-margin coastlines? Why do they look so different?

11. What would be the major changes in the world's coastal zones if sea level were to rise?

Consider and Respond

1. Explain why the coastal regions of the world are considered the most dynamic systems of Earth's environments.

2. What are some of the major ways humans change coastlines? Explain how this human activity may interfere with natural coastal processes.

3. Assume you are the geographer in charge of coastal zone management for a rapidly growing coastal community on the Atlantic or Gulf coast of the United States. Describe the major coastal landforms. What natural hazards would you have to plan for to protect the community?

4. Assume you are a geographer in charge of coastal zone management for a rapidly growing coastal community on the Pacific coast of the United States. Describe the major coastal landforms. What natural hazards would you have to plan for to protect the community?

CHAPTER
23

DYNAMIC PHYSICAL GEOGRAPHY

CHAPTER PREVIEW

▶ Physical geography offers an integrative framework for thinking about and understanding the way Earth operates environmentally and as a life support system.

What have we learned in this course about Earth as a dynamic system? How does geography contribute to our understanding of the Earth system? How is the human exploration of space related to our understanding of the Earth system?

▶ Earth is an oasis in space, a life support system of great environmental diversity and richness.

What basic characteristics of Earth combine to provide a suitable habitat for life? Why does Earth's surface vary in its ability to sustain a human population? What factors provide limits to the amount of life that Earth can sustain?

▶ Planet Earth is a complex system composed of many major and minor subsystems.

What major subsystems of Earth interact to provide the diversity found on our planet's surface? How are thresholds related to the operation of Earth systems?

▶ Geography is a modern discipline that involves spatial and environmental problem-solving, generally supported by computer-assisted technologies.

What are some of the computer-assisted and space age technologies that are particularly applicable to geographic problem solving? Why will the application of computer and space age technologies be important to geographers in the 21st century? In terms of studying the Earth system, how can computers help us to "dissect" problems and "reassemble" them?

▶ The perspective of physical geography employs some combination of six major geographical themes that were discussed throughout this book. Applying these themes gives us a geographical perspective of the world.

What are these themes, and how is each related to geography as a discipline? What are some examples of the use of each of these themes that were discussed earlier in the book? How does physical geography relate to you and your existence—what is meant by "Geography for Life," the title of the national standards in geography education?

▲ The first image taken from a spacecraft that includes both Earth and the moon surrounded by the darkness of space was imaged by NASA's *Voyager* satellite.

Our journey through the essentials of physical geography began with an astronaut's view of Earth from orbit. From this vantage point and beyond, the space scientist is able to observe the Earth system, as a whole, from outside its boundaries. The perspective is striking—a blue planetary oasis surrounded by the vast darkness of space.

> **S**uddenly from behind the rim of the Moon . . . there emerges a sparkling blue and white jewel, a light, delicate, sky-blue sphere laced with slowly swirling veils of white, rising like a small pearl in a thick sea of black mystery. It takes more than a moment to fully realize this is Earth . . . home.
>
> **Edgar Mitchell**
> Lunar astronaut, 1971

The space program has yielded a critical learning experience for humankind. Looking outward to space has resulted in a fresh, inward look at Earth. Exploring space has focused human efforts and new technologies on *examining and understanding our planet, how it operates, how it functions as a habitat for life, and its place in the universe.* Acquiring new knowledge about our own planet is, in fact, a primary justification for studying the rest of the solar system and the universe beyond. As your course in physical geography comes to an end, how much of this new knowledge have you acquired? How has your perception of your planetary home improved? Have your attitudes toward Earth environments changed? In the years to come, what should you remember about physical geography? What important knowledge, ideas, theories, and skills should you retain? This final chapter in your textbook is written to help you think about the role of physical geography in your future.

You have learned that physical geography involves studying spatial variations in terrestrial, aquatic, biotic, and atmospheric environments, the factors responsible for this variability, and the dynamic processes that operate the Earth system.

These factors, along with the integrative nature of geography, provide a perspective and a body of knowledge that contribute greatly to our understanding of Earth. A knowledge of physical geography and how the Earth system works sharpens the observational skills of people as they examine the features of our planet. Increasing one's understanding also fosters a stronger appreciation of our planet's complexity and beauty. These are a few of the reasons why astronauts receive training in physical geography, helping them to be skilled observers as they gather data and photograph Earth features from space.

At the dawn of a new millennium, humanity marks the beginning of an unprecedented cooperation among the space agencies of the United States and many other nations in planning to construct and launch a truly international space station (Fig. 23.1). The crews' missions will involve studying a variety of environmental aspects of our planet—and scientific equipment will be sent up for that purpose—

Figure 23.1 *The International Space Station (ISS), being planned through the cooperation of the United States and 15 other countries, will be the largest structure ever placed in orbit by humans. Modular in construction and put together in space, the ISS will be the size of two football fields and weigh 500 tons. Its mission will be to support long-term study of the Earth by crews of scientist-astronauts.* • **What kinds of environmental problems can be studied from orbit in the space station?**

but safety, survival, and logistical support for the human occupants are overriding concerns. Many factors had to be considered in designing living quarters that would be hospitable for humans during a long-term habitation in space, and, in terms of its function as a human habitat, this space station will be an orbiting microcosm of the Earth system. As scientists and aerospace engineers dealt with the difficult task of creating an artificial life-support system for humans in space, they could not help but appreciate the wonder of the intricacies and complexities of Earth's environments.

The space station will operate on solar energy, much like Earth but with a critical difference: this space capsule cannot function over the long-term as a closed system. Direct links to Earth (bringing in food, water, and medical supplies, and returning refuse) will be continually necessary to support the well-being of the crew. Long-term sustenance of human life in space without these necessary connections to Earth remains the subject of science fiction.

A Special Planet

Is Earth unique? It has been argued that, in its ability to support life, Earth may not be unique in the expanse of space. This opinion is based on the idea that somewhere, because there are so many billions of stars in the universe, distant solar systems may contain planets that support living organisms. Further, this reasoning suggests that, because of the vastness of space, the existence of life as complex as that in Earth's biosphere is likely. There are two problems with these contentions, however. First, just because we know that space is vast, that fact alone does not prove that life exists elsewhere. When using the scientific method, the burden of proof is on the person who proposes an idea, or hypothesis; it is not the responsibility of others to prove that the idea is invalid. Second, even our most distant space probes have yet to discover the existence of any other planets capable of supporting life as we know it. Human and scientific curiosity, however, continue to drive a quest to answer the intriguing and compelling question of whether life exists elsewhere in the universe.

An eminent scientist provides us with an interesting reaction to images of Earth like the one from space on page 593, and to what they suggest about our planet. He contrasts Earth with the moon, a lifeless satellite without an atmosphere, hydrosphere, or biosphere, although its surface is composed of similar lithospheric materials and it receives energy from the sun (Fig. 23.2).

Figure 23.2 Earthrise as seen from lunar orbit on the first Apollo mission to the Moon. The lunar lander is in the foreground, orbiting above the lifeless surface of the moon. The striking blue Earth, shrouded in clouds, is in stark contrast to the ashen-gray lunar surface.

Viewed from the distance of the moon, the astonishing thing about the Earth, catching the breath, is that it is alive. The photographs show the dry, pounded surface of the moon in the foreground, dead as an old bone. Aloft, floating green beneath the moist, gleaming membrane of bright blue sky, is the rising Earth, the only exuberant thing in this part of the cosmos. If you could look long enough, you would see the swirling of the great drifts of white clouds, covering and uncovering the half-hidden masses of land. If you had been looking for a very long, geologic time, you could have seen the continents themselves in motion, drifting apart on their crustal plates, held aloft by the fire beneath.

The Lives of a Cell
Thomas Lewis

An important consideration in determining whether or not our planet is unique, at least among other features of the known universe, is the nature of the Earth system. Other than receiving inputs of energy from the sun, Earth is a completely self-sustaining life-support system. Now that we have gained a knowledge of physical geography, we should be well prepared to appreciate the conditions that make life possible here on Earth.

The environmental richness and the nurturing characteristics that make our planet suitable for life are the long-term results of an amazing combination of interrelated and interacting factors. Earth orbits at just the right distance from the sun, a star that radiates the right amount and wavelengths of energy to provide light for photosynthesis and vision. Our planet absorbs that light, and converts it into thermal energy to heat Earth's surface, atmosphere, and hydrosphere within a suitable temperature range to permit life to exist. Rotating on its axis, most of Earth receives a daily dose of direct solar energy throughout the year, and its speed of rotation does not allow days to be too hot or nights to be too cold, at least on most of Earth's surface.

Gases in the atmosphere allow much *insolation* to penetrate while providing *insulation* by absorbing and retaining heat energy and by blocking its loss to space. Without this result of the greenhouse effect, our oceans would freeze over, shutting down the hydrologic cycle. Our planet's surface is covered by water and terrestrial materials that effectively store solar energy, which can be heat. The inclination of Earth's axis, combined

with our annual revolution, cause the Northern and Southern Hemispheres to face both toward and away from the sun and cause direct solar rays to annually track over a wide tropical belt. These factors give us the seasons, and act to moderate air, land, and water temperatures in extreme latitudes.

Our planet has an adequate mass and a resulting gravitational force to hold an atmosphere in place. Earth also possesses the unique characteristic of abundant, life-sustaining water. Yet landmasses project above sea level because of crustal dynamics expressed as plate tectonics, volcanism, solid tectonic activity, and isostasy. Without these processes, our planet's surface would be completely covered by oceans. Differences of relief on the land caused by uplift or subsidence, and by erosion or deposition, provide a variety of elevational environments for the biosphere and interact with the atmosphere and hydrosphere. Crustal activity and lithospheric materials combine with processes related to the hydrosphere, atmosphere, and biosphere to produce an intriguing variety of landforms and landscapes. Minerals in the crust undergo weathering to provide soil and nutrients for plant life.

The biosphere is exceedingly complex, with organisms occupying diverse niches in virtually every environment in the planet's hydrosphere and near-surface lithosphere (Fig. 23.3). Earth is a complex system of interrelated features, processes, organisms, and regions. In our vast universe, the existence of this complex suite of environmental variables occurring together on even one solitary planet is indeed a miracle. Given this situation, any future discovery of life in other parts of the universe would still not diminish the miraculous nature of Earth.

So, Earth—the life-support system for humanity and the rest of the biosphere—is at the very least a special place. The variation in environmental conditions that support life on Earth is tremendous. Every ecosystem offers a different set of challenges and advantages to the success or survival of indigenous life-forms, as well as to human sustenance. Geographic transitions and regional boundaries exist in a nearly unlimited terrestrial variety, and their expression in the landscape ranges from the obvious to the subtle—food for thought for legions of geographers in the millennia to come.

We have learned that the physical geography of Earth's surface varies greatly in its ability to directly sustain a human population. Seventy-one percent of Earth is covered by water. The remaining 29 percent is land—continents and is-

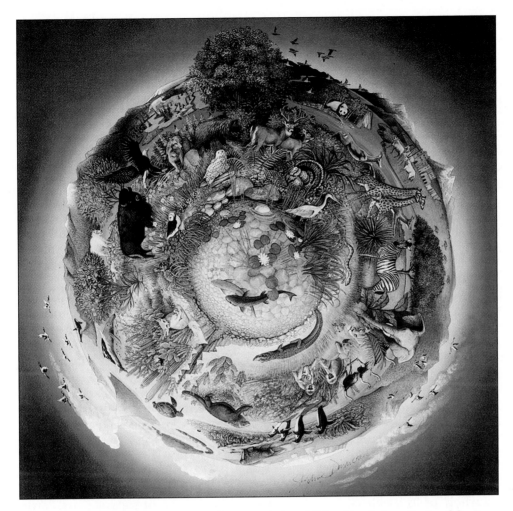

Figure 23.3 *The Smithsonian Institution commissioned this whimsical poster to celebrate Earth Day in 1990. Earth Day is a time when people reflect on the wonder, diversity, and fragility of Earth environments and consider how humans can help to maintain the planet as a habitat for life. All four of the major Earth subsystems are represented in this illustration.* • **Why is an awareness of Earth as a habitat for life important and how does physical geography relate to this awareness?**

lands. Of this *land,* a considerable amount is taken up by areas that are unsuitable for human habitation, generally because of extreme environmental conditions. These locations include virtually uninhabitable deserts, glaciers, and tundra, although these "unsuitable" areas of Earth are not "wastelands" (an unfortunate word that should be used sparingly) because they contribute to or support in many ways the environment of our planet as well as human existence. Some other lands are also excluded as potential living space for the human population because they are heavily forested or too mountainous for most land uses. We even set aside land as parks and wilderness areas for the enjoyment of future generations. The majority of the human popula-

tion lives on the lands that remain, and there is strong competition for areas that have the greatest utility or that are attractive for some other practical reason.

Thus, planetary resources and acceptable living space are finite, while the human population continues its exponential growth. Dealing with population growth is difficult and complex, but no matter what the answer to that challenge may be, gathering information about our planet and its environments will always play a key role in sustaining human existence. Advancing our knowledge of how Earth and its environments operate is critical to understanding the life-support system for humankind and the rest of the biosphere. It is through stewardship, care, and maintenance of our planet's environments that humans will function as the *crew* of Spaceship Earth, rather than as *passengers*.

Complexities of Planet Earth

How should we think about our planet—as a collection of environmental features and regions, as spatial bits and pieces, or as a unified whole, the Earth system? In *Operating Manual for Spaceship Earth*, discussed in the opening chapter of this text, Buckminster Fuller noted how strange it is that people seem to be moving toward increasingly narrow academic and career specializations. He contended that the desire " . . . to understand all and put everything together" is a basic human trait. Categories and classification schemes are really human inventions. We use them to divide our seamless, complex world into meaningful segments, to organize these parts in ways that our brains find logical, and to reduce complicated features to a level of simplicity with which we mentally can deal. While we classify and subdivide parts of the Earth system in order to increase our understanding, we also learn much by focusing on features that do not seem to fit our schemes, and on the overlap that often exists between categories.

To respond to our original question about thinking about Earth as a whole or as a collection of parts, let us reexamine the four "spheres," or major subsystems, of Earth: the atmosphere, lithosphere, hydrosphere, and biosphere. To a certain extent they operate as individual systems, but they are also well integrated and strongly interrelated—one might even say separable but also inseparable. Imagine Earth without one of these four major subsystems; each is a critical and essential part of the whole (Fig. 23.4).

Too many instances of overlap exist between these physical domains to permit a full inventory in this chapter, but consider the following examples. The lithosphere incorporates portions of the hydrosphere in the forms of soil moisture and groundwater. It interacts with the biosphere through organic activity in the soil, and with the atmosphere as rocks weather and soil develops. The atmosphere contains water vapor—part of the hydrosphere—and volcanic gases, volcanic ash, and soil particles, all of lithospheric origin. Pollen, mold spores, bacteria, and smoke from burning vegetation are also in the air and are products of the biosphere. The biosphere, in all of its abundance and variety, would not exist without the involvement and support of land, air, and water. Water, a unique ingredient of our planet, plays the critical role in processes that characterize each of the other spheres in the Earth system.

Whether or not we consider our planet as a whole or as a collection of interacting parts, Earth

Figure 23.4 This remarkably clear image of Earth reveals global and regional information about all of the four major subsystems of Earth. The distributions of landmasses and oceans are visible in this Atlantic Ocean–centered image, as well as cloud patterns associated with the polar fronts and the Intertropical Convergence Zone. The vegetation boundary in Africa between the Sahara and the Sahel can also be seen.

is further complicated by its ever-changing nature; it is a dynamic system. Fortunately, it is also robust and resilient, in that negative feedback in its subsystems acts to maintain balance. Studies of natural systems on Earth indicate that they are able to accommodate much change without impacts that might be catastrophically detrimental to a living population. However, Earth systems do have limits (thresholds) to how much change can occur without changing the environment. If these threshold conditions are exceeded or are not met, major problems can develop. Increasing this possibility is the fact that human activities appear to be exerting ever-increasing pressures on the "normal" function of environmental systems. These activities may be planned environmental intervention, such as flood control structures, or

they may be the inadvertent effects of human activities, such as the increasing release of greenhouse gases into the atmosphere. A valid question concerns the degree of human impact that can be accommodated by the naturally operating processes that are in place regulating the environment (Fig. 23.5).

Physical Geography, Computers, and Space-Age Technology

We have now reached the end of a century and the beginning of a new one. In the early 1900s, geography in the colleges and universities was mainly studied by those who planned to teach the subject in the schools (Fig. 23.6). Wide parts of our planet were unmapped in any detail, and we knew nothing (or virtually nothing) about the deep ocean floors, or of plate tectonics, the jet stream, air masses, weather fronts, the ozone layer, acid rain, or El Niño. Gaining information about Earth from space was the subject of science fiction, and the age of flight, which would provide an ability to view our landscapes from above, was only beginning. Early in the twentieth century, the study of Earth was mainly limited to exploration, field observation, surveying, mapping, and map interpretation. Communication, travel, and data collection, all essential elements in the study of physical geography, were slower and generally much more difficult than they are today.

During the past century, particularly over the last 25 years, our ability and effectiveness in

Figure 23.5 Madagascar has suffered severe environmental degradation as a result of deforestation associated with a growing population and the cutting and burning of forests to clear lands for agriculture. Most of the island was covered with trees at the beginning of the twentieth century, but today only about 10 percent of the forest remains. Soil erosion on the barren slopes is intense as heavy tropical rains carve gullies into hillsides and choke rivers with sediment.

Figure 23.6 When compared to the photographs and images in this textbook, this artistic lithograph from Matthew Fontaine Maury's 1900 textbook on physical geography illustrates how technology has affected the study of physical geography in the last 100 years. Today, we have much more access to data, information, and images than people did a hundred years ago. Maury's textbook is highly respected for its time.

studying Earth have dramatically increased, primarily through the application of technology. As it has been demonstrated throughout your textbook, physical geographers today employ many different tools and methods to gather data and information, conduct analyses, and graphically display the results. New equipment, and ways to apply these tools, continue to become available at an ever-increasing rate. Most of these technologies are not exclusive to the discipline of geography, but through a "spatial science" perspective, knowledge gained in studying Earth with these techniques becomes uniquely geographical in nature.

Contemporary environmental problems involve so many variables that several approaches to information gathering and data analysis tend to be used, rather than a single method. Because they are supported by a variety of data, infor-

Figure 23.7 *Originally designed for military purposes, the global positioning system is being integrated into many civilian uses, such as this direction-finding system in an automobile. GPS is revolutionizing location finding and navigation on Earth, as it can provide a precise location of the receiver and track its movements on a computer map display.* • **What other kinds of civilian applications could be assisted by GPS technology?**

mation, and imagery that have been gathered and examined, multiple lines of reasoning and analysis can be applied to check one form of evidence against another until we are reasonably certain that our conclusions are correct.

The methods that geographers apply in studying Earth features may be simple, straightforward, and time-honored—for example, direct observation, map interpretation, and field measurement. Other techniques involve applying high-tech, computer-assisted methods to solve geographic and environmental problems, particularly those which include complex spatial dimensions and the interactions of several processes.

It is interesting to consider that while the increase of technology has contributed to many problems that our environment faces, new technologies are also critical to addressing these problems. These technologies are much like "power tools" for studying Earth, because in the hands of skilled users, they make the task easier and more effective and allow users to accomplish or learn more in a shorter time. Having a good, basic knowledge of geography and the expertise to apply appropriate tools and techniques is not only useful in academic studies but also important in a variety of career fields. No technology or method is useful without the knowledge, training, and ability that are required to apply it effectively.

In Chapter 2, the role of space-age technology, such as the global positioning system (GPS), geographic information systems (GIS), remote sensing, and computer-assisted cartography, was discussed in some detail. However, it would be worthwhile to consider further how these modern tools of physical geography will be a part of your future. For example, it is likely that GPS receivers, which are rapidly decreasing in price, will someday replace magnetic compasses as the standard tool for direction finding and navigation on Earth. Newly developed cartographic systems are being combined with GPS locational data to show—automatically and directly on a computerized map—where a receiver is located. These systems are now available for automobiles and could completely replace the traditional road map (Fig. 23.7).

A GIS, which can integrate, store, access, and display spatial data of all types, has become an essential component in many aspects of problem solving in business, industry, and the public sector (Fig. 23.8). Countless critical decisions affecting your life in the years to come will be better

(Text continues on page 602.)

JULIE SMITH

High School Education: *Antelope Valley High School, Lancaster, California, 1990*

University Education: *Bachelor of Arts, Geography/Environmental Studies, June 1994, California State University, San Bernardino*

Early Interests: *Camping, outdoor sports, environmental issues, animals*

Prior Work Experience: *Internship, Aerial Information Systems*

Julie Smith is currently a teacher at La Mesa Middle School in Santa Clarita, California. She is also taking classes toward obtaining a teaching certificate and Master's degree at the University of LaVerne. Before entering the teaching profession, Julie recently held a series of positions at companies that specialized in using geographic data and technology. Julie describes how the collection and use of data in GPS and GIS technology affect everyday life.

As an undergraduate at California State University, San Bernardino, I worked as an intern for Aerial Information Systems doing manual land-use mapping for city and county agencies. The kinds of tools we used were definitely low-tech—colored pencils and light desks. We mapped areas by taking two aerial photos and viewing them through a stereoscope, which made the photographs appear in 3-D. We would then identify the buildings in the photos, trace them onto Mylar sheets and code them. This kind of information is vital to city planners who need to know, for budgetary and other reasons, how land is being used.

My first geography-related job after graduation was working for a company called Navigation Technologies in Sunnyvale, California. NavTech specialized in using technology to turn database information into navigational systems for cars. As an Update Geographer, it was my responsibility to take data that was collected in the field and enter it into the GIS system in our office. This data could include street names and locations, new construction—anything about the location that was relevant to identifying it. We might also retrieve information from government agencies that had their own mapping data. The company pulled this information together, organized it, and compiled it on a CD-ROM that consumers could purchase each year as a subscription.

I spent a year at NavTech before accepting a position as a Field Data Capture Coordinator in the Los Angeles field office of ETAK, another company headquartered in the San Francisco Bay area that produced navigational systems. There, I supervised ten people who used GPS units to collect data on specific geographic locations. The GPS units were similar to laptop computers, with touch screens. The units themselves are fairly compact and they transmit information about position—longitude and latitude—to a group of 24 satellites in space. The satellite then sends back an 'address' for the location.

Over the course of my three years at ETAK, I was promoted to Source Acquisitions Specialist. In that job I sought out new contacts or vendors who could either provide me with data or who were building their own navigational units. Finally, I became the West Region Manager at ETAK, overseeing the Los Angeles field office, as well as staff at the Phoenix and San Francisco offices. When ETAK was bought out by another company, I decided to reevaluate my career options, which led me to consider teaching. The middle school curriculum does not include geography, but I'd like to be able to teach at the high school level or beyond when I receive my certificate and Master's degree.

I've already seen in my previous work experience that both GPS and GIS technologies are becoming a big part of everyday life. The technology is changing rapidly. In three years at ETAK, we basically went from paper to computer in all of our work.

The uses for GPS and GIS are expanding—many people who hike or mountain bike use handheld GPS units to keep track of where they are, and all cars within the next five years will be equipped with navigation systems.

Job opportunities in GPS and GIS are increasing as well. Everyone from government agencies to oil companies has an interest in the development of these technologies. But even specialized, high-tech companies like ETAK look for geography majors who have the basic tools for and interest in this kind of work. As our need to understand the world and how it functions becomes more crucial, GPS and GIS technologies will continue to come to the forefront and be more accessible to consumers.

Figure 23.8 Geographic Information Systems (GIS) facilitate the storage, retrieval, and display of spatial information and data. Like any technology, however, the most important component is a knowledgeable and skilled person to apply the GIS. A strong background in geography and spatial analysis is essential to effectively use GIS, understand the spatial nature of environments, and solve environmental problems.

informed through the availability of a GIS. Similarly, the future use of remotely sensed imagery will improve our ability to monitor and protect our environment, will provide earlier and more accurate predictions of environmental change, and will help to make Earth a better and safer place in which to live.

Perhaps most significant of all will be the future role of the computer in physical geography. The computer has already revolutionized cartography; maps, graphs, and charts that once took days or weeks to construct by manual drafting methods can now be completed in hours or even minutes (Fig. 23.9). Knowledge of geography, the nature of maps, and basic cartographic principles is still essential to producing an acceptable map, but merging that knowledge with computer skills assures that future maps can be more accurate, more economical, more easily revised, and more readily designed to help solve a specific problem.

Aided by the computer, the "information explosion" of the twentieth century will expand in force throughout the twenty-first century. Geographic information about the nature of places will become even more readily available on the Internet and the World Wide Web. Both will offer nearly unlimited opportunities for instantaneous, convenient, and inexpensive worldwide

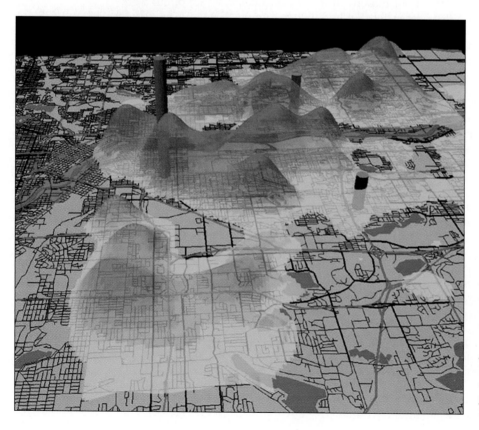

Figure 23.9 Computer-assisted cartography and GIS, combined with digital elevation data and satellite imagery can be used to easily generate complex maps that would have been impossible, or exceedingly difficult to produce without such technology.

communication. It is hoped that this opportunity will lead to better international understanding and a strong feeling of community among the crew of Spaceship Earth.

Finally, computer applications will become increasingly important to physical geographers as they face the ever more complex and multifaceted problems of the future. Through modeling, simulation, and prediction, computers allow geographers to "dissect" the Earth system by breaking it down into its component variables. As physical geographers have in the past, their counterparts in the future will analyze how these variables function, either individually or in conjunction with other variables, and then reassemble them, subsystem by subsystem, until Earth is "whole" again.

Themes in Physical Geography—Revisited

The first chapter of this book outlined six themes in physical geography that provide a guide to asking informed questions about the Earth system. Throughout the book we have encountered these themes because they are prominent in topics concerning physical geography. We now have the opportunity to reexamine these themes, recall how they provided a framework for the study of physical geography in earlier chapters of your textbook, and consider how you might use them as the physical geographer does: to structure your future thoughts about the environment.

Location

You have learned that location is an essential part of spatial communication. Having a reference system or a "global address" to describe a location or to find features is the first step in the study of geography. Location also provides the framework for an understanding of the world in which you live, and, if you have learned only one lesson from your brief exposure to geography, it should be that there is inestimable value in knowing where things are located in respect to political units or boundaries and major physical features. Location is the language of geography, whether it is absolute location in terms of a grid system, or relative location in terms of proximity or distance from other features or landmarks. Recall how important location has been in previous discussions of weather elements, climate types, environmental hazards, tectonic plate boundaries, catastrophic natural events, and ma-

jor landforms. Locational information provides the spatial context that is relevant in every aspect of geography and it inevitably leads the geographer to question *why* things are located where they are.

Characteristics of Places

The environmental diversity of places on Earth is one of our planet's most compelling characteristics. Describing the physical characteristics of a place was once the *goal* of many geographical studies, but today that task usually represents only the *beginning* of our efforts to understand a site, locale, or region. Description based on detailed observation and an inventory of relevant environmental features is still important, but being able to explain the factors that have contributed to the nature of a place is the general intent of many studies in physical geography today (Fig. 23.10). The range of environments on Earth varies greatly according to location, elevation, and position on a continent (or in the ocean). Perhaps most important to recall, unequal distribution of energy in various forms helps to explain the variability on Earth throughout the hydrosphere, biosphere, lithosphere, and atmosphere.

Once the characteristics of places are known, geographers often group places with similar characteristics into regions. We have encountered many examples of physical regions, such as climatic or vegetation regions (Mediterranean climates, tropical rainforests), geomorphological or landform regions (Rocky Mountains, Great Plains), and hydrologic or oceanic regions (drainage basins, Indian Ocean, abyssal plains). The physical and conceptual boundaries that divide such natural regions tend to be fuzzy rather than sharp. The edges of most physical regions on Earth tend to be zones of transition rather than readily defined lines.

Knowledge about the physical characteristics of places and regions is important also because environmental concerns are generally linked to the nature of a location or area. Every environment on Earth presents a set of advantages and limitations to both human existence and the existence of other life-forms.

Spatial Distributions and Spatial Patterns

Our study has indicated that physical geographers are especially interested in the general extent of, and causal factors associated with, the

Figure 23.10 The way that an environment looks is often geographically distinctive—a function of its physical characteristics, its location on Earth, and a variety of other factors we have learned in this course. This photograph shows a tropical savanna area of the East African rift valley in Kenya—a tropical-wet and dry region of faulting and volcanic activity. • **What factors of location and characteristics of place have combined to give this area its distinctive appearance?**

distributions of natural features and areas. You have encountered numerous examples of distributions of environmental phenomena in your textbook. Maps have provided the visual medium for portraying these distributions, either using points to show locations of phenomena or indicating regions where the phenomena may be found.

Whatever phenomena a map of distribution shows—whether they are natural features, categories, or environmental features (earthquakes, vegetation types, precipitation, tectonic plates, deserts, ocean currents, loess deposits, atmospheric pressure, glaciation, and so forth)—the physical geographer attempts to answer two important questions: What environmental processes explain the distribution, and what can be learned from the spatial pattern revealed on the map? The answers to either question may involve complex explanations that draw on the accumulated knowledge and theory of several related science disciplines and is often influenced by the geographer who designed the map. As an example, an adequate explanation of the distribution revealed by the map of world climates that appears in Chapter 10 required no fewer than five earlier chapters, dealing with insolation, temperature, pressure, wind, atmospheric moisture, and precipitation. Much of the content in those chapters is also the subject matter of physics and chemistry. In addition, because the authors chose to use a modified Köppen classification for the map, the climate types and boundaries shown are closely associated with biology.

When studying maps of distribution, the physical geographer seeks to identify patterns; in the future, you should be alert to their appearance as well as you encounter maps in newspa-

pers, magazines, or on the Internet. You should ask yourself if a spatial distribution is random, clustered, or uniform. Good examples of spatial patterns can be seen in a satellite image of Earth at night (Fig. 23.11), which shows an approximation of the distribution of the human population (although some cultures use more electrical lighting than others). On the image shown, random means that there is no pattern; clustered means that large numbers of people exist closely together, as seen in the clustering of light in urban areas; and uniform means that there is a regular distribution, as seen by the even spacing of light that occurs in rural areas influenced by the public lands survey system in the West and Midwest regions of the United States.

Analyzing patterns of distribution often helps us explain the reasons behind the location of a particular category of phenomena. Does the distributional pattern of one feature seem to be related to the distribution of another feature, region, process, or environmental condition? An example can be seen on a world map of the distribution of volcanic regions (see Fig. 15.8). The distribution pattern of volcanic regions reveals a remarkable correlation with the distribution pattern of earthquake epicenters (see Fig. 15.24), and both spatial patterns exhibit unmistakable relationships with the pattern of tectonic plate boundaries (see Fig. 14.14).

Spatial Interactions

The interactions between places and regions as well as the interconnections that link them are nearly boundless. We have learned about feedback operations and systems analysis, concepts that help us to better understand not only com-

Figure 23.11 *A close approximation of the distribution of the world's population is shown in this satellite image of the Earth at night. This image is a mosaic of scenes taken at about midnight local time for each time zone, and shows electric lights and fires, such as those from agricultural burning and gas flares from oil fields.* • **How would a scene like this look if it had been taken 100 years ago, and how do you think it will look 100 years from now?**

plex environments, but also how the variables interact with each other as part of a system or subsystem. Earlier in this chapter we discussed how the hydrosphere, atmosphere, lithosphere, and biosphere are linked together to form the Earth system. Changes in one of these subsystems is generally accompanied by changes in at least one (if not all) of the others.

At one time humans thought that places on the planet operated in virtual isolation from environmental events that occurred in distant locations, but the more we learn about the Earth system, the better we understand how such occurrences, even in places remote to us, can affect our local environments. Volcanic eruptions can put ash and dust in the atmosphere, affecting weather over wide parts of Earth. El Niño, once thought to be a condition mainly affecting the west coast of South America, now is known to be associated with global weather conditions. The deforestation process occurring in the tropical rainforests and other forests is now thought to have an impact on global climates. Some interrelationships may involve amazing distances; a researcher has recently suggested that soil nutrients are supplied to the rainforests of the Ama-

zon Basin through dust carried by winds over the Atlantic from the Sahara Desert, in Africa (Fig. 23.12). It is important to remember that, in its environmental interactions and interrelationships, nature knows no boundaries.

Ever-Changing Earth

Our planet, as we have learned, is an ever-changing dynamic system. In fact, it is often said that the only thing we can predict with certainty is that change will occur. That is easy to understand, but we should also realize that the nearly infinite number of changes taking place on Earth also operate at different rates and over different time spans. Certain changes in the Earth system can be directly witnessed, such as volcanic eruptions, passage of weather systems, seasonal changes in vegetation, the results of a wildfire, the flooding of rivers, coastal erosion, and many others affecting all four of the major Earth subsystems. Changes in the Earth system can also operate at a rate that requires scientific study and measurement to monitor. Examples of this type of change include gradual slippage along a fault, shifts in the boundaries between climatic zones,

Figure 23.12 *A dust storm blows a cloud of sediment and soil particles from a desert region over the ocean. Some of the particles will settle out on the ocean to be deposited as bottom sediments. Fine dust particles can also be suspended by winds and carried thousands of kilometers from their source region, until they settle out of the atmosphere or are washed out by precipitation. Recent research suggests that such storms can provide nutrients to the rainforests of the Amazon Basin across the Atlantic from the Sahara. This relatively small dust storm is about 64 × 112 km (40 × 70 miles) in size.* • **What do trans-oceanic dust storms suggest about the Earth as a system?**

vegetation succession from meadow to forest, long-term changes in sea level, glacial advance or recession, and revegetation of the devastated zone adjacent to Washington's Mount St. Helens (see Fig. 15.3c).

Human activity can make it difficult to fully understand changes taking place in the environment. Superimposed on naturally occurring changes in the Earth system are human modifications that complicate the processes involved. For example, most scientists agree that global warming is occurring, and many believe that human impacts on the atmosphere play a role in this climatic shift, particularly as a result of the increasing release of greenhouse gases. As carbon dioxide levels rise because of human activity, so does the global average temperature. But the problem is not as simple as that, because, as we noted in Chapter 9, many factors influence climate change and most involve components of the atmosphere, hydrosphere, biosphere, and lithosphere. Thus, climate can change "naturally," or be a result of carbon dioxide emissions (from automobiles, or factories and other human

sources), or maybe a result of a combination of the two.

Whatever the causes of environmental change, it is most important to realize that change in one part of the Earth system can produce changes in other parts. For example, if the trend in global warming were to continue so that climate became warm enough for the ice caps to melt, Earth would undergo myriad other changes. The temperature and salinity of the oceans would be altered. Sea level would rise and the map of the world would be different. The coastlines would migrate inland, drowning coastal lowlands and shrinking continental landmasses. Climatic changes would cause a shift in Earth's vegetation zones, and animals adapted to certain environments would migrate with these shifts. Rivers would alter the positions of their deltas and regrade their channels to a new base level.

A point to consider is that these changes, as radical as they seem, would be accommodated by adjustments in the variables comprising the Earth system. Further, whether the changes would be "good or bad" would depend to a great extent on geography. The impact of the changes would differ from one part of Earth to another. If you lived on a lowland coast, the change would likely be bad, as your home would be threatened by a rising ocean. However, if you lived in Russia or Canada, the growing season would likely become significantly longer, something that most residents would consider advantageous.

Human Interactions with the Environment

Through their life activities, humans are becoming an increasing factor in changing Earth's surface. There are two major reasons for this: First,

Figure 23.13 *Humans change the environment to fit a particular need. An aqueduct (concrete-lined, artificial stream channel) transports water from the mountains to arid and semi-arid regions of southern California. Without this human-induced transfer of water, many arid regions could not support either a large population or intensive agriculture.* • **What environmental problems may be created through the transfer of water from wet regions to arid regions?**

our ability to affect Earth environments has increased over time with technological advances and with the growth in human population. Secondly, much environmental change is cumulative and/or not easily reversible. Given enough time the amount of total change increases. For example, once a place is urbanized, it will not likely revert back to a natural condition, at least for as long as humans maintain the artificial environment. There are few, if any, parts of the Earth system that have not been affected in some way by human-induced modification.

Today we understand that every environment presents both challenges and advantages to the organisms living there. *Humans have an impact on the environment and the environment also has an impact on humans.* Humans have an ability, although not an unlimited ability, to adapt the environments to fit their needs (Fig. 23.13), but humans will never completely control the environment. While some of the environmental problems that we face can be solved, humans must adapt to and live with the environment as they interact with the Earth system. The more we learn about the Earth system, the more we should appreciate that we must work within the "rules" that govern how natural systems operate or risk creating or aggravating environmental problems.

Scientists do not seek to impose their needs and wants on Nature, but instead humbly interrogate Nature and take seriously what they find.

The Demon-Haunted World
Carl Sagan

Physical Geography and You—Revisited

A pertinent question remains: How does knowledge of physical geography relate to me? The real answer lies in your nature as an individual and your personal interests, but you should consider several possibilities. The phrase "Geography for Life," used in the title of the Geography National Standards, is meant to convey the development of a lasting, lifelong awareness of and appreciation for geography as the spatial science. This awareness may grow into a career, or perhaps continue as a part of the broadening experience of a liberal education. In any case, physical geography is a part of our everyday life. An understanding of how the Earth system operates contributes to our appreciation of the planet and helps us to become responsible citizens.

In addition to those in teaching careers, many geographers are employed by federal, state and local government agencies, corporations, businesses, and consulting firms. Virtually any commercial or governmental activity that deals with maps, spatial information, and environmental data could benefit from the services of a well-trained geographer. Physical geographers offer skills and technologies for gathering data and information that are important to making informed decisions concerning all aspects of the environment. The Career Vision Series, featuring a selection of physical geographers who are involved in a variety of career fields, illustrates the point that expertise in physical geography, with its tools and methods, has much relevance in today's world.

At the end of the first chapter, it was stated that after learning about physical geography, you should "see Earth differently." Basically, this is a function of awareness and appreciation of the elements of the Earth system and their spatial, or *geographic,* nature. This awareness can also often come from really *examining* your environment and the elements that make up the landscape that you see (Fig. 23.14). For many people, this is an enriching experience and one that attracts them to physical geography whether their involvement is career-oriented or avocational.

The *Titanic* and Physical Geography

In Chapter 1 the location of the wreck of the *Titanic* was used as an example of absolute and relative location, which are two ways of expressing the position of any feature on our planet. In several ways the example of the *Titanic* provides also an excellent case study in the significance of physical geography. The Global Positioning System (GPS) played a key role in the pinpointing of the wreck's location by Dr. Robert Ballard's discovery team, which was supported by the National Geographic Society (Fig. 23.15). One problem was finding the ship's exact location given the featureless (in terms of landmarks) surface of the ocean's floor. Dr. Ballard said that finding the location of a ship on the bottom of the ocean under miles of seawater, using a tethered, robotic submarine-camera system, was like "lowering a penny on a thread from the top of the Empire State Building to its base and getting it to land on a toothpick." But once the wreck was located, the GPS guaranteed that the *Titanic* never would be lost again.

We recognize that location is the basic theme of physical geography, which leads us to ask why the wreck is located where it is, at a specific latitude and longitude on the ocean floor of the North Atlantic. Much of the story, of course, concerns human error—trying to cross the Atlantic

Figure 23.14 How have physical geographic and environmental factors influenced this landscape? Look at this photograph of an area in the Colorado Rockies. What aspects of this environment do you recognize in this landscape that make it look the way it does? • What can you explain about this photograph in terms of the Earth system and its four major subsystems? Can you examine this photo with the "educated eye" of the geographer that you learned about in Chapter 1?

Figure 23.15 Space age technology and equipment along with much searching was involved in Robert Ballard's expedition to find the wreck of the Titanic. *Global Positioning Systems played a key role in locating and mapping the site in detail. Ballard's expedition was supported by the National Geographic Society.*

Ocean in record time on the maiden voyage of a ship that was thought to be unsinkable and immune to the environmental conditions under which it would operate. This last error of calculation led to another tragic decision—specifically, that because of this "immunity to nature" adequate lifeboats were unnecessary. But there are also some important geographical factors to the story that deal not only with *location,* but also with the *physical characteristics* of place and other themes of physical geography.

First, we might look at a world map and ask why the *Titanic* met its fate in its voyage from England to New York City at the particular latitude it did. The answer is that the ship, in its attempt to set a record, was sailing a great circle route (Fig. 23.16), which is the shortest distance between two points on Earth. On many world maps, as we know, a great circle does not appear as a straight line, which we normally associate with the shortest distance, because of the distortion present in most of these maps, particularly in the high latitudes. This information yields examples of the importance of location and how locations are displayed on maps, and gives us a part of the answer to our question, Why there?

Secondly, we need to consider some physical characteristics of place. Everyone knows that the *Titanic* hit an iceberg, but why were icebergs there? Icebergs calve off of glaciers in Greenland to the north of where the *Titanic* met its fate, and drift southward, unfortunately into the heavily used shipping lane between Europe and North America. Since icebergs float, because ice is

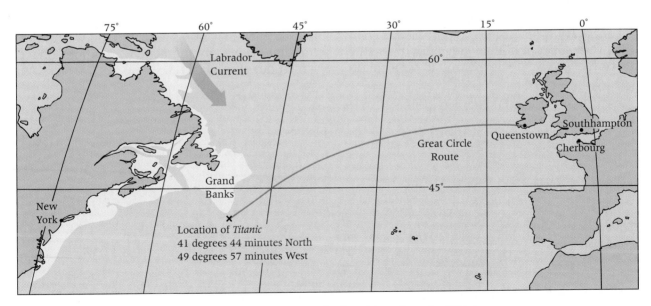

Figure 23.16 This map shows the latitudinal and longitudinal position where the Titanic rests on the North Atlantic Ocean floor, about 3810 meters (12,500 feet) below the surface. The great circle route is shown along with a few of the physical geographic factors that contributed to the demise of the ship, such as the Labrador current which delivers icebergs southward each spring into one of the busiest shipping lanes in the world.

slightly lighter than seawater, only a small part of them projects above the sea surface. The reason the iceberg was there in the path of the *Titanic* was that it was adrift in the Labrador Current, which flows southward between Greenland and Baffin Island. Icebergs are common in the waters off Newfoundland, Canada, and, since the *Titanic* sank, they are counted and monitored annually—250 per year is the average number. The icebergs eventually shrink in size, melting away as they encounter warmer waters to the south.

Another factor related to the physical characteristics of place is the climate of that part of the world—more specifically, the weather conditions on that fateful April night in 1912. The north Atlantic off the east coast of Canada includes a region called the Grand Banks, one of the foggiest places on Earth. The fog is advection fog, caused by the cooling of the warm, moist air that follows the Gulf Stream, a warm current flowing poleward along the east coast of North America. As this warm air moves over the cold water of the Labrador Current, dense fog forms, a condition that lasts much of the year in the Grand Banks and nearby parts of the north Atlantic. So, did the *Titanic* hit an iceberg that its crew could not see because of foggy conditions? Actually, no, because on that night the air was cold and clear; high pressure, with its typical calm conditions, dominated, the sea was smooth, and visibility should have been good. However, there was no moon that evening, making the clear skies and the calm ocean particularly dark.

These atmospheric and oceanic factors combined with errors of human judgment to seal the fate of the *Titanic*. If the moon had been full in the clear night, icebergs would have been easier to spot, perhaps early enough to avoid a collision. Unfortunately, the icebergs were obscured by the cover of darkness in a moonless sky, a situation compounded by the pressure on the crew to maintain speed. It is ironic that, if the night had been foggy—as it typically tends to be in the Atlantic near the Grand Banks—the *Titanic* probably would have had to reduce its speed, thus perhaps avoiding collision with the iceberg with enough force to sink the ship. So, not only did human error contribute to this notorious event, but the geographical factors associated with location, the physical characteristics of place, and the other themes of physical geography played key roles in the *Titanic*'s story.

Epilogue

As a part of a general education, knowledge of physical geography and how the Earth system operates can deepen our understanding of events that involve human interaction with the environment. Physical geography can also focus our awareness and understanding of the Earth system and how it operates as a function of strictly natural processes. In both cases, knowledge of our physical environment enriches our existence on planet Earth and helps us to be informed, active participants in our finite life-support system rather than just idle passengers on the planet. Unlike those on board the *Titanic*, we know that, as crew members of Spaceship Earth, we are not immune to the environmental conditions under which our ship operates.

Credits

Chapter 1
Opener NASA; **1.1** NASA Goddard Space Flight Center; **1.2 (a)** Olaf Soot/Tony Stone Images; **1.2 (b)** Schafer and Hill/Tony Stone Images; **1.2 (d)** Gary Braasch/Tony Stone Images; **1.3** Lee Malis/Gamma Liaison; **1.4 (a)** R. Gabler, **(b)** R. Sager; **1.5 (a, b)** U.S. National Park Service, Air Quality Division; **1.6 (a)** Oxford Cartographers/Tony Stone Worldwide, **(b)** WAGDA and the University of Washington Map Collection, **(c)** NASA, **(d)** EPA, South Florida Water Management Division; **1.9** NASA; **1.11** Donald Marshall Collection; **1.12 (a)** R. Gabler, **(b, c)** EPA, South Florida Water Management Division; **1.14** Kee Chang, Chicago Association of Commerce and Industry; **1.15** R. Gabler; **The Environment, p. 16** Martin Bond/Science Photo Library, **p. 17** John Freeman/Gamma Liaison

Chapter 2
Opener NASA; **2.1** © De Sazo/Photo Researchers, Inc.; **2.2 (a and b)** NASA; **2.6** R. Sager; **2.8** After U.S. Navy Oceanographic Office, No. 5192; **2.12** Grant Heilman/Grant Heilman Photography; **2.13** Magellan Systems Corporation; **2.19** Western Illinois University Cartography Laboratory; **2.24** After Hydrologic Chart 1706; **2.25** Courtesy William Westerhold, Illinois Institute for Rural Affairs; **2.26** USGS; **2.28** ESRI; **2.30** Courtesy ESRI; **2.31** Space Imaging; **2.32** U.S.D.A.; **2.33 (a, b)** USGS; **2.34** NASA; **2.35 (a)** courtesy of the NOAA, National Weather Service, Austin/San Antonio, Texas, **(b)** NASA; **2.36** U.S. Geologic Survey; **Map Interpretation p. 50** USGS, **p. 51** Aerial Eye, Inc., Irvine, California

Chapter 3
Opener J. Lotter Gurling/Tom Stack & Associates; **3.1** Jerry Schad/Science Source/Photo Researchers, Inc.; **3.3 (a–d)** NASA; **3.4 (a–e)** NASA; **3.5** NASA

Chapter 4
Opener R. Gabler; **4.17** R. Sager

Chapter 5
Opener © Laura Wight/Peter Arnold, Inc.

Chapter 6
Opener NASA/Peter Arnold, Inc.; **6.6** NASA

Chapter 7
Opener R. Gabler; **7.7** R. Sager; **7.11 (a)** Gregory Scott/Photo Researchers, Inc., **(b)** Steve McCutcheon/Visuals Unlimited, **(c) and (d)** Mark A. Schneider/Visuals Unlimited, **(e)** Bruce Berg/Visuals Unlimited, **(f)** John D. Cunningham/Visuals Unlimited, **(g)** Lincoln Nutting/Photo Researchers, Inc., **(h)** William J. Weber/Visuals Unlimited, **(i)** YVA Momatiuk and John Eastcott/Photo Researchers Inc., **(j)** Martin Miller/Visuals Unlimited, **(k)** Mark A. Schneider/Visuals Unlimited; **7.14 (b and c)** R. Gabler

Chapter 8
Opener R. Gabler; **8.1** After Trewartha; **8.10** Science VU/Visuals Unlimited; **8.11** Jan Halaska/Photo Researchers, Inc.; **8.13** National Weather Service, Davenport, IA; **8.15** Cameron Davidson/Tony Stone Images; **Map Interpretation, p. 198** NOAA

Chapter 9
Opener R. Gabler; **9.4** Norman Meek; **9.10** Courtesy Arizona State University Departments of Geography and Computer Science; **9.17 (a)** National Center for Atmospheric Research, **(b)** NASA/Goddard Institute for Space Studies, **(c)** United Kingdom Meteorological Office

Chapter 10
Opener R. Gabler; **10.1 (all)** R. Gabler; **10.3 (all)** R. Gabler; **10.7** R. Sager; **10.8** R. Gabler; **10.10** R. Gabler; **10.11 (a)** David Watanabe/Tony Stone Images, **(b)** H. Armstrong Roberts/Tony Stone Images; **10.13** R. Gabler; **10.14** Bildarchiv Okapia/Photo Researchers; **10.15** Donald Marshall Collection; **10.18–10.19** R. Gabler; **10.21** David Schultz/Tony Stone Images; **The Environment, p. 254** Dan Guarvich/Photo Researchers, **p. 255** Gregory Dimijian/Photo Researchers

Chapter 11
Opener R. Gabler; **11.3–11.5** R. Gabler; **11.7** R. Gabler; **11.8** Herman Frass/Tony Stone Images; **11.9** Norman Meek; **11.10** Kenneth Murray/Photo Researchers; **11.13** George Herben/Alaska Stock Images; **11.14–11.15** R. Gabler; **11.21** Courtesy Doug Miller; **11.22 (a,c,d)** R. Gabler, **(b)** © David Lorenz Winston; **11.23** R. Gabler; **11.25** GeorgGerster/Comstock; **11.27** R. Gabler; **11.30** Ron Levy/Gamma Liaison; **11.32** R. Gabler

Chapter 12

Opener R. Gabler; **12.1** R. Gabler; **12.2** R. Gabler; **12.10–12.13** R. Gabler; **12.14** David Butler; **12.17** R. Gabler; **12.18** Courtesy Sheila Brazier; **12.19 (all)** R. Gabler; **12.20** R. Gabler; **12.21** Donald Marshall Collection; **12.22** R. Gabler; **12.23** David Butler; **12.24** Peter Lamberti/Tony Stone Images; **12.25** Randy Brandon/Alaska Stock Images; **12.26–12.27** R. Gabler; **12.28** NASA; **The Environment (all)** R. Gabler

Chapter 13

Opener R. Gabler; **13.6 (a–h)** Marbut Memorial Slide Collection, courtesy Soil Science Society of America, **(i)** Larry Wilding, Soil and Crop Sciences Department, Texas A&M University, College Station, TX, **(j–l)** Marbut Memorial Slide Collection, courtesy Soil Science Society of America; **13.7** R. Gabler; **13.10** William E. Ferguson/William E. Ferguson Nature Photography; **13.12** Norman Meek; **13.13** Norman Meek; **13.14** R. Gabler; **13.16** R. Gabler; **13.17** Jerry Irwin/Photo Researchers

Chapter 14

Opener R. J. Sager; **14.3 (all)** R. Sager; **14.4** R. Sager; **14.5** R. Sager; **14.6 (a)** Courtesy Sheila Brazier, **(b)** R. Gabler; **14.7** R. Sager; **14.8** R. Sager; **14.9** R. Sager; **14.10** Courtesy Sheila Brazier; **14.11** R. Sager; **14.15** NOAA/National Geophysical Data Center, Boulder, CO; **14.17** NASA, Johnson Space Center, Houston, TX

Chapter 15

Opener Duncan Livingston/The News Tribune; **15.1** J. D. Griggs/USGS; **15.2 (a)** Courtesy of Sheila Brazier, **(b)** R. Sager; **15.3 (a)** Jim Hughes/U.S.D.A. Forest Service, **(b)** Austin Post/USGS, **(c)** C. Crisafulli; **15.4 (a and b)** USGS; **15.5** Alan L. Mayo/Geophoto Publishing Co.; **15.6** R. Sager; **15.7 (a)** Raven Maps and Images, **(b)** R. Sager; **15.9** Jeff Gnass; **15.10** Science Graphics, Inc./Ward's Natural Science Establishment, Inc.; **15.12** R. Sager; **15.13** Courtesy Sheila Brazier; **15.16** R. Sager; **15.17** R. Gabler; **15.19** R. Sager; **15.20** R. Gabler; **15.21** Kevin Schafer/Allstock; **15.22** Courtesy Sheila Brazier; **15.23 (a and b)**

Patrick Roberts/Sygma; **Map Interpretation pg. 404–405** USGS; **The Environment pg. 418** Bob Riha/Gamma Liason, **pg. 419** Les Stone/SYGMA

Chapter 16

Opener R.J. Sager; **16.3** Austin Post, USGS; **16.5** R. Sager; **16.6** R. Sager; **16.7** R. Gabler; **16.8** National Park Service; **16.9** B. Bradley, NOAA, National Geophysical Data Center; **16.10** R. Sager; **16.11** Sheila Brazier; **16.12-16.16** R. Sager; **16.18** NASA; **16.19** B. Bradley, NOAA, National Geophysical Data Center; **16.21** B. Bradley, NOAA, National Geophysical Data Center; **16.22** B. Bradley, NOAA, National Geophysical Data Center; **16.23** Courtesy of Dr. Gerald F. Wieczorek, USGS; **16.25** © Doug Morton, USGS; **The Environment pg. 439 (top)** T.J. Casadevall/USGS, **pg. 439 (bottom)** Austin Post, USGS

Chapter 17

Opener Jeffrey Alford/Asia Access; **17.09** © St. Petersburg Times; **17.10** Jeffrey Alford/Asia Access; **17.11** Tom Till Photography; **17.12** Thomas Brase/Tony Stone Images; **17.13** R. Gabler

Chapter 18

Opener R. Gabler; **18.1 (a)** National Air & Space Museum, **(b)** NASA; **18.3 (c)** J. Petersen; **18.5** Courtesy Sheila Brazier; **18.6** Courtesy Sheila Brazier; **18.8** Austin Post, USGS; **18.10** R. Sager; **18.11** R. Sager; **18.13** NASA, Jet Propulsion Laboratory; **18.14** Nick Decker, Missouri Department of Natural Resources; **18.16 (a)** Nigel Press/Tony Stone Images, **(b)** NASA, **(c)** NOAA; **18.17** Alan L. Mayo, GeoPhoto Publishing Co.; **18.18 (b)** Courtesy Sheila Brazier; **18.22** U.S. Army Corps of Engineers, Portland District; **18.23** R. Sager; **18.24** James Petersen; **18.25** NASA; **18.26** R. Sager; **18.27** R. Gabler; **18.29** R. Sager; **18.31** Alan L. Mayo, GeoPhoto Publishing Co.; **18.32** R. Sager; **18.33** Courtesy Sheila Brazier; **18.34** NASA, Aerospace Education Services Program (AESP); **18.36** David Ball/Allstock; **Map Interpretation pg. 483** NASA, **pg. 482** NASA

Chapter 19

Opener–19.3 Austin Post, U.S. Geologic Survey; **19.5** Austin Post, U.S. Geologic Survey; **19.6 (a and b)** R. Sager; **19.7** Austin Post, U.S. Geologic Survey; **19.9** Austin Post, USGS; **19.10 (a)** R. Gabler, **(b-d)** R. Sager; **19.13 (a)** Alan L. Mayo, GeoPhoto Publishing Company, **(b)** William Felger/Grant Heilman Photography; **19.14 (a)** NASA; **19.15** USGS/Science Photo Library/Photo Researchers, Inc.; **19.16** U.S. Coast Guard; **19.19** John S. Shelton; **19.21 (a)** R. Gabler, **(b)** Henry Kyllingstad/Photo Researchers, Inc.; **19.23** Spaceshots, Inc.; **Map Interpretation pg. 525** R. Sager, **pg. 524** NOAA

Chapter 20

Opener Robert Van Der Hilst/Tony Stone Images; **20.2** R. Sager; **20.3** Courtesy Sheila Brazier; **20.4** Robert J. Sager; **20.5 (a) and (b)** R. Sager; **20.7** Robert J. Sager; **20.8** David B. Loope, University of Nebraska, Lincoln; **20.10** Gary J. James/Biological Photo Service; **20.11** NASA; **20.12** © Jane Sanders; **20.13** © Michael Andrews/Earth Scenes; **20.14** Vicksburg Convention and Visitors Bureau; **20.15** Greg Probst/Tony Stone Images; **Map Interpretation pg. 540** Kalmback Publishing Company

Chapter 21

Opener Warren Bolster/Tony Stone Images; **21.1** Rod Catanach, Woods Hole Oceanographic Institution; **21.1 (a and b)** *The Floor of the Oceans Map,* © Marie Tharp; **21.3** Geological Data Center, Scripps Institution of Oceanography; **21.5** Friends of Monterey Bay, (408) 336-5342; **21.7** Courtesy Aluminum Company of America; **21.8** Ocean Drilling Program, Texas A&M University; **21.9** Sunbeam Instruments; **21.10** Robert J. Sager; **21.11** Rod Catanach, Woods Oceanographic Institution; **21.12 (a)** University of Hawaii, School of Ocean and Earth Science and Technology Transfer and Economic Development, **(b)** A. Malahoff, Hawaii Undersea Research Laboratory; **21.13 (a)** David Hiser/Tony Stone Images, **(b)** Paul Chesley/Tony Stone Images, **(c)** Tahiti Tourism Board; **21.16 (a and b)** Nova Scotia Tourism and Culture; **21.19** U.S. Coast Guard; **The**

Environment (a) and (b) NOAA, National Geophysical Data Center

Chapter 22
Opener R. Sager; **22.1** R. Sager; **22.4 (b), (c), (d), and (e)** R. Sager; **22.6** R. Sager; **22.7 (all)** R. Sager; **22.8** R. Sager; **22.9 (b–d)** R. Sager; **22.10** NASA; **22.11** Austin Post, USGS; **Map Interpretation pg. 588** NASA; **Map Interpretation pg. 589** NASA

Chapter 23
Opener Goddard Space Flight Center, NASA; **23.1** NASA; **23.2** NASA; **23.3** © Suzanne Duranceau; **23.4** NASA; **23.5** © Frans Lanting/Minden Pictures; **23.7** © 1998 Hertz System, Inc. Hertz is a registered service mark and trademark of Hertz System, Inc.; **23.8** Courtesy of Environmetnal Systems Research Institute, Inc.; **23.9** Courtesy of ESRI; **23.10** Nicholas Parfitt/Tony Stone Images; **23.11**

© 1994 Hansen Planetarium Publications, Salt Lake City, Utah, and W.T. Sullivan III, University of Washington, Seattle; **23.12** NASA, Johnson Space Center, Houston, TX; **23.13** Mark Wagner/Tony Stone Images; **23.14** Paul Chesley/Tony Stone Images; **23.15** Emory Kristof/National Geographic Image Collection

Appendix

Köppen Climate Classification

The key to understanding any system of classification is found by personally practicing use of the system. This is one reason why the Consider and Respond review sections of both Chapters 10 and 11 are based on the classification of data from sites selected throughout the world. Correctly classifying these sites in the modified Köppen system may seem complicated at first, but you will find that, after applying the system to a few of the sample locations, the determination of a correct letter symbol and associated climate name for any other site data should be routine. In addition, this will be the case whether the site data is in your textbook or provided by your instructor from some other source.

Before you begin, you should take the time to familiarize yourself with Table 1. You will note that there are precise definitions in regard to temperature or precipitation that identify a site as one of the five major climate categories in

TABLE 1 Simplified Köppen Classification of Climates

FIRST LETTER	SECOND LETTER	THIRD LETTER
E Warmest month less than 10°C (50°F) POLAR CLIMATES ET—Tundra EF—Ice-cap	T Warmest month between 10°C (50°F) and 0°C (32°F)	NO THIRD LETTER (with polar climates) SUMMERLESS
	F Warmest month below 0°C (32°F)	
B Arid or semiarid climates ARID CLIMATES BS—Steppe BW—Desert	S Semiarid climate (see Graph 1)	h Mean annual temperature greater than 18°C (64.4°F)
	W Arid climate (See Graph 1)	k Mean annual temperature less than 18°C (64.4°F)

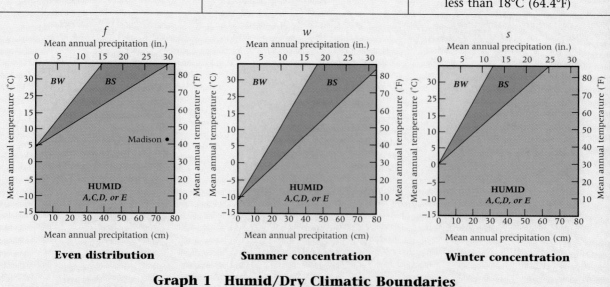

Even distribution · Summer concentration · Winter concentration

Graph 1 Humid/Dry Climatic Boundaries

the Köppen system (A, Tropical; B, Arid; C, Mesothermal; D, Microthermal; and E, Polar). Furthermore, you will note that the additional letters, which are required to complete the letter symbol and identify the actual climate type, also have precise definitions or are determined by the use of graphs that distinguish one climate type from another. In other words, Table 1 is all you need to classify a site if monthly and annual means of precipitation and temperature are available. Table 1 should be used in a systematic fashion to determine first the major climate category and, once that is determined, the second and third letter symbols (if needed) that complete the classification. As you begin to classify, it is strongly recommended that you use the following procedure. (After a few examples you may find you can omit some steps with a glance at the statistics.)

Step 1. Ask: Is this a polar climate (E)? Is the warmest month less than 10°C (50°F)? If so, is the warmest month between 10°C (50°F) and 0°C (32°F) (ET) or below 0°C (32°F) (EF)? If not, move on to:

Step 2. Ask: Is there a seasonal concentration of precipitation? The answer is determined by examining the monthly precipitation data for the driest and wettest summer and winter months for the site. Take careful note of temperature data as well, because you must take into consideration whether the site is located in the Northern or Southern Hemisphere. (April to September are summer months in the Northern Hemisphere but winter months in the Southern Hemisphere. Similarly, the Northern Hemisphere winter months of October to March are summer south of the equator.) As the table indicates, a site has a

TABLE 1   Simplified Köppen Classification of Climates (continued)

FIRST LETTER	SECOND LETTER	THIRD LETTER
A Coolest month greater than 18°C (64.4°F) TROPICAL CLIMATES Am—Tropical monsoon Aw—Tropical savanna Af—Tropical rainforest	f Driest month has at least 6 cm (2.4 in.) of precipitation m Seasonally, excessively moist (see Graph 2) w Dry winter, wet summer (see Graph 2)	NO THIRD LETTER (with tropical climates) WINTERLESS

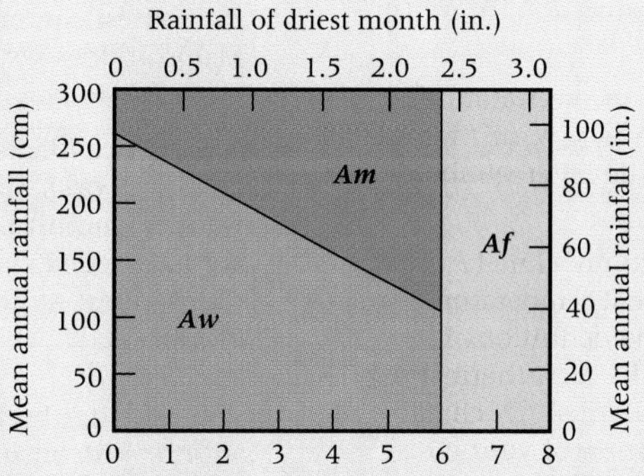

Graph 2   Rainfall of driest month (cm)

C Coldest month between 18°C (64.4°F) and 0°C (32°F); at least one month over 10°C (50°F) MESOTHERMAL CLIMATES Csa, Csb—Mediterranean Cfa, Cwa—Humid Subtropical Cfb, Cfc—Marine West Coast	s (DRY SUMMER) Driest month in the summer half of the year, with less than 3 cm (1.2 in.) of precipitation and less than 1/3 of the wettest winter month	a Warmest month above 22°C (71.6°F)
	w (DRY WINTER) Driest month in the winter half of the year, with less than 1/10 the precipitation of the wettest summer month	b Warmest month below 22°C (71.6°F), with at least four months above 10°C (50°F)
D Coldest month less than 0°C (32°F); at least one month over 10°C (50°F) MICROTHERMAL CLIMATES Dfa, Dwa—Humid Continental, Hot Summer Dfb, Dwb—Humid Continental, Mild Summer Dfc, Dwc, Dfc, Dwc—Subarctic		c Warmest month below 22°C (71.6°F), with one to three months above 10°C (50°F)
	f (ALWAYS MOIST) Does not meet conditions for s or w above	d Same as c, but coldest month is below −38°C (−36.4°F)

dry summer *(s)* if the driest month in summer has less than 3 cm (1.2 in.) of precipitation and less than 1/3 of the precipitation of the wettest winter month. It has a dry winter *(w)* if the driest month in winter has less than 1/10 the precipitation of the wettest summer month. If the site has neither a dry summer nor a dry winter, it is classified as having an even distribution of precipitation *(f)*. Move on to:

Step 3:   Ask: Is this an arid climate *(B)*? You must use one of the small graphs (included in Graph 1) to decide. Based on your answer in Step 2, select one of the small graphs and compare mean annual temperature with mean annual precipitation. The graph will indicate whether the site is an arid *(B)* climate or not. If it is, the graph will indicate which one *(BW* or *BS)*. You should further classify the site by adding *h* if the mean annual temperature is above 18°C (64.4°F) and *k* if it is below. If the site is neither *BW* nor *BS*, it is a humid climate *(A, C,* or *D)*. Move on to:

Step 4a.   Ask: Is this a tropical climate *(A)*? The site has a tropical climate if the temperature of the coolest month is higher than 18°C (64.4°F). If so, use Graph 2 in Table 1 to determine which tropical climate the site represents. (Note that there are no additional lowercase letters required.) If not, move on to:

Step 4b.   Ask: Which major middle-latitude climate group does this site represent, mesothermal *(C)* or microthermal *(D)*? If the temperature of the coldest month is between 18°C (64.4°F) and 0°C (32°F), the site has a mesothermal climate. If it is below 0°C (32°F), it has a microthermal climate. Once you have answered the question, move on to:

Step 5.   Ask: What was the distribution of precipitation? This was determined back in Step 2. Add *s, w,* or *f* for a *C* climate, or *w* or *f* for a *D* climate, to the letter symbol for the climate. Then, move on to:

Step 6.   Ask: What letter should I add to my climate symbol to express the details of seasonal temperature for the site? Refer again to Table 1 and the definitions for the letter symbols. Add *a, b,* or *c* for the mesothermal *(C)* climates or *a, b, c,* or *d* for the microthermal *(D)* climates and you have completed the classification of your climate. However, note that you may not have come this far because you might have completed your classification at Steps 1, 3, or 4a.

We should now be ready to try out the use of Table 1, following the steps we have recommended. Data for Madison, Wisconsin, is presented below for our example.

Step 1.   First, we must determine whether or not our site has an *E* climate. Since Madison has several months averaging above 10°C, it does not have an *E* climate.

Step 2.   Second, we must determine if there is a seasonal concentration of precipitation. Since Madison is driest in winter, we compare the 2.5 cm of February precipitation with the precipitation of June (1/10 of 11.0 cm, or 1.1 cm) and conclude that Madison has neither a dry summer nor a dry winter, but instead has an even distribution of precipitation *(f)*. (Note: The 2.5 cm of February precipitation is not less than 1/10 (1.1 cm) of June precipitation.

Step 3.   Next we assess, through the use of Graphs 1*(f)*, 1*(w)*, or 1*(s)*, whether our site is an arid climate *(BW, BS)* or a humid climate *(A, C,* or *D)*. Since we have previously determined that Madison has an even distribution of precipitation, we will use Graph 1*(f)*. Based upon Madison's mean annual precipitation (77.0 cm) and mean annual temperature (7°C), we conclude that Madison is a humid climate *(A, C,* or *D)*.

Step 4.   Now we must assess which humid climate type Madison falls under. Since the coldest month (−8°C) is below 18°C, Madison does *not* have an *A* climate. Although the warmest month (21°C) is above 10°C, the coldest month (−8°C) is not between 0° and 18°C, so Madison does *not* have a *C* climate. Since the warmest month (21°C) is above 10°C, and the coldest month is below 0°C, Madison *does* have a *D* climate.

Step 5.   Since Madison has a *D* climate, the second letter will be *w* or *f*. Since precipitation in the driest month of winter (2.5 cm) is not less than 1/10 of the amount of the wettest summer month (1/10 × 11.0 cm = 1.1 cm), Madison does *not* have a *Dw* climate. Madison, therefore, has a *Df* climate.

Step 6.   Since Madison is a *Df* climate, the third letter will be *a, b, c,* or *d*. Since the average temperature of the warmest month (21°C) is not above 22°C, Madison does *not* have a *Dfa* climate. Since the average temperature of the warmest month is below 22°C, with at least 4 months above 10°C, Madison is a *Dfb* climate.

If you have not been able to follow the explanation for why Madison, Wisconsin, is classified *Dfb* (a site with a humid continental, mild summer climate), try the example again using only Table 1 before reading again the explanation at each step in the procedure. If you still are having difficulty, do not hesitate to ask your instructor for assistance. Stick with it; classification can be fun once you begin getting the correct answers. In addition, you will be better prepared for the real challenge of explaining why a site has a particular climate.

	J	F	M	A	M	J	J	A	S	O	N	D	Year
T(°C)	−8	−7	−1	7	13	19	21	21	16	10	2	−6	7
P(cm)	3.3	2.5	4.8	6.9	8.6	11.0	9.6	7.9	8.6	5.6	4.8	3.8	77.0

Glossary

ablation loss of glacial ice and snow cover by means of melting, evaporation, and sublimation.

abrasion (corrasion) erosion of a stream by the grinding and rolling of rock particles and boulders carried by a stream, or by wave action, wind, or glacial ice.

absolute humidity mass of water vapor present per unit volume of air, expressed as grams per cubic meter, or grains per cubic foot.

abyssal plains deep low-relief ocean floor, generally covered with a blanket of marine sediments.

accretion the growth of continents by the adding of large pieces of their borders through collision.

acid rain rain with a pH value of less than 5.6, the pH of natural rain; often linked to the pollution associated with the burning of fossil fuels.

adiabatic heating and cooling change of temperature *within* a gas because of compression (resulting in heating) or expansion (resulting in cooling); no heat is added or subtracted from outside.

advection horizontal heat transfer within the atmosphere; air masses moved horizontally, usually by wind.

advection fog fog produced by the movement of warm moist air across a cold sea or land surface.

air mass large portion of the atmosphere, sometimes subcontinental in size, that may move over Earth's surface as a distinct, relatively homogeneous entity.

air mass analysis explanation of weather phenomena by a study of the actions and interactions of major portions of the atmosphere.

albedo proportion of solar radiation reflected back from a surface, expressed as a percentage of radiation received on that surface.

Aleutian low center of low atmospheric pressure in the area of the Aleutian Islands, especially persistent in winter.

alluvial fan fan-shaped aggradational feature where a stream emerges from a mountain channel onto a flat plain and deposits material, generally found in arid regions.

alluvium fragmented Earth materials deposited by a river or stream.

alpine glacier moving glacial ice accumulated in high sheltered mountain valleys; also called *valley glacier*.

altithermal an interval of time about 7,000 years ago when the climate was hotter than it is today.

altitude heights of points above Earth's surface.

analemma a diagram that shows the declination of the sun throughout the year.

angle of inclination tilt of Earth's polar axis at an angle of $23\frac{1}{2}°$ from the vertical to the plane of the ecliptic.

angle of repose (rest) angle of slope created by a bed of loose sand, gravel, or rock.

annual range of temperature difference between the mean daily temperatures for the warmest and coolest months of the year.

Antarctic circle parallel of latitude at $66\frac{1}{2}°$S; the northern limit of the zone in the Southern Hemisphere that experiences a 24-hour period of sunlight and a 24-hour period of darkness at least once a year.

anticline upfold in a wave of crustal folding.

anticyclone an area of high atmospheric pressure, also known as a *high*.

aphelion position of Earth's orbit at farthest distance from the sun during each earth revolution.

aquiclude rock layer that restricts flow and storage of groundwater; it is impermeable and nonporous.

aquifer rock layer that is a container and transmitter of groundwater; it is both porous and permeable.

Arctic circle parallel of latitude at $66\frac{1}{2}°$N; the southern limit of the zone in the Northern Hemisphere that experiences a 24-hour period of sunlight and a 24-hour period of darkness at least once a year.

arête jagged, sawtooth spine or wall of rock separating two expanding cirque basins.

artesian well groundwater that flows toward the surface under its own pressure.

asthenosphere thick, plastic layer within Earth's mantle that theoretically flows in response to convection and instigates the surface movement of tectonic plates.

atmosphere blanket of air, composed of various gases, that envelops Earth.

atmospheric air pressure (barometric pressure) force per unit area that the atmosphere exerts on any surface at a particular elevation.

atmospheric controls geographic features that affect climate and weather patterns; e.g., distance from the ocean, wind direction, altitude.

atmospheric disturbance refers to variation in

the secondary circulation of the atmosphere that cannot correctly be classified as a storm; e.g., front, air mass.

atmospheric effect the absorption of long-wave Earth radiation by water vapor, carbon dioxide, and dust in the atmosphere so that Earth temperatures are moderated.

atmospheric elements components of weather; e.g., temperature, precipitation, pressure, wind.

atoll ring of coral reefs and islands encircling a lagoon, with no inner island.

autotroph organism which, because it is capable of photosynthesis, is at the foundation of a food web and is considered a basic producer.

axis an imaginary line between the geographic North Pole and South Pole, around which the planet rotates.

Azores high *see* Bermuda high.

backing wind shift change in wind direction counterclockwise around the compass; e.g., from east to northeast, to north, to northwest.

badlands region of rugged, barren topography with sharp ridges and ravines; caused by gully erosion of soft materials.

bajada continuous series of alluvial fans forming a gently sloping, low-relief area along the base of a mountain range.

barchan crescent-shaped sand dune with tips that point downwind.

barometer instrument for measuring atmospheric pressure.

barrier island long, narrow, wave-built island separated from the mainland by a lagoon, formed on low-relief coastlines.

basalt a dark-colored fine-grained extrusive igneous rock generally associated with the oceanic crust and oceanic volcanoes.

base level elevation below which a river or stream cannot erode; although sea level is the ultimate base level, basins or lakes may be local base levels.

batholith largest of the deep-seated igneous masses generally known as plutons.

beach coastal region of unconsolidated sediments between the low tide line and the upper limit of wave action.

bedding plain boundary between different sedimentary strata marking a change in deposits.

bedrock solid rock of Earth's crust that underlies soil and other unconsolidated materials.

benthos the ocean bottom and the plants and animals that live on the sea floor.

Bermuda high persistent, high atmospheric pressure center located in the subtropics of the north Atlantic Ocean.

biomass amount of living material or standing crop in an ecosystem or at a particular trophic level within an ecosystem.

biome one of Earth's major terrestrial ecosystems, classified by the vegetation types that dominate the plant communities within the ecosystem.

biosphere the life-forms, human, animal, or plant, of Earth that form one of the major Earth subsystems.

bolson desert basin, surrounded by mountains, with no drainage outlet.

bora cold downslope wind in Yugoslavia (*see* katabatic wind).

boreal forest (taiga) coniferous forest dominated by spruce, fir, and pine found growing in subarctic conditions around the world north of the 50th parallel of latitude.

braided stream stream channel with multiple subchannels that form a braided pattern flowing through alluvial deposits.

butte isolated erosional remnant of a tableland with a flat summit, often bordered by steep-sided escarpments. Buttes are usually found in arid regions of flat-lying sediments and are smaller than mesas.

calcification soil-forming process of subhumid and semiarid climates. Soil types in the mollisol order, the typical end products of the process, are characterized by little leaching or eluviation and by the accumulation of both humus and mineral bases (especially calcium carbonate, $CaCO_3$).

caldera collapsed summit area of a stratovolcano, thought to be caused by the expulsion or withdrawal of supporting magma.

calíche hardened layers of lime ($CaCO_3$) deposited at the surface of a soil by evaporating capillary water.

calving the formation of icebergs by a mass of ice breaking away from the snout of a glacier at its junction with the sea or a lake.

campos region of characteristic tropical savanna vegetation in Brazil, located primarily in the Amazon Basin bordering the tropical rainforest.

Canadian high high atmospheric pressure area that tends to develop over the central North American continent in winter.

capacity the maximum amount of water vapor that can be contained in a given quantity of air at a given temperature.

capillary water soil water that clings to soil peds and individual soil particles as a result of surface tension. Capillary water moves in all directions through the soil from areas of surplus water to areas of deficit.

carbonate a mineral group characterized by carbon's ability to form complex compounds of organic and inorganic origins.

carnivore animal that eats only other animals.

cartography the science of map-making.

catastrophism once-popular theory that all Earth's landforms developed in a relatively short time in a catastrophic fashion.

Celsius (or centigrade) scale temperature scale in which 0° is the freezing point of water and 100° its boiling point at standard sea level pressure.

centrifugal force force that pulls a rotating object away from the center of rotation.

chaparral sclerophyllous woodland vegetation found growing in the Mediterranean climate of the western United States; these seasonal, drought-resistant plants are low-growing, with small, hard-surfaced leaves and deep, water-probing roots.

chinook dry warm wind on the eastern slopes of the Rocky Mountains (see foehn wind).

cinder cone volcano formed primarily from the expulsion of cinders, ash, and other solid rock fragments.

circle of illumination line dividing the sunlit (day) hemisphere from the shaded (night) hemisphere; experienced by individuals on Earth's surface as sunrise and/or sunset.

cirque deep, sometimes steep-sided amphitheater formed at the head of an alpine valley by glacial ice erosion.

cirque glacier glacial ice limited to a cirque basin and not entering the alpine valley itself.

cirrus high, detached clouds consisting of ice particles. Cirrus clouds are white and feathery or fibrous in appearance.

classification process of systematically arranging phenomena into groups, classes, or categories based on some established criteria.

clastic rock sedimentary rock formed by the compaction and cementation of pre-existing rock debris.

climate accumulated and averaged weather patterns of a locality or region; the full description is based upon long-term statistics and includes extremes or deviations from the norm.

climatology scientific study of climates of Earth and their distribution.

climax community the final step in the succession of plant communities that occupy a specific location.

climograph graphic means of giving information on mean monthly temperature and rainfall for a select location or station.

closed system system in which no substantial amount of energy and/or materials can cross its boundaries.

cloud mass of suspended water droplets (or at high altitudes, ice particles) in air above ground level.

cold front leading edge of a relatively cooler, denser air mass that advances upon a warmer, less dense air mass.

comet a small body of icy and dusty matter that revolves about the sun. When a comet comes near the sun, some of its material vaporizes, forming a large head and often a tail.

composite cone (stratovolcano) volcano formed from alternating layers of lava and pyroclastic materials generally known for violent eruptions.

condensation process by which a vapor is converted to a liquid during which energy is released in the form of latent heat.

condensation nuclei minute particles in the atmosphere (e.g., dust, smoke, pollen, sea salt) on which condensation can take place.

conduction transfer of heat within a body or between adjacent matter by means of internal molecular movement.

conformal map projection a map projection that maintains the true shape of small areas on Earth's surface.

connate water groundwater trapped in the pore spaces of sedimentary rock at the time it was first deposited; water locked out of the hydrologic cycle in sedimentary rocks.

continental crust the less dense (av. 2.7 gm/cm^3) portion of Earth's crust that underlies all the continents.

Continental Divide line of separation dividing runoff between the Pacific and Atlantic Oceans. In North America it generally follows the crest of the Rocky Mountains.

continental drift theory proposed by Alfred Wegener stating that the continents joined, broke apart, and moved on Earth's surface, it was later replaced by the theory of plate tectonics.

continental ice sheet thick ice mass that covers a major portion of a continent and buries all but the highest mountain peaks; it usually flows from one or more areas of accumulation outward in all directions.

continental islands islands that are geologically part of a continent and are usually located on the continental shelf.

continentality the distance a particular place is located in respect to a large body of water: the greater the distance, the greater the continentality.

continental rise gently sloping depositional surface at the base of the continental slope.

continental shelf gently sloping submarine surface extending from the coast to the steep continental slope.

continental shields ancient crystalline rock cores of the continents.

continental slope steeply sloping submarine surface that is seaward of the continental shelf.

contour interval vertical distance represented by two adjacent contour lines on a topographic map.

contour line line on a map connecting points that are the same elevation above mean sea level.

contour map (topographic map) map that uses contour lines to show differences in elevation (topography).

convection process by which a circulation is produced within an air mass or fluid body (heated material rises, cooled material sinks); also, in tectonic plate theory, the method whereby heat is transferred to Earth's surface from deep within the mantle.

convectional precipitation precipitation resulting from condensation of water vapor in an air mass that is rising convectionally as it is heated from below.

convergent wind circulation pressure-and-wind system where the airflow is inward toward the center, where pressure is lowest.

coral reef ridge of limestone built up by accumulation of skeletal remains of tiny sea animals.

core extremely hot and dense, innermost portion of Earth's interior; the molten outer core is 2400 km (1500 mi) thick; the solid inner core is 1120 km (700 mi) thick.

Coriolis effect effect of Earth's rotation on horizontally moving bodies, such as wind and ocean currents; such bodies tend to be deflected to the right in the Northern Hemisphere and to the left in the Southern Hemisphere.

corridor a relatively linear feature cutting across a mosaic (ecosystem supporting a particular plant community) caused by nature (a stream) or by humans (a road or powerline).

coulee snaking, steep-sided channel cut through lava formations by glacial meltwater.

creep slow downslope movement of soil and regolith caused by the pull of gravity, also refers to slow fault zone displacement.

crevasse stress crack commonly found along the margins and at the snout or terminus of a glacier.

crust relatively thin, approximately 8–64 km (5–40 mi) deep, low-density surface layer of Earth.

cumulus globular clouds, usually with a horizontal base and strong vertical development.

cyclone center of low atmospheric pressure, also known as a *low*.

cyclonic precipitation *see* frontal precipitation.

debris flow channelled movement of Earth material mixed with water usually following drainage; a *mud-flow* is a rapidly moving debris flow with the consistency of mud.

decomposer organism that promotes decay by feeding on dead plant and animal material and returns mineral nutrients to the soil or water in a form that plants can utilize.

deflation surface erosion and removal of fine Earth materials by the wind.

delta depositional landform where a river flows into a still body of water, such as a sea or lake.

dendritic term used to describe a drainage pattern that is treelike with tributaries joining the main stream at acute angles.

deposition accumulation of Earth materials at a new site after they have been dropped by the transporting agents: water, wind, or glacial ice.

desert pavement desert surface accumulation of pebbles and stones, the finer materials having been removed by wind and/or water erosion.

detritivore animal that feeds on dead plant and animal material.

dew tiny droplets of water on ground surfaces, grass blades, or solid objects. Dew is formed by condensation when air at the surface reaches the dew point.

dew point the temperature at which an air mass becomes saturated; any further cooling will cause condensation of water vapor in the air.

differential weathering (and erosion) process whereby different types of rock weather (and erode) at varying speeds due to differing resistance to the weathering (and erosion) processes; such differing resistance often produces distinctive landform features.

dike igneous intrusion that forms a vertical rock mass after molten material has been forced through a crustal fracture and cools at right angles to flat-lying rock layers.

dip the angle that a stratum of rocks or a fault makes with the horizontal plane.

discharge (stream discharge) rate of stream flow; measured as the volume of water flowing past a cross section of a stream per unit of time (cubic meters or cubic feet per second).

distributary branching stream that flows away from the main stream, common on deltas; the opposite of a tributary.

diurnal (daily) range of temperature difference between the highest and lowest temperatures of the day (usually recorded hourly).

divergent wind circulation pressure-and-wind system where the airflow is outward away from the center, where pressure is highest.

divide line of separation between drainage basins; generally follows high ground or ridge lines.

doldrums zone of low pressure and calms along the equator.

drainage basin (watershed) total land surface area drained by a stream system.

drainage wind *see* katabatic wind.

drift all material deposited by glacier; includes both unsorted and unstratified material and sorted debris deposited by meltwater.

drizzle fine mist or haze of very small water droplets with a barely perceptible falling motion.

drumlin streamlined, elongated hill composed of glacial drift. Drumlins are usually found in swarms, with as many as 100 or more clustered together; their elongated shapes indicate the direction of ice flow.

dry adiabatic rate rate at which a rising mass of air is cooled by expansion when no condensation is occurring (10°C/1000 m or 5.6°F/1000 ft).

dune (sand dune) mound of sand-sized materials deposited and shaped by the wind.

dynamic equilibrium constantly changing relationship among the variables of a system, which produces a balance between the amounts of energy and/or materials that enter a system and the amounts that leave.

earthflow linear movement downslope of moist, clay-rich soil and regolith, usually exhibiting a tongue-like shape.

earthquake series of vibrations or shock waves set in motion by sudden movement along a fault.

Earth system set of interrelated components or variables (e.g., atmosphere, lithosphere, biosphere, hydrosphere), which interact and function together to make up Earth as it is currently constituted.

easterly wave trough-shaped, weak, low-pressure cell that progresses slowly from east to west in the tradewind belt of the tropics; this type of disturbance sometimes develops into a tropical hurricane.

eccentricity cycle the change in Earth's orbit from slightly elliptical to more circular, and back to its earlier shape every 100,000 years.

ecological niche combination of role and habitat as represented by a particular species in an ecosystem.

ecology science that studies the interactions between organisms and their environment.

ecosystem community of organisms functioning together in an interdependent relationship with the environment which they occupy.

ecotone transition zone of varied natural vegetation occupying the boundary between two adjacent and differing plant communities.

effective precipitation actual precipitation available to supply plants and soil with usable moisture; does not take into consideration storm runoff or evaporation.

El Niño warm countercurrent that influences the central and eastern Pacific.

elevation vertical distance from mean sea level to a point or object on Earth's surface.

ellipsoid of rotation a rotating, near-sphere with an elliptical (oval-shaped), rather than pure circular, plane or cross section.

eluviation removal by gravitational water of fine soil components from the surface layer (*A* horizon) of the soil.

empirical classification classification process based on statistical, physical, or observable characteristics of phenomena; it ignores the causes or theory behind their occurrence.

end moraine accumulation of rocks and fine glacial material at the terminus or snout of a glacier.

environment surroundings, whether of man or of any other living organism; includes physical, social, and cultural conditions that affect the development of that organism.

eolian (aeolian) referring to the work of wind; associated with wind erosion, transportation, and deposition.

epicenter point on Earth's surface directly above the focus of an earthquake.

epipedon surface soil layer that possesses specific characteristics essential to the identification of soils in the National Resources Conservation Service System (Examples of epipedons may be found in Table 13.1.)

equal-area map projection a map projection on which any given areas of Earth's surface are shown in correct proportional sizes on the map.

equator great circle of Earth midway between the poles; the zero degree parallel of latitude that divides Earth into the Northern and Southern Hemispheres.

equatorial low zone of low atmospheric pressure centered more or less over the equator where heated air is rising. (*See also* doldrums.)

equilibrium state of balance between the interconnected components of an organized whole.

equinox one of two times each year (approximately March 21 and September 23) when the position of the noon sun is overhead (and its vertical rays strike) at the equator; all over Earth, day and night are of equal length.

erg desert region of active sand dunes, most common in the Sahara.

erosion removal of Earth materials by water, wind, or glacial ice.

erratic large rock or boulder transported and deposited by a glacier above bedrock of different composition.

esker narrow, winding ridge composed of glaciofluvial gravels; believed to have been formed by streams of meltwater flowing in tunnels of a stagnant ice sheet, or on a melting glacial surface.

estuary coastal waters where salt and fresh water mix.

evaporation process by which a liquid is converted to the gaseous (or vapor) state by the addition of latent heat.

evaporite mineral salts that are soluble in water and accumulate when water evaporates.

evapotranspiration combined water loss to the atmosphere from ground and water surfaces by evaporation and, from plants, by transpiration.

exfoliation progressive breaking off of concentric slabs or sheets from the exposed portions of massive rocks due to weathering.

exotic stream (or river) stream that originates in a humid region and has sufficient water volume to flow across a desert region.

extratropical disturbance convergence of cold polar and warm subtropical air masses over the middle latitudes.

extrusive rock igneous rock that was erupted and solidified on Earth's surface.

Fahrenheit scale temperature scale in which 32° is the freezing point of water, and 212° its boiling point, at standard sea level pressure.

faulting movement of adjacent crustal blocks along joints, or fracture planes, in bedrock.

fault scarp (escarpment) the steep cliff or exposed face of a fault where one crustal block has been displaced vertically relative to another.

feedback sequence of changes in the elements of a system, which ultimately affects the element that was initially altered to begin the sequence.

fetch distance over open water that winds blow without interruption.

firn compact granular snow formed by partial melting and refreezing due to overlying layers of snow.

firn line boundary between the zones of ablation and accumulation on a glacier, representing the equilibrium point between net snowfall and ablation.

fissure an extensive crack or break in rocks which may allow lava to be extruded.

fjord deep, glacial trough along the coast invaded by the sea after the removal of the glacier.

fluvial term used to describe landform processes associated with the work of streams and rivers.

focus point within Earth's crust where an earthquake originates.

foehn wind warm, dry, downslope wind on lee of

mountain range, caused by adiabatic heating of descending air.

fog mass of suspended water droplets within the atmosphere that is in contact with the ground.

folding the wrinkling of the Earth's crust due to compressional forces.

foliation process whereby metamorphic rocks tend to develop parallel banding or platy structures during formation.

front sloping boundary or contact surface between air masses with different properties of temperature, moisture content, density, and atmospheric pressure.

frontal lifting lifting or rising of warmer, lighter air above cooler, denser air along a frontal boundary.

frontal precipitation precipitation resulting from condensation of water vapor in an air mass that is rising over another mass along a front.

frost frozen condensation that occurs when air at ground level is cooled to a dew point of 0°C (32°F) or below; also any temperature near or below freezing that threatens sensitive plants.

frost wedging breaking apart of bedrock by the expansive power of water freezing, melting, and refreezing in joints, cracks, and crevices.

galactic movement movement of the solar system within the Milky Way galaxy.

galaxy a large assemblage of stars; a typical galaxy contains millions to hundreds of billions of stars.

galeria forest jungle-like vegetation extending along and over streams in tropical forest regions.

gap an area within the territory occupied by a plant community when the climax vegetation has been destroyed or damaged by some natural process, such as a hurricane, forest fire, or landslide.

General Circulation Model (GCM) complex computer simulations based on the relationships of selected variables within the Earth system that are used in attempts to predict future climates.

genetic classification classification process based on the causes, theory, or origins of phenomena; generally ignoring their statistical, physical, or observable characteristics.

Geographic Information Systems complex computer programs that combine the features of automated (computer) cartography and database management to produce new data to solve spatial problems.

geography study of Earth phenomena; includes an analysis of distributional patterns and interrelationships among these phenomena.

geomorphology the study of the origin and development of landforms.

geostrophic winds upper-level winds in which the Coriolis effect and pressure gradient are balanced, resulting in a wind flowing parallel to the isobars.

giant planets the four largest planets—Jupiter, Saturn, Uranus, and Neptune.

glaciofluvial deposit sorted glacial drift deposited by meltwater.

glaciolacustrine deposit sorted glacial drift deposited by meltwater in lakes associated with the margins of glaciers.

glaze translucent coating of ice that develops when rain strikes a freezing surface.

gleization soil-forming process of poorly drained areas in cold, wet climates. The resulting soils have a heavy surface layer of humus with a water-saturated clay horizon directly beneath.

Global Positioning System (GPS) GPS uses satellites and computers to compute positions anywhere on Earth to within a few centimeters of their true location.

gnomonic projection planar projection with greatly distorted land and water areas; valuable for navigation because all great circles on the projection appear as straight lines.

graben depressed landform or crustal trough that develops when the crust between two parallel faults is lowered relative to blocks on either side.

gradational processes processes that derive their energy indirectly from the sun and directly from Earth gravitation and serve to wear down, fill in, and level off Earth's surface.

graded stream stream where slope and channel size provide velocity just sufficient to transport the load supplied by the drainage basin; a theoretical balanced state averaged over a period of many years.

granite a coarse-grained intrusive igneous rock generally associated with continental crust.

gravitational water meteoric water that passes through the soil under the influence of gravitation.

gravity the mutual attraction of bodies or particles.

great circle any circle formed by a full circumference of the globe; the plane of a great circle passes through the center of the globe.

greenhouse effect warming of the atmosphere that occurs because short-wave solar radiation heats the planet's surface, but the loss of long-wave heat radiation is hindered by the release of gases associated with human activity (e.g., CO_2).

Greenwich mean time (G.M.T.) time at zero degrees longitude used as the base time for Earth's 24 time zones; also called Universal Time or Zulu Time.

ground-inversion fog see radiation fog.

ground moraine glacial till deposited on Earth's surface beneath a melting glacier.

groundwater (underground water) all subsurface water, especially in the zone of saturation.

guyot flat-topped seamount thought to be formed by the slow subsidence of a volcanic island.

habitat location within an ecosystem occupied by a particular organism.

hail form of precipitation consisting of pellets or balls of ice with a concentric layered structure usually associated with the strong convection of cumulonimbus clouds.

hamada desert plains covered with boulders or featuring large expanses of exposed bedrock.

hanging valley tributary trough that enters a main glaciated valley at a level high above the valley floor.

hardpan dense, compacted, clay-rich layer occasionally found in the subsoil (*B* horizon) that is an end product of excessive illuviation.

heat energy budget relationship between solar energy input, storage, and output within the Earth system.

heat island mass of warmer air overlying urban areas.

herbivore animal that eats only living plant material.

heterotroph organism that is incapable of producing its own food and that must survive by consuming other organisms.

high *see* anticyclone.

Holocene the most recent time interval of warm, relatively stable climate that began with the retreat of major glaciers about 10,000 years ago.

horizon the visual boundary between Earth and sky; *see also* soil horizon.

horn pyramid-like peak created where three or more expanding cirques meet at a mountain summit.

horst raised landform that develops when the crust between two parallel faults is uplifted relative to blocks on either side.

humidity amount of water vapor in an air mass at a given time.

humus organic matter found in the surface soil layers that is in various stages of decomposition as a result of bacterial action.

hurricane severe tropical cyclone of great size with nearly concentric isobars. Its torrential rains and high-velocity winds create unusually high seas and extensive coastal flooding; also called willy-willies, tropical cyclones, baguios, and typhoons.

hydration attachment of water molecules to molecules of other elements or compounds without chemical change.

hydrologic cycle circulation of water within the Earth system, from evaporation to condensation, precipitation, runoff, storage, and re-evaporation back into the atmosphere.

hydrolysis union of water with other substances involving chemical change and the formation of new compounds.

hydrosphere major Earth subsystem consisting of the waters of Earth, including oceans, ice, freshwater bodies, groundwater, and water within the atmosphere and biomass.

ice age period of Earth history when much of the Earth's surface was covered with massive continental glaciers. The most recent ice age is referred to as the Pleistocene Epoch.

iceberg free-floating mass of glacier broken off by melting, tidal, and wave action.

ice cap small ice sheet found in highland areas that usually covers all but the highest mountain peaks.

ice fall portion of a glacier moving over and down a steep slope, creating a rigid white cascade, crisscrossed with deep crevasses.

Icelandic Low center of low atmospheric pressure located in the north Atlantic, especially persistent in winter.

ice-marginal lake temporary lake formed by the disruption of meltwater drainage by deposition along a glacial margin, usually in the area of an end moraine.

ice sheet mass of glacial ice thousands of feet thick that is of continental proportions and covers all but the highest points of land. The sheet usually flows from one or more areas of accumulation outward in all directions.

ice shelf large flat-topped plate of ice from the Antarctic ice cap, which overlies Antarctic waters and is a source of icebergs.

igneous rock one of the three major rock types: formed from the cooling and solidification of molten Earth material.

illuviation deposition of fine soil components in the subsoil (*B* horizon) by gravitational water.

inner core the innermost portion of the Earth's core that forms the center of the Earth and is considered to be a dense hot solid mass of iron.

inselberg remnant residual hill rising above an arid or semiarid plain; produced by stream erosion of a former mountainous area.

insolation incoming solar radiation, i.e., energy received from the sun.

instability condition of air when it is warmer than the surrounding atmosphere and is buoyant with a tendency to rise; the lapse rate of the surrounding atmosphere is greater than that of *unstable* air.

interglacial warmer period between glacial advances, during which continental ice sheets and many valley glaciers retreat and disappear or are greatly reduced in size.

intermittent stream stream that flows part of the time, usually only during, and shortly after, a rainy period.

International Date Line line roughly along the 180-degree meridian, where each day begins and ends; it is always a day later west of the line than east of the line.

Intertropical Convergence Zone (ITC) zone of low pressure and calms along the equator, where air carried by the trade winds from both sides of the equator converges and is forced to rise.

intrusive rock igneous rock that was cooled and crystallized beneath the Earth's surface.

inversion *see* temperature inversion.

isarithm line on a map that connects all points of the same numerical value, such as isotherms, isobars, and isobaths.

island arc curved row of volcanic islands along a deep oceanic trench; found near tectonic plate boundaries where subduction occurs.

isobar line drawn on a map to connect all points with the same atmospheric pressure.

isostasy   theory which holds that Earth's crust *floats* in hydrostatic equilibrium in the denser plastic layer of the mantle.

isotherm   line drawn on a map to connect all points with the same temperature.

jet stream   high-velocity upper-air current with speeds of 120–640 kph (75–250 mph).

joints   cracks or systems of cracks revealing lines of weakness in bedrock.

kame   conical hill composed of sorted glaciofluvial deposits; presumed to have formed in contact with glacial ice when sediment accumulated in ice pits, crevasses, and among jumbles of detached ice blocks.

kame terraces   landform resulting from accumulation of glaciofluvial sand and gravel along the margin of a glacier occupying a valley in an area of hilly relief.

karst   unique landforms developed as a result of the dissolving of limestone by groundwater.

katabatic wind   downslope flow of cold, dense air that has accumulated in a high mountain valley or over an elevated plateau or ice cap.

kettle hole   water-filled pit formed by the melting of a remnant ice block left buried in drift after the retreat of a glacier.

Köppen system   climate classification based on monthly and annual averages of temperature and precipitation; boundaries between climate classes are designed so that climate types coincide with vegetation regions.

laccolith   massive igneous intrusion that bows overlying rock layers upwards in a domal fashion as it forces its way toward the surface.

lahar   rapid form of mass movement involving mudflows from volcanic materials.

land breeze   air flow at night from the land toward the sea, caused by the movement of air from a zone of higher pressure associated with cooler nighttime temperatures over the land.

landslide   mass of Earth material, including all loose debris and often portions of bedrock, moving as a unit rapidly downslope.

lapse rate   *see* normal lapse rate.

latent heat of condensation   energy release in the form of heat, as water is converted from the gaseous (vapor) to the liquid state.

lateral moraine   moraine deposited along the side margin of an alpine glacier or lobe of a continental ice sheet.

laterite   iron, aluminum, and manganese rich layer in the subsoil (*B* horizon) that can be an end product of laterization in the wet-dry tropics (tropical savanna climate).

laterization   soil-forming process of hot, wet climates. Oxisols, the typical end product of the process, are characterized by the presence of little or no humus, the removal of soluble and most fine soil components, and the heavy accumulation of iron and aluminum compounds.

latitude   angular distance (distance measured in degrees) north or south of the equator.

lava   molten Earth material expelled at the surface from volcanoes or fissures. From this material extrusive igneous rock is formed.

leaching   removal by gravitational water of soluble inorganic soil components from the surface layers of the soil.

leeward   located on the side facing away from the wind.

levee   natural raised alluvial bank along margins of a river on a floodplain; artificial levees may be constructed along river banks for flood control.

liana   woody vine found in tropical forests that roots in the forest floor but uses trees for support as it grows upward toward available sunshine.

life support system   interacting and interdependent units (e.g., oxygen cycle, nitrogen cycle) that together provide an environment within which life can exist.

lightning   visible electrical discharge produced within a thunderstorm.

light year   the distance light travels in one year—6 trillion miles.

lithification   the combined processes of compaction and cementation that transform clastic sediments into sedimentary rocks.

lithosphere   solid crust of Earth that forms one of the major Earth subsystems. In a more technical definition related to tectonic plate theory, the lithosphere consists of Earth's crust and the uppermost rigid zone of the mantle, which is divided into individual plates that move independently on the plastic material of the asthenosphere.

Little Ice Age   an especially cold interval of time during the early 14th century that had major impacts on civilizations in the Northern Hemisphere.

llanos   region of characteristic tropical savanna vegetation in Venezuela, located primarily in the plains of the Orinoco River.

loam soil   soil with a texture in which none of the three soil grades (sand, silt, or clay) predominate over the others.

loess   wind-deposited silt; usually transported in dust storms and derived from arid or glaciated regions.

longitude   angular distance (distance measured in degrees) east or west of the prime meridian.

longshore current   current flowing parallel to the shore within the surf zone, produced by waves breaking at an angle to the shore.

long-wave radiation   electromagnetic radiation emitted by Earth in the form of waves more than 4.0 micrometers in amplitude, which includes heat reradiated by Earth's surface.

low   *see* cyclone.

magma   melt or molten Earth material, situated beneath Earth's surface, from which plutonic and intrusive igneous rock is formed.

magnetic declination   horizontal angle between geographic north and magnetic north.

mantle   moderately dense, relatively thick (2885 km/1800 mi) middle layer of Earth's interior that separates the crust from the outer core.

maquis   sclerophyllous woodland and plant community, similar to North American chaparral; can be found growing throughout the Mediterranean region.

mass   a measure of the total amount of matter in a body.

mass wasting (mass movement)   movement of surface materials downslope as a result of Earth gravitation.

matrix   the dominant area of a mosaic (ecosystem supporting a particular plant community) where the major plant in the community is concentrated.

medial moraine   central moraine in a large valley glacier; formed when two smaller valley glaciers come together to form the larger glacier and their interior lateral moraines merge.

Mercator projection   mathematically produced, conformal map projection showing true compass bearings as straight lines.

mercury barometer   instrument measuring atmospheric pressure by balancing it against a column of mercury.

meridian   one half of a great circle on the globe connecting all points of equal longitude; all meridians connect the North and South Poles.

mesa   flat-topped erosional remnant of a tableland characteristic of arid regions with flat-lying sediments; typically bordered by steep-sided escarpments and may cover large areas.

mesopause   upper limit of mesophere, separating it from the thermosphere.

mesosphere   layer of atmosphere above the stratosphere; characterized by temperatures that decrease regularly with altitude.

metamorphic rock   one of the three major rock types; formed from other rock within the crust by change induced by heat and pressure.

meteor   the luminous phenomenon observed when a small piece of solid matter enters Earth's atmosphere and burns up.

meteorology   study of the patterns and causes associated with short-term changes in the elements of the atmosphere.

microclimate   climate associated with a small area at or near Earth's surface; the area may range from a few inches to several miles in size.

microplate terrane   material added to continents as they collide with smaller areas of distinct geology such as volcanic islands or continent fragments.

millibar   unit of measurement for atmospheric pressure; one millibar equals a force of 1000 dynes per square centimeter; 1013.2 millibars is standard sea level pressure.

mineral   naturally occurring inorganic substance that possesses fairly definite physical characteristics and unique chemical composition.

mistral   cold downslope wind in southern France (*see* katabatic wind).

Mohorovicic discontinuity (Moho)   interface between Earth's crust and the more dense mantle.

monadnock   erosional remnant of more resistant rock on a plain of old age; associated with a theoretical cycle of erosion in humid lands.

monsoon   seasonal wind that reverses direction during the year in response to a reversal of pressure over a large landmass. The classic monsoons of Southeast Asia blow onshore in response to low pressure over Eurasia in summer and offshore in response to high pressure in winter.

moraine   unsorted glacial drift deposited beneath and along the margins of a glacier.

mosaic   a plant community and the ecosystem upon which it is based, viewed as a landscape of interlocking parts by ecologists.

mountain breeze   air flow downslope from mountains toward valleys during the night.

muskeg   poorly drained vegetation-rich marshes or swamps usually overlying permafrost areas of polar climatic regions.

natural resource   any element, material, or organism existing in nature that may be useful to humans.

natural vegetation   vegetation that has been allowed to develop naturally without obvious interference from or modification by humans.

nekton   marine organisms that swim freely in the oceans.

nimbus   term used in cloud description to indicate precipitation; thus cumulonimbus is a cumulus cloud from which rain is falling.

normal lapse rate   decrease in temperature with altitude under normal atmospheric conditions; approximately 6.5°C/1000 m (3.6°F/1000 ft).

northeast trades   *see* trade winds.

obliquity cycle   the change in the tilt of the Earth's axis relative to the plane of the ecliptic over a 41,000 year period.

occluded front   boundary between a rapidly advancing cold air mass and an uplifted warm air mass cut off from Earth's surface; denotes the last stage of a midlatitude cyclone.

ocean current   horizontal movement of ocean water, usually in response to major patterns of atmospheric circulation.

oceanic crust   the denser (av. 3.0 gm/cm^3) portion of the Earth's crust that underlies the ocean basins.

oceanic islands   volcanic islands that rise from the deep ocean floor.

oceanic ridge (midocean ridge)   linear seismic mountain range that interconnects through all the major oceans; it is where new molten crustal material rises through the oceanic crust.

oceanic trench (trench)   long, narrow depression on the sea floor usually associated with an island arc. Trenches mark the deepest portions of the oceans and are associated with subduction of oceanic crust.

omnivore   animal that can feed on both plants and other animals.

open system   system in which energy and/or materials can freely cross its boundaries.

orographic precipitation   precipitation resulting from condensation of water vapor in an air mass that is forced to rise over a mountain range or other raised landform.

outcrop   bedrock exposed at Earth's surface with no overlying regolith or soil.

outer core   the upper portion of the Earth's core; considered to be composed of molten iron liquefied by the Earth's internal heat.

outwash   glacial drift deposited beyond an end moraine by glacial meltwater.

outwash plain   extensive, relatively smooth plain covered with sorted deposits carried forward by the meltwater from an ice sheet.

oxbow lake   crescent-shaped lake or pond formed on a river floodplain in an abandoned meander channel.

oxidation   chemical union of oxygen with other elements to form new chemical compounds.

oxide   a mineral group composed of oxygen combining with other Earth elements, especially metallics.

oxygen-isotope analysis   a dating method used to reconstruct climate history; it is based on the varying evaporation rates of different oxygen isotopes and the changing ratio between the isotopes revealed in foraminifera fossils.

ozone   gas with a molecule consisting of three atoms of oxygen, (O_3); forms a layer in the upper atmosphere that serves to screen out ultraviolet radiation harmful at the earth's surface.

Pacific high   persistent cell of high atmospheric pressure located in the subtropics of the North Pacific Ocean.

paleogeography   the study of past geographic environments, based on climatic and geologic evidence.

parallel   circle on the globe connecting all points of equal latitude.

parallelism   tendency of Earth's polar axis to remain parallel to itself at all positions in its orbit around the sun.

parent material   residual (derived from bedrock directly beneath) or transported (by water, wind, or ice) mineral matter from which soil is formed.

patch   a gap or area within a matrix (territory occupied by a dominant plant community) where the dominant vegetation is not supported due to natural causes.

paternoster lakes   chain of lakes connected by a post-glacial stream occupying the trough of a glaciated mountain valley.

ped   soil aggregate or mass of individual mineral particles with a distinctive shape that characterizes a soil's structure.

pediment   gently sloping bedrock surface, usually covered with fluvial gravels, located at the base of a stream-eroded mountain range in an arid region.

pediplain   desert plain of pediments and alluvial fans; the presumed final erosion stage in an arid region.

peneplain   theoretical plain of extreme old age; the last stage in a cycle of erosion, reached when a landmass has been reduced to near base level by stream erosion in a humid region.

perihelion   position of Earth at closest distance to sun during each Earth revolution.

permafrost   permanently frozen layer of subsoil and underlying rock found in midlatitude subarctic and polar climates where the season is too short for summer thaw to penetrate more than a few feet below ground level.

permeability   characteristic of soil or bedrock that determines the ease with which water moves through Earth material.

pH scale   scale from 0 to 14 that describes the acidity or alkalinity of a substance and which is based on a measurement of hydrogen ions; pH values below 7 indicate acidic conditions; pH values above 7 indicate alkaline conditions.

photosynthesis   the process by which carbohydrates (sugars and starches) are manufactured in plant cells; requires carbon dioxide, water, light, and chlorophyll (the green color in plants).

piedmont glacier   glacier that forms where two or more valley glaciers coalesce to cover lower lands at the base of a mountainous region.

plane of the ecliptic   plane of Earth's orbit about the sun and the apparent annual path of the sun along the stars.

planet   any of the nine largest bodies revolving about the sun, or any similar bodies that may orbit other stars.

plankton   passively drifting or weakly swimming marine organisms, including both phytoplankton (plants) and zooplankton (animals).

plant community   variety of individual plants living in harmony with each other and the surrounding physical environment.

plate tectonics   theory that superseded continental drift and is based on the idea that the lithosphere is composed of a number of segments or *plates* that move independently of one another, at varying speeds, over Earth's surface.

playa   dry lake bed in a desert basin.

Pleistocene   the name given to the most recent "ice age" or period of Earth history experiencing cycles of continental glaciation; it commenced approximately 2.4 million years ago.

plucking   *see* quarrying.

plug dome   a steep-sided, explosive type of volcano with its central vent or vents plugged by the rapid congealing of its highly acidic lava.

pluton   extensive mass of igneous rock formed by the cooling of magma deep within the Earth's crust.

pluvial   rainy time period, usually pertaining to glacial periods when deserts were wetter than at present.

podzolization   soil-forming process of humid climates with long cold winter seasons. Spodosols, the typical end product of the process, are characterized by the surface accumulation of raw humus, strong acidity, and the leaching or eluviation of soluble bases and iron and aluminum compounds.

polar easterlies   easterly surface winds that move out from the polar highs toward the subpolar lows.

polar front   shifting boundary between cold polar air and warm subtropical air, located within the middle latitudes and strongly influenced by the polar jet stream.

polar highs   high-pressure systems located near the poles where air is settling and diverging.

polar jet stream   high-velocity air current within the upper air westerlies.

pollution   alteration of the physical, chemical, or biological balance of the environment that has adverse effects on the normal functioning of all lifeforms, including humans.

porosity   characteristic of soil or bedrock that relates to the amount of pore space between individual peds or soil and rock particles and which determines the water storage capacity of Earth material.

potential evapotranspiration   hypothetical rate of evapotranspiration if at all times there is a more than adequate amount of soil water for growing plants.

prairie   almost treeless tall grasslands in middle latitudes.

precession cycle   changes in the time (date) of the year that perihelion occurs; the date is determined on the basis of a major period 23,000 years in length and a secondary period 19,000 years in length.

precipitation   water in liquid or solid form that falls from the atmosphere and reaches Earth's surface.

pressure belts   zones of high or low pressure that tend to circle Earth parallel to the equator in a theoretical model of world atmospheric pressure.

pressure gradient   rate of change of atmospheric pressure horizontally with distance, measured along a line perpendicular to the isobars on a map of pressure distribution.

prevailing wind   direction from which the wind for a particular location blows during the greatest proportion of the time.

prime meridian (Greenwich meridian)   half of a great circle that connects the North and South Poles and marks zero degrees longitude. By international agreement the meridian passes through the Royal Observatory at Greenwich, England.

productivity   rate at which new organic material is created at a particular trophic level.

pyroclastic material   solid rock material (cinders, ash, and rock fragments) thrown into the air by a volcanic eruption.

quarrying (plucking)   process whereby active glaciers break away and carry forward weathered and fractured bedrock.

radiation   emission of waves that transmit energy through space. (*See also* short-wave radiation and long-wave radiation.)

radiation fog   fog produced by cooling of air in contact with a cold ground surface.

rain   falling droplets of liquid water.

rain shadow   dry, leeward side of a mountain range, resulting from the adiabatic warming of descending air.

recessional moraine   end moraine deposited behind the terminal moraine, marking pauses in the retreat of a valley glacier or ice sheet.

reg   desert surface of gravel and pebbles with finer materials removed; common to large areas in the Sahara.

regolith   weathered surface materials that usually cover bedrock.

relative humidity   ratio between the amount of water vapor in air of a given temperature and the maximum amount of water vapor that the air could hold at that temperature, if saturated; usually expressed as a percentage.

remote sensing   mechanical collection of information about the environment from a distance, usually from aircraft or spacecraft, e.g., photography, radar, infrared.

remote sensing devices   variety of techniques by which information about Earth can be gathered from great heights, typically from very high-flying aircraft or spacecraft.

revolution (Earth)   motion of Earth along a path, or orbit, around the sun. One complete revolution requires approximately $365\frac{1}{4}$ days and determines an Earth year.

rhumb line   line of true compass bearing (heading).

ria   coastline with many narrow bays mainly due to submerged river valleys.

ribbon falls   high, narrow waterfall dropping from a hanging glacial valley.

rift valley   major lowland that forms in a graben or down-faulted crustal block.

rime   ice crystals formed along the windward side of tree branches, airplane wings, etc., under conditions of supercooling.

rip current   strong, narrow surface current flowing away from shore. It is produced by the return flow of water piled up near shore by incoming waves.

roche moutonnée   bedrock hill subjected to intense glacial abrasion on its upstream side, with some plucking evident on the downstream side.

rockfall   nearly vertical drop of individual rocks or a small rock mass caused by the pull of gravity on steep slopes.

rock flour   rock fragments finely ground between the base of a glacier and the underlying bedrock surface.

rockslide   rapid downslope movement of huge masses of bedrock.

Rossby waves   horizontal undulations in the flow of the upper air winds of the middle and upper latitudes.

rotation (Earth)   turning of Earth on its polar axis;

one complete rotation requires 24 hours and determines one Earth day.

runoff   flow of water from the land surface, generally in the form of streams and rivers.

salinity   the amount of dissolved solids in seawater.

salinization   soil-forming process of low-lying areas in desert regions; the resulting soils are characterized by a high concentration of soluble salts as a result of the evaporation of surface water.

saltation   the transportation by running water or wind of particles too large to be carried in suspension; the particles are bounced along on the surface or stream bed by repeated lifting and deposition.

Santa Ana   very dry foehn wind occurring in southern California. (*See also* foehn wind.)

saturation (saturated air)   point at which sufficient cooling has occurred so that an air mass contains the maximum amount of water vapor it can hold. Further cooling produces condensation of excess water vapor.

savanna   tropical vegetation consisting primarily of coarse grasses, often associated with scattered low-growing trees or patches of bare ground.

scale   ratio between distance as measured on Earth and the same distance as measured on a map, globe, or other representation of Earth.

sclerophyllous   vegetation type commonly associated with the Mediterranean climate; characterized by tough surfaces, deep roots, and thick, shiny leaves that resist moisture loss.

sea breeze   air flow by day from the sea toward the land; caused by the movement of air toward a zone of lower pressure associated with higher daytime temperatures over the land.

sea floor spreading   movement of oceanic crust in opposite directions away from the midocean ridges, associated with the formation of new crust at the ridges and subduction of old crust at ocean margins.

seamount   submarine volcanic peak rising from the deep ocean floor.

sedimentary rock   one of the three major rock types; formed by the accumulation, compaction, and cementation of fragmented Earth materials, organic remains, or chemical precipitates.

seismograph   scientific instrument utilized to read the passage of vibratory earthquake and shock waves.

selva   characteristic tropical rainforest comprised of multistoried, broad-leaf evergreen trees with significant development of lianas and relatively little undergrowth.

sextant   navigation instrument used to determine latitude by star and sun positions.

shield volcano   gentle-sloped volcano formed by the cooling and accumulation of successive fluid lava flows extruded from a central vent or system of vents.

short-wave radiation   radiation energy emitted by the sun in the form of waves of less than 4.0 micrometers (1 micrometer equals one ten-thousandth

of a centimeter); includes X-rays, gamma rays, ultraviolet rays, and visible light waves.

Siberian high   intensively developed center of high atmospheric pressure located in northern central Asia in winter.

silicate   the largest mineral group, composed of oxygen and silica and forming most of the Earth's crust.

sill   igneous intrusion that forms a horizontal rock mass after molten material has been forced between rock layers and subsequently cools.

sinkhole   circular surface depression produced by the dissolving of limestone by groundwater.

slash-and-burn agriculture   also called swidden or shifting cultivation; typical subsistence agriculture of primitive societies in the tropical rainforest. Trees are cut, the smaller residue is burned, and crops are planted between the larger trees or stumps before rapid deterioration of the soil forces a move to a new area.

sleet   form of precipitation produced when raindrops freeze as they fall through a layer of cold air; may also, locally, refer to a mixture of rain and snow.

slope aspect   direction a mountain slope faces in respect to the sun's rays.

slump   mass of soil and regolith that slips or collapses downslope with a backward rotation.

small circle   any circle that is not a full circumference of the globe. The plane of a small circle does not pass through the center of the globe.

smog   combination of chemical pollutants and particulate matter in the lower atmosphere, typically over urban-industrial areas.

snow   precipitation in the form of ice crystals.

snow line   elevation in mountain regions above which summer melting is insufficient to prevent the accumulation of permanent snow or ice.

soil grade   classification of soil texture by particle size: clay (less than 0.002 mm), silt (0.002–0.05 mm), and sand (0.05–2.0 mm) are soil grades.

soil horizon   distinct soil layer characteristic of vertical zonation in soils; horizons are distinguished by their general appearances and their specific chemical and physical properties.

soil profile   vertical cross section of a soil that displays the various horizons or soil layers that characterize it; used for classification.

soil survey   a publication of the United States Soil Survey Division of the Natural Resources Conservation Service which includes maps showing the distribution of soil within a given area, usually a county.

soil taxonomy   the classification and naming of soils.

solar constant   rate at which insolation is received just outside Earth's atmosphere on a surface at right angles to the incoming radiation.

solar energy   *see* insolation.

solar system   the system of the sun and the planets, their satellites, comets, meteoroids, and other objects revolving around the sun.

solid tectonic processes   those processes that distort the solid Earth crust by bending, folding, warping, or fracturing (faulting).

solifluction   slow movement or flow of water-saturated soil and regolith downslope due to gravity; causes characteristic lobes on slopes in permafrost areas where only the top few feet of Earth material thaws in summer and drainage is poor.

solstice   one of two times each year when the position of the noon sun is overhead at its farthest distance from the equator; this occurs when the sun is overhead at the Tropic of Cancer (about June 22) and the Tropic of Capricorn (about December 22).

source region   nearly homogeneous surface of land or ocean over which an air mass acquires its temperature and humidity characteristics.

southeast trades   *see* trade winds.

southern oscillation   the systematic variation in atmospheric pressure between the eastern and western Pacific Ocean.

specific humidity   mass of water vapor present per unit mass of air, expressed as grams per kilogram of moist air.

spit   beach feature attached to the mainland and built partially across a bay or inlet by the depositional action of longshore currents.

spring   any surface outflow of groundwater, generally where the water table intersects the ground surface.

squall line   narrow line of rapidly advancing storm clouds, strong winds, and heavy precipitation; usually develops in front of a fast-moving cold front.

stability   condition of air when it is cooler than the surrounding atmosphere and resists the tendency to rise; the lapse rate of the surrounding atmosphere is less than that of *stable* air.

stationary front   frontal system between air masses of nearly equal strength; produces stagnation over one location for an extended period of time.

steppe   middle-latitude semiarid vegetation, treeless and dominated by short bunch grasses.

stock   individual deep-seated igneous mass of limited size; it is often associated with other igneous masses known generally as plutons.

storm   local atmospheric disturbance often associated with rain, hail, snow, sleet, lightning, or strong winds.

storm surge   rise in sea level due to wind and reduced air pressure during a hurricane or other severe storm.

storm track   path frequently traveled by a cyclonic storm as it moves in a generally eastward direction from its point of origin.

strata (stratification)   distinct layers or beds of sedimentary rock.

stratus   uniform layer of low sheetlike clouds, frequently grayish in appearance.

stratopause   upper limit of stratosphere, separating it from the mesosphere.

stratosphere   layer of atmosphere lying above the troposphere and below the mesosphere, characterized by fairly constant temperatures and ozone concentration.

stratovolcano   *see* composite cone.

stream load   amount of material transported by a stream at a given instant; includes bed load, suspended load, and dissolved load.

striations   gouges, grooves, and scratches produced in bedrock by rock fragments and boulders imbedded in a glacier.

strike   the compass direction taken by a rock stratum or fault plane, which is at right angles to their dip.

subduction   process associated with plate tectonic theory whereby an oceanic crustal plate is forced downward into the mantle beneath a lighter continental plate when the two converge.

submarine canyon   steep-sided erosional valley cut into the continental shelf or continental slope.

subpolar lows   east/west trending belts or cells of low atmospheric pressure located in the upper middle latitudes.

subsurface horizon   buried soil layer that possesses specific characteristics essential to the identification of soils in the National Resources Conservation Service System. (Example of subsurface horizons may be found in Table 13.1.)

subtropical highs   cells of high atmospheric pressure centered over the eastern portions of the oceans in the vicinity of 30°N and 30°S latitude; source of the westerlies poleward and the trades equatorward.

subtropical jet stream   high-velocity air current flowing above the sinking air of the subtropical high-pressure cells; most prominent in the winter season.

succession   progression of natural vegetation from one plant community to the next until a final stage of equilibrium has been reached with the natural environment.

surface of discontinuity   three-dimensional surface with length, width, and height separating two different air masses; also referred to as a *front*.

surge (glacial)   sudden shift downslope of glacial ice, possibly caused by a reduction of basal friction with underlying bedrock.

swell   regular longer-period sea wave traveling a significant distance from the area where it was generated by the wind.

syncline   trough or downfold in a wave of crustal folding.

system   group of interacting and interdependent units that together form an organized whole.

taiga   term used to describe the northern coniferous forest of subarctic regions on the Eurasian landmass.

taku   cold downslope wind in Alaska. (*See also* katabatic wind.)

talus (talus cone)   rock debris in a cone-shaped deposit at the base of a steep slope or escarpment; usually a result of frost wedging and individual rockfalls with debris accumulating at the angle of rest.

tarn   mountain lake in a glacial cirque.

tectonic processes processes that derive their energy from within Earth's interior and serve to create landforms by elevating, disrupting, and roughening Earth's surface.

temperature degree of heat or cold and its measurement.

temperature gradient rate of change of temperature with distance in any direction from a given point; refers to rate of change horizontally; a vertical temperature gradient is referred to as the *lapse rate*.

temperature inversion reverse of the normal pattern of vertical distribution of air temperature; in the case of inversion, temperature *increases* rather than decreases with increasing altitude.

terminal moraine end moraine that marks the farthest advance of an alpine glacier or ice sheet.

terra rosa characteristic calcium-rich (developed over limestone bedrock) red-brown soils of the climatic regions surrounding the Mediterranean Sea.

terrestrial planets the four closest planets to the sun—Mercury, Venus, Earth, and Mars.

thermocline vertical zone of ocean water where there is a sharp change in temperature with depth.

thermosphere highest layer of atmosphere extending from the mesopause to outer space.

Thornthwaite system climate classification based on moisture availability and of greatest use at the local level; climate types are distinguished by examining and comparing potential and actual evapotransportation.

thunder sound produced by the rapidly expanding, heated air along the channel of a lightning discharge.

thunderstorm intense convectional storm characterized by thunder and lightning, short in duration and often accompanied by heavy rain, hail, and strong winds.

tide periodic rise and fall of sea level in response to the gravitational interaction of the moon, sun, and Earth.

till unsorted glacial drift, characterized by variation in size of deposit from clay particles to boulders.

tombolo wave depositional beach feature connecting an island to the mainland.

tornado small, intense, funnel-shaped cyclonic storm of very low pressure, violent updrafts, and converging winds of enormous velocity.

trade winds consistent surface winds blowing in low latitudes from the subtropical highs toward the intertropical convergence zone; labeled north-east trades in the Northern Hemisphere and south-east trades in the Southern Hemisphere.

transpiration transfer of moisture from living plants to the atmosphere by the emission of water vapor, primarily from leaf pores.

transportation movement of Earth materials from one site to another as a result of the transporting power of water, wind, or glacial ice.

travertine calcium carbonate (limestone) deposits resulting from the evaporation in caves or caverns and near surface openings of groundwater saturated with lime.

tree line elevation in mountain regions above which cold temperatures and wind stress prohibit tree growth.

trophic level number of feeding steps that a given organism is removed from the autotrophs (e.g., green plant—first level, herbivore—second level, carnivore—third level, etc.).

trophic structure organization of an ecosystem based on the feeding patterns of the organisms that comprise the ecosystem.

Tropic of Cancer parallel of latitude at $23\frac{1}{2}°$N; the northern limit of the migration of the sun's vertical rays throughout the year.

Tropic of Capricorn parallel of latitude at $23\frac{1}{2}°$S; the southern limit to the migration of the sun's vertical rays throughout the year.

tropopause boundary between the troposphere and stratosphere.

troposphere lowest layer of the atmosphere, exhibiting a steady decrease in temperature with increasing altitude and containing virtually all atmospheric dust and water vapor.

trough elongated area or "belt" of low atmospheric pressure; also glacial trough, a U-shaped valley carved by a glacier.

tsunami ocean wave produced by submarine earthquake, volcanic eruption, or landslide; not noticeable in deep ocean waters, but building to dangerous heights in shallow waters.

tundra treeless vegetation of polar regions and very high mountains, consisting of mosses, lichens, and low-growing shrubs and flowering plants.

turbidity current submarine flow of sediment-laden water.

unconformity an interruption in the sequence of deposition mainly due to erosion.

uniformitarianism widely accepted theory that Earth's landforms have developed over exceedingly long periods of time as a result of processes that may be observed in the present landscape.

upper air westerlies system of westerly winds in the upper atmosphere, flowing in latitudes poleward of 20°.

upwelling upward movement of colder, nutrient rich, subsurface ocean water, replacing surface water that is pushed away from shore by winds.

valley breeze air flow upslope from the valleys toward the mountains during the day.

valley glacier *see* alpine glacier.

valley train outwash deposit from glacial meltwater, resembling an alluvial fan confined by valley walls.

variable one of a set of objects and/or characteristics of objects, which are interrelated in such a way that they function together as a system.

varve a pairing of organic-rich summer sediments and organic-poor winter sediments found in exposed lake beds; because each pair represents one year of time, counting varves is useful as a dating technique for recent Earth history.

veering wind shift   the change in wind direction clockwise around the compass; e.g., east to southeast to south, to southwest, to west, and northwest.

ventifact   wind-fashioned rock produced by wind abrasion (sandblasting).

volcanism   the upward movement of molten material (magma) and its cooling above Earth's surface.

warm front   leading edge of a relatively warmer, less dense air mass advancing upon a cooler, denser air mass.

warping   broad and general uplift or settling of Earth's crust with little or no local distortion.

wash (arroyo, wadi, barranca)   generally steep-walled channel of an intermittent stream in an arid region; the stream bed is characteristically choked with coarse alluvium.

water budget   relationship between evaporation, condensation, and storage of water within the Earth system.

water table   upper limit of the zone of saturation below which all pore spaces are filled with water.

water vapor   water in its gaseous form.

wave-cut bench   gently sloping surface produced by wave erosion at the base of a sea cliff.

wave refraction   bending of waves as they approach a shore, aligning themselves with the bottom contours of the surf zone.

weather   atmospheric conditions, at a given time, in a specific location.

weathering   physical (mechanical) fragmentation and chemical decomposition of rocks and minerals in Earth's crust.

westerlies   surface winds flowing from the polar portions of the subtropical highs, carrying fronts, storms, and variable weather conditions from west to east through the middle latitudes.

wet adiabatic rate   rate at which a rising mass of air is cooled by expansion when condensation is taking place. The rate varies but averages 5°C/1000 m (3.2°F/1000 ft).

wind   air in motion from areas of higher pressure to areas of lower pressure; movement is generally horizontal, relative to the ground surface.

windward   location on the side that faces toward the wind and is therefore exposed or unprotected; usually refers to mountain and island locations.

xerophyte   vegetation type that has genetically evolved to withstand the extended periods of drought common to arid regions.

yazoo stream   a stream tributary that flows parallel to the main stream for a considerable distance before joining it.

zone of ablation   lower portion of a glacier, below the firn line, where melting, evaporation, and sublimation exceed net snow accumulation.

zone of accumulation   subsoil or *B* horizon of a soil, characterized by deposition or illuviation of soil components by gravitational water; also the upper portion of a glacier, above the firn line, where net snow accumulation exceeds the melting, evaporation, and sublimation of snowfall.

zone of aeration   upper groundwater zone above the water table where pore spaces may be alternately filled with air or water.

zone of depletion   top layer, or *A* horizon, of a soil, characterized by the removal of soluble and insoluble soil components through leaching and eluviation by gravitational water.

zone of saturation   zone immediately below the water table, where all pore spaces in soil and rock are filled with groundwater.

Index